SI Base Units

Base Quantity	Name of Unit	Symbol
Length	meter	m
Mass	kilogram	kg
Time	second	s
Electrical current	ampere	A
Temperature	kelvin	K
Amount of substance	mole	mol
Luminous intensity	candela	cd

Derived Units in the SI System

Physical Quantity	Name	Symbol	Units
Energy	Joule	J	$kg\ m^2\ s^{-2}$
Force	Newton	N	$kg\ m\ s^{-2}$
Power	Watt	W	$kg\ m^2\ s^{-3}$
Electric charge	Coulomb	C	$A\ s$
Electrical resistance	Ohm	Ω	$kg\ m^2\ s^{-3}\ A^{-2}$
Electrical potential difference	Volt	V	$kg\ m^2\ s^{-3}\ A^{-1}$
Electrical capacity	Farad	F	$A^2\ s^4\ kg^{-1}\ m^{-2}$
Frequency	Hertz	Hz	s^{-1}

Some Commonly Used Non-SI Units

Unit	Quantity	Symbol	Conversion Factor
Angstrom	Length	Å	$1\ \text{Å} = 10^{-10}\ m = 100\ pm$
Calorie	Energy	cal	$1\ cal = 4.184\ J$
Debye	Dipole moment	D	$1\ D = 3.3356 \times 10^{-30}\ C\ m$
Gauss	Magnetic field	G	$1\ G = 10^{-4}\ T$
Liter	Volume	L	$1\ L = 10^{-3}\ m^3$

Prefixes Used with SI Units

Prefix	Symbol	Meaning
Tera-	T	$1,000,000,000,000$, or 10^{12}
Giga-	G	$1,000,000,000$, or 10^9
Mega-	M	$1,000,000$, or 10^6
Kilo-	k	$1,000$, or 10^3
Deci-	d	$1/10$, or 10^{-1}
Centi-	c	$1/100$, or 10^{-2}
Milli-	m	$1/1,000$, or 10^{-3}
Micro-	μ	$1/1,000,000$, or 10^{-6}
Nano-	n	$1/1,000,000,000$, or 10^{-9}
Pico-	p	$1/1,000,000,000,000$, or 10^{-12}

PHYSICAL CHEMISTRY
for the Biosciences

PHYSICAL CHEMISTRY
for the Biosciences

Raymond Chang
WILLIAMS COLLEGE

University Science Books
Mill Valley, California

University Science Books
www.uscibooks.com

Production Manager: *Cecile Joyner*
Manuscript Editor: *John Murdzek*
Designer: *Robert Ishi*
Illustrators: *John and Judy Waller*
Compositor: *Asco Typesetters*
Printer & Binder: *Edwards Brothers, Inc.*

This book is printed on acid-free paper.

Library of Congress Cataloging-in-Publication Data

Chang, Raymond.
    Physical chemistry for the biosciences / Raymond Chang
        p.   cm.
    Includes bibliographical references (p.).
    ISBN 1-891389-33-5 (alk. paper)
      1. Biochemistry. I. Title.
    QH345.C425   2004
    572—dc22   2004049612

Printed in the United States of America
10 9 8 7 6 5 4 3

Contents

Preface   xiii

CHAPTER **1**   Introduction   1

1.1   Nature of Physical Chemistry   1
1.2   Units   2
1.3   Atomic Mass, Molecular Mass, and the Chemical Mole   5

CHAPTER **2**   Properties of Gases   7

2.1   Some Definitions   7
2.2   An Operational Definition of Temperature   8
2.3   Ideal Gases   8
   •   Boyle's law   8   •   Charles' and Gay-Lussac's Law   9   •   Avogadro's Law   10   •   The Ideal-Gas Equation   11   •   Dalton's Law of Partial Pressures   12
2.4   Real Gases   14
   •   The van der Waals Equation   15   •   The Virial Equation of State   16
2.5   Condensation of Gases and the Critical State   18
2.6   Kinetic Theory of Gases   21
   •   The Model   21   •   Pressure of a Gas   21   •   Kinetic Energy and Temperature   24
2.7   The Maxwell Distribution Laws   25
2.8   Molecular Collisions and the Mean Free Path   28
2.9   Graham's Laws of Diffusion and Effusion   30
Problems   32

CHAPTER **3**   The First Law of Thermodynamics   39

3.1   Work and Heat   39
   •   Work   39   •   Heat   43
3.2   The First Law of Thermodynamics   44
   •   Enthalpy   46   •   A Comparison of ΔU with ΔH 47
3.3   Heat Capacities   49
   •   Constant-Volume and Constant-Pressure Heat Capacities   50
   •   Molecular Interpretation of Heat Capacity   51   •   A Comparison of C_V with C_P 53
3.4   Gas Expansion   55
   •   Isothermal Expansions   55   •   Adiabatic Expansions   56
3.5   Calorimetry   59
   •   Constant-Volume Calorimetry   59   •   Constant-Pressure Calorimetry   61
   •   Differential Scanning Calorimetry   62
3.6   Thermochemistry   64
   •   Standard Enthalpy of Formation   64   •   Dependence of Enthalpy of Reaction on Temperature   68
3.7   Bond Energies and Bond Enthalpies   70
   •   Bond Enthalpy and Bond Dissociation Enthalpy   71
Problems   75

CHAPTER **4**   The Second Law of Thermodynamics   81

4.1   Spontaneous Processes   81
4.2   Entropy   83
   · Statistical Definition of Entropy   83   ·   Thermodynamic Definition of
   Entropy   83   ·   The Carnot Heat Engine and Thermodynamic Efficiency   87
4.3   The Second Law of Thermodynamics   88
4.4   Entropy Changes   90
   · Entropy Change Due to Mixing of Ideal Gases   90   ·   Entropy Change
   Due to Phase Transitions   91   ·   Entropy Change Due to Heating   92
4.5   The Third Law of Thermodynamics   95
   · Third Law or Absolute Entropies   96   ·   Entropy of Chemical
   Reactions   97   ·   The Meaning of Entropy   98
4.6   Gibbs Energy   101
   · The Meaning of Gibbs Energy   103
4.7   Standard Molar Gibbs Energy of Formation $(\Delta_f \bar{G}^\circ)$ 105
4.8   Dependence of Gibbs Energy on Temperature and Pressure   107
   · Dependence of G on Temperature   107   ·   Dependence of G on
   Pressure   108
4.9   Phase Equilibria   110
   · The Clapeyron and the Clausius-Clapeyron Equations   112   ·   Phase
   Diagrams   115   ·   The Phase Rule   117
4.10 Thermodynamics of Rubber Elasticity   117
Problems   121

CHAPTER **5**   Solutions   127

5.1   Concentration Units   127
   · Percent by Weight   127   ·   Mole Fraction   128   ·   Molarity (M)   128
   · Molality (m)   128
5.2   Partial Molar Quantities   129
   · Partial Molar Volume   129   ·   Partial Molar Gibbs Energy   131
   · The Meaning of Chemical Potential   131
5.3   The Thermodynamics of Mixing   132
5.4   Binary Mixtures of Volatile Liquids   134
   · Raoult's Law   135   ·   Henry's Law   137
5.5   Real Solutions   139
   · The Solvent Component   139   ·   The Solute Component   141
5.6   Colligative Properties   142
   · Vapor-Pressure Lowering   143   ·   Boiling-Point Elevation   143
   · Freezing-Point Depression   146   ·   Osmotic Pressure   148
5.7   Electrolyte Solutions   154
   · A Molecular View of the Electrolyte Solution Process   154
   · Thermodynamics of Ions in Solution   157   ·   Enthalpy, Entropy, and
   Gibbs Energy of Formation of Ions in Solution   159
5.8   Ionic Activity   160
   Debye-Hückel Theory of Electrolytes   164   ·   The Salting-In and
   Salting-Out Effects   167
5.9   Colligative Properties of Electrolyte Solutions   170
   · The Donnan Effect   172
5.10 Biological Membranes   175
   · Membrane Transport   177
Appendix 5.1   Notes on Electrostatics   182
Problems   186

CHAPTER **6** Chemical Equilibrium 193

6.1 Chemical Equilibrium in Gaseous Systems 193
· Ideal Gases 193 · A Closer Look at Equation 6.7 197
· A Comparison of $\Delta_r G°$ with $\Delta_r G$ 199 · Real Gases 200
6.2 Reactions in Solutions 201
6.3 Heterogeneous Equilibria 203
6.4 The Influence of Temperature, Pressure, and Catalysts on the Equilibrium Constant 205
· The Effect of Temperature 205 · The Effect of Pressure 208
· The Effect of a Catalyst 208
6.5 Binding of Ligands and Metal Ions to Macromolecules 209
· One Binding Site per Macromolecule 210 · n Equivalent Binding Sites per Macromolecule 211 · Experimental Studies of Binding Equilibria 213
6.6 Bioenergetics 217
· The Standard State in Biochemistry 218 · ATP—The Currency of Energy 220 · Principles of Coupled Reactions 222 · Glycolysis 223
· Some Limitations of Thermodynamics in Biology 228
Problems 230

CHAPTER **7** Electrochemistry 235

7.1 Electrochemical Cells 235
7.2 Single Electrode Potentials 236
7.3 Thermodynamics of Electrochemical Cells 238
· The Nernst Equation 242 · Temperature Dependence of EMF 244
7.4 Types of Electrochemical Cells 245
· Concentration Cells 245 · Fuel Cells 245
7.5 Applications of EMF Measurements 246
· Determination of Activity Coefficients 246 · Determination of pH 247
7.6 Biological Oxidation 248
· The Chemiosmotic Theory of Oxidative Phosphorylation 252
7.7 Membrane Potential 255
· The Goldman Equation 258 · The Action Potential 258
Problems 262

CHAPTER **8** Acids and Bases 267

8.1 Definitions of Acids and Bases 267
8.2 Acid-Base Properties of Water 268
· pH—A Measure of Acidity 269
8.3 Dissociation of Acids and Bases 270
· The Relationship Between the Dissociation Constant of an Acid and Its Conjugate Base 275 · Salt Hydrolysis 275
8.4 Diprotic and Polyprotic Acids 276
8.5 Buffer Solutions 280
· The Effect of Ionic Strength and Temperature on Buffer Solutions 283
· Preparing a Buffer Solution with a Specific pH 284
· Buffer Capacity 285
8.6 Acid-Base Titrations 286
· Acid-Base Indicators 287
8.7 Amino Acids 289
· Dissociation of Amino Acids 290 · The Isoelectric Point (pI) 291
· Titration of Proteins 292
8.8 Maintaining the pH of Blood 293

Appendix 8.1   A More Exact Treatment of Acid-Base Equilibria   298
Problems   305

CHAPTER **9**   Chemical Kinetics   311

9.1   Reaction Rates   311
9.2   Reaction Order   312
· Zero-Order Reactions   313 · First-Order Reactions   314
· Second-Order Reactions   318 · Determination of Reaction Order   323
9.3   Molecularity of a Reaction   324
· Unimolecular Reactions   325 · Bimolecular Reactions   327
· Termolecular Reactions   327
9.4   More Complex Reactions   328
· Reversible Reactions   328 · Consecutive Reactions   330
· Chain Reactions   332
9.5   The Effect of Temperature on Reaction Rates   332
· The Arrhenius Equation   333
9.6   Potential Energy Surfaces   335
9.7   Theories of Reaction Rates   336
· Collision Theory   336 · Transition-State Theory   338
· Thermodynamic Formulation of Transition-State Theory   340
9.8   Isotope Effects in Chemical Reactions   343
9.9   Reactions in Solution   346
9.10 Fast Reactions in Solution   347
· The Flow Method   349 · The Relaxation Method   349
9.11 Oscillating Reactions   353
Problems   356

CHAPTER **10**   Enzyme Kinetics   363

10.1   General Principles of Catalysis   363
· Enzyme Catalysis   364
10.2   The Equations of Enzyme Kinetics   367
· Michaelis-Menten Kinetics   367 · Steady-State Kinetics   368
· The Significance of K_M and V_{max} 370
10.3   Chymotrypsin: A Case Study   372
10.4   Multisubstrate Systems   375
· The Sequential Mechanism   376 · The Nonsequential or "Ping-Pong"
Mechanism   376
10.5   Enzyme Inhibition   377
· Reversible Inhibition   377 · Irreversible Inhibition   384
10.6   Allosteric Interactions   385
· Oxygen Binding to Myoglobin and Hemoglobin   385 · The Hill
Equation   387 · The Concerted Model   390 · The Sequential
Model   391 · Conformational Changes in Hemoglobin Induced by
Oxygen Binding   392
10.7   The Effect of pH on Enzyme Kinetics   393
Problems   398

CHAPTER **11**   Quantum Mechanics and Atomic Structure   401

11.1   The Wave Theory of Light   401
11.2   Planck's Quantum Theory   403
11.3   The Photoelectric Effect   405

11.4   Bohr's Theory of the Hydrogen Emission Spectrum   407
11.5   de Broglie's Postulate   410
11.6   The Heisenberg Uncertainty Principle   414
11.7   The Schrödinger Wave Equation   416
11.8   Particle in a One-Dimensional Box   418
   •   Electronic Spectra of Polyenes   423
11.9   Quantum-Mechanical Tunneling   424
11.10 The Schrödinger Wave Equation for the Hydrogen Atom   426
11.11 Many-Electron Atoms and the Periodic Table   432
   •   Electronic Configurations   433   •   Variations in Periodic
   Properties   437
Problems   441

CHAPTER **12**   The Chemical Bond   447

12.1   Lewis Structures   447
12.2   Valence Bond Theory   448
12.3   Hybridization of Atomic Orbitals   450
12.4   Electronegativity and Dipole Moment   455
   •   Electronegativity   455   •   Dipole Moment   456
12.5   Molecular Orbital Theory   458
12.6   Diatomic Molecules   460
   •   Homonuclear Diatomic Molecules of the Second-Period Elements   460
   •   Heteronuclear Diatomic Molecules of the Second-Period Elements   463
12.7   Resonance and Electron Delocalization   465
   •   The Peptide Bond   467
12.8   Coordination Compounds   469
   •   Crystal Field Theory   470   •   Molecular Orbital Theory   475
   •   Valence Bond Theory   476
12.9   Coordination Compounds in Biological Systems   477
   •   Iron   477   •   Copper   480   •   Cobalt, Manganese, and Nickel   480
   •   Zinc   481   •   Toxic Heavy Metals   482
Problems   485

CHAPTER **13**   Intermolecular Forces   489

13.1   Intermolecular Interactions   489
13.2   The Ionic Bond   490
13.3   Types of Intermolecular Forces   492
   •   Dipole-Dipole Interaction   492   •   Ion-Dipole Interaction   494
   •   Ion-Induced Dipole and Dipole-Induced Dipole Interactions   495
   •   Dispersion, or London, Forces   497   •   Repulsive and Total
   Interactions   498   •   The Role of Dispersion Forces in Sickle-Cell
   Anemia   500
13.4   Hydrogen Bonding   502
13.5   The Structure and Properties of Water   505
   •   The Structure of Ice   505   •   The Structure of Water   506
   •   Some Physiochemical Properties of Water   507
13.6   Hydrophobic Interaction   508
Problems   511

CHAPTER **14**   Spectroscopy   513

14.1   Vocabulary   513
   •   Absorption and Emission   513   •   Units   513   •   Regions of the

Spectrum   514   ·   Line Width   514   ·   Resolution   517
·   Intensity   518   ·   Selection Rules   519   ·   Signal-to-Noise Ratio   521
·   The Beer-Lambert Law   521

14.2   Microwave Spectroscopy   522
14.3   Infrared Spectroscopy   527
·   Simultaneous Vibrational and Rotational Transitions   532
14.4   Electronic Spectroscopy   534
·   Organic Molecules   535   ·   Transition Metal Complexes   537
·   Molecules that Undergo Charge-Transfer Interactions   537
·   Application of the Beer-Lambert Law   538
14.5   Nuclear Magnetic Resonance   539
·   The Boltzmann Distribution   542   ·   Chemical Shifts   542
·   Spin-Spin Coupling   544   ·   NMR and Rate Processes   545
·   NMR of Nuclei Other than ^{1}H   546   ·   Fourier-Transform NMR   547
·   Magnetic Resonance Imaging (MRI)   551
14.6   Electron Spin Resonance   552
14.7   Fluorescence and Phosphorescence   554
·   Fluorescence   554   ·   Phosphorescence   556
14.8   Lasers   557
·   Properties and Applications of Laser Light   560
14.9   Optical Rotatory Dispersion and Circular Dichroism   562
·   Molecular Symmetry and Optical Activity   562   ·   Polarized Light and
Optical Rotation   563   ·   Optical Rotatory Dispersion (ORD) and Circular
Dichroism (CD)   566
Problems   570

CHAPTER **15**   Photochemistry and Photobiology   575

15.1   Introduction   575
·   Thermal Versus Photochemical Reactions   575   ·   Primary Versus
Secondary Processes   576   ·   Quantum Yields   576   ·   Measurement of
Light Intensity   578   ·   Action Spectrum   579
15.2   Photosynthesis   580
·   The Chloroplast   580   ·   Chlorophyll and Other Pigment
Molecules   581   ·   The Reaction Center   581
·   Photosystems I and II   583   ·   Dark Reactions   586
15.3   Vision   586
·   Structure of Rhodopsin   588   ·   Mechanism of Vision   588
·   Rotation About the C=C Bond   589
15.4   Biological Effects of Radiation   591
·   Sunlight and Skin Cancer   591   ·   Photomedicine   592
Problems   598

CHAPTER **16**   Macromolecules   599

16.1   Methods for Determining Size, Shape, and Molar Mass of
Macromolecules   599
·   Molar Mass of Macromolecules   599   ·   Sedimentation in the
Ultracentrifuge   600   ·   Viscosity   607   ·   Electrophoresis   608
16.2   Structure of Synthetic Polymers   613
·   Configuration and Conformation   613   ·   The Random-Walk
Model   614
16.3   Structure of Proteins and DNA   616
·   Proteins   616   ·   DNA   621

16.4   Protein Stability   624
      · Hydrophobic Interaction   625 · Denaturation   626 · Protein
      Folding   629
Problems   635

Appendix 1   Review of Mathematics   639
Appendix 2   Thermodynamic Data   651
Glossary   655
Answers to Even-Numbered Computational Problems   665
Index   669

Preface

Physical Chemistry for the Biosciences is intended for use in a one-semester introductory course in physical chemistry. Most students enrolled in this course have taken general chemistry, organic chemistry, and a year of physics and calculus. Only basic skills of differential and integral calculus are required for understanding the equations. For premedical students, this text will form the basis for taking courses like physiology and pharmacology in medical school. For those intending to pursue graduate study in the biological sciences, the materials presented here will serve as an introduction to topics in biophysical chemistry courses, where more advanced texts such as those by Gennis, van Holde, and Cantor & Schimmel are used.

My aim is to emphasize on understanding physical concepts and their applications to chemical and biological systems rather than on precise mathematical development or on actual experimental details. To keep the text at a reasonable length, I have had to make choices of what to omit. All the basic and essential topics in physical chemistry like thermodynamics, chemical kinetics, and bonding are treated in detail. Additional chapters on spectroscopy, photochemistry and photobiology, and macromolecules can also be covered if time permits. Each chapter has an extensive reference list of texts and journal articles. A color tint is used to show the key equations. The end-of-chapter problems (about 900 in all) are arranged according to topics in each chapter. The Additional Problems section contains more challenging and multi-concept problems. A Solutions Manual written by Helen Leung and Mark Marshall containing full solutions to all even-numbered problems is available.

It is a pleasure to thank the following people who provided helpful comments and suggestions: Christopher Barrett (McGill University), Ron Christensen (Bowdoin College), Kirsten Eberth (The Royal Danish School of Pharmacy), Raymond Esquerra (San Francisco State University), Gary Lorigan (Miami University), Robert O'Brien (Portland State University), Keith Orrell (University of Exeter), and Karen Singmaster (San Jose State University). I also thank Bruce Armbruster and Kathy Armbruster for general assistance, Cecile Joyner for expertly supervising the production, Bob Ishi for his functional and tasteful design, John Murdzek for a meticulous job of copyediting, and John Waller and Judy Waller for their pleasing and effective illustrations. Finally, my special thanks go to Jane Ellis, who supervised the project from beginning to end and took care of all the details big and small.

Raymond Chang
Williamstown, Massachusetts

PHYSICAL CHEMISTRY
for the Biosciences

CHAPTER 1

Introduction

1.1 Nature of Physical Chemistry

Physical chemistry can be described as a set of characteristically quantitative approaches to the study of chemical problems. A physical chemist seeks to predict and/ or explain chemical events using certain models and postulates.

Because the problems encountered in physical chemistry are diversified and often complex, they require a number of different approaches. For example, in the study of thermodynamics and rates of chemical reactions, we employ a phenomenological, macroscopic approach. But a microscopic, molecular approach is necessary to understand the kinetic behavior of molecules and reaction mechanisms. Ideally, we study all phenomena at the molecular level, because that is where change occurs. In fact, our knowledge of atoms and molecules is neither extensive nor thorough enough to permit this type of investigation in all cases, and we sometimes have to settle for a good, semiquantitative understanding. It is useful to keep in mind the scope and limitations of a given approach.

The principles of physical chemistry can be applied to the study of any chemical system. For example, let us consider how we use these principles to understand the binding of dioxygen (O_2) to hemoglobin. This system is one of the most important biochemical reactions and is probably the most extensively studied. Hemoglobin, a protein molecule with a molar mass of about 65,000 g, contains four subunits, made up of two α chains (141 amino acids each) and two β chains (146 amino acids each). Each chain contains a heme group to which an oxygen molecule can bind. The main functions of hemoglobin are to carry oxygen in the blood from the lungs to the tissues, where it transfers the oxygen molecules to myoglobin, and to transport carbon dioxide from the tissues back to the lungs. Myoglobin, which possesses only one polypeptide chain (153 amino acids) and one heme group, stores oxygen for metabolic processes.

A detailed understanding of the three-dimensional structure of a protein molecule, such as hemoglobin, is perhaps the most crucial key to revealing the secrets of its functions. Physical chemistry provides us with a number of techniques, including spectroscopy and X-ray diffraction, for such structural determination.

Another question for which we use physical chemistry principles concerns the binding of oxygen to hemoglobin. To understand how oxygen and other molecules, such as carbon monoxide, bind to the heme group, we need to investigate the coordination chemistry of transition-metal ions in general and complexes of iron in particular. For example, it is important to know which orbitals are involved in the iron–ligand complex and the reasons the binding constant for CO is some 200 times stronger than that for O_2. Knowledge of the molecular orbitals involved will also help explain hemoglobin's spectroscopic properties, including the purple color of venous blood (deoxyhemoglobin) and the red color of arterial blood (oxyhemoglobin).

A very important phenomenon is the cooperative nature of binding of oxygen to hemoglobin. Scientists noticed many years ago that oxygen molecules did not bind to the four heme groups independently; rather, the presence of the first molecule facilitates the binding of the second, and so on. Similarly, when the first oxygen molecule is released from a fully oxygenated hemoglobin, the remaining molecules come off with increasing ease. The biological function of cooperative binding is to increase the efficiency of the transport and release of oxygen. The kinetic and thermodynamic details of this phenomenon have been successfully accounted for by current theories based on *allosteric interaction*, which is the long-range interaction between spatially distant ligand-binding sites mediated by the structure of a protein molecule.

The function and efficiency of most proteins and enzymes depend critically on the pH. Hemoglobin is no exception. The CO_2–O_2 transport process in blood is buffered by the bicarbonate–carbonic acid system. Being amphoteric, that is, possessing the ability to act both as an acid and as a base, hemoglobin itself can act as a buffer. This process is an acid–base equilibrium reaction.

Finally, we may raise the following question: Of the numerous possible structures that a molecule this size can assume, why is only one predominant structure observed for hemoglobin? We must realize that in addition to the normal chemical bonds, many other types of molecular interaction, such as electrostatic forces, hydrogen bonding, and van der Waals forces, exist. In principle, a macromolecule can fold in many different ways; the native conformation represents the minimum Gibbs energy structure. The specificity in binding depends precisely on the environment at and near the active site, an environment that is maintained by the rest of the three-dimensional molecule. To appreciate how delicate the balance of these forces can be in some cases, consider the replacement of a glutamic acid by valine in the β chains of hemoglobin:

$$HOOC-(CH_2)_2-\overset{\overset{+}{N}H_3}{\underset{H}{C}}-COO^- \qquad \overset{CH_3}{\underset{CH_3}{C}}H-\overset{\overset{+}{N}H_3}{\underset{H}{C}}-COO^-$$

Glutamic acid   Valine

This seemingly small alteration is sufficient to produce a significant conformational change—an increase in the attraction between protein molecules, resulting in polymerization. The insoluble polymers that form distort red blood cells into a sickle shape, causing the symptoms of the disease sickle-cell anemia.

All these phenomena can be understood, at least in theory, by applying the principles of physical chemistry. Obviously, very different approaches are necessary for a thorough investigation of the chemistry of hemoglobin—or photosynthesis or atmospheric chemistry, for that matter. The point is that the principles of physical chemistry provide a foundation for the study of many exciting chemical and biochemical phenomena.

1.2 Units

Before we proceed with the study of physical chemistry, let us review the units chemists use for quantitative measurements.

For many years, scientists recorded measurements in *metric units*, which are related decimally, that is, by powers of 10. In 1960, however, the General Conference

Table 1.1
SI Base Units

Base Quantity	Name of Unit	Symbol
Length	meter	m
Mass	kilogram	kg
Time	second	s
Electrical current	ampere	A
Temperature	kelvin	K
Amount of substance	mole	mol
Luminous intensity	candela	cd

of Weights and Measures, the international authority on units, proposed a revised metric system called the *International System of Units* (abbreviated SI). The advantage of the SI system is that many of its units can be derived from natural constants. For example, the SI system defines meter (m) as the length equal to 1,650,763.73 wavelengths of radiation corresponding to a particular electronic transition from the *6d* to the *5p* orbital in krypton. The unit of time, the second, is equivalent to 9,192,631,770 cycles of the radiation associated with a certain electronic transition of the cesium atom. In contrast, the fundamental unit of mass, the kilogram (kg), is defined in terms of an artifact, not in terms of a naturally occurring phenomenon. One kilogram is the mass of a platinum–iridium alloy cylinder kept by the International Bureau of Weights and Measures in Sevres, France.

Table 1.1 gives the seven SI base units. Note that in SI units, temperature is given as K without the degree symbol, and the unit is plural—for example, 300 kelvins (300 K). A number of physical quantities can be derived from the list in Table 1.1. We shall discuss only a few of them here (see the inside front cover of the book).

Force

The unit of force in the SI system is the *newton* (N) (after the English physicist Sir Isaac Newton, 1642–1726), defined as the force required to give a mass of 1 kg an acceleration of 1 m s^{-2}; that is,

$$1 \text{ N} = 1 \text{ kg m s}^{-2}$$

It is interesting to note that one newton is approximately equal to the gravitational pull on an apple.

Pressure

Pressure is defined as

$$\text{pressure} = \frac{\text{force}}{\text{area}}$$

The SI unit of pressure is the *pascal* (Pa) (after the French mathematician and physicist Blaise Pascal, 1623–1662), where

$$1 \text{ Pa} = 1 \text{ N m}^{-2}$$

The following relations are exact:

$$1 \text{ bar} = 1 \times 10^5 \text{ Pa} = 100 \text{ kPa}$$

$$1 \text{ atm} = 1.01325 \times 10^5 \text{ Pa} = 101.325 \text{ kPa}$$

$$1 \text{ atm} = 1.01325 \text{ bar}$$

$$1 \text{ atm} = 760 \text{ torr}$$

The torr is named after the Italian mathematician Evangelista Torricelli (1608–1674). The standard atmosphere (1 atm) is used to define the normal melting point and boiling point of substances, and the bar is used to define standard states in physical chemistry. We shall use all of these units in this text.

Pressure is sometimes expressed in millimeters of mercury (mmHg): 1 mmHg is the pressure exerted by a column of mercury 1 mm high when its density is 13.5951 g cm^{-3} and the acceleration due to gravity is 980.67 cm s^{-2}. The relation between mmHg and torr is 1 mmHg = 1 torr.

One instrument that measures atmospheric pressure is the barometer. A simple barometer can be constructed by filling a long glass tube, closed at one end, with mercury, and then carefully inverting the tube in a dish of mercury, making sure that no air enters the tube. Some mercury will flow down into the dish, creating a vacuum at the top (Figure 1.1). The weight of the mercury column remaining in the tube is supported by atmospheric pressure acting on the surface of the mercury in the dish.

The device used to measure the pressure of gases other than the atmosphere is called a manometer. Its principle of operation is similar to that of a barometer. There are two types of manometers (Figure 1.2): the closed-tube manometer (Figure 1.2a) is normally used to measure pressures lower than atmospheric pressure, and the open-tube manometer (Figure 1.2b) is more suited for measuring pressures equal to or greater than atmospheric pressure.

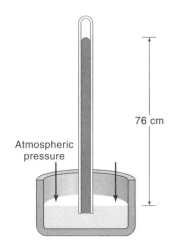

76 cm

Atmospheric pressure

Figure 1.1
A barometer for measuring atmospheric pressure. Above the mercury in the tube is a vacuum. The column of mercury is supported by atmospheric pressure.

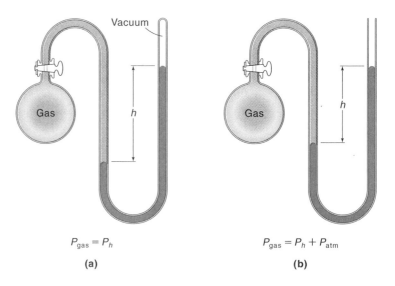

Vacuum

Gas

h

$P_{gas} = P_h$

(a)

Gas

h

$P_{gas} = P_h + P_{atm}$

(b)

Figure 1.2
Two types of manometers used to measure gas pressures. (a) Gas pressure is less than atmospheric pressure. (b) Gas pressure is greater than atmospheric pressure.

Energy

The SI unit of energy is the joule (J) (after the English physicist James Prescott Joule, 1818–1889). Because energy is the ability to do work and work is force times distance, we have

$$1\ J = 1\ N\ m$$

Some chemists have continued to use the non-SI unit of energy, calorie (cal), where 1 cal = 4.184 J.

1.3 Atomic Mass, Molecular Mass, and the Chemical Mole

By international agreement, an atom of the carbon-12 isotope, which has six protons and six neutrons, has a mass of exactly 12 atomic mass units (amu). One atomic mass unit is defined as a mass equal to exactly one-twelfth the mass of one carbon-12 atom. Experiments have shown that a hydrogen atom is only 8.400% as massive as the standard carbon-12 atom. Thus, the atomic mass of hydrogen must be $0.08400 \times 12 = 1.008$ amu. Similar experiments show that the atomic mass of oxygen is 16.00 amu and that of iron is 55.85 amu.

One atomic mass unit is also called the dalton.

When you look up the atomic mass of carbon in a table such as the one on the inside front cover of this book, you will find it listed as 12.01 amu rather than 12.00 amu. The reason for the difference is that most naturally occurring elements (including carbon) have more than one isotope. This means that when we measure the atomic mass of an element, we must generally settle for the average mass of the naturally occurring mixture of isotopes. For example, the natural abundances of carbon-12 and carbon-13 are 98.90% and 1.10%, respectively. The atomic mass of carbon-13 has been determined to be 13.00335 amu. Thus, the average atomic mass of carbon can be calculated as follows:

$$\text{average atomic mass of carbon} = (0.9890)(12.0000\ \text{amu})$$
$$+ (0.0110)(13.00335\ \text{amu})$$
$$= 12.01\ \text{amu}$$

Because there are many more carbon-12 isotopes than carbon-13 isotopes, the average atomic mass is much closer to 12 amu than 13 amu. Such an average is called the *weighted* average.

If we know the atomic masses of the component atoms, we can calculate the mass of a molecule. Thus, the molecular mass of H_2O is

$$2(1.008\ \text{amu}) + 16.00\ \text{amu} = 18.02\ \text{amu}$$

A *mole* (abbreviated mol) is the amount of substance that contains as many atoms, molecules, ions, or any other entities as there are atoms in exactly 12 g of carbon-12. It has been determined experimentally that the number of atoms in one mole of carbon-12 is 6.0221367×10^{23}. This number is called *Avogadro's number* (after the Italian physicist and mathematician Amedeo Avogadro, 1776–1856). Avogadro's number has no units, but dividing this number by one mole gives us Avogadro's constant (N_A), where

$$N_A = 6.0221367 \times 10^{23}\ \text{mol}^{-1}$$

For most purposes, N_A can be taken as 6.022×10^{23} mol^{-1}. The following examples indicate the number and kind of particles in one mole of any substance.

1. One mole of helium atoms contains 6.022×10^{23} He atoms.

2. One mole of water molecules contains 6.022×10^{23} H_2O molecules, or $2 \times (6.022 \times 10^{23})$ H atoms and 6.022×10^{23} O atoms.

3. One mole of NaCl contains 6.022×10^{23} NaCl units, or 6.022×10^{23} Na$^+$ ions and 6.022×10^{23} Cl$^-$ ions.

The *molar mass* of a substance is the mass in grams or kilograms of 1 mole of the substance. Thus, the molar mass of atomic hydrogen is 1.008 g mol^{-1}, of molecular hydrogen 2.016 g mol^{-1}, of hemoglobin 65,000 g mol^{-1}. In many calculations, molar masses are more conveniently expressed as kg mol^{-1}.

Suggestions for Further Reading

The following standard texts are useful references. The physical and biophysical texts contain many mathematical derivations of equations and provide experimental details for a number of topics covered in this book. The biochemistry texts provide the necessary background for the biological examples used in this book.

PHYSICAL CHEMISTRY
General
1. Alberty, R. A. and R. J. Silbey, *Physical Chemistry*, 3rd ed., John Wiley & Sons, Inc., New York, 2001.
2. Atkins, P. W. *Physical Chemistry*, 7th ed., W. H. Freeman and Company, New York, 2002.
3. Chang, R. *Physical Chemistry for the Chemical and Biological Sciences*, 3rd ed., University Science Books, Sausalito, CA, 2000.
4. Laidler, K. J., J. H. Meiser, and B. C. Sanctuary, *Physical Chemistry*, 4th ed., Houghton Mifflin Company, Boston, 2002.
5. Levine, I. N. *Physical Chemistry*, 5th ed., McGraw-Hill, Inc., New York, 2002.
6. McQuarrie, D. A. and J. D. Simon, *Physical Chemistry*, University Science Books, Sausalito, CA, 1997.
7. Noggle, J. H. *Physical Chemistry*, 3rd ed., Harper-Collins College Publishers, New York, 1996.
8. Winn, J. S. *Physical Chemistry*, HarperCollins College Publishers, New York, 1995.

Historical Development of Physical Chemistry
"One Hundred Years of Physical Chemistry," E. B. Wilson, Jr., *Am. Sci.* **74**, 70 (1986).
Laidler, K. J. *The World of Physical Chemistry*, Oxford University Press, New York, 1993.

Cobb, C. *Magic, Mayhem, and Mavericks: The Spirited History of Physical Chemistry*, Prometheus Books, Amherst, NY, 2002.

BIOPHYSICAL CHEMISTRY
9. Bergethon, P. R. *The Physical Basis of Biochemistry*, Springer-Verlag, New York, 1998.
10. Bergethon, P. R. and E. R. Simons, *Biophysical Chemistry: Molecules to Membranes*, Springer-Verlag, New York, 1990.
11. Cantor, C. R. and P. R. Schimmel, *Biophysical Chemistry*, W. H. Freeman and Company, San Francisco, CA, 1980.
12. Freifelder, D. *Physical Biochemistry*, 2nd ed., W. H. Freeman, New York, 1982.
13. van Holde, K. *Physical Biochemistry*, 2nd ed., Prentice Hall, Inc., Englewood Cliffs, NJ, 1984.
14. van Holde, K., W. C. Johnson, and P. S. Ho, *Principles of Physical Biochemistry*, Prentice Hall, Upper Saddle River, NJ, 1998.

BIOCHEMISTRY
15. Lehninger, A. L., D. C. Nelson, and M. M. Cox, *Principles of Biochemistry*, 3rd ed., W. H. Freeman, New York, 2000.
16. Mathews, C. K. and K. E. van Holde, *Biochemistry*, The Benjamin/Cummings Publishing Company, Inc., Menlo Park, CA, 1996.
17. Berg, J. M., J. T. Tymoczko, and L. Stryer, *Biochemistry*, 5th ed., W. H. Freeman and Company, New York, 2002.
18. Voet, D. and J. G. Voet, *Biochemistry*, 3rd ed., John Wiley & Sons, New York, 2004.

CHAPTER 2

Properties of Gases

The study of the behavior of gases has given rise to numerous chemical and physical theories. In many ways, the gaseous state is the easiest to investigate. In this chapter, we examine several gas laws based on experimental observations, introduce the concept of temperature, and discuss the kinetic theory of gases.

2.1 Some Basic Definitions

Before we discuss the gas laws, it is useful to define a few basic terms that will be used throughout the book. We often speak of the *system* in reference to a particular part of the universe in which we are interested. Thus, a system could be a collection of oxygen molecules in a container, a NaCl solution, a tennis ball, or a Siamese cat. Having defined a system, we call the rest of the universe the *surroundings*. There are three types of systems (Figure 2.1). An *open system* is one that can exchange both mass and energy with its surroundings. A *closed system* is one that does not exchange mass with its surroundings but can exchange energy. An *isolated system* is one that can exchange neither mass nor energy with its surroundings. To completely define a system, we need to understand certain experimental variables, such as pressure, volume, temperature, and composition, that collectively describe the *state of the system*.

A system is separated from the surroundings by definite boundaries such as walls or surfaces.

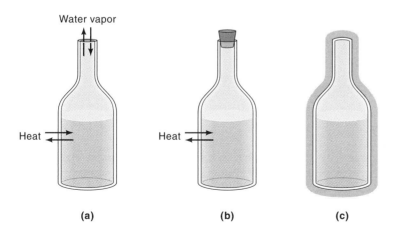

Figure 2.1
(a) An open system allows the exchange of both mass and energy; (b) a closed system allows the exchange of energy but not mass; and (c) an isolated system allows exchange of neither mass nor energy.

Most of the properties of matter may be divided into two classes: *extensive properties* and *intensive properties*. Consider, for example, two beakers containing the same amounts of water at the same temperature. If we combine these two systems by pouring the water from one beaker to the other, we find that the volume of the water is doubled and so is its mass. On the other hand, the temperature and the density of the water do not change. Properties whose values are directly proportional to the amount of the material present in the system are called extensive properties; those that do not depend on the amount are called intensive properties. Extensive properties include mass, area, volume, energy, and electrical charge. As mentioned, temperature and density are both intensive properties; so are pressure and electrical potential.

2.2 An Operational Definition of Temperature

Temperature is a very important quantity in many branches of science, and not surprisingly, it can be defined in different ways. Daily experience tells us that temperature is a measure of coldness and hotness, but for our purposes we need a more precise operational definition. Consider the following system of a container of gas A. The walls of the container are flexible so that its volume can expand and contract. This is a closed system that allows heat but not mass to flow into and out of the container. The initial pressure (P) and volume (V) are P_A and V_A. Now we bring the container in contact with a similar container of gas B at P_B and V_B. Heat exchange will take place until thermal equilibrium is reached. At equilibrium, the pressure and volume of A and B will be altered to P'_A, V'_A and P'_B, V'_B. It is possible to remove container A temporarily, readjust its pressure and volume to P''_A and V''_A, and still have A in thermal equilibrium with B at P'_B and V'_B. In fact, an infinite set of such values $(P'_A, V'_A), (P''_A, V''_A), (P'''_A, V'''_A), \ldots$ can be obtained that will satisfy the equilibrium conditions. Figure 2.2 shows a plot of these points.

For all these states of A to be in thermal equilibrium with B, they must have the same value of the variable we call temperature. It follows that if two systems are in thermal equilibrium with a third system, they must also be in thermal equilibrium with each other. This statement is generally known as the *zeroth law of thermodynamics*. The curve in Figure 2.2 is the locus of all the points that represent the states that can be in thermal equilibrium with system B. Such a curve is called an *isotherm*, or "same temperature." At another temperature, a different isotherm is obtained.

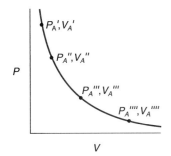

Figure 2.2
Plot of pressure versus volume at constant temperature for a given amount of a gas. Such a graph is called an isotherm.

2.3 Ideal Gases

In this section we shall briefly examine the various gas laws formulated for the behavior of an ideal gas.

Boyle's Law

In a 1662 study of the physical behavior of gases, the English chemist Robert Boyle (1627–1691) found that the volume (V) of a given amount of gas at constant temperature is inversely proportional to its pressure (P):

$$V \propto \frac{1}{P}$$

or

$$PV = \text{constant} \qquad (2.1)$$

Equation 2.1 is known as *Boyle's law*. A plot of P versus V at a given temperature gives a hyperbola, which is the isotherm illustrated in Figure 2.2.

Boyle's law is used to predict the pressure of a gas when its volume changes and vice versa. Letting the initial values of pressure and volume be P_1 and V_1 and the final values of pressure and volume be P_2 and V_2, we have

$$P_1 V_1 = P_2 V_2 \quad \text{(constant } n \text{ and temperature)} \tag{2.2}$$

where n is the number of moles of the gas present.

Charles' and Gay-Lussac's Law

Boyle's law depends on the amount of gas and the temperature of the system remaining constant. But suppose the temperature changes. How does a change in temperature affect the volume and pressure of a gas? Let us first look at the effect of temperature on the volume of a gas. The earliest investigators of this relationship were the French physicists Jacques Alexandre Charles (1746–1823) and Joseph Louis Gay-Lussac (1778–1850). Their studies showed that, at constant pressure, the volume of a gas sample expands when heated and contracts when cooled. The quantitative relations involved in changes in gas temperature and volume turn out to be remarkably consistent. For example, we observe an interesting phenomenon when we study the temperature–volume relationship at various pressures. At any given pressure, the plot of volume versus temperature yields a straight line. By extending the line to zero volume, we find the intercept on the temperature axis to be $-273.15°C$. At any other pressure, we obtain a different straight line for the volume–temperature plot, but we get the *same* zero-volume temperature intercept at $-273.15°C$ (Figure 2.3).

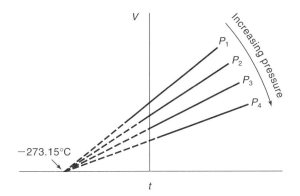

Figure 2.3
Plots of the volume of a given amount of gas versus temperature (t) at different pressures. All gases ultimately condense if they are cooled to low enough temperatures. When these lines are extrapolated, they all converge at the point representing zero volume and a temperature of $-273.15°C$.

(In practice, we can measure the volume of a gas over only a limited temperature range, because all gases condense at low temperatures to form liquids.)

In 1848, the Scottish mathematician and physicist William Thomson (Lord Kelvin, 1824–1907) realized the significance of this phenomenon. He identified $-273.15°C$ as *absolute zero*, which is theoretically the lowest attainable temperature. Then he set up an *absolute temperature scale*, now called the kelvin temperature scale, with absolute zero as the starting point. On the kelvin scale, one kelvin (K) is equal *in magnitude* to one degree Celsius. The only difference between the absolute temperature scale and the Celsius scale is that the zero position is shifted. The relation between the two scales is

$$T/\text{K} = t/°\text{C} + 273.15$$

Dividing the symbol by the unit gives us a pure number. Thus, if $T = 298$ K, then $T/\text{K} = 298$.

Important points on the two scales match up as follows:

	Kelvin Scale	Celsius Scale
Absolute zero	0 K	−273.15°C
Freezing point of water	273.15 K	0°C
Boiling point of water	373.15 K	100°C

The normal freezing point and normal boiling point are measured at 1 atm pressure.

In most cases, we shall use 273 instead of 273.15 as the term relating the two scales. By convention, we use T to denote absolute (kelvin) temperature and t to indicate temperature on the Celsius scale. As we shall soon see, the absolute zero of temperature has major theoretical significance; absolute temperatures must be used in gas law problems and thermodynamic calculations.

At constant pressure, the volume of a given amount of gas is directly proportional to the absolute temperature:

$$V \propto T$$

or

$$\frac{V}{T} = \text{constant} \tag{2.3}$$

Equation 2.3 is known as *Charles' law*, or the *law of Charles and Gay-Lussac*. An alternative form of Charles' law relates the pressure of a given amount of gas to its temperature at constant volume:

$$P \propto T$$

or

$$\frac{P}{T} = \text{constant} \tag{2.4}$$

Equations 2.3 and 2.4 permit us to relate the volume–temperature and pressure–temperature values of a gas in states 1 and 2 as follows:

$$\frac{V_1}{T_1} = \frac{V_2}{T_2} \quad \text{(constant } n \text{ and } P) \tag{2.5}$$

$$\frac{P_1}{T_1} = \frac{P_2}{T_2} \quad \text{(constant } n \text{ and } V) \tag{2.6}$$

A practical consequence of Equation 2.6 is that automobile tire pressure should be checked only when the car has been idle for a while. After a long drive, tires become quite hot and the air pressure in them rises.

Avogadro's Law

Another important gas law was formulated by Amedeo Avogadro in 1811. He proposed that equal volumes of gases at the same temperature and pressure contain the same number of molecules. This concept means that

$$V \propto n$$

or

$$\frac{V}{n} = \text{constant} \quad (\text{constant } T \text{ and } P) \tag{2.7}$$

Equation 2.7 is known as *Avogadro's law*.

The Ideal-Gas Equation

According to Equations 2.1, 2.3, and 2.7, the volume of a gas depends on the pressure, temperature, and number of moles as follows:

$$V \propto \frac{1}{P} \quad (\text{constant } T \text{ and } n) \quad (\text{Boyle's law})$$

$$V \propto T \quad (\text{constant } P \text{ and } n) \quad (\text{Charles' law})$$

$$V \propto n \quad (\text{constant } T \text{ and } P) \quad (\text{Avogadro's law})$$

Therefore, V must be proportional to the product of these three terms, that is,

$$V \propto \frac{nT}{P}$$

$$V = R\frac{nT}{P}$$

or

$$PV = nRT \tag{2.8}$$

where R, a proportionality constant, is the *gas constant*. Equation 2.8 is called the *ideal-gas equation*. The ideal-gas equation is an example of an *equation of state*, which provides the mathematical relationships among the properties that define the state of the system, such as P, T, and V.

The value of R can be obtained as follows. Experimentally, it is found that 1 mole of an ideal gas occupies 22.414 L at 1 atm and 273.15 K (a condition known as *standard temperature and pressure*, or STP). Thus,

$$R = \frac{(1 \text{ atm})(22.414 \text{ L})}{(1 \text{ mol})(273.15 \text{ K})} = 0.08206 \text{ L atm K}^{-1} \text{ mol}^{-1}$$

To express R in units of J K^{-1} mol^{-1}, we use the conversion factors

$$1 \text{ atm} = 1.01325 \times 10^5 \text{ Pa}$$

$$1 \text{ L} = 1 \times 10^{-3} \text{ m}^3$$

and obtain

$$R = \frac{(1.01325 \times 10^5 \text{ N m}^{-2})(22.414 \times 10^{-3} \text{ m}^3)}{(1 \text{ mol})(273.15 \text{ K})}$$

$$= 8.314 \text{ N m K}^{-1} \text{ mol}^{-1}$$

$$= 8.314 \text{ J K}^{-1} \text{ mol}^{-1} \quad (1 \text{ J} = 1 \text{ N m})$$

From the two values of R, we can write

$$0.08206 \text{ L atm K}^{-1} \text{ mol}^{-1} = 8.314 \text{ J K}^{-1} \text{ mol}^{-1}$$

or

$$1 \text{ L atm} = 101.3 \text{ J}$$

and

$$1 \text{ J} = 9.87 \times 10^{-3} \text{ L atm}$$

Example 2.1

Air entering the lungs ends up in tiny sacs called alveoli, from which oxygen diffuses into the blood. The average radius of the alveoli is 0.0050 cm, and the air inside contains 14 mole percent oxygen. Assuming that the pressure in the alveoli is 1.0 atm and the temperature is 37°C, calculate the number of oxygen molecules in one of the alveoli.

ANSWER

The volume of one alveolus is

$$V = \frac{4}{3}\pi r^3 = \frac{4}{3}\pi(0.0050 \text{ cm})^3$$

$$= 5.2 \times 10^{-7} \text{ cm}^3 = 5.2 \times 10^{-10} \text{ L} \quad (1 \text{ L} = 10^3 \text{ cm}^3)$$

The number of moles of air in one alveolus is given by

$$n = \frac{PV}{RT} = \frac{(1.0 \text{ atm})(5.2 \times 10^{-10} \text{ L})}{(0.08206 \text{ L atm K}^{-1} \text{ mol}^{-1})(310 \text{ K})} = 2.0 \times 10^{-11} \text{ mol}$$

Because the air inside the alveolus is 14% oxygen, the number of oxygen molecules is

$$2.0 \times 10^{-11} \text{ mol air} \times \frac{14\% \text{ O}_2}{100\% \text{ air}} \times \frac{6.022 \times 10^{23} \text{ O}_2 \text{ molecules}}{1 \text{ mol O}_2}$$

$$= 1.7 \times 10^{12} \text{ O}_2 \text{ molecules}$$

Dalton's Law of Partial Pressures

So far, we have discussed the pressure–volume–temperature behavior of a pure gas. Frequently, however, we work with mixtures of gases. For example, a chemist researching the depletion of ozone in the atmosphere must deal with several gaseous components. For a system containing two or more different gases, the total pressure (P_T) is the sum of the individual pressures that each gas would exert if it were alone and occupied the same volume. Thus,

$$P_T = P_1 + P_2 + \cdots$$

$$P_T = \sum_i P_i \tag{2.9}$$

where $P_1, P_2, \ldots$ are the individual or *partial* pressures of components $1, 2, \ldots$ and $\sum$ is the summation sign. Equation 2.9 is known as *Dalton's law of partial pressures* (after the English chemist and mathematician John Dalton, 1766–1844).

Consider a system containing two gases (1 and 2) at temperature T and volume V. The partial pressures of the gases are P_1 and P_2, respectively. From Equation 2.8,

$$P_1 V = n_1 RT \quad \text{or} \quad P_1 = \frac{n_1 RT}{V}$$

$$P_2 V = n_2 RT \quad \text{or} \quad P_2 = \frac{n_2 RT}{V}$$

where n_1 and n_2 are the numbers of moles of the two gases. According to Dalton's law,

$$P_T = P_1 + P_2$$
$$= n_1 \frac{RT}{V} + n_2 \frac{RT}{V}$$
$$= (n_1 + n_2) \frac{RT}{V}$$

Dividing the partial pressures by the total pressure and rearranging, we get

$$P_1 = \frac{n_1}{n_1 + n_2} P_T = x_1 P_T$$

and

$$P_2 = \frac{n_2}{n_1 + n_2} P_T = x_2 P_T$$

where x_1 and x_2 are the mole fractions of gases 1 and 2. A mole fraction, defined as the ratio of the number of moles of one gas to the total number of moles of all gases present, is a dimensionless quantity. Furthermore, by definition, the sum of all the mole fractions in a mixture must be unity, that is

$$\sum_i x_i = 1 \tag{2.10}$$

In general, in a mixture of gases the partial pressure of the ith component, P_i, is given by

$$P_i = x_i P_T \tag{2.11}$$

How are partial pressures determined? A manometer can measure only the total pressure of a gaseous mixture. To obtain partial pressures, we need to know the mole fractions of the components. The most direct method of measuring partial pressures is using a mass spectrometer. The relative intensities of the peaks in a mass spectrum are directly proportional to the amounts, and hence to the mole fractions, of the gases present.

The gas laws played a key role in the development of atomic theory, and many practical illustrations of them appear in everyday life. The two brief examples below are particularly important to scuba divers. Seawater has a slightly higher density than fresh water—approximately 1.03 g mL^{-1} compared with 1.00 g mL^{-1}. The pressure exerted by a 33-ft (10-m) column of seawater is equivalent to 1 atm pressure. What would happen if a diver were to rise to the surface rather quickly, holding his breath? If the ascent started at 40 ft under water, the decrease in pressure from this depth to the surface would be (40 ft/33 ft) × 1 atm, or 1.2 atm. Assuming constant temperature, when the diver reached the surface, the volume of air trapped in his lungs would have increased by a factor of (1 + 1.2) atm/1 atm, or 2.2 times! This sudden expansion of air could damage or rupture the membranes of his lungs, seriously injuring or killing the diver.

Dalton's law has a direct application to scuba diving. The partial pressure of oxygen in air is approximately 0.2 atm. Because oxygen is essential for our survival, we may have a hard time believing that it could be harmful to breathe in more than normal. In fact, the toxicity of oxygen is well documented.* Physiologically, our bodies function best when the partial pressure of oxygen is 0.2 atm. For this reason, the composition of the air in a scuba tank is adjusted when the diver is submerged. For example, at a depth where the total pressure (hydrostatic plus atmospheric) is 4 atm, the oxygen content in the air supply should be reduced to 5% by volume to maintain the optimal partial pressure (0.05 × 4 atm = 0.2 atm). At a greater depth, the oxygen content must be even lower. Although nitrogen would seem to be the obvious choice for mixing with oxygen in a scuba tank because it is the major component of air, it is not the best choice. When the partial pressure of nitrogen exceeds 1 atm, a sufficient amount will dissolve in the blood to cause *nitrogen narcosis*. Symptoms of this condition, which resembles alcohol intoxication, include light-headedness and impaired judgment. Divers suffering from nitrogen narcosis have been known to do strange things, such as dancing on the sea floor and chasing sharks. For this reason, helium is usually employed to dilute oxygen in diving tanks. Helium, an inert gas, is much less soluble in blood than nitrogen, and it does not produce narcotic effects.

2.4 Real Gases

The ideal-gas equation holds only for gases that have the following properties: (1) the gas molecules possess negligible volume, and (2) there is no interaction, attractive or repulsive, among the molecules. Obviously, no such gases exist. Nevertheless, Equation 2.8 is quite useful for many gases at high temperatures or moderately low pressures (≤ 10 atm).

When a gas is being compressed, the molecules are brought closer to one another, and the gas will deviate appreciably from ideal behavior. One way to measure the deviation from ideality is to plot the *compressibility factor* (Z) of a gas versus pressure. Starting with the ideal-gas equation, we write

$$PV = nRT$$

or

---

*At partial pressures above 2 atm, oxygen becomes toxic enough to produce convulsions and coma. Years ago, newborn infants placed in oxygen tents often developed *retrolental fibroplasia*, damage of the retinal tissues by excess oxygen. This damage usually resulted in partial or total blindness.

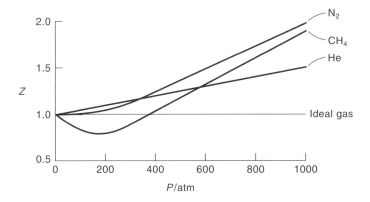

Figure 2.4
Plot of the compressibility factor versus pressure for real gases and an ideal gas at 273 K. Note that for an ideal gas $Z = 1$, no matter how great the pressure.

$$Z = \frac{PV}{nRT} = \frac{P\bar{V}}{RT} \qquad (2.12)$$

where $\bar{V}$ is the molar volume of the gas (V/n) or the volume of 1 mole of the gas at the specified temperature and pressure. For an ideal gas, $Z = 1$ for any value of P at a given T. However, as Figure 2.4 shows, the compressibility factors for real gases exhibit fairly divergent dependence on pressure. At low pressures, the compressibility factors of most gases are close to unity. In fact, in the limit of P approaching zero, we have $Z = 1$ for all gases. This finding is expected because all real gases behave ideally at low pressures. As pressure increases, some gases have $Z < 1$, which means that they are easier to compress than an ideal gas. Then, as pressure increases further, all gases have $Z > 1$. Over this region, the gases are harder to compress than an ideal gas. These behaviors are consistent with our understanding of intermolecular forces. In general, attractive forces are long-range forces, whereas repulsive forces operate only within a short range (more on this topic in Chapter 13). When molecules are far apart (for example, at low pressures), the predominant intermolecular interaction is attraction. As the distance of separation between molecules decreases, the repulsive interaction among molecules becomes more significant.

Over the years, considerable effort has gone into modifying the ideal-gas equation for real gases. Of the numerous such equations proposed, we shall consider two: the van der Waals equation and the virial equation of state.

The van der Waals Equation

The *van der Waals equation of state* (after the Dutch physicist Johannes Diderik van der Waals, 1837–1923) attempts to account for the finite volume of individual molecules in a nonideal gas and the attractive forces between them.

$$\left(P + \frac{an^2}{V^2}\right)(V - nb) = nRT \qquad (2.13)$$

The pressure exerted by the individual molecules on the walls of the container depends on both the frequency of molecular collisions with the walls and the momentum imparted by the molecules to the walls. Both contributions are diminished by the attractive intermolecular forces (Figure 2.5). In each case, the reduction in pressure

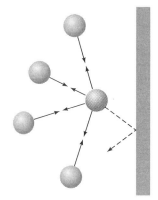

Figure 2.5
Effect of intermolecular forces on the pressure exerted by a gas. The speed of a molecule that is moving toward the container wall (red sphere) is reduced by the attractive forces exerted by its neighbors (gray spheres). Consequently, the impact this molecule makes with the wall is not as great as it would be if no intermolecular forces were present. In general, the measured gas pressure is lower than the pressure the gas would exert if it behaved ideally.

Table 2.1
van der Waals Constants and Boiling Points of Some Substances

Substance	a/atm $\cdot$ L$^2 \cdot$ mol^{-2}	b/L $\cdot$ mol^{-1}	Boiling Point/K
He	0.0341	0.0237	4.2
Ne	0.214	0.0174	27.2
Ar	1.34	0.0322	87.3
H$_2$	0.240	0.0264	20.3
N$_2$	1.35	0.0386	77.4
O$_2$	1.34	0.0312	90.2
CO	1.45	0.0395	83.2
CO$_2$	3.60	0.0427	195.2
CH$_4$	2.26	0.0430	109.2
H$_2$O	5.47	0.0305	373.15
NH$_3$	4.25	0.0379	239.8

depends on the number of molecules present or the density of the gas, n/V, so that

$$\text{reduction in pressure due to attractive forces} \propto \left(\frac{n}{V}\right)\left(\frac{n}{V}\right)$$

$$= a\frac{n^2}{V^2}$$

where a is a proportionality constant.

Note that in Equation 2.13, P is the *measured* pressure of the gas and $(P + an^2/V^2)$ would be the pressure of the gas if there were no intermolecular forces present. Because an^2/V^2 must have units of pressure, a is expressed as atm L^2 mol^{-2}. To allow for the finite volume of molecules, we replace V in the ideal-gas equation with $(V - nb)$, where nb represents the total effective volume of n moles of the gas. Thus, nb must have the unit of volume, and b has the units L mol^{-1}. Both a and b are constants characteristic of the gas under study. Table 2.1 lists the values of a and b for several gases. The value of a is related to the magnitude of attractive forces. Using the boiling point as a measure of the strength of intermolecular forces (the higher the boiling point, the stronger the intermolecular forces), we see that there is a rough correlation between the values of a and the boiling points of these substances. The quantity b is more difficult to interpret. Although b is proportional to the size of the molecule, the correlation is not always straightforward. For example, the value of b for helium is 0.0237 L mol^{-1} and that for neon is 0.0174 L mol^{-1}. Based on these values, we might expect that helium is larger than neon, which we know is not true.

The van der Waals equation is valid over a wider range of pressure and temperature than is the ideal-gas equation. Furthermore, it provides a molecular interpretation for the equation of state. At very high pressures and low temperatures, the van der Waals equation also becomes unreliable.

The Virial Equation of State

Another way of representing gas nonideality is the *virial equation of state*. In this relationship, the compressibility factor is expressed as a series expansion in inverse powers of molar volume $\overline{V}$:

$$Z = 1 + \frac{B}{\overline{V}} + \frac{C}{\overline{V}^2} + \frac{D}{\overline{V}^3} + \cdots \tag{2.14}$$

where B, C, D, are called the second, third, fourth, ... virial coefficients. The first virial coefficient is 1. The second and higher virial coefficients are all temperature dependent. For a given gas, they are evaluated from the P–V–T data of the gas by a curve-fitting procedure using a computer. For an ideal gas, the second and higher virial coefficients are zero and Equation 2.14 becomes Equation 2.8.

An alternate form of the virial equation is given by a series expansion of the compressibility factor in terms of the pressure, P:

$$Z = 1 + B'P + C'P^2 + D'P^3 + \cdots \qquad (2.15)$$

Because P and V are related, it is not surprising that relationships exist between B and B', C and C', and so on. In each equation, the values of the coefficients decrease rapidly. For example, in Equation 2.15, the magnitude of the coefficients is such that $B' \gg C' \gg D'$ so that at pressures between zero and 10 atm, say, we need to include only the second term provided the temperature is not very low:

$$Z = 1 + B'P \qquad (2.16)$$

Equations 2.13 and 2.14 or 2.15 exemplify two rather different approaches. The van der Waals equation accounts for the nonideality of gases by correcting for the finite molecular volume and intermolecular forces. Although these corrections do result in a definite improvement over the ideal gas equation, Equation 2.13 is still an approximate equation. The reason is that our present knowledge of intermolecular forces is insufficient to quantitatively explain macroscopic behavior. Of course, we could further improve this equation by adding more corrective terms; indeed, numerous other equations of state have been proposed since van der Waals first presented his analysis. On the other hand, Equation 2.14 is accurate for real gases, but it does not provide us with any direct molecular interpretation. The nonideality of the gas is accounted for mathematically by a series expansion in which the coefficients $B, C, \ldots$ can be determined experimentally. These coefficients do not have any physical meaning, although they can be related to intermolecular forces in an indirect way. Thus, our choice in this case is between an approximate equation that gives us some physical insight or an equation that describes the gas behavior accurately (if the coefficients are known), but tells us nothing about molecular behavior.

Example 2.2

Calculate the molar volume of methane at 300 K and 100 atm, given that the second virial coefficient (B) of methane is -0.042 L mol^{-1}. Compare your result with that obtained using the ideal-gas equation.

ANSWER

From Equation 2.14, neglecting terms containing $C, D, \ldots$,

$$Z = 1 + \frac{B}{\overline{V}}$$

$$= 1 + \frac{BP}{RT}$$

$$= 1 + \frac{(-0.042 \text{ L mol}^{-1})(100 \text{ atm})}{(0.08206 \text{ L atm K}^{-1} \text{ mol}^{-1})(300 \text{ K})}$$

$$= 1 - 0.17 = 0.83$$

$$\bar{V} = \frac{ZRT}{P}$$

$$= \frac{(0.83)(0.08206 \text{ L atm K}^{-1} \text{ mol}^{-1})(300 \text{ K})}{100 \text{ atm}}$$

$$= 0.20 \text{ L mol}^{-1}$$

For an ideal gas,

$$\bar{V} = \frac{RT}{P}$$

$$= \frac{(0.08206 \text{ L atm K}^{-1} \text{ mol}^{-1})(300 \text{ K})}{100 \text{ atm}}$$

$$= 0.25 \text{ L mol}^{-1}$$

COMMENT

At 100 atm and 300 K, methane is more compressible than an ideal gas ($Z = 0.83$ compared with $Z = 1$) due to the attractive intermolecular forces between the CH_4 molecules.

2.5 Condensation of Gases and the Critical State

The condensation of gas to liquid is a familiar phenomenon. The first quantitative study of the pressure–volume relationship of this process was made in 1869 by the Irish chemist Thomas Andrews (1813–1885). He measured the volume of a given amount of carbon dioxide as a function of pressure at various temperatures and obtained a series of isotherms like those shown in Figure 2.6. At high temperatures, the curves are roughly hyperbolic, indicating that the gas obeys Boyle's law. As the temperature is lowered, deviations become evident, and a drastically different behavior is observed at T_4. Moving along the isotherm from right to left, we see that

Figure 2.6
Isotherms of carbon dioxide at various temperatures (temperature increases from T_1 to T_7). The critical temperature is T_5. Above this temperature carbon dioxide cannot be liquefied no matter how great the pressure.

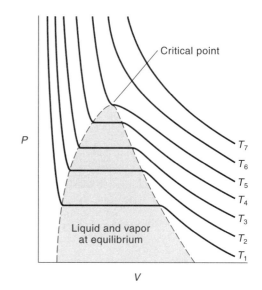

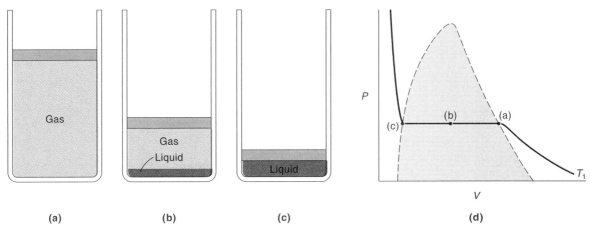

Figure 2.7
The liquefaction of carbon dioxide at T_1 (see Figure 2.6). At (a), the first drop of the liquid appears. From (b) to (c) the gas is gradually and completely converted to liquid at constant pressure. Beyond (c) the volume decreases only slightly with increasing pressure because liquids are highly incompressible. As temperature increases, the horizontal line becomes shorter until it becomes a point at T_5, the critical temperature.

although the volume of the gas decreases with pressure, the product PV is no longer a constant (because the curve is no longer a hyperbola). Increasing the pressure further, we reach a point that is the intersection between the isotherm and the dashed curve on the right. If we could observe this process, we would note the formation of liquid carbon dioxide at this pressure. With the pressure held constant, the volume continues to decrease (as more vapor is converted to liquid) until all the vapor has been condensed. Beyond this point (the intersection between the horizontal line and the dashed curve on the left), the system is entirely liquid, and any further increase in pressure will result in only a very small decrease in volume, because liquids are much less compressible than gases (Figure 2.7).

The pressure corresponding to the horizontal line (the region in which vapor and liquid coexist) is called the *equilibrium vapor pressure* or simply the *vapor pressure* of the liquid at the temperature of the experiment. The length of the horizontal line decreases with increasing temperature. At a particular temperature (T_5 in this case), the isotherm is tangential to the dashed curve and only one phase (the gas phase) is present. The horizontal line is now at a point known as the *critical point*. The corresponding temperature, pressure, and volume at this point are called the critical temperature (T_c), critical pressure (P_c), and critical volume (V_c), respectively. The *critical temperature* is the temperature above which no condensation can occur no matter how great the pressure. The critical constants of several gases are listed in Table 2.2.* Note that the critical volume is usually expressed as a molar quantity, called the molar critical volume ($\bar{V}_c$), given by V_c/n, where n is the number of moles of the substance present.

The phenomenon of condensation and the existence of a critical temperature are direct consequences of the nonideal behavior of gases. After all, if molecules did not attract one another, no condensation would occur and if molecules had no volume, then we would not be able to observe liquids. As mentioned earlier, the nature of molecular interaction is such that the force among molecules is attractive when they

---

*The van der Waals constants of a gas (see Equation 2.13) can be obtained from its critical constants. For mathematical details see the physical chemistry texts listed in Chapter 1.

Table 2.2
Critical Constants of Some Substances

Substance	P_c/atm	$\bar{V}_c$/L·mol^{-1}	T_c/K
He	2.25	0.0578	5.2
Ne	26.2	0.0417	44.4
Ar	49.3	0.0753	151.0
H_2	12.8	0.0650	32.9
N_2	33.6	0.0901	126.1
O_2	50.8	0.0764	154.6
CO	34.5	0.0931	132.9
CO_2	73.0	0.0957	304.2
CH_4	45.4	0.0990	190.2
H_2O	217.7	0.0560	647.6
NH_3	109.8	0.0724	405.3
SF_6	37.6	0.2052	318.7

are relatively far apart, but as they get closer to one another (for example, a liquid under pressure) this force becomes repulsive, because of electrostatic repulsions between nuclei and between electrons. In general, the attractive force reaches a maximum at a certain finite intermolecular distance. At temperatures below T_c, it is possible to compress the gas and bring the molecules within this attractive range, where condensation will occur. Above T_c, the kinetic energy of the gas molecules is such that they will always be able to break away from this attraction and no condensation can take place. Figure 2.8 shows the critical phenomenon of sulfur hexafluoride (SF_6).

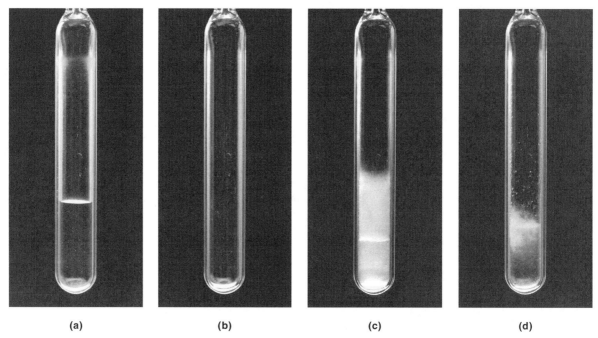

(a) (b) (c) (d)

Figure 2.8
The critical phenomenon of sulfur hexafluoride ($T_c = 45.5°C$; $P_c = 37.6$ atm). (a) Below the critical temperature, a clear liquid phase is visible. (b) Above the critical temperature, the liquid phase disappears. (c) The substance is cooled just below its critical temperature. (d) Finally, the liquid phase reappears.

In recent years, there has been much interest in the practical applications of *supercritical fluids* (SCF), that is, of the state of matters above the critical temperature. One of the most studied SCFs is carbon dioxide. Under appropriate conditions of temperature and pressure, SCF CO_2 can be used as a solvent for removing caffeine from raw coffee beans and cooking oil from potato chips to produce crisp, oil-free chips. It is also being used in environmental cleanups because it dissolves chlorinated hydrocarbons. SCFs of CO_2, NH_3, and certain hydrocarbons such as hexane and heptane are used in chromatography. SCF CO_2 has been shown to be an effective carrier medium for substances such as antibiotics and hormones, which are unstable at the high temperatures required for normal chromatographic separations.

2.6 Kinetic Theory of Gases

The study of gas laws exemplifies the phenomenological, macroscopic approach to physical chemistry. Equations describing the gas laws are relatively simple, and experimental data are readily accessible. Yet studying gas laws gives us no real physical insight into processes that occur at the molecular level. Although the van der Waals equation attempts to account for nonideal behavior in terms of intermolecular interactions, it does so in a rather vague manner. It does not answer such questions as: How is the pressure of a gas related to the motion of individual molecules, and why do gases expand when heated at constant pressure? The next logical step, then, is to try to explain the behavior of gases in terms of the dynamics of molecular motion. To interpret the properties of gas molecules in a more quantitative manner, we turn to the kinetic theory of gases.

The Model

Any time we try to develop a theory to account for experimental observations, we must first define our system. If we do not understand all the properties of a system, as is usually the case, we must make a number of assumptions. Our *model for the kinetic theory of gases* is based on the following assumptions:

1. A gas is made up of a great number of atoms or molecules, separated by distances that are large compared to their size.

2. The molecules have mass, but their volume is negligibly small.

3. The molecules are constantly in random motion.

4. Collisions among molecules and between molecules and the walls of the container are *elastic*; that is, kinetic energy may be transferred from one molecule to another, but it is not converted to other forms of energy.

5. There is no interaction, attractive or repulsive, between the molecules.

Assumptions 2 and 5 should be familiar from our discussion of ideal gases. The difference between the ideal-gas laws and the kinetic theory of gases is that for the latter, we shall use the foregoing assumptions in an explicit manner to derive expressions for macroscopic properties, such as pressure and temperature, in terms of the motion of individual molecules.

Pressure of a Gas

Using the model for the kinetic theory of gases, we can derive an expression for the pressure of a gas in terms of its molecular properties. Consider an ideal gas

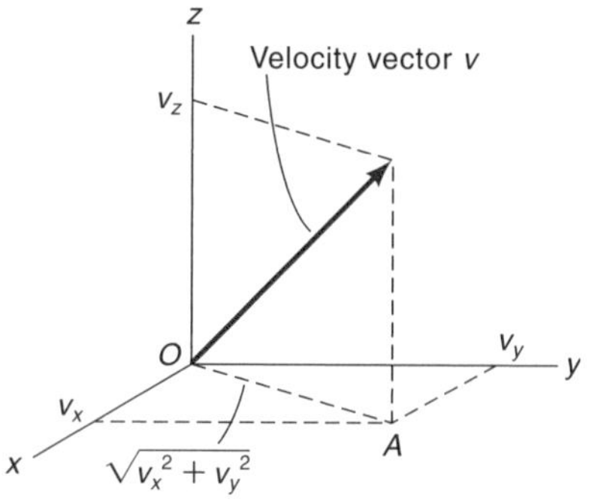

Figure 2.9
Velocity vector v and its components along the x, y, and z directions.

made up of N molecules, each of mass m, confined in a cubic box of length l. At any instant, the molecular motion inside the container is completely random. Let us analyze the motion of a particular molecule with velocity v. Because velocity is a *vector* quantity—it has both magnitude and direction—v can be resolved into three mutually perpendicular components v_x, v_y, and v_z. These three components give the rates at which the molecule is moving along the x, y, and z directions, respectively; v is simply the resultant velocity (Figure 2.9). The projection of the velocity vector on the xy plane is $\overline{OA}$, which, according to Pythagoras' theorem, is given by

$$\overline{OA}^2 = v_x^2 + v_y^2$$

Similarly,

$$v^2 = \overline{OA}^2 + v_z^2$$
$$= v_x^2 + v_y^2 + v_z^2 \tag{2.17}$$

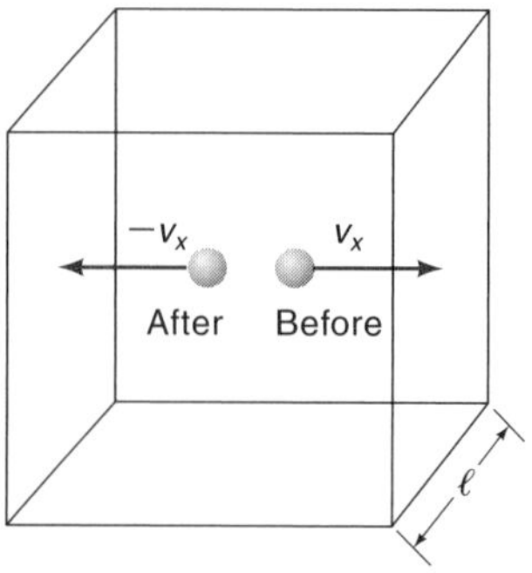

Figure 2.10
Change in velocity upon collision of a molecule moving with v_x with the wall of the container.

Let us for the moment consider the motion of a molecule only along the x direction. Figure 2.10 shows the changes that take place when the molecule collides with the wall of the container (the yz plane) with velocity component v_x. Because the collision is elastic, the velocity after collision is the same as before but opposite in direction. The momentum of the molecule is mv_x, where m is its mass, so that the *change* in momentum is given by

$$mv_x - m(-v_x) = 2mv_x$$

The sign of v_x is positive when the molecule moves from left to right and negative when it moves in the opposite direction. Immediately after the collision, the molecule will take time l/v_x to collide with the other wall, and in time $2l/v_x$ the molecule will strike the same wall again.* Thus, the frequency of collision between the molecule and a given wall (that is, the number of collisions per unit time) is $v_x/2l$, and the change in momentum per unit time is $(2mv_x)(v_x/2l)$, or mv_x^2/l. According to Newton's second law of motion,

$$\text{force} = \text{mass} \times \text{acceleration}$$
$$= \text{mass} \times \text{distance} \times \text{time}^{-2}$$
$$= \text{momentum time}^{-1}$$

---

*We assume that the molecule does not collide with other molecules along the way. A more rigorous treatment including molecular collision gives exactly the same result.

Therefore, the force exerted by one molecule on one wall as a result of the collision is mv_x^2/l, and the total force due to N molecules is Nmv_x^2/l. Because pressure is force/area and area is l^2, we can now express the total pressure exerted on one wall as

$$P = \frac{F}{A}$$

$$= \frac{Nmv_x^2}{l(l^2)} = \frac{Nmv_x^2}{V}$$

or

$$PV = Nmv_x^2 \tag{2.18}$$

where V is the volume of the cube (equal to l^3). When we are dealing with a large collection of molecules (for example, when N is on the order of 6×10^{23}), there is a tremendous spread of molecular velocities. It is more appropriate, therefore, to replace v_x^2 in Equation 2.18 with the mean or average quantity, $\overline{v_x^2}$. Referring to Equation 2.17, we see that the relation between the average of the square of the velocity components and the average of the square of the velocity, $\overline{v^2}$, is still

$$\overline{v^2} = \overline{v_x^2} + \overline{v_y^2} + \overline{v_z^2}$$

The quantity $\overline{v^2}$ is called the *mean-square velocity*, defined as

$$\overline{v^2} = \frac{v_1^2 + v_2^2 + \cdots + v_N^2}{N} \tag{2.19}$$

When N is a large number, it is correct to assume that molecular motions along the x, y, and z directions are equally probable. This means that

$$\overline{v_x^2} = \overline{v_y^2} = \overline{v_z^2} = \frac{\overline{v^2}}{3}$$

and Equation 2.18 can now be written as

$$P = \frac{Nm\overline{v^2}}{3V}$$

Multiplying the top and bottom by 2 and recalling that the kinetic energy of the molecule E_{trans} is given by $\frac{1}{2}mv^2$ (where the subscript *trans* denotes translational motion; that is, motion through space of the whole molecule), we obtain

$$P = \frac{2N}{3V}\left(\frac{1}{2}m\overline{v^2}\right) = \frac{2N}{3V}\overline{E}_{\text{trans}} \tag{2.20}$$

This is the pressure exerted by N molecules on one wall. The same result can be obtained regardless of the direction (x, y, or z) we describe for the molecular motion. We see that the pressure is directly proportional to the average kinetic energy or, more explicitly, to the mean-square velocity of the molecule. The physical meaning of this dependence is that the larger the velocity, the more frequent the collisions and the greater the change in momentum. Thus, these two independent terms give us the quantity $\overline{v^2}$ in the kinetic theory expression for the pressure.

Kinetic Energy and Temperature

Let us compare Equation 2.20 with the ideal-gas equation (Equation 2.8):

$$PV = nRT$$

$$= \frac{N}{N_A} RT$$

or

$$P = \frac{NRT}{N_A V} \tag{2.21}$$

where N_A is the Avogadro constant. Combining the pressures in Equations 2.20 and 2.21, we get

$$\frac{2}{3} \frac{N}{V} \bar{E}_{trans} = \frac{N}{N_A} \frac{RT}{V}$$

or

$$\bar{E}_{trans} = \frac{3}{2} \frac{RT}{N_A} = \frac{3}{2} k_B T \tag{2.22}$$

The kinetic energy of 1 mole of the gas is given by $(3/2)N_A k_B T = (3/2)RT$.

where $R = k_B N_A$ and k_B is the Boltzmann constant, equal to 1.380658×10^{-23} J K^{-1} [after the Austrian physicist Ludwig Eduard Boltzmann (1844–1906)]. (In most calculations, we shall round k_B to 1.381×10^{-23} J K^{-1}.) We see that the mean kinetic energy of one molecule is proportional to absolute temperature.

The significance of Equation 2.22 is that it provides an explanation for the temperature of a gas in terms of molecular motion. For this reason, random molecular motion is sometimes referred to as *thermal motion*. It is important to keep in mind that the kinetic theory is a *statistical* treatment of our model; hence, it is meaningless to associate temperature with the kinetic energy of just a few molecules. Equation 2.22 also tells us that whenever two ideal gases are at the same temperature T, they must have the *same* average kinetic energy. The reason is that $\bar{E}_{trans}$ in Equation 2.22 is independent of molecular properties such as size or molar mass or amount of the gas present, as long as N is a large number.

It is easy to see that $\overline{v^2}$ would be a very difficult quantity to measure, if indeed it could be measured at all. To do so, we would need to measure each individual velocity, square it, and then take the average (see Equation 2.19). Fortunately, $\overline{v^2}$ can be obtained quite directly from other quantities. From Equation 2.22, we write

$$\frac{1}{2}m\overline{v^2} = \frac{3}{2} \frac{RT}{N_A} = \frac{3}{2} k_B T$$

so that

Note that k_B refers to one molecule and R refers to one mole of such molecules; $k_B = R/N_A$.

$$\overline{v^2} = \frac{3RT}{mN_A} = \frac{3k_B T}{m}$$

or

$$\sqrt{\overline{v^2}} = v_{rms} = \sqrt{\frac{3RT}{\mathscr{M}}} = \sqrt{\frac{3k_B T}{m}} \quad (\mathscr{M} = mN_A) \tag{2.23}$$

where v_{rms} is the *root-mean-square velocity** and m is the mass (in kg) of one molecule; $\mathscr{M}$ is the molar mass (in kg mol^{-1}). Note that v_{rms} is directly proportional to the square root of temperature and inversely proportional to the square root of molar mass of the molecule. Therefore, the heavier the molecule, the slower its motion.

2.7 The Maxwell Distribution Laws

The root-mean-square velocity gives us an average measure that is very useful in the study of a large number of molecules. When we are studying, say, one mole of a gas, it is impossible to know the velocity of each individual molecule for two reasons. First, the number of molecules is so huge that there is no way we can follow all their motions. Second, although molecular motion is a well-defined quantity, we cannot measure its velocity exactly. Therefore, rather than concerning ourselves with individual molecular velocities, we ask this question: For a given system at some known temperature, how many molecules are moving at velocities between v and $v + \Delta v$ at any moment? Or, how many molecules in a macroscopic gas sample have velocities, say, between 306.5 m s^{-1} and 306.6 m s^{-1} at any moment?

Because the total number of molecules is very large, there is a continuous spread, or *distribution*, of velocities as a result of collisions. We can therefore make the velocity range Δv smaller and smaller, and in the limit it becomes dv. This fact has great significance, because it enables us to replace the summation sign with the integral sign in calculating the number of molecules whose velocities fall between v and $v + dv$. Mathematically speaking, it is easier to integrate than to sum a large series. This distribution-of-velocities approach was first employed by the Scottish physicist James Clerk Maxwell (1831–1879) in 1860 and later refined by Boltzmann. They showed that for a system containing N ideal gas molecules at thermal equilibrium with its surroundings, the fraction of molecules dN/N moving at velocities between v_x and $v_x + dv_x$ along the x direction is given by

$$\frac{dN}{N} = \left(\frac{m}{2\pi k_B T}\right)^{1/2} e^{-mv_x^2/2k_B T} \, dv_x \tag{2.24}$$

where m is the mass of the molecule, k_B the Boltzmann constant, and T the absolute temperature.

As mentioned earlier, velocity is a vector quantity. In many cases, we need to deal only with the speed of molecules (c), which is a scalar quantity; that is, it has magnitude but no directional properties. The fraction of molecules dN/N moving between speeds c and $c + dc$ is given by

$$\frac{dN}{N} = 4\pi c^2 \left(\frac{m}{2\pi k_B T}\right)^{3/2} e^{-mc^2/2k_B T} \, dc$$

$$= f(c)dc \tag{2.25}$$

where $f(c)$, the *Maxwell speed distribution function*, is given by

$$f(c) = 4\pi c^2 \left(\frac{m}{2\pi k_B T}\right)^{3/2} e^{-mc^2/2k_B T} \tag{2.26}$$

---

* Because velocity is a vector quantity, the average molecular velocity, $\bar{v}$, must be zero; there are just as many molecules moving in the positive direction as there are in the negative direction. On the other hand, v_{rms} is a scalar quantity; that is, it has magnitude but no direction.

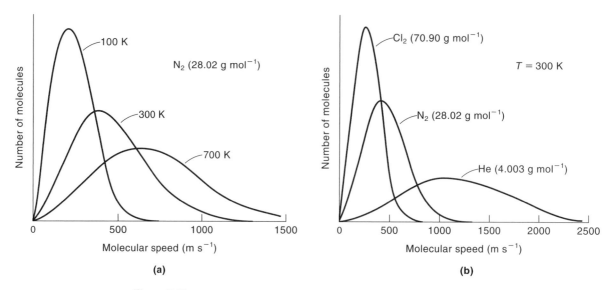

Figure 2.11
(a) The distribution of speeds for nitrogen gas at three different temperatures. At higher temperatures, more molecules are moving at faster speeds. (b) The distribution of speeds for three gases at 300 K. At a given temperature, the light molecules are moving faster, on the average.

Figure 2.11 shows the dependence of the speed distribution curve on temperature and molar mass. At any given temperature, the general shape of a distribution curve can be explained as follows. Initially, at small c values, the c^2 term in Equation 2.25 dominates so $f(c)$ increases with increasing c. At larger values of c, the term $e^{-mc^2/2k_BT}$ becomes more important. These two opposing terms cause the curve to reach a maximum beyond which it decreases roughly exponentially with increasing c. The speed corresponding to the maximum value of $f(c)$ is called the *most probable speed*, c_{mp}, because it is the speed of the largest number of molecules.

Figure 2.11a shows how the shape of the distribution curve is influenced by temperature. At low temperatures the distribution has a rather narrow range. As the temperature increases, the curve becomes flatter, meaning that there are now more fast-moving molecules. This temperature dependence of the distribution curve has important implications in chemical reaction rates. As we shall see in Chapter 9, in order to react, a molecule must possess a minimum amount of energy, called *activation energy*. At low temperatures the number of fast-moving molecules is small; hence most reactions proceed at a slow rate. Raising the temperature increases the number of energetic molecules and causes the reaction rate to increase. In Figure 2.11b we see that heavier gases have a narrower range of speed distribution than lighter gases at the same temperature. This is to be expected considering that heavier gases move slower, on the average, than lighter gases. The validity of the Maxwell speed distribution has been verified experimentally.

The usefulness of the Maxwell speed distribution function is that it enables us to calculate average quantities. In fact, we can obtain three related expressions for speed called the most probable speed (c_{mp}), as defined above, average speed ($\bar{c}$), which is defined as the sum of the speeds of all the molecules divided by the number of molecules, and root-mean-square speed (c_{rms}):*

---

* For the derivation of c_{mp}, see Problem 2.66; for the derivation of $\bar{c}$, see the physical chemistry texts listed in Chapter 1. Because the square of the average velocity is a scalar quantity, it follows that $\overline{v^2} = \overline{c^2}$; hence $v_{rms} = c_{rms}$.

$$c_{mp} = \sqrt{\frac{2RT}{\mathcal{M}}} \tag{2.27}$$

$$\bar{c} = \sqrt{\frac{8RT}{\pi \mathcal{M}}} \tag{2.28}$$

$$c_{rms} = \sqrt{\frac{3RT}{\mathcal{M}}} \tag{2.29}$$

Example 2.3

Calculate the values of c_{mp}, $\bar{c}$, and c_{rms} for O_2 at 300 K.

ANSWER

The constants are

$$R = 8.314 \text{ J K}^{-1} \text{ mol}^{-1} \quad T = 300 \text{ K}$$
$$\mathcal{M} = 0.03200 \text{ kg mol}^{-1}$$

The most probable speed is given by

$$c_{mp} = \sqrt{\frac{2 \times 8.314 \text{ J K}^{-1} \text{ mol}^{-1} \times 300 \text{ K}}{0.03200 \text{ kg mol}^{-1}}}$$

$$= \sqrt{1.56 \times 10^5 \text{ J kg}^{-1}}$$

$$= \sqrt{1.56 \times 10^5 \text{ m}^2 \text{ s}^{-2}}$$

$$= 395 \text{ m s}^{-1}$$

Similarly, we can show that

$$\bar{c} = \sqrt{\frac{8RT}{\pi \mathcal{M}}} = 446 \text{ m s}^{-1}$$

and

$$c_{rms} = \sqrt{\frac{3RT}{\mathcal{M}}} = 484 \text{ m s}^{-1}$$

COMMENT

The calculation shows, and indeed it is generally true, that $c_{rms} > \bar{c} > c_{mp}$. That c_{mp} is the smallest of the three speeds is due to the asymmetry of the curve (see Figure 2.11). The reason c_{rms} is greater than $\bar{c}$ is that the squaring process in Equation 2.19 is weighted toward larger values of c.

Finally, note that both the $\bar{c}$ and c_{rms} values of N_2 and O_2 are close to the speed of sound in air. Sound waves are pressure waves. The propagation of these waves is directly related to the movement of molecules and hence to their speeds.

2.8 Molecular Collisions and the Mean Free Path

Now that we have an explicit expression for the average speed, $\bar{c}$, we can use it to study some dynamic processes involving gases. We know that the speed of a molecule is not constant but changes frequently as a result of collisions. Therefore, the question we ask is: How often do molecules collide with one another? The collision frequency depends on the density of the gas and the molecular speed, and therefore on the temperature of the system. In the kinetic theory model, we assume each molecule to be a hard sphere of diameter d. A molecular collision is one in which the separation between the two spheres (measured from each center) is d.

Let us consider the motion of a particular molecule. A simple approach is to assume that at a given instant, all molecules except this one are standing still. In time t, this molecule moves a distance $\bar{c}t$ (where $\bar{c}$ is the average speed) and sweeps out a collision tube that has a cross-sectional area πd^2 (Figure 2.12). The volume of the

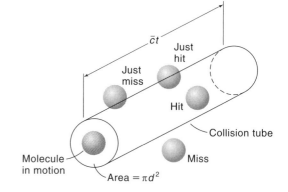

Figure 2.12
The collision cross section and the collision tube.
Any molecule whose center lies within or touches the
tube will collide with the moving molecule (red sphere).

cylinder is $(\pi d^2)(\bar{c}t)$. Any molecule whose center lies within this cylinder will collide with the moving molecule. If there are altogether N molecules in volume V, then the number density of the gas is N/V, the number of collisions in time t is $\pi d^2 \bar{c}t(N/V)$, and the number of collisions per unit time, or the *collision frequency*, Z_1, is $\pi d^2 \bar{c}(N/V)$. The expression for the collision frequency needs a correction. If we assume that the rest of the molecules are not frozen in position, we should replace $\bar{c}$ with the average *relative* speed. Figure 2.13 shows three different collisions for two molecules. The relative speed for the case shown in Figure 2.13c is $\sqrt{2}\bar{c}$, so that

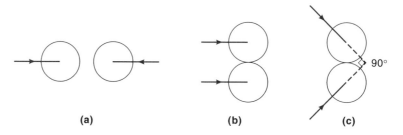

(a)   (b)   (c)

Figure 2.13
Three different approaches for two colliding molecules. The situations shown in (a) and (b) represent the two extreme cases, while that shown in (c) may be taken as the "average" case for molecular encounter.

$$Z_1 = \sqrt{2}\pi d^2 \bar{c} \left(\frac{N}{V} \right) \quad \text{collisions s}^{-1} \tag{2.30}$$

This is the number of collisions a *single* molecule makes in one second. Because there are N molecules in volume V and each makes Z_1 collisions per second, the total number of binary collisions, or collisions between two molecules, per unit volume per unit time, Z_{11}, is given by

$$
\begin{aligned}
Z_{11} &= \frac{1}{2} Z_1 \left(\frac{N}{V} \right) \\
&= \frac{\sqrt{2}}{2} \pi d^2 \bar{c} \left(\frac{N}{V} \right)^2 \quad \text{collisions m}^{-3}\,\text{s}^{-1}
\end{aligned}
\tag{2.31}
$$

The factor $1/2$ is introduced in Equation 2.31 to ensure that we are counting each collision between two molecules only once. The probability of three or more molecules colliding at once is very small except at high pressures. Because the rate of a chemical reaction generally depends on how often reacting molecules come in contact with one another, Equation 2.31 is quite important in gas-phase chemical kinetics. We shall return to this equation in Chapter 9.

A quantity closely related to the collision number is the average distance traveled by a molecule between successive collisions. This distance, called the *mean free path*, λ (Figure 2.14), is defined as

$$\lambda = (\text{average speed}) \times (\text{average time between collisions})$$

Because the average time between collisions is the reciprocal of the collision frequency, we have

$$\lambda = \frac{\bar{c}}{Z_1} = \frac{\bar{c}}{\sqrt{2}\pi d^2 \bar{c}(N/V)} = \frac{1}{\sqrt{2}\pi d^2(N/V)} \tag{2.32}$$

Notice that the mean free path is inversely proportional to the number density of the gas (N/V). This behavior is reasonable because in a dense gas, a molecule makes more collisions per unit time and hence travels a shorter distance between successive collisions. The mean free path can also be expressed in terms of the gas pressure. Assuming ideal behavior,

$$P = \frac{nRT}{V}$$

$$= \frac{(N/N_A)RT}{V}$$

$$\frac{N}{V} = \frac{PN_A}{RT}$$

Equation 2.32 can now be written as

$$\lambda = \frac{RT}{\sqrt{2}\pi d^2 P N_A} \tag{2.33}$$

Figure 2.14
The distances traveled by a molecule between successive collisions. The average of these distances is called the mean free path.

Example 2.4

The concentration of dry air at 1.00 atm and 298 K is about 2.5×10^{19} molecules cm^{-3}. Assuming that air contains only nitrogen molecules, calculate the collision frequency, the binary collision number, and the mean free path of nitrogen molecules under these conditions. The collision diameter of nitrogen is 3.75 Å. (1 Å $= 10^{-8}$ cm.)

ANSWER

Our first step is to calculate the average speed of nitrogen. From Equation 2.28, we find $\bar{c} = 4.8 \times 10^2$ m s^{-1}. The collision frequency is given by

$$Z_1 = \sqrt{2}\pi(3.75 \times 10^{-8} \text{ cm})^2(4.8 \times 10^4 \text{ cm s}^{-1})(2.5 \times 10^{19} \text{ molecules cm}^{-3})$$

$$= 7.5 \times 10^9 \text{ collisions s}^{-1}$$

Note that we have replaced the unit "molecules" with "collisions" because, in the derivation of Z_1, every molecule in the collision volume represents a collision. The binary collision number is

$$Z_{11} = \frac{Z_1}{2}\left(\frac{N}{V}\right)$$

$$= \frac{(7.5 \times 10^9 \text{ collisions s}^{-1})}{2} \times 2.5 \times 10^{19} \text{ molecules cm}^{-3}$$

$$= 9.4 \times 10^{28} \text{ collisions cm}^{-3} \text{ s}^{-1}$$

Again, we converted molecules to collisions in calculating the total number of binary collisions. Finally, the mean free path is given by

$$\lambda = \frac{\bar{c}}{Z_1} = \frac{4.8 \times 10^4 \text{ cm s}^{-1}}{7.5 \times 10^9 \text{ collisions s}^{-1}}$$

$$= 6.4 \times 10^{-6} \text{ cm collision}^{-1}$$

$$= 640 \text{ Å collision}^{-1}$$

COMMENT

It is usually sufficient to express mean free path in terms of distance alone rather than distance per collision. Thus, in this example, the mean free path of nitrogen is 640 Å, or 6.4×10^{-6} cm.

2.9 Graham's Laws of Diffusion and Effusion

Perhaps without thinking about it, we witness molecular motion on a daily basis. The scent of perfume and the shrinking of an inflated helium rubber balloon are examples of diffusion and effusion, respectively. We can apply the kinetic theory of gases to both processes.

The phenomenon of gas diffusion offers direct evidence of molecular motion. Were it not for diffusion, there would be no perfume industry, and skunks would be just another cute, furry species. Removing a partition separating two different gases in a container quickly leads to a complete mixing of molecules. These are spontaneous processes for which we shall discuss the thermodynamic basis in Chapter 4. During effusion, a gas travels from a high-pressure region to a low-pressure one

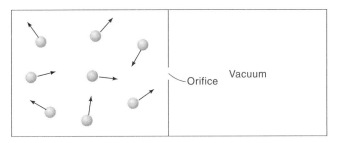

Figure 2.15
An effusion process in which molecules move through an opening (orifice) into an evacuated region. The conditions for effusion are that the mean free path of the molecules is large compared to the size of the opening and the wall containing the opening is thin so that no molecular collisions occur during the exit. Also, the pressure in the right chamber must be low enough so as not to obstruct the molecular movement through the hole.

through a pinhole or orifice (Figure 2.15). For effusion to occur, the mean free path of the molecules must be large compared with the diameter of the orifice. This ensures that a molecule is unlikely to collide with another molecule when it reaches the opening but will pass right through it. It follows then that the number of molecules passing through the orifice is equal to the number that would normally strike an area of wall equal to the area of the hole.

Although the basic molecular mechanisms for diffusion and effusion are quite different (the former involves *bulk* flow, whereas the latter involves *molecular* flow), these two phenomena obey laws of the same form. Both laws were discovered by the Scottish chemist Thomas Graham (1805–1869), the law of diffusion in 1831 and the law of effusion in 1864. These laws state that under the same conditions of temperature and pressure, the rates of diffusion (or effusion) of gases are inversely proportional to the square roots of their molar masses. Thus, for two gases 1 and 2, we have

$$\frac{r_1}{r_2} = \sqrt{\frac{\mathcal{M}_2}{\mathcal{M}_1}} \qquad (2.34)$$

where r_1 and r_2 are the rates of diffusion (or effusion) of the two gases.

Suggestions for Further Reading

BOOKS
Hirschfelder, J. O., C. F. Curtiss, and R. B. Bird, *The Molecular Theory of Gases and Liquids*, John Wiley & Sons, New York, 1954.
Hildebrand, J. H. *An Introduction to Molecular Kinetic Theory*, Chapman & Hall, London, 1963 (Van Nostrand Reinhold Company, New York).
Walton, A. J. *The Three Phases of Matter*, 2nd ed., Oxford University Press, New York, 1983.
Tabor, D. *Gases, Liquids, and Solids*, 3rd ed., Cambridge University Press, New York, 1996.

ARTICLES
Gas Laws and Equations of State
"The van der Waals Gas Equation," F. S. Swinbourne, *J. Chem. Educ.* **32**, 366 (1955).
"A Simple Model for van der Waals," S. S. Winter, *J. Chem. Educ.* **33**, 459 (1959).
"Comparisons of Equations of State in Effectively Describing *PVT* Relations," J. B. Ott, J. R. Goales, and H. T. Hall, *J. Chem. Educ.* **48**, 515 (1971).
"Scuba Diving and the Gas Laws," E. D. Cooke, *J. Chem. Educ.* **50**, 425 (1973).

"Derivation of the Ideal Gas Law," S. Levine, *J. Chem. Educ.* **62**, 399 (1985).

"The Ideal Gas Law at the Center of the Sun," D. B. Clark, *J. Chem. Educ.* **66**, 826 (1989).

"The Many Faces of van der Waals's Equation of State," J. G. Eberhart, *J. Chem. Educ.* **66**, 906 (1989).

"Does a One-Molecule Gas Obey Boyle's Law?" G. Rhodes, *J. Chem. Educ.* **69**, 16 (1992).

"Equations of State," M. Ross in *Encyclopedia of Applied Physics*, G. L. Trigg, Ed., VCH Publishers, New York, 1993, Vol. 6, p. 291.

"Interpretation of the Second Virial Coefficient," J. Wisniak, *J. Chem. Educ.* **76**, 671 (1999).

The Critical State

"The Critical Temperature: A Necessary Consequence of Gas Nonideality," F. L. Pilar, *J. Chem. Educ.* **44**, 284 (1967).

"Supercritical Fluids: Liquid, Gas, Both, or Neither? A Different Approach," E. F. Meyer and T. P. Meyer, *J. Chem. Educ.* **63**, 463 (1986).

"Past, Present, and Possible Future Applications of Supercritical Fluid Extraction Technology," C. L. Phelps, N. G. Smart, and C. M. Wai, *J. Chem. Educ.* **73**, 1163 (1996).

Kinetic Theory of Gases

"Kinetic Energies of Gas Molecules," J. C. Aherne, *J. Chem. Educ.* **42**, 655 (1965).

"Kinetic Theory, Temperature, and Equilibrium," D. K. Carpenter, *J. Chem. Educ.* **43**, 332 (1966).

"Graham's Laws of Diffusion and Effusion," E. A. Mason and B. Kronstadt, *J. Chem. Educ.* **44**, 740 (1967).

"The Cabin Atmosphere in Manned Space Vehicles," W. H. Bowman and R. M. Lawrence, *J. Chem. Educ.* **48**, 152 (1971).

"The Assumption of Elastic Collisions in Elementary Gas Kinetic Theory," B. Rice and C. J. G. Raw, *J. Chem. Educ.* **51**, 139 (1974).

"Velocity and Energy Distribution in Gases," B. A. Morrow and D. F. Tessier, *J. Chem. Educ.* **59**, 193 (1982).

"Applications of Maxwell-Boltzmann Distribution Diagrams," G. D. Peckham and I. J. McNaught, *J. Chem. Educ.* **69**, 554 (1992).

"Misuse of Graham's Laws," S. J. Hawkes, *J. Chem. Educ.* **70**, 836 (1993).

"Graham's Law and Perpetuation of Error," S. J. Hawkes, *J. Chem. Educ.* **74**, 1069 (1997).

General

"The Lung," J. H. Comroe, *Sci. Am.* February 1966.

"The Invention of the Balloon and the Birth of Modern Chemistry," A. F. Scott, *Sci. Am.* January 1984.

"Temperature, Cool but Quick," S. M. Cohen, *J. Chem. Educ.* **63**, 1038 (1986).

"Mountain Sickness," C. S. Houston, *Sci. Am.* October 1992.

Problems

Ideal Gases

2.1 Classify each of the following properties as intensive or extensive: force, pressure (P), volume (V), temperature (T), mass, density, molar mass, molar volume ($\bar{V}$).

2.2 Some gases, such as NO_2 and NF_2, do not obey Boyle's law at any pressure. Explain.

2.3 An ideal gas originally at 0.85 atm and 66°C was allowed to expand until its final volume, pressure, and temperature were 94 mL, 0.60 atm, and 45°C, respectively. What was its initial volume?

2.4 Some ballpoint pens have a small hole in the main body of the pen. What is the purpose of this hole?

2.5 Starting with the ideal-gas equation, show how you can calculate the molar mass of a gas from a knowledge of its density.

2.6 At STP (standard temperature and pressure), 0.280 L of a gas weighs 0.400 g. Calculate the molar mass of the gas.

2.7 Ozone molecules in the stratosphere absorb much of the harmful radiation from the sun. Typically, the temperature and partial pressure of ozone in the strato-sphere are 250 K and 1.0×10^{-3} atm, respectively. How many ozone molecules are present in 1.0 L of air under these conditions? Assume ideal-gas behavior.

2.8 Calculate the density of HBr in g L^{-1} at 733 mmHg and 46°C. Assume ideal-gas behavior.

2.9 Dissolving 3.00 g of an impure sample of $CaCO_3$ in an excess of HCl acid produced 0.656 L of CO_2 (measured at 20°C and 792 mmHg). Calculate the percent by mass of $CaCO_3$ in the sample.

2.10 The saturated vapor pressure of mercury is 0.0020 mmHg at 300 K and the density of air at 300 K is 1.18 g L^{-1}. **(a)** Calculate the concentration of mercury vapor in air in mol L^{-1}. **(b)** What is the number of parts per million (ppm) by mass of mercury in air?

2.11 A very flexible balloon with a volume of 1.2 L at 1.0 atm and 300 K is allowed to rise to the stratosphere, where the temperature and pressure are 250 K and 3.0×10^{-3} atm, respectively. What is the final volume of the balloon? Assume ideal-gas behavior.

2.12 Sodium bicarbonate ($NaHCO_3$) is called baking soda because when heated, it releases carbon dioxide gas, which causes cookies, doughnuts, and bread to rise during baking. **(a)** Calculate the volume (in liters) of CO_2 produced by heating 5.0 g of $NaHCO_3$ at 180°C and 1.3 atm. **(b)** Ammonium bicarbonate (NH_4HCO_3) has also been used as a leavening agent. Suggest one advantage and one disadvantage of using NH_4HCO_3 instead of $NaHCO_3$ for baking.

2.13 A common, non-SI unit for pressure is pounds per square inch (psi). Show that 1 atm = 14.7 psi. An automobile tire is inflated to 28.0 psi gauge pressure when cold, at 18°C. **(a)** What will the pressure be if the tire is heated to 32°C by driving the car? **(b)** What percentage of the air in the tire would have to be let out to reduce the pressure to the original 28.0 psi? Assume that the volume of the tire remains constant with temperature. (A tire gauge measures not the pressure of the air inside but its excess over the external pressure, which is 14.7 psi.)

2.14 (a) What volume of air at 1.0 atm and 22°C is needed to fill a 0.98-L bicycle tire to a pressure of 5.0 atm at the same temperature? (Note that 5.0 atm is the gauge pressure, which is the difference between the pressure in the tire and atmospheric pressure. Initially, the gauge pressure in the tire was 0 atm.) **(b)** What is the total pressure in the tire when the gauge reads 5.0 atm? **(c)** The tire is pumped with a hand pump full of air at 1.0 atm; compressing the gas in the cylinder adds all the air in the pump to the air in the tire. If the volume of the pump is 33% of the tire's volume, what is the gauge pressure in the tire after 3 full strokes of the pump?

2.15 A student breaks a thermometer and spills most of the mercury (Hg) onto the floor of a laboratory that measures 15.2 m long, 6.6 m wide, and 2.4 m high. **(a)** Calculate the mass of mercury vapor (in grams) in the room at 20°C. **(b)** Does the concentration of mercury vapor exceed the air quality regulation of 0.050 mg Hg m^{-3} of air? **(c)** One way to treat small quantities of spilled mercury is to spray powdered sulfur over the metal. Suggest a physical and a chemical reason for this treatment. The vapor pressure of mercury at 20°C is 1.7×1.0^{-6} atm.

2.16 Nitrogen forms several gaseous oxides. One of them has a density of 1.27 g L^{-1} measured at 764 mmHg and 150°C. Write the formula of the compound.

2.17 Nitrogen dioxide (NO_2) cannot be obtained in a pure form in the gas phase because it exists as a mixture of NO_2 and N_2O_4. At 25°C and 0.98 atm, the density of this gas mixture is 2.7 g L^{-1}. What is the partial pressure of each gas?

2.18 An ultra-high-vacuum pump can reduce the pressure of air from 1.0 atm to 1.0×10^{-12} mmHg. Calculate the number of air molecules in a liter at this pressure and 298 K. Compare your results with the number of molecules in 1.0 L at 1.0 atm and 298 K. Assume ideal-gas behavior.

2.19 An air bubble with a radius of 1.5 cm at the bottom of a lake where the temperature is 8.4°C and the pressure is 2.8 atm rises to the surface, where the temperature is 25.0°C and the pressure is 1.0 atm. Calculate the radius of the bubble when it reaches the surface. Assume ideal-gas behavior. [*Hint:* The volume of a sphere is given by $(4/3)\pi r^3$, where r is the radius.]

2.20 The density of dry air at 1.00 atm and 34.4°C is 1.15 g L^{-1}. Calculate the composition of air (percent by mass) assuming that it contains only nitrogen and oxygen and behaves like an ideal gas. (*Hint:* First calculate the "molar mass" of air, then the mole fractions, and then the mass fractions of O_2 and N_2.)

2.21 A gas that evolved during the fermentation of glucose has a volume of 0.78 L when measured at 20.1°C and 1.0 atm. What was the volume of this gas at the fermentation temperature of 36.5°C? Assume ideal-gas behavior.

2.22 Two bulbs of volumes V_A and V_B are connected by a stopcock. The number of moles of gases in the bulbs are n_A and n_B, and initially the gases are at the same pressure, P, and temperature, T. Show that the final pressure of the system, after the stopcock has been opened, equals P. Assume ideal-gas behavior.

2.23 The composition of dry air at sea level is 78.03% N_2, 20.99% O_2, and 0.033% CO_2 by volume. **(a)** Calculate the average molar mass of this air sample. **(b)** Calculate the partial pressures of N_2, O_2, and CO_2 in atm. (At constant temperature and pressure, the volume of a gas is directly proportional to the number of moles of the gas.)

2.24 A mixture containing nitrogen and hydrogen weighs 3.50 g and occupies a volume of 7.46 L at 300 K and 1.00 atm. Calculate the mass percent of these two gases. Assume ideal-gas behavior.

2.25 The relative humidity in a closed room with a volume of 645.2 m^3 is 87.6% at 300 K, and the vapor pressure of water at 300 K is 0.0313 atm. Calculate the mass of water in the air. [*Hint:* The relative humidity is defined as $(P/P_s) \times 100\%$, where P and P_s are the partial pressure and saturated partial pressure of water vapor, respectively.]

2.26 Death by suffocation in a sealed container is normally caused not by oxygen deficiency but by CO_2 poisoning, which occurs at about 7% CO_2 by volume. For what length of time would it be safe to be in a sealed room $10 \times 10 \times 20$ ft? [*Source:* "Eco-Chem," J. A. Campbell, *J. Chem. Educ.* **49**, 538 (1972).]

2.27 A flask contains a mixture of two ideal gases, A and B. Show graphically how the total pressure of the system depends on the amount of A present; that is, plot the total pressure versus the mole fraction of A. Do the same for B on the same graph. The total number of moles of A and B is constant.

2.28 A mixture of helium and neon gases is collected over water at 28.0°C and 745 mmHg. If the partial pressure of helium is 368 mmHg, what is the partial pressure of neon? (*Note:* The vapor pressure of water at 28°C is 28.3 mmHg.)

2.29 If the barometric pressure falls in one part of the world, it must rise somewhere else. Explain why.

2.30 A piece of sodium metal reacts completely with water as follows:

$$2Na(s) + 2H_2O(l) \rightarrow 2NaOH(aq) + H_2(g)$$

The hydrogen gas generated is collected over water at 25.0°C. The volume of the gas is 246 mL measured at 1.00 atm. Calculate the number of grams of sodium used in the reaction. (*Note:* The vapor pressure of water at 25°C is 0.0313 atm.)

2.31 A sample of zinc metal reacts completely with an excess of hydrochloric acid:

$$Zn(s) + 2HCl(aq) \rightarrow ZnCl_2(aq) + H_2(g)$$

The hydrogen gas produced is collected over water at 25.0°C. The volume of the gas is 7.80 L, and the pressure is 0.980 atm. Calculate the amount of zinc metal in grams consumed in the reaction. (*Note:* The vapor pressure of water at 25°C is 23.8 mmHg.)

2.32 Helium is mixed with oxygen gas for deep sea divers. Calculate the percent by volume of oxygen gas in the mixture if the diver has to submerge to a depth where the total pressure is 4.2 atm. The partial pressure of oxygen is maintained at 0.20 atm at this depth.

2.33 A sample of ammonia (NH_3) gas is completely decomposed to nitrogen and hydrogen gases over heated iron wool. If the total pressure is 866 mmHg, calculate the partial pressures of N_2 and H_2.

2.34 The partial pressure of carbon dioxide in air varies with the seasons. Would you expect the partial pressure in the Northern Hemisphere to be higher in the summer or winter? Explain.

2.35 A healthy adult exhales about 5.0×10^2 mL of a gaseous mixture with each breath. Calculate the number of molecules present in this volume at 37°C and 1.1 atm. List the major components of this gaseous mixture.

2.36 Describe how you would measure, by either chemical or physical means (other than mass spectrometry), the partial pressures of a mixture of gases: **(a)** CO_2 and H_2, **(b)** He and N_2.

2.37 The gas laws are vitally important to scuba divers. The pressure exerted by 33 ft of seawater is equivalent to 1 atm pressure. **(a)** A diver ascends quickly to the surface of the water from a depth of 36 ft without exhaling gas from his lungs. By what factor would the volume of his lungs increase by the time he reaches the surface? Assume that the temperature is constant. **(b)** The partial pressure of oxygen in air is about 0.20 atm. (Air is 20% oxygen by volume.) In deep-sea diving, the composition of air the diver breathes must be changed to maintain this partial pressure. What must the oxygen content (in percent by volume) be when the total pressure exerted on the diver is 4.0 atm?

2.38 A 1.00-L bulb and a 1.50-L bulb, connected by a stopcock, are filled, respectively, with argon at 0.75 atm and helium at 1.20 atm at the same temperature. Calculate the total pressure and the partial pressures of each gas after the stopcock has been opened and the mole fraction of each gas. Assume ideal-gas behavior.

2.39 A mixture of helium and neon weighing 5.50 g occupies a volume of 6.80 L at 300 K and 1.00 atm. Calculate the composition of the mixture in mass percent.

Nonideal Gases

2.40 Suggest two demonstrations to show that gases do not behave ideally.

2.41 Which of the following combinations of conditions most influences a gas to behave ideally: **(a)** low pressure and low temperature, **(b)** low pressure and high temperature, **(c)** high pressure and high temperature, and **(d)** high pressure and low temperature.

2.42 The van der Waals constants of a gas can be obtained from its critical constants, where $a = (27R^2T_c^2/64P_c)$ and $b = (RT_c/8P_c)$. Given that $T_c = 562$ K and $P_c = 48.0$ atm for benzene, calculate its a and b values.

2.43 Using the data shown in Table 2.1, calculate the pressure exerted by 2.500 moles of carbon dioxide confined in a volume of 1.000 L at 450 K. Compare the pressure with that calculated assuming ideal behavior.

2.44 Without referring to a table, select from the following list the gas that has the largest value of b in the van der Waals equation: CH_4, O_2, H_2O, CCl_4, Ne.

2.45 Referring to Figure 2.4, we see that for He the plot has a positive slope even at low pressures. Explain this behavior.

2.46 At 300 K, the virial coefficients (B) of N_2 and CH_4 are -4.2 cm^3 mol^{-1} and -15 cm^3 mol^{-1}, respectively. Which gas behaves more ideally at this temperature?

2.47 Calculate the molar volume of carbon dioxide at 400 K and 30 atm, given that the second virial coefficient (B) of CO_2 is -0.0605 L mol^{-1}. Compare your result with that obtained using the ideal-gas equation.

2.48 Consider the virial equation $Z = 1 + B'P + C'P^2$, which describes the behavior of a gas at a certain temperature. From the following plot of Z versus P, deduce the signs of B' and C' ($< 0, = 0, > 0$).

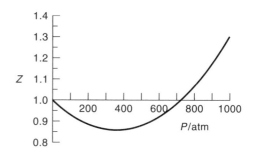

Kinetic Theory of Gases

2.49 Apply the kinetic theory of gases to explain Boyle's law, Charles' law, and Dalton's law.

2.50 Is temperature a microscopic or macroscopic concept? Explain.

2.51 In applying the kinetic molecular theory to gases, we have assumed that the walls of the container are elastic for molecular collisions. Actually, whether these collisions are elastic or inelastic makes no difference as long as the walls are at the same temperature as the gas. Explain.

2.52 If 2.0×10^{23} argon (Ar) atoms strike 4.0 cm^2 of wall per second at a 90° angle to the wall when moving with a speed of 45,000 cm s^{-1}, what pressure (in atm) do they exert on the wall?

2.53 A square box contains He at 25°C. If the atoms are colliding with the walls perpendicularly (at 90°) at the rate of 4.0×10^{22} times per second, calculate the force and the pressure exerted on the wall given that the area of the wall is 100 cm^2 and the speed of the atoms is 600 m s^{-1}.

2.54 Calculate the average translational kinetic energy for a N_2 molecule and for 1 mole of N_2 at 20°C.

2.55 To what temperature must He atoms be cooled so that they have the same v_{rms} as O_2 at 25°C?

2.56 The c_{rms} of CH_4 is 846 m s^{-1}. What is the temperature of the gas?

2.57 Calculate the value of the c_{rms} of ozone molecules in the stratosphere, where the temperature is 250 K.

2.58 At what temperature will He atoms have the same c_{rms} value as N_2 molecules at 25°C? Solve this problem without calculating the value of c_{rms} for N_2.

Maxwell Speed Distribution

2.59 List the conditions used for deriving the Maxwell speed distribution.

2.60 Plot the speed distribution function for **(a)** He, O_2, and UF_6 at the same temperature, and **(b)** CO_2 at 300 K and 1000 K.

2.61 Account for the maximum in the Maxwell speed distribution curve (Figure 2.11) by plotting the following two curves on the same graph: (1) c^2 versus c and (2) $e^{-mc^2/2k_BT}$ versus c. Use neon (Ne) at 300 K for the plot in (2).

2.62 A N_2 molecule at 20°C is released at sea level to travel upward. Assuming that the temperature is constant and that the molecule does not collide with other molecules, how far would it travel (in meters) before coming to rest? Do the same calculation for a He atom. [*Hint:* To calculate the altitude, h, the molecule will travel, equate its kinetic energy with the potential energy, mgh, where m is the mass and g the acceleration due to gravity (9.81 m s^{-2}).]

2.63 The speeds of 12 particles (in cm s^{-1}) are 0.5, 1.5, 1.8, 1.8, 1.8, 1.8, 2.0, 2.5, 2.5, 3.0, 3.5, and 4.0. Find **(a)** the average speed, **(b)** the root-mean-square speed, and **(c)** the most probable speed of these particles. Explain your results.

2.64 At a certain temperature, the speeds of six gaseous molecules in a container are 2.0 m s^{-1}, 2.2 m s^{-1}, 2.6 m s^{-1}, 2.7 m s^{-1}, 3.3 m s^{-1}, and 3.5 m s^{-1}. Calculate the root-mean-square speed and the average speed of the molecules. These two average values are close to each other, but the root-mean-square value is always the larger of the two. Why?

2.65 The following diagram shows the Maxwell speed distribution curves for a certain ideal gas at two different temperatures (T_1 and T_2). Calculate the value of T_2.

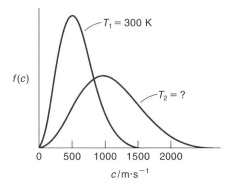

2.66 Derive an expression for c_{mp}. [*Hint:* Differentiate $f(c)$ with respect to c in Equation 2.26 and set the result to zero.]

2.67 Calculate the values of c_{rms}, c_{mp}, and $\bar{c}$ for argon at 298 K.

2.68 Calculate the value of c_{mp} for C_2H_6 at 25°C. What is the ratio of the number of molecules with a speed of 989 m s^{-1} to the number of molecules with this value of c_{mp}?

Molecular Collisions and the Mean Free Path

2.69 Considering the magnitude of molecular speeds, explain why it takes so long (on the order of minutes) to detect the odor of ammonia when someone opens a bottle of concentrated ammonia at the other end of a laboratory bench.

2.70 How does the mean free path of a gas depend on **(a)** the temperature at constant volume, **(b)** the density, **(c)** the pressure at constant temperature, **(d)** the volume at constant temperature, and **(e)** the size of molecules?

2.71 A bag containing 20 marbles is being shaken vigorously. Calculate the mean free path of the marbles if the volume of the bag is 850 cm^3. The diameter of each marble is 1.0 cm.

2.72 Calculate the mean free path and the binary number of collisions per liter per second between HI molecules at 300 K and 1.00 atm. The collision diameter of the HI molecules may be taken to be 5.10 Å. Assume ideal-gas behavior.

2.73 Ultra-high-vacuum experiments are routinely performed at a total pressure of 1.0×10^{-10} torr. Calculate the mean free path of N_2 molecules at 350 K under these conditions. The collision diameter of N_2 is 3.75 Å.

2.74 Suppose that helium atoms in a sealed container all start with the same speed, 2.74×10^4 cm s^{-1}. The atoms are then allowed to collide with one another until the Maxwell distribution is established. What is the temperature of the gas at equilibrium? Assume that there is no heat exchange between the gas and its surroundings.

2.75 Compare the collision number and the mean free path for air molecules at **(a)** sea level ($T = 300$ K and density $= 1.2$ g L^{-1}) and **(b)** in the stratosphere ($T = 250$ K and density $= 5.0 \times 10^{-3}$ g L^{-1}). The molar mass of air may be taken as 29.0 g, and the collision diameter is 3.72 Å.

2.76 Calculate the values of Z_1 and Z_{11} for mercury (Hg) vapor at 40°C, both at $P = 1.0$ atm and at $P = 0.10$ atm. How do these two quantities depend on pressure? The collision diameter of Hg is 4.26 Å.

Gas Diffusion and Effusion

2.77 Derive Equation 2.34 from Equation 2.23.

2.78 An inflammable gas is generated in marsh lands and sewage by a certain anaerobic bacterium. A pure sample of this gas was found to effuse through an orifice in 12.6 min. Under identical conditions of temperature and pressure, oxygen takes 17.8 min to effuse through the same orifice. Calculate the molar mass of the gas, and suggest what this gas might be.

2.79 Nickel forms a gaseous compound of the formula $Ni(CO)_x$. What is the value of x given the fact that under the same conditions of temperature and pressure, methane (CH_4) effuses 3.3 times faster than the compound?

2.80 In 2.00 min, 29.7 mL of He effuse through a small hole. Under the same conditions of temperature and pressure, 10.0 mL of a mixture of CO and CO_2 effuse through the hole in the same amount of time. Calculate the percent composition by volume of the mixture.

2.81 Uranium-235 can be separated from uranium-238 by the effusion process involving UF_6. Assuming a 50:50 mixture at the start, what is the percentage of enrichment after a single stage of separation?

2.82 An equimolar mixture of H_2 and D_2 effuses through an orifice at a certain temperature. Calculate the composition (in mole fractions) of the gas that passes through the orifice. The molar mass of deuterium is 2.014 g mol^{-1}.

2.83 The rate (r_{eff}) at which molecules confined to a volume V effuse through an orifice of area A is given by $(1/4)nN_A\bar{c}A/V$, where n is the number of moles of the gas. An automobile tire of volume 30.0 L and pressure 1,500 torr is punctured as it runs over a sharp nail. **(a)** Calculate the effusion rate if the diameter of the hole is 1.0 mm. **(b)** How long would it take to lose half of the air in the tire through effusion? Assume a constant effusion rate and constant volume. The molar mass of air is 29.0 g, and the temperature is 32.0°C.

Additional Problems

2.84 A barometer with a cross-sectional area of 1.00 cm^2 at sea level measures a pressure of 76.0 cm of mercury. The pressure exerted by this column of mercury is equal to the pressure exerted by all the air on 1 cm^2 of Earth's surface. Given that the density of mercury is 13.6 g cm^{-3} and the average radius of Earth is 6371 km, calculate the total mass of Earth's atmosphere in kilograms. (*Hint:* The surface area of a sphere is $4\pi r^2$, where r is the radius of the sphere.)

2.85 It has been said that every breath we take, on average, contains molecules once exhaled by Wolfgang Amadeus Mozart (1756–1791). The following calculations demonstrate the validity of this statement. **(a)** Calculate the total number of molecules in the atmosphere. (*Hint:* Use the result from Problem 2.84 and 29.0 g mol^{-1} as the molar mass of air.) **(b)** Assuming the volume of every breath (inhale or exhale) is 500 mL, calculate the number of molecules exhaled in each breath at 37°C, which is the body temperature. **(c)** If Mozart's life span was exactly 35 years, how many molecules did he exhale in that period (given that an average person breathes 12 times per minute)? **(d)** Calculate the fraction of molecules in the atmosphere that were exhaled by Mozart. How many of Mozart's molecules do we inhale with each breath of air? Round your answer to one significant digit. **(e)** List three important assumptions in these calculations.

2.86 A stockroom supervisor measured the contents of a partially filled 25.0-gallon acetone drum on a day when the temperature was 18.0°C and the atmospheric pressure was 750 mmHg, and found that 15.4 gallons of the solvent remained. After tightly sealing the drum, an assistant dropped the drum while carrying it upstairs to the organic laboratory. The drum was dented and its internal volume was decreased to 20.4 gallons. What is the total pressure inside the drum after the accident? The vapor pressure of acetone at 18.0°C is 400 mmHg. (*Hint:* At the time the drum was sealed, the pressure inside the drum, which is equal to the sum of the pressures of air and acetone, was equal to the atmospheric pressure.)

2.87 A relation known as the barometric formula is useful for estimating the change in atmospheric pressure with altitude. **(a)** Starting with the knowledge that atmospheric pressure decreases with altitude, we have $dP = -\rho g \, dh$, where ρ is the density of air, g is the acceleration due to gravity (9.81 m s^{-2}), and P and h are the pressure and height, respectively. Assuming ideal-gas behavior and constant temperature, show that the pressure P at height h is related to the pressure at sea level P_0 ($h = 0$) by $P = P_0 e^{-g\mathcal{M}h/RT}$. (*Hint:* For an ideal gas, $\rho = P\mathcal{M}/RT$, where $\mathcal{M}$ is the molar mass.) (b) Calculate the atmospheric pressure at a height of 5.0 km, assuming the temperature is constant at 5.0°C, given that the average molar mass of air is 29.0 g mol^{-1}.

2.88 In terms of the hard-sphere gas model, molecules are assumed to possess finite volume, but there is no interaction among the molecules. (a) Compare the P–V isotherm for an ideal gas and that for a hard-sphere gas. (b) Let b be the effective volume of the gas. Write an equation of state for this gas. (c) From this equation, derive an expression for $Z = P\bar{V}/RT$ for the hard-sphere gas and make a plot of Z versus P for two values of T (T_1 and T_2, $T_2 > T_1$). Be sure to indicate the value of the intercepts on the Z axis. (d) Plot Z versus T for fixed P for an ideal gas and for the hard-sphere gas.

2.89 One way to gain a physical understanding of b in the van der Waals equation is to calculate the "excluded volume." Assume that the distance of closest approach between two similar spherical molecules is the sum of their radii ($2r$). (a) Calculate the volume around each molecule into which the center of another molecule cannot penetrate. (b) From your result in (a), calculate the excluded volume for one mole of molecules, which is the constant b. How does this compare with the sum of the volumes of 1 mole of the same molecules?

2.90 You may have witnessed a demonstration in which a burning candle standing in water is covered by an upturned glass. The candle goes out and the water rises in the glass. The explanation usually given for this phenomenon is that the oxygen in the glass is consumed by combustion, leading to a decrease in volume and hence the rise in the water level. However, the loss of oxygen is only a minor consideration. (a) Using $C_{12}H_{26}$ as the formula for paraffin wax, write a balanced equation for the combustion. Based on the nature of the products, show that the predicted rise in water level due to the removal of oxygen is far less than the observed change. (b) Devise a chemical process that would allow you to measure the volume of oxygen in the trapped air. (*Hint:* Use steel wool.) (c) What is the main reason for the water rising in the glass after the flame is extinguished?

2.91 Express the van der Waals equation in the form of Equation 2.14. Derive relationships between the van der Waals constants (a and b) and the virial coefficients (B, C, and D), given that

$$\frac{1}{1-x} = 1 + x + x^2 + x^3 + \cdots \quad |x| < 1$$

2.92 The Boyle temperature is the temperature at which the coefficient B is zero. Therefore, a real gas behaves like an ideal gas at this temperature. (a) Give a physical interpretation of this behavior. (b) Using your result for B for the van der Waals equation in Problem 2.91, calculate the Boyle temperature for argon, given that $a = 1.345$ atm L^2 mol^{-2} and $b = 3.22 \times 10^{-2}$ L mol^{-1}.

2.93 Estimate the distance (in Å) between molecules of water vapor at 100°C and 1.0 atm. Assume ideal-gas behavior. Repeat the calculation for liquid water at 100°C, given that the density of water at 100°C is 0.96 g cm^{-3}. Comment on your results. (The diameter of a H_2O molecule is approximately 3 Å. 1 Å $= 10^{-8}$ cm.)

2.94 The following apparatus can be used to measure atomic and molecular speed. A beam of metal atoms is directed at a rotating cylinder in a vacuum. A small opening in the cylinder allows the atoms to strike a target area. Because the cylinder is rotating, atoms traveling at different speeds will strike the target at different positions. In time, a layer of the metal will deposit on the target area, and the variation in its thickness is found to correspond to Maxwell's speed distribution. In one experiment, it is found that at 850°C, some bismuth (Bi) atoms struck the target at a point 2.80 cm from the spot directly opposite the slit. The diameter of the cylinder is 15.0 cm, and it is rotating at 130 revolutions per second. (a) Calculate the speed (m s^{-1}) at which the target is moving. (*Hint:* The circumference of a circle is given by $2\pi r$, where r is the radius.) (b) Calculate the time (in seconds) it takes for the target to travel 2.80 cm. (c) Determine the speed of the Bi atoms. Compare your result in (c) with the c_{rms} value for Bi at 850°C. Comment on the difference.

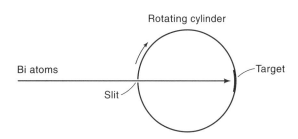

2.95 The escape velocity, v, from Earth's gravitational field is given by $(2GM/r)^{1/2}$, where G is the universal gravitational constant (6.67×10^{-11} m^3 kg^{-1} s^{-2}), M is the mass of Earth (6.0×10^{24} kg), and r is the distance from the center of Earth to the object, in meters. Compare the average speeds of He and N_2 molecules in the thermosphere (altitude about 100 km, $T = 250$ K). Which of the two molecules will have a greater tendency to escape? The radius of Earth is 6.4×10^6 m.

2.96 Calculate the ratio of the number of O_3 molecules with a speed of 1300 m s^{-1} at 360 K to the number with that speed at 293 K.

2.97 Calculate the collision frequency for 1.0 mole of krypton (Kr) at equilibrium at 300 K and 1.0 atm pressure. Which of the following alterations increases the collision frequency more: (a) doubling the temperature at constant pressure or (b) doubling the pressure at constant temperature? (*Hint:* The collision diameter of Kr is 4.16 Å.)

2.98 Apply your knowledge of the kinetic theory of gases to the following situations. **(a)** Two flasks of volumes V_1 and V_2 (where $V_2 > V_1$) contain the same number of helium atoms at the same temperature. **(i)** Compare the root-mean-square (rms) speeds and average kinetic energies of the helium (He) atoms in the flasks. **(ii)** Compare the frequency and the force with which the He atoms collide with the walls of their containers. **(b)** Equal numbers of He atoms are placed in two flasks of the same volume at temperatures T_1 and T_2 (where $T_2 > T_1$). **(i)** Compare the rms speeds of the atoms in the two flasks. **(ii)** Compare the frequency and the force with which the He atoms collide with the walls of their containers. **(c)** Equal numbers of He and neon (Ne) atoms are placed in two flasks of the same volume. The temperature of both gases is 74°C. Comment on the validity of the following statements: **(i)** The rms speed of He is equal to that of Ne. **(ii)** The average kinetic energies of the two gases are equal. **(iii)** The rms speed of each He atom is 1.47×10^3 m s^{-1}.

2.99 Consider 1 mole each of gaseous He and N_2 at the same temperature and pressure. State which gas (if any) has the greater value for: **(a)** $\bar{c}$, **(b)** c_{rms}, **(c)** $\bar{E}_{trans}$, **(d)** Z_1, **(e)** Z_{11}, **(f)** density, and **(g)** mean free path. The diameter of N_2 is 1.7 times that of He.

2.100 The root-mean-square velocity of a certain gaseous oxide is 493 m s^{-1} at 20°C. What is the molecular formula of the compound?

2.101 Calculate the mean kinetic energy ($\bar{E}_{trans}$) in joules of the following molecules at 350 K: **(a)** He, **(b)** CO_2, and **(c)** UF_6. Explain your results.

2.102 A sample of neon gas is heated from 300 K to 390 K. Calculate the percent increase in its kinetic energy.

2.103 A CO_2 fire extinguisher is located on the outside of a building in Massachusetts. During the winter months, one can hear a sloshing sound when the extinguisher is gently shaken. In the summertime, there is often no sound when it is shaken. Explain. Assume that the extinguisher has no leaks and that it has not been used.

CHAPTER 3

The First Law of Thermodynamics

Thermodynamics is the science of heat and temperature and, in particular, of the laws governing the conversion of thermal energy into mechanical, electrical, or other forms of energy. It is a central branch of science that has important applications in chemistry, physics, biology, and engineering. What makes thermodynamics such a powerful tool? It is a completely logical discipline and can be applied without any sophisticated mathematical techniques. The immense practical value of thermodynamics lies in the fact that it systematizes the information obtained from experiments performed on systems and enables us to draw conclusions, without further experimentation, about other aspects of the same systems and about similar aspects of other systems. It allows us to predict whether a certain reaction will proceed and what the maximum yield might be.

Thermodynamics is a macroscopic science concerning such properties as pressure, temperature, and volume. Unlike quantum mechanics, thermodynamics is not based on a specific molecular model, and therefore it is unaffected by our changing concepts of atoms and molecules. Indeed, the major foundations of thermodynamics were laid long before detailed atomic theories became available. This fact is one of its major strengths. On the negative side, equations derived from laws of thermodynamics do not provide us with a molecular interpretation of complex phenomena. Furthermore, although thermodynamics helps us predict the direction and extent of chemical reactions, it tells us nothing about the *rate* of a process; that issue is addressed by chemical kinetics, the topic of Chapter 9.

This chapter introduces the first law of thermodynamics and discusses some examples of thermochemistry.

3.1 Work and Heat

In this section, we shall study two concepts that form the basis of the first law of thermodynamics: work and heat.

Work

In classical mechanics, *work* is defined as force times distance. In thermodynamics, work becomes a more subtle concept; it encompasses a broader range of processes, including surface work, electrical work, work of magnetization, and so on. Let us consider a particularly useful example of a system doing work—the expansion of a gas. A sample of a gas is placed in a cylinder fitted with a weightless and frictionless piston. We assume the temperature of the system is kept constant at *T*. The gas is

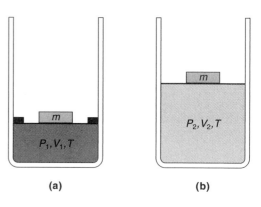

Figure 3.1
An isothermal expansion of a gas.
(a) Initial state. (b) Final state.

allowed to expand from its initial state—P_1, V_1, T—to P_2, V_2, T as shown in Figure 3.1. We assume no atmospheric pressure is present, so the gas is expanding only against the weight of an object of mass m placed on the piston. The work done (w) in lifting the mass from the initial height, h_1, to the final height, h_2, is given by

$$w = -\text{force} \times \text{distance}$$

$$= -\text{mass} \times \text{acceleration} \times \text{distance}$$

$$= -mg(h_2 - h_1)$$

$$= -mg\Delta h \tag{3.1}$$

where g is the acceleration (9.81 m s^{-2}) due to gravity and $\Delta h = h_2 - h_1$. Because m is in kilograms (kg) and h in meters (m), we see that w has the unit of energy (J). The minus sign in Equation 3.1 has the following meaning: in an expansion process, $h_2 > h_1$ and w is negative. This notation follows the convention that when a system does work on its surroundings, the work performed is a negative quantity. In a compression process, $h_2 < h_1$, so work is done on the system, and w is positive.

The external, opposing pressure, P_{ex}, acting on the gas is equal to force/area, so that

$$P_{ex} = \frac{mg}{A}$$

or

$$w = -P_{ex}A\Delta h = -P_{ex}(V_2 - V_1)$$

$$\boxed{w = -P_{ex}\Delta V} \tag{3.2}$$

where A is the area of the piston, and the product $A\Delta h$ gives the change in volume. Equation 3.2 shows that the amount of work done during expansion depends on the value of P_{ex}. Depending on experimental conditions, the amount of work performed by a gas during expansion from V_1 to V_2 at T can vary considerably from one case to another. In one extreme, the gas is expanding against a vacuum (for example, if the mass m is removed from the piston). Because $P_{ex} = 0$, the work done, $-P_{ex}\Delta V$, is also zero. A more common arrangement is to have some mass resting on the piston so that the gas is expanding against a *constant* external pressure. As we saw earlier, the amount of work performed by the gas in this case is $-P_{ex}\Delta V$, where $P_{ex} \neq 0$. Note that as the gas expands, the pressure of the gas, P_{in}, decreases constantly. For the gas to expand, however, we must have $P_{in} > P_{ex}$ at every stage of expansion. For example, if initially $P_{in} = 5$ atm and the gas is expanding against a constant external pressure of 1 atm ($P_{ex} = 1$ atm) at constant temperature T, then the piston will finally come to a halt when P_{in} decreases to exactly 1 atm.

Is it possible to have the gas perform a greater amount of work for the same increase in volume? Suppose we have an infinite number of identical weights exerting a total pressure of 5 atm on the piston. Because $P_{in} = P_{ex}$, the system is at mechanical equilibrium. Removing one weight will decrease the external pressure by an infinitesimal amount so that $P_{in} > P_{ex}$ and the gas will very slightly expand until P_{in} is again equal to P_{ex}. When the second weight is removed, the gas expands a bit further and so on until enough weights have been lifted from the piston to decrease the external pressure to 1 atm. At this point, we have completed the expansion process as before. How do we calculate the amount of work done in this case? At every stage of expansion (that is, each time one weight is lifted), the infinitesimal amount of work done is given by $-P_{ex}\, dV$, where dV is the infinitesimal increase in volume. The total work done in expanding from V_1 to V_2 is therefore

$$w = -\int_{V_1}^{V_2} P_{ex}\, dV \tag{3.3}$$

Because P_{ex} is no longer a constant value, the integral cannot be evaluated in this form. We note, however, that at every instant, P_{in} is only infinitesimally greater than P_{ex}, that is,

$$P_{in} - P_{ex} = dP$$

so that we can rewrite Equation 3.3 as

$$w = -\int_{V_1}^{V_2} (P_{in} - dP)\, dV$$

Realizing that $dPdV$ is a product of two infinitesimal quantities, we have $dPdV \approx 0$, and we can write

$$w = -\int_{V_1}^{V_2} P_{in}\, dV \tag{3.4}$$

Equation 3.4 is a more manageable form, because P_{in} is the pressure of the system (that is, the gas), and we can express it in terms of a particular equation of state. For an ideal gas,

$$P_{in} = \frac{nRT}{V}$$

so that

$$w = -\int_{V_1}^{V_2} \frac{nRT}{V}\, dV$$

$$w = -nRT \ln \frac{V_2}{V_1} = -nRT \ln \frac{P_1}{P_2} \tag{3.5}$$

because $P_1 V_1 = P_2 V_2$ (at constant n and T).

Equation 3.5 looks quite different from our earlier expression for work done ($-P_{ex}\Delta V$), and in fact it represents the *maximum* amount of work of expansion from V_1 to V_2. The reason for this result is not difficult to see. Because work in expansion is performed against external pressure, we can maximize the work done by adjusting the external pressure so that it is only infinitesimally smaller than the internal pressure at every stage, as described above. Under these conditions, expansion is a *reversible* process. By reversible, we mean that if we increase the external pressure by an infinitesimal amount, dP, we can bring the expansion to a stop. A further increase in P_{ex} by dP would actually result in compression. Thus, a reversible process is one in which the system is always infinitesimally close to equilibrium.

A truly reversible process would take an infinite amount of time to complete, and therefore it can never really be done. We could set up a system so that the gas does expand very slowly and try to approach reversibility, but actually attaining it is impossible. In the laboratory, we must work with *real* processes that are always irreversible. The reason we are interested in a reversible process is that it enables us to calculate the maximum amount of work that could possibly be extracted from a process. This quantity is important in estimating the efficiency of chemical and biological processes, as we shall see in Chapter 4.

It would take an infinite amount of time to remove an infinite number of weights from the piston at the rate of one weight at a time.

Example 3.1

A quantity of 0.850 mole of an ideal gas initially at a pressure of 15.0 atm and 300 K is allowed to expand isothermally until its final pressure is 1.00 atm. Calculate the value of the work done if the expansion is carried out (a) against a vacuum, (b) against a constant external pressure of 1.00 atm, and (c) reversibly.

ANSWER

(a) Because $P_{ex} = 0$, $-P_{ex}\Delta V = 0$, so that no work is performed in this case.
(b) Here the external, opposing pressure is 1.00 atm, so work will be done in expansion. The initial and final volumes can be obtained from the ideal gas equation:

$$V_1 = \frac{nRT}{P_1} \quad V_2 = \frac{nRT}{P_2}$$

Furthermore, the final pressure of the gas is equal to the external pressure, so $P_{ex} = P_2$. From Equation 3.2, we write

$$w = -P_2(V_2 - V_1)$$

$$= -nRTP_2\left(\frac{1}{P_2} - \frac{1}{P_1}\right)$$

Recall that 1 L atm = 101.3 J.

$$= -(0.850 \text{ mol})(0.08206 \text{ L atm K}^{-1} \text{ mol}^{-1})(300 \text{ K})(1.00 \text{ atm})$$

$$\times \left(\frac{1}{1.00 \text{ atm}} - \frac{1}{15.0 \text{ atm}}\right)$$

$$= -19.5 \text{ L atm} = -1.98 \times 10^3 \text{ J}$$

(c) For an isothermal, reversible process, the work done is given by Equation 3.5:

$$w = -nRT \ln \frac{V_2}{V_1}$$

$$= -nRT \ln \frac{P_1}{P_2}$$

$$= -(0.850 \text{ mol})(8.314 \text{ J K}^{-1} \text{ mol}^{-1})(300 \text{ K}) \ln \frac{15}{1}$$

$$= -5.74 \times 10^3 \text{ J}$$

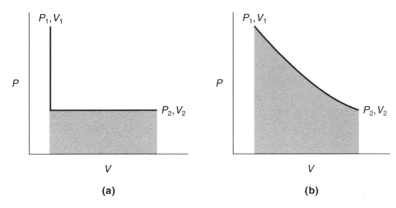

Figure 3.2
Isothermal gas expansion from P_1, V_1 to P_2, V_2. (a) An irreversible process. Note that P_2 is the constant external pressure. (b) A reversible process. In each case the shaded area represents the work done during expansion. The reversible process does the most work.

As we can see, the reversible process produces the most work. Figure 3.2 shows graphically the work done for cases (b) and (c) in Example 3.1. In an irreversible process (Figure 3.2a), the amount of work done is given by $P_2(V_2 - V_1)$, which is the area under the curve. For a reversible process, the amount of work is also given by the area under the curve (Figure 3.2b); because the external pressure is no longer held constant, however, the area is considerably greater.

From the foregoing discussion, we can draw several conclusions about work. First, work should be thought of as a mode of energy transfer. Gas expands because there is a pressure difference. When the internal and external pressure are equalized, the word *work* is no longer applicable. Second, the amount of work done depends on how the process is carried out—that is, the *path* (for example, reversible versus irreversible)—even though the initial and final states are the same in each case. Thus, work is not a *state function*, a property that is determined by the state of the system, and we cannot say that a system has, within itself, so much work or work content.

An important property of state functions is that when the state of a system is altered, a change in any state function depends only on the initial and final states of the system, not on how the change is accomplished. Let us assume that the change involves the expansion of a gas from an initial volume V_1 (2 L) to a final volume V_2 (4 L) at constant temperature. The change or the increase in volume is given by

$$\Delta V = V_2 - V_1$$
$$= 4\,L - 2\,L = 2\,L$$

The change can be brought about in many ways. We can let the gas expand directly from 2 L to 4 L as described above, or first allow it to expand to 6 L and then compress the volume down to 4 L, and so on. No matter how we carry out the process, the change in volume is always 2 L. Similarly, we can show that pressure and temperature, like volume, are state functions.

Heat

Heat is the transfer of energy between two bodies that are at different temperatures. Like work, heat appears only at the boundary of the system and is defined by a process. Energy is transferred from a hotter object to a colder one because there is a

Although *heat* already implies energy transfer, it is customary to speak of heat flow, heat absorbed, and heat released.

temperature difference. When the temperatures of the two objects are equal, the word *heat* is no longer applicable. Heat is not a property of a system and is not a state function. It is therefore path dependent. Suppose that we raise the temperature of 100.0 g of water initially at 20.0°C and 1 atm to 30.0°C and 1 atm. What is the heat transfer for this process? We do not know the answer because the process is not specified. One way to raise the temperature is to heat the water using a Bunsen burner or electrically using an immersion heater. The heat change, q (transferred from the surroundings to the system), is given by

$$q = ms\Delta T$$
$$= (100.0 \text{ g})(4.184 \text{ J g}^{-1} \text{ K}^{-1})(10.0 \text{ K})$$
$$= 4184 \text{ J}$$

where s is the specific heat of water. Alternatively, we can bring about the temperature increase by doing mechanical work on the system; for example, by stirring the water with a magnetic stirring bar until the desired temperature is reached as a result of friction. The heat transfer in this case is zero. Or, we could first raise the temperature of water from 20.0°C to 25.0°C by direct heating and then stir the bar to bring it up to 30.0°C. In this case, q is somewhere between zero and 4184 J. Clearly, then, an infinite number of ways are available to increase the temperature of the system by the same amount, but the heat change always depends on the path of the process.

In conclusion, work and heat are not functions of state. They are measures of energy transfer, and changes in these quantities are path dependent. The conversion factor between the *thermochemical calorie* and the joule, which is the mechanical equivalent of heat, is 1 cal = 4.184 J.

3.2 The First Law of Thermodynamics

The *first law of thermodynamics* states that energy can be converted from one form to another but cannot be created or destroyed. Put another way, this law says that the total energy of the universe is a constant. In general, we can divide the energy of the universe into two parts:

$$E_{univ} = E_{sys} + E_{surr}$$

where the subscripts denote the universe, system, and surroundings, respectively. For any given process, the changes in energies are

$$\Delta E_{univ} = \Delta E_{sys} + \Delta E_{surr} = 0$$

or

$$\Delta E_{sys} = -\Delta E_{surr}$$

Thus, if one system undergoes an energy change, ΔE_{sys}, the rest of the universe, or the surroundings, must undergo a change in energy that is equal in magnitude but opposite in sign; energy gained in one place must have been lost somewhere else. Furthermore, because energy can be changed from one form to another, the energy lost by one system can be gained by another system in a different form. For example, the energy lost by burning oil in a power plant may ultimately turn up in our homes as electrical energy, heat, light, and so on.

In chemistry, we are normally interested in the energy changes associated with the system, not with the surroundings. We have seen that because heat and work are not state functions, it is meaningless to ask how much heat or work a system possesses. On the other hand, the internal energy of a system is a state function, because it depends only on the thermodynamic parameters of the state, such as temperature, pressure, and composition. Note that the adjective *internal* implies that other kinds of energy may be associated with the system. For example, the whole system may be in motion and therefore possess kinetic energy (KE). The system may also possess potential energy (PE). Thus, the total energy of the system, E_{total}, is given by

$$E_{total} = KE + PE + U$$

where U denotes internal energy. This internal energy consists of translational, rotational, vibrational, electronic, and nuclear energies of the molecules, as well as intermolecular interactions. In most cases we shall consider, the system will be at rest, and external fields (for example, electric or magnetic fields) will not be present. Thus, both KE and PE are zero and $E_{total} = U$. As mentioned earlier, thermodynamics is not based on a particular model; therefore, we have no need to know the exact nature of U. In fact, we normally have no way to calculate this quantity accurately. All we are interested in, as you will see below, are methods for measuring the *change* in U for a process. For simplicity, we shall frequently refer to internal energy simply as energy and write its change, ΔU, as

$$\Delta U = U_2 - U_1$$

where U_2 and U_1 are the internal energies of the system in the final and initial states, respectively.

Energy differs from both heat and work in that it always changes by the same amount in going from one state to another, regardless of the nature of the path. Mathematically, the first law of thermodynamics can be expressed as

$$\Delta U = q + w \tag{3.6}$$

We assume all work done is of the *P–V* type.

or, for an infinitesimal change,

$$dU = đq + đw \tag{3.7}$$

Equations 3.6 and 3.7 tell us that the change in the internal energy of a system in a given process is the sum of the heat exchange, q, between the system and its surroundings and the work done, w, on (or by) the system. The sign conventions for q and w are summarized in Table 3.1. Note that we have deliberately omitted the Δ

Table 3.1
Sign Conventions for Work and Heat

Process	Sign
Work done by the system on the surroundings	−
Work done on the system by the surroundings	+
Heat absorbed by the system from the surroundings (endothermic process)	+
Heat absorbed by the surroundings from the system (exothermic process)	−

sign for q and w, because this notation represents the difference between the final and initial states and is therefore not applicable to heat and work, which are not state functions. Similarly, although dU is an *exact differential* (see Appendix 1), that is, an integral of the type $\int_1^2 dU$ is independent of the path, the $đ$ notation reminds us that $đq$ and $đw$ are *inexact differentials* and therefore are path dependent. In this text, we shall use capital letters for thermodynamic quantities (such as U, P, T, and V) that are state functions, and lowercase letters for thermodynamic quantities (such as q and w) that are not.

As a simple illustration of Equation 3.6, consider the heating of a gas in a closed container. Because the volume of the gas is constant, no expansion work can be done so that $w = 0$ and we can write

$$\Delta U = q_V + w$$
$$= q_V \tag{3.8}$$

where the subscript V denotes that this is a constant-volume process. Thus, the increase in the energy of the gas equals the heat absorbed by the gas from the surroundings. Equation 3.8 may seem strange at first: ΔU is equated to the heat absorbed, but heat, as we said earlier, is not a state function. However, we have restricted ourselves to a particular process or path—that is, one that takes place at constant volume; hence, q_V can have only one value under the given conditions.

Enthalpy

In the laboratory, most chemical and physical processes are carried out under constant pressure (that is, atmospheric pressure) rather than constant volume conditions. Consider a gas undergoing an irreversible expansion against a constant external pressure P, so that $w = -P\Delta V$. Equation 3.6 becomes

$$\Delta U = q + w$$
$$= q_P - P\Delta V$$

or

$$U_2 - U_1 = q_P - P(V_2 - V_1)$$

where the subscript P reminds us that this is a constant-pressure process. Rearrangement of the equation above gives

$$q_P = (U_2 + PV_2) - (U_1 + PV_1) \tag{3.9}$$

We define a function, called *enthalpy* (H), as follows

$$H = U + PV \tag{3.10}$$

where U, P, and V are the energy, pressure, and volume of the system. All the terms in Equation 3.10 are functions of state; H has the units of energy. From Equation 3.10, we can write the change in H as

$$\Delta H = H_2 - H_1 = (U_2 + P_2 V_2) - (U_1 + P_1 V_1)$$

Setting $P_2 = P_1 = P$ for a constant-pressure process, we obtain, by comparison with

Equation 3.9,

$$\Delta H = (U_2 + PV_2) - (U_1 + PV_1) = q_P$$

Again, we have restricted the change to a specific path—this time at constant pressure—so that the heat change, q_P, can be equated directly to the change in the state function, H.

In general, when a system undergoes a change from state 1 to state 2, the change in enthalpy is given by

$$\Delta H = \Delta U + \Delta(PV)$$
$$= \Delta U + P\Delta V + V\Delta P + \Delta P\Delta V \qquad (3.11)$$

This equation applies if neither pressure nor volume is kept constant. The last term, $\Delta P\Delta V$, is not negligible.* Recall that both P and V in Equation 3.11 refer to the system. If the change is carried out, say, at constant pressure, and if the pressure exerted by the system on the surroundings (P_{in}) is equal to the pressure exerted by the surroundings on the system (P_{ext})—that is,

$$P_{in} = P_{ext} = P$$

then we have $\Delta P = 0$, and Equation 3.11 now becomes

$$\Delta H = \Delta U + P\Delta V \qquad (3.12)$$

Similarly, for an infinitesimal change

$$dH = dU + P\,dV \qquad (3.13)$$

A Comparison of ΔU with ΔH

What is the difference between ΔU and ΔH? Both terms represent the change in energy, but their values differ because the conditions are not the same. Consider the following situation. The heat evolved when 2 moles of sodium react with water,

$$2Na(s) + 2H_2O(l) \rightarrow 2NaOH(aq) + H_2(g)$$

is 367.5 kJ. Because the reaction takes place at constant pressure, $q_P = \Delta H = -367.5$ kJ. To calculate the change in internal energy, from Equation 3.12, we write

$$\Delta U = \Delta H - P\Delta V$$

If we assume the temperature to be 25°C and ignore the small change in the volume of solution, we can show that the volume of 1 mole of H_2 generated at 1 atm is 24.5 L, so that $-P\Delta V = -24.5$ L atm or -2.5 kJ. Finally,

$$\Delta U = -367.5 \text{ kJ} - 2.5 \text{ kJ}$$
$$= -370.0 \text{ kJ}$$

---

*Note that $\Delta(PV)$, which represents the change in PV from state 1 to state 2, can be written as $[(P + \Delta P)(V + \Delta V) - PV] = P\Delta V + V\Delta P + \Delta P\Delta V$. For an infinitesimal change, we would write $dH = dU + P\,dV + V\,dP + dP\,dV$. Because $dPdV$ is the product of two infinitesimal quantities, we would ignore it and have $dH = dU + P\,dV + V\,dP$.

This calculation shows that ΔU and ΔH are slightly different. The reason ΔH is smaller than ΔU in this case is that some of the internal energy released is used to do gas expansion work (the H_2 generated has to push the air back), so less heat is evolved. In general, the difference between ΔH and ΔU in reactions involving gases is $\Delta(PV)$ or $\Delta(nRT) = RT\Delta n$ (if T is constant), where Δn is the change in the number of moles of gases, that is,

$$\Delta n = n_{\text{products}} - n_{\text{reactants}}$$

At constant temperature we have $\Delta H = \Delta U + RT\Delta n$. For the above reaction, $\Delta n = 1$ mol. Thus, at $T = 298$ K, $RT\Delta n$ is approximately 2.5 kJ, which is a small but not negligible quantity in accurate work. On the other hand, for chemical reactions occurring in the condensed phases (liquids and solids), ΔV is usually a small number (≤ 0.1 L per mole of reactant converted to product) so that $P\Delta V = 0.1$ L atm, or 10 J, which can be neglected in comparison with ΔU and ΔH. Thus, changes in enthalpy and energy in reactions not involving gases or in cases for which $\Delta n = 0$ are one and the same for all practical purposes.

Example 3.2

Compare the difference between ΔH and ΔU for the following physical changes: (a) 1 mol ice $\rightarrow$ 1 mol water at 273 K and 1 atm and (b) 1 mol water $\rightarrow$ 1 mol steam at 373 K and 1 atm. The molar volumes of ice and water at 273 K are 0.0196 L mol^{-1} and 0.0180 L mol^{-1}, respectively, and the molar volumes of water and steam at 373 K are 0.0188 L mol^{-1} and 30.61 L mol^{-1}, respectively.

ANSWER

Both cases are constant-pressure processes:

$$\Delta H = \Delta U + \Delta(PV) = \Delta U + P\Delta V$$

or

$$\Delta H - \Delta U = P\Delta V$$

(a) The change in molar volume when ice melts is

$$\Delta V = \bar{V}(l) - \bar{V}(s)$$
$$= (0.0180 - 0.0196) \text{ L mol}^{-1}$$
$$= -0.0016 \text{ L mol}^{-1}$$

Hence,

$$P\Delta V = (1 \text{ atm})(-0.0016 \text{ L mol}^{-1})$$
$$= -0.0016 \text{ L atm mol}^{-1}$$
$$= -0.16 \text{ J mol}^{-1}$$

(b) The change in molar volume when water boils is

$$\Delta V = \bar{V}(g) - \bar{V}(l)$$
$$= (30.61 - 0.0188) \text{ L mol}^{-1}$$
$$= 30.59 \text{ L mol}^{-1}$$

Hence,

$$P\Delta V = (1\text{ atm})(30.59\text{ L mol}^{-1})$$
$$= 30.59\text{ L atm mol}^{-1}$$
$$= 3100\text{ J mol}^{-1}$$

COMMENT

This example clearly shows that $(\Delta H - \Delta U)$ is negligibly small for condensed phases but can be quite appreciable if the process involves gases. Further, in (a), $\Delta U > \Delta H$; that is, the increase in the internal energy of the system is greater than the heat absorbed by the system because when ice melts, there is a decrease in volume. Consequently, work is done on the system by the surroundings. The opposite situation holds for (b) because in this case, steam is doing work on the surroundings.

3.3 Heat Capacities

In this section we shall study a thermodynamic quantity, called heat capacity, that enables us to measure energy changes (ΔU and ΔH) of a system as a result of changes in its temperature.

When heat is added to a substance, its temperature will rise. This fact we know well. But just how much the temperature will rise depends on (1) the amount of heat delivered (q), (2) the amount of the substance present (m), (3) the specific heat (s), which is determined by the chemical nature and physical state of the substance, and (4) the conditions under which heat is added to the substance. The temperature rise (ΔT) for a given amount of a substance is related to heat added by the equation

For this discussion of heat capacity, we assume no phase change.

$$q = ms\Delta T$$
$$= C\Delta T$$

or

$$C = \frac{q}{\Delta T} \tag{3.14}$$

where C, a proportionality constant, is called the heat capacity. (The specific heat of a substance is the energy required to raise the temperature of 1 g of the substance by $1°C$ or 1 K. It has the units of J g^{-1} K^{-1}. Here we have $C = ms$, where m is in grams and hence C has the units of J K^{-1}.)

Because the increase in temperature depends on the amount of substance present, it is often convenient to speak of the heat capacity of 1 mole of a substance or molar heat capacity $\bar{C}$, where

$$\bar{C} = \frac{C}{n} = \frac{q}{n\Delta T} \tag{3.15}$$

where n is the number of moles of the substance present and $\bar{C}$ has the units of J K^{-1} mol^{-1}. Note that C is an extensive property but $\bar{C}$, like all molar quantities, is intensive.

Constant-Volume and Constant-Pressure Heat Capacities

Heat capacity is a directly measurable quantity. Knowing the amount of the substance present, the heat added, and the temperature rise, we can readily calculate $\bar{C}$ using Equation 3.15. However, it turns out that the value we calculate also depends on how the heating process is carried out. Although many different conditions can be realized in practice, we shall consider only two important cases here: constant volume and constant pressure. We have already seen in Section 3.2 that for a constant-volume process, the heat absorbed by the system equals the increase in internal energy; that is, $\Delta U = q_V$. Hence the heat capacity at constant volume, C_V, of a given amount of substance is given by

$$C_V = \frac{q_V}{\Delta T} = \frac{\Delta U}{\Delta T}$$

or, expressed in partial derivatives (see Appendix 1)

$$C_V = \left(\frac{\partial U}{\partial T}\right)_V \tag{3.16}$$

so that

$$dU = C_V \, dT \tag{3.17}$$

We have seen that for a constant-pressure process we have $\Delta H = q_P$, so that the heat capacity at constant pressure is

$$C_P = \frac{q_P}{\Delta T} = \frac{\Delta H}{\Delta T}$$

or, expressed in partial derivatives

$$C_P = \left(\frac{\partial H}{\partial T}\right)_P \tag{3.18}$$

so that

$$dH = C_P \, dT \tag{3.19}$$

From the definitions of C_V and C_P we can calculate ΔU and ΔH for processes carried out under constant-volume or constant-pressure conditions. Integrating Equations 3.17 and 3.19 between temperatures T_1 and T_2, we obtain

$$\Delta U = \int_{T_1}^{T_2} C_V \, dT = C_V(T_2 - T_1) = C_V \Delta T = n\bar{C}_V \Delta T \tag{3.20}$$

$$\Delta H = \int_{T_1}^{T_2} C_P \, dT = C_P(T_2 - T_1) = C_P \Delta T = n\bar{C}_P \Delta T \tag{3.21}$$

where n is the number of moles of the substance present and $C_V = n\bar{C}_V$ and $C_P = n\bar{C}_P$. We have assumed that both C_V and C_P are independent of temperature. This is not true, however. Studies of the temperature dependence of constant-pressure heat capacity for many substances, for example, show that it can be represented by an equation $C_P = a + bT$, where a and b are constants for a given substance over a particular temperature range. Such an expression must be used in Equation 3.21 in accurate work. A similar equation applies to C_V. However, if the temperature change in a process is small, say 50 K or less, we can often treat C_V and C_P as if they were temperature independent.

Molecular Interpretation of Heat Capacity

Let us for now focus on gases, assuming ideal behavior. In Section 2.5 we saw that the translational kinetic energy of 1 mole of a gas is $\frac{3}{2}RT$; therefore, the molar heat capacity $\bar{C}_V$ is given by

Table 3.2
Constant-Volume Molar Heat Capacities of Gases at 298 K

$$\bar{C}_V = \left(\frac{\partial U}{\partial T}\right)_V = \left[\frac{\partial\left(\frac{3}{2}RT\right)}{\partial T}\right]_V = \frac{3}{2}R = 12.47 \text{ J K}^{-1} \text{ mol}^{-1}$$

Gas	$\bar{C}_V/\text{J} \cdot \text{K}^{-1} \cdot \text{mol}^-$
He	12.47
Ne	12.47
Ar	12.47
H_2	20.50
N_2	20.50
O_2	21.05
CO_2	28.82
H_2O	25.23
SO_2	31.51

Table 3.2 shows the measured molar heat capacities for several gases. The agreement is excellent for monatomic gases (that is, the noble gases), but considerable discrepancies are found for molecules. To see why the heat capacities are larger than 12.47 J K^{-1} mol^{-1} for molecules, we need to use quantum mechanics. Molecules, unlike atoms, can have rotational and vibrational motions in addition to translational motion—that is, motion through space of the whole molecule (Figure 3.3).

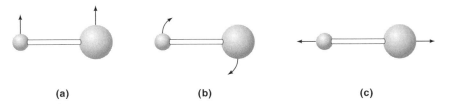

(a)   **(b)**   **(c)**

Figure 3.3
(a) Translational, (b) rotational, and (c) vibrational motion of a diatomic molecule, such as HCl.

According to quantum mechanics, the electronic, vibrational, and rotational energies of a molecule are quantized (further discussed in Chapters 11 and 14). That is, different molecular energy levels are associated with each type of motion, as shown in Figure 3.4. Note that the spacing between successive electronic energy levels is much

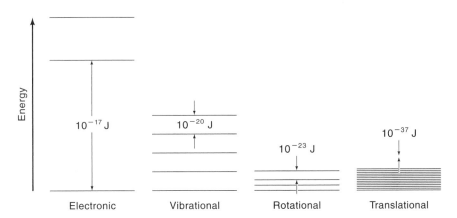

Figure 3.4
Energy levels associated with translational, rotational, vibrational, and electronic motions.

larger than that between the vibrational energy levels, which in turn is much larger than that between the rotational energy levels. The spacing between successive translational energy levels is so small that the levels practically merge into a continuum of energy. In fact, for most practical purposes, they can be treated as a continuum. Translational motion, then, is treated as a classical rather than a quantum mechanical phenomenon because its energy can vary continuously.

What do these energy levels have to do with heat capacities? When a gas sample absorbs heat from the surroundings, the energy is used to promote various kinds of motion. In this sense, the term *heat capacity* really means *energy capacity* because its value tells us the capacity of the system to store energy. Energy may be stored partly in rotational motion—the molecules may be promoted to a higher rotational energy level (that is, they will rotate faster), or it may be stored partly in vibrational motion. In each case, the molecules are promoted to higher energy levels.

Figure 3.4 suggests that it is much easier to excite a molecule to a higher rotational energy level than to a higher vibrational or electronic energy level, and this is indeed the case. Quantitatively, the ratio of the populations in any two energy levels E_2 and E_1 at temperature T is given by the *Boltzmann distribution law*:

$$\frac{N_2}{N_1} = e^{-\Delta E/k_B T} \tag{3.22}$$

where N_2 and N_1 are the number of molecules in E_2 and E_1, respectively, $\Delta E = E_2 - E_1$, and k_B is the Boltzmann constant (1.381×10^{-23} J K^{-1}). Equation 3.22 tells us that for a system at thermal equilibrium at a finite temperature, $N_2/N_1 < 1$, which means that the number of molecules in the upper level is *always* less than that in the lower level (Figure 3.5).

We can make some simple estimates using Equation 3.22. For translational motion, the spacing between adjacent energy levels ΔE is about 10^{-37} J, so that $\Delta E/k_B T$ at 298 K is

$$\frac{10^{-37} \text{ J}}{(1.381 \times 10^{-23} \text{ J K}^{-1})(298 \text{ K})} = 2.4 \times 10^{-17}$$

This number is so small that the exponential term on the right-hand side of Equation 3.22 is essentially unity. Thus, the number of molecules in a higher energy level is the same as in the one below it. The physical meaning of this result is that the kinetic energy is not quantized and a molecule can absorb any arbitrary amount of energy to increase its kinetic motion.

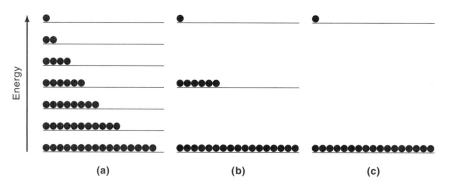

Figure 3.5
Qualitative illustration of the Boltzmann distribution law at some finite temperature T for three different types of energy levels. Note that if the energy spacing is large compared to $k_B T$, the molecules will crowd into the lowest energy level.

With rotational motion we find that ΔE is also small compared to the $k_B T$ term; therefore, the ratio N_2/N_1 is close to (although smaller than) unity. This means that the molecules are distributed fairly evenly among the rotational energy levels. The difference between rotational and translational motions is that only the energies of the former are quantized.

The situation is quite different when we consider vibrational motion. Here the spacing between levels is quite large (that is, $\Delta E \gg k_B T$), so that the ratio N_2/N_1 is much smaller than 1. Thus, at 298 K, most of the molecules are in the lowest vibrational energy level and only a small fraction of them are in the higher levels. Finally, because the spacing between electronic energy levels is very large, almost all the molecules are found in the lowest electronic energy level at room temperature.

From this discussion, we see that at room temperature both the translational and rotational motions contribute to the heat capacity of molecules and hence the $\bar{C}_V$ values are greater than 12.47 J K^{-1} mol^{-1}. As temperature increases, vibrational motions also begin to contribute to $\bar{C}_V$. Therefore, heat capacity increases with temperature. (In most cases we can ignore electronic motion's contribution to heat capacity except at very high temperatures.)

A Comparison of C_V with C_P

In general, C_V and C_P for a given substance are not equal to each other. The reason is that work has to be done on the surroundings in a constant-pressure process, so that *more* heat is required to raise the temperature by a definite amount in a constant-pressure process than in a constant-volume process. It follows, therefore, that $C_P > C_V$. This is true mainly for gases. The volume of a liquid or solid does not change appreciably with temperature, so the work done as it expands is quite small. For most purposes, therefore, C_V and C_P are practically the same for condensed phases.

Our next step is to see how C_P differs from C_V for an ideal gas. We start by writing

$$H = U + PV = U + nRT$$

For an infinitesimal change in temperature, dT, the change in enthalpy for a given amount of an ideal gas is

$$dH = dU + d(nRT)$$
$$= dU + nR\,dT$$

Substituting $dH = C_P\,dT$ and $dU = C_V\,dT$ into the above equation, we get

$$C_P\,dT = C_V\,dT + nR\,dT$$
$$C_P = C_V + nR \tag{3.23}$$
$$C_P - C_V = nR$$

or

$$\bar{C}_P - \bar{C}_V = R \tag{3.24}$$

Thus, for an ideal gas, the molar constant-pressure heat capacity is greater than the molar constant-volume heat capacity by R, the gas constant. Appendix 2 lists the $\bar{C}_P$ values of many substances.

Example 3.3

Calculate the values of ΔU and ΔH for the heating of 55.40 g of xenon from 300 K to 400 K. Assume ideal-gas behavior and that the heat capacities at constant volume and constant pressure are independent of temperature.

ANSWER

Xenon is a monatomic gas. Earlier we saw that $\bar{C}_V = \frac{3}{2}R = 12.47$ J K^{-1} mol^{-1}. Thus, from Equation 3.24, we have $\bar{C}_P = \frac{3}{2}R + R = \frac{5}{2}R = 20.79$ J K^{-1} mol^{-1}. The quantity 55.40 g of Xe corresponds to 0.4219 mole. From Equations 3.21 and 3.22,

$$\Delta U = n\bar{C}_V \Delta T$$
$$= (0.4219 \text{ mol})(12.47 \text{ J K}^{-1} \text{ mol}^{-1})(400 - 300) \text{ K}$$
$$= 526 \text{ J}$$
$$\Delta H = n\bar{C}_P \Delta T$$
$$= (0.4219 \text{ mol})(20.79 \text{ J K}^{-1} \text{ mol}^{-1})(400 - 300) \text{ K}$$
$$= 877 \text{ J}$$

Example 3.4

The molar heat capacity of oxygen at constant pressure is given by $(25.7 + 0.0130T)$ J K^{-1} mol^{-1}. Calculate the enthalpy change when 1.46 moles of O_2 are heated from 298 K to 367 K.

ANSWER

From Equation 3.21

$$\Delta H = \int_{T_1}^{T_2} n\bar{C}_P \, dT = \int_{298 \text{ K}}^{367 \text{ K}} (1.46 \text{ mol})(25.7 + 0.0130T) \text{ J K}^{-1} \text{ mol}^{-1} \, dT$$
$$= (1.46 \text{ mol})\left(25.7T + \frac{0.0130T^2}{2}\right)_{298 \text{ K}}^{367 \text{ K}} \text{ J K}^{-1} \text{ mol}^{-1}$$
$$= 3.02 \times 10^3 \text{ J}$$

Heat Capacity and Hypothermia

Hypothermia describes the condition when the body mechanisms for producing and conserving heat are exceeded by exposure to severe cold. As warm-blooded animals, our body temperature is maintained at around 37°C. The human body is about 70 percent water by mass, and the high heat capacity of water ensures that normally there is only a slight fluctuation in body temperature. An ambient temperature of 25°C (often described as room temperature) feels warm to us because air has a small specific heat (about 1 J g^{-1} °C^{-1}) and a low density. Consequently, little heat is lost from the body to the surrounding air. The situation is drastically different if one's body is immersed in water at the same temperature. A rough estimate shows that for the same increase in temperature of the surrounding fluid (air or water), the heat lost by the body is nearly 3,000 times greater in the case of water. A mild case of hypothermia occurs when the body temperature is reduced to 35°C, and in severe cases the body temperature may be as low as 28°C. Victims of hypothermia from mild to severe cases exhibit symptoms ranging from shivering (a way to generate heat), muscle rigidity, abnormal heart rhythms, and eventually, to death. Once the body temperature has been significantly lowered, the metabolic rate is slowed to the point that the body cannot recover without an external heat source.

The above discussion shows that hypothermia occurs much more quickly when a person falls into a frozen lake than when exposed to cold air temperature. Sometimes a child survives the trauma of hypothermia caused by immersion in ice water even after breathing has stopped for as long as 30 minutes. When the icy water is inhaled, it enters the lungs and spreads quickly through the relatively short blood stream. The chilled blood cools the brain and reduces the cells' need for oxygen.

3.4 Gas Expansions

The expansion of a gas is a simple process to which we can apply what we have learned so far about the first law of thermodynamics. Although gas expansion does not have much chemical significance, it enables us to use some of the equations derived in previous sections to calculate changes in thermodynamic quantities. We shall assume ideal behavior and consider two special cases: isothermal expansion and adiabatic expansion.

Isothermal Expansions

An *isothermal* process is one in which the temperature is held constant. The work done in isothermal reversible and irreversible expansions was discussed in some detail in Section 3.1 and will not be repeated here. Instead we shall look at changes in heat, internal energy, and enthalpy during such a process.

Because temperature does not change in an isothermal process, the change in energy is zero; that is, $\Delta U = 0$. This follows from the fact that ideal gas molecules neither attract nor repel each other. Consequently, their total energy is independent of the distance of separation between them and therefore the volume. Mathematically, this relationship is expressed as

$$\left(\frac{\partial U}{\partial V}\right)_T = 0$$

This partial derivative tells us that the change in internal energy of the system with respect to volume is zero at constant temperature. For an isothermal process, then, Equation 3.6 becomes

$$\Delta U = q + w = 0$$

or

$$q = -w$$

In an isothermal expansion, the heat absorbed by the gas equals the work done by the ideal gas on its surroundings. Referring to Example 3.1, we see that the heat absorbed by an ideal gas when it expands from a pressure of 15 atm to 1 atm at 300 K is zero in (a), 1980 J in (b), and 5740 J in (c). Because maximum work is performed in a reversible process, it is not surprising that the heat absorbed is greatest for (c).

Finally, we also wish to calculate the enthalpy change for such an isothermal process. Starting from

$$\Delta H = \Delta U + \Delta(PV)$$

we have $\Delta U = 0$, as mentioned above, and because PV is constant at constant T and n (Boyle's law), $\Delta(PV) = 0$, so that $\Delta H = 0$. Alternatively, we could write $\Delta(PV) = \Delta(nRT)$. Because the temperature is unchanged and no chemical reaction occurs, both n and T are constant and $\Delta(nRT) = 0$; hence $\Delta H = 0$.

Adiabatic Expansions

Referring to Figure 3.1, suppose we now isolate the cylinder thermally from its surroundings so there is no heat exchange during the expansion. This means that $q = 0$ and the process is *adiabatic*. (The word adiabatic means no heat exchange with the surroundings.) Consequently, there will be a temperature drop and T will no longer be a constant. We consider two cases here.

Reversible Adiabatic Expansion. Let us first suppose that the expansion is reversible. The two questions we ask are: What is the P–V relationship for the initial and final states, and how much work is done in the expansion?

For an infinitesimal adiabatic expansion, the first law takes the form

$$dU = đq + đw$$

$$= đw = -P\,dV = -\frac{nRT}{V}dV$$

or

$$\frac{dU}{nT} = -R\frac{dV}{V}$$

Note that $đq = 0$ and we have replaced the external, opposing pressure with the internal pressure of the gas, because it is a reversible process. Substituting $dU = C_V\,dT$ into the above equation, we obtain

$$\frac{C_V\,dT}{nT} = \bar{C}_V\frac{dT}{T} = -R\frac{dV}{V} \tag{3.25}$$

Integration of Equation 3.25 between the initial and final states gives

$$\int_{T_1}^{T_2} \bar{C}_V\frac{dT}{T} = -R\int_{V_1}^{V_2}\frac{dV}{V}$$

$$\bar{C}_V \ln\frac{T_2}{T_1} = R\ln\frac{V_1}{V_2}$$

(We assume $\bar{C}_V$ to be temperature independent.) Because $\bar{C}_P - \bar{C}_V = R$ for an ideal gas, we write

$$\bar{C}_V \ln\frac{T_2}{T_1} = (\bar{C}_P - \bar{C}_V)\ln\frac{V_1}{V_2}$$

Dividing by $\bar{C}_V$ on both sides, we obtain

$$\ln\frac{T_2}{T_1} = \left(\frac{\bar{C}_P}{\bar{C}_V} - 1\right)\ln\frac{V_1}{V_2}$$

$$= (\gamma - 1)\ln\frac{V_1}{V_2} = \ln\left(\frac{V_1}{V_2}\right)^{\gamma-1}$$

where γ is called the *heat capacity ratio*, given by

$$\gamma = \frac{\bar{C}_P}{\bar{C}_V} \tag{3.26}$$

For a monatomic gas, $\bar{C}_V = \frac{3}{2}R$ and $\bar{C}_P = \frac{5}{2}R$, so that $\gamma = \frac{5}{3}$ or 1.67. Finally, we arrive at the following useful results:

$$\left(\frac{V_1}{V_2}\right)^{\gamma-1} = \frac{T_2}{T_1} = \frac{P_2 V_2}{P_1 V_1} \quad \left(\frac{P_1 V_1}{T_1} = \frac{P_2 V_2}{T_2}\right)$$

or

$$\left(\frac{V_1}{V_2}\right)^{\gamma} = \frac{P_2}{P_1}$$

Thus, for an adiabatic process, the P–V relationship becomes

$$P_1 V_1^{\gamma} = P_2 V_2^{\gamma} \tag{3.27}$$

It is useful to keep in mind the conditions under which this equation was derived: (1) It applies to an ideal gas, and (2) it applies to a reversible adiabatic change. Equation 3.27 differs from Boyle's law $(P_1 V_1 = P_2 V_2)$ in the exponent γ, because temperature is *not* kept constant in an adiabatic expansion.

The work done in an adiabatic process is given by

$$w = \int_1^2 dU = \Delta U = \int_{T_1}^{T_2} C_V\, dT$$

$$= C_V(T_2 - T_1)$$

$$w = n\bar{C}_V(T_2 - T_1) \tag{3.28}$$

where $T_2 < T_1$. Because the gas expands, the internal energy of the system decreases.

Appearance of the quantity $\bar{C}_V$ in Equation 3.28 may seem strange because the volume is not held constant. However, adiabatic expansion (from P_1, V_1, T_1 to P_2, V_2, T_2) can be thought of as a two-step process, as shown in Figure 3.6. First, the

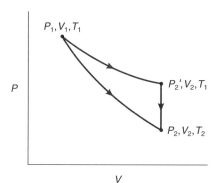

Figure 3.6
Because U is a state function, ΔU is the same whether the change of a gas from P_1, V_1, T_1 to P_2, V_2, T_2 occurs directly or indirectly.

gas is isothermally expanded from P_1V_1 to $P_2'V_2$ at T_1. Because the temperature is constant, $\Delta U = 0$. Next the gas is cooled at constant volume from T_1 to T_2 and its pressure drops from P_2' to P_2. In this case, $\Delta U = n\bar{C}_V(T_2 - T_1)$, which is Equation 3.28. The fact that U is a state function enables us to analyze the process by employing a different path.

Example 3.5

A quantity of 0.850 mole of a monatomic ideal gas initially at a pressure of 15.0 atm and 300 K is allowed to expand until its final pressure is 1.00 atm (see Example 3.1). Calculate the work done if the expansion is carried out adiabatically and reversibly.

ANSWER

Our first task is to calculate the final temperature, T_2. This is done in three steps. First, we need to evaluate V_1, given by $V_1 = nRT_1/P_1$.

$$V_1 = \frac{(0.850 \text{ mol})(0.08206 \text{ L atm K}^{-1} \text{ mol}^{-1})(300 \text{ K})}{15.0 \text{ atm}}$$

$$= 1.40 \text{ L}$$

Next, we calculate the value of V_2 using the following relation:

$$P_1 V_1^{\gamma} = P_2 V_2^{\gamma}$$

$$V_2 = \left(\frac{P_1}{P_2}\right)^{1/\gamma} V_1 = \left(\frac{15.0}{1.00}\right)^{3/5}(1.40 \text{ L}) = 7.1 \text{ L}$$

Finally, we have $P_2 V_2 = nRT_2$, or

$$T_2 = \frac{P_2 V_2}{nR} = \frac{(1.00 \text{ atm})(7.1 \text{ L})}{(0.850 \text{ mol})(0.08206 \text{ L atm K}^{-1} \text{ mol}^{-1})}$$

$$= 102 \text{ K}$$

Hence,

$$\Delta U = w = n\bar{C}_V(T_2 - T_1)$$

$$= (0.850 \text{ mol})(12.47 \text{ J K}^{-1} \text{ mol}^{-1})(102 - 300) \text{ K}$$

$$= -2.1 \times 10^3 \text{ J}$$

This comparison shows that not all reversible processes do the same amount of work.

Examples 3.1 and 3.5 show that less work is performed in a reversible adiabatic expansion than in a reversible isothermal expansion. In the latter case, heat is absorbed from the surroundings to make up for the work done by the gas, but this does not occur in an adiabatic process so that the temperature drops. Plots of reversible isothermal and adiabatic expansions are shown in Figure 3.7.

Irreversible Adiabatic Expansion. Finally, consider what happens in an irreversible adiabatic expansion. Suppose we start with an ideal gas at P_1, V_1, and T_1, and P_2 is the constant external pressure. The final volume and temperature of the gas are V_2 and T_2. Again, $q = 0$ so that

$$\Delta U = n\bar{C}_V(T_2 - T_1) = w = -P_2(V_2 - V_1) \tag{3.29}$$

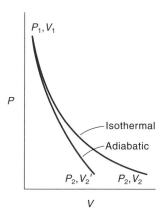

Figure 3.7
Pressure versus volume plots of an adiabatic, reversible and an isothermal, reversible expansion of an ideal gas. In each case, the work done in expansion is represented by the area under the curve. Note that the curve is steeper for the adiabatic expansion because γ is greater than one. Consequently, the area is smaller than the isothermal case.

Furthermore, from the ideal gas equation we write

$$V_1 = \frac{nRT_1}{P_1} \quad \text{and} \quad V_2 = \frac{nRT_2}{P_2}$$

Substituting the expressions for V_1 and V_2 in Equation 3.29, we obtain

$$n\bar{C}_V(T_2 - T_1) = -P_2\left(\frac{nRT_2}{P_2} - \frac{nRT_1}{P_1}\right) \tag{3.30}$$

Thus, knowing the initial conditions and P_2, we can solve for T_2 and hence the work done (see Problem 3.35).

The decrease in temperature, or the cooling effect that occurs in an adiabatic expansion, has some interesting practical consequences. A familiar example is the formation of fog when the caps of soft drinks or corks of champagne bottles are removed. Initially, the bottles are pressurized with carbon dioxide and air, and the space above the liquid is saturated with water vapor. When the cap is removed, the gases inside rush out. The process takes place so rapidly that the expansion of the gases can be compared to adiabatic expansion. As a result, the temperature drops and water vapor condenses to form the observed fog.

Liquefaction of gases is based on the same principle. Normally, the steps are: (1) compress a gas isothermally, (2) let the compressed gas expand adiabatically, (3) recompress the cooled gas isothermally, and so on, until the gas condenses to liquid.

3.5 Calorimetry

Calorimetry is the measurement of heat changes. In the laboratory, heat changes in physical and chemical processes are measured with a calorimeter, an apparatus designed specifically for this purpose. There are many types of calorimeters, depending on the purpose of the experiment. We shall consider three types in this section.

Constant-Volume Calorimetry

Heat of combustion is usually measured in a constant-volume adiabatic bomb calorimeter (Figure 3.8). It is a tightly sealed, heavy-walled, stainless steel container, which, together with the water, are thermally isolated from the surroundings. The substance under investigation is placed inside the container, which is filled with oxygen at about 30 atm. The combustion is started by an electrical discharge through a

Adiabatic means no heat exchange between the calorimeter and the surroundings and *bomb* denotes the explosive nature of the reaction (on a small scale).

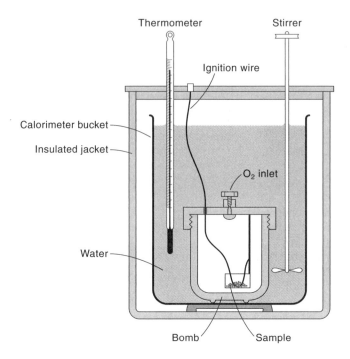

Figure 3.8
Schematic diagram of a constant-volume bomb calorimeter.

pair of wires that are in contact with the substance. The heat released by the reaction can be measured by registering the rise in the temperature of the water filling the inner jacket of the calorimeter. To determine the heat of combustion, we need to know the heat capacity of the calorimeter. The combined heat capacities of the water and the bomb are first determined by burning a compound of accurately known heat of combustion. As shown earlier, in a constant-volume process the heat produced is equal to the change in internal energy of the system:

$$\Delta U = q_V + w$$
$$= q_V - P\Delta V$$
$$= q_V$$

Example 3.6

When a 0.5122-g sample of naphthalene ($C_{10}H_8$) was burned in a constant-volume bomb calorimeter, the temperature of the water in the inner jacket (see Figure 3.8) rose from 20.17°C to 24.08°C. If the effective heat capacity (C_V) of the bomb calorimeter plus water is 5267.8 J K^{-1}, calculate ΔU and ΔH for the combustion of naphthalene in kJ mol^{-1}.

ANSWER

The reaction is

$$C_{10}H_8(s) + 12O_2(g) \rightarrow 10CO_2(g) + 4H_2O(l)$$

The amount of heat evolved is given by

$$C_V \Delta T = (5267.8 \text{ J K}^{-1})(3.91 \text{ K}) = 20.60 \text{ kJ}$$

From the molar mass of naphthalene (128.2 g), we write

$$q_V = \Delta U = -\frac{(20.60 \text{ kJ})(128.2 \text{ g mol}^{-1})}{0.5122 \text{ g}} = -5156 \text{ kJ mol}^{-1}$$

The negative sign indicates that the reaction is exothermic.

To calculate ΔH, we start with $\Delta H = \Delta U + \Delta(PV)$. When all reactants and products are in condensed phases, $\Delta(PV)$ is negligible in comparison with ΔH and ΔU. When gases are involved, $\Delta(PV)$ cannot be ignored. Assuming ideal gas behavior, we have $\Delta(PV) = \Delta(nRT) = RT\Delta n$, where Δn is the change in the number of moles of gas in the reaction. Note that T refers to the *initial* temperature here because we are comparing reactants and products under the same conditions. For our reaction, $\Delta n = (10 - 12) \text{ mol} = -2 \text{ mol}$ so that

$$\Delta H = \Delta U + RT\Delta n$$

$$= -5156 \text{ kJ mol}^{-1} + \frac{(8.314 \text{ J K}^{-1} \text{ mol}^{-1})(293.32 \text{ K})(-2)}{1000 \text{ J/kJ}}$$

$$= -5161 \text{ kJ mol}^{-1}$$

COMMENT

(1) The difference between ΔU and ΔH is quite small for this reaction. The reason is that $\Delta(PV)$ (which in this case is equal to $RT\Delta n$) is small compared to ΔU or ΔH. Because we assumed ideal gas behavior (we ignored the volume change of condensed phases), ΔU has the same value ($-5156 \text{ kJ mol}^{-1}$) whether the process occurs at constant V or at constant P because the internal energy is independent of pressure or volume. Similarly, $\Delta H = -5161 \text{ kJ mol}^{-1}$ whether the process is carried out at constant V or at constant P. The heat change q, however, is $-5156 \text{ kJ mol}^{-1}$ at constant V and $-5161 \text{ kJ mol}^{-1}$ at constant P, because it is path dependent. (2) In our calculation we ignored the heat capacities of the products (water and carbon dioxide) because the amounts of these substances are small compared to the bomb calorimeter itself. This omission does not introduce serious errors.

Constant-Pressure Calorimetry

For many physical processes (such as phase transitions and dissolutions and dilutions) and chemical reactions (for example, acid–base neutralization), which take place under atmospheric conditions, the heat change is equal to the enthalpy change; that is, $q_P = \Delta H$. Figure 3.9 shows a crude constant-pressure calorimeter, made of two Styrofoam coffee cups. To determine ΔH, we need to know the heat capacity of the calorimeter (C_P) and the temperature change. A more refined calorimeter employs a Dewar flask and a thermocouple to monitor the temperature. As in the case of constant-volume calorimeter, no heat exchange is assumed to take place between the calorimeter and the surroundings during the experiment.

Suppose we wish to measure the heat of hydrolysis (ΔH) of adenosine 5′-triphosphate (ATP) to give adenosine 5′-diphosphate (ADP), a key reaction in many biological processes:

$$\text{ATP} + \text{H}_2\text{O} \rightarrow \text{ADP} + \text{P}_i$$

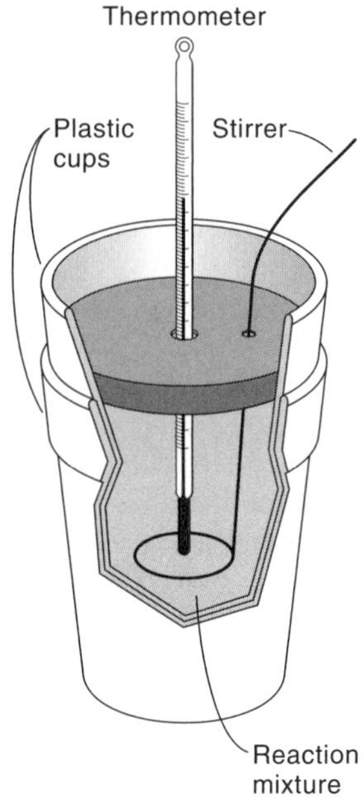

Figure 3.9
A constant-pressure calorimeter made of two Styrofoam coffee cups.
The outer cup helps to insulate the reacting mixture from the surroundings.
Two solutions of known volume containing the reactants at the same
temperature are carefully mixed in the calorimeter. The heat produced or
absorbed by the reaction can be determined from the temperature change,
the quantities of the solutions used, and the heat capacity of the calorimeter.

where P_i denotes the inorganic phosphate group. We start by placing a solution of known concentration of ATP in the calorimeter. The reaction is then initiated by adding a small amount of the enzyme adenosine triphosphatase (ATPase), which catalyzes the hydrolysis reaction. From the rise in temperature, ΔT, we have

$$\Delta H = C_P \Delta T$$

In reality, ΔH depends on many factors such as pH and the nature of counter ions. A typical value is $\Delta H \approx -30$ kJ per mole of ATP hydrolyzed to ADP.

Differential Scanning Calorimetry

Differential scanning calorimetry (DSC) is a powerful technique that enables us to study the energetics of biopolymers (for example, proteins and nucleic acids). Suppose we are interested in the thermodynamics of protein denaturation. In its physiologically functioning state, a protein molecule has a unique three-dimensional structure held together by various intra- and intermolecular forces. But this delicately balanced structure can be disrupted by various chemical reagents (called denaturants), changes in pH, and temperature, which cause the protein to unfold. When this happens, the protein has lost its biological function and is said to be denatured. Most proteins denature at elevated temperatures, sometimes only a few degrees higher than those at which they function.

To measure the enthalpy change of protein denaturation (ΔH_d), we use a differential scanning calorimeter (shown schematically in Figure 3.10). The sample cell containing a protein in a buffer solution and a reference cell containing only the buffer solution are slowly heated electrically and the temperature rise in both cells is kept the same. Because the protein solution has a greater heat capacity, additional electric current is required to maintain its temperature rise. Figure 3.11a shows a plot of specific heat capacity, which is the difference between the heat capacity of the protein solution and the reference solution, versus temperature. Initially the curve rises slowly. As the protein begins to denature, there is a large absorption of heat because the process is highly endothermic. The temperature corresponding to the peak is called the

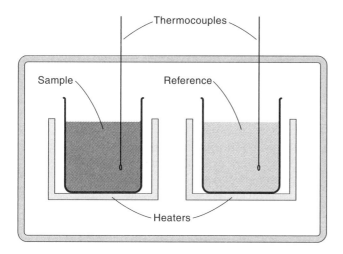

Figure 3.10
Schematic diagram of a differential scanning calorimeter. The sample and reference solutions are heated slowly. A feedback electronic circuit (via the thermocouples) is used to add additional heat to the sample solution so that the two solutions are maintained at the same temperature throughout the experiment.

melting temperature (T_m). When unfolding is complete, the specific heat capacity again rises slowly, although at an elevated level because the denatured protein has a greater heat capacity than the native one. The shaded area gives the enthalpy of denaturation; that is,

$$\Delta H_d = \int_{T_1}^{T_2} C_P \, dT$$

The mechanism of protein denaturation for many small proteins (molar mass $\leq$ 40,000 g) is that of a two-state model. That is, we can describe the process as the

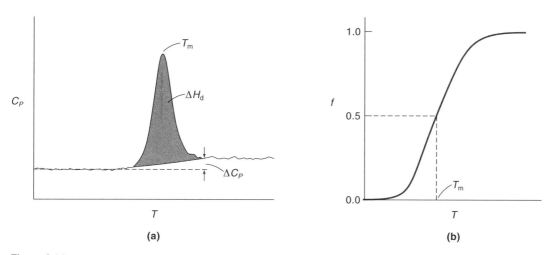

(a)   **(b)**

Figure 3.11
(a) A typical DSC thermogram. ΔH_d is the enthalpy of denaturation and T_m is the temperature at which half of the protein is denatured. ΔC_P is the heat capacity of the denatured protein solution minus the heat capacity of the native protein solution. (b) The fraction (f) of denatured protein as a function of temperature.

transition between the native protein (N) and denatured protein (D) as follows (Figure 3.12):

$$N \rightleftharpoons D$$

Near the denaturation temperature, when one non-covalent bond breaks, all such bonds break simultaneously in an all-or-none fashion. The more cooperative the process, the narrower the peak. (A familiar example of a two-state cooperative transition is that between ice and water at 0°C.) The melting temerature T_m is defined as the temperature at which $[N] = [D]$, or that the protein is half denatured (Figure 3.11b). Studies show that the enthalpy of denaturation, ΔH_d, falls between 200 kJ mol^{-1} and 800 kJ mol^{-1} for most proteins. As a general rule, the more stable the protein, the greater are the T_m and ΔH_d values. Note that in such a calorimetric measurement, we do not need to know the molecular structure or composition of the protein in determining ΔH_d. By the same token, when we say that one protein is more stable than another because it has a higher T_m and a larger ΔH_d value, we have no idea why this is so in the absence of X-ray diffraction and spectroscopic data.

Figure 3.12
Schematic diagram of the native (top) and denatured (bottom) forms of a protein.

3.6 Thermochemistry

In this section we shall apply the first law of thermodynamics to thermochemistry, the study of energy changes in chemical reactions. Calorimetry provides us with the experimental approach; here we shall derive the necessary equations for dealing with such processes.

Standard Enthalpy of Formation

Chemical reactions almost always involve changes in heat. The *heat of reaction* can be defined as the heat change in the transformation of reactants at some temperature and pressure to products at the same temperature and pressure. For a constant-pressure process, the heat of reaction, q_P, is equal to the enthalpy change of the reaction, $\Delta_r H$, where the subscript r denotes reaction. An *exothermic reaction* is a process that gives off heat to its surroundings and for which $\Delta_r H$ is negative; for an *endothermic reaction*, $\Delta_r H$ is positive because the process absorbs heat from the surroundings.

Consider the following reaction:

$$\mathrm{C(graphite)} + \mathrm{O_2}(g) \rightarrow \mathrm{CO_2}(g)$$

In quoting $\Delta_r H$ values we must specify the stoichiometric equation. For simplicity, we shall express $\Delta_r H°$ as kJ mol^{-1}.

When 1 mole of graphite is burned in an excess of oxygen at 1 bar and 298 K to form 1 mole of carbon dioxide at the same temperature and pressure, 393.5 kJ of heat is given off.* The enthalpy change for this process is called the *standard enthalpy of reaction*, denoted $\Delta_r H°$. It has the units of kJ (mol reaction)$^{-1}$. One mole of reaction is when the appropriate numbers of moles of substances (as specified by the stoichiometric coefficients) on the left side are converted to the substances on the right side of the equation. We express $\Delta_r H°$ for the combustion of graphite as -393.5 kJ mol^{-1}, which is defined as the enthalpy change when the reactants in their standard states are converted to the product in its standard state. The standard state is defined as follows: For a pure solid or liquid, it is the state at a pressure $P = 1$ bar (see Section

The standard state is defined only in terms of pressure.

---

*The temperature during combustion is much higher than 298 K, but we are measuring the total heat change from reactants at 1 bar and 298 K to product at 1 bar and 298 K. Therefore, the heat given off when the product cools to 298 K is part of the enthalpy of reaction.

1.2) and some temperature T. For a pure gas, the standard state refers to the hypothetical ideal gas at a pressure of 1 bar and some temperature of interest. The symbol for a standard state is a zero superscript.

In general, the standard enthalpy change of a chemical reaction can be thought of as the total enthalpy of the products minus the total enthalpy of the reactants:

$$\Delta_r H^\circ = \Sigma v \overline{H}^\circ (\text{products}) - \Sigma v \overline{H}^\circ (\text{reactants})$$

where $\overline{H}^\circ$ is the standard molar enthalpy and v is the stoichiometric coefficient. The units of $\overline{H}^\circ$ are kJ mol^{-1}, and v is a number without units. For the hypothetical reaction

$$a\text{A} + b\text{B} \rightarrow c\text{C} + d\text{D}$$

the standard enthalpy of reaction is given by

$$\Delta_r H^\circ = c\overline{H}^\circ(\text{C}) + d\overline{H}^\circ(\text{D}) - a\overline{H}^\circ(\text{A}) - b\overline{H}^\circ(\text{B})$$

We cannot measure the *absolute* values of molar enthalpy of any substance, however. To circumvent this dilemma, we use *standard molar enthalpy of formation* ($\Delta_f \overline{H}^\circ$) of reactants and products, where the subscript f denotes formation. The standard molar enthalpy of formation is an intensive quantity. It is the enthalpy change when 1 mole of a compound is formed from its constituent elements at 1 bar and 298 K. The standard enthalpy change for the above reaction can now be written as

$$\Delta_r H^\circ = c\Delta_f \overline{H}^\circ(\text{C}) + d\Delta_f \overline{H}^\circ(\text{D}) - a\Delta_f \overline{H}^\circ(\text{A}) - b\Delta_f \overline{H}^\circ(\text{B}) \qquad (3.31)$$

In general, we write

$$\Delta_r H^\circ = \Sigma v \Delta_f \overline{H}^\circ (\text{products}) - \Sigma v \Delta_f \overline{H}^\circ (\text{reactants}) \qquad (3.32)$$

For the formation of CO_2 shown above, we express the standard enthalpy of reaction as

$$\Delta_r H^\circ = \Delta_f \overline{H}^\circ(CO_2) - \Delta_f \overline{H}^\circ(\text{graphite}) - \Delta_f \overline{H}^\circ(O_2)$$
$$= -393.5 \text{ kJ mol}^{-1}$$

By convention, we arbitrarily assign a value of zero to $\Delta_f \overline{H}^\circ$ for elements in their most stable allotropic forms at a particular temperature. If we choose 298 K, then

$$\Delta_f \overline{H}^\circ(O_2) = 0$$
$$\Delta_f \overline{H}^\circ(\text{graphite}) = 0$$

Allotropes are two or more forms of the same element that differ in physical and chemical properties.

because O_2 and graphite are the stable allotropic forms of oxygen and carbon at this temperature. Now neither ozone nor diamond is the more stable allotropic form at 1 bar and 298 K, and so we have

$$\Delta_f \overline{H}^\circ(O_3) \neq 0 \quad \text{and} \quad \Delta_f \overline{H}^\circ(\text{diamond}) \neq 0$$

The standard enthalpy for the combustion of graphite can now be written as

$$\Delta_r H^\circ = \Delta_f \overline{H}^\circ(CO_2) = -393.5 \text{ kJ mol}^{-1}$$

This procedure is analogous to choosing the sea level at zero meters for measuring terrestrial altitudes.

Thus, the standard molar enthalpy of formation for CO_2 is equal to the standard enthalpy of reaction.

There is no mystery in the assignment of a zero value to $\Delta_f \overline{H}^\circ$ for the elements. As mentioned above, we cannot determine the absolute value of the enthalpy of a substance. Only values *relative* to an arbitrary reference can be given. In thermodynamics, we are primarily interested in the changes of H. Although any arbitrarily assigned value of $\Delta_f \overline{H}^\circ$ for an element would work, zero makes calculations simpler. The importance of the standard molar enthalpies of formation is that once we know their values, we can calculate the standard enthalpies of reaction. The $\Delta_f \overline{H}^\circ$ values are obtained by either the direct method or the indirect method, described below.

The Direct Method. This method of measuring $\Delta_f \overline{H}^\circ$ works for compounds that can be readily synthesized from their elements. The formation of CO_2 from graphite and O_2 is such an example. Other compounds that can be directly synthesized from their elements are SF_6, P_4O_{10}, and CS_2. The equations representing their syntheses are

$$S(\text{rhombic}) + 3F_2(g) \rightarrow SF_6(g) \qquad \Delta_r H^\circ = -1209 \text{ kJ mol}^{-1}$$

$$4P(\text{white}) + 5O_2(g) \rightarrow P_4O_{10}(s) \qquad \Delta_r H^\circ = -2984.0 \text{ kJ mol}^{-1}$$

$$C(\text{graphite}) + 2S(\text{rhombic}) \rightarrow CS_2(l) \qquad \Delta_r H^\circ = 89.7 \text{ kJ mol}^{-1}$$

Note that S(rhombic) and P(white) are the most stable allotropes of sulfur and phosphorus, respectively, at 1 bar and 298 K, so their $\Delta_f \overline{H}^\circ$ values are zero. As for CO_2, the standard enthalpy of reaction ($\Delta_r H^\circ$) for the three reactions shown is equal to $\Delta_f \overline{H}^\circ$ for the compound in each case.

The Indirect Method. Most compounds cannot be synthesized directly from their elements. In some cases, the reaction proceeds too slowly or not at all, or side reactions produce compounds other than the desired product. In these cases, the value of $\Delta_f \overline{H}^\circ$ can be determined by an indirect approach, which is based on Hess's law. *Hess's law* (after the Swiss chemist Germain Henri Hess, 1802–1850) can be stated as follows: When reactants are converted to products, the change in enthalpy is the same whether the reaction takes place in one step or in a series of steps. In other words, if we can break down a reaction into a series of reactions for which the value of $\Delta_r H^\circ$ can be measured, we can calculate the $\Delta_r H^\circ$ value for the overall reaction. The logic of Hess's law is that because enthalpy is a state function, its change is path independent.

A simple analogy for Hess's law is as follows. Suppose you go from the first floor to the sixth floor of a building by elevator. The gain in your gravitational potential energy (which corresponds to the enthalpy change for the overall process) is the same whether you go directly there or stop at each floor on your way up (breaking the reaction into a series of steps).

Let us apply Hess's law to find the value of $\Delta_f \overline{H}^\circ$ for carbon monoxide. We might represent the synthesis of CO from its elements as

$$C(\text{graphite}) + \tfrac{1}{2}O_2(g) \rightarrow CO(g)$$

We cannot burn graphite in oxygen without also forming some CO_2, however, so this approach will not work. To circumvent this difficulty, we can carry out the following two separate reactions, which do go to completion:

(1) $C(\text{graphite}) + O_2(g) \rightarrow CO_2(g)$ $\Delta_r H^\circ = -393.5 \text{ kJ mol}^{-1}$

(2) $CO(g) + \frac{1}{2}O_2(g) \rightarrow CO_2(g)$ $\Delta_r H^\circ = -283.0 \text{ kJ mol}^{-1}$

First, we reverse equation 2 to get

(3) $CO_2(g) \rightarrow CO(g) + \frac{1}{2}O_2(g)$ $\Delta_r H^\circ = +283.0 \text{ kJ mol}^{-1}$ When we reverse an equation, $\Delta_r H^\circ$ changes sign.

Because chemical equations can be added and subtracted just like algebraic equations, we carry out the operation $(1) + (3)$ and obtain

(4) $C(\text{graphite}) + \frac{1}{2}O_2(g) \rightarrow CO(g)$ $\Delta_r H^\circ = -110.5 \text{ kJ mol}^{-1}$

Thus, $\Delta_f \overline{H}^\circ(CO) = -110.5 \text{ kJ mol}^{-1}$. Looking back, we see that the overall reaction is the formation of CO_2 (reaction 1), which can be broken down into two parts (reactions 2 and 4). Figure 3.13 shows the overall scheme of our procedure.

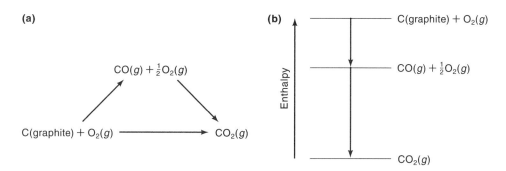

Figure 3.13
(a) The formation of CO_2 from graphite and O_2 can be broken into two steps. (b) The enthalpy change for the overall reaction is equal to the sum of the enthalpy changes for the two steps.

The general rule in applying Hess's law is that we should arrange a series of chemical equations (corresponding to a series of steps) in such a way that, when added together, all species cancel except for the reactants and product that appear in the overall reaction. This means that we want the elements on the left and the compound of interest on the right of the arrow. To achieve this goal, we often need to multiply some or all of the equations representing the individual steps by the appropriate coefficients.

Table 3.3 lists the $\Delta_f \overline{H}^\circ$ values for a number of common elements and inorganic and organic compounds. (A more extensive listing is given in Appendix 2.) Note that compounds differ not only in the magnitude of $\Delta_f \overline{H}^\circ$ but in the sign as well. Water and other compounds that have negative $\Delta_f \overline{H}^\circ$ values lie "downhill" on the enthalpy scale relative to their constituent elements (Figure 3.14). These compounds tend to be more stable than those that have positive $\Delta_f \overline{H}^\circ$ values. The reason is that energy has to be supplied to the former to decompose them into the elements, while the latter decompose with the evolution of heat.

Table 3.3
Standard Molar Enthalpies of Formation at 298 K and 1 Bar for Some Inorganic and Organic Substances

Substance	$\Delta_f \overline{H}°/\text{kJ} \cdot \text{mol}^{-1}$	Substance	$\Delta_f \overline{H}°/\text{kJ} \cdot \text{mol}^{-1}$
C(graphite)	0	$CH_4(g)$	−74.85
C(diamond)	1.90	$C_2H_6(g)$	−84.7
CO(g)	−110.5	$C_3H_8(g)$	−103.8
$CO_2(g)$	−393.5	$C_2H_2(g)$	226.6
HF(g)	−268.6	$C_2H_4(g)$	52.3
HCl(g)	−92.3	$C_6H_6(l)$	49.04
HBr(g)	−36.4	$CH_3OH(l)$	−238.7
HI(g)	26.48	$C_2H_5OH(l)$	−277.0
$H_2O(g)$	−241.8	$CH_3CHO(l)$	−192.3
$H_2O(l)$	−285.8	HCOOH(l)	−424.7
$NH_3(g)$	−46.3	$CH_3COOH(l)$	−484.2
NO(g)	90.4	$C_6H_{12}O_6(s)$	−1274.5
$NO_2(g)$	33.9	$C_{12}H_{22}O_{11}(s)$	−2221.7
$N_2O_4(g)$	9.7		
$N_2O(g)$	81.56		
$O_3(g)$	142.7		
$SO_2(g)$	−296.1		
$SO_3(g)$	−395.2		

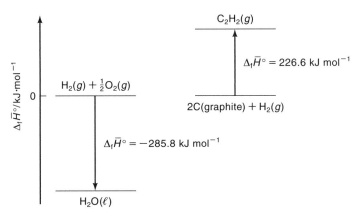

Figure 3.14
Two representative compounds with negative and positive $\Delta_f \overline{H}°$ values.

Dependence of Enthalpy of Reaction on Temperature

Suppose you have measured the standard enthalpy of a reaction at a certain temperature, say 298 K, and want to know its value at 350 K. One way to find out is to repeat the measurement at the higher temperature. Fortunately, we also can obtain the desired quantity from tabulated thermodynamic data without doing another experiment. For any reaction, the change in enthalpy at a particular temperature is

$$\Delta_r H = \Sigma H(\text{products}) - \Sigma H(\text{reactants})$$

To see how the enthalpy of reaction ($\Delta_r H$) itself changes with temperature, we differentiate this equation with respect to temperature at constant pressure as follows

$$\left(\frac{\partial \Delta_r H}{\partial T}\right)_P = \left[\frac{\partial \Sigma H(\text{products})}{\partial T}\right]_P - \left[\frac{\partial \Sigma H(\text{reactants})}{\partial T}\right]_P$$

$$= \Sigma C_P(\text{products}) - \Sigma C_P(\text{reactants})$$

$$= \Delta C_P \qquad (3.33)$$

because $(\partial H/\partial T)_P = C_P$. Integration of Equation 3.33 gives

$$\int_1^2 d\Delta_r H = \Delta_r H_2 - \Delta_r H_1 = \int_{T_1}^{T_2} \Delta C_P \, dT = \Delta C_P(T_2 - T_1) \qquad (3.34)$$

where $\Delta_r H_1$ and $\Delta_r H_2$ are the enthalpies of reaction at T_1 and T_2, respectively. Equation 3.34 is known as *Kirchhoff's law* (after the German physicist Gustav-Robert Kirchhoff, 1824–1887). This law says that the difference between the enthalpies of a reaction at two different temperatures is just the difference in the enthalpies of heating the products and reactants from T_1 to T_2 (Figure 3.15). Note that in deriving this equation, we have assumed the C_P values are all independent of temperature. Otherwise, they must be expressed as a function of T in the integration, as mentioned in Section 3.3.

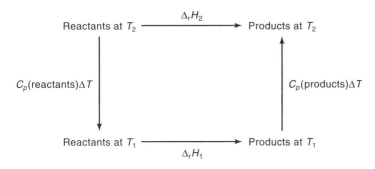

Figure 3.15
Schematic diagram showing Kirchhoff's law (Equation 3.34). The enthalpy change is $\Delta_r H_2 = \Delta_r H_1 + \Delta C_P(T_2 - T_1)$, where ΔC_P is the difference in heat capacities between products and reactants.

Example 3.7

The standard enthalpy change for the reaction

$$3O_2(g) \rightarrow 2O_3(g)$$

is given by $\Delta_r H^\circ = 285.4$ kJ mol^{-1} at 298 K and 1 bar. Calculate the value of $\Delta_r H^\circ$ at 380 K. Assume that the $\bar{C}_P$ values are all independent of temperature.

ANSWER

In Appendix 2, we find the molar heat capacities at constant pressure for O_2 and O_3 to be 29.4 J K^{-1} mol^{-1} and 38.2 J K^{-1} mol^{-1}, respectively. From Equation 3.34

$$\Delta_r H^\circ_{380} - \Delta_r H^\circ_{298} = \frac{[(2)38.2 - (3)29.4]\text{ J K}^{-1}\text{ mol}^{-1}}{(1000 \text{ J/kJ})} \times (380 - 298) \text{ K}$$

$$= -0.97 \text{ kJ mol}^{-1}$$

$$\Delta_r H^\circ_{380} = (285.4 - 0.97) \text{ kJ mol}^{-1}$$

$$= 284.4 \text{ kJ mol}^{-1}$$

3.7 Bond Energies and Bond Enthalpies

Because chemical reactions involve the breaking and making of chemical bonds in the reactant and product molecules, a proper understanding of the thermochemical nature of reactions clearly requires a detailed knowledge of bond energies. Bond energy is the energy required to break a bond between two atoms. Consider the dissociation of 1 mole of H_2 molecules at 298 K and 1 bar:

$$H_2(g) \rightarrow 2H(g) \qquad\qquad \Delta_r H^\circ = 436.4 \text{ kJ mol}^{-1}$$

Assigning the energy of the H–H bond a value of 436.4 kJ mol^{-1} might be tempting, but the situation is more complicated. What is measured is actually the bond enthalpy of H_2, not its bond energy. To understand the difference between these two quantities, consider first what we mean by bond energy.

Figure 3.16 shows the *potential-energy curve* of the H_2 molecule. Let us start by asking how the molecule is formed. At first, the two hydrogen atoms are far apart and exert no influence on each other. As the distance of separation is shortened, both Coulombic attraction (between electron and nucleus) and Coulombic repulsion (between electron and electron and nucleus and nucleus) begin to affect each atom. Because attraction outweighs repulsion, the potential energy of the system decreases with decreasing distance of separation. This process continues until the net attraction force reaches a maximum, leading to the formation of a hydrogen molecule. Further shortening of the distance increases the repulsion, and the potential rises steeply. The

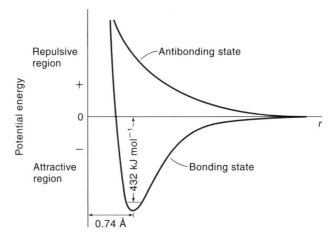

Figure 3.16
Potential-energy curve for a diatomic molecule. The short horizontal line represents the lowest vibrational energy level (the zero-point energy) of the molecule (for example, H_2). The intercepts of this line with the curve shows the maximum and minimum bond lengths during a vibration.

reference state (zero potential energy) corresponds to the case of two infinitely separated H atoms. Potential energy is a negative quantity for the bound state (that is, H_2), and energy in the form of heat is given off as a result of the bond formation.

The most important features in Figure 3.16 are the minimum point on the potential-energy curve for the bonding state, which represents the most stable state for the molecule, and the corresponding distance of separation, called the *equilibrium distance*. However, molecules are constantly executing vibrational motions that persist even at absolute zero. Furthermore, the energies associated with vibration, like the energies of an electron in an atom, are quantized. The lowest vibrational energy is not zero but equal to $\frac{1}{2}h\nu$, called the *zero-point energy*, where ν is the frequency of vibration of H_2. Consequently, the two hydrogen atoms cannot be held rigidly in the molecule, as is the case when a molecule is situated at the minimum point. Instead, the lowest vibrational state for H_2 is represented by the horizontal line. The intercepts between this line and the potential-energy curve represent the two extreme bond lengths during the course of a vibration. We can still speak of equilibrium distance in this case, although technically it is the average of the two extreme bond lengths. The bond energy of H_2 is the vertical distance from lowest vibrational energy level to the reference state of zero potential energy.

The measured enthalpy change ($436.4 \text{ kJ mol}^{-1}$) should not be identified with the bond energy of H_2 for two reasons. First, upon dissociation, the number of moles of gas doubles, and hence gas-expansion work is done on the surroundings. The enthalpy change (ΔH) is not equal to the internal energy change (ΔU), which is the bond energy, but is related to it by the equation

$$\Delta H = \Delta U + P\Delta V$$

Second, the hydrogen molecules have vibrational, rotational, and translational energy before they dissociate, whereas the hydrogen atoms have only translational energy. Thus, the total kinetic energy of the reactant differs from that of the product. Although these kinetic energies are not relevant to the bond energy, their difference is unavoidably incorporated in the $\Delta_r H^\circ$ value. Thus, despite the fact that bond energy has a firmer theoretical basis, for practical reasons we shall use bond enthalpies to help us study energy changes of chemical reactions.

Bond Enthalpy and Bond Dissociation Enthalpy

With respect to diatomic molecules such as H_2 and the following examples,

$$N_2(g) \rightarrow 2N(g) \qquad \Delta_r H^\circ = 941.4 \text{ kJ mol}^{-1}$$
$$HCl(g) \rightarrow H(g) + Cl(g) \qquad \Delta_r H^\circ = 430.9 \text{ kJ mol}^{-1}$$

bond enthalpy has a special significance because there is only one bond in each molecule, so that the enthalpy change can be assigned unequivocally to that bond. For this reason, we shall use the term *bond dissociation enthalpy* for diatomic molecules. Polyatomic molecules are not so straightforward. Measurements show that the energy needed to break the first O–H bond in H_2O, for example, is different from that needed to break the second O–H bond:

$$H_2O(g) \rightarrow H(g) + OH(g) \qquad \Delta_r H^\circ = 502 \text{ kJ mol}^{-1}$$
$$OH(g) \rightarrow H(g) + O(g) \qquad \Delta_r H^\circ = 427 \text{ kJ mol}^{-1}$$

In each case, an O–H bond is broken, but the first step is more endothermic than the second. The difference between the two $\Delta_r H^\circ$ values suggests that the second O–H

Table 3.4
Average Bond Enthalpies/kJ · mol^{-1a}

Bond	Bond Enthalpy	Bond	Bond Enthalpy
H–H	436.4	C–S	255
H–N	393	C=S	477
H–O	460	N–N	393
H–S	368	N=N	418
H–P	326	N≡N	941.4
H–F	568.2	N–O	176
H–Cl	430.9	N–P	209
H–Br	366.1	O–O	142
H–I	298.3	O=O	498.8
C–H	414	O–P	502
C–C	347	O=S	469
C=C	619	P–P	197
C≡C	812	P=P	490
C–N	276	S–S	268
C=N	615	S=S	351
C≡N	891	F–F	150.6
C–O	351	Cl–Cl	242.7
C=O^b	724	Br–Br	192.5
C–P	264	I–I	151.0

^{a} Bond enthalpies for diatomic molecules have more significant figures than those for polyatomic molecules because they are directly measurable quantities and are not averaged over many compounds as for polyatomic molecules.
^{b} The C=O bond enthalpy in CO_2 is 799 kJ mol^{-1}.

bond itself undergoes change, because the chemical environment has been altered. If we were to study the O–H breaking process in other compounds, such as H_2O_2, CH_3OH, and so on, we would find still other $\Delta_r H°$ values. Thus, for polyatomic molecules, we can speak only of the *average* bond enthalpy of a particular bond. For example, we can measure the bond enthalpy of the O–H bond in 10 different polyatomic molecules and obtain the average O–H bond enthalpy by dividing the sum of the bond enthalpies by 10. When we use the term *bond enthalpy*, then, it is understood that we are referring to an average quantity, whereas *bond dissociation enthalpy* means a precisely measured value. Table 3.4 lists the bond enthalpies of a number of common chemical bonds. As you can see, triple bonds are stronger than double bonds, which, in turn, are stronger than single bonds.

The usefulness of bond enthalpies is that they enable us to estimate $\Delta_r H°$ values when precise thermochemical data (that is, $\Delta_f H°$ values) are not available. Because energy is required to break chemical bonds and chemical bond formation is accompanied by a release of heat, we can estimate $\Delta_r H°$ values by counting the total number of bonds broken and formed in the reaction and recording all the corresponding energy changes. The enthalpy of reaction in the *gas phase* is given by

$$\Delta_r H° = \Sigma BE(\text{reactants}) - \Sigma BE(\text{products})$$

$$= \text{total energy input} - \text{total energy released} \tag{3.35}$$

where BE stands for average bond enthalpy. As written, Equation 3.35 takes care of the sign convention for $\Delta_r H°$. If the total energy input is greater than the total energy

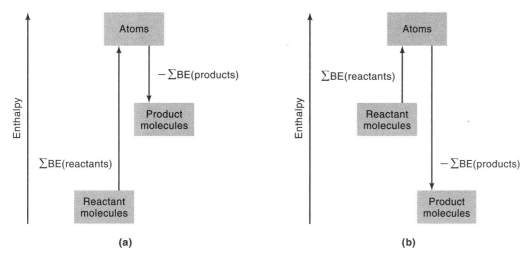

Figure 3.17
Bond enthalpy changes in (a) an endothermic reaction and (b) an exothermic reaction.

released, the $\Delta_r H°$ value is positive and the reaction is endothermic. Conversely, if more energy is released than absorbed, the $\Delta_r H°$ value is negative and the reaction is exothermic (Figure 3.17). If reactants and products are all diatomic molecules, then Equation 3.35 will yield accurate results because the bond dissociation enthalpies of diatomic molecules are accurately known. If some or all of the reactants and products are polyatomic molecules, Equation 3.35 will yield only approximate results because the bond enthalpies for calculation will be average values.

Example 3.8

Estimate the enthalpy of combustion for methane

$$CH_4(g) + 2O_2(g) \rightarrow CO_2(g) + 2H_2O(g)$$

at 298 K and 1 bar using the bond enthalpies in Table 3.4. Compare your result with that calculated from the enthalpies of formation of products and reactants.

ANSWER

The first step is to count the number of bonds broken and the number of bonds formed. This is best done by creating a table:

Type of bonds broken	Number of bonds broken	Bond enthalpy kJ mol^{-1}	Enthalpy change kJ mol^{-1}
C–H	4	414	1656
O=O	2	498.8	997.6

Type of bonds formed	Number of bonds formed	Bond enthalpy kJ mol^{-1}	Enthalpy change kJ mol^{-1}
C=O	2	799	1598
O–H	4	460	1840

From Equation 3.35,

$$\Delta_r H^\circ = [(1656 \text{ kJ mol}^{-1} + 997.6 \text{ kJ mol}^{-1}) - (1598 \text{ kJ mol}^{-1} + 1840 \text{ kJ mol}^{-1})]$$

$$= -784.4 \text{ kJ mol}^{-1}$$

To calculate the $\Delta_r H^\circ$ value using Equation 3.32, we obtain the $\Delta_f H^\circ$ values from Table 3.3 and write

$$\Delta_r H^\circ = [\Delta_f H^\circ(CO_2) + 2\Delta_f H^\circ(H_2O)] - [\Delta_f H^\circ(CH_4) + 2\Delta_f H^\circ(O_2)]$$

$$= [-393.5 \text{ kJ mol}^{-1} + 2(-241.8 \text{ kJ mol}^{-1})] - (-74.85 \text{ kJ mol}^{-1})$$

$$= -802.3 \text{ kJ mol}^{-1}$$

COMMENT

The agreement between the estimated $\Delta_r H^\circ$ value using bond enthalpies and the actual $\Delta_r H^\circ$ value is fairly good in this case. In general, the more exothermic (or endothermic) the reaction, the better the agreement. If the actual $\Delta_r H^\circ$ value is a small positive or negative quantity, then the value obtained from bond enthalpies becomes unreliable. Such values may even give the wrong sign for the reaction.

Suggestions for Further Reading

BOOKS

Edsall, J. T. and H. Gutfreund, *Biothermodynamics*, John Wiley & Sons, New York, 1983.

Rock, P. A. *Chemical Thermodynamics*, University Science Books, Sausalito, CA, 1983.

Klotz, I. M. and R. M. Rosenberg, *Chemical Thermodynamics: Basic Theory and Methods*, 5th ed., John Wiley & Sons, New York, 1994.

McQuarrie, D. A. and J. D. Simon, *Molecular Thermodynamics*, University Science Books, Sausalito, CA, 1999.

ARTICLES

Work and Heat

"What Is Heat?" F. J. Dyson, *Sci. Am.* September 1954.

"The Definition of Heat," T. B. Tripp, *J. Chem. Educ.* **53**, 782 (1976).

"Heat, Work, and Metabolism," J. N. Spencer, *J. Chem. Educ.* **62**, 571 (1985).

"General Definitions of Work and Heat in Thermodynamic Processes," E. A. Gislason and N. C. Craig, *J. Chem. Educ.* **64**, 670 (1987).

First Law of Thermodynamics

"Simplification of Some Thermochemical Calculations," E. R. Boyko and J. F. Belliveau, *J. Chem. Educ.* **67**, 743 (1990).

"Understanding the Language: Problem Solving and the First Law of Thermodynamics," M. Hamby, *J. Chem. Educ.* **67**, 923 (1990).

Heat Capacity

"Heat Capacity and the Equipartition Theorem," J. B. Dence, *J. Chem. Educ.* **49**, 798 (1972).

"Heat Capacity, Body Temperature, and Hypothermia," D. R. Kimbrough, *J. Chem. Educ.* **75**, 48 (1998).

Thermochemistry

"Standard Enthalpies of Formation of Ions in Solution," T. Solomon, *J. Chem. Educ.* **68**, 41 (1991).

"Bond Energies and Enthalpies," R. S. Treptow, *J. Chem. Educ.* **72**, 497 (1995).

"Thermochemistry," P. A. G. O'Hare in *Encyclopedia of Applied Physics*, G. L. Trigg, Ed., VCH Publishers, New York, 1997, Vol. 21, p. 265.

General

"The Use and Misuse of the Laws of Thermodynamics," M. L. McGlashan, *J. Chem. Educ.* **43**, 226 (1966).

"The Scope and Limitations of Thermodynamics," K. G. Denbigh, *Chem. Brit.* **4**, 339 (1968).

"Perpetual Motion Machines," S. W. Angrist, *Sci. Am.* January 1968.

"The Thermostat of Vertebrate Animals," H. C. Heller, L. L. Crawshaw, and H. T. Hammel, *Sci. Am.* August 1978.

"Conversion of Standard (1 atm) Thermodynamic Data to the New Standard-State Pressure, 1 bar (10^5 Pa)," *Bull. Chem. Thermodynamics* **25**, 523 (1982).

"Conversion of Standard Thermodynamic Data to the New Standard-State Pressure," R. D. Freeman, *J. Chem. Educ.* **62**, 681 (1985).

"Student Misconceptions in Thermodynamics," M. F. Granville, *J. Chem. Educ.* **62**, 847 (1985).

"Power From the Sea," T. R. Penney and P. Bharathan, *Sci. Am.* January 1987.

"Why There's Frost on the Pumpkin," W. H. Corkern and L. H. Holmes, Jr., *J. Chem. Educ.* **68**, 825 (1991).

"The Thermodynamics of Drunk Driving," R. Q. Thompson, *J. Chem. Educ.* **74**, 532 (1997).

"How Thermodynamic Data and Equilibrium Constants Changed When the Standard-State Pressure Became 1 Bar," R. S. Treptow, *J. Chem. Educ.* **76**, 212 (1999).

Problems

Work and Heat

3.1 Explain the term *state function*. Which of the following are state functions? P, V, T, w, q.

3.2 What is heat? How does heat differ from thermal energy? Under what condition is heat transferred from one system to another?

3.3 Show that 1 L atm = 101.3 J.

3.4 A 7.24-g sample of ethane occupies 4.65 L at 294 K. **(a)** Calculate the work done when the gas expands isothermally against a constant external pressure of 0.500 atm until its volume is 6.87 L. **(b)** Calculate the work done if the same expansion occurs reversibly.

3.5 A 19.2-g quantity of dry ice (solid carbon dioxide) is allowed to sublime (evaporate) in an apparatus like the one shown in Figure 3.1. Calculate the expansion work done against a constant external pressure of 0.995 atm and at a constant temperature of 22°C. Assume that the initial volume of dry ice is negligible and that CO_2 behaves like an ideal gas.

3.6 Calculate the work done by the reaction

$$Zn(s) + H_2SO_4(aq) \rightarrow ZnSO_4(aq) + H_2(g)$$

when 1 mole of hydrogen gas is collected at 273 K and 1.0 atm. (Neglect volume changes other than the change in gas volume.)

First Law of Thermodynamics

3.7 A truck traveling 60 kilometers per hour is brought to a complete stop at a traffic light. Does this change in velocity violate the law of conservation of energy?

3.8 Some driver's test manuals state that the stopping distance quadruples as the velocity doubles. Justify this statement by using mechanics and thermodynamic arguments.

3.9 Provide a first law analysis for each of the following cases: **(a)** When a bicycle tire is inflated with a hand pump, the temperature inside rises. You can feel the warming effect at the valve stem. **(b)** Artificial snow is made by quickly releasing a mixture of compressed air and water vapor at about 20 atm from a snow-making machine to the surroundings.

3.10 An ideal gas is compressed isothermally by a force of 85 newtons acting through 0.24 meter. Calculate the values of ΔU and q.

3.11 Calculate the internal energy of 2 moles of argon gas (assuming ideal behavior) at 298 K. Suggest two ways to increase its internal energy by 10 J.

3.12 A thermos bottle containing milk is shaken vigorously. Consider the milk as the system. **(a)** Will the temperature rise as a result of the shaking? **(b)** Has heat been added to the system? **(c)** Has work been done on the system? **(d)** Has the system's internal energy changed?

3.13 A 1.00-mole sample of ammonia at 14.0 atm and 25°C in a cylinder fitted with a movable piston expands against a constant external pressure of 1.00 atm. At equilibrium, the pressure and volume of the gas are 1.00 atm and 23.5 L, respectively. **(a)** Calculate the final temperature of the sample. **(b)** Calculate the values of q, w, and ΔU for the process.

3.14 An ideal gas is compressed isothermally from 2.0 atm and 2.0 L to 4.0 atm and 1.0 L. Calculate the values of ΔU and ΔH if the process is carried out **(a)** reversibly and **(b)** irreversibly.

3.15 Explain the energy changes at the molecular level when liquid acetone is converted to vapor at its boiling point.

3.16 A piece of potassium metal is added to water in a beaker. The reaction that takes place is

$$2K(s) + 2H_2O(l) \rightarrow 2KOH(aq) + H_2(g)$$

Predict the signs of $w, q, \Delta U$, and ΔH.

3.17 At 373.15 K and 1 atm, the molar volume of liquid water and steam are 1.88×10^{-5} m^3 and 3.06×10^{-2} m^3, respectively. Given that the heat of vaporization of water is 40.79 kJ mol^{-1}, calculate the values of ΔH and ΔU for 1 mole in the following process:

$$H_2O(l, 373.15 \text{ K}, 1 \text{ atm}) \rightarrow H_2O(g, 373.15 \text{ K}, 1 \text{ atm})$$

3.18 Consider a cyclic process involving a gas. If the pressure of the gas varies during the process but returns to the original value at the end, is it correct to write $\Delta H = q_P$?

3.19 Calculate the value of ΔH when the temperature of 1 mole of a monatomic gas is increased from 25°C to 300°C.

3.20 One mole of an ideal gas undergoes an isothermal expansion at 300 K from 1.00 atm to a final pressure while performing 200 J of expansion work. Calculate the final pressure of the gas if the external pressure is 0.20 atm.

Heat Capacities

3.21 A 6.22-kg piece of copper metal is heated from 20.5°C to 324.3°C. Given that the specific heat of Cu is 0.385 J g^{-1} °C^{-1}, calculate the heat absorbed (in kJ) by the metal.

3.22 A 10.0-g sheet of gold with a temperature of 18.0°C is laid flat on a sheet of iron that weighs 20.0 g and has a temperature of 55.6°C. Given that the specific heats of Au and Fe are 0.129 J g^{-1} °C^{-1} and 0.444 J g^{-1} °C^{-1}, respectively, what is the final temperature of the combined metals? Assume that no heat is lost to the surroundings. (*Hint:* The heat gained by the gold must be equal to the heat lost by the iron.)

3.23 It takes 330 joules of energy to raise the temperature of 24.6 g of benzene from 21.0°C to 28.7°C at constant pressure. What is the molar heat capacity of benzene at constant pressure?

3.24 The molar heat of vaporization for water is 44.01 kJ mol^{-1} at 298 K and 40.79 kJ mol^{-1} at 373 K. Give a qualitative explanation of the difference in these two values.

3.25 The constant-pressure molar heat capacity of nitrogen is given by the expression

$$\bar{C}_P = (27.0 + 5.90 \times 10^{-3} \ T$$
$$- 0.34 \times 10^{-6} \ T^2) \ \text{J K}^{-1} \ \text{mol}^{-1}$$

Calculate the value of ΔH for heating 1 mole of nitrogen from 25.0°C to 125°C.

3.26 The heat capacity ratio (γ) of an ideal gas is 1.38. What are its $\bar{C}_V$ and $\bar{C}_P$ values?

3.27 One way to measure the heat capacity ratio (γ) of a gas is to measure the speed of sound in the gas (c), which is given by

$$c = \left(\frac{\gamma RT}{\mathcal{M}}\right)^{1/2}$$

where $\mathcal{M}$ is the molar mass of the gas. Calculate the speed of sound in helium at 25°C.

3.28 Which of the following gases has the largest $\bar{C}_V$ value at 298 K? He, N_2, CCl_4, HCl.

3.29 (a) For most efficient use, refrigerator freezer compartments should be fully packed with food. What is the thermochemical basis for this recommendation? **(b)** Starting at the same temperature, tea and coffee remain hot longer in a thermal flask than soup. Explain.

3.30 In the nineteenth century, two scientists named Dulong and Petit noticed that the product of the molar mass of a solid element and its specific heat is approximately 25 J °C^{-1}. This observation, now called Dulong and Petit's law, was used to estimate the specific heat of metals. Verify the law for aluminum (0.900 J g^{-1} °C^{-1}), copper (0.385 J g^{-1} °C^{-1}), and iron (0.444 J g^{-1} °C^{-1}). The law does not apply to one of the metals. Which one is it? Why?

Gas Expansion

3.31 The following diagram represents the P–V changes of a gas. Write an expression for the total work done.

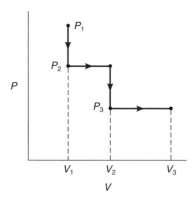

3.32 The equation of state for a certain gas is given by $P[(V/n) - b] = RT$. Obtain an expression for the maximum work done by the gas in a reversible isothermal expansion from V_1 to V_2.

3.33 Calculate the values of $q, w, \Delta U$, and ΔH for the reversible adiabatic expansion of 1 mole of a monatomic ideal gas from 5.00 m^3 to 25.0 m^3. The temperature of the gas is initially 298 K.

3.34 A quantity of 0.27 mole of neon is confined in a container at 2.50 atm and 298 K and then allowed to expand adiabatically under two different conditions: **(a)** reversibly to 1.00 atm and **(b)** against a constant pressure of 1.00 atm. Calculate the final temperature in each case.

3.35 One mole of an ideal monatomic gas initially at 300 K and a pressure of 15.0 atm expands to a final pressure of 1.00 atm. The expansion can occur via any one of four different paths: **(a)** isothermal and reversible, **(b)** isothermal and irreversible, **(c)** adiabatic and reversible, and **(d)** adiabatic and irreversible. In irreversible processes, the expansion occurs against an external pressure of 1.00 atm. For each case, calculate the values of $q, w, \Delta U$, and ΔH.

Calorimetry

3.36 A 0.1375-g sample of magnesium is burned in a constant-volume bomb calorimeter that has a heat capacity of 1769 J °C^{-1}. The calorimeter contains exactly 300 g of water, and the temperature increases by 1.126°C. Calculate the heat given off by the burning magnesium, in kJ g^{-1} and in kJ mol^{-1}.

3.37 The enthalpy of combustion of benzoic acid (C$_6$H$_5$COOH) is commonly used as the standard for calibrating constant-volume bomb calorimeters; its value has been accurately determined to be −3226.7 kJ mol^{-1}. **(a)** When 0.9862 g of benzoic acid was oxidized, the temperature rose from 21.84°C to 25.67°C. What is the heat capacity of the calorimeter? **(b)** In a separate experiment, 0.4654 g of α-D-glucose (C$_6$H$_{12}$O$_6$) was oxidized in the same calorimeter, and the temperature rose from 21.22°C to 22.28°C. Calculate the enthalpy of combustion of glucose, the value of $\Delta_r U$ for the combustion, and the molar enthalpy of formation of glucose.

3.38 A quantity of 2.00 × 10^2 mL of 0.862 M HCl is mixed with 2.00 × 10^2 mL of 0.431 M Ba(OH)$_2$ in a constant-pressure calorimeter that has a heat capacity of 453 J °C^{-1}. The initial temperature of the HCl and Ba(OH)$_2$ solutions is the same at 20.48°C. For the process

$$H^+(aq) + OH^-(aq) \rightarrow H_2O(l)$$

the heat of neutralization is −56.2 kJ mol^{-1}. What is the final temperature of the mixed solution?

3.39 When 1.034 g of naphthalene (C$_{10}$H$_8$) are completely burned in a constant-volume bomb calorimeter at 298 K, 41.56 kJ of heat is evolved. Calculate the values of $\Delta_r U$ and $\Delta_r H$ for the reaction.

Thermochemistry

3.40 Consider the following reaction:

$$2CH_3OH(l) + 3O_2(g) \rightarrow 4H_2O(l) + 2CO_2(g)$$
$$\Delta_r H^\circ = -1452.8 \text{ kJ mol}^{-1}$$

What is the value of $\Delta_r H^\circ$ if **(a)** the equation is multiplied throughout by 2, **(b)** the direction of the reaction is reversed so that the products become the reactants and vice versa, and **(c)** water vapor instead of liquid water is the product?

3.41 Which of the following standard enthalpy of formation values is not zero at 25°C? Na(s), Ne(g), CH$_4$(g), S$_8$(s), Hg(l), H(g).

3.42 The standard enthalpies of formation of ions in aqueous solution are obtained by arbitrarily assigning a value of zero to H$^+$ ions; that is, $\Delta_f \bar{H}^\circ[H^+(aq)] = 0$. **(a)** For the following reaction,

$$HCl(g) \rightarrow H^+(aq) + Cl^-(aq)$$
$$\Delta_r H^\circ = -74.9 \text{ kJ mol}^{-1}$$

calculate the value of $\Delta_f \bar{H}^\circ$ for the Cl$^-$ ions. **(b)** The standard enthalpy of neutralization between a HCl solution and a NaOH solution is found to be −56.2 kJ mol^{-1}. Calculate the standard enthalpy of formation of the hydroxide ion at 25°C.

3.43 Determine the amount of heat (in kJ) given off when 1.26 × 10^4 g of ammonia is produced according to the equation

$$N_2(g) + 3H_2(g) \rightarrow 2NH_3(g)$$
$$\Delta_r H^\circ = -92.6 \text{ kJ mol}^{-1}$$

Assume the reaction takes place under standard-state conditions at 25°C.

3.44 When 2.00 g of hydrazine decomposed under constant-pressure conditions, 7.00 kJ of heat were transferred to the surroundings:

$$3N_2H_4(l) \rightarrow 4NH_3(g) + N_2(g)$$

What is the $\Delta_r H^\circ$ value for the reaction?

3.45 Consider the reaction

$$N_2(g) + 3H_2(g) \rightarrow 2NH_3(g)$$
$$\Delta_r H^\circ = -92.6 \text{ kJ mol}^{-1}$$

If 2.0 moles of N$_2$ react with 6.0 moles of H$_2$ to form NH$_3$, calculate the work done (in joules) against a pressure of 1.0 atm at 25°C. What is the value of $\Delta_r U$ for this reaction? Assume the reaction goes to completion.

3.46 The standard enthalpies of combustion of fumaric acid and maleic acid (to form carbon dioxide and water) are −1336.0 kJ mol^{-1} and −1359.2 kJ mol^{-1}, respectively. Calculate the enthalpy of the following isomerization process:

Maleic acid   Fumaric acid

3.47 From the reaction

$$C_{10}H_8(s) + 12O_2(g) \rightarrow 10CO_2(g) + 4H_2O(l)$$
$$\Delta_r H^\circ = -5153.0 \text{ kJ mol}^{-1}$$

and the enthalpies of formation of CO$_2$ and H$_2$O (see Appendix 2), calculate the enthalpy of formation of naphthalene (C$_{10}$H$_8$).

3.48 The standard molar enthalpy of formation of molecular oxygen at 298 K is zero. What is its value at 315 K? (*Hint:* Look up the $\bar{C}_P$ value in Appendix 2.)

3.49 Which of the following substances has a nonzero $\Delta_f \bar{H}^\circ$ value at 25°C? Fe(s), I$_2$(l), H$_2$(g), Hg(l), O$_2$(g), C(graphite).

3.50 The hydrogenation for ethylene is

$$C_2H_4(g) + H_2(g) \rightarrow C_2H_6(g)$$

Calculate the change in the enthalpy of hydrogenation from 298 K to 398 K. The $\bar{C}_P^\circ$ values are: C_2H_4: 43.6 J K^{-1} mol^{-1} and C_2H_6: 52.7 J K^{-1} mol^{-1}.

3.51 Use the data in Appendix 2 to calculate the value of $\Delta_r H^\circ$ for the following reaction at 298 K:

$$N_2O_4(g) \rightarrow 2NO_2(g)$$

What is its value at 350 K? State any assumptions used in your calculation.

3.52 Calculate the standard enthalpy of formation for diamond, given that

$$C(\text{graphite}) + O_2(g) \rightarrow CO_2(g)$$
$$\Delta_r H^\circ = -393.5 \text{ kJ mol}^{-1}$$
$$C(\text{diamond}) + O_2(g) \rightarrow CO_2(g)$$
$$\Delta_r H^\circ = -395.4 \text{ kJ mol}^{-1}$$

3.53 Photosynthesis produces glucose, $C_6H_{12}O_6$, and oxygen from carbon dioxide and water:

$$6CO_2 + 6H_2O \rightarrow C_6H_{12}O_6 + 6O_2$$

(a) How would you determine the $\Delta_r H^\circ$ value for this reaction experimentally? **(b)** Solar radiation produces approximately 7.0×10^{14} kg glucose a year on Earth. What is the corresponding change in the $\Delta_r H^\circ$ value?

3.54 From the following heats of combustion,

$$CH_3OH(l) + \tfrac{3}{2}O_2(g) \rightarrow CO_2(g) + 2H_2O(l)$$
$$\Delta_r H^\circ = -726.4 \text{ kJ mol}^{-1}$$
$$C(\text{graphite}) + O_2(g) \rightarrow CO_2(g)$$
$$\Delta_r H^\circ = -393.5 \text{ kJ mol}^{-1}$$
$$H_2(g) + \tfrac{1}{2}O_2(g) \rightarrow H_2O(l)$$
$$\Delta_r H^\circ = -285.8 \text{ kJ mol}^{-1}$$

calculate the enthalpy of formation of methanol (CH_3OH) from its elements:

$$C(\text{graphite}) + 2H_2(g) + \tfrac{1}{2}O_2(g) \rightarrow CH_3OH(l)$$

3.55 The standard enthalpy change for the following reaction is 436.4 kJ mol^{-1}:

$$H_2(g) \rightarrow H(g) + H(g)$$

Calculate the standard enthalpy of formation of atomic hydrogen (H).

3.56 Calculate the difference between the values of $\Delta_r H^\circ$ and $\Delta_r U^\circ$ for the oxidation of α-D-glucose at 298 K:

$$C_6H_{12}O_6(s) + 6O_2(g) \rightarrow 6CO_2(g) + 6H_2O(l)$$

3.57 Alcoholic fermentation is the process in which carbohydrates are broken down into ethanol and carbon dioxide. The reaction is very complex and involves a number of enzyme-catalyzed steps. The overall change is

$$C_6H_{12}O_6(s) \rightarrow 2C_2H_5OH(l) + 2CO_2(g)$$

Calculate the standard enthalpy change for this reaction, assuming that the carbohydrate is α-D-glucose.

Bond Enthalpy

3.58 (a) Explain why the bond enthalpy of a molecule is always defined in terms of a gas-phase reaction. **(b)** The bond dissociation enthalpy of F_2 is 150.6 kJ mol^{-1}. Calculate the value of $\Delta_f \bar{H}^\circ$ for F(g).

3.59 From the molar enthalpy of vaporization of water at 373 K and the bond dissociation enthalpies of H_2 and O_2 (see Table 3.4), calculate the average O–H bond enthalpy in water, given that

$$H_2(g) + \tfrac{1}{2}O_2(g) \rightarrow H_2O(l)$$
$$\Delta_r H^\circ = -285.8 \text{ kJ mol}^{-1}$$

3.60 Use the bond enthalpy values in Table 3.4 to calculate the enthalpy of combustion for ethane,

$$2C_2H_6(g) + 7O_2(g) \rightarrow 4CO_2(g) + 6H_2O(l)$$

Compare your result with that calculated from the enthalpy of formation values of the products and reactants listed in Appendix 2.

Additional Problems

3.61 A 2.10-mole sample of crystalline acetic acid, initially at 17.0°C, is allowed to melt at 17.0°C and is then heated to 118.1°C (its normal boiling point) at 1.00 atm. The sample is allowed to vaporize at 118.1°C and is then rapidly quenched to 17.0°C, so that it recrystallizes. Calculate the value of $\Delta_r H^\circ$ for the total process as described.

3.62 Predict whether the values of $q, w, \Delta U$, and ΔH are positive, zero, or negative for each of the following processes: **(a)** melting of ice at 1 atm and 273 K, **(b)** melting of solid cyclohexane at 1 atm and the normal melting point, **(c)** reversible isothermal expansion of an ideal gas, and **(d)** reversible adiabatic expansion of an ideal gas.

3.63 Einstein's special relativity equation is $E = mc^2$, where E is energy, m is mass, and c is the velocity of light. Does this equation invalidate the law of conservation of energy, and hence the first law of thermodynamics?

3.64 The convention of arbitrarily assigning a zero enthalpy value to all the (most stable) elements in the standard state and (usually) 298 K is a convenient way of dealing with the enthalpy changes of chemical processes. This convention does not apply to one kind of process, however. What process is it? Why?

3.65 Two moles of an ideal gas are compressed isothermally at 298 K from 1.00 atm to 200 atm. Calculate the values of $q, w, \Delta U$, and ΔH for the process if it is carried out reversibly.

3.66 The fuel value of hamburger is approximately 3.6 kcal g^{-1}. If a person eats 1 pound of hamburger for lunch and if none of the energy is stored in his body, estimate the amount of water that would have to be lost in perspiration to keep his body temperature constant. (1 lb = 454 g.)

3.67 A quantity of 4.50 g of CaC_2 is reacted with an excess of water at 298 K and atmospheric pressure:

$$CaC_2(s) + 2H_2O(l) \rightarrow Ca(OH)_2(aq) + C_2H_2(g)$$

Calculate the work done in joules by the acetylene gas against the atmospheric pressure.

3.68 An oxyacetylene flame is often used in the welding of metals. Estimate the flame temperature produced by the reaction

$$2C_2H_2(g) + 5O_2(g) \rightarrow 4CO_2(g) + 2H_2O(g)$$

Assume the heat generated from this reaction is all used to heat the products. (*Hint:* First calculate the value of $\Delta_r H^\circ$ for the reaction. Next, look up the heat capacities of the products. Assume the heat capacities are temperature independent.)

3.69 The $\Delta_f \bar{H}^\circ$ values listed in Appendix 2 all refer to 1 bar and 298 K. Suppose that a student wants to set up a new table of $\Delta_f \bar{H}^\circ$ values at 1 bar and 273 K. Show how she should proceed on the conversion, using acetone as an example.

3.70 The enthalpies of hydrogenation of ethylene and benzene have been determined at 298 K:

$$C_2H_4(g) + H_2(g) \rightarrow C_2H_6(g)$$
$$\Delta_r H^\circ = -132 \text{ kJ mol}^{-1}$$
$$C_6H_6(g) + 3H_2(g) \rightarrow C_6H_{12}(g)$$
$$\Delta_r H^\circ = -246 \text{ kJ mol}^{-1}$$

What would be the enthalpy of hydrogenation for benzene if it contained three isolated, unconjugated double bonds? How would you account for the difference between the calculated value based on this assumption and the measured value?

3.71 The molar enthalpies of fusion and vaporization of water are 6.01 kJ mol^{-1} and 44.01 kJ mol^{-1} (at 298 K), respectively. From these values, estimate the molar enthalpy of sublimation of ice.

3.72 The standard enthalpy of formation at 298 K of $HF(aq)$ is -320.1 kJ mol^{-1}; $OH^-(aq)$, -229.6 kJ mol^{-1}; $F^-(aq)$, -329.11 kJ mol^{-1}; and $H_2O(l)$, -285.84 kJ mol^{-1}. **(a)** Calculate the enthalpy of neutralization of $HF(aq)$,

$$HF(aq) + OH^-(aq) \rightarrow F^-(aq) + H_2O(l)$$

(b) Using the value of -55.83 kJ mol^{-1} as the enthalpy change from the reaction

$$H^+(aq) + OH^-(aq) \rightarrow H_2O(l)$$

calculate the enthalpy change for the dissociation

$$HF(aq) \rightarrow H^+(aq) + F^-(aq)$$

3.73 It was stated in the chapter that for reactions in condensed phases, the difference between the values of $\Delta_r H$ and $\Delta_r U$ is usually negligibly small. This statement holds for processes carried out under atmospheric conditions. For certain geochemical processes, however, the external pressures may be so great that $\Delta_r H$ and $\Delta_r U$ values can differ by a significant amount. A well-known example is the slow conversion of graphite to diamond under Earth's surface. Calculate the value of the quantity $(\Delta_r H - \Delta_r U)$ for the conversion of 1 mole of graphite to 1 mole of diamond at a pressure of 50,000 atm. The densities of graphite and diamond are 2.25 g cm^{-3} and 3.52 g cm^{-3}, respectively.

3.74 Metabolic activity in the human body releases approximately 1.0×10^4 kJ of heat per day. Assuming the body is 50 kg of water, how fast would the body temperature rise if it were an isolated system? How much water must the body eliminate as perspiration to maintain the normal body temperature (98.6°F)? Comment on your results. The heat of vaporization of water may be taken as 2.41 kJ g^{-1}.

3.75 An ideal gas in a cylinder fitted with a movable piston is adiabatically compressed from V_1 to V_2. As a result, the temperature of the gas rises. Explain what causes the temperature of the gas to rise.

3.76 Calculate the fraction of the enthalpy of vaporization of water used for the expansion of steam at its normal boiling point.

3.77 The combustion of what volume of ethane (C_2H_6), measured at 23.0°C and 752 mmHg, would be required to heat 855 g of water from 25.0°C to 98.0°C?

3.78 Calculate the internal energy of a Goodyear blimp filled with helium gas at 1.2×10^5 Pa (compared to the empty blimp). The volume of the inflated blimp is 5.5×10^3 m^3. If all the energy were used to heat 10.0

tons of copper at 21°C, calculate the final temperature of the metal. (*Hint:* 1 ton = 9.072×10^5 g.)

3.79 Without referring to the chapter, state the conditions for each of the following equations:
(a) $\Delta H = \Delta U + P\Delta V$, **(b)** $C_P = C_V + nR$, **(c)** $\gamma = 5/3$, **(d)** $P_1 V_1^{\gamma} = P_2 V_2^{\gamma}$, **(e)** $w = n\bar{C}_V(T_2 - T_1)$, **(f)** $w = -P\Delta V$, **(g)** $w = -nRT \ln(V_2/V_1)$, and **(h)** $dH = dq$.

3.80 An ideal gas is isothermally compressed from P_1, V_1 to P_2, V_2. Under what conditions would the work done be a minimum? A maximum? Write the expressions for minimum and maximum work done for this process. Explain your reasoning.

3.81 Construct a table with the headings $q, w, \Delta U$, and ΔH. For each of the following processes, deduce whether each of the quantities listed is positive $(+)$, negative $(-)$, or zero (0). **(a)** Freezing of acetone at 1 atm and its normal melting point. **(b)** Irreversible isothermal expansion of an ideal gas. **(c)** Adiabatic compression of an ideal gas. **(d)** Reaction of sodium with water. **(e)** Boiling of liquid ammonia at its normal boiling point. **(f)** Irreversible adiabatic expansion of a gas against an external pressure. **(g)** Reversible isothermal compression of an ideal gas. **(h)** Heating of a gas at constant volume. **(i)** Freezing of water at 0°C.

3.82 State whether each of the following statements is true or false: **(a)** $\Delta U \approx \Delta H$ except for gases or high-pressure processes. **(b)** In gas compression, a reversible process does maximum work. **(c)** ΔU is a state function. **(d)** $\Delta U = q + w$ for an open system. **(e)** C_V is temperature independent for gases. **(f)** The internal energy of a real gas depends only on temperature.

3.83 Show that $(\partial C_V/\partial V)_T = 0$ for an ideal gas.

3.84 Derive an expression for the work done during the isothermal, reversible expansion of a van der Waals gas. Account physically for the way in which the coefficients a and b appear in the final expression. [*Hint:* You need to apply the Taylor series expansion:

$$\ln(1 - x) = -x - \frac{x^2}{2} \cdots \quad \text{for } |x| \ll 1$$

to the expression $\ln(V - nb)$. Recall that the a term represents attraction and the b term repulsion.]

3.85 Show that for the adiabatic reversible expansion of an ideal gas,

$$T_1^{C_V/R} V_1 = T_2^{C_V/R} V_2$$

3.86 One mole of ammonia initially at 5°C is placed in contact with 3 moles of helium initially at 90°C. Given that $\bar{C}_V$ for ammonia is $3R$, if the process is carried out at constant total volume, what is the final temperature of the gases?

3.87 The typical energy differences between successive rotational, vibrational, and electronic energy levels are 5.0×10^{-22} J, 0.50×10^{-19} J, and 1.0×10^{-17} J, respectively. Calculate the ratios of the numbers of molecules in the two adjacent energy levels (higher to lower) in each case at 298 K.

3.88 The first excited electronic energy level of the helium atom is 3.13×10^{-18} J above the ground level. Estimate the temperature at which the electronic motion will begin to make a significant contribution to the heat capacity. That is, at what temperature will the ratio of the population of the first excited state to the ground state be 5.0%?

3.89 Calculate the total translational kinetic energy of the air molecules in a spherical balloon of radius 43.0 cm at 24°C and 1.2 atm. Is this enough energy to heat 200 mL of water from 20°C to 90°C for a cup of tea? The density of water is 1.0 g cm^{-3}, and its specific heat is 4.184 J g^{-1} °C^{-1}.

3.90 From your knowledge of heat capacity, explain why hot, humid air is more uncomfortable than hot, dry air and cold, damp air is more uncomfortable than cold, dry air.

3.91 A hemoglobin molecule (molar mass = 65,000 g) can bind up to four oxygen molecules. In a certain experiment a 0.085-L solution containing 6.0 g of deoxygenated hemoglobin was reacted with an excess of oxygen in a constant-pressure calorimeter of negligible heat capacity. Calculate the enthalpy of reaction per mole of oxygen bound if the temperature rose by 0.044°C. Assume the solution is dilute so that the specific heat of the solution is equal to that of water.

3.92 Give an interpretation for the following DSC thermogram for the thermal denaturation of a protein.

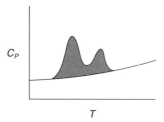

CHAPTER 4

The Second Law of Thermodynamics

As we saw in Chapter 3, the first law of thermodynamics specifies that energy can be neither created nor destroyed, but flows from one part of the universe to another or is converted from one form into another. The total amount of energy in the universe remains constant. Despite its immense value in the study of the energetics of chemical reactions, the first law does have a major limitation: it cannot predict the direction of change. It helps us to do the bookkeeping of energy balance, such as the energy input, heat released, work done, and so forth, but it says nothing about whether a particular process can indeed occur. For this kind of information we must turn to the second law of thermodynamics.

In this chapter we introduce a new thermodynamic function, called entropy (S), which is central to both the second law and the third law of thermodynamics. We shall see that changes in entropy, ΔS, provide the necessary criterion for predicting the direction of any process. To help us focus on the system and on specific practical conditions, we shall develop a function that forms the basis of chemical thermodynamics: Gibbs energy (G).

4.1 Spontaneous Processes

A lump of sugar dissolves in a cup of coffee, an ice cube melts in your hand, and a struck match burns in air; we witness so many of these *spontaneous* processes in everyday life that it is almost impossible to list them all. The interesting aspect of a spontaneous process is that the reverse process never happens under the same set of conditions. Ice melts at 20°C and 1 atm, but water at the same temperature and pressure will not spontaneously turn into ice. A leaf lying on the ground will not rise into the air on its own and return to the branch from which it came. Viewed backwards, a movie of a baseball smashing a window to pieces is funny because everyone knows that such a process is impossible. But why? Surely we can demonstrate that any of the changes just described (and countless more) can occur in *either* direction in accord with the first law of thermodynamics; yet, in fact, each process occurs in only one direction. After many observations, we can conclude that processes occurring spontaneously in one direction cannot also take place spontaneously in the opposite direction; otherwise, nothing would ever happen (Figure 4.1).

Why can't the reverse of a spontaneous process occur by itself? Consider a rubber ball held at some distance above the floor. When the ball is released, it falls. The impact between the ball and the floor causes the ball to bounce upward, and when it has reached a certain height, it repeats its downward motion. In the process of falling, the potential energy of the ball is converted to kinetic energy. Experience tells us that

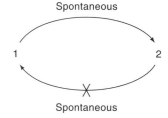

Figure 4.1
If the change from state 1 to 2 occurs spontaneously, then the reverse step, that is, 2 to 1, cannot also be a spontaneous process.

81

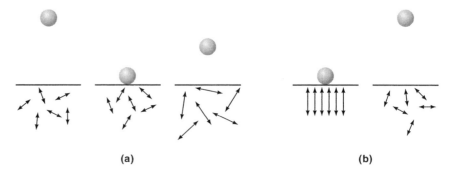

Figure 4.2
(a) A spontaneous process. A falling ball strikes the floor and loses some of its kinetic energy to the molecules in the floor. As a result, the ball does not bounce quite as high and the floor heats up a little. The length of arrows indicates the amplitude of molecular vibration. (b) An impossible event. A ball resting on the floor cannot spontaneously rise into the air by absorbing thermal energy from the floor.

after every bounce, the ball does not rise quite as high as before. The reason is that the collision between the ball and the floor is inelastic, so that upon each impact some of the ball's kinetic energy is dissipated among the molecules in the floor. After each bounce, the floor becomes a little bit hotter.* This intake of energy increases the rotational and vibrational motions of the molecules in the floor (Figure 4.2a). Eventually, the ball comes to a complete rest because its kinetic energy is totally lost to the floor. To describe this process in another way, we say that the original potential energy of the ball, through its conversion to kinetic energy, is degraded into heat.

Now let us consider what would be necessary for the reverse process to occur on its own; that is, a ball sitting on the floor spontaneously rises to a certain height in the air by absorbing heat from the floor. Such a process will not violate the first law. If the mass of the ball is m, and the height above the floor to which it rises is h, we have

$$\text{energy extracted from the floor} = mgh$$

where g is acceleration due to gravity. The thermal energy of the floor is random molecular motion. To impart an amount of energy large enough to raise the ball from the floor, most of the molecules would have to line up under the ball and vibrate in phase with one another, as shown in Figure 4.2b. At the instant the ball leaves the floor, all the atoms in these molecules must be moving upward for proper energy transfer. It is conceivable for 2 million molecules to execute this kind of synchronized motion, but because of the magnitude of energy transfer, the number of molecules involved would have to be on the order of Avogadro's number, or 6×10^{23}. Given the random nature of molecular motion, this is such an improbable event that it is virtually impossible. Indeed, no one has ever witnessed the spontaneous rising of a ball from the floor, and we can safely conclude that no one ever will.

Thinking about the improbability of a ball spontaneously rising upward from the floor helps us understand the nature of many spontaneous processes. Consider the familiar example of a gas in a cylinder fitted with a movable piston. If the pressure of the gas is greater than the external pressure, then the gas will expand until the internal and external pressures are equal. This is a spontaneous process. What would it take for the gas to contract spontaneously? Most of the gas molecules would have to move away from the piston and toward other parts of the cylinder at the same

---

*Actually, the temperatures of the ball and of the surrounding air also rise slightly after each impact. But here, we are concerned only with what happens to the floor.

time. Now, at any given moment many molecules are indeed doing this, but we will never find 6×10^{23} molecules engaged in unidirectional motion because molecular translational motion is totally random. By the same token, a metal bar at a uniform temperature will not suddenly become hotter at one end and colder at the other. To establish this temperature gradient, the thermal motion resulting from collisions between randomly vibrating atoms would have to decrease at one end and rise at the other—a highly improbable event.

Let us look at this problem from a different angle and ask what changes accompany a spontaneous process. Logically, we can assume that all spontaneous processes occur in such a way as to decrease the energy of the system. This assumption helps us explain why things fall downward, why springs unwind, and so forth. But a change of energy alone is not enough to predict whether a process will be spontaneous. For example, in Chapter 3, we saw that the expansion of an ideal gas against a vacuum does not result in a change in its internal energy. Yet the process is spontaneous. When ice melts spontaneously at 20°C to form water, the internal energy of the system actually *increases*. In fact, many endothermic physical and chemical processes are spontaneous. If energy change cannot be used to indicate the direction of a spontaneous process, then we need another thermodynamic function to help us. This function turns out to be entropy (S).

4.2 Entropy

Our discussion of spontaneous processes is based on macroscopic events. In trying to understand spontaneous processes, we should focus our attention on the *statistical* behavior of a very large number of molecules, not on the motion of just a few of them. In this section, we derive a statistical definition of entropy and then define entropy in terms of thermodynamic quantities.

Statistical Definition of Entropy

For a cylinder containing helium atoms, depicted in Figure 4.3, the probability of finding any one He atom in V_2, the entire volume of the cylinder, is 1, because all He atoms are known to be inside the cylinder. On the other hand, the probability of finding a helium atom in half of the volume of the cylinder, V_1, is only $\frac{1}{2}$. If the number of He atoms is increased to 2, the probability of finding both of them in V_2 is still 1, but that of finding both of them in V_1 becomes $\left(\frac{1}{2}\right)\left(\frac{1}{2}\right)$, or $\frac{1}{4}$.* Because $\frac{1}{4}$ is an appreciable quantity, finding both He atoms in the same region at a given time would not be surprising. We can see, however, that as the number of He atoms increases,

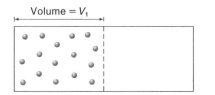
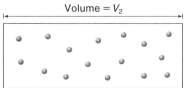

Figure 4.3
Schematic diagram showing N molecules occupying volumes V_1 and V_2 of a container.

---

*The probability of both events occurring is a *product* of the probabilities of two independent events. We assume that the He gas behaves ideally so that the presence of one He atom in V_1 does not affect the presence of another He atom in the same volume in any way.

the probability of finding all of them in V_1 becomes progressively smaller:

$$W = \left(\tfrac{1}{2}\right)\left(\tfrac{1}{2}\right)\left(\tfrac{1}{2}\right)\cdots\cdots$$
$$= \left(\tfrac{1}{2}\right)^N$$

where N is the total number of atoms present. If $N = 100$, we have

$$W = \left(\tfrac{1}{2}\right)^{100} = 8 \times 10^{-31}$$

Put in perspective, this probability is less than that for the production of Shakespeare's complete works 15 quadrillion times in succession without a single error by a tribe of wild monkeys randomly pounding on computer keyboards.

If N is of the order of 6×10^{23}, the probability becomes $\left(\tfrac{1}{2}\right)^{6 \times 10^{23}}$, a quantity so small that for all practical purposes it can be regarded as zero. From the results of these simple calculations come a most important message. If initially we had compressed all the He atoms into V_1 and allowed the gas to expand on its own, we would find that eventually the atoms would be evenly distributed over the entire volume, V_2, because this situation corresponds to the most probable state. Thus, the direction of spontaneous change is from a situation in which the gas is in V_1 to one in which it is in V_2, or from a state with low probability of occurring to one of maximum probability.

Now that we know how to predict the direction of a spontaneous change in terms of probabilities of the initial and final states, it may seem appropriate to treat entropy as being directly proportional to probability as $S = k_B W$, where k_B is a proportionality constant. But this expression is invalid for the following reason. Entropy, like U and H, is an extensive property. Consequently, doubling the number of molecules would lead to a twofold increase in the entropy of the system. As we just saw, however, probability is proportional to the volume raised to the number of molecules, that is, $W \propto V^N$.* Therefore, changing from one to two molecules gives us W^2. Thus, the increases in entropy (from S to $2S$) and probability (from W to W^2) are not related to each other as predicted by the simple equation given above. A way out of this dilemma is to express entropy as a natural logarithmic function of probability as follows:

$$S = k_B \ln W \tag{4.1}$$

This equation tells us that as W increases to W^2, S increases to $2S$ because $\ln W^2 = 2 \ln W$. Equation 4.1 is known as the Boltzmann equation, and k_B is the Boltzmann constant, given by 1.381×10^{-23} J K^{-1}. Because the quantity $\ln W$ is dimensionless, the units of entropy are therefore J K^{-1}.

Equation 4.1 enables us to calculate changes in entropy when a system changes from an initial state, 1, to a final state, 2. The entropies of the system in these two states are given by

$$S_1 = k_B \ln W_1$$
$$S_2 = k_B \ln W_2$$

Ludwig Boltzmann's gravestone in Vienna, Austria is inscribed with his famous equation. (Photo courtesy of John Simon, reprinted with permission.)

Because entropy is a state function (it depends only on the probability of a state occurring and not on the manner in which the state is created), the change in entropy, ΔS, for the $1 \rightarrow 2$ process is

---

*Because $W \propto V$, we have $W = CV$, where C is a proportionality constant. Referring to the situation shown in Figure 4.3, this constant is given by $1/V_2$. Thus, the probability of finding a He atom in volume V_1 is given by $W = (1/V_2)(V_1) = (V_1/V_2) = \left(\tfrac{1}{2}\right)$, because $V_1 = V_2/2$. In general, the probability of finding N particles in volume V is given by the product of the individual probabilities; that is, $W = (CV)^N$, so $W \propto V^N$.

$$\Delta S = S_2 - S_1 = k_B \ln \frac{W_2}{W_1} \qquad (4.2)$$

Equation 4.2 can be used to calculate the entropy change when an ideal gas expands isothermally from V_1 to V_2. As we saw earlier, for N molecules, the probabilities W_1 and W_2 are related to the volumes V_1 and V_2 as follows:

$$W_1 = (CV_1)^N$$
$$W_2 = (CV_2)^N$$

Substituting these relations in Equation 4.2, we obtain

$$\Delta S = k_B \ln \frac{(CV_2)^N}{(CV_1)^N} = k_B \ln \left(\frac{V_2}{V_1}\right)^N$$

The Boltzmann constant k_B is given by R/N_A, where R is the gas constant and N_A is the Avogadro constant. Therefore, we write

$$\Delta S = \frac{N}{N_A} R \ln \frac{V_2}{V_1} = nR \ln \frac{V_2}{V_1} \qquad (4.3)$$

where n is the number of moles of the gas present. Remember that Equation 4.3 holds only for an isothermal expansion because the entropy of a system is also affected by changes in temperature. Furthermore, we do not have to specify the manner in which the expansion was brought about (that is, reversible or irreversible), because S is a state function.

Example 4.1

Calculate the entropy change when 2.0 moles of an ideal gas are allowed to expand isothermally from an initial volume of 1.5 L to 2.4 L. Estimate the probability that the gas will contract spontaneously from the final volume to the initial one.

ANSWER

From Equation 4.3, we write

$$\Delta S = (2.0 \text{ mol})(8.314 \text{ J K}^{-1} \text{ mol}^{-1}) \ln \frac{2.4 \text{ L}}{1.5 \text{ L}}$$

$$= 7.8 \text{ J K}^{-1}$$

To estimate the probability for spontaneous contraction, we note that this process must be accompanied by a *decrease* in entropy equal to -7.8 J K^{-1}. Because the process is now defined as $2 \rightarrow 1$, we have, from Equation 4.2,

$$\Delta S = k_B \ln \frac{W_1}{W_2}$$

$$-7.8 \text{ J K}^{-1} = (1.381 \times 10^{-23} \text{ J K}^{-1}) \ln \frac{W_1}{W_2}$$

$$\ln \frac{W_1}{W_2} = -5.7 \times 10^{23}$$

or

$$\frac{W_1}{W_2} = e^{-5.7 \times 10^{23}}$$

This exceedingly small ratio means that the probability of state 1 occurring is so much smaller than that of state 2 that there is virtually no possibility for the process to occur by itself. This result does not mean, of course, that the gas cannot be compressed from 2.4 L to 1.5 L, but it must be done with the aid of an external force.

Thermodynamic Definition of Entropy

Equation 4.1 is a statistical formulation of entropy; defining entropy in terms of probability provides us with a molecular interpretation. In general, however, this equation is not used for calculating changes in entropy. Calculating the value of W for complex systems, those in which chemical reactions occur, for example, is too difficult. Entropy changes can be conveniently measured from changes of other thermodynamic quantities, such as ΔH. In Section 3.5, we saw that the heat absorbed by an ideal gas in an isothermal, reversible expansion is given by

$$q_{rev} = -w_{rev}$$

$$q_{rev} = nRT \ln \frac{V_2}{V_1}$$

or

$$\frac{q_{rev}}{T} = nR \ln \frac{V_2}{V_1}$$

Because the right side of the equation above is equal to ΔS (see Equation 4.3), we have

$$\Delta S = \frac{q_{rev}}{T} \tag{4.4}$$

In words, Equation 4.4 says that the entropy change of a system in a reversible process is given by the heat absorbed divided by the temperature at which the process occurs. For an infinitesimal process, we can write

$$dS = \frac{dq_{rev}}{T} \tag{4.5}$$

Both Equations 4.4 and 4.5 are the thermodynamic definition of entropy. Although these equations were derived for the expansion of gases, they are applicable to any type of process at constant temperature. Note that the definition holds only for a reversible process, as the subscript *rev* indicates. Although S is a path-independent state function, q is not, so we must specify the reversible path in defining entropy. If the expansion were irreversible, then the work done by the gas on the surroundings would be less, and so would be the heat absorbed by the gas from the surroundings; that is, $q_{irrev} < q_{rev}$. Although the entropy change would be the same (that is, $\Delta S_{rev} = \Delta S_{irrev} = \Delta S$), we would have $\Delta S > q_{irrev}/T$. We shall return to this point in the next section.

The Carnot Heat Engine and Thermodynamic Efficiency

A heat engine converts heat to mechanical work. Heat engines play an essential role in our technological society; they include the now almost obsolete steam locomotives, steam turbines that generate electricity, and the internal combustion engines in automobiles. In 1824 a French engineer named Sadi Carnot (1796–1832) presented an analysis of the efficiency of heat engines, which laid the foundation of the second law of thermodynamics. The Carnot heat engine is an idealized model for the operation of any heat engine. For our purpose, it can be represented by an ideal gas in a cylinder fitted with a movable, frictionless piston that allows P–V work to be done on and by the gas.

Figure 4.4 shows the relationships of the engine to its thermal and mechanical surroundings. Like all machines, a heat engine works in a cyclic process. By drawing heat from a heat source, the gas expands and does work on the surroundings. It then undergoes a compression and discharges some of its heat to a cold reservoir. Finally, it is restored to the original state to repeat the process. The efficiency of a heat engine is defined as the ratio of output to input, or

$$\text{efficiency} = \frac{\text{net work done by heat engine}}{\text{heat absorbed by engine}} \tag{4.6}$$

For an engine working under reversible conditions, Carnot showed that*

$$\text{efficiency} = \frac{T_2 - T_1}{T_2} = 1 - \frac{T_1}{T_2} \tag{4.7}$$

where T_2 and T_1 are the temperatures of the heat source and cold reservoir, respectively. Because the engine is assumed to perform reversibly, the efficiency given in Equation 4.7 is the *maximum* value that can be obtained. In practice T_2 cannot be infinite and T_1 cannot be zero; therefore, we see that the efficiency can never be 1 or 100%. For example, at a power plant, superheated steam at about 560°C (833 K) is used to drive a turbine for electricity generation. The steam is discharged to a cooling tower at 38°C (311 K). From Equation 4.7 we have

$$\text{efficiency} = \frac{833\ \text{K} - 311\ \text{K}}{833\ \text{K}} = 0.63 \text{ or } 63\%$$

In reality, engines do not operate reversibly and when we include frictional loss and other complicating factors, the efficiency of a heat engine is considerably less than the theoretically estimated value. Most steam turbines operate at less than 40% efficiency.

The significance of Carnot's result is that heat cannot be totally converted to work—part of it is always discharged to the surroundings. In our example of the steam engine, as in other cases involving heat engines, the heat discharged to the cold reservoir is often called waste heat. Imagine constructing another heat engine to utilize the discharged heat to do work. Such a heat engine with $T_2 = 311$ K and $T_1 = 298$ K (the ambient temperature) would have a dismally low efficiency of 4.2% even under ideal conditions. Therefore, this heat is not captured and ends up only in enhancing the motions of air molecules (mostly translational and rotational). In contrast, consider generating electricity at a hydroelectric plant by releasing water

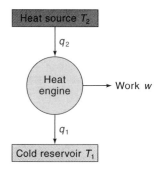

Figure 4.4
A heat engine draws heat from a heat source to do work on the surroundings and discharges some of the heat to a cold reservoir.

---

* See any physical chemistry text listed on p. 6 for the derivation of Equation 4.7.

from a dam. In principle, all the gravitational potential energy of water can be converted to mechanical work (that is, to drive a turbine) because this process does not involve heat and is therefore not subject to the thermodynamic limitations in energy conversion. Compared to other forms of energy, heat is low grade. Alternately, we say that energy is degraded when it is converted to heat.

This discussion of thermodynamic efficiency helps us understand the nature of many chemical and biological processes. In photosynthesis, for example, plants capture the radiant energy from the sun to make complex molecules for growth and function. To do so, plants have evolved molecules like chlorophyll that can absorb concentrated photon energy in the visible region to do chemical work. But if the high-energy portion of the radiation is allowed to degrade to heat, it will be too dispersed to be of any use. The efficiency of a hypothetical engine based on heated leaves as the heat source would be too low for any meaningful biosynthesis.

4.3 The Second Law of Thermodynamics

So far, our discussion of entropy changes has focused on the system. For a proper understanding of entropy, we must also examine what happens to the surroundings. Because of its size and the amount of material it contains, the surroundings can be thought of as an infinitely large reservoir. Therefore, the exchange of heat and work between a system and its surroundings alters the properties of the surroundings by only an infinitesimal amount. Because infinitesimal changes are characteristic of reversible processes, it follows that *any* process has the same effect on the surroundings as a reversible process. Thus, regardless of whether a process is reversible or irreversible with respect to the system, we can write the heat change in the surroundings as

$$(dq_{surr})_{rev} = (dq_{surr})_{irrev} = dq_{surr}$$

For this reason, we shall not bother to specify the path for dq_{surr}. The change in entropy of the surroundings is

$$dS_{surr} = \frac{dq_{surr}}{T_{surr}}$$

and for a finite isothermal process—that is, for a process that can be studied in the laboratory,

$$\Delta S_{surr} = \frac{q_{surr}}{T_{surr}}$$

Returning to the isothermal expansion of an ideal gas, we saw earlier that the heat absorbed from the surroundings during a reversible process is $nRT_{sys}\ln(V_2/V_1)$, where T_{sys} is the temperature of the system. Because the system is at thermal equilibrium with its surroundings throughout the process, $T_{sys} = T_{surr} = T$. The heat lost by the surroundings to the system is therefore $-nRT\ln(V_2/V_1)$, and the corresponding change in entropy is

$$\Delta S_{surr} = \frac{q_{surr}}{T}$$

The total change in the entropy of the universe (system plus surroundings), ΔS_{univ}, is given by

$$\Delta S_{univ} = \Delta S_{sys} + \Delta S_{surr}$$

$$= \frac{q_{sys}}{T} + \frac{q_{surr}}{T}$$

$$= \frac{nRT \ln(V_2/V_1)}{T} + \frac{[-nRT \ln(V_2/V_1)]}{T} = 0$$

Thus, for a reversible process, the total change in the entropy of the universe is equal to zero.

Now let us consider what happens if the expansion is irreversible. In the extreme case, we can assume that the gas is expanding against a vacuum. Again, the change in the entropy of the system is given by $\Delta S_{sys} = nR \ln(V_2/V_1)$ because S is a state function. Because no work is done in this process, however, no heat is exchanged between the system and the surroundings. Therefore, we have $q_{surr} = 0$ and $\Delta S_{surr} = 0$. The change in the entropy of the universe is now given by

$$\Delta S_{univ} = \Delta S_{sys} + \Delta S_{surr}$$

$$= nR \ln(V_2/V_1) > 0$$

Combining these two expressions for ΔS_{univ}, we obtain

$$\Delta S_{univ} = \Delta S_{sys} + \Delta S_{surr} \geq 0 \qquad (4.8)$$

where the equality sign applies to a reversible process and the greater-than ($>$) sign applies to an irreversible (that is, spontaneous) process. Equation 4.8 is the mathematical statement of the *second law of thermodynamics*. In words, the second law may be stated as follows: *The entropy of an isolated system increases in an irreversible process and remains unchanged in a reversible process. It can never decrease.* Thus, either ΔS_{sys} or ΔS_{surr} can be a negative quantity for a particular process, but their sum can never be less than zero.

Just thinking about entropy increases its value in the universe.

Example 4.2

A quantity of 0.50 mole of an ideal gas at 20°C expands isothermally against a constant pressure of 2.0 atm from 1.0 L to 5.0 L. Calculate the values of ΔS_{sys}, ΔS_{surr}, and ΔS_{univ}.

ANSWER

From the initial conditions, we can show that the pressure of the gas is 12 atm. First we calculate the value of ΔS_{sys}. Noting that the process is isothermal and ΔS_{sys} is the same whether the process is reversible or irreversible, we write, from Equation 4.3,

$$\Delta S_{sys} = nR \ln \frac{V_2}{V_1}$$

$$= (0.50 \text{ mol})(8.314 \text{ J K}^{-1} \text{ mol}^{-1}) \ln \frac{5.0 \text{ L}}{1.0 \text{ L}}$$

$$= 6.7 \text{ J K}^{-1}$$

To calculate the value of ΔS_{surr}, we first determine the work done in the irreversible gas expansion

$$w = -P\Delta V$$
$$= -(2.0 \text{ atm})(5.0 - 1.0) \text{ L}$$
$$= -8.0 \text{ L atm}$$
$$= -810 \text{ J} \quad (1 \text{ L atm} = 101.3 \text{ J})$$

Because $\Delta U = 0$, $q = -w = +810$ J. The heat lost by the surroundings must then be -810 J. The change in entropy of the surroundings is given by

$$\Delta S_{surr} = \frac{q_{surr}}{T}$$
$$= \frac{-810 \text{ J}}{293 \text{ K}}$$
$$= -2.8 \text{ J K}^{-1}$$

Finally, from Equation 4.8,

$$\Delta S_{univ} = 6.7 \text{ J K}^{-1} - 2.8 \text{ J K}^{-1}$$
$$= 3.9 \text{ J K}^{-1}$$

COMMENT

The result shows that the process is spontaneous, which is what we would expect given the initial pressure of the gas.

4.4 Entropy Changes

Having learned the statistical and thermodynamic definitions of entropy and examined the second law of thermodynamics, we are now ready to study how various processes affect the entropy of the system. We have already seen that the entropy change for the reversible, isothermal expansion of an ideal gas is given by $nR \ln(V_2/V_1)$. In this section, we shall consider several other examples of entropy changes.

Entropy Change Due to Mixing of Ideal Gases

Figure 4.5 shows a container in which n_A moles of ideal gas A at T, P, and V_A are separated by a partition from n_B moles of ideal gas B at T, P, and V_B. When the partition is removed, the gases mix spontaneously and the entropy of the system increases. To calculate the entropy of mixing, $\Delta_{mix} S$, we can treat the process as two separate, isothermal gas expansions.

$$\text{For gas A} \quad \Delta S_A = n_A R \ln \frac{V_A + V_B}{V_A}$$

$$\text{For gas B} \quad \Delta S_B = n_B R \ln \frac{V_A + V_B}{V_B}$$

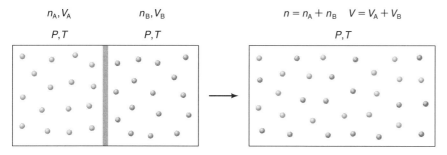

Figure 4.5
The mixing of two ideal gases at the same temperature and pressure leads to an increase in entropy.

Therefore,

$$\Delta_{mix}S = \Delta S_A + \Delta S_B = n_A R \ln \frac{V_A + V_B}{V_A} + n_B R \ln \frac{V_A + V_B}{V_B}$$

According to Avogadro's law, volume is directly proportional to the number of moles of the gas at constant T and P, and so the above equation can be written as

$$\Delta_{mix}S = n_A R \ln \frac{n_A + n_B}{n_A} + n_B R \ln \frac{n_A + n_B}{n_B}$$

$$= -n_A R \ln \frac{n_A}{n_A + n_B} - n_B R \ln \frac{n_B}{n_A + n_B}$$

$$= -n_A R \ln x_A - n_B R \ln x_B$$

$$\Delta_{mix}S = -R(n_A \ln x_A + n_B \ln x_B) \qquad (4.9)$$

where x_A and x_B are the mole fractions of A and B, respectively. Because $x < 1$, it follows that $\ln x < 0$ and that the right side of Equation 4.9 is a positive quantity, which is consistent with the spontaneous nature of the process.*

Entropy Change Due to Phase Transitions

The melting of ice is a familiar phase change. At 0°C and 1 atm, ice and water are in equilibrium. Under these conditions, heat is absorbed reversibly by the ice during the melting process. Furthermore, because this is a constant-pressure process, the heat absorbed is equal to the enthalpy change of the system, so that $q_{rev} = \Delta_{fus}H$, where $\Delta_{fus}H$ is called the *heat* or *enthalpy of fusion*. Note that because H is a state function, it is no longer necessary to specify the path, and the melting process need not be carried out reversibly. The entropy of fusion, $\Delta_{fus}S$, is given by

$$\Delta_{fus}S = \frac{\Delta_{fus}H}{T_f} \qquad (4.10)$$

Equations 4.10 and 4.11 hold for an isothermal process.

---

* Because A and B are ideal gases, there are no intermolecular forces between the molecules, so no heat change results from the mixing. Consequently, the change in entropy of the surroundings is zero, and the direction of the process depends solely on the change in entropy of the system.

where T_f is the fusion or melting point (273 K for ice). Similarly, we can write the entropy of vaporization, $\Delta_{vap}S$, as

$$\Delta_{vap}S = \frac{\Delta_{vap}H}{T_b} \qquad (4.11)$$

where $\Delta_{vap}H$ and T_b are the enthalpy of vaporization and boiling point of the liquid, respectively.

Entropy Change Due to Heating

When the temperature of a system is raised from T_1 to T_2, its entropy also increases. We can calculate this entropy increase as follows. Let S_1 and S_2 be the entropies of the system in states 1 and 2 (characterized by T_1 and T_2). If heat is transferred reversibly to the system, then the increase in entropy for an infinitesimal amount of heat transfer is given by Equation 4.5:

$$dS = \frac{dq_{rev}}{T}$$

The entropy at T_2 is given by

$$S_2 = S_1 + \int_{T_1}^{T_2} \frac{dq_{rev}}{T}$$

If we define this to be a constant-pressure process, as is usually the case, then $dq_{rev} = dH$, so that

$$S_2 = S_1 + \int_{T_1}^{T_2} \frac{dH}{T}$$

From Equation 3.19, we have $dH = C_P\, dT$, and so we can write

Remember that $\int \frac{dx}{x} = \int d\ln x.$
$$S_2 = S_1 + \int_{T_1}^{T_2} \frac{C_P}{T}\, dT = S_1 + \int_{T_1}^{T_2} C_P\, d\ln T \qquad (4.12)$$

If the temperature range is small, we can assume that C_P is independent of temperature. Then Equation 4.12 becomes

$$S_2 = S_1 + C_P \ln \frac{T_2}{T_1} \qquad (4.13)$$

and the increase in entropy, ΔS, as a result of heating is

$$\Delta S = S_2 - S_1 = C_P \ln \frac{T_2}{T_1}$$

$$\Delta S = n\bar{C}_P \ln \frac{T_2}{T_1} \qquad (4.14)$$

Example 4.3

At constant pressure, 200 g of water is heated from 10°C to 20°C. Calculate the increase in entropy for this process. The molar heat capacity of water at constant pressure is 75.3 J K^{-1} mol^{-1}.

ANSWER

The number of moles of water present is 200 g/18.02 g mol^{-1} = 11.1 mol. The increase in entropy, according to Equation 4.14, is given by

$$\Delta S = (11.1 \text{ mol})(75.3 \text{ J K}^{-1} \text{ mol}^{-1}) \ln \frac{293 \text{ K}}{283 \text{ K}}$$

$$= 29.0 \text{ J K}^{-1}$$

COMMENT

For this calculation, we have assumed that $\bar{C}_P$ is independent of temperature and that water does not expand when heated, so that no work is done.

Suppose that the heating of water in Example 4.3 had been carried out irreversibly (as is the case in practice), say with a Bunsen burner. What would be the increase in entropy? We note that regardless of the path, the initial and final states are the same; that is, 200 g of water heated from 10°C to 20°C. Therefore, the integral on the right side of Equation 4.12 gives ΔS for the irreversible heating. This conclusion follows from the fact that ΔS depends only on T_1 and T_2 and not on the path. Thus, ΔS for this process is 29.0 J K^{-1}, whether the heating is done reversibly or irreversibly.

Example 4.4

Supercooled water is liquid water that has been cooled below its normal freezing point. This state is thermodynamically unstable and tends to freeze into ice spontaneously. Suppose we have 2.0 moles of supercooled water turning into ice at −10°C and 1.0 atm. Calculate the values of ΔS_{sys}, ΔS_{surr}, and ΔS_{univ} for this process. The $\bar{C}_P$ values of water and ice for the temperature range between 0 and −10°C are 75.3 J K^{-1} mol^{-1} and 37.7 J K^{-1} mol^{-1}, respectively, and the molar heat of fusion of water is 6.01 kJ mol^{-1}.

ANSWER

First, we note that a change in phase is reversible only at the temperature at which the two phases are at equilibrium. Because supercooled water at −10°C and ice at −10°C are not at equilibrium, the freezing process is not reversible. To calculate the value of ΔS_{sys}, we devise a series of reversible steps by which supercooled water at −10°C is converted to ice at −10°C (Figure 4.6).

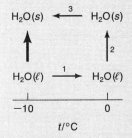

Figure 4.6
The spontaneous freezing of the supercooled water (thick arrow) at −10°C can be broken down into three reversible paths (1, 2, and 3).

Step 1. Reversible heating of supercooled water at $-10°C$ to $0°C$:

$$\underset{-10°C}{H_2O(l)} \rightarrow \underset{0°C}{H_2O(l)}$$

From Equation 4.14,

$$\Delta S_1 = (2.0 \text{ mol})(75.3 \text{ J K}^{-1} \text{ mol}^{-1}) \ln \frac{273 \text{ K}}{263 \text{ K}}$$

$$= 5.6 \text{ J K}^{-1}$$

Step 2. Water freezes into ice at $0°C$:

$$\underset{0°C}{H_2O(l)} \rightarrow \underset{0°C}{H_2O(s)}$$

According to Equation 4.10, the molar entropy of fusion is given by

$$\Delta_{\text{fus}}\bar{S} = \frac{\Delta_{\text{fus}}\bar{H}}{T_{\text{f}}}$$

Freezing is an exo-thermic process.

So, for the freezing of 2.0 moles of water we reverse the sign of $\Delta_{\text{fus}}\bar{H}$ and write

$$\Delta S_2 = (2.0 \text{ mol}) \frac{-6.01 \times 10^3 \text{ J mol}^{-1}}{273 \text{ K}}$$

$$= -44.0 \text{ J K}^{-1}$$

Step 3. Reversible cooling of ice from $0°C$ to $-10°C$:

$$\underset{0°C}{H_2O(s)} \rightarrow \underset{-10°C}{H_2O(s)}$$

Again from Equation 4.14,

$$\Delta S_3 = (2.0 \text{ mol})(37.7 \text{ J K}^{-1} \text{ mol}^{-1}) \ln \frac{263 \text{ K}}{273 \text{ K}}$$

$$= -2.8 \text{ J K}^{-1}$$

Finally,

$$\Delta S_{\text{sys}} = \Delta S_1 + \Delta S_2 + \Delta S_3$$

$$= (5.6 - 44.0 - 2.8) \text{ J K}^{-1}$$

$$= -41.2 \text{ J K}^{-1}$$

To calculate the value of ΔS_{surr}, we first determine the heat change in the surroundings for each of the above steps.

Step 1. Heat gained by the supercooled water equals heat lost by the surroundings, given by

$$(q_{\text{surr}})_1 = -n\bar{C}_P \Delta T$$

$$= -(2.0 \text{ mol})(75.3 \text{ J K}^{-1} \text{ mol}^{-1})(10 \text{ K})$$

$$= -1.5 \times 10^3 \text{ J}$$

Step 2. When the water freezes at 0°C, heat is given off to the surroundings:

$$(q_{surr})_2 = (2.0 \text{ mol})(6010 \text{ J mol}^{-1})$$
$$= 1.2 \times 10^4 \text{ J}$$

Step 3. Cooling ice from 0°C to −10°C releases heat to the surroundings equal to

$$(q_{surr})_3 = (2.0 \text{ mol})(37.7 \text{ J K}^{-1} \text{ mol}^{-1})(10 \text{ K})$$
$$= 754 \text{ J}$$

The total heat change is given by

$$(q_{surr})_{total} = (-1.5 \times 10^3 + 1.2 \times 10^4 + 754) \text{ J}$$
$$= 1.1 \times 10^4 \text{ J}$$

and the change in entropy at −10°C is

$$\Delta S_{surr} = \frac{1.1 \times 10^4 \text{ J}}{263 \text{ K}}$$
$$= 41.8 \text{ J K}^{-1}$$

Finally,

$$\Delta S_{univ} = \Delta S_{sys} + \Delta S_{surr}$$
$$= -41.2 \text{ J K}^{-1} + 41.8 \text{ J K}^{-1}$$
$$= 0.6 \text{ J K}^{-1}$$

COMMENT

The result ($\Delta S_{univ} > 0$) confirms the statement that supercooled water is unstable and will spontaneously freeze on standing. Note that in this process, the entropy of the system decreases because water is converted to ice. However, the heat released to the surroundings results in an increase in the value of ΔS_{surr} that is greater (in magnitude) than ΔS_{sys}, so the ΔS_{univ} value is positive.

4.5 The Third Law of Thermodynamics

In thermodynamics, we are normally interested only in changes in properties, such as ΔU and ΔH. Although we have no way of measuring the absolute values of internal energy and enthalpy, we can find the absolute entropy of a substance. As written, Equation 4.12 enables us to measure the change in entropy over a suitable temperature range between T_1 and T_2. Suppose we set the lower temperature to absolute zero (that is, $T_1 = 0$ K), and call the upper temperature T. Equation 4.12 now becomes

$$\Delta S = S_T - S_0 = \int_0^T \frac{C_P}{T} dT \tag{4.15}$$

Because entropy is an extensive property, its value at any temperature T is equal to the sum of the contributions from 0 K to the specified temperature. We can measure heat capacity as a function of temperature and calculate entropy changes, including phase transitions, if any, to evaluate the integral in Equation 4.15. This approach presents two obstacles, however. First, what is the entropy of a substance at absolute zero; that is, what is S_0? Second, how do we account for the part of the contribution to the total entropy that lies between absolute zero and the lowest temperature at which measurements are feasible?

According to the Boltzmann equation (Equation 4.1), entropy is related to the probability of a certain state occurring. We can also use W to denote the number of *microstates* of a macroscopic system. The meaning of microstate in this context can be illustrated by a hypothetical, perfect crystalline substance with no impurities or crystal defects. Such a crystal can have only one particular arrangement of atoms or molecules (that is, only one microstate). Consequently, $W = 1$ and

$$S = k_B \ln W = k_B \ln 1 = 0$$

This principle is known as the *third law of thermodynamics*, which states that *every substance has a finite positive entropy, but at the absolute zero of temperature the entropy may become zero, and it does in the case of a pure, perfect crystalline substance.* Mathematically, the third law can be expressed as

$$\lim_{T \to 0\,K} S = 0 \quad \text{(perfect crystalline substance)}$$

At temperatures above absolute zero, thermal motion contributes to the entropy of the substance so that its entropy is no longer zero, even if it is pure and remains perfectly crystalline. The significance of the third law is that it enables us to calculate the absolute values of entropies, discussed below.

Third-Law or Absolute Entropies

The third law of thermodynamics enables us to measure the entropy of a substance at temperature T. For a perfect crystalline substance, $S_0 = 0$, so Equation 4.15 becomes

Contributions to S due to phase transitions must also be included in Equation 4.16.

$$S_T = \int_0^T \frac{C_P}{T} dT = \int_0^T C_P \, d\ln T \tag{4.16}$$

Now we can measure the heat capacity over the desired temperature range. For very low temperatures (≤ 15 K), for which such measurements are difficult to carry out, we can use Debye's theory (after the Dutch–American physicist Peter Debye, 1884–1966) of heat capacity:

$$C_P = aT^3 \tag{4.17}$$

where a is a constant for a given substance. The entropy change over this small temperature range is

$$\Delta S = \int_0^T \frac{aT^3}{T} dT = \int_0^T aT^2 \, dT$$

Note that Equation 4.17 is applicable only near absolute zero. In applying Equation 4.16, we must recall that it holds only for a perfectly ordered substance at 0 K. Figure 4.7 shows a plot of $S°$ versus temperature for substances in general, where the superscript denotes the standard state.

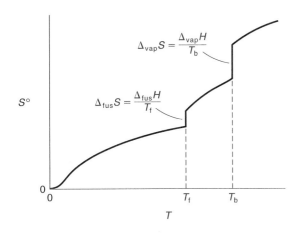

Figure 4.7
The increase in entropy of a perfect crystalline substance from absolute zero to its gaseous state at some temperature. Note the contributions to the $S°$ value due to phase transitions (melting and boiling).

Entropy values calculated by using Equation 4.16 are called third-law or absolute entropies because these values are not based on some reference state. Table 4.1 lists the absolute standard molar entropy values of a number of common elements and compounds at 1 bar and 298 K. More data are given in Appendix 2. Note that because these are absolute values, we omit the Δ sign and the subscript f for $\bar{S}°$, but we retain them for the molar standard enthalpies of formation ($\Delta_f \bar{H}°$).

For convenience, we shall often omit "absolute" in referring to the standard molar entropy values.

Table 4.1
Standard Molar Entropies at 298 K and 1 Bar for Some Inorganic and Organic Substances

Substance	$\bar{S}°/\text{J} \cdot \text{K}^{-1} \cdot \text{mol}^{-1}$	Substance	$\bar{S}°/\text{J} \cdot \text{K}^{-1} \cdot \text{mol}^{-1}$
C(graphite)	5.7	$CH_4(g)$	186.2
C(diamond)	2.4	$C_2H_6(g)$	229.5
$CO(g)$	197.9	$C_3H_8(g)$	269.9
$CO_2(g)$	213.6	$C_2H_2(g)$	200.8
$HF(g)$	173.5	$C_2H_4(g)$	219.5
$HCl(g)$	186.5	$C_6H_6(l)$	172.8
$HBr(g)$	198.7	$CH_3OH(l)$	126.8
$HI(g)$	206.3	$C_2H_5OH(l)$	161.0
$H_2O(g)$	188.7	$CH_3CHO(l)$	160.2
$H_2O(l)$	69.9	$HCOOH(l)$	129.0
$NH_3(g)$	192.5	$CH_3COOH(l)$	159.8
$NO(g)$	210.6	$C_6H_{12}O_6(s)$	210.3
$NO_2(g)$	240.5	$C_{12}H_{22}O_{11}(s)$	360.2
$N_2O_4(g)$	304.3		
$N_2O(g)$	220.0		
$O_2(g)$	205.0		
$O_3(g)$	237.7		
$SO_2(g)$	248.5		

Entropy of Chemical Reactions

We are now ready to calculate the entropy change that occurs in a chemical reaction. As for the enthalpy of reaction (see Equation 3.32), the change in entropy for the hypothetical reaction

$$a\text{A} + b\text{B} \rightarrow c\text{C} + d\text{D}$$

is given by the equation

$$\Delta_r S^\circ = c\bar{S}^\circ(\text{C}) + d\bar{S}^\circ(\text{D}) - a\bar{S}^\circ(\text{A}) - b\bar{S}^\circ(\text{B})$$

$$\boxed{\Delta_r S^\circ = \Sigma v \bar{S}^\circ(\text{products}) - \Sigma v \bar{S}^\circ(\text{reactants})} \tag{4.18}$$

where v represents the stoichiometric coefficient.

Example 4.5

Calculate $\Delta S_{sys}, \Delta S_{surr}$, and ΔS_{univ} for the synthesis of ammonia at 25°C:

$$\text{N}_2(g) + 3\text{H}_2(g) \rightarrow 2\text{NH}_3(g) \quad \Delta_r H^\circ = -92.6 \text{ kJ mol}^{-1}$$

ANSWER

First we calculate ΔS_{sys}. Using Equation 4.18 and the data in Appendix 2, we write

$$\begin{aligned}
\Delta S_{sys} &= 2\bar{S}^\circ(\text{NH}_3) - [\bar{S}^\circ(\text{N}_2) + 3\bar{S}^\circ(\text{H}_2)] \\
&= (2)(192.5 \text{ J K}^{-1} \text{ mol}^{-1}) - [191.6 \text{ J K}^{-1} \text{ mol}^{-1} + (3)(130.6 \text{ J K}^{-1} \text{ mol}^{-1})] \\
&= -198.4 \text{ J K}^{-1} \text{ mol}^{-1}
\end{aligned}$$

To calculate ΔS_{surr}, we note that the system is in thermal equilibrium with the surroundings. Because $\Delta H_{surr} = -\Delta H_{sys}$, ΔS_{surr} is given by

$$\begin{aligned}
\Delta S_{surr} &= \frac{\Delta H_{surr}}{T} \\
&= \frac{-(-92.6 \times 1000) \text{ J mol}^{-1}}{298 \text{ K}} = 311 \text{ J K}^{-1} \text{ mol}^{-1}
\end{aligned}$$

The change in entropy for the universe is

$$\begin{aligned}
\Delta S_{univ} &= \Delta S_{sys} + \Delta S_{surr} \\
&= -198 \text{ J K}^{-1} \text{ mol}^{-1} + 311 \text{ J K}^{-1} \text{ mol}^{-1} \\
&= 113 \text{ J K}^{-1} \text{ mol}^{-1}
\end{aligned}$$

COMMENT

Because the ΔS_{univ} value is positive, we predict that the reaction is spontaneous at 25°C. Recall that just because a reaction is spontaneous does not mean that it will occur at an observable rate. The synthesis of ammonia is, in fact, extremely slow at room temperature because it has a large activation energy. Thermodynamics can tell us whether a reaction will occur spontaneously under specific conditions, but it does not say how fast it will occur. Reaction rates are the subject of chemical kinetics (Chapter 9).

The Meaning of Entropy

At this point we have defined entropy statistically and thermodynamically. With the third law of thermodynamics it is possible to determine the absolute entropy of substances. We have seen some examples of entropy changes in physical processes and chemical reactions. But what is entropy, really?

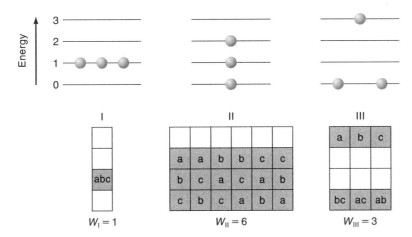

Figure 4.8
Arrangement of three molecules among energy levels with a total energy of three units.

Frequently entropy is described as a measure of disorder or randomness. The higher the disorder, the greater the entropy of the system. While useful, these terms must be used with caution because they are subjective concepts.* On the other hand, relating entropy to probability makes more sense because probability is a quantitative concept. Earlier we saw how gas expansion was viewed in terms of probability. In a spontaneous process, a system goes from a less probable state to a more probable one. The corresponding change in entropy is calculated using the Boltzmann equation (Equation 4.1). The quantity W was identified with probability but in general it should be interpreted as the number of microscopic states or microstates that corresponds to a given macrostate.

To clarify the difference between microstate and macrostate, consider a system comprising three identical, noninteracting molecules distributed over energy levels and the total energy of the system is restricted to three units. How many different ways can this distribution be accomplished? Although the molecules are identical, they can be distinguished from one another by their locations (for example, if they occupy different lattice points in a crystal). We see that there are ten ways (ten microstates) to distribute the molecules that make up three distinct *distributions* (three macrostates), designated I, II, and III, as shown in Figure 4.8. Not all the macrostates are equally probable—macrostate II is six times more probable as I and twice as probable as III. Based on this analysis, we conclude that the probability of occurence of a particular distribution (state) depends on the number of ways (microstates) in which the distribution can be achieved. Three molecules do not constitute a macroscopic system, but as the number of molecules increases (and hence also the total energy of the system), we find that there will be one macrostate with many more microstates than the other distributions. When the number approaches Avogadro's number, say, the most probable macrostate will have such an overwhelming number of microstates compared with all other macrostates that we will always find the system in this macrostate (Figure 4.9).

From this discussion we can say the following about entropy. Entropy is related to the distribution or spread of energy among the available molecular energy levels. At thermal equilibrium we always find the system in the most probable macrostate, which has the largest number of microstates and the most probable distribution of

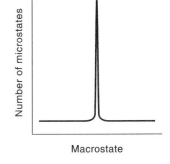

Figure 4.9
For an Avogadro's number of molecules, the most probable macrostate has an overwhelmingly large number of microstates compared to other macrostates.

---

* See D. F. Styer, *Am. J. Phys.* **68**, 1090 (2000) and F. L. Lambert, *J. Chem. Educ.* **79**, 187 (2002).

energy. The greater the number of energy levels that have significant occupation, the larger the entropy. It follows, therefore, that the entropy of a system is a maximum at equilibrium because W itself is a maximum. Equation 4.1 is not used to calculate entropy, however, because in general we do not know what W is. As mentioned earlier, entropy values are usually determined by calorimetric methods. Nevertheless, the molecular interpretation enables us to gain a better understanding of the nature of entropy and its changes. We consider a few examples below.

Isothermal Gas Expansion. In an expansion, the gas molecules move in a larger volume. As we shall see in Chapter 11 (p. 419), the translational kinetic energy of a molecule is quantized and the energy of any particular level is *inversely* proportional to the dimension of the container. It follows, therefore, that in the larger volume the levels become more closely spaced and hence more accessible for distributing energy. Consequently, more energy levels will be occupied, resulting in an increased number of microstates corresponding to the most probable macrostate and hence an increase in entropy.

Isothermal Mixing of Gases. The mixing of two gases at constant temperature can be treated as two separate gas expansions. Again, we predict an increase in entropy.

Heating. When the temperature of a substance is raised, the energy input is used to promote the molecular motions (translational, rotational, and vibrational) from the low-lying levels to higher ones. The result is an increase in the occupancy among the molecular energy levels and hence the number of microstates. Consequently, there will be an increase in entropy. This is what happens for heating at constant volume. If heating is carried out at constant pressure, there will be an additional contribution to entropy due to expansion. The difference between constant volume and constant pressure conditions is significant only if the substance is a gas.

Phase Transitions. In a solid the atoms or molecules are confined to fixed positions and the number of microstates is small. Upon melting, these atoms or molecules can occupy many more positions as they move away from the lattice points. Consequently, the number of microstates increases because there are now many more ways to arrange the particles. Therefore, we predict this "order → disorder" phase transition to result in an increase in entropy because the number of microstates has increased. Similarly, we predict the vaporization process will also lead to an increase in the entropy of the system. The increase will be considerably greater than that for melting, however, because molecules in the gas phase occupy much more space and therefore there are far more microstates than in the liquid phase.

Chemical Reactions. Referring to Example 4.5, we see that the synthesis of ammonia from nitrogen and hydrogen results in a net loss of two moles of gases per reaction unit. The decrease in molecular motions is reflected in fewer microstates so we would expect to see a decrease in the entropy of the system. Because the reaction is exothermic, the heat released energizes the motions of the surrounding air molecules. The increase in the microstates of the air molecules leads to an increase in the entropy of the surroundings, which outweighs the decrease in the entropy of the system so the reaction is spontaneous. Keep in mind that prediction of entropy changes becomes less certain for reactions involving condensed phases or in cases where there are no changes in the number of gaseous components.

4.6 Gibbs Energy

With the first law of thermodynamics to take care of energy balance and the second law to help us decide which processes can occur spontaneously, we might reasonably expect that we have enough thermodynamic quantities to deal with any situation. Although this expectation is true in principle, in practice, the equations we have derived so far are not the most convenient to apply. For example, to use the second law (Equation 4.8), we must calculate the entropy change in both the system and the surroundings. Because we are generally interested only in what happens in the system and are not concerned with events in the surroundings, it would be simpler if we could establish criteria for equilibrium and spontaneity in terms of the change in a certain thermodynamic function of the system and not of the entire universe, as for ΔS_{univ}.

Consider a system in thermal equilibrium with its surroundings at temperature T. A process occurring in the system results in the transfer of an infinitesimal amount of heat, dq, from the system to the surroundings. Thus, we have $-dq_{\text{sys}} = dq_{\text{surr}}$. The total change in entropy, according to Equation 4.8, is

$$dS_{\text{univ}} = dS_{\text{sys}} + dS_{\text{surr}} \geq 0$$

$$= dS_{\text{sys}} + \frac{dq_{\text{surr}}}{T} \geq 0$$

$$= dS_{\text{sys}} - \frac{dq_{\text{sys}}}{T} \geq 0$$

Note that every quantity on the right side of the above equation refers to the system. If the process takes place at constant pressure, then $dq_{\text{sys}} = dH_{\text{sys}}$, or

$$dS_{\text{sys}} - \frac{dH_{\text{sys}}}{T} \geq 0$$

Multiplying the equation above by $-T$, we obtain

$$dH_{\text{sys}} - T\, dS_{\text{sys}} \leq 0$$

The reversal of the inequality sign follows from the fact that if $x > 0$ then $-x < 0$.

We now define a function, called *Gibbs energy** (after the American physicist Josiah Willard Gibbs, 1839–1903), G, as

$$G = H - TS \tag{4.19}$$

From Equation 4.19, we see that because H, T, and S are all state functions, G is also a state function. Further, like enthalpy, G has the units of energy.

At constant temperature, the change in the Gibbs energy of the system in an infinitesimal process is given by

$$dG_{\text{sys}} = dH_{\text{sys}} - T\, dS_{\text{sys}}$$

---

* Gibbs energy was previously called *Gibbs free energy* or just *free energy*. However, IUPAC (the International Union of Pure and Applied Chemistry) has recommended that the *free* be dropped. The same recommendation applies to Helmholtz energy, to be discussed shortly.

We can apply dG_{sys} as a criterion for equilibrium and spontaneity as follows:

$$dG_{sys} \leq 0 \qquad (4.20)$$

where the $<$ sign denotes a spontaneous process and the equality sign denotes equilibrium at constant temperature and pressure.

Unless otherwise indicated, from now on we shall consider only the system in our discussion of Gibbs energy changes. For this reason, the subscript *sys* will be omitted for simplicity. For a finite isothermal process $1 \rightarrow 2$, the change of Gibbs energy is given by

$$\Delta G = \Delta H - T\Delta S \qquad (4.21)$$

and the conditions of equilibrium and spontaneity at constant temperature and pressure are given by

$$\Delta G = G_2 - G_1 = 0 \quad \text{system at equilibrium}$$

$$\Delta G = G_2 - G_1 < 0 \quad \text{spontaneous process from 1 to 2}$$

If ΔG is negative, the process is said to be *exergonic* (from the Greek word for "work producing"); if positive, the process is *endergonic* (work consuming). Note that pressure must be constant to set $q = \Delta H$, and temperature must be constant to derive Equation 4.21. In general, we can replace q with ΔH only if pressure is constant *throughout* the process. Because G is a state function, however, ΔG is independent of path. Therefore, Equation 4.21 applies to any process as long as the temperature and pressure are the same in the initial and final states.

Gibbs energy is useful because it incorporates both enthalpy and entropy. In some reactions, the enthalpy and entropy contributions reinforce each other. For example, if ΔH is negative (an exothermic reaction) and ΔS is positive, then $(\Delta H - T\Delta S)$ or ΔG is a negative quantity, and the process is favored from left to right. In other reactions, enthalpy and entropy may work against each other; that is, ΔH and $(-T\Delta S)$ have different signs. In such cases, the sign of ΔG is determined by the *magnitudes* of ΔH and $T\Delta S$. If $|\Delta H| \gg |T\Delta S|$, then the reaction is said to be enthalpy-driven because the sign of ΔG is predominantly determined by ΔH. Conversely, if $|T\Delta S| \gg |\Delta H|$, then the process is entropy-driven. Table 4.2 shows how positive and negative values of ΔH and ΔS affect ΔG at different temperatures.

A similar thermodynamic function can be derived for processes in which the temperature and volume are kept constant. *Helmholtz energy* (after the German physiologist and physicist Hermann Ludwig Helmholtz, 1821–1894), A, is defined as

$$A = U - TS \qquad (4.22)$$

where all the terms refer to the system. Like G, A is a state function and has the units of energy. Following the same procedure described above for Gibbs energy, we can show that at constant temperature and volume, the criteria for equilibrium and spontaneity are given by

Constant-volume processes are less common in biological systems.

$$dA_{sys} \leq 0 \qquad (4.23)$$

Omitting the subscript for system, we have, for a finite process at constant temperature,

$$\Delta A = \Delta U - T\Delta S \qquad (4.24)$$

Table 4.2
Factors Affecting ΔG of a Reaction[a]

ΔH	ΔS	ΔG	Example
+	+	Positive at low temperatures; negative at high temperatures. Reaction spontaneous in the forward direction at high temperatures and spontaneous in the reverse direction at low temperatures.	$2HgO(s) \rightarrow 2Hg(l) + O_2(g)$
+	−	Positive at all temperatures. Reaction spontaneous in the reverse direction at all temperatures.	$3O_2(g) \rightarrow 2O_3(g)$
−	+	Negative at all temperatures. Reaction spontaneous in the forward direction at all temperatures.	$2H_2O_2(l) \rightarrow 2H_2O(l) + O_2(g)$
−	−	Negative at low temperatures; positive at high temperatures. Reaction spontaneous at low temperatures; tends to reverse at high temperatures.	$NH_3(g) + HCl(g) \rightarrow NH_4Cl(s)$

[a] Assuming both ΔH and ΔS are independent of temperature.

The Meaning of Gibbs Energy

Equation 4.20 provides us with an extremely useful criterion for dealing with the direction of spontaneous changes and the nature of chemical and physical equilibria. In addition, it also enables us to determine the amount of work that can be done in a given process.

To show the relationship between the change in Gibbs energy and work, we start with the definition of G:

$$G = H - TS$$

For an infinitesimal process,

$$dG = dH - T\,dS - S\,dT$$

Now, because

$$H = U + PV$$

$$dH = dU + P\,dV + V\,dP$$

According to the first law of thermodynamics,

$$dU = \bar{d}q + \bar{d}w$$

and

$$dU = \bar{d}q - P\,dV$$

For a reversible process,

$$dq_{\text{rev}} = T\,dS$$

so that

$$dU = T\,dS - P\,dV \tag{4.25}$$

and

$$dH = (T\,dS - P\,dV) + P\,dV + V\,dP$$
$$= T\,dS + V\,dP$$

Finally, we have

$$dG = (T\,dS + V\,dP) - T\,dS - S\,dT$$

$$dG = V\,dP - S\,dT \tag{4.26}$$

Equation 4.25 incorporates the first and second laws, whereas Equation 4.26 shows how G depends on pressure and temperature. Both are important, fundamental equations of thermodynamics.

Equation 4.26 holds for a process in which only expansion work occurs. If, in addition to expansion work, another type of work is done, we must take that into account. For example, for a redox reaction in an electrochemical cell that generates electrons and does electrical work (w_{el}), Equation 4.25 is modified to be

$$dU = T\,dS - P\,dV + dw_{el}$$

and therefore

$$dG = V\,dP - S\,dT + dw_{el}$$

where the subscript el denotes electrical. At constant P and T, we have

$$dG = dw_{el,rev}$$

and for a finite change

$$\Delta G = w_{el,rev} = w_{el,max} \tag{4.27}$$

This derivation shows that ΔG is the maximum nonexpansion work we can obtain for a process at constant P and T. We shall make use of Equation 4.27 when we discuss electrochemistry in Chapter 7.

Example 4.6

In a fuel cell, natural gases such as methane undergo the same redox reaction as in the combustion process to produce carbon dioxide and water and generate electricity (see Section 7.4). Calculate the maximum electrical work that can be obtained from 1 mole of methane at 25°C.

ANSWER

The reaction is

$$CH_4(g) + 2O_2(g) \rightarrow CO_2(g) + 2H_2O(l)$$

From the $\Delta_f \bar{H}°$ and $\bar{S}°$ values in Appendix 2, we find that $\Delta_r H = -890.3 \text{ kJ mol}^{-1}$ and $\Delta_r S = -242.8 \text{ J K}^{-1} \text{ mol}^{-1}$. Therefore, from Equation 4.21

$$\Delta_r G = -890.3 \text{ kJ mol}^{-1} - 298 \text{ K} \left(\frac{-242.8 \text{ J K}^{-1} \text{ mol}^{-1}}{1000 \text{ J/kJ}} \right)$$

$$= -818.0 \text{ kJ mol}^{-1}$$

From Equation 4.27 we write

$$w_{\text{el, max}} = -818.0 \text{ kJ mol}^{-1}$$

Thus, the maximum electrical work the system can do on the surroundings is equal to 818.0 kJ per mole of CH_4 reacted.

COMMENT

Two points of interest: First, because the reaction results in a decrease in entropy, the electrical work done is *less* than the heat generated. Second, if the enthalpy of combustion were used to do work in a heat engine, then the efficiency of the heat-to-work conversion would be limited by Equation 4.7. In principle, 100% of the Gibbs energy released in a fuel cell can be converted to work because the cell is not a heat engine and therefore is not governed by the second law restrictions.

4.7 Standard Molar Gibbs Energy of Formation ($\Delta_f \bar{G}°$)

As for enthalpy, we cannot measure the absolute value of Gibbs $\bar{G}°$ energies, and so for convenience, we assign a value of zero to the standard molar Gibbs energy of formation of an element in its most stable allotropic form at 1 bar and 298 K. Again using the combustion of graphite as an example (see Section 3.6):

$$C(\text{graphite}) + O_2(g) \rightarrow CO_2(g)$$

If the reaction is carried out with reactants at 1 bar being converted to products at 1 bar, then the standard Gibbs energy change, $\Delta_r G°$, for the reaction is

$$\Delta_r G° = \Delta_f \bar{G}°(CO_2) - \Delta_f \bar{G}°(\text{graphite}) - \Delta_f \bar{G}°(O_2)$$

$$= \Delta_f \bar{G}°(CO_2)$$

or

$$\Delta_f \bar{G}°(CO_2) = \Delta_r G°$$

because the $\Delta_f \bar{G}°$ values for both graphite and O_2 are zero. To determine the value of $\Delta_r G°$, we use Equation 4.21

$$\Delta_r G° = \Delta_r H° - T\Delta_r S°$$

In Chapter 3 (p. 65), we saw that $\Delta_r H° = -393.5 \text{ kJ mol}^{-1}$. To find the value of

$\Delta_r S^\circ$, we use Equation 4.18 and the data in Appendix 2:

$$\Delta_r S^\circ = \bar{S}^\circ(CO_2) - \bar{S}^\circ(\text{graphite}) - \bar{S}^\circ(O_2)$$
$$= (213.6 - 5.7 - 205.0)\ \text{J K}^{-1}\ \text{mol}^{-1}$$
$$= 2.9\ \text{J K}^{-1}\ \text{mol}^{-1}$$

Thus,

$$\Delta_r G^\circ = -393.5\ \text{kJ mol}^{-1} - 298\ \text{K}\left(\frac{2.9\ \text{J K}^{-1}\ \text{mol}^{-1}}{1000\ \text{J/kJ}}\right)$$
$$= -394.4\ \text{kJ mol}^{-1}$$

Finally, we arrive at the result:

$$\Delta_f \bar{G}^\circ(CO_2) = -394.4\ \text{kJ mol}^{-1}$$

In this manner, we can determine the $\Delta_f \bar{G}^\circ$ values of most substances. Table 4.3 lists the $\Delta_f \bar{G}^\circ$ values for a number of common inorganic and organic substances (a more extensive listing is given in Appendix 2).

In general, $\Delta_r G^\circ$ for a reaction of the type

$$a\text{A} + b\text{B} \rightarrow c\text{C} + d\text{D}$$

is given by

$$\Delta_r G^\circ = c\Delta_f \bar{G}^\circ(C) + d\Delta_f \bar{G}^\circ(D) - a\Delta_f \bar{G}^\circ(A) - b\Delta_f \bar{G}^\circ(B)$$

$$\Delta_r G^\circ = \Sigma v\Delta_f \bar{G}^\circ(\text{products}) - \Sigma v\Delta_f \bar{G}^\circ(\text{reactants}) \tag{4.28}$$

Table 4.3
Standard Molar Gibbs Energies of Formation at 1 Bar and 298 K for Some Inorganic and Organic Substances

Substance	$\Delta_f \bar{G}^\circ/\text{kJ}\cdot\text{mol}^{-1}$	Substance	$\Delta_f \bar{G}^\circ/\text{kJ}\cdot\text{mol}^{-1}$
C(graphite)	0	$CH_4(g)$	−50.79
C(diamond)	2.87	$C_2H_6(g)$	−32.9
$CO(g)$	−137.3	$C_3H_8(g)$	−23.49
$CO_2(g)$	−394.4	$C_2H_2(g)$	209.2
$HF(g)$	−270.7	$C_2H_4(g)$	68.12
$HCl(g)$	−95.3	$C_6H_6(l)$	124.5
$HBr(g)$	−53.45	$CH_3OH(l)$	−166.3
$HI(g)$	1.7	$C_2H_5OH(l)$	−174.2
$H_2O(g)$	−228.6	$CH_3CHO(l)$	−128.1
$H_2O(l)$	−237.2	$HCOOH(l)$	−361.4
$NH_3(g)$	−16.6	$CH_3COOH(l)$	−389.9
$NO(g)$	86.7	$C_6H_{12}O_6(s)$	−910.6
$NO_2(g)$	51.84	$C_{12}H_{22}O_{11}(s)$	−1544.3
$N_2O_4(g)$	98.29		
$N_2O(g)$	103.6		
$O_3(g)$	163.4		
$SO_2(g)$	−300.4		
$SO_3(g)$	−370.4		

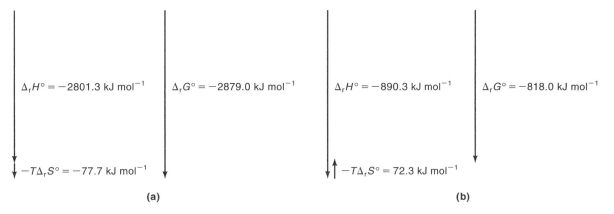

Figure 4.10
Vector diagrams show the changes of $\Delta_r H^\circ$, $-T\Delta_r S^\circ$, and $\Delta_r G^\circ$ at 298 K for the combustion of (a) glucose and (b) methane.

where v is the stoichiometric coefficient. In later chapters, we shall see that $\Delta_r G^\circ$ can also be obtained from the equilibrium constant and electrochemical measurements.

Because the Gibbs energy change is made up of two parts—enthalpy and temperature times entropy—comparing their contributions to $\Delta_r G^\circ$ in a process is instructive. Consider the combustions of methane and glucose:

$$CH_4(g) + 2O_2(g) \rightarrow CO_2(g) + 2H_2O(l)$$

$$C_6H_{12}O_6(s) + 6O_2(g) \rightarrow 6CO_2(g) + 6H_2O(l)$$

Following the same procedure as that used for the combustion of graphite to form carbon dioxide shown above, we obtain the following data:

$C_6H_{12}O_6$	$\Delta_r H^\circ = -2801.3$ kJ mol^{-1}	$-T\Delta_r S^\circ = -77.7$ kJ mol^{-1}	$\Delta_r G^\circ = -2879.0$ kJ mol^{-1}
CH_4	$\Delta_r H^\circ = -890.3$ kJ mol^{-1}	$-T\Delta_r S^\circ = 72.3$ kJ mol^{-1}	$\Delta_r G^\circ = -818.0$ kJ mol^{-1}

Figure 4.10 compares the changes for each reaction on a vector diagram.

4.8 Dependence of Gibbs Energy on Temperature and Pressure

Because Gibbs energy plays such a central role in chemical thermodynamics, understanding its properties is important. Equation 4.26 shows that it is a function of both pressure and temperature. Here, we shall see how the value of G changes with each of these variables and derive expressions for ΔG for a particular process under these conditions.

Dependence of G on Temperature

We start with Equation 4.26:

$$dG = V\, dP - S\, dT$$

At constant pressure, this equation becomes

$$dG = -S\, dT$$

so that the variation of G with respect to T at constant pressure is given by

$$\left(\frac{\partial G}{\partial T}\right)_P = -S \qquad (4.29)$$

Equation 4.19 now becomes

$$G = H + T\left(\frac{\partial G}{\partial T}\right)_P$$

Dividing the above equation by T^2 and rearranging, we obtain

$$-\frac{G}{T^2} + \frac{1}{T}\left(\frac{\partial G}{\partial T}\right)_P = -\frac{H}{T^2}$$

The left side of the above equation is the partial derivative of G/T with respect to T; that is,

$$\left[\frac{\partial\left(\frac{G}{T}\right)}{\partial T}\right]_P = -\frac{G}{T^2} + \frac{1}{T}\left(\frac{\partial G}{\partial T}\right)_P$$

Therefore,

$$\left[\frac{\partial\left(\frac{G}{T}\right)}{\partial T}\right]_P = -\frac{H}{T^2} \qquad (4.30)$$

Equation 4.30 is known as the Gibbs–Helmholtz equation. When applied to a finite process, G and H become ΔG and ΔH so that the equation becomes

$$\left[\frac{\partial\left(\frac{\Delta G}{T}\right)}{\partial T}\right]_P = -\frac{\Delta H}{T^2} \qquad (4.31)$$

Equation 4.31 is important because it relates the temperature dependence of the Gibbs energy change, and hence the position of equilibrium, to the enthalpy change. We shall return to this equation in Chapter 6.

Dependence of G on Pressure

To see how the Gibbs energy depends on pressure, we again employ Equation 4.26. At constant temperature,

$$dG = V\,dP$$

or

$$\left(\frac{\partial G}{\partial P}\right)_T = V \qquad (4.32)$$

Because volume must be a positive quantity, Equation 4.32 says that the Gibbs energy of a system always increases with pressure at constant temperature. We are interested in how the value of G increases when the pressure of the system increases from P_1 to P_2. We can write the change in G, ΔG, as the system goes from state 1 to state 2 as

$$\Delta G = \int_1^2 dG = G_2 - G_1 = \int_{P_1}^{P_2} V \, dP$$

For an ideal gas, $V = nRT/P$, so that

$$\Delta G = G_2 - G_1 = \int_{P_1}^{P_2} \frac{nRT}{P} \, dP$$

$$\Delta G = nRT \ln \frac{P_2}{P_1} \qquad (4.33)$$

If we set $P_1 = 1$ bar (the standard state), we can replace G_1 with the symbol for the standard state, G°, G_2 by G, and P_2 by P. Equation 4.33 now becomes

$$G = G^\circ + nRT \ln \frac{P}{1 \text{ bar}}$$

Expressed in molar quantities,

$$\bar{G} = \bar{G}^\circ + RT \ln \frac{P}{1 \text{ bar}} \qquad (4.34)$$

where $\bar{G}$ depends on both temperature and pressure, and $\bar{G}^\circ$ is a function of temperature only. Equation 4.34 relates the molar Gibbs energy of an ideal gas to its pressure. Later, we shall see a similar equation relating the Gibbs energy of a substance to its concentration in a mixture.

Example 4.7

A 0.590-mol sample of an ideal gas initially at 300 K and 1.50 bar is compressed isothermally to a final pressure of 6.90 bar. Calculate the change in Gibbs energy for this process.

ANSWER

From Equation 4.33, we write

$$P_1 = 1.50 \text{ bar} \qquad P_2 = 6.90 \text{ bar}$$

so that

$$\Delta G = (0.590 \text{ mol})(8.314 \text{ J K}^{-1} \text{ mol}^{-1})(300 \text{ K}) \ln \frac{6.90 \text{ bar}}{1.50 \text{ bar}}$$

$$= 2.25 \times 10^3 \text{ J}$$

Thus far, we have focused on gases in discussing the dependence of G on pressure. Because the volume of a liquid or a solid is practically independent of applied pressure, we write

$$G_2 - G_1 = \int_{P_1}^{P_2} V \, dP$$
$$= V(P_2 - P_1) = V \Delta P$$

or

$$G_2 = G_1 + V \Delta P$$

The volume, V, is treated as a constant and may be taken outside the integral. In general, the Gibbs energies of liquids and solids are much less dependent on pressure so that the variation of G with P can be ignored, except when dealing with geological processes in Earth's interior or specially created high-pressure conditions in the laboratory.

4.9 Phase Equilibria

In this section, we shall see how Gibbs energy can be applied to the study of phase equilibria. A *phase* is a homogeneous part of a system that is in contact with other parts of the system but separated from them by a well-defined boundary. Examples of phase equilibria are physical processes such as freezing and boiling. In Chapter 6, we shall apply Gibbs energy to the study of chemical equilibria. Our discussion here is restricted to one-component systems.

Consider that at some temperature and pressure, two phases, say solid and liquid, of a one-component system are in equilibrium. How do we formulate this condition? We might be tempted to equate the Gibbs energies as follows:

$$G_{\text{solid}} = G_{\text{liquid}}$$

But this formulation will not hold, for it is possible to have a small ice cube floating in an ocean of water at 0°C, and yet the Gibbs energy of water is much larger than that of the ice cube. Instead, we must insist that the Gibbs energy *per mole* (or the molar Gibbs energy) of the substance, an intensive property, be the same in both phases at equilibrium because intensive quantities are independent of the amount present:

$$\bar{G}_{\text{solid}} = \bar{G}_{\text{liquid}}$$

If external conditions (temperature or pressure) were altered so that $\bar{G}_{\text{solid}} > \bar{G}_{\text{liquid}}$, then some solid would melt because

$$\Delta G = \bar{G}_{\text{liquid}} - \bar{G}_{\text{solid}} < 0$$

On the other hand, if $\bar{G}_{\text{solid}} < \bar{G}_{\text{liquid}}$, then some liquid would freeze spontaneously.

Next, let us see how the molar Gibbs energies of solid, liquid, and vapor depend on temperature and pressure. Equation 4.29 expressed in molar quantities becomes

$$\left(\frac{\partial \bar{G}}{\partial T} \right)_P = -\bar{S}$$

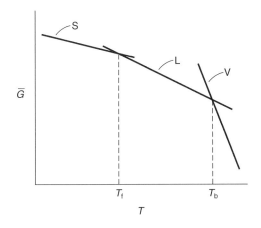

Figure 4.11
Dependence of molar Gibbs energy on temperature for the gas, liquid, and solid phases of a substance at constant pressure. The phase with the lowest $\bar{G}$ is the most stable phase at that temperature. The intercept of the gas and liquid lines gives the boiling point (T_b) and that between the liquid and solid lines gives the melting point (T_f).

Because the entropy of a substance in any phase is always positive, a plot of $\bar{G}$ versus T at constant pressure gives us a line with a negative slope. For the three phases of a single substance, we have*

$$\left(\frac{\partial \bar{G}_{\text{solid}}}{\partial T}\right)_P = -\bar{S}_{\text{solid}} \qquad \left(\frac{\partial \bar{G}_{\text{liquid}}}{\partial T}\right)_P = -\bar{S}_{\text{liquid}} \qquad \left(\frac{\partial \bar{G}_{\text{vap}}}{\partial T}\right)_P = -\bar{S}_{\text{vap}}$$

At any temperature, the molar entropies of a substance decrease in the order

$$\bar{S}_{\text{vap}} \gg \bar{S}_{\text{liquid}} > \bar{S}_{\text{solid}}$$

These differences are reflected in the slopes of lines shown in Figure 4.11. At high temperatures, the vapor phase is the most stable, because it has the lowest molar Gibbs energy. As temperature decreases, however, liquid becomes the stable phase, and finally, at even lower temperatures, solid becomes the most stable phase. The intercept between the vapor and liquid lines is the point at which these two phases are in equilibrium—that is, $\bar{G}_{\text{vap}} = \bar{G}_{\text{liquid}}$. The corresponding temperature is T_b, the boiling point. Similarly, solid and liquid coexist in equilibrium at the temperature T_f, the melting (or fusion) point.

How does an increase in pressure affect the phase equilibria? In the previous section, we saw that the Gibbs energy of a substance always increases with pressure (see Equation 4.32). Further, for a given change in pressure, the increase is greatest for vapors, much less for liquids and solids. This result follows from Equation 4.32, expressed in molar quantities:

$$\left(\frac{\partial \bar{G}}{\partial P}\right)_T = \bar{V}$$

The molar volume of a vapor is normally about a thousand times greater than that for a liquid or a solid.

Figure 4.12 shows the increases in the value of $\bar{G}$ for the three phases as the pressure increases from P_1 to P_2. We see that both T_f and T_b shift to higher values, but the shift in T_b is greater because of the larger increase in the value of $\bar{G}$ for the

---

* Although we use the terms *gas* and *vapor* interchangeably, strictly speaking, there is a difference. A gas is a substance that is normally in the gaseous state at ordinary temperatures and pressures; a vapor is the gaseous form of any substance that is a liquid or a solid at normal temperatures and pressures. Thus, at 25°C and 1 atm, we speak of water vapor and oxygen gas.

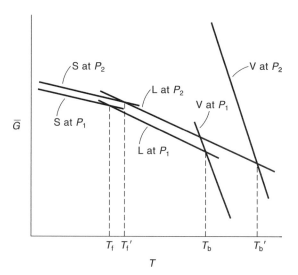

Figure 4.12
Pressure dependence of molar Gibbs energy. For the majority of substances (water being the important exception), an increase in pressure leads to an increase in both the melting point and the boiling point. (Here we have $P_2 > P_1$.)

vapor. Thus, in general, an increase in external pressure will raise both the melting point and boiling point of a substance. Although not shown in Figure 4.12, the reverse also holds true; that is, decreasing the pressure will lower both the melting point and boiling point. Keep in mind that our conclusion about the effect of pressure on melting point is based on the assumption that the molar volume of liquid is greater than that of solid. This assumption is true of most, but not all, substances. A key exception is water. The molar volume of ice is actually greater than that of liquid water, accounting for the fact that ice floats on water. For water, then, an increase in pressure will *lower* the melting point. More will be said about this property of water later.

The Clapeyron and the Clausius–Clapeyron Equations

We shall now derive some useful, general relations for the quantitative understanding of phase equilibria. Consider a substance that exists in two phases, α and β. The condition for equilibrium at constant temperature and pressure is that

$$\bar{G}_\alpha = \bar{G}_\beta$$

so that

$$d\bar{G}_\alpha = d\bar{G}_\beta$$

To establish the relationship of dT to dP in the change that links these two phases, we have, from Equation 4.26,

$$d\bar{G}_\alpha = \bar{V}_\alpha \, dP - \bar{S}_\alpha \, dT = d\bar{G}_\beta = \bar{V}_\beta \, dP - \bar{S}_\beta \, dT$$

$$(\bar{S}_\beta - \bar{S}_\alpha)dT = (\bar{V}_\beta - \bar{V}_\alpha)dP$$

or

$$\frac{dP}{dT} = \frac{\Delta \bar{S}}{\Delta \bar{V}}$$

where $\Delta \bar{V}$ and $\Delta \bar{S}$ are the change in molar volume and molar entropies for the $\alpha \rightarrow \beta$

phase transition, respectively. Because $\Delta \bar{S} = \Delta \bar{H}/T$ at equilibrium, the above equation becomes

$$\frac{dP}{dT} = \frac{\Delta \bar{H}}{T \Delta \bar{V}} \tag{4.35}$$

where T is the phase transition temperature (it may be the melting point or the boiling point or any other temperature at which the two phases can coexist in equilibrium). Equation 4.35 is known as the Clapeyron equation (after the French engineer Benoit-Paul-Émile Clapeyron, 1799–1864). This simple expression gives us the ratio of the change in pressure to the change in temperature in terms of some readily measurable quantities, such as molar volume and molar enthalpy change for the process. It applies to fusion, vaporization, and sublimation, as well as to equilibria between two allotropic forms, such as graphite and diamond.

The Clapeyron equation can be expressed in a convenient approximate form for vaporization and sublimation equilibria. In these cases, the molar volume of the vapor is so much greater than that for the condensed phase, we can write

$$\Delta_{vap} \bar{V} = \bar{V}_{vap} - \bar{V}_{condensed} \approx \bar{V}_{vap}$$

Further, if we assume ideal-gas behavior, then

$$\Delta_{vap} \bar{V} \approx \bar{V}_{vap} = \frac{RT}{P}$$

Substitution for $\Delta_{vap}\bar{V}$ in Equation 4.35 yields

$$\frac{dP}{dT} = \frac{P\Delta_{vap}\bar{H}}{RT^2}$$

or

$$\frac{dP}{P} = d\ln P = \frac{\Delta_{vap}\bar{H}\,dT}{RT^2} \tag{4.36}$$

Equation 4.36 is known as the Clausius–Clapeyron equation (after Clapeyron and the German physicist Rudolf Julius Clausius, 1822–1888). Integrating Equation 4.36 between limits of P_1, T_1 and P_2, T_2, we obtain

$$\int_{P_1}^{P_2} d\ln P = \ln\frac{P_2}{P_1} = \frac{\Delta_{vap}\bar{H}}{R}\int_{T_1}^{T_2}\frac{dT}{T^2} = -\frac{\Delta_{vap}\bar{H}}{R}\left(\frac{1}{T_2} - \frac{1}{T_1}\right)$$

or

$$\ln\frac{P_2}{P_1} = \frac{\Delta_{vap}\bar{H}}{R}\frac{(T_2 - T_1)}{T_1 T_2} \tag{4.37}$$

We assume that $\Delta_{vap}\bar{H}$ is independent of temperature. If we had carried out an indefinite integral (integration without the limits), we could express $\ln P$ as a function of temperature as follows:

$$\ln P = -\frac{\Delta_{vap}\bar{H}}{RT} + \text{constant} \tag{4.38}$$

Thus, a plot of $\ln P$ versus $1/T$ gives a straight line whose slope (which is negative) is equal to $-\Delta_{\text{vap}}\bar{H}/R$.

Example 4.8

The following data show the variation of the vapor pressure of water as a function of temperature:

P/mmHg	17.54	31.82	55.32	92.51	149.38	233.7
$t/°C$	20	30	40	50	60	70

Determine the molar enthalpy of vaporization for water.

ANSWER

We need Equation 4.38. The first step is to convert the data into a suitable form for plotting:

$\ln P$	2.865	3.460	4.013	4.527	5.007	5.454
K/T	3.41×10^{-3}	3.30×10^{-3}	3.19×10^{-3}	3.10×10^{-3}	3.00×10^{-3}	2.92×10^{-3}
$10^3\ K/T$	3.41	3.30	3.19	3.10	3.00	2.92

Figure 4.13 shows the plot of $\ln P$ versus $1/T$. From the measured slope, we have

$$-5090\ \text{K} = -\frac{\Delta_{\text{vap}}\bar{H}}{R}$$

or

$$\Delta_{\text{vap}}\bar{H} = (8.314\ \text{J K}^{-1}\ \text{mol}^{-1})(5090\ \text{K})$$
$$= 42.3\ \text{kJ mol}^{-1}$$

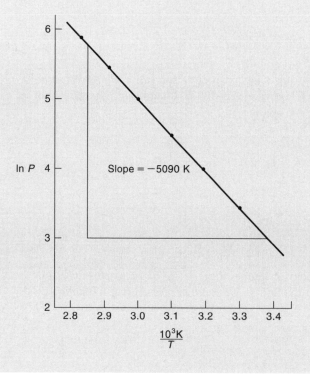

Figure 4.13
Plot of $\ln P$ versus $1/T$ to determine $\Delta_{\text{vap}}\bar{H}$ of water. Note that the same slope is obtained whether we express the pressure as mmHg or atm.

Phase Diagrams

At this point, we are ready to examine the phase equilibria of some familiar systems. The conditions at which a system exists as a solid, liquid, or vapor are conveniently summarized in a *phase diagram*, which is a plot of pressure versus temperature. We shall consider the phase equilibria of water and carbon dioxide.

Water. Figure 4.14 shows the phase diagram of water, where S, L, and V represent regions in which only one phase (solid, liquid, or vapor) can exist. Along any one curve, however, the two corresponding phases can coexist. The slope of any curve is given by dP/dT. For example, the curve separating regions L and V shows how the vapor pressure of water varies with respect to temperature. At 373.15 K, its vapor pressure is 1 atm, and these conditions mark the normal boiling point of water. Note that the L–V curve stops abruptly at the critical point, beyond which the liquid phase cannot exist. The normal freezing point of water (or melting point of ice) is similarly defined by the S–L curve at 1 atm, which is 273.15 K. Finally, all three phases can coexist at only one point called the triple point; for water, the triple point is at $T = 273.16$ K and $P = 0.006$ atm.

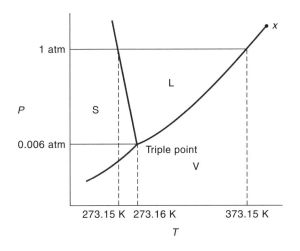

Figure 4.14
Phase diagram of water. Note that the solid–liquid curve has a negative slope. The liquid–vapor curve stops at x, the critical point (647.6 K and 219.5 atm).

Example 4.9

Calculate the slope of the S–L curve at 273.15 K in atm K^{-1}, given that $\Delta_{fus}\overline{H} = 6.01$ kJ mol^{-1}, $\overline{V}_L = 0.0180$ L mol^{-1}, and $\overline{V}_S = 0.0196$ L mol^{-1}.

ANSWER

We need the Clapeyron equation (Equation 4.35):

$$\frac{dP}{dT} = \frac{\Delta_{fus}\overline{H}}{T_f \Delta_{fus}\overline{V}}$$

Using the conversion factor $1\ \text{J} = 9.87 \times 10^{-3}\ \text{L atm}$, we obtain

$$\frac{dP}{dT} = \frac{(6010\ \text{J mol}^{-1})(9.87 \times 10^{-3}\ \text{L atm J}^{-1})}{(273.15\ \text{K})(0.0180 - 0.0196)\ \text{L mol}^{-1}}$$

$$= -136\ \text{atm K}^{-1}$$

COMMENT

(1) Because the molar volume of liquid water is smaller than that for ice, the slope is negative, as shown in Figure 4.14. Furthermore, because the quantity $(\bar{V}_L - \bar{V}_S)$ is small, the slope is also quite steep. (2) An interesting result is obtained by calculating the quantity dT/dP, which gives the change (decrease) in melting point as a function of pressure. We find that $dT/dP = -7.35 \times 10^{-3}\ \text{K atm}^{-1}$, which means that the melting point of ice decreases by $7.35 \times 10^{-3}\ \text{K}$ whenever the pressure increases by 1 atm. This effect helps make ice skating possible. The weight of a skater exerts considerable pressure on the ice (of the order of 500 atm) because of the small area of the blades. As ice melts, the film of water formed between skates and ice acts as a lubricant to facilitate movement over the ice. However, more detailed studies indicate that the frictional heat generated between the skates and ice is the main reason for ice melting.

Figure 4.16
At 1 atm, solid carbon dioxide cannot melt; it can only sublime.

Carbon Dioxide. Figure 4.15 shows the phase diagram for carbon dioxide. The main difference between this diagram and that for water is that the S–L curve for CO_2 has a positive slope. This follows from the fact that because $\bar{V}_{liq} > \bar{V}_{solid}$, the quantity on the right side of Equation 4.35 is positive and therefore so is dP/dT. Note that liquid CO_2 is unstable at pressures lower than 5 atm. For this reason, solid CO_2 is called "dry ice"—under atmospheric conditions it does not melt; it can only sublime. Furthermore, it looks like ice (Figure 4.16). Liquid CO_2 does exist at room temperature, but it is normally confined to a metal cylinder under a pressure of 67 atm!

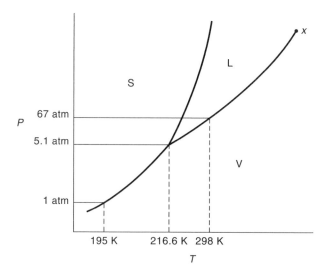

Figure 4.15
Phase diagram of carbon dioxide. Note that the solid–liquid curve has a positive slope. This is true of most substances. The liquid–vapor curve stops at x, the critical point (304.2 K and 73.0 atm).

The Phase Rule

To conclude our discussion of phase equilibria, let us consider a useful rule that was derived by Gibbs*:

$$f = c - p + 2 \qquad (4.39)$$

where c is the number of components and p is the number of phases present in a system. The *degree of freedom*, f, gives the number of intensive variables (pressure, temperature, and composition) that can be changed independently without disturbing the number of phases in equilibrium. For example, in a single-component, single-phase system ($c = 1$, $p = 1$), say, a gas in a container, the pressure and temperature of the gas may be changed independently without changing the number of phases, so $f = 2$, or the system has two degrees of freedom.

Now let us apply the phase rule to water ($c = 1$). Figure 4.14 shows that in the pure phase region (S, L, or V) we have $p = 1$ and $f = 2$, meaning that the pressure can be varied independently of temperature (two degrees of freedom). Along each of the S–L, L–V, or S–V boundaries, however, $p = 2$, and $f = 1$. Thus, for every value of P, there can be only one specific value of T and vice versa (one degree of freedom). Finally, at the triple point, $p = 3$ and $f = 0$ (no degree of freedom). Under these conditions, the system is totally fixed, and no variation of either the pressure or the temperature is possible. Such a system is said to be *invariant* and is represented by a point in a plot of pressure versus temperature.

4.10 Thermodynamics of Rubber Elasticity

In this section we shall see an application of thermodynamics functions to a system other than gases—the familiar rubber band.

Natural rubber is poly-*cis*-isoprene and has the following repeating monomeric unit:

$$\left(\begin{array}{c} CH_3 \quad\quad H \\ \diagdown \quad\quad \diagup \\ C=C \\ \diagup \quad\quad \diagdown \\ -CH_2 \quad\quad CH_2- \end{array} \right)_n$$

where n is in the hundreds. The characteristic property of rubber is its elasticity. It can be stretched up to 10 times its length, and, if released, will return to its original size. This behavior is due to the flexibility of rubber's long-chain molecules. In the bulk state, rubber is a tangle of polymeric chains, and if the external force is strong enough, individual chains will slip past one another, causing the rubber to lose most of its elasticity. In 1839, the American chemist Charles Goodyear (1800–1860) discovered that natural rubber could be cross-linked with sulfur to prevent chain slippage in a process called *vulcanization*. As Figure 4.17 shows, rubber in the unstretched state has many possible conformations and hence a greater entropy than the stretched state, which has relatively few conformations and a lower entropy.

When a rubber band is stretched elastically by a force, f, the work done, dw, is given by two terms:

$$dw = f\,dl - P\,dV \qquad (4.40)$$

The difference in signs between $f\,dl$ and $P\,dV$ arises because whereas a positive dV denotes work done by the system, a positive dl denotes work done on the system.

---

* For derivation of the phase rule, see the physical chemistry texts listed on p. 6.

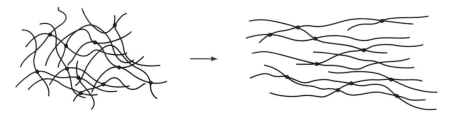

Figure 4.17
Unstretched rubber (left) has many more conformations than stretched rubber (right). The long chains of vulcanized rubber molecules are held together by sulfur linkages to prevent slippage.

The first term is force times the extension. The second term is small, however, and can usually be ignored. (The rubber band becomes thinner when stretched, but it also gets longer so that the change in volume, dV, is negligible.) If the rubber band is stretched slowly, the restoring force is equal to the applied force at every stage and we can therefore assume the process to be reversible and write

$$dw_{rev} = f \, dl \tag{4.41}$$

Treating this as a constant volume and temperature process, we start with the definition of Helmholtz energy (Equation 4.22). For an infinitesimal change, the equation takes the form

$$dA = dU - T \, dS \tag{4.42}$$

For a reversible process, $dq_{rev} = T \, dS$ so that Equation 4.42 becomes

$$dA = dU - dq_{rev}$$

From the first law of thermodynamics, $dU = dq_{rev} + dw_{rev}$, so that

$$dA = dq_{rev} + dw_{rev} - dq_{rev}$$
$$= dw_{rev} \tag{4.43}$$

Combining Equations 4.41 and 4.43 we write

$$dA = f \, dl \tag{4.44}$$

We can now express the restoring force in terms of the Helmholtz energy as

$$f = \left(\frac{\partial A}{\partial l} \right)_T \tag{4.45}$$

From Equation 4.22,

$$A = U - TS$$

we find the variation of A with extension l at constant temperature to be

$$\left(\frac{\partial A}{\partial l} \right)_T = \left(\frac{\partial U}{\partial l} \right)_T - T \left(\frac{\partial S}{\partial l} \right)_T \tag{4.46}$$

Substituting Equation 4.46 into 4.45 we obtain

$$f = \left(\frac{\partial U}{\partial l}\right)_T - T\left(\frac{\partial S}{\partial l}\right)_T \qquad (4.47)$$

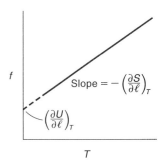

Equation 4.47 shows that there are two contributions to the restoring force—one from energy change with extension and the other from entropy change.

The restoring force of a stretched rubber band can be readily measured.* Figure 4.18 shows a plot of the restoring force as a function of temperature. Note that the line has a positive slope, which means that $(\partial S/\partial l)_T$ is negative. This is consistent with the notion that the polymer molecules become less tangled (fewer microstates) in the stretched state, leading to a decrease in entropy. Experimental results also show that the $(\partial U/\partial l)_T$ term (the intercept on the y axis) is 5–10 times smaller than the $(\partial S/\partial l)_T$ term. The reason is that intermolecular forces between hydrocarbon molecules are relatively small so that the internal energy of the rubber band does not vary appreciably with extension. Therefore, the predominant contribution to the restoring force is entropy, not energy. When a stretched rubber band snaps back to its original position, the process is largely driven by an increase in entropy!

Finally, it is interesting to note the analogy between the stretching of a rubber band and the compression of a gas. If the rubber and the gas behave ideally, then

Figure 4.18
Plot of restoring force in a rubber band versus temperature.

$$\left(\frac{\partial U}{\partial l}\right)_T = 0 \text{ (rubber)} \quad \text{and} \quad \left(\frac{\partial U}{\partial V}\right)_T = 0 \text{ (gas)}$$

Ideal behavior for rubber means that intermolecular forces are independent of conformations, whereas no intermolecular forces are present in an ideal gas. Similarly, the entropy of a rubber band decreases on stretching at constant temperature just as the entropy of a gas decreases when it is compressed isothermally.

* See J. P. Byrne, *J. Chem. Educ.* **71**, 531 (1994).

Suggestions for Further Reading

BOOKS

Atkins, P. W. *The Second Law*, Scientific American Books, New York, 1984.

Bent, H. A. *The Second Law*, Oxford University Press, New York, 1965.

Edsall, J. T. and H. Gutfreund, *Biothermodynamics*, John Wiley & Sons, New York, 1983.

Klotz, I. M. and R. M. Rosenberg, *Chemical Thermodynamics: Basic Theory and Methods*, 6th ed., John Wiley & Sons, New York, 2000.

McQuarrie, D. A. and J. D. Simon, *Molecular Thermodynamics*, University Science Books, Sausalito, CA, 1999.

Rock, P. A. *Chemical Thermodynamics*, University Science Books, Sausalito, CA, 1983.

von Baeyer, H. C. *Warmth Disperses and Time Passes*, Random House, New York, 1998.

ARTICLES

Entropy and the Second Law of Thermodynamics

"The Second Law of Thermodynamics," H. A. Bent, *J. Chem. Educ.* **39**, 491 (1962).

"States, Indistinguishability, and the Formula $S = k \ln W$ in Thermodynamics," J. Braunstein, *J. Chem. Educ.* **46**, 719 (1969).

"Temperature-Entropy Diagrams," A. Wood, *J. Chem. Educ.* **47**, 285 (1970).

"The Arrow of Time," D. Layzer, *Sci. Am.* December 1975.

"Negative Absolute Temperature," W. G. Proctor, *Sci. Am.* August 1978.

"Reversibility and Returnability," J. A. Campbell, *J. Chem. Educ.* **57**, 345 (1980).

"Heat-Fall and Entropy," J. P. Lowe, *J. Chem. Educ.* **59**, 353 (1982).

"Entropy and Unavailable Energy," J. N. Spencer and E. S. Holmboe, *J. Chem. Educ.* **60**, 1018 (1983).

"A Simple Method for Showing that Entropy is a Function of State," P. Djurdjevic and I. Gutman, *J. Chem. Educ.* **65**, 399 (1985).

"Entropy: Conceptual Disorder," J. P. Lowe, *J. Chem. Educ.* **65**, 403 (1988).

"Entropy Analyses of Four Familiar Processes," N. C. Craig, *J. Chem. Educ.* **65**, 760 (1988).

"Order and Disorder and Entropies of Fusion," D. F. R. Gilson, *J. Chem. Educ.* **69**, 23 (1992).

"Periodic Trends for the Entropy of Elements," T. Thoms, *J. Chem. Educ.* **72**, 16 (1995).

"Thermodynamics and Spontaneity," R. S. Ochs, *J. Chem. Educ.* **73**, 952 (1996).

"Entropy Diagrams," N. C. Craig, *J. Chem. Educ.* **73**, 716 (1996).

"Shuffled Cards, Messy Desks, and Disorderly Dorm Rooms—Examples of Entropy Increase? Nonsense!" F. L. Lambert, *J. Chem. Educ.* **76**, 1385 (1999).

"Entropy, Disorder, and Freezing," B. B. Laird, *J. Chem. Educ.* **76**, 1388 (1999).

"Inside into Entropy," D. F. Styer, *Am. J. Phys.* **68**, 1090 (2000).

"Stories to Make Thermodynamics and Related Subjects More Palatable," L. S. Bartell, *J. Chem. Educ.* **78**, 1059 (2001).

"Entropy Explained: The Origin of Some Simple Trends," L. A. Watson and O. Eisenstein, *J. Chem. Educ.* **79**, 1269 (2002).

The Third Law of Thermodynamics

"Ice," L. K. Runnels, *Sci. Am.* December 1966.

"The Third Law of Thermodynamics and the Residual Entropy of Ice," M. M. Julian, F. H. Stillinger, and R. R. Festa, *J. Chem. Educ.* **60**, 65 (1983).

Phase Equilibria

"The Triple Point of Water," F. L. Swinton, *J. Chem. Educ.* **44**, 541 (1967).

"Reappearing Phases," J. Walker and C. A. Vanse, *Sci. Am.* May 1987.

"Subtleties of Phenomena Involving Ice-Water Equilibria," L. F. Loucks, *J. Chem. Educ.* **63**, 115 (1986). Also see *J. Chem. Educ.* **65**, 186 (1988).

"Supercritical Phase Transitions at Very High Pressure," K. M. Scholsky, *J. Chem. Educ.* **66**, 989 (1989).

"The Direct Relation Between Altitude and Boiling Point," B. L. Earl, *J. Chem. Educ.* **67**, 45 (1990).

"Phase Diagrams of One-Component Systems," G. D. Peckham and I. J. McNaught, *J. Chem. Educ.* **70**, 560 (1993).

"Melting Below Zero," J. S. Wettlaufer and J. G. Dash, *Sci. Am.* February 2000.

General

"The Synthesis of Diamond," H. Hall, *J. Chem. Educ.* **38**, 484 (1961).

"Chance," A. J. Ayer, *Sci. Am.* October 1965.

"The Use and Misuse of the Laws of Thermodynamics," M. L. McGlashan, *J. Chem. Educ.* **43**, 226 (1966).

"Maxwell's Demon," W. Ehrenberg, *Sci. Am.* November 1967.

"The Scope and Limitations of Thermodynamics," K. G. Denbigh, *Chem. Brit.* **4**, 339 (1968).

"Thermodynamics of Hard Molecules," L. K. Runnels, *J. Chem. Educ.* **47**, 742 (1970).

"The Thermodynamic Transformation of Organic Chemistry," D. E. Stull, *Am. Sci.* **54**, 734 (1971).

"Introduction to the Thermodynamics of Biopolymer Growth," C. Kittel, *Am. J. Phys.* **40**, 60 (1972).

"High Pressure Synthetic Chemistry," A. P. Hagen, *J. Chem. Educ.* **55**, 620 (1978).

"Conversion of Standard Thermodynamic Data to the New Standard-State Pressure," R. D. Freeman, *J. Chem. Educ.* **62**, 681 (1985).

"Student Misconceptions in Thermodynamics," M. F. Granville, *J. Chem. Educ.* **62**, 847 (1985).

"Demons, Engines, and the Second Law," C. H. Bennett, *Sci. Am.* November 1987.

"The True Meaning of Isothermal," D. Fain, *J. Chem. Educ.* **65**, 187 (1988).

"The Conversion of Chemical Energy," D. J. Wink, *J. Chem. Educ.* **69**, 109 (1992).

"The Thermodynamics of Home-Made Ice Cream," D. L. Gibbon, K. Kennedy, N. Reading, and M. Quierox, *J. Chem. Educ.* **69**, 658 (1992).

"The Gibbs Function Controversy," S. E. Wood and R. B. Battino, *J. Chem. Educ.* **73**, 408 (1996).

"How Thermodynamic Data and Equilibrium Constants Changed When the Standard-State Pressure Became 1 Bar," R. S. Treptow, *J. Chem. Educ.* **76**, 212 (1999).

Problems

The Second Law of Thermodynamics and Entropy Changes

4.1 Comment on the statement: "Even thinking about entropy increases its value in the universe."

4.2 One of the many statements of the second law of thermodynamics is: Heat cannot flow from a colder body to a warmer one without external aid. Assume two systems, 1 and 2, at T_1 and T_2 $(T_2 > T_1)$. Show that if a quantity of heat q did flow spontaneously from 1 to 2, the process would result in a decrease in entropy of the universe. (You may assume that the heat flows very slowly so that the process can be regarded as reversible. Assume also that the loss of heat by system 1 and the gain of heat by system 2 do not affect T_1 and T_2.)

4.3 A ship sailing in the Indian Ocean takes the warmer surface water at 32°C to run a heat engine that powers the ship and discharges the used water back to the surface of the sea. Does this scheme violate the second law of thermodynamics? If so, what change would you implement to make it work?

4.4 Molecules of a gas at any temperature T above the absolute zero are in constant motion. Does this "perpetual motion" violate the laws of thermodynamics?

4.5 According to the second law of thermodynamics, the entropy of an irreversible process in an isolated system must always increase. On the other hand, it is well known that the entropy of living systems remains small. (For example, the synthesis of highly complex protein molecules from individual amino acids is a process that leads to a decrease in entropy.) Is the second law invalid for living systems? Explain.

4.6 On a hot summer day, a person tries to cool himself by opening the door of a refrigerator. Is this a wise action, thermodynamically speaking?

4.7 The molar heat of vaporization of ethanol is 39.3 kJ mol^{-1}, and the boiling point of ethanol is 78.3°C. Calculate the value of $\Delta_{vap}S$ for the vaporization of 0.50 mole of ethanol.

4.8 Calculate the values of $\Delta U, \Delta H,$ and ΔS for the following process:

$$\begin{array}{cc} \text{1 mole of liquid water} & \text{1 mole of steam} \\ \text{at 25°C and 1 atm} & \xrightarrow{} \quad \text{at 100°C and 1 atm} \end{array}$$

The molar heat of vaporization of water at 373 K is 40.79 kJ mol^{-1}, and the molar heat capacity of water is 75.3 J K^{-1} mol^{-1}. Assume the molar heat capacity to be temperature independent and ideal-gas behavior.

4.9 Calculate the value of ΔS in heating 3.5 moles of a monatomic ideal gas from 50°C to 77°C at constant pressure.

4.10 A quantity of 6.0 moles of an ideal gas is reversibly heated at constant volume from 17°C to 35°C. Calculate the entropy change. What would be the value of ΔS if the heating were carried out irreversibly?

4.11 One mole of an ideal gas is first, heated at constant pressure from T to $3T$ and second, cooled back to T at constant volume. **(a)** Derive an expression for ΔS for the overall process. **(b)** Show that the overall process is equivalent to an isothermal expansion of the gas at T from V to $3V$, where V is the original volume. **(c)** Show that the value of ΔS for the process in **(a)** is the same as that in **(b)**.

4.12 A quantity of 35.0 g of water at 25.0°C (called A) is mixed with 160.0 g of water at 86.0°C (called B). **(a)** Calculate the final temperature of the system, assuming that the mixing is carried out adiabatically. **(b)** Calculate the entropy change of A, B, and the entire system.

4.13 The heat capacity of chlorine gas is given by

$$\bar{C}_P = (31.0 + 0.008\,T) \text{ J K}^{-1} \text{ mol}^{-1}$$

Calculate the entropy change when 2 moles of gas are heated from 300 K to 400 K at constant pressure.

4.14 A sample of neon (Ne) gas initially at 20°C and 1.0 atm is expanded from 1.2 L to 2.6 L and simultaneously heated to 40°C. Calculate the entropy change for the process.

4.15 One of the early experiments in the development of the atomic bomb was to demonstrate that ^{235}U and not ^{238}U is the fissionable isotope. A mass spectrometer was employed to separate ^{235}UF$_6$ from ^{238}UF$_6$. Calculate the value of ΔS for the separation of 100 mg of the mixture of gas, given that the natural abundances of ^{235}U and ^{238}U are 0.72% and 99.28%, respectively, and that of ^{19}F is 100%.

4.16 One mole of an ideal gas at 298 K expands isothermally from 1.0 L to 2.0 L **(a)** reversibly and **(b)** against a constant external pressure of 12.2 atm. Calculate the values of $\Delta S_{sys}, \Delta S_{surr},$ and ΔS_{univ} in both cases. Are your results consistent with the nature of the processes?

4.17 The absolute molar entropies of O_2 and N_2 are 205 J K^{-1} mol^{-1} and 192 J K^{-1} mol^{-1}, respectively, at 25°C. What is the entropy of a mixture made up of 2.4 moles of O_2 and 9.2 moles of N_2 at the same temperature and pressure?

4.18 A quantity of 0.54 mole of steam initially at 350°C and 2.4 atm undergoes a cyclic process for which $q = -74$ J. Calculate the value of ΔS for the process.

4.19 Predict whether the entropy change is positive or negative for each of the following reactions at 298 K:

(a) $4Fe(s) + 3O_2(g) \rightarrow 2Fe_2O_3(s)$

(b) $O(g) + O(g) \rightarrow O_2(g)$

(c) $NH_4Cl(s) \rightarrow NH_3(g) + HCl(g)$

(d) $H_2(g) + Cl_2(g) \rightarrow 2HCl(g)$

4.20 Use the data in Appendix 2 to calculate the values of $\Delta_r S°$ of the reactions listed in the previous problem.

4.21 A quantity of 0.35 mole of an ideal gas initially at 15.6°C is expanded from 1.2 L to 7.4 L. Calculate the values of $w, q, \Delta U$, and ΔS if the process is carried out **(a)** isothermally and reversibly, and **(b)** isothermally and irreversibly against an external pressure of 1.0 atm.

4.22 One mole of an ideal gas is isothermally expanded from 5.0 L to 10 L at 300 K. Compare the entropy changes for the system, surroundings, and the universe if the process is carried out **(a)** reversibly, and **(b)** irreversibly against an external pressure of 2.0 atm.

4.23 The heat capacity of hydrogen may be represented by

$$\bar{C}_P = (1.554 + 0.0022T) \text{ J K}^{-1} \text{ mol}^{-1}$$

Calculate the entropy changes for the system, surroundings, and the universe for the **(a)** reversible heating, and **(b)** irreversible heating of 1.0 mole of hydrogen from 300 K to 600 K. [*Hint:* In **(b)**, assume the surroundings to be at 600 K.]

4.24 Consider the reaction

$$N_2(g) + O_2(g) \rightarrow NO(g)$$

Calculate the values of $\Delta_r S°$ for the reaction mixture, surroundings, and the universe at 298 K. Why is your result reassuring to Earth's inhabitants?

The Third Law of Thermodynamics

4.25 The $\Delta_f \bar{H}°$ values can be negative, zero, or positive, but the $\bar{S}°$ values can be only zero or positive. Explain.

4.26 Choose the substance with the greater molar entropy in each of the following pairs: **(a)** $H_2O(l)$, $H_2O(g)$, **(b)** $NaCl(s)$, $CaCl_2(s)$, **(c)** N_2 (0.1 atm), N_2 (1 atm), **(d)** C (diamond), C (graphite), **(e)** $O_2(g)$, $O_3(g)$, **(f)** ethanol (C_2H_5OH), dimethly ether (C_2H_6O), **(g)** $N_2O_4(g)$, $2NO_2(g)$, and **(h)** Fe(s) at 298 K, Fe(s) at 398 K. (Unless otherwise stated, assume the temperature is 298 K.)

4.27 Explain why the value of $\bar{S}°$ (graphite) is greater than that of $\bar{S}°$ (diamond) at 298 K (see Appendix 2). Would this inequality hold at 0 K?

Gibbs Energy

4.28 A quantity of 0.35 mole of an ideal gas initially at 15.6°C is expanded from 1.2 L to 7.4 L. Calculate the values of $w, q, \Delta U, \Delta H, \Delta S$, and ΔG if the process is carried out **(a)** isothermally and reversibly, and **(b)** isothermally and irreversibly against an external pressure of 1.0 atm.

4.29 At one time, the domestic gas used for cooking, called "water gas," was prepared as follows:

$$H_2O(g) + C(\text{graphite}) \rightarrow CO(g) + H_2(g)$$

From the thermodynamic quantities listed in Appendix 2, predict at what temperature the reaction will favor the formation of products. Assume $\Delta_r H°$ and $\Delta_r S°$ are temperature independent.

4.30 Use the values listed in Appendix 2 to calculate the value of $\Delta_r G°$ for the following alcohol fermentation:

$$\alpha\text{-D-glucose}(aq) \rightarrow 2C_2H_5OH(l) + 2CO_2(g)$$

$(\Delta_f \bar{G}°[\alpha\text{-D-glucose}(aq)] = -914.5 \text{ kJ mol}^{-1})$

4.31 As an approximation, we can assume that proteins exist either in the native (or physiologically functioning) state or the denatured state. The standard molar enthalpy and entropy of the denaturation of a certain protein are 512 kJ mol^{-1} and 1.60 kJ K^{-1} mol^{-1}, respectively. Comment on the signs and magnitudes of these quantities, and calculate the temperature at which the denaturation becomes spontaneous.

4.32 Certain bacteria in the soil obtain the necessary energy for growth by oxidizing nitrite to nitrate:

$$2NO_2^-(aq) + O_2(g) \rightarrow 2NO_3^-(aq)$$

Given that the standard Gibbs energies of formation of NO_2^- and NO_3^- are -34.6 kJ mol^{-1} and -110.5 kJ mol^{-1}, respectively, calculate the amount of Gibbs energy released when 1 mole of NO_2^- is oxidized to 1 mole of NO_3^-.

4.33 Consider the synthesis of urea according to the equation

$$CO_2(g) + 2NH_3(g) \rightarrow (NH_2)_2CO(s) + H_2O(l)$$

From the data listed in Appendix 2, calculate the value of $\Delta_r G°$ for the reaction at 298 K. Assuming ideal-gas behavior, calculate the value of $\Delta_r G$ for the reaction at a pressure of 10.0 bar. The $\Delta_f \bar{G}°$ of urea is -197.15 kJ mol^{-1}.

4.34 This problem involves the synthesis of diamond from graphite:

$$C(\text{graphite}) \rightarrow C(\text{diamond})$$

(a) Calculate the values of $\Delta_r H°$ and $\Delta_r S°$ for the reaction. Will the conversion be favored at 25°C or any other temperature? **(b)** From density measurements, the molar volume of graphite is found to be 2.1 cm^3 greater than that of diamond. Can the conversion of graphite to diamond be brought about at 25°C by applying pressure on graphite? If so, estimate the pressure at which the process becomes spontaneous. [*Hint:* Starting from Equation 4.32, derive the equation $\Delta G = (\bar{V}_{\text{diamond}} - \bar{V}_{\text{graphite}})\Delta P$ for a constant-temperature process. Next, calculate the ΔP value that would lead to the necessary decrease in Gibbs energy.]

4.35 A student placed 1 g of each of three compounds A, B, and C in a container and found that no change had occurred after one week. Offer possible explanations for the lack of reaction. Assume that A, B, and C are totally miscible liquids.

4.36 Predict the signs of $\Delta H, \Delta S$, and ΔG of the system for the following processes at 1 atm: **(a)** ammonia melts at $-60°C$, **(b)** ammonia melts at $-77.7°C$, and **(c)** ammonia melts at $-100°C$. (The normal melting point of ammonia is $-77.7°C$.)

4.37 Crystallization of sodium acetate from a supersaturated solution occurs spontaneously. What can you deduce about the signs of ΔS and ΔH?

4.38 A student looked up the $\Delta_f \bar{G}°, \Delta_f \bar{H}°$, and $\bar{S}°$ values for CO_2 in Appendix 2. Plugging these values into Equation 4.21, he found that $\Delta_f \bar{G}° \neq \Delta_f \bar{H}° - T\bar{S}°$ at 298 K. What is wrong with his approach?

4.39 A certain reaction is spontaneous at $72°C$. If the enthalpy change for the reaction is 19 kJ, what is the *minimum* value of $\Delta_r S$ (in joules per kelvin) for the reaction?

4.40 A certain reaction is known to have a $\Delta_r G°$ value of -122 kJ. Will the reaction necessarily occur if the reactants are mixed together?

Phase Equilibria

4.41 The vapor pressure of mercury at various temperatures has been determined as follows:

T/K	$P/mmHg$
323	0.0127
353	0.0888
393.5	0.7457
413	1.845
433	4.189

Calculate the value of $\Delta_{vap}\bar{H}$ for mercury.

4.42 The pressure exerted on ice by a 60.0-kg skater is about 300 atm. Calculate the depression in freezing point. The molar volumes are $\bar{V}_L = 0.0180$ L mol^{-1} and $\bar{V}_S = 0.0196$ L mol^{-1}.

4.43 Use the phase diagram of water (Figure 4.14) to predict the direction for the following changes: **(a)** at the triple point of water, temperature is lowered at constant pressure, and **(b)** somewhere along the S–L curve of water, pressure is increased at constant temperature.

4.44 Use the phase diagram of water (Figure 4.14) to predict the dependence of the freezing and boiling points of water on pressure.

4.45 Consider the following system at equilibrium

$$CaCO_3(s) \rightleftharpoons CaO(s) + CO_2(g)$$

How many phases are present?

4.46 Below is a rough sketch of the phase diagram of carbon. **(a)** How many triple points are there, and what are the phases that can coexist at each triple point? **(b)** Which has a higher density, graphite or diamond? **(c)** Synthetic diamond can be made from graphite. Using the phase diagram, how would you go about making diamond?

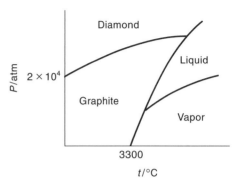

4.47 What is wrong with the following phase diagram for a one-component system?

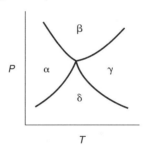

4.48 The plot in Figure 4.13 is no longer linear at high temperatures. Explain.

4.49 Pike's Peak in Colorado is approximately 4,300 m above sea level ($0°C$). What is the boiling point of water at the summit? (*Hint:* See Problem 2.87. The molar mass of air is 29.0 g mol^{-1}, and $\Delta_{vap}\bar{H}$ for water is 40.79 kJ mol^{-1}.)

4.50 The normal boiling point of ethanol is $78.3°C$, and its molar enthalpy of vaporization is 39.3 kJ mol^{-1}. What is its vapor pressure at $30°C$?

Additional Problems

4.51 Entropy has sometimes been described as "time's arrow" because it is the property that determines the forward direction of time. Explain.

4.52 State the condition(s) under which the following equations can be applied: **(a)** $\Delta S = \Delta H/T$, **(b)** $S_0 = 0$, **(c)** $dS = C_P \, dT/T$, and **(d)** $dS = dq/T$.

4.53 Without referring to any table, predict whether the entropy change is positive, nearly zero, or negative for each of the following reactions:

(a) $N_2(g) + O_2(g) \rightarrow 2NO(g)$

(b) $2Mg(s) + O_2(g) \rightarrow 2MgO(s)$

(c) $2H_2O_2(l) \rightarrow 2H_2O(l) + O_2(g)$

(d) $H_2(g) + CO_2(g) \rightarrow H_2O(g) + CO(g)$

4.54 Calculate the entropy change when neon at 25°C and 1.0 atm in a container of volume 0.780 L is allowed to expand to 1.25 L and is simultaneously heated to 85°C. Assume ideal behavior. (*Hint:* Because S is a state function, you can first calculate the value of ΔS for expansion and then calculate the value of ΔS for heating at constant final volume.)

4.55 Photosynthesis makes use of photons of visible light to bring about chemical changes. Explain why heat energy in the form of infrared photons is ineffective for photosynthesis.

4.56 One mole of an ideal monatomic gas is compressed from 2.0 atm to 6.0 atm while being cooled from 400 K to 300 K. Calculate the values of $\Delta U, \Delta H,$ and ΔS for the process.

4.57 The three laws of thermodynamics are sometimes stated colloquially as follows: First law: You cannot get something for nothing; Second law: The best you can do is break even; Third law: You cannot break even. Provide a scientific basis for each of these statements. (*Hint:* One consequence of the third law is that it is impossible to attain the absolute zero of temperature.)

4.58 Use the following data to determine the normal boiling point, in kelvins, of mercury. What assumptions must you make to do the calculation?

$$Hg(l): \quad \Delta_f \bar{H}° = 0 \text{ (by definition)}$$
$$\bar{S}° = 77.4 \text{ J K}^{-1} \text{ mol}^{-1}$$
$$Hg(g): \quad \Delta_f \bar{H}° = 60.78 \text{ kJ mol}^{-1}$$
$$\bar{S}° = 174.7 \text{ J K}^{-1} \text{ mol}^{-1}$$

4.59 Trouton's rule states that the ratio of the molar enthalpy of vaporization of a liquid to its boiling point in kelvins is approximately 90 J K^{-1} mol^{-1}. **(a)** Use the following data to show that this is the case and explain why Trouton's rule holds true.

	$t_{bp}/°C$	$\Delta_{vap}\bar{H}/kJ \cdot mol^{-1}$
Benzene	80.1	31.0
Hexane	68.7	30.8
Mercury	357	59.0
Toluene	110.6	35.2

(b) Trouton's rule does not hold for ethanol ($t_{bp} = 78.3°C$, $\Delta_{vap}\bar{H} = 39.3$ kJ mol^{-1}) and water ($t_{bp} = 100°C$, $\Delta_{vap}\bar{H} = 40.79$ kJ mol^{-1}). Explain. **(c)** The ratio in **(a)** is considerably smaller for liquid HF. Why?

4.60 Give a detailed example of each of the following, with an explanation: **(a)** a thermodynamically spontaneous process; **(b)** a process that would violate the first

law of thermodynamics; **(c)** a process that would violate the second law of thermodynamics; **(d)** an irreversible process; and **(e)** an equilibrium process.

4.61 In the reversible adiabatic expansion of an ideal gas, there are two contributions to entropy changes: the expansion of the gas and the cooling of the gas. Show that these two contributions are equal in magnitude but opposite in sign. Show also that for an irreversible adiabatic gas expansion, these two contributions are no longer equal in magnitude. Predict the sign of ΔS.

4.62 Superheated water is water heated above 100°C without boiling. As for supercooled water (see Example 4.4), superheated water is thermodynamically unstable. Calculate the values of $\Delta S_{sys}, \Delta S_{surr},$ and ΔS_{univ} when 1.5 moles of superheated water at 110°C and 1.0 atm are converted to steam at the same temperature and pressure. (The molar enthalpy of vaporization of water is 40.79 kJ mol^{-1}, and the molar heat capacities of water and steam in the temperature range 100–110°C are 75.5 J K^{-1} mol^{-1} and 34.4 J K^{-1} mol^{-1}, respectively.)

4.63 Toluene (C_7H_8) has a dipole moment, whereas benzene (C_6H_6) is nonpolar:

		CH₃
m.pt.	5.5°C	−95°C
b.pt.	80.1°C	110.6°C

Explain why, contrary to our expectation, benzene melts at a much higher temperature than toluene. Why is the boiling point of toluene higher than that of benzene?

4.64 Give the conditions under which each of the following equations may be applied: **(a)** $dA \leq 0$ (for equilibrium and spontaneity), **(b)** $dG \leq 0$ (for equilibrium and spontaneity), **(c)** $\ln\dfrac{P_2}{P_1} = \dfrac{\Delta\bar{H}}{R}\dfrac{(T_2 - T_1)}{T_1 T_2}$, and **(d)** $\Delta G = nRT \ln\dfrac{P_2}{P_1}$.

4.65 When ammonium nitrate is dissolved in water, the solution becomes colder. What conclusion can you draw about $\Delta S°$ for the process?

4.66 Protein molecules are polypeptide chains made up of amino acids. In their physiologically functioning or native state, these chains fold in a unique manner such that the nonpolar groups of the amino acids are usually buried in the interior region of the proteins, where there is little or no contact with water. When a protein denatures, the chain unfolds so that these nonpolar groups are exposed to water. A useful estimate of the changes of the thermodynamic quantities as a result of denaturation is to consider the transfer of a hydrocarbon such as methane (a nonpolar substance) from an inert solvent (such as benzene or carbon tetrachloride) to the aqueous environment:

(a) $CH_4(\text{inert solvent}) \rightarrow CH_4(g)$

(b) $CH_4(g) \rightarrow CH_4(aq)$

If the values of ΔH° and ΔG° are approximately 2.0 kJ mol^{-1} and -14.5 kJ mol^{-1}, respectively, for (a) and -13.5 kJ mol^{-1} and 26.5 kJ mol^{-1}, respectively, for (b), calculate the values of ΔH° and ΔG° for the transfer of 1 mole of CH_4 according to the equation

$$CH_4(\text{inert solvent}) \rightarrow CH_4(aq)$$

Comment on your results. Assume $T = 298$ K.

4.67 Find a rubber band that is about 0.5 cm wide. Quickly stretch the rubber band and then press it against your lips. You will feel a slight warming effect. Next, reverse the process. Stretch a rubber band and hold it in position for a few seconds. Then quickly release the tension and press the rubber band against your lips. This time you will feel a slight cooling effect. Present a thermodynamic analysis of this behavior.

4.68 A rubber band under tension will contract when heated. Explain.

4.69 Hydrogenation reactions are facilitated by the use of a transition metal catalyst, such as Ni or Pt. Predict the signs of $\Delta_r H, \Delta_r S,$ and $\Delta_r G$ when hydrogen gas is adsorbed onto the surface of nickel metal.

4.70 A sample of supercooled water freezes at -10°C. What are the signs of $\Delta H, \Delta S,$ and ΔG for this process? All the changes refer to the system.

4.71 The boiling point of benzene is 80.1°C. Estimate (a) its $\Delta_{vap}\bar{H}$ value and (b) its vapor pressure at 74°C. (*Hint:* See Problem 4.59.)

4.72 A chemist has synthesized a hydrocarbon compound (C_xH_y). Briefly describe what measurements are needed to determine the values of $\Delta_f\bar{H}^\circ, \bar{S}^\circ,$ and $\Delta_f\bar{G}^\circ$ of the compound.

4.73 A closed, 7.8-L flask contains 1.0 g of water. At what temperature will half of the water be in the vapor phase? (*Hint:* Look up the vapor pressures of water in the inside back cover.)

4.74 A person heated water in a closed bottle in a microwave oven for tea. After removing the bottle from the oven, she added a tea bag to the hot water. To her surprise, the water started to boil violently. Explain what happened.

4.75 Consider the reversible, isothermal compression of 0.45 mole of helium gas from 0.50 atm and 22 L to 1.0 atm at 25°C. (a) Calculate the values of $w, \Delta U, \Delta H, \Delta S,$ and ΔG for the process. (b) Can you use the sign of ΔG to predict whether the process is spontaneous? Explain. (c) What is the maximum work that can be done for the compression process? Assume ideal-gas behavior.

4.76 The molar entropy of argon (Ar) is given by

$$\bar{S}^\circ = (36.4 + 20.8 \ln T) \text{ J K}^{-1} \text{ mol}^{-1}$$

Calculate the change in Gibbs energy when 1.0 mole of Ar is heated at constant pressure from 20°C to 60°C. (*Hint:* Use the relation $\int \ln x \, dx = x \ln x - x$.)

4.77 In Section 4.2 we saw that the probability of finding all 100 helium atoms in half of the cylinder is 8×10^{-31} (see Figure 4.3). Assuming that the age of the universe is 13 billion years, calculate the time in seconds during which this event can be observed.

4.78 Comment on the analogy sometimes used to relate a student's dormitory room becoming disorderly and untidy to an increase in entropy.

4.79 Use Equation 4.1 to account for the fact that carbon monoxide has a residual entropy (that is, an entropy at absolute zero) of 4.2 J K^{-1} mol^{-1}.

4.80 In a DSC experiment (see p. 62), the melting temperature (T_m) of a certain protein is found to be 46°C and the enthalpy of denaturation is 382 kJ mol^{-1}. Estimate the entropy of denaturation assuming that the denaturation is a two-state process; that is, native protein $\rightleftharpoons$ denatured protein. The single polypeptide protein chain has 122 amino acids. Calculate the entropy of denaturation per amino acid. Comment on your result.

CHAPTER 5

Solutions

The study of solutions is of great importance because most of the interesting and useful chemical and biological processes occur in liquid solutions. Generally, a solution is defined as a homogeneous mixture of two or more components that form a single phase. Most solutions are liquids, although gas solutions (for example, air) and solid solutions (for example, solder) also exist. This chapter starts with the thermodynamic study of ideal and nonideal solutions of nonelectrolytes—solutions that do not contain ionic species—and the colligative properties of these solutions.

Because all biological and many chemical systems are aqueous solutions containing various ions, we shall also study the properties of electrolyte solutions. The stability of biomacromolecules and the rates of many biochemical reactions very much depend on the type and concentration of ions present. It is important to have a clear understanding of the behavior of ions in solution. Finally, we shall look briefly at biological membranes and membrane transport.

5.1 Concentration Units

Any quantitative study of solutions requires that we know the amount of solute dissolved in a solvent or the concentration of the solution. Chemists employ several different concentration units, each one having advantages and limitations. The use of the solution generally determines how we express its concentration. In this section, we shall define four concentration units: percent by weight, mole fraction, molarity, and molality.

Percent by Weight

The percent by weight (also called percent by mass) of a solute in a solution is defined as

$$\text{percent by weight} = \frac{\text{weight of solute}}{\text{weight of solute} + \text{weight of solvent}} \times 100\%$$

$$= \frac{\text{weight of solute}}{\text{weight of solution}} \times 100\% \tag{5.1}$$

Mole Fraction (*x*)

The concept of mole fraction was introduced in Section 2.7. We define the mole fraction of a component *i* of a solution, x_i, as

$$x_i = \frac{\text{number of moles of component } i}{\text{number of moles of all components}}$$

$$= \frac{n_i}{\sum_i n_i} \tag{5.2}$$

The mole fraction has no units.

Molarity (*M*)

Molarity is defined as the number of moles of solute dissolved in 1 liter of solution; that is,

$$\text{molarity} = \frac{\text{number of moles of solute}}{\text{L solution}} \tag{5.3}$$

Thus, molarity has the units moles per liter (mol L^{-1}). By convention, we use square brackets [ ] to represent molarity.

Molality (*m*)

Molality is defined as the number of moles of solute dissolved in 1 kg (1000 g) of solvent; that is,

$$\text{molality} = \frac{\text{number of moles of solute}}{\text{weight of solvent in kg}} \tag{5.4}$$

Thus, molality has the units of moles per kg of solvent (mol kg^{-1}).

We shall now compare the usefulness of these four concentration terms. Percent by weight has the advantage that we do not need to know the molar mass of the solute. This unit is useful to biochemists, who frequently work with macromolecules either of unknown molar mass or of unknown purity. (A common unit for protein and DNA solutions is mg mL^{-1}, or mg per milliliter.) Furthermore, the percent by weight of a solute in a solution is independent of temperature, because it is defined in terms of weight. Mole fractions are useful for calculating partial pressures of gases (see Section 2.7) and in the study of vapor pressures of solutions (to be introduced later). Molarity is one of the most commonly employed concentration units. The advantage of using molarity is that it is generally easier to measure the volume of a solution using precisely calibrated volumetric flasks than to weigh the solvent. Its main drawback is that it is temperature dependent, because the volume of a solution usually increases with increasing temperature. Another drawback is that molarity does not tell us the amount of solvent present. Molality, on the other hand, is temperature independent because it is a ratio of the number of moles of solute to the weight of the solvent. For this reason, molality is the preferred concentration unit in studies that involve changes in temperature, as in some of the colligative properties of solutions (see Section 5.6).

5.2 Partial Molar Quantities

The extensive properties of a one-component system at a constant temperature and pressure depend only on the amount of the system present. For example, the volume of water depends on the quantity of water present. If the volume is expressed as a molar quantity, however, it is an intensive property. Thus, the molar volume of water at 1 atm and 298 K is 0.018 L mol^{-1}, no matter how little or how much water is present. For solutions, the criteria are different. A solution, by definition, contains at least two components. The extensive properties of a solution depend on temperature, pressure, and the composition of the solution. In discussing the properties of any solution, we cannot employ molar quantities; instead, we must use *partial molar quantities*. Perhaps the easiest partial molar quantity to understand is *partial molar volume*, described below.

Partial Molar Volume

The molar volumes of water and ethanol at 298 K are 0.018 L and 0.058 L, respectively. If we mix half a mole of each liquid, we might expect the combined volume to be the sum of 0.018 L/2 and 0.058 L/2, or 0.038 L. Instead, we find the volume to be only 0.036 L. The shrinkage of the volume is the result of unequal intermolecular interaction between unlike molecules. Because the forces of attraction between water and ethanol molecules are greater than those between water molecules and between ethanol molecules, the total volume is less than the sum of the individual volumes. If the intermolecular forces are weaker, then expansion will occur and the final volume will be greater than the sum of individual volumes. Only if the interactions between like and unlike molecules are the same will the volumes be additive. If the final volume is equal to the sum of the separate volumes, the solution is called an *ideal* solution. Figure 5.1 shows the total volume of a water–ethanol solution as a function of their mole fractions. In a real (nonideal) solution, the partial molar volume of each component is affected by the presence of the other components.

At constant temperature and pressure, the volume of a solution is a function of the number of moles of different substances present; that is,

$$V = V(n_1, n_2, \dots)$$

For a two-component system the total differential, dV, is given by

$$dV = \left(\frac{\partial V}{\partial n_1}\right)_{T,P,n_2} dn_1 + \left(\frac{\partial V}{\partial n_2}\right)_{T,P,n_1} dn_2$$

$$dV = \bar{V}_1\, dn_1 + \bar{V}_2\, dn_2 \tag{5.5}$$

where $\bar{V}_1$ and $\bar{V}_2$ are the partial molar volumes of components 1 and 2, respectively. The partial molar volume $\bar{V}_1$, for example, tells us the rate of change in volume with number of moles of component 1, at constant T, P, and component 2. Alternatively, $\bar{V}_1$ can be viewed as the increase in volume resulting from the addition of 1 mole of component 1 to a very large volume of solution so that its concentration remains unchanged. The quantity $\bar{V}_2$ can be similarly interpreted. Equation 5.5 can be integrated to give

$$V = n_1 \bar{V}_1 + n_2 \bar{V}_2 \tag{5.6}$$

Financially, this shrinkage in volume has a detrimental effect on bartenders.

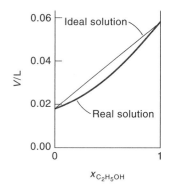

Figure 5.1
Total volume of a water–ethanol mixture as a function of the mole fraction of ethanol. At any concentration the sum of the number of moles is 1. The straight line represents the variation of volume with mole fraction for an ideal solution. The curve represents the actual variation. Note that at $x_{C_2H_5OH} = 0$, the volume corresponds to that of the molar volume of water, and at $x_{C_2H_5OH} = 1$, V is the molar volume of ethanol.

This equation enables us to calculate the volume of the solution by summing the products of the number of moles and the partial molar volume of each component (see Problem 5.63).

Figure 5.2 suggests a way of measuring partial molar volumes. Consider a solution composed of substances 1 and 2. To measure $\overline{V}_2$, we prepare a series of solutions at certain T and P, all of which contain a fixed number of moles of component 1 (that is, n_1 is fixed) but different amounts of n_2. When we plot the measured volumes,

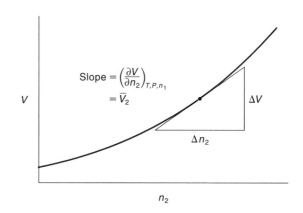

Figure 5.2
Determination of partial molar volume. The volume of a two-component solution is measured as a function of the number of moles n_2 of component 2. The slope at a particular value of n_2 gives the partial molar volume $\overline{V}_2$ at that concentration while holding temperature, pressure, and number of moles of component 1 constant.

V, of the solutions against n_2, the slope of the curve at a particular composition of 2 gives $\overline{V}_2$ for that composition. Once $\overline{V}_2$ has been measured, $\overline{V}_1$ at the same composition can be calculated using Equation 5.6:

$$\overline{V}_1 = \frac{V - n_2\overline{V}_2}{n_1}$$

Figure 5.3 shows the partial molar volumes of ethanol and water in an ethanol–water solution. Note that whenever the partial molar volume of one component rises, that of the other component falls. This relationship is a characteristic of *all* partial molar quantities.

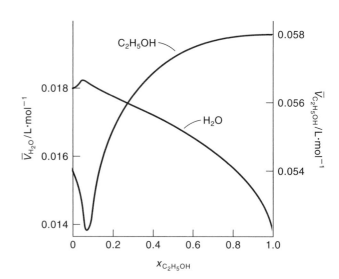

Figure 5.3
The partial molar volumes of water and ethanol as a function of the mole fraction of ethanol. Note the different vertical scales for water (left) and ethanol (right).

Partial Molar Gibbs Energy

Partial molar quantities permit us to express the total extensive properties, such as volume, energy, enthalpy, and Gibbs energy, of a solution of any composition. The partial molar Gibbs energy of the ith component in solution $\bar{G}_i$ is given by

$$\bar{G}_i = \left(\frac{\partial G}{\partial n_i}\right)_{T,P,n_j} \tag{5.7}$$

where n_j represents the number of moles of all other components present. Again, we can think of $\bar{G}_i$ as the coefficient that gives the increase in the Gibbs energy of the solution upon the addition of 1 mole of component i at constant temperature and pressure to a large amount of solution of specified concentration. Partial molar Gibbs energy is also called the *chemical potential* (μ), so we can write

$$\bar{G}_i = \mu_i \tag{5.8}$$

The expression for the total Gibbs energy of a two-component solution is similar to Equation 5.6 for volumes:

$$G = n_1\mu_1 + n_2\mu_2 \tag{5.9}$$

The Meaning of Chemical Potential

The chemical potential provides a criterion for equilibrium and spontaneity for a multicomponent system, just as molar Gibbs energy does for a single-component system. Consider the transfer of dn_i moles of component i from some initial state A, where its chemical potential is μ_i^A, to some final state B, where its chemical potential is μ_i^B. For a process carried out at constant temperature and pressure, the change in the Gibbs energy, dG, is given by

$$dG = \mu_i^B \, dn_i - \mu_i^A \, dn_i$$
$$= (\mu_i^B - \mu_i^A)dn_i$$

If $\mu_i^B < \mu_i^A$, $dG < 0$, and transfer of dn_i moles from A to B will be a spontaneous process; if $\mu_i^B > \mu_i^A$, $dG > 0$, and the process will be spontaneous from B to A. As we shall see later, the transfer can be from one phase to another or from one state of chemical combination to another. The transfer can be transport by diffusion, evaporation, sublimation, condensation, crystallization, solution formation, or chemical reaction. Regardless of the nature of the process, in each case the transfer proceeds from a higher μ_i value to a lower μ_i value. This characteristic explains the name *chemical potential*. In mechanics, the direction of spontaneous change always takes the system from a higher potential-energy state to a lower one. In thermodynamics, the situation is not quite so simple because we have to consider both energy and entropy factors. Nevertheless, we know that at constant temperature and pressure, the direction of a spontaneous change is always toward a decrease in the system's Gibbs energy. Thus, the role Gibbs energy plays in thermodynamics is analogous to that of potential energy in mechanics. This is the reason that molar Gibbs energy or, more commonly, partial molar Gibbs energy, is called the chemical potential.

5.3 The Thermodynamics of Mixing

The formation of solutions is governed by the principles of thermodynamics. In this section, we shall discuss the changes in thermodynamic quantities that result from mixing. In particular, we shall focus on gases.

Equation 5.9 gives the dependence of the Gibbs energy of a system on its composition. The spontaneous mixing of gases is accompanied by a change in composition; consequently, the system's Gibbs energy decreases. In Section 4.8, we obtained an expression for the molar Gibbs energy of an ideal gas (Equation 4.34):

$$\bar{G} = \bar{G}^\circ + RT \ln \frac{P}{1 \text{ bar}}$$

In a mixture of ideal gases, the chemical potential of the ith component is given by

$$\mu_i = \mu_i^\circ + RT \ln \frac{P_i}{1 \text{ bar}} \tag{5.10}$$

where P_i is the partial pressure of component i in the mixture and μ_i° is the standard chemical potential of component i when its partial pressure is 1 bar. Now consider the mixing of n_1 moles of gas 1 at temperature T and pressure P with n_2 moles of gas 2 at the same T and P. Before mixing, the total Gibbs energy of the system is given by Equation 5.9, where chemical potentials are the same as molar Gibbs energies,

$$G = n_1 \bar{G}_1 + n_2 \bar{G}_2$$

$$G_{\text{initial}} = n_1(\mu_1^\circ + RT \ln P) + n_2(\mu_2^\circ + RT \ln P)$$

For simplicity, we omit the term "1 bar." Note that the resulting P values are dimensionless.

After mixing, the gases exert partial pressures P_1 and P_2, where $P_1 + P_2 = P$, and the Gibbs energy is*

$$G_{\text{final}} = n_1(\mu_1^\circ + RT \ln P_1) + n_2(\mu_2^\circ + RT \ln P_2)$$

The Gibbs energy of mixing, $\Delta_{\text{mix}} G$, is given by

$$\Delta_{\text{mix}} G = G_{\text{final}} - G_{\text{initial}}$$

$$= n_1 RT \ln \frac{P_1}{P} + n_2 RT \ln \frac{P_2}{P}$$

$$= n_1 RT \ln x_1 + n_2 RT \ln x_2$$

where $P_1 = x_1 P$ and $P_2 = x_2 P$, and x_1 and x_2 are the mole fractions of 1 and 2, respectively. (The standard chemical potential, μ°, is the same in the pure state and in the mixture.) Further, from the relations

$$x_1 = \frac{n_1}{n_1 + n_2} = \frac{n_1}{n} \quad \text{and} \quad x_2 = \frac{n_2}{n_1 + n_2} = \frac{n_2}{n}$$

where n is the total number of moles, we have

---

*Note that $P_1 + P_2 = P$ only if there is no change in volume as a result of mixing; that is, $\Delta_{\text{mix}} V = 0$. This condition holds for ideal solutions.

$$\Delta_{mix}G = nRT(x_1 \ln x_1 + x_2 \ln x_2) \tag{5.11}$$

Because both x_1 and x_2 are less than unity, $\ln x_1$ and $\ln x_2$ are negative quantities, and hence so is $\Delta_{mix}G$. This result is consistent with our expectation that the mixing of gases is a spontaneous process at constant T and P.

Now we can calculate other thermodynamic quantities of mixing. From Equation 4.29, recall that at constant pressure

$$\left(\frac{\partial G}{\partial T}\right)_P = -S$$

Thus, the entropy of mixing is obtained by differentiating Equation 5.11 with respect to temperature at constant pressure:

$$\left(\frac{\partial \Delta_{mix}G}{\partial T}\right)_P = nR(x_1 \ln x_1 + x_2 \ln x_2)$$

$$= -\Delta_{mix}S$$

or

$$\Delta_{mix}S = -nR(x_1 \ln x_1 + x_2 \ln x_2) \tag{5.12}$$

This result is equivalent to Equation 4.9. The minus sign in Equation 5.12 makes $\Delta_{mix}S$ a positive quantity, in accord with a spontaneous process. The enthalpy of mixing is given by rearranging Equation 4.21:

$$\Delta_{mix}H = \Delta_{mix}G + T\Delta_{mix}S$$

$$= 0$$

This result is not surprising, because molecules of ideal gases do not interact with one another, so no heat is absorbed or produced as a result of mixing. Figure 5.4 shows the plots of $\Delta_{mix}G$, $T\Delta_{mix}S$, and $\Delta_{mix}H$ for a two-component system as a function of composition. Note that both the maximum (for $T\Delta_{mix}S$) and the minimum (for $\Delta_{mix}G$) occur at $x_1 = 0.5$. This result means that we achieve the maximum number of microstates by mixing equimolar amounts of gases and that the Gibbs energy of mixing reaches a minimum at this point (see Problem 5.65).

Reversing the process for a two-component solution of equal mole fractions leads to an increase in Gibbs energy and a decrease in entropy of the system, so energy must be supplied to the system from the surroundings. Initially, at $x_1 \approx x_2$, the $\Delta_{min}G$ and $T\Delta_{mix}S$ curves are fairly flat (see Figure 5.4), and separation can be carried out easily. However, as the solution becomes progressively richer in one component, say 1, the curves become very steep. Then, a considerable amount of energy input is needed to separate component 2 from 1. This difficulty is encountered, for example, in the attempt to clean up a lake contaminated by small amounts of undesirable chemicals. The same consideration applies to the purification of compounds. Preparing most compounds in 95% purity is relatively easy, but much more effort is required to attain 99% or higher purity, which is needed, for example, for the silicon crystals used in solid-state electronics.

As another example, let us explore the possibility of mining gold from the oceans. Estimates are that there is approximately 4×10^{-12} g of gold/mL of seawater. This amount may not seem like much, but when we multiply it by the total volume of ocean water, 1.5×10^{21} L, we find the amount of gold present to be

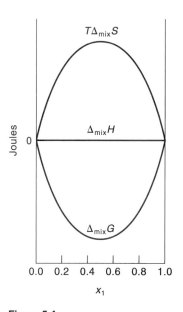

Figure 5.4
Plots of $T\Delta_{mix}S$, $\Delta_{mix}H$, and $\Delta_{mix}G$ as a function of x_1 for the mixing of two components to form an ideal solution.

6×10^{12} g or 7 million tons, which should satisfy anybody. Unfortunately, not only is the concentration of gold in seawater very low, but gold is also just one of some 60 different elements in the ocean. Separating one pure component initially present in a very low concentration in seawater (that is, starting at the steep portions of the curves in Figure 5.4) would be a very formidable (and expensive) undertaking indeed.

Example 5.1

Calculate the Gibbs energy and entropy of mixing 1.6 moles of argon at 1 atm and 25°C with 2.6 moles of nitrogen at 1 atm and 25°C. Assume ideal behavior.

ANSWER

The mole fractions of argon and neon are

$$x_{Ar} = \frac{1.6}{1.6 + 2.6} = 0.38 \qquad x_{N_2} = \frac{2.6}{1.6 + 2.6} = 0.62$$

From Equation 5.11,

$$\Delta_{mix} G = (4.2 \text{ mol})(8.314 \text{ J K}^{-1} \text{ mol}^{-1})(298 \text{ K})[(0.38) \ln 0.38 + (0.62) \ln 0.62]$$
$$= -6.9 \text{ kJ}$$

Because $\Delta_{mix} S = -\Delta_{mix} G / T$, we write

$$\Delta_{mix} S = -\frac{-6.9 \times 10^3 \text{ J}}{298 \text{ K}}$$
$$= 23 \text{ J K}^{-1}$$

COMMENT

In this example, the gases are at the same temperature and pressure when they are mixed. If the initial pressures of the gases differ, then there will be two contributions to $\Delta_{mix} G$: the mixing itself and the changes in pressure. Problem 5.66 illustrates this situation.

5.4 Binary Mixtures of Volatile Liquids

The results obtained in Section 5.3 for mixtures of gases also apply to ideal liquid solutions. For the study of solutions, we need to know how to express the chemical potential of each component. We shall consider a solution containing two volatile liquids (that is, liquids with easily measurable vapor pressures).

Let us start with a liquid in equilibrium with its vapor in a closed container. Because the system is at equilibrium, the chemical potentials of the liquid phase and the vapor phase must be the same; that is,

$$\mu^*(l) = \mu^*(g)$$

where the asterisk denotes a pure component. Further, from the expression for $\mu^*(g)$

for an ideal gas, we can write[†]

$$\mu^*(l) = \mu^*(g) = \mu^\circ(g) + RT \ln \frac{P^*}{1 \text{ bar}} \tag{5.13}$$

where $\mu^\circ(g)$ is the standard chemical potential at $P^* = 1$ bar. For a two-component solution at equilibrium with its vapor, the chemical potential of each component is still the same in the two phases. Thus, for component 1 we have

$$\mu_1(l) = \mu_1(g) = \mu_1^\circ(g) + RT \ln \frac{P_1}{1 \text{ bar}} \tag{5.14}$$

where P_1 is the partial pressure. Now the standard chemical potential of component 1 is the same in the pure state and in the solution; that is, $\mu^\circ(g) = \mu_1^\circ(g)$ and $\mu_1^\circ(g) = \mu_1^*(l) - RT \ln(P_1^*/1 \text{ bar})$. Combining Equations 5.13 and 5.14 we get

$$\mu_1(l) = \mu_1^\circ(g) + RT \ln \frac{P_1}{1 \text{ bar}}$$

$$= \mu_1^*(l) - RT \ln \frac{P_1^*}{1 \text{ bar}} + RT \ln \frac{P_1}{1 \text{ bar}}$$

$$= \mu_1^*(l) + RT \ln \frac{P_1}{P_1^*} \tag{5.15}$$

Thus, the chemical potential of component 1 in solution is expressed in terms of the chemical potential of the liquid in the pure state and the natural log of the ratio of the vapor pressures of the liquid in solution to that in the pure state.

Raoult's Law

The French chemist François Marie Raoult (1830–1901) found that for some solutions, the ratio P_1/P_1^* in Equation 5.15 is equal to the mole fraction of component 1; that is,

$$\frac{P_1}{P_1^*} = x_1$$

or

$$P_1 = x_1 P_1^* \tag{5.16}$$

Equation 5.16 is known as *Raoult's law*, which states that the vapor pressure of a component of a solution is equal to the product of its mole fraction and the vapor pressure of the pure liquid. Substituting Equation 5.16 into Equation 5.15, we obtain

$$\mu_1(l) = \mu_1^*(l) + RT \ln x_1 \tag{5.17}$$

We see that in a pure liquid ($x_1 = 1$ and $\ln x_1 = 0$), $\mu_1(l) = \mu_1^*(l)$. Solutions that obey Raoult's law are called *ideal solutions*. An example of a nearly ideal solution is the

---

[†] This equation follows from Equation 4.34. For a pure component, the chemical potential is equal to the molar Gibbs energy.

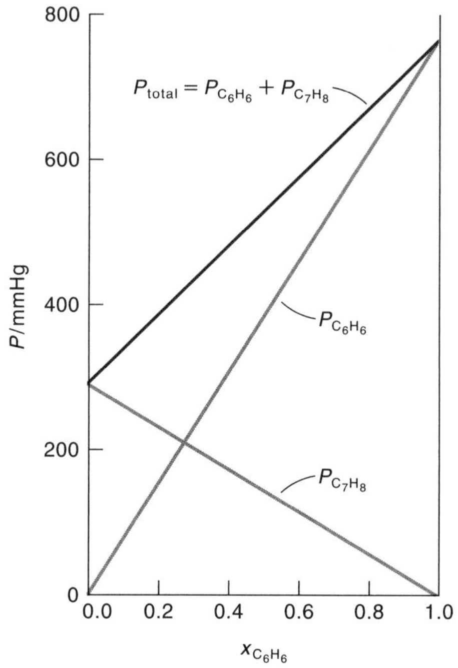

800

$P_{total} = P_{C_6H_6} + P_{C_7H_8}$

600

P/mmHg

400

$P_{C_6H_6}$

200

$P_{C_7H_8}$

0

0.0   0.2   0.4   0.6   0.8   1.0

$x_{C_6H_6}$

Figure 5.5
Total vapor pressure of the benzene–toluene mixture as a function of the benzene mole fraction at 80.1°C. The lighter lines represent the partial pressures of the two components.

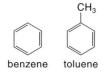

benzene   toluene

benzene–toluene system. Figure 5.5 shows a plot of the vapor pressures versus the mole fraction of benzene.

Example 5.2

Liquids A and B form an ideal solution. At 45°C, the vapor pressures of pure A and pure B are 66 torr and 88 torr, respectively. Calculate the composition of the vapor in equilibrium with a solution containing 36 mole percent A at this temperature.

ANSWER

Because $x_A = 0.36$ and $x_B = 1 - 0.36 = 0.64$, we have, according to Raoult's law

$$P_A = x_A P_A^* = 0.36(66 \text{ torr}) = 23.8 \text{ torr}$$
$$P_B = x_B P_B^* = 0.64(88 \text{ torr}) = 56.3 \text{ torr}$$

The total vapor pressure, P_T, is given by

$$P_T = P_A + P_B = 23.8 \text{ torr} + 56.3 \text{ torr} = 80.1 \text{ torr}$$

Finally, the mole fractions of A and B in the vapor phase, x_A^v and x_B^v, are given by

$$x_A^v = \frac{23.8 \text{ torr}}{80.1 \text{ torr}} = 0.30$$

and

$$x_B^v = \frac{56.3 \text{ torr}}{80.1 \text{ torr}} = 0.70$$

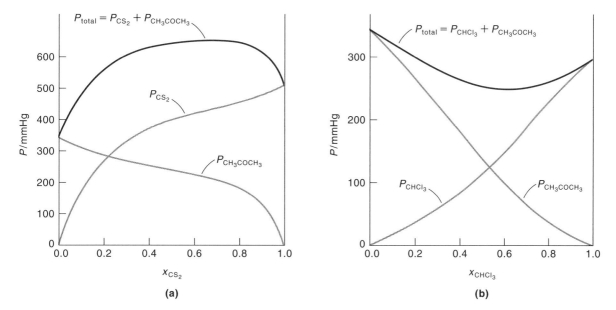

Figure 5.6
Nonideal solutions. (a) Positive deviation from Raoult's law: carbon-disulfide–acetone system at 35.2°C. (b) Negative deviation: chloroform–acetone system at 35.2°C. (From *The Solubility of Nonelectrolytes* by J. Hildebrand and R. Scott. © 1950 by Litton Educational Publishing, Inc. Reprinted by permission of Van Nostrand Reinhold Company.)

In an ideal solution, all intermolecular forces are equal, whether the molecules are alike or not. The benzene–toluene system approximates this requirement because benzene and toluene molecules have similar shapes and electronic structures. For an ideal solution, we have both $\Delta_{mix}H = 0$ and $\Delta_{mix}V = 0$. Most solutions do not behave ideally, however. Figure 5.6 shows the positive and negative deviations from Raoult's law. The positive deviation (Figure 5.6a) corresponds to the case in which the intermolecular forces between unlike molecules are weaker than those between like molecules, and there is a greater tendency for these molecules to leave the solution than in the case of an ideal solution. Consequently, the vapor pressure of the solution is greater than the sum of the vapor pressures for an ideal solution. Just the opposite holds for a negative deviation from Raoult's law (Figure 5.6b). In this case, unlike molecules attract each other more strongly than they do their own kind, and the vapor pressure of the solution is less than the sum of the vapor pressures for an ideal solution.

Henry's Law

When one solution component is present in excess (this component is called the solvent), its vapor pressure is quite accurately described by Equation 5.16. The regions where Raoult's law is applicable are shown for the carbon disulfide–acetone system in Figure 5.7. In contrast, the vapor pressure of the component present in a small amount (this component is called the solute) does not vary with the composition of the solution, as predicted by Equation 5.16. Still, the vapor pressure of the solute varies with concentration in a linear manner:

$$P_2 = Kx_2 \qquad (5.18)$$

There is no sharp distinction between solvent and solute. Where applicable, we shall call component 1 the solvent and component 2 the solute.

Equation 5.18 is known as *Henry's law* (after the English chemist William Henry, 1775–1836), where K, the Henry's law constant, is in atm or torr. Henry's law relates

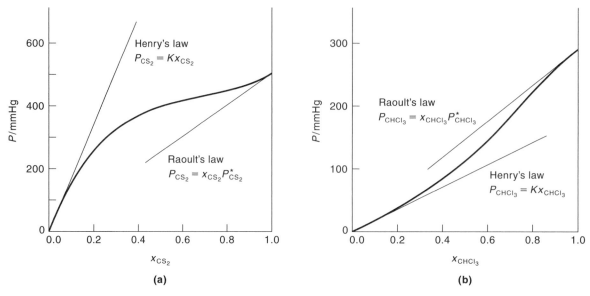

Figure 5.7
Diagrams showing regions over which Raoult's law and Henry's law are applicable for a two-component system (see Figure 5.6). Part (a) shows positive deviation and part (b) shows negative deviation. In each case the Henry's law constants can be obtained from the intercepts on the y (pressure) axis.

the mole fraction of the solute to its partial (vapor) pressure. Alternatively, Henry's law can be expressed as

$$P_2 = K'm \tag{5.19}$$

where m is the molality of the solution and the constant K' now has the units atm mol^{-1} kg of the solvent. Table 5.1 lists the values of K and K' for several gases in water at 298 K.

Henry's law is normally associated with solutions of gases in liquids, although it is equally applicable to solutions containing nongaseous volatile solutes. It has great practical importance in chemical and biological systems and therefore merits further discussion. The effervescence observed when a soft drink or champagne bottle is opened is a nice demonstration of the decrease in gas—mostly CO_2—solubility as its partial pressure is lowered. The emboli (gas bubbles in the bloodstream) suffered by deep-sea divers who rise to the surface too rapidly also illustrate Henry's law. At a

Table 5.1
Henry's Law Constants for Some Gases in Water at 298 K

Gas	K/torr	K'/atm $\cdot$ mol^{-1} $\cdot$ kg H_2O
H_2	5.54×10^7	1311
He	1.12×10^8	2649
Ar	2.80×10^7	662
N_2	6.80×10^7	1610
O_2	3.27×10^7	773
CO_2	1.24×10^6	29.3
H_2S	4.27×10^5	10.1

point some 40 m below the surface of seawater, the total pressure is about 6 atm. The solubility of N_2 in the blood plasma is 0.8×6 atm/1610 atm mol^{-1} kg H_2O, or 3.0×10^{-3} mol (kg H_2O)$^{-1}$—six times the solubility at sea level. If the diver swims upward too rapidly, dissolved nitrogen gas will start boiling off. The mildest result is dizziness; the most serious, death.[†] Because helium is less soluble in the blood plasma than nitrogen is, it is the preferred gas for diluting oxygen gas for use in deep-sea diving tanks.

There are several types of deviations from Henry's law. First, as mentioned earlier, the law holds only for dilute solutions. Second, if the dissolved gas interacts chemically with the solvent, then the solubility can be greatly enhanced. Gases such as CO_2, H_2S, NH_3, and HCl all have high solubilities in water because they react with the solvent. The third type of deviation is illustrated by the dissolution of oxygen in blood. Normally, oxygen is only sparingly soluble in water (see Table 5.1), but its solubility increases dramatically if the solution contains hemoglobin or myoglobin. The nature of oxygen binding to the heme group in these molecules will be discussed further in later chapters.

Example 5.3

Calculate the molal solubility of carbon dioxide in water at 298 K and a CO_2 pressure of 3.3×10^{-4} atm, which corresponds to the partial pressure of CO_2 in air.

ANSWER

We use Equation 5.19 and the data in Table 5.1:

$$m = \frac{P_{CO_2}}{K'}$$

$$= \frac{3.3 \times 10^{-4} \text{ atm}}{29.3 \text{ atm mol}^{-1} \text{ kg } H_2O} = 1.12 \times 10^{-5} \text{ mol (kg } H_2O)^{-1}$$

COMMENT

Carbon dioxide dissolved in water is converted to carbonic acid, which causes water that is exposed to air for a long period of time to become acidic.

5.5 Real Solutions

As pointed out in Section 5.4, most solutions do not behave ideally. One problem that immediately arises in dealing with nonideal solutions is how to write the chemical potentials for the solvent and solute components.

The Solvent Component

Let us look at the solvent component first. As we saw earlier, the chemical potential of the solvent in an ideal solution is given by (see Equation 5.17)

$$\mu_1(l) = \mu_1^*(l) + RT \ln x_1$$

[†] For other interesting illustrations of Henry's law, see T. C. Loose, *J. Chem. Educ.* **48**, 154 (1971); W. J. Ebel, *J. Chem. Educ.* **50**, 559 (1973); and E. D. Cook, *J. Chem. Educ.* **50**, 425 (1973).

where $x_1 = P_1/P_1^*$ and P_1^* is the equilibrium vapor pressure of pure component 1 at T. The standard state is the pure liquid and is attained when $x_1 = 1$. For a nonideal solution, we write

$$\mu_1(l) = \mu_1^*(l) + RT \ln a_1 \tag{5.20}$$

where a_1 is the *activity* of the solvent. Nonideality is the consequence of unequal intermolecular forces between solvent–solvent and solvent–solute molecules. Therefore, the extent of nonideality depends on the composition of solution, and the activity of the solvent plays the role of "effective" concentration. The solvent's activity can be expressed in terms of vapor pressure as

$$a_1 = \frac{P_1}{P_1^*} \tag{5.21}$$

where P_1 is the partial vapor pressure of component 1 over the (nonideal) solution. Activity is related to concentration (mole fraction) as follows:

$$a_1 = \gamma_1 x_1 \tag{5.22}$$

where γ_1 is the *activity coefficient*. Equation 5.20 can now be written as

$$\mu_1(l) = \mu_1^*(l) + RT \ln \gamma_1 + RT \ln x_1 \tag{5.23}$$

The value of γ_1 is a measure of the deviation from ideality. In the limiting case, where $x_1 \rightarrow 1$, $\gamma_1 \rightarrow 1$ and activity and the mole fraction are identical. This condition also holds for an ideal solution at all concentrations.

Equation 5.21 provides a way of obtaining the activity of the solvent. By measuring P_1 of the solvent vapor over a range of concentrations, we can calculate the value of a_1 at each concentration if P_1^* is known.[†]

[†] To obtain the value of P_1, we must measure the total pressure, P, and also analyze the composition of the mixture. Then we can calculate partial pressure P_1 using Dalton's law; that is, $P_1 = x_1^v P$, where x_1^v is the mole fraction of the solvent in the vapor phase.

Example 5.4

The vapor pressure of water in a 6.00 m urea solution is 5.501×10^{-3} atm at 273 K. Calculate the activity and activity coefficient of water. The vapor pressure of pure water is 6.025×10^{-3} atm at this temperature.

ANSWER

To calculate the activity of water we use Equation 5.21:

$$a_1 = \frac{5.501 \times 10^{-3} \text{ atm}}{6.025 \times 10^{-3} \text{ atm}} = 0.913$$

To calculate the activity coefficient, we must first determine the mole fraction of water in the solution. Because the number of moles of water in 1 kg of the solvent is (1000 g/ 18.02 g mol^{-1}) or 55.50 mol, we have

$$x_1 = \frac{m_1}{m_1 + m_2}$$

$$= \frac{55.50}{55.50 + 6.00} = 0.902$$

Finally, from Equation 5.22,

$$\gamma_1 = \frac{a_1}{x_1}$$

$$= \frac{0.913}{0.902} = 1.012$$

The Solute Component

We now come to the solute. Ideal solutions in which both components obey Raoult's law over the entire concentration range are rare. If a nonideal solution is dilute, and there is no chemical interaction, then the solvent obeys Raoult's law and the solute obeys Henry's law.[†] Such solutions are sometimes called "ideal dilute solutions." If the solution were ideal, the chemical potential of the solute is also given by Raoult's law:

$$\mu_2(l) = \mu_2^*(l) + RT \ln x_2$$

$$= \mu_2^*(l) + RT \ln \frac{P_2}{P_2^*}$$

In an ideal dilute solution, Henry's law applies. This is, $P_2 = Kx_2$, so that

$$\mu_2(l) = \mu_2^*(l) + RT \ln \frac{K}{P_2^*} + RT \ln x_2$$

$$\mu_2(l) = \mu_2^\circ(l) + RT \ln x_2 \tag{5.24}$$

where $\mu_2^\circ(l) = \mu_2^*(l) + RT \ln(K/P_2^*)$. Although Equation 5.24 seems to take the same form as Equation 5.17, there is an important difference, which lies in the choice of standard state. According to Equation 5.24, the standard state is defined as the pure solute, attained by setting $x_2 = 1$. Equation 5.24 holds only for dilute solutions, however. How can these two conditions be met simultaneously? The simple way out of this dilemma is to recognize that standard states are often hypothetical states, not physically realizable. Thus, the standard state of the solute defined by Equation 5.24 is the hypothetical pure component 2 with a vapor pressure equal to K (when $x_2 = 1$, $P_2 = K$). In a sense, this is an "infinite dilution state of unit mole fraction"; that is, it is infinitely dilute with respect to component 1, the solvent, with the solute at unit mole fraction. For nonideal solutions in general (beyond the dilute solution limit), Equation 5.24 is modified to

$$\mu_2(l) = \mu_2^\circ(l) + RT \ln a_2 \tag{5.25}$$

[†] For ideal solutions, Raoult's law and Henry's law become identical; that is, $P_2 = Kx_2 = P_2^*x_2$.

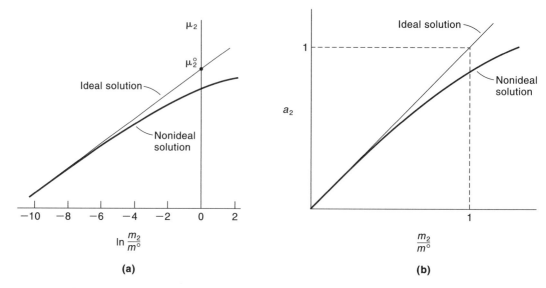

Figure 5.8
(a) Chemical potential of a solute plotted against the logarithm of molality for a nonideal solution. (b) Activity of a solute as a function of molality for a nonideal solution. The standard state is at $m_2/m^\circ = 1$.

where a_2 is the activity of the solute. As in the case of the solvent component, we have $a_2 = \gamma_2 x_2$, where γ_2 is the activity coefficient of the solute. Here, we have $a_2 \to x_2$ or $\gamma_2 \to 1$ as $x_2 \to 0$. Henry's law is now given by

$$P_2 = Ka_2 \tag{5.26}$$

Concentrations are usually expressed in molalities (or molarities) instead of mole fractions. In molality, Equation 5.24 takes the form

$$\mu_2(l) = \mu_2^\circ(l) + RT \ln \frac{m_2}{m^\circ} \tag{5.27}$$

where $m^\circ = 1$ mol kg^{-1} so that the ratio m_2/m° is dimensionless. Here, the standard state is defined as a state at unit molality but in which the solution is behaving ideally. Again, this standard state is a hypothetical state, not attainable in practice (Figure 5.8). For nonideal solutions, Equation 5.27 is rewritten as

$$\mu_2(l) = \mu_2^\circ(l) + RT \ln a_2 \tag{5.28}$$

where $a_2 = \gamma_2(m_2/m^\circ)$. In the limiting case of $m_2 \to 0$, we have $a_2 \to m_2/m^\circ$ or $\gamma_2 \to 1$ (see Figure 5.8b).

Keep in mind that although Equations 5.24 and 5.27 were derived using Henry's law, they are applicable to any solute, whether or not it is volatile. These expressions are useful in discussing the colligative properties of solutions (see Section 5.6), and, as we shall see in Chapter 6, in deriving the equilibrium constant.

5.6 Colligative Properties

General properties of solutions include vapor-pressure lowering, boiling-point elevation, freezing-point depression, and osmotic pressure. These properties are commonly referred to as *colligative*, or *collective*, *properties* because they are bound together through their common origin. Colligative properties depend only on the number of solute molecules present, not on the size or molar mass of the molecules. To derive equations describing these phenomena, we shall make two important assumptions: (1) The solutions are ideal dilute, so that the solvent obeys Raoult's law and (2) the solutions contain nonelectrolytes. As usual, we shall consider only a two-component system.

Vapor-Pressure Lowering

Consider a solution that contains a solvent 1 and a *nonvolatile* solute 2, such as a solution of sucrose in water. Because the solution is ideal dilute, Raoult's law applies:

$$P_1 = x_1 P_1^*$$

Because $x_1 = 1 - x_2$, the equation above becomes

$$P_1 = (1 - x_2)P_1^*$$

Rearranging this equation gives

$$P_1^* - P_1 = \Delta P = x_2 P_1^* \tag{5.29}$$

where ΔP, the decrease in vapor pressure from that of the pure solvent, is directly proportional to the mole fraction of the solute.

Why does the vapor pressure of a solution fall in the presence of a solute? It is tempting to suggest that it is because of the modification of intermolecular forces. But this cannot be so, because vapor-pressure lowering occurs even in ideal solutions, in which there is no difference between solute–solvent and solvent–solvent interactions. A more convincing explanation is provided by the entropy effect. When a solvent evaporates, the entropy of the universe increases, because the entropy of any substance in the gaseous state is greater than that in the liquid state (at the same temperature). As we saw in Section 5.3, the solution process itself is accompanied by an increase in entropy. Because the solution has a greater entropy than the pure solvent, it therefore has a smaller driving force for evaporation. In other words, evaporation from a solution will result in a smaller increase in entropy than the case for a pure solvent. Consequently, the solvent has less of a tendency to leave the solution, and the solution will have a lower vapor pressure than the pure solvent.

Boiling-Point Elevation

The boiling point of a solution is the temperature at which its vapor pressure is equal to the external pressure. The previous discussion might lead you to expect that because the addition of a nonvolatile solute lowers the vapor pressure, it should also raise the boiling point of a solution. This effect is indeed the case.

For a solution containing a *nonvolatile* solute, the boiling-point elevation originates in the change in the chemical potential of the solvent due to the presence of the solute. From Equation 5.17, we can see that the chemical potential of the solvent in a solution is less than the chemical potential of the pure solvent by an amount equal to $RT \ln x_1$. How this change affects the boiling point of the solution can be seen from Figure 5.9. The solid lines refer to the pure solvent. Because the solute is nonvolatile, it does not vaporize; therefore, the curve for the vapor phase is the same as that for the pure vapor. On the other hand, because the liquid contains a solute, the chemical potential of the solvent decreases (see the dashed curve). The points where the curve

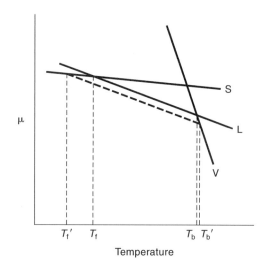

Figure 5.9
Plot of chemical potentials versus temperature to illustrate colligative properties. The dashed red line denotes the solution phase. T_b and T_b' are the boiling points of the solvent and solution, and T_f and T_f' are the freezing points of the solvent and solution, respectively.

for the vapor intersects the curves for the liquids (pure and solution) correspond to the boiling points of the pure solvent and the solution, respectively. We see that the boiling point of the solution (T_b') is higher than that of the pure solvent (T_b).

We now turn to a quantitative treatment of the boiling-point-elevation phenomenon. At the boiling point, the solvent vapor is in equilibrium with the solvent in solution, so that

$$\mu_1(g) = \mu_1(l) = \mu_1^*(l) + RT \ln x_1$$

or

$$\Delta\mu_1 = \mu_1(g) - \mu_1^*(l) = RT \ln x_1 \tag{5.30}$$

where $\Delta\mu_1$ is the Gibbs energy change associated with the evaporation of 1 mole of solvent from the solution at temperature T, its boiling point. Thus, we can write $\Delta\mu_1 = \Delta_{vap}\overline{G}$. Dividing Equation 5.30 by T, we obtain

$$\frac{\Delta\overline{G}_{vap}}{T} = \frac{\mu_1(g) - \mu_1^*(l)}{T} = R \ln x_1$$

From the Gibbs–Helmholz equation (Equation 4.31), we write

$$\frac{d(\Delta G/T)}{dT} = -\frac{\Delta H}{T^2} \quad (\text{at constant } P)$$

or

$$\frac{d(\Delta\overline{G}_{vap}/T)}{dT} = \frac{-\Delta_{vap}\overline{H}}{T^2} = R\frac{d(\ln x_1)}{dT}$$

where $\Delta_{vap}\overline{H}$ is the molar enthalpy of vaporization of the solvent from the solution. Because the solution is dilute, $\Delta_{vap}\overline{H}$ is taken to be the same as the molar enthalpy of vaporization of the pure solvent. Rearranging the last equation gives

$$d \ln x_1 = \frac{-\Delta_{vap}\overline{H}}{RT^2} dT \tag{5.31}$$

To find the relationship between x_1 and T, we integrate Equation 5.31 between the limits T_b' and T_b, the boiling points of the solution and pure solvent, respectively. Because the mole fraction of the solvent is x_1 at T_b' and 1 at T_b, we write

$$\int_{\ln 1}^{\ln x_1} d \ln x_1 = \int_{T_b}^{T_b'} \frac{-\Delta_{vap}\overline{H}}{RT^2} dT$$

or

$$\begin{aligned}
\ln x_1 &= \frac{\Delta_{vap}\overline{H}}{R}\left(\frac{1}{T_b'} - \frac{1}{T_b}\right) \\
&= \frac{-\Delta_{vap}\overline{H}}{R}\left(\frac{T_b' - T_b}{T_b'T_b}\right) \\
&= \frac{-\Delta_{vap}\overline{H}}{R}\frac{\Delta T}{T_b^2}
\end{aligned} \qquad (5.32)$$

where $\Delta T = T_b' - T_b$. Two assumptions were used to obtain Equation 5.32, both of which are based on the fact that T_b' and T_b differ only by a small amount (a few degrees). First, we assumed $\Delta_{vap}\overline{H}$ to be temperature independent and second, $T_b' \approx T_b$, so that $T_b'T_b \approx T_b^2$.

Equation 5.32 gives the elevation of boiling point, ΔT, in terms of the concentration of the solvent (x_1). By custom, however, we express the concentration in terms of the amount of solute present, so we write

$$\ln x_1 = \ln(1 - x_2) = \frac{-\Delta_{vap}\overline{H}}{R}\frac{\Delta T}{T_b^2}$$

where*

$$\begin{aligned}
\ln(1 - x_2) &= -x_2 - \frac{x_2^2}{2} - \frac{x_2^3}{3} \cdots \\
&= -x_2 \quad (x_2 \ll 1)
\end{aligned}$$

We now have

$$\Delta T = \frac{RT_b^2}{\Delta_{vap}\overline{H}}x_2$$

To convert the mole fraction x_2 into a more practical concentration unit, such as molality (m_2), we write

$$x_2 = \frac{n_2}{n_1 + n_2} \approx \frac{n_2}{n_1} = \frac{n_2}{w_1/\mathcal{M}_1} \quad (n_1 \gg n_2)$$

where w_1 is the mass of the solvent in kg and $\mathcal{M}_1$ is the molar mass of the solvent in kg mol^{-1}, respectively. Because n_2/w_1 gives the molality of the solution, m_2, it follows that $x_2 = \mathcal{M}_1 m_2$ and thus

$$\Delta T = \frac{RT_b^2 \mathcal{M}_1}{\Delta_{vap}\overline{H}}m_2 \qquad (5.33)$$

---

*This series expansion is known as Maclaurin's theorem. You can verify this relationship by employing a small numerical value for x_2 (≤ 0.2).

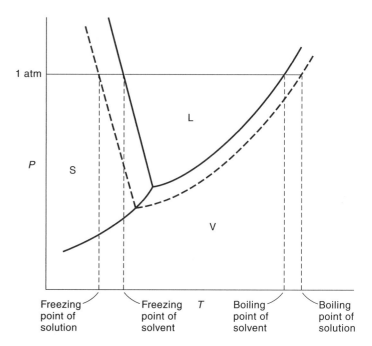

Figure 5.10
Phase diagrams of pure water (solid red lines) and of water in an aqueous solution containing a nonvolatile solute (dashed red lines).

Note that all the quantities in the first term on the right of Equation 5.33 are constants for a given solvent, and so we have

$$K_b = \frac{RT_b^2 \mathscr{M}_1}{\Delta_{vap}\overline{H}} \tag{5.34}$$

where K_b is called the *molal boiling-point-elevation constant*. The units of K_b are K mol^{-1} kg. Finally,

$$\Delta T = K_b m_2 \tag{5.35}$$

The advantage of using molality, as mentioned in Section 5.1, is that it is independent of temperature and thus is suitable for boiling-point-elevation studies.

Figure 5.10 shows the phase diagrams of pure water and an aqueous solution. Upon the addition of a nonvolatile solute, the vapor pressure of the solution decreases at every temperature. Consequently, the boiling point of the solution at 1 atm will be greater than 373.15 K.

Freezing-Point Depression

A nonchemist may be forever unaware of the boiling-point-elevation phenomenon, but any casual observer living in a cold climate witnesses an illustration of freezing-point depression: ice on winter roads and sidewalks melts readily when sprinkled with salt.* This method of thawing depresses the freezing point of water.

The thermodynamic analysis of freezing-point depression is similar to that of boiling-point elevation. If we assume that when a solution freezes, the solid that separates from the solution contains only the solvent, then the curve for the chemical

---

*The salt employed is usually sodium chloride, which attacks cement and is harmful to many plants. Also see "Freezing Ice Cream and Making Caramel Topping," J. O. Olson and L. H. Bowman, *J. Chem. Educ.* **53**, 49 (1976).

potential of the solid does not change (see Figure 5.9). Consequently, the solid curve for the solid and the dashed curve for the solvent in solution now intersect at a point (T_f') *below* the freezing point of the pure solvent (T_f). By following exactly the same procedure as that for the boiling-point elevation, we can show that the drop in freezing point ΔT (that is, $T_f - T_f'$, where T_f and T_f' are the freezing points of the pure solvent and solution, respectively) is

$$\Delta T = K_f m_2 \tag{5.36}$$

where K_f is the *molal freezing-point-depression constant* given by

$$K_f = \frac{RT_f^2 \mathscr{M}_1}{\Delta_{fus}\bar{H}} \tag{5.37}$$

where $\Delta_{fus}\bar{H}$ is the molar enthalpy of fusion of the solvent.

The freezing-point-depression phenomenon can also be understood by studying Figure 5.10. At 1 atm, the freezing point of solution lies at the intersection point of the dashed curve (between the solid and liquid phases) and the horizontal line at 1 atm. It is interesting that whereas the solute must be nonvolatile in the boiling-point-elevation case,* no such restriction applies to lowering the freezing point. A proof of this statement is the use of ethanol (b.p. = 351.65 K) as an antifreeze.

Both Equations 5.35 and 5.36 can be used to determine the molar mass of a solute. In general, the freezing-point-depression experiment is much easier to carry out. Table 5.2 lists the K_b and K_f values for several common solvents.

Table 5.2
Molal Boiling-Point-Elevation and Molal Freezing-Point-Depression Constants of Some Common Solvents

Solvent	$K_b/K \cdot mol^{-1} \cdot kg$	$K_f/K \cdot mol^{-1} \cdot kg$
H_2O	0.51	1.86
C_2H_5OH	1.22	—
C_6H_6	2.53	5.12
$CHCl_3$	3.63	—
CH_3COOH	2.93	3.90
CCl_4	5.03	—

---

* A volatile solute component can actually lower the boiling point of the solution. For example, 95% by volume ethanol in water boils at 351.3 K.

Example 5.5

For a solution of 45.20 g of sucrose ($C_{12}H_{22}O_{11}$) in 316.0 g of water, calculate (a) the boiling point, and (b) the freezing point.

ANSWER

(a) Boiling point: $K_b = 0.51$ K mol^{-1} kg, and the molality of the solution is given by

$$m_2 = \frac{(45.20 \text{ g})(1000 \text{ g/1 kg})}{(342.3 \text{ g mol}^{-1})(316.0 \text{ g})} = 0.418 \text{ mol kg}^{-1}$$

From Equation 5.35,

$$\Delta T = (0.51 \text{ K mol}^{-1} \text{ kg})(0.418 \text{ mol kg}^{-1})$$
$$= 0.21 \text{ K}$$

Thus, the solution will boil at $(373.15 + 0.21)$ K, or 373.36 K.
(b) Freezing point: From Equation 5.36,

$$\Delta T = (1.86 \text{ K mol}^{-1} \text{ kg})(0.418 \text{ mol kg}^{-1})$$
$$= 0.78 \text{ K}$$

Thus, the solution will freeze at $(273.15 - 0.78)$ K, or 272.37 K.

COMMENT

For aqueous solutions of equal concentrations, the depression in freezing point is always greater than the corresponding elevation in boiling point. The reason can be seen by comparing the following two expressions from Equations 5.34 and 5.37:

$$K_b = \frac{RT_b^2 \mathcal{M}_1}{\Delta_{vap}\bar{H}} \qquad K_f = \frac{RT_f^2 \mathcal{M}_1}{\Delta_{fus}\bar{H}}$$

Although $T_b > T_f$, $\Delta_{vap}\bar{H}$ for water is 40.79 kJ mol^{-1}, whereas $\Delta_{fus}\bar{H}$ is only 6.01 kJ mol^{-1}. The large value of $\Delta_{vap}\bar{H}$ in the denominator is what causes K_b and hence ΔT to be smaller.

The freezing-point-depression phenomenon has many examples in everyday life and in biological systems. As mentioned above, salts, such as sodium chloride and calcium chloride, are used to melt ice on roads and sidewalks. The organic compound ethylene glycol [$CH_2(OH)CH_2(OH)$] is the common automobile antifreeze. It is also employed to de-ice airplanes. In recent years, there has been much interest in understanding how certain species of fish manage to survive in the ice-cold waters of the polar oceans. The freezing point of seawater is approximately $-1.9°C$, which is the temperature of seawater surrounding an iceberg. A depression in freezing point of 1.9 degrees corresponds to a concentration of one molal, which is much too high for proper physiological function; for example, it alters osmotic balance (see the section below on Osmotic Pressure). Besides dissolved salts and other substances that can lower the freezing point colligatively, a special class of proteins resides in the blood of polar fishes that has some kind of protective effect. These proteins contain both amino acid and sugar units and are called glycoproteins. The concentration of glycoproteins in the fishes' blood is quite low (approximately 4×10^{-4} m), so their action cannot be explained by colligative properties. The belief is that the glycoproteins have the ability to adsorb onto the surface of each tiny ice crystal as soon as it begins to form, thus preventing it from growing to a size that would cause biological damage. Consequently, the freezing point of blood in these fishes is below $-2°C$.

Osmotic Pressure

The phenomenon of *osmosis* is illustrated in Figure 5.11. The left compartment of the apparatus contains pure solvent; the right compartment contains a solution. The two compartments are separated by a *semipermeable membrane* (for example, a cellophane membrane), one that permits the solvent molecules to pass through but

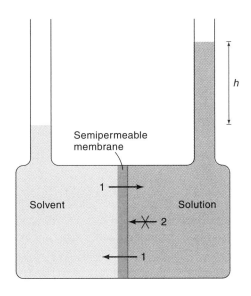

Figure 5.11
Apparatus demonstrating the
osmotic pressure phenomenon.

does not permit the movement of solute molecules from right to left. Practically speaking, then, this system has two different phases. At equilibrium, the height of the solution in the tube on the right is greater than that of the pure solvent in the left tube by h. This excess hydrostatic pressure is called the *osmotic pressure*. We can now derive an expression for osmotic pressure as follows.

Let μ_1^L and μ_1^R be the chemical potential of the solvent in the left and right compartments, respectively. Initially, before equilibrium is established, we have

$$\mu_1^L = \mu_1^* + RT \ln x_1$$
$$= \mu_1^* \quad (x_1 = 1)$$

and

$$\mu_1^R = \mu_1^* + RT \ln x_1 \quad (x_1 < 1)$$

Thus,

$$\mu_1^L = \mu_1^* > \mu_1^R = \mu_1^* + RT \ln x_1$$

Note that μ_1^L is the same as the standard chemical potential for the pure solvent, μ_1^*, and the inequality sign denotes that $RT \ln x_1$ is a negative quantity. Consequently, more solvent molecules, on the average, will move from left to right across the membrane. The process is spontaneous because the dilution of the solution in the right compartment by solvent leads to a decrease in the Gibbs energy and an increase in entropy. Equilibrium is finally reached when the flow of solvent is exactly balanced by the hydrostatic pressure difference in the two side tubes. This extra pressure increases the chemical potential of the solvent in solution, μ_1^R. From Equation 4.32, we know that

$$\left(\frac{\partial G}{\partial P}\right)_T = V$$

We can write a similar equation for the variation of the chemical potential with pressure at constant temperature. Thus, for the solvent component in the right compartment,

$$\left(\frac{\partial \mu_1^R}{\partial P}\right)_T = \overline{V}_1 \qquad (5.38)$$

where $\overline{V}_1$ is the partial molar volume of the solvent. For a dilute solution, $\overline{V}_1$ is approximately equal to $\overline{V}$, the molar volume of the pure solvent. The increase in the chemical potential of the solvent in the solution compartment $(\Delta\mu_1^R)$ when the pressure increases from P, the external atmospheric pressure, to $(P + \pi)$ is given by

$$\Delta\mu_1^R = \int_P^{P+\pi} \overline{V} \, dP = \overline{V}\pi$$

Note that $\overline{V}$ is treated as a constant because the volume of a liquid changes little with pressure. The Greek letter π represents the osmotic pressure. The term *osmotic pressure of a solution* refers to the pressure that must be applied to the solution to increase the chemical potential of the solvent to the value of its pure liquid under atmospheric pressure.

At equilibrium, the following relations must hold:

$$\mu_1^L = \mu_1^R = \mu_1^* + RT \ln x_1 + \pi\overline{V}$$

Because $\mu_1^L = \mu_1^*$, we have

$$\pi\overline{V} = -RT \ln x_1 \qquad (5.39)$$

To relate π to the concentration of the solute, we take the following steps. From the procedure employed for boiling-point elevation (p. 144):

$$-\ln x_1 = -\ln(1 - x_2)$$
$$= x_2 \quad (x_2 \ll 1)$$

Furthermore,

$$x_2 = \frac{n_2}{n_1 + n_2} \approx \frac{n_2}{n_1} \quad (n_1 \gg n_2)$$

where n_1 and n_2 are the number of moles of solvent and solute, respectively. Equation 5.39 now becomes

$$\pi\overline{V} = RTx_2$$
$$= RT\left(\frac{n_2}{n_1}\right) \qquad (5.40)$$

Substituting $\overline{V} = V/n_1$ into Equation 5.40, we get

$$\pi V = n_2 RT \qquad (5.41)$$

If V is in liters, then

Equation 5.42 also applies to two similar solutions that have different concentrations. In this case, M is the difference in concentrations between the solutions.

$$\pi = \frac{n_2}{V} RT$$

$$\pi = MRT \qquad (5.42)$$

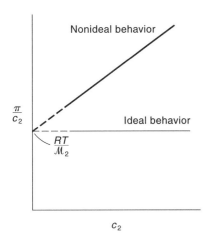

Figure 5.12
Determination of molar mass of a solute by osmotic pressure measurement for an ideal and a nonideal solution. The intercept on the y axis (as $c_2 \to 0$) gives the correct value for the molar mass.

where M is the molarity of the solution. Note that molarity is a convenient concentration unit here, because osmotic pressure measurements are normally made at constant temperature. Alternatively, we can rewrite Equation 5.42 as

$$\pi = \frac{c_2}{\mathscr{M}_2} RT \qquad (5.43)$$

or

$$\frac{\pi}{c_2} = \frac{RT}{\mathscr{M}_2} \qquad (5.44)$$

where c_2 is the concentration of the solute in g L^{-1} of the solution and $\mathscr{M}_2$ is the molar mass of the solute in g mol^{-1}. Equation 5.44 provides a way to determine molar masses of compounds from osmotic pressure measurements.

Equation 5.44 is derived by assuming ideal behavior, so it is desirable to measure π at several different concentrations and extrapolate to zero concentration for molar mass determination (Figure 5.12). For a nonideal solution, the osmotic pressure at *any* concentration, c_2, is given by

$$\frac{\pi}{c_2} = \frac{RT}{\mathscr{M}_2} \left(1 + Bc_2 + Cc_2^2 + Dc_2^3 + \cdots \right) \qquad (5.45)$$

Compare Equation 5.45 with Equation 2.14.

where B, C, and D are called the second, third, and fourth virial coefficients, respectively. The magnitude of the virial coefficients is such that $B \gg C \gg D$. In dilute solutions, we need be concerned only with the second virial coefficient. For an ideal solution, the second and higher virial coefficients are all equal to zero, so Equation 5.45 reduces to Equation 5.44.

Even though osmosis is a well-studied phenomenon, the mechanism involved is not always clearly understood. In some cases, a semipermeable membrane may act as a molecular sieve, allowing smaller solvent molecules to pass through while blocking larger solute molecules. In other cases, osmosis may be caused by the higher solubility of the solvent in the membrane than the solute. Each system must be studied individually. The previous discussion illustrates both the usefulness and limitation of thermodynamics. We have derived a convenient equation relating the molar mass of the solute to an experimentally measurable quantity—the osmotic pressure—simply in terms of the chemical potential difference. Because thermodynamics is not based on any specific model, however, Equation 5.44 tells us nothing about the mechanism of osmosis.

Example 5.6

Consider the following arrangement, in which a solution containing 20 g of hemoglobin in 1 liter of the solution is placed in the right compartment, and pure water is placed in the left compartment (see Figure 5.11). At equilibrium, the height of the water in the right column is 77.8 mm in excess of the height of the solution in the left column. What is the molar mass of hemoglobin? The temperature of the system is constant at 298 K.

ANSWER

To determine the molar mass of hemoglobin, we first need to calculate the osmotic pressure of the solution. We start by writing

$$\text{pressure} = \frac{\text{force}}{\text{area}}$$

$$\text{pressure} = \frac{Ah\rho g}{A} = h\rho g$$

where A is the area of the cross section of the tube, h the excess liquid height in the right column, ρ the density of the solution, and g the acceleration due to gravity. The constants are

$$h = 0.0778 \text{ m}$$
$$g = 9.81 \text{ m s}^{-2}$$
$$\rho = 1 \times 10^3 \text{ kg m}^{-3}$$

(We have assumed that the density of the dilute solution is the same as that of water.) The osmotic pressure in pascals ($N m^{-2}$) is given by

$$\pi = 0.0778 \text{ m} \times 1 \times 10^3 \text{ kg m}^{-3} \times 9.81 \text{ m s}^{-2}$$
$$= 763 \text{ kg m}^{-1} \text{ s}^{-2}$$
$$= 763 \text{ N m}^{-2}$$

From Equation 5.44,

$$\mathcal{M}_2 = \frac{c_2}{\pi} RT$$

Using the conversion factors $1 \text{ g} = 1 \times 10^{-3} \text{ kg}$ and $1 \text{ L} = 1 \times 10^{-3} \text{ m}^3$, we have

$$\mathcal{M}_2 = \frac{(20 \text{ kg m}^{-3})(8.314 \text{ J K}^{-1} \text{ mol}^{-1})(298 \text{ K})}{763 \text{ N m}^{-2}}$$
$$= 65 \text{ kg mol}^{-1}$$

Example 5.6 shows that osmotic pressure measurement is a more sensitive method to determine molar mass than the boiling-point-elevation and freezing-point-depression techniques, because 7.8 cm is an easily measurable height. On the other hand, the same solution will lead to an elevation in boiling point of approximately 1.6×10^{-4}°C and a depression in freezing point of 5.8×10^{-4}°C, which are too small to measure accurately. Most proteins are less soluble than hemoglobin. Nevertheless, their molar masses can often be determined by osmotic pressure measurements. In

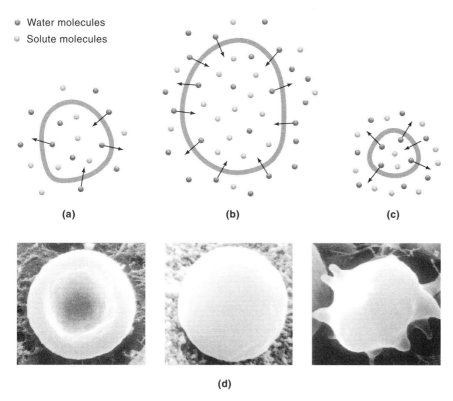

- Water molecules
- Solute molecules

(a)   (b)   (c)

(d)

Figure 5.13
A cell in (a) an isotonic solution, (b) a hypotonic solution, and (c) a hypertonic solution. The cell remains unchanged in (a), swells in (b), and shrinks in (c). (d) From left to right, a red blood cell is shown in an isotonic solution, a hypotonic solution, and a hypertonic solution. (Part d Copyright David Phillips/PhotoResearchers Inc. Reprinted with permission.)

Chapter 16, we shall discuss other useful techniques for determining the molar mass of macromolecules.

Many examples of the osmotic-pressure phenomenon are found in chemical and biological systems. If two solutions are of equal concentration, and hence have the same osmotic pressure, they are said to be *isotonic*. For two solutions of unequal osmotic pressures, the more concentrated solution is said to be *hypertonic*, and the less concentrated solution is said to be *hypotonic* (Figure 5.13). To study the contents of red blood cells, which are protected from the outside environment by a semipermeable membrane, biochemists employ a technique called *hemolysis*. They place the red blood cells in a hypotonic solution, which causes water to move into the cell. The cells swell and eventually burst, releasing hemoglobin and other protein molecules. When a cell is placed in a hypertonic solution, on the other hand, the intracellular water tends to move out of the cell by osmosis to the more concentrated, surrounding solution. This process, known as *crenation*, causes the cell to shrink and eventually cease functioning.

The mammalian kidney is a particularly effective osmotic device. Its main function is to remove metabolic waste products and other impurities from the bloodstream to the urine outside through a semipermeable membrane. Biologically important ions (such as Na^+ and Cl^-) lost in this manner are then actively pumped back into the blood through the same membrane (see Section 5.10). The loss of water through the kidney is controlled by the antidiuretic hormone (ADH), which is secreted into the blood by the hypothalamus and posterior pituitary gland. When little or no ADH is secreted, large amounts of water (perhaps 10 times normal) pass into the urine each

Crenation helps to prolong the shelf life of jams when exposed to air.

day. On the other hand, when large quantities of ADH are present in the blood, the permeability of water through the membrane decreases so that the volume of urine formed may be as little as one-half the normal amount. Thus, the kidney–ADH combination controls the rate of loss of both water and other small waste molecules.

The chemical potential of water within the body fluids of freshwater fishes is lower than that in their environment, so they are able to draw in water by osmosis through their gill membranes. Surplus water is excreted as urine. An opposite process occurs for the marine teleost fishes. They lose body water to the more concentrated environment by osmosis across the gill membranes. To balance the loss, they drink seawater.

Osmotic pressure is also the major mechanism for water rising upward in plants. The leaves of trees constantly lose water to their surroundings, a process called *transpiration*, so the solute concentrations in leaf fluids increase. Water is then pushed up through the trunks and branches by osmotic pressure, which, to reach the tops of the tallest trees, can be as high as 10 to 15 atm.* Leaf movement is an interesting phenomenon that may also be related to osmotic pressure. The belief is that some processes can increase salt concentration in leaf cells in the presence of light. Osmotic pressure rises and cells become enlarged and turgid, causing the leaves to orient toward light.

Reverse Osmosis. A related phenomenon to osmosis is called *reverse osmosis*. If we apply pressure greater than the equilibrium osmotic pressure to the solution compartment shown in Figure 5.11, pure solvent will flow from the solution to the solvent compartment. This reversal of the osmotic process results in the unmixing of the solution components. An important application of reverse osmosis is the desalination of water. Several techniques discussed in this chapter are suitable, at least in principle, for obtaining pure water from the sea. For example, either distilling or freezing seawater would achieve the goal. However, these processes involve a phase change from liquid to vapor or liquid to solid and so would require considerable energy input to maintain. Reverse osmosis is more appealing, for it does not involve a phase change and is economically sound for large amounts of water.† Seawater, which is approximately 0.7 M in NaCl, has an estimated osmotic pressure of 30 atm. For a 50% recovery of pure water from the sea, an additional 60 atm would have to be applied on the seawater-side compartment to cause reverse osmosis. The success of large-scale desalination depends on the selection of a suitable membrane that is permeable to water but not to dissolved salts and that can withstand the high pressure over long periods of time.

5.7 Electrolyte Solutions

Having studied the general properties of nonelectrolyte solutions, we now turn our attention to electrolyte solutions. An electrolyte is a substance that, when dissolved in a solvent (usually water), produces a solution that will conduct electricity. An electrolyte can be an acid, a base, or a salt.

A Molecular View of the Electrolyte Solution Process

Why does NaCl dissolve in water and not in benzene? We know that NaCl is a stable compound in which the Na^+ and Cl^- ions are held together by electrostatic

---

* *See* "Entropy Makes Water Run Uphill—in Trees," P. E. Steveson, *J. Chem. Educ.* **48**, 837 (1971).

† *See* "Desalination of Water by Reverse Osmosis," C. E. Hecht, *J. Chem. Educ.* **44**, 53 (1967).

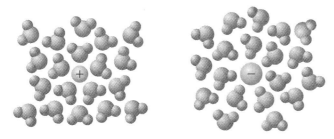

Figure 5.14
Hydration of a cation and an anion. In general, each cation and anion has a specific number of water molecules associated with it in the hydration sphere.

forces in the crystal lattice. In order for NaCl to enter the aqueous environment, the strong attractive forces must somehow be overcome. The dissolution of NaCl in water presents two questions: How do the ions interact with water molecules and how do they interact with one another?

Water is a good solvent for ionic compounds because it is a polar molecule and therefore can stabilize the ions through ion–dipole interaction that results in hydration. Generally, smaller ions can be hydrated more effectively than larger ions. A small ion contains a more concentrated charge, which leads to greater electrostatic interaction with the polar water molecules.* Figure 5.14 shows a schematic diagram of hydration. Because a different number of water molecules surrounds each type of ion, we speak of the *hydration number* of an ion. This number is directly proportional to the charge and inversely proportional to the size of the ion. Note that water in the "hydration sphere" and bulk water molecules have different properties,[†] which can be distinguished by spectroscopic techniques such as nuclear magnetic resonance. There is a dynamic equilibrium between the two types of molecules. Depending on the ion, the mean lifetime of a H_2O molecule in the hydration sphere can vary tremendously. For example, consider the mean lifetime of H_2O in the hydration sphere for the following ions: Br^-, 10^{-11} s; Na^+, 10^{-9} s; Cu^{2+}, 10^{-7} s; Fe^{3+}, 10^{-5} s; Al^{3+}, 7 s; and Cr^{3+}, 1.5×10^5 s, or 42 h.

The ion–dipole interaction (see Chapter 13) between dissolved ions and water molecules can affect several bulk properties of water. Small and/or multicharged ions such as Li^+, Na^+, Mg^{2+}, Al^{3+}, Er^{3+}, OH^-, and F^- are often called *structure-making ions*. The strong electric fields exerted by these ions can polarize water molecules, producing additional order beyond the first hydration layer. This interaction increases the solution's viscosity. On the other hand, large monovalent ions such as K^+, Rb^+, Cs^+, NH_4^+, Cl^-, NO_3^-, and ClO_4^- are *structure-breaking ions*. Because of their diffuse surface charges and hence weak electric fields, these ions are unable to polarize water molecules beyond the first layer of hydration. Consequently, the viscosities of solutions containing these ions are usually lower than that of pure water.

The effective radii of hydrated ions in solution can be appreciably greater than their crystal or ionic radii. For example, the radii of the hydrated Li^+, Na^+, and K^+ ions are estimated to be 3.66 Å, 2.80 Å, and 1.87 Å, respectively, although the ionic radii actually increase from Li^+ to K^+.

We now turn to the other question raised earlier: How do ions interact with one another? According to Coulomb's law (after the French physicist Charles Augustin

---

* According to electrostatic theory, the electric field at the surface of a charged sphere of radius r is proportional to ze/r^2, where z is the number of charges and e is the electronic charge.
[†] Water molecules in the hydration sphere of an ion do not exhibit individual translational motion. They move with the ion as a whole.

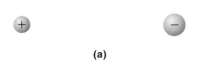

(a)

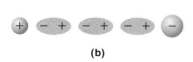

(b)

Figure 5.15
(a) Separation of a cation and an anion in vacuum. (b) Separation of the same ions in water. The alignment of polar water molecules is exaggerated. Because of thermal motion, the polar molecules are only partially aligned. Nevertheless, this arrangement reduces the electric field and hence the attractive force between the ions.

de Coulomb, 1736–1806), the force (F) between Na^+ and Cl^- ions in a vacuum is given by

$$F = \frac{q_{Na^+} q_{Cl^-}}{4\pi\varepsilon_0 r^2} \tag{5.46}$$

where ε_0 is the *permittivity of the vacuum* (8.854×10^{-12} C^2 N^{-1} m^{-2}), q_{Na^+} and q_{Cl^-} are the charges on the ions, and r is the distance of separation. The factor $4\pi\varepsilon_0$ is present as a result of using SI units so that F is expressed in newtons. In the polar medium of water, as Figure 5.15 shows, the dipolar molecules align themselves with their positive ends facing the negative charge and the negative ends facing the positive charge. This arrangement reduces the effective charge at the positive and negative charge centers by a factor of $1/\varepsilon$, where ε is the *dielectric constant* of the medium (see Appendix 5.1 on p. 182). Therefore, in any medium other than a vacuum, Equation 5.46 takes the form

$$F = \frac{q_{Na^+} q_{Cl^-}}{4\pi\varepsilon_0 \varepsilon r^2} \tag{5.47}$$

Table 5.3 lists the dielectric constants of several solvents. Keep in mind that ε always decreases with increasing temperature. For example, at 343 K, the dielectric constant of water is reduced to about 64. It is the large dielectric constant of water that reduces the attractive force between the Na^+ and Cl^- ions and allows them to separate in solution.

The dielectric constant of a solvent also determines the "structure" of ions in solution. To maintain electrical neutrality in solution, an anion must be near a cation, and vice versa. Depending on the proximity of these two ions, we can think of them either as "free" ions or as "ion pairs." Each free ion is surrounded by at least one and perhaps several layers of water molecules. In an ion pair, the cation and anion are close to each other, and few or no solvent molecules are between them. Generally, free ions and ion pairs are thermodynamically distinguishable species that have quite different chemical reactivities. For dilute 1:1 aqueous electrolyte solutions, such as

Table 5.3
Dielectric Constant of Some Pure Liquids at 298 K

Liquid	Dielectric Constant, ε^a
H_2SO_4	101
H_2O	78.54
$(CH_3)_2SO$ (dimethylsulfoxide)	49
$C_3H_8O_3$ (glycerol)	42.5
CH_3NO_2 (nitromethane)	38.6
$HOCH_2CH_2OH$ (ethylene glycol)	37.7
CH_3CN (acetonitrile)	36.2
CH_3OH	32.6
C_2H_5OH	24.3
CH_3COCH_3	20.7
CH_3COOH	6.2
C_6H_6	4.6
$C_2H_5OC_2H_5$	4.3
CS_2	2.6

^{a} The dielectric constant is a dimensionless quantity.

NaCl, ions are believed to be in the free-ion form. In higher-valence electrolytes, such as $CaCl_2$ and Na_2SO_4, the formation of ion pairs is indicated by conductance measurements, for a neutral ion pair cannot conduct electricity. Two opposing factors determine whether we have free ions or ion pairs in solution: the potential energy of attraction between the cation and anion, and the kinetic or thermal energy, of the order of $k_B T$, for individual ions.

We can now understand easily why NaCl does not dissolve in benzene. A nonpolar molecule, benzene does not solvate Na^+ and Cl^- ions effectively. In addition, its small dielectric constant means that the cations and anions will have little tendency to enter the solution as separate ions.

Example 5.7

Calculate the force in newtons between a pair of Na^+ and Cl^- ions separated by exactly 1 nm (10 Å) in (a) a vacuum and (b) water at 25°C. The charges on the Na^+ and Cl^- ions are 1.602×10^{-19} C and -1.602×10^{-19} C, respectively.

ANSWER

From Equations 5.46 and 5.47 and Table 5.3, we proceed as follows:

(a)
$$F = \frac{(1.602 \times 10^{-19}\ \text{C})(-1.602 \times 10^{-19}\ \text{C})}{4\pi(8.854 \times 10^{-12}\ \text{C}^2\ \text{N}^{-1}\ \text{m}^{-2})(1 \times 10^{-9}\ \text{m})^2}$$
$$= -2.31 \times 10^{-10}\ \text{N}$$

(b)
$$F = \frac{(1.602 \times 10^{-19}\ \text{C})(-1.602 \times 10^{-19}\ \text{C})}{4\pi(8.854 \times 10^{-12}\ \text{C}^2\ \text{N}^{-1}\ \text{m}^{-2})(78.54)(1 \times 10^{-9}\ \text{m})^2}$$
$$= -2.94 \times 10^{-12}\ \text{N}$$

COMMENT

As expected, the attractive force between the ions is reduced by a factor of about 80 from vacuum to the aqueous environment. The negative sign convention for F denotes attraction.

Thermodynamics of Ions in Solution

In this section, we shall briefly examine the thermodynamic parameters of the solution process involving ionic compounds and the thermodynamic functions of the formation of ions in aqueous solution.

The constant-pressure dissolution of NaCl can be represented by

The enthalpy change for process 1 corresponds to the energy required to separate the ions from the crystal lattice to an infinite distance. This energy is called the *lattice energy* (U_0). The enthalpy change for process 3 is the enthalpy of solution, $\Delta_{soln} H$, which is the heat absorbed or released when NaCl dissolves in a large amount of water. The heat of hydration, $\Delta_{\text{hydr}}H$, for process 2 is given by Hess's law:

$$\Delta_{\text{hydr}}H = \Delta_{\text{soln}}H - U_0$$

The quantity $\Delta_{\text{soln}}H$ is experimentally measurable; the value of U_0 can be estimated if the structure of the lattice is known. For NaCl, we have $U_0 = 787$ kJ mol^{-1} and $\Delta_{\text{soln}}H = 3.8$ kJ mol^{-1}, so that

$$\Delta_{\text{hydr}}H = 3.8 - 787 = -783 \text{ kJ mol}^{-1}$$

Thus, the hydration of Na$^+$ and Cl$^-$ ions by water releases a large amount of heat.

The enthalpy of hydration obtained above comes from both ions together. We often want to know the value of individual ions. In reality, we cannot study them separately, but their values can be obtained as follows. The hydration enthalpy for the process

$$H^+(g) \rightarrow H^+(aq)$$

has been reliably estimated by theoretical methods as 1089 kJ mol^{-1}. Using this value as a starting point, we can calculate the $\Delta_{\text{hydr}}H$ values for individual anions such as F$^-$, Cl$^-$, Br$^-$, and I$^-$ (from data on HF, HCl, HBr, and HI), and in turn obtain $\Delta_{\text{hydr}}H$ values for Li$^+$, Na$^+$, K$^+$, and other cations (from data on alkali metal halides). Table 5.4 lists the standard $\Delta_{\text{hydr}}H$ values for a number of ions. All the $\Delta_{\text{hydr}}H$ values are negative because the hydration of a gaseous ion is an exothermic process. Furthermore, there is a rough correlation between ionic charge/radius and hydration enthalpy. The values of $\Delta_{\text{hydr}}H$ are larger (more negative) for smaller ions than for large ions of the same charge. A smaller ion has a more concentrated charge and can interact more strongly with water molecules. Ions bearing higher charges also have larger $\Delta_{\text{hydr}}H$ values.

Table 5.4
Thermodynamic Values for the Hydration of Gaseous Ions at 298 K

Ion	$-\Delta_{\text{hydr}}H^\circ$ kJ·mol^{-1}	$-\Delta_{\text{hydr}}S^\circ$ J·K^{-1}·mol^{-1}	Ionic Radius/Å
H$^+$	1089[a]	132[a]	—
Li$^+$	520	119	0.60
Na$^+$	405	89	0.95
K$^+$	314	51	1.33
Ag$^+$	468	94	1.26
Mg^{2+}	1926	268	0.65
Ca^{2+}	1579	209	0.99
Ba^{2+}	1309	159	1.35
Mn^{2+}	1832	243	0.80
Fe^{2+}	1950	272	0.76
Cu^{2+}	2092	259	0.72
Fe^{3+}	4355	460	0.64
F$^-$	506	151	1.36
Cl$^-$	378	96	1.81
Br$^-$	348	80	1.95
I$^-$	308	60	2.16

[a] This is a theoretical estimate.

The other quantity of interest is the entropy of hydration, $\Delta_{hydr}S$. The hydration process results in considerable ordering of water molecules around each ion, which reduces various molecular motions and hence the number of microstates so that $\Delta_{hydr}S$ also is a negative quantity. As Table 5.4 shows, the variation in standard $\Delta_{hydr}S$ with ionic radius closely corresponds to that for $\Delta_{hydr}H$. Finally, note that there are two contributions to the entropy of solution, $\Delta_{soln}S$. The first is the hydration process, which results in a decrease in entropy. The other is the entropy gained when the solid breaks up into ions, which can move freely in solution. The sign of $\Delta_{soln}S$ depends on the magnitudes of these opposing factors.

Enthalpy, Entropy, and Gibbs Energy of Formation of Ions in Solution

Because ions cannot be studied separately, we cannot measure the standard molar enthalpy of formation, $\Delta_f\overline{H}^\circ$, of an individual ion. To get around this problem, we arbitrarily assign a zero value to the formation of the hydrogen ion—that is, $\Delta_f\overline{H}^\circ[H^+(aq)] = 0$—and then evaluate the $\Delta_f\overline{H}^\circ$ values of other ions relative to this scale. Consider the following reaction:

$$\tfrac{1}{2}H_2(g) + \tfrac{1}{2}Cl_2(g) \rightarrow H^+(aq) + Cl^-(aq) \qquad \Delta_rH^\circ = -167.2 \text{ kJ mol}^{-1}$$

The standard enthalpy of the reaction, which is an experimentally measurable quantity, can be expressed as

$$\Delta_rH^\circ = \Delta_f\overline{H}^\circ[H^+(aq)] + \Delta_f\overline{H}^\circ[Cl^-(aq)] - \left(\tfrac{1}{2}\right)(0) - \left(\tfrac{1}{2}\right)(0)$$

so that

$$\Delta_rH^\circ = \Delta_f\overline{H}^\circ[Cl^-(aq)]$$

or

$$\Delta_f\overline{H}^\circ[Cl^-(aq)] = -167.2 \text{ kJ mol}^{-1}$$

Once the value of $\Delta_f\overline{H}^\circ[Cl^-(aq)]$ has been determined, we can measure the $\Delta_r\overline{H}^\circ$ of the reaction

$$Na(s) + \tfrac{1}{2}Cl_2(g) \rightarrow Na^+(aq) + Cl^-(aq)$$

from which we can determine the value of $\Delta_f\overline{H}^\circ[Na^+(aq)]$ and so on.

Table 5.5 lists the $\Delta_f\overline{H}^\circ$ values of some common cations and anions. Two points are worth noting about this table. First, for aqueous solutions, the standard state at 298 K is a hypothetical state defined as the ideal solution of unit molality at 1 bar pressure, in which the activity of the solute (the ion) is unity. The ion thus has the properties it would possess in an infinitely dilute solution, in which interactions between the ions are negligible. Second, all the $\Delta_f\overline{H}^\circ$ values are *relative* values based on the $\Delta_f\overline{H}^\circ[H^+(aq)] = 0$ scale.

We can determine the standard molar Gibbs energy of formation of ions and standard molar entropy of ions at 298 K in a similar fashion; that is, by arbitrarily assigning zero values to $\Delta_f\overline{G}^\circ[H^+(aq)]$ and $\overline{S}^\circ[H^+(aq)]$. These values are also listed in Table 5.5. Because the entropy values of ions in aqueous solution are relative to that of the H^+ ion, they may be either positive or negative. For example, the entropy of $Ca^{2+}(aq)$ is -55.23 J K^{-1} mol^{-1}, and that of $NO_3^-(aq)$ is 146.4 J K^{-1} mol^{-1}. The magnitude and sign of these entropies are influenced by the extent to which they can order the water molecules around themselves in solution, compared with $H^+(aq)$.

Table 5.5
Thermodynamic Data for Aqueous Ions at 1 bar and 298 K

	$\Delta_f \overline{H}°/\text{kJ} \cdot \text{mol}^{-1}$	$\Delta_f \overline{G}°/\text{kJ} \cdot \text{mol}^{-1}$	$\overline{S}°/\text{J} \cdot \text{K}^{-1} \cdot \text{mol}^{-1}$
H^+	0	0	0
Li^+	−278.5	−293.8	14.23
Na^+	−239.7	−261.9	60.25
K^+	−251.2	−282.3	102.5
Mg^{2+}	−462.0	−456.0	−138.1
Ca^{2+}	−543.0	−553.0	−55.23
Fe^{2+}	−87.9	−84.9	−137.7
Zn^{2+}	−152.4	−147.2	−112.1
Fe^{3+}	−47.7	−4.7	−293.3
OH^-	−229.9	−157.3	−10.54
F^-	−329.1	−276.5	−13.8
Cl^-	−167.2	−131.2	56.5
Br^-	−120.9	−102.8	80.71
I^-	−55.9	−51.7	109.37
CO_3^{2-}	−676.3	−528.1	−53.14
NO_3^-	−206.6	−110.5	146.4
PO_4^{3-}	−1284.1	−1025.6	−217.6

Small, highly charged ions have negative entropy values, whereas large, singly charged ions have positive entropy values.

Example 5.8

Use the standard enthalpy of the reaction

$$Na(s) + \tfrac{1}{2}Cl_2(g) \rightarrow Na^+(aq) + Cl^-(aq) \quad \Delta_r H° = -406.9 \text{ kJ mol}^{-1}$$

to calculate the value of $\Delta_f \overline{H}°[Na^+(aq)]$, given that $\Delta_f \overline{H}°[Cl^-(aq)] = -167.2 \text{ kJ mol}^{-1}$.

ANSWER
The standard enthalpy of reaction is given by

$$\Delta_r H° = \Delta_f H°[Na^+(aq)] + \Delta_f H°[Cl^-(aq)] - (0) - \left(\tfrac{1}{2}\right)(0)$$
$$-406.9 \text{ kJ mol}^{-1} = \Delta_f \overline{H}°[Na^+(aq)] - 167.2 \text{ kJ mol}^{-1}$$

so

$$\Delta_f \overline{H}°[Na^+(aq)] = -239.7 \text{ kJ mol}^{-1}$$

5.8 Ionic Activity

Our next task is to learn to write chemical potentials of electrolytes in solution. First, we shall discuss ideal electrolyte solutions in which the concentrations are expressed on the molality scale.

For an ideal NaCl solution, the chemical potential, μ_{NaCl}, is given by

$$\mu_{\text{NaCl}} = \mu_{\text{Na}^+} + \mu_{\text{Cl}^-} \tag{5.48}$$

Because cations and anions cannot be studied individually, μ_{Na^+} and μ_{Cl^-} are not measurable. Nevertheless, we can express the chemical potentials of the cation and anion as

$$\mu_{\text{Na}^+} = \mu^{\circ}_{\text{Na}^+} + RT \ln m_{\text{Na}^+}$$

$$\mu_{\text{Cl}^-} = \mu^{\circ}_{\text{Cl}^-} + RT \ln m_{\text{Cl}^-}$$

Each m term is divided by m°, where $m^{\circ} = 1$ mol kg^{-1}, so the logarithmic term is dimensionless.

where $\mu^{\circ}_{\text{Na}^+}$ and $\mu^{\circ}_{\text{Cl}^-}$ are the standard chemical potentials of the ions. Equation 5.48 can now be written as

$$\mu_{\text{NaCl}} = \mu^{\circ}_{\text{NaCl}} + RT \ln m_{\text{Na}^+} m_{\text{Cl}^-}$$

where

$$\mu^{\circ}_{\text{NaCl}} = \mu^{\circ}_{\text{Na}^+} + \mu^{\circ}_{\text{Cl}^-}$$

In general, a salt with the formula $M_{\nu_+} X_{\nu_-}$ dissociates as follows:

$$M_{\nu_+} X_{\nu_-} \rightleftharpoons \nu_+ M^{z+} + \nu_- X^{z-}$$

where ν_+ and ν_- are the numbers of cations and anions per unit and z_+ and z_- are the numbers of charges on the cation and anion, respectively. For NaCl, $\nu_+ = \nu_- = 1$, $z_+ = +1$, and $z_- = -1$. For CaCl$_2$, $\nu_+ = 1$, $\nu_- = 2$, $z_+ = +2$, and $z_- = -1$. The chemical potential is given by

$$\mu = \nu_+ \mu_+ + \nu_- \mu_- \tag{5.49}$$

where

$$\mu_+ = \mu^{\circ}_+ + RT \ln m_+$$

and

$$\mu_- = \mu^{\circ}_- + RT \ln m_-$$

The molalities of the cation and anion are related to the molality of the salt orginally dissolved in solution, m, as follows:

$$m_+ = \nu_+ m \qquad m_- = \nu_- m$$

Substituting the expressions for μ_+ and μ_- into Equation 5.49 yields

$$\mu = (\nu_+ \mu^{\circ}_+ + \nu_- \mu^{\circ}_-) + RT \ln m_+^{\nu_+} m_-^{\nu_-} \tag{5.50}$$

We define *mean ionic molality* $(m_\pm)$ as a geometric mean (see Appendix 1) of the individual ionic molalities

$$m_\pm = (m_+^{\nu_+} m_-^{\nu_-})^{1/\nu} \tag{5.51}$$

where $v = v_+ + v_-$, and Equation 5.50 becomes

$$\mu = (v_+\mu_+^\circ + v_-\mu_-^\circ) + vRT \ln m_\pm \tag{5.52}$$

Mean ionic molality can also be expressed in terms of the molality of the solution, m. Because $m_+ = v_+m$ and $m_- = v_-m$, we have

$$m_\pm = [(v_+m)^{v_+}(v_-m)^{v_-}]^{1/v}$$
$$= m[(v_+^{v_+})(v_-^{v_-})]^{1/v} \tag{5.53}$$

Example 5.9

Write the expression for the chemical potential of $Mg_3(PO_4)_2$ in terms of the molality of the solution.

ANSWER

For $Mg_3(PO_4)_2$, we have $v_+ = 3$, $v_- = 2$, and $v = 5$. The mean ionic molality is

$$m_\pm = (m_+^3 m_-^2)^{1/5}$$

and the chemical potential is given by

$$\mu_{Mg_3(PO_4)_2} = \mu_{Mg_3(PO_4)_2}^\circ + 5RT \ln m_\pm$$

From Equation 5.51,

$$m_\pm = m(3^3 \times 2^2)^{1/5}$$
$$= 2.55\, m$$

so that

$$\mu_{Mg_3(PO_4)_2} = \mu_{Mg_3(PO_4)_2}^\circ + 5RT \ln 2.55\, m$$

Unlike nonelectrolyte solutions, most electrolyte solutions behave nonideally. The reason is as follows. The intermolecular forces between uncharged species generally depend on $1/r^7$, where r is the distance of separation; a 0.1-m nonelectrolyte solution is considered ideal for most practical purposes. But Coulomb's law has a $1/r^2$ dependence (Figure 5.16). This dependence means that even in quite dilute solutions (for example, 0.05 m), the electrostatic forces exerted by ions on one another are enough to cause a deviation from ideal behavior. Thus, in the vast majority of cases, we must replace molality with activity. By analogy to the mean ionic molality, we define the *mean ionic activity* ($a_\pm$) as

$$a_\pm = (a_+^{v_+} a_-^{v_-})^{1/v} \tag{5.54}$$

where a_+ and a_- are the activities of the cation and anion, respectively. The mean ionic activity and mean ionic molality are related by the *mean ionic activity coefficient*, $\gamma_\pm$; that is,

$$a_\pm = \gamma_\pm m_\pm \tag{5.55}$$

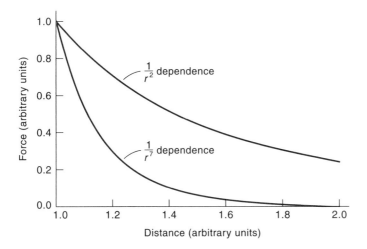

Figure 5.16
Comparison of dependence of attractive forces on distance r: electrostatic forces that exist between ions $(1/r^2)$ and van der Waals forces that exist between molecules $(1/r^7)$.

where

$$\gamma_\pm = (\gamma_+^{v_+} \gamma_-^{v_-})^{1/v} \tag{5.56}$$

The chemical potential of a nonideal electrolyte solution is given by

$$
\begin{aligned}
\mu &= (v_+\mu_+^\circ + v_-\mu_-^\circ) + vRT \ln a_\pm \\
&= (v_+\mu_+^\circ + v_-\mu_-^\circ) + RT \ln a_\pm^v \\
&= (v_+\mu_+^\circ + v_-\mu_-^\circ) + RT \ln a
\end{aligned}
\tag{5.57}
$$

where the activity of the electrolyte, a, is related to its mean ionic activity by

$$a = a_\pm^v$$

Experimental values of $\gamma_\pm$ can be obtained from freezing-point-depression and osmotic-pressure measurements* or electrochemical studies (see Chapter 7). Hence, the value of $a_\pm$ can be calculated from Equation 5.55. In the limiting case of infinite dilution $(m \to 0)$, we have

$$\lim_{m \to 0} \gamma_\pm = 1$$

Figure 5.17 shows the plots of $\gamma_\pm$ versus m for several electrolytes. At very low concentrations, $\gamma_\pm$ approaches unity for all types of electrolytes. As the concentrations of electrolytes increase, deviations from ideality occur. The variation of $\gamma_\pm$ with concentration for dilute solutions can be explained by the Debye–Hückel theory, to be discussed next.

---

*Interested readers should consult the standard physical chemistry texts listed in Chapter 1 (p. 6) for details of $\gamma_\pm$ measurements.

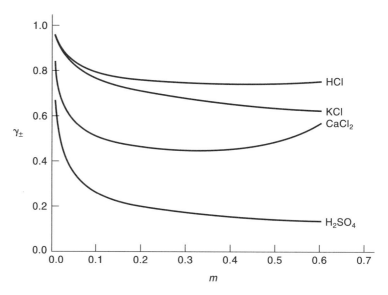

Figure 5.17
Plots of mean activity coefficient, $\gamma_{\pm}$, versus molality, m, for several electrolytes. At infinite dilution $(m \to 0)$, the mean activity coefficient approaches unity.

Example 5.10

Write expressions for the activities (a) of KCl, Na_2CrO_4, and $Al_2(SO_4)_3$ in terms of their molalities and mean ionic activity coefficients.

ANSWER

We need the relations $a = a_{\pm}^{\nu} = (\gamma_{\pm}m_{\pm})^{\nu}$.

KCl: $\nu = 1 + 1 = 2; \ m_{\pm} = (m^2)^{1/2} = m$

                 Therefore, $a_{KCl} = m^2\gamma_{\pm}^2$

Na_2CrO_4: $\nu = 2 + 1 = 3; \ m_{\pm} = [(2m)^2(m)]^{1/3} = 4^{1/3}m$

                 Therefore, $a_{Na_2CrO_4} = 4m^3\gamma_{\pm}^3$

$Al_2(SO_4)_3$: $\nu = 2 + 3 = 5; \ m_{\pm} = [(2m)^2(3m)^3]^{1/5} = 108^{1/5}m$

                 Therefore, $a_{Al_2(SO_4)_3} = 108m^5\gamma_{\pm}^5$

Debye–Hückel Theory of Electrolytes

Our treatment of deviations from ideality by electrolyte solutions has been empirical: Using the ionic activities obtained from the activity coefficient and the known concentration, we calculate chemical potential, the equilibrium constant, and other properties. Missing in this approach is a physical interpretation of ionic behavior in solution. In 1923, Debye and the German chemist Walter Karl Hückel (1895–1980) put forward a quantitative theory that has greatly advanced our knowledge of electrolyte solutions. Based on a rather simple model, the Debye–Hückel theory enables us to calculate the value of $\gamma_{\pm}$ from the properties of the solution.

The mathematical details of Debye's and Hückel's treatment are too complex to present here. (Interested readers should consult the standard physical chemistry

Figure 5.18
(a) Simplified presentation of an ionic atmosphere surrounding a cation in solution. (b) In a conductance measurement, the movement of a cation toward the cathode is retarded by the electric field exerted by the ionic atmosphere left behind.

texts listed in Chapter 1.) Instead, we shall discuss the underlying assumptions and final results. Debye and Hückel began by assuming the following: (1) electrolytes are completely dissociated into ions in solution; (2) the solutions are dilute, with a concentration of 0.01 m or lower; and (3) on average, each ion is surrounded by ions of opposite charge, forming an *ionic atmosphere* (Figure 5.18a). Working from these assumptions, Debye and Hückel calculated the average electric potential at each ion caused by the presence of other ions in the ionic atmosphere. The Gibbs energy of the ions was then related to the activity coefficient of the individual ion. Because neither γ_+ nor γ_- could be measured directly, the final result is expressed in terms of the mean ionic activity coefficient of the electrolyte as follows:

$$\log \gamma_{\pm} = -\frac{1.824 \times 10^6}{(\varepsilon T)^{3/2}} |z_+ z_-| \sqrt{I} \qquad (5.58)$$

where the | | signs denote the magnitude but not the signs of the product $z_+ z_-$. Thus for $CuSO_4$, we have $z_+ = 2$ and $z_- = -2$, but $|z_+ z_-| = 4$. The quantity I, called the *ionic strength*, is defined as follows:

$$I = \frac{1}{2} \sum_i m_i z_i^2 \qquad (5.59)$$

In Equation 5.58 and all subsequent equations involving mean ionic activity coefficients, the I term is divided by $m°$ so that the ionic strength becomes a dimensionless quantity.

where m_i and z_i are the molality and the charge of the ith ion in the electrolyte, respectively. This quantity was first introduced by the American chemist Gilbert Newton Lewis (1875–1946), who noted that nonideality observed in electrolyte solutions primarily stems from the *total* concentration of charges present rather than from the chemical nature of the individual ionic species. Equation 5.59 enables us to express the ionic concentrations for all types of electrolytes on a common basis so that we need not sort out the charges on the individual ions. Because most studies are carried out in water at 298 K (that is, $\varepsilon = 78.54$, and $T = 298$ K), Equation 5.58 becomes

$$\log \gamma_{\pm} = -0.509 |z_+ z_-| \sqrt{I} \qquad (5.60)$$

This equation is known as the *Debye–Hückel limiting law*, as is Equation 5.58.

Example 5.11

Calculate the mean activity coefficient $(\gamma_\pm)$ of a 0.010 m aqueous solution of $CuSO_4$ at 298 K.

ANSWER

The ionic strength of the solution is given by Equation 5.59:

$$I = \tfrac{1}{2}[(0.010 \ m) \times 2^2 + (0.010 \ m) \times (-2)^2]$$
$$= 0.040 \ m$$

From Equation 5.60,

$$\log \gamma_\pm = -0.509(2 \times 2)\sqrt{0.040}$$
$$= -0.407$$

or

$$\gamma_\pm = 0.392$$

Experimentally, $\gamma_\pm$ is found to be 0.41 at the same concentration.

Two points are worth noting in applying Equation 5.60. First, in a solution containing several electrolytes, *all* the ions in solution contribute to the ionic strength, but z_+ and z_- refer only to the ionic charges of the particular electrolyte for which $\gamma_\pm$ is being calculated. Second, Equation 5.60 can be used to calculate the ionic activity coefficient of individual cations or anions. Thus, for the ith ion, we write

$$\log \gamma_i = -0.509 z_i^2 \sqrt{I} \tag{5.61}$$

where z_i is the charge of the ion. The γ_+ and γ_- values calculated this way are related to $\gamma_\pm$ according to Equation 5.56.

Figure 5.19 shows calculated and measured values of $\log \gamma_\pm$ at various ionic strengths. We can see Equation 5.60 holds quite well for dilute solutions but must be modified to account for the drastic deviations that occur at high concentrations of electrolytes. Several improvements and modifications have been applied to this equation for treating more concentrated solutions.

The generally good agreement between experimentally determined $\gamma_\pm$ values and those calculated using the Debye–Hückel theory provides strong support for the existence of an ionic atmosphere in solution. The model can be tested by taking a conductance measurement in a very strong electric field. In reality, ions do not move in a straight line toward the electrodes in a conductance cell but move along a zigzag path. Microscopically, the solvent is not a continuous medium. Each ion actually "jumps" from one solvent hole to another, and as the ion moves across the solution, its ionic atmosphere is repeatedly being destroyed and formed again. The formation of an ionic atmosphere does not occur instantaneously but requires a finite amount of time, called the *relaxation time*, which is approximately 10^{-7} s in a 0.01-m solution. Under normal conditions of conductance measurement, the velocity of an ion is sufficiently slow so that the electrostatic force exerted by the atmosphere on the ion tends to retard its motion and hence to decrease the conductance (see Figure 5.18b).

The Debye–Hückel theory has been described as applicable to solutions so dilute that they are unkindly called slightly contaminated distilled water.

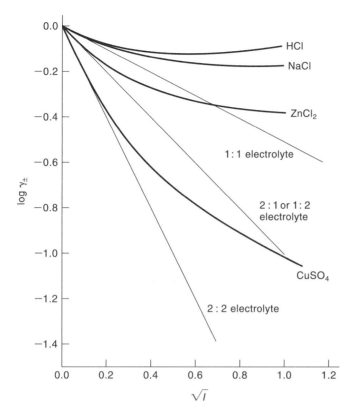

Figure 5.19
Plots of $\log \gamma_{\pm}$ versus $\sqrt{I}$ for several electrolytes. The straight lines are those predicted by Equation 5.60.

If the conductance measurement were carried out at a very strong electric field (approximately 2×10^5 V cm^{-1}), the ionic velocity would be approximately 10 cm s^{-1}. The radius of the ionic atmosphere in a 0.01-m solution is approximately 5 Å, or 5×10^{-8} cm, so that the time required for the ion to move out of the atmosphere is 5×10^{-8} cm/(10 cm s^{-1}), or 5×10^{-9} s, which is considerably shorter than the relaxation time. Consequently, the ion can move through the solution free of the retarding influence of the ionic atmosphere. The free movement leads to a marked increase in conductance. This phenomenon is called the *Wien effect*, after the German physicist Wilhelm Wien (1864–1928), who first performed the experiment in 1927. The Wien effect is one of the strongest pieces of evidence for the existence of an ionic atmosphere.

The Salting-In and Salting-Out Effects

The Debye–Hückel limiting law can be applied to study the solubility of proteins. The solubility of a protein in an aqueous solution depends on the temperature, pH, dielectric constant, ionic strength, and other characteristics of the medium. In this section, however, we shall focus on the influence of ionic strength.

Let us first investigate the effect of ionic strength on the solubility of an inorganic compound, AgCl. The solubility equilibrium is

$$AgCl(s) \rightleftharpoons Ag^+(aq) + Cl^-(aq)$$

The *thermodynamic* solubility product for the process, K_{sp}°, is given by

$$K_{sp}^{\circ} = a_{Ag^+} a_{Cl^-}$$

The ionic activities are related to ionic concentrations as follows:

$$a_+ = \gamma_+ m_+ \quad \text{and} \quad a_- = \gamma_- m_-$$

so that

$$K_{sp}^{\circ} = \gamma_{Ag^+} m_{Ag^+} \gamma_{Cl^-} m_{Cl^-}$$
$$= \gamma_{Ag^+} \gamma_{Cl^-} K_{sp}$$

where $K_{sp} = m_{Ag^+} m_{Cl^-}$ is the *apparent* solubility product. The difference between the thermodynamic and apparent solubility products is as follows. As we can see, the apparent solubility product is expressed in molalities (or some other concentration unit). We can readily calculate this quantity if we know the amount of AgCl dissolved in a known amount of water to produce a saturated solution. Because of electrostatic forces, however, the dissolved ions are under the influence of their immediate neighbors. Consequently, the actual or effective number of ions is not the same as that calculated from the concentration of the solution. For example, if a cation forms a tight ion pair with an anion, then the actual number of species in solution, from a thermodynamic perspective, is one and not, as we would expect, two. This is the reason for replacing concentration with activity, which is the effective concentration. Thus, the thermodynamic solubility product represents the true value of the solubility product, which generally differs from the apparent solubility product. Because

$$\gamma_{Ag^+} \gamma_{Cl^-} = \gamma_{\pm}^2$$

we write

$$K_{sp}^{\circ} = \gamma_{\pm}^2 K_{sp}$$

Taking the logarithm of both sides and rearranging, we obtain

$$-\log \gamma_{\pm} = \log \left(\frac{K_{sp}}{K_{sp}^{\circ}}\right)^{1/2} = 0.509|z_+ z_-|\sqrt{I}$$

The last equality in the above equation is the Debye–Hückel limiting law. The solubility product can be directly related to the solubility (S) itself; for a 1:1 electrolyte,

$$(K_{sp})^{1/2} = S \quad \text{and} \quad (K_{sp}^{\circ})^{1/2} = S^{\circ}$$

For dilute aqueous solutions, molality is approximately equal to molarity.

where S and S° are the apparent and thermodynamic solubilities in mol L^{-1}. Finally, we obtain the following equation relating the solubility of an electrolyte to the ionic strength of the solution:

$$\log \frac{S}{S^{\circ}} = 0.509|z_+ z_-|\sqrt{I} \tag{5.62}$$

Note that the value of S° can be determined by plotting $\log S$ versus $\sqrt{I}$. The intercept on the $\log S$ axis ($I = 0$) gives $\log S^{\circ}$, and hence S°.

If AgCl is dissolved in pure water, its solubility (S) is 1.3×10^{-5} mol L^{-1}. If it is dissolved in a KNO_3 solution, according to Equation 5.62, its solubility is greater because of the solution's increase in ionic strength. In a KNO_3 solution, the ionic strength is a sum of two concentrations, one from AgCl and the other from KNO_3. The increase in solubility, caused by the increase in ionic strength, is called the *salting-in effect*.

Equation 5.62 holds up only to a certain value of ionic strength. As the ionic strength of a solution increases further, it must be replaced by the following expression:

$$\log \frac{S}{S^\circ} = -K'I \tag{5.63}$$

where K' is a positive constant whose value depends on the nature of the solute and on the electrolyte present. The larger the solute molecule, the greater is the value of K'. Equation 5.63 tells us that the ratio of the solubilities in the region of high ionic strength actually decreases with I (note the negative sign). The decrease in solubility with increasing ionic strength of the solution is called the *salting-out effect*. This phenomenon can be explained in terms of hydration. Recall that hydration is the process that stabilizes ions in solution. At high salt concentrations, the availability of water molecules decreases, and so the solubility of ionic compounds also decreases. The salting-out effect is particularly noticeable with proteins, whose solubility in water is sensitive to ionic strength because of their large surface areas. Combining Equations 5.62 and 5.63, we have the approximate equation

$$\log \frac{S}{S^\circ} = 0.509|z_+z_-|\sqrt{I} - K'I \tag{5.64}$$

Equation 5.64 is applicable over a wider range of ionic strengths.

Figure 5.20 shows how the ionic strength of various inorganic salts affects the solubility of horse hemoglobin. As we can see, the protein exhibits a salting-in region

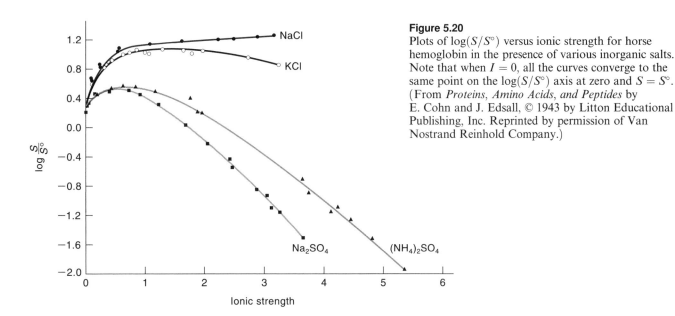

Figure 5.20
Plots of $\log(S/S^\circ)$ versus ionic strength for horse hemoglobin in the presence of various inorganic salts. Note that when $I = 0$, all the curves converge to the same point on the $\log(S/S^\circ)$ axis at zero and $S = S^\circ$. (From *Proteins, Amino Acids, and Peptides* by E. Cohn and J. Edsall, © 1943 by Litton Educational Publishing, Inc. Reprinted by permission of Van Nostrand Reinhold Company.)

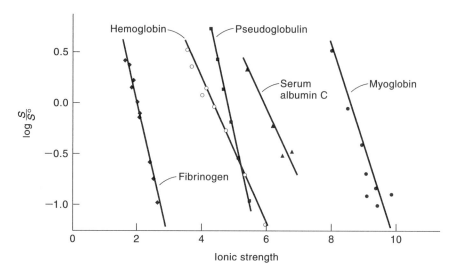

Figure 5.21
Salting-out phenomenon for several proteins in aqueous ammonium sulfate. [Reprinted with permission from E. J. Cohn, *Chem. Rev.* **19**, 241 (1936). Copyright 1936 American Chemical Society.]

at low ionic strengths.* As I increases, the curve goes through a maximum and eventually the slope becomes negative, indicating that the solubility decreases with increasing ionic strength. In this region, the second term in Equation 5.64 predominates. This trend is most pronounced for salts such as Na_2SO_4 and $(NH_4)_2SO_4$.

The practical value of the salting-out effect is that it enables us to precipitate proteins from solutions. In addition, the effect can also be used to purify proteins. Figure 5.21 shows the range of the salting-out phenomenon for several proteins in the presence of ammonium sulfate. Although the solubility of proteins is sensitive to the degree of hydration, the strength of binding of water molecules is not the same for all proteins. The relative solubility of different proteins at a particular ionic strength provides a means for selective precipitation. The point is that although higher ionic strengths are needed to salt out proteins, precipitation occurs over a small range of ionic strength, providing sharp separations.

5.9 Colligative Properties of Electrolyte Solutions

The colligative properties of an electrolyte solution are influenced by the number of ions present in solution. For example, we expect the aqueous freezing-point depression caused by a 0.01 m solution of NaCl to be twice that effected by a 0.01 m sucrose solution, assuming complete dissociation of the former. With incompletely dissociated salts or in situations in which ion pair formation occurs, the relationship is more complicated. Let us define a factor i, called the van't Hoff factor [after the Dutch chemist Jacobus Hendricus van't Hoff (1852–1911)], as follows:

$$i = \frac{\text{actual number of particles in solution at equilibrium}}{\text{number of particles in solution before dissociation}} \qquad (5.65)$$

---

* When I is less than unity, $\sqrt{I} > I$. Thus, at low ionic strengths, the first term in Equation 5.64 predominates.

If a solution contains N units of a weak electrolyte and if α is the degree of dissociation,

$$\begin{array}{cccc}
M_{v_+}X_{v_-} & \rightleftharpoons & v_+M^{z+} & + \ v_-X^{z-} \\
N(1-\alpha) & & Nv_+\alpha & Nv_-\alpha
\end{array}$$

there will be $N(1-\alpha)$ undissociated units and $(Nv_+\alpha + Nv_-\alpha)$, or $Nv\alpha$ ions in solution at equilibrium, where $v = v_+ + v_-$. We can now write the van't Hoff factor as

$$i = \frac{N(1-\alpha) + Nv\alpha}{N} = 1 - \alpha + v\alpha$$

and

$$\alpha = \frac{i-1}{v-1} \qquad (5.66)$$

For strong electrolytes, i is approximately equal to the number of ions formed from each unit of the electrolyte; for example, $i \approx 2$ for NaCl and $CuSO_4$; $i \approx 3$ for K_2SO_4 and $BaCl_2$, and so on. The value of i decreases with increasing concentration of the solution, which is attributed to the formation of ion pairs.

The presence of ion pairs will also affect the colligative properties because it decreases the number of free particles in solution. In general, formation of ion pairs is most pronounced between highly charged cations and anions and in media of low dielectric constants. In an aqueous solution of $Ca(NO_3)_2$, for example, the Ca^{2+} and NO_3^- ions form ion pairs as follows:

$$Ca^{2+}(aq) + NO_3^-(aq) \rightleftharpoons Ca(NO_3)^+(aq)$$

The equilibrium constants for such ion pairing are not accurately known, however, thus making calculations of colligative properties of electrolyte solutions difficult. A recent study* showed that deviation from colligative properties for many electrolyte solutions is not the result of ion pair formation, but of hydration. In an electrolyte solution the cations and anions can tie up large numbers of water molecules in their hydration spheres, thus reducing the number of free water molecules in the bulk solvent. Deviations disappear when the correct number of water molecules in the hydration sphere is subtracted from the total water solvent molecules in calculating the concentrations (molality or molarity) of the solution.

---

*A. A. Zavitsas, *J. Phys. Chem.* **105**, 7805 (2002).

Example 5.12

The osmotic pressures of a 0.01 m solution of a monoprotic acid HA and a 0.01 m sucrose solution at 298 K are 0.306 atm and 0.224 atm, respectively. Calculate the van't Hoff factor and the degree of dissociation for HA. Assume ideal behavior and no ion pair formation.

ANSWER

As far as osmotic pressure measurements are concerned, the main difference between HA and sucrose is that only HA can dissociate into ions (H^+ and A^-). Otherwise, equal concentrations of HA and sucrose solutions would have the same osmotic pressure. Because the osmotic pressure of a solution is directly proportional to the number of particles present, the ratio of the osmotic pressure for HA (with dissociation) to that of sucrose (no dissociation) must therefore be equal to the van't Hoff factor; that is,

$$i = \frac{0.306 \text{ atm}}{0.224 \text{ atm}} = 1.37$$

For HA, $v_+ = 1$ and $v_- = 1$, so $v = 2$. From Equation 5.66,

$$\alpha = \frac{1.37 - 1}{2 - 1} = 0.37$$

The acid is 37% dissociated.

Finally, the equations used to determine the colligative properties of non-electrolyte solutions (Equations 5.35, 5.36, and 5.43) must be modified for electrolyte solutions as follows:

$$\Delta T = K_b(im_2) \tag{5.67}$$

$$\Delta T = K_f(im_2) \tag{5.68}$$

$$\pi = \frac{RT(ic_2)}{\mathcal{M}_2} \tag{5.69}$$

We assume ideal behavior and no ion pair formation.

The Donnan Effect

The Donnan effect (after the British chemist Frederick George Donnan, 1870–1956) has its starting point in the treatment of osmotic pressure. It describes the uneven distribution at equilibrium of small diffusible ions on the two sides of a membrane that is freely permeable to these ions but impermeable to macromolecular ions, in the presence of a macromolecular electrolyte on one side of the membrane.

Suppose that a cell is separated into two parts by a semipermeable membrane that allows the diffusion of water and small ions but not protein molecules. Let us consider the following two cases.

Case 1. The protein solution is placed in the left compartment and water is placed in the right compartment. Proteins are ampholytes—that is, they possess both acidic and basic properties. Depending on the pH of the medium, a protein (P) can exist as an anion, a cation, or a neutral species. Let us assume that in this situation the protein is an anion bearing z-charges and Na^+ is the counterion. In solution it dissociates as follows:

$$Na_z^+ P^{z-} \rightarrow z Na^+ + P^{z-}$$

The osmotic pressure of the solution (π_1), which depends on the number of particles present, is given by

$$\pi_1 = (z+1)cRT \tag{5.70}$$

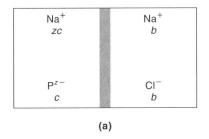

Figure 5.22
Schematic representation of the Donnan effect. (a) Before diffusion has begun. (b) At equilibrium. The membrane separating the left and right compartments is permeable to all but the P^{z-} ions.

where c is the concentration (molarity) of the protein solution. Because z is typically of the order of 30, using this arrangement to determine the molar mass of the protein yields a value that is only $\frac{1}{31}$ of the true value.

Case 2. Again the protein solution is placed in the left compartment, but a NaCl solution is placed in the right compartment (Figure 5.22). The requirement that the chemical potential of a component be the same throughout the system applies to the NaCl as well as to the water. To attain equilibrium, some of the NaCl will move from the right to the left compartment. We can calculate the actual amount of NaCl that is transported. The initial molar concentration of $Na_z^+ P^{z-}$ is c, and that of NaCl is b. At equilibrium, the concentrations are

$$[P^{z-}]^L = c \qquad [Na^+]^L = (zc + x) \qquad [Cl^-]^L = x$$

and

$$[Na^+]^R = (b - x) \qquad [Cl^-]^R = (b - x)$$

where L and R denote the left and right compartment, respectively, and x is the amount of NaCl transported from right to left.

At equilibrium $(\mu_{NaCl})^L = (\mu_{NaCl})^R$ so that

$$(\mu^\circ + RT \ln a_\pm)_{NaCl}^L = (\mu^\circ + RT \ln a_\pm)_{NaCl}^R$$

Because μ°, the standard chemical potential, is the same on both sides, we obtain

$$(a_\pm)_{NaCl}^L = (a_\pm)_{NaCl}^R$$

From Equation 5.54,

$$(a_{Na^+} a_{Cl^-})^L = (a_{Na^+} a_{Cl^-})^R$$

If the solutions are dilute, the ionic activities may be replaced by the corresponding concentrations; that is, $a_{Na^+} = [Na^+]$ and $a_{Cl^-} = [Cl^-]$. Hence,

$$([Na^+][Cl^-])^L = ([Na^+][Cl^-])^R$$

or

$$(zc + x)(x) = (b - x)(b - x)$$

Solving for x, we obtain

$$x = \frac{b^2}{zc + 2b} \tag{5.71}$$

The osmotic pressure (π_2), which is proportional to the difference in solute concentration between the two sides, is now given by

$$\pi_2 = \underbrace{[(c + zc + x + x)}_{\text{left compartment}} - \underbrace{(b - x + b - x)]RT}_{\text{right compartment}}$$

or

$$\pi_2 = (c + zc - 2b + 4x)RT$$

Substituting for x (Equation 5.71), we obtain

$$\pi_2 = \left(c + zc - 2b + \frac{4b^2}{zc + 2b}\right)RT$$

$$= \frac{zc^2 + 2cb + z^2c^2}{zc + 2b}RT \tag{5.72}$$

Equation (5.72) was derived assuming no change in either the pH or the volume of the solutions. Two limiting cases follow.

Case 1. If $b \ll zc$ (the salt concentration is much less than the protein concentration), then

$$\pi_2 = \frac{zc^2 + z^2c^2}{zc}RT = (zc + c)RT$$

$$= (z + 1)cRT$$

$$= \pi_1$$

Case 2. If $b \gg z^2c$ (the salt concentration is much greater than the protein concentration),* then

$$\pi_2 = \frac{2cb}{2b}RT = cRT$$

In this limiting case, the osmotic pressure approaches that of the solution in which the protein bears no net charge. In effect, the added salt reduces (and at high enough salt concentrations, eliminates) the Donnan effect. Under these conditions, the molar mass determined by osmotic pressure measurement would correspond closely to the true value.

An alternative approach to eliminate the Donnan effect is to choose a pH at which the protein has no net charge, called the *isoelectric point* (see Chapter 8). At this pH, the distribution of any type of diffusible ions will always be the same in both compartments. This method is difficult in practice because most proteins are least soluble at their isoelectric points.

---

*In practice, $c \leq 1 \times 10^{-4}$ M, $z \leq 30$, so that $z^2c \approx 0.1$ M. Thus, in order for this limiting case to hold, the concentration of the added salt should be about 1 M.

We have simplified the discussion of the Donnan effect by assuming no change in either pH or volume of the solution. Also, for the sake of simplicity, we have used a common singly charged ion, Na^+. Equations for more complicated situations can be derived using the same principles. The Donnan effect is essential to understanding the distribution of ions across the membranes of living organisms. A particularly important case is the distribution of bicarbonate and chloride ions between plasma and red blood cells, discussed in Section 8.8. In other cases, such as the nerve cells, the Donnan effect cannot be easily applied because of the active transport phenomenon in which ions are transported across the membrane against a concentration gradient (see p. 178).

5.10 Biological Membranes

In this section we shall consider the structure and function of biological membranes. In particular, we shall concentrate on the transport of ions across these membranes.

Cell membranes are composed of two kinds of molecules: lipids and proteins. Lipids, such as fats and waxes, are insoluble in water but soluble in many organic solvents. There are three types of membrane lipids: *phospholipids, cholesterol,* and *glycolipids*. We shall consider only the phospholipids here. One of the most common phospholipids found in cellular membranes is phosphatidic acid, shown in Figure 5.23. Membrane lipids are unique in that one end of the molecule contains a polar group that is hydrophilic ("water-loving"), and the other end is a long hydrocarbon chain that is hydrophobic ("water-fearing"). The lipids form a bilayer (about 60 Å in thickness), arranged so their polar groups constitute the top and bottom surfaces of the membrane and the nonpolar groups are buried in the interior region, which has a dielectric constant of about 3.

Figure 5.24 shows a widely accepted model of cell membrane structure called the *fluid mosaic model*. Protein molecules may lie at or near the inner or outer membrane surface, or they may partially or totally penetrate the membrane. The extent of interaction between the proteins and the lipids depends on the types of intermolecular forces and thermodynamic considerations. Generally, cell membranes have

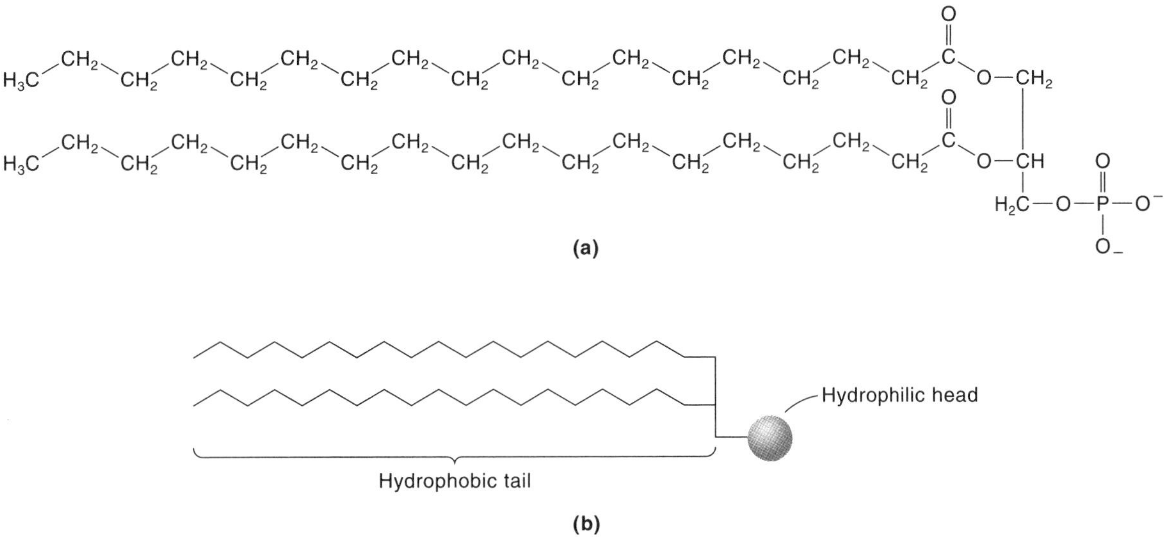

(a)

(b)

Figure 5.23
(a) Structure of phosphatidic acid (a phospholipid). (b) Simplified phosphatidic acid structure.

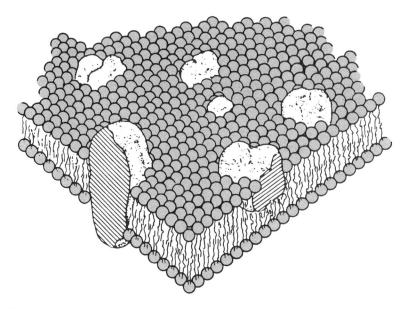

Figure 5.24
The fluid mosaic model of cell membrane structure. The large bodies are protein molecules that are embedded to varying degrees in the lipid bilayer. [From S. J. Singer and G. L. Nicolson, *Science* **175**, 723 (1972). Copyright 1972 by the American Association for the Advancement of Science.]

great physical strength and high electrical insulating properties. They are not rigid structures, however. On the contrary, many membrane proteins and lipids are constantly in motion (Figure 5.25). Fluorescence probes have shown that membrane proteins can normally diffuse laterally a distance of about several hundred angstroms in 1 minute. The phospholipid molecules, because of their smaller size, can diffuse much more quickly.

Membrane proteins have multiple functions: to act as receptors, to act as enzymes (for example, in the synthesis of important molecules, such as adenosine triphosphate (ATP) at the mitochondrial membrane, or in the initiation of photosynthesis at the chloroplast membrane), and to act as carriers for ions and other molecules across the membrane.

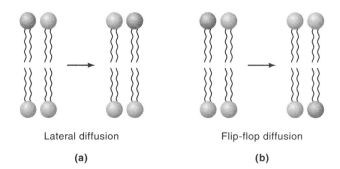

Lateral diffusion

(a)

Flip-flop diffusion

(b)

Figure 5.25
Lateral diffusion (a) of lipids is much more rapid than flip-flop diffusion (b). Membrane proteins can only undergo lateral diffusion.

Membrane Transport

One of the key functions of cell membranes is to act as a permeability barrier. In general, membranes selectively control the passage of substances from one region to another; for example, they permit nutrients to enter the interior of the cell and expel waste products from metabolism to the outside. Membrane transport, the movement of substances across membranes, relies on membrane permeability and membrane transport mechanisms.

Cell membranes are freely permeable to water, carbon dioxide, and oxygen, but much less so to ions, polar molecules, and other substances. Generally speaking, small molecules pass through the membrane more readily than larger ones and most membranes are impermeable to large molecules, such as proteins. Because the interior region of the lipid bilayer is hydrophobic, we would expect the membrane to be more permeable to uncharged species and nonpolar molecules than to ions and polar molecules. This is indeed the case, although water is a conspicuous exception. The membrane permeability of ions in general is very low. Their movement across the membrane must therefore be aided by special mechanisms to be discussed shortly.

In order for many essential biological processes to take place, the ionic and molecular composition of the internal aqueous phase of the cell must be appreciably different from its external environment. For example, the concentration of potassium ions is some 35 times higher in erythrocytes (red blood cells) than in the extracellular blood plasma. The reverse is true for sodium ions—the extracellular sodium ion concentration is about 15 times that of the intracellular fluid. We would expect that, given enough time, the concentrations of the same types of ions would eventually become equal in the intra- and extracellular fluids. That this is not the case means that processes other than normal diffusion must be sustaining the differences in ionic concentrations. We shall briefly consider the three types of membrane transport mechanisms.

Simple Diffusion. A number of substances pass through the membrane by simple diffusion; that is, they pass from one region to another containing a lower concentration of the substance. There may be discrete pores (thought to be formed by the membrane-bound proteins) in the membrane structure through which small molecules pass, thereby circumventing the hydrophobic lipid bilayer. This mechanism not only accounts for the rapid rate at which oxygen and carbon dioxide move through the membrane, but also explains why polar molecules such as water penetrate the membrane with ease. The pores can be envisioned as protein-lined channels that penetrate the lipid bilayer. Because of the fluidity of the membrane structure, these pores are continually being destroyed and created. The rate of molecular movement across the membrane is directly proportional to the concentration gradient across the membrane. It is known that there are water-specific membrane-channel proteins, called aquaporins, that allow the passage of water molecules at a high rate (about 10^9 molecules per second), which is essential for maintaining osmotic balance in cells. Computer simulation shows that water molecules diffuse through the channel in a highly cooperative fashion via hydrogen bonds with one another and with the atoms on the protein backbone that forms the channel wall. Ions (for example, H_3O^+) cannot pass through because they are repelled by the local electrostatic fields along the channel.

Facilitated Diffusion. Facilitated diffusion also involves the movement of molecules from a region of higher concentration to one having a lower concentration, but it differs from simple diffusion in that it is a carrier-assisted transport process. In facilitated diffusion, the substance to be transported is complexed with a carrier

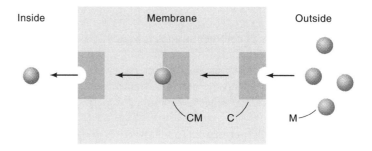

Figure 5.26
Representation of a facilitated diffusion. A metabolite (M) binds to a carrier molecule (C) at the outer face of the membrane. The complex (CM) diffuses through the membrane and dissociates at the inner surface, releasing M into the cell interior.

molecule (Figure 5.26). The carrier molecule is free to move back and forth across the membrane. At the outer face of the membrane (that is, the surface in contact with extracellular fluid), the carrier can bind to the substance, and the resulting complex diffuses across the membrane from a high- to a low-concentration region, where it dissociates into the carrier and the free substance. Each carrier has one or more specific binding sites for the substance it transports. After releasing its cargo, the carrier molecule diffuses back to the other side of the membrane and the process is repeated. Facilitated diffusion continues as long as there is a concentration gradient across the membrane for the transported substance. A different example of facilitated diffusion is the transport of glucose molecules into the erythrocyte. Glucose by itself is quite insoluble in the lipid bilayer, so the rate of simple diffusion of these molecules would be too slow to sustain metabolic processes. Instead, glucose enters the erthrocyte through a protein complex called glucose permease, which forms a hydrophilic path through the hydrophobic center of the membrane.

As mentioned, membrane transport by simple diffusion and facilitated diffusion are spontaneous processes. Consider the transfer of 1 mole of a solute X from a higher concentration region (α) to a lower one (β). From Equation 5.27 we write the chemical potential of X as

$$\mu_X = \mu_X^\circ + RT \ln[X]$$

where μ_X° is the standard chemical potential and we have used concentration instead of activity. The change in Gibbs energy in terms of the difference in chemical potentials is

$$\Delta G = (\mu_X)_\beta - (\mu_X)_\alpha = RT \ln[X]_\beta - RT \ln[X]_\alpha$$

$$= RT \ln \frac{[X]_\beta}{[X]_\alpha}$$

(Note that μ_X° is a constant and therefore cancels in the subtraction.) Because $[X]_\beta < [X]_\alpha$, $\Delta G < 0$ and the process will continue until the concentrations in the two regions become equal and $\Delta G = 0$.

Active Transport. Unlike simple diffusion and facilitated diffusion, active transport involves the movement of substances across the membrane *against* a concentration gradient. Thermodynamics tells us that this is a nonspontaneous process; therefore, energy must be supplied from an external source in order for active transport to occur. Earlier we mentioned the unequal concentrations of K^+ and Na^+ ions in the intra- and extracellular fluids. Maintaining these differences in ionic concentration requires active transport. Normal diffusion processes transport Na^+ ions from the exterior to the interior of the cell and move K^+ ions in the opposite direction. At the same time, Na^+ ions are constantly "pumped" out of the cell while K^+ ions are pumped in. The word *pump* is used to describe the movement of ions against concentration gradients. A steady state is reached when the flow of the ions of one type in one direction by active transport is balanced by the "leaking" (that is, diffusion) of the ions of the same type in the opposite direction.

The mechanism of active transport is fairly well understood. We know, for example, that the active transport of Na^+ and the active transport of K^+ ions across the membrane are linked together. Thus, the active transport system is often referred to as the *sodium–potassium pump*. As in facilitated diffusion, a carrier molecule is needed for active transport. Furthermore, the molecule must be such that an energy supply can be directly coupled to it. This carrier molecule is an enzyme called *sodium–potassium ATPase* (adenosinetriphosphatase). The energy provider is adenosine triphosphate (ATP), which hydrolyzes to adenosine diphosphate (ADP) and inorganic phosphate:

$$ATP^{4-} + H_2O \rightarrow ADP^{3-} + HPO_4^{2-} + H^+ \qquad \Delta_r G^{\circ} \approx -30 \text{ kJ mol}^{-1}$$

This spontaneous process (note the large decrease in the standard Gibbs energy) is coupled to sodium–potassium ATPase to drive the energetically unfavorable process of moving ions against a concentration gradient (Figure 5.27).

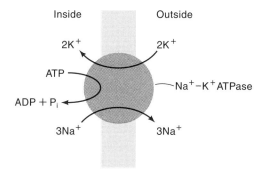

Figure 5.27
The Na^+–K^+ ATPase is primarily responsible for setting and maintaining the intracellular concentrations of Na^+ and K^+ ions, which it does by moving 3 Na^+ ions out of the cell for every 2 K^+ ions it moves in. Consequently, a potential is also established across the membrane.

We can estimate the increase in Gibbs energy for the process of transporting against a concentration gradient as follows. Suppose that 1 mole of a certain compound X is transported from the interior of the cell, where its concentration is 1.0×10^{-4} *M*, to the outside, where its concentration is 1.0×10^{-3} *M*. Because transport occurs against a concentration gradient, we have the following inequality in chemical potentials:

$$(\mu_X)_{ex} > (\mu_X)_{in}$$

where *in* and *ex* denote interior and exterior, respectively. At 37°C, the change in Gibbs energy for this process is given by

$$\Delta G = (\mu_X)_{ex} - (\mu_X)_{in} = RT \ln \frac{[X]_{ex}}{[X]_{in}}$$

$$= (8.314 \text{ J K}^{-1} \text{ mol}^{-1})(310 \text{ K}) \ln \frac{1.0 \times 10^{-3} \text{ } M}{1.0 \times 10^{-4} \text{ } M}$$

$$= 5.9 \text{ kJ mol}^{-1}$$

If the substance transported carries a charge—for example, X^{z+}, where X denotes the ion and $z+$ its charge—there is an additional effect, namely, the electrical potential established by the unequal concentrations of the ions across the membrane. Consequently, the Gibbs energy change is given by

$$\Delta G = RT \ln \frac{[X^{z+}]_{ex}}{[X^{z+}]_{in}} + zF\Delta V \tag{5.73}$$

where F is the faraday constant (96,500 C mol^{-1}) and ΔV is the difference in electrical potential (in volts) across the membrane. Consider the case in which Na$^+$ ions are being carried from the inside of the cell to the outside. To obtain an estimate for ΔG, let us use the following typical data: $[Na^+]_{ex}/[Na^+]_{in} = 10$, $\Delta V = 80$ mV or 8.0×10^{-2} V, and $T = 37$°C or 310 K. Thus,

$$\Delta G = (8.314 \text{ J K}^{-1} \text{ mol}^{-1})(310 \text{ K}) \ln 10 + (1)(96,500 \text{ C mol}^{-1})(8.0 \times 10^{-2} \text{ V})$$

$$= 13.7 \text{ kJ mol}^{-1}$$

where 1 J = 1 C × 1 V. In this case the active transport becomes more nonspontaneous by the positive electrical potential due to the Na$^+$ ions. As mentioned earlier, this nonspontaneous process is made possible by the hydrolysis of ATP.

In recent years, a great deal of effort has been devoted to understand the structure and function of ion channels—the protein-based gatekeepers governing the cellular influx and outflow of ions. Much of the knowledge gained has come from the X-ray studies of the membrane-bound proteins that form these channels. Consider, for example, the potassium channels. In animals, these channels govern nerve cell signaling, heart beating rhythm, and the release of insulin. Experiments show that the channels are formed by a complex made up of four protein subunits. One of the early mysteries about these channels was how they can favor the larger K$^+$ ions (ionic radius: 1.33 Å) over the smaller Na$^+$ ions (ionic radius: 0.98 Å) in a ratio of about 1000 to 1. When an ion enters such a channel, it is stripped of its hydrated sphere that normally surrounds it when dissolved in water. The size of K$^+$ ions fits nicely inside the oxygen rich tunnel structure. The oxygen atoms (from the amino acids) take the place of water molecules in the favorable ion–dipole interaction that exists in the hydration shell. The Na$^+$ ions, on the other hand, are too small for this interaction and generally remain outside the channel. Another area of intense study is the mechanism of *gating* of ion channels. Depending on the chemical and/or physical environment, some channels can open and close very quickly, on the order of microseconds. Certain channels open only in response to changes in voltage differences across the surrounding cell membranes. These so-called voltage-gated ion channels are of particular significance in nerve conduction (see Section 7.6).

Finally, we note that the ionic permeability across cell membranes can be altered by the naturally occurring antibiotics such as valinomycin and nonactin. These mol-

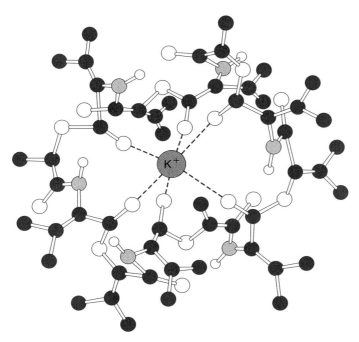

Figure 5.28
The valinomycin–K^+ complex has a nonpolar periphery that allows it to penetrate cell membranes. The color codes are: filled black circles, C; filled gray circles, N; open circles, O; small open circles, H.

ecules have a macrocyclic structure, with a cavity in the center and a hydrophobic exterior (Figure 5.28). They can form complexes with alkali metal ions (usually in a 1:1 ratio). Because of their external hydrophobic nature, the antibiotic–metal complexes can pass through cell membranes. Moreover, the complex formation is quite specific for individual ions, which must be of the appropriate size to fit into the cavity. For example, the formation constant for the valinomycin–K^+ complex is about 1000 times greater than that of the valinomycin–Na^+ complex. The increase in K^+ ion permeability disturbs the delicate intra- and extracellular concentration balance, and can kill the targeted bacterial cells. Gramicidin A, another type of natural antibiotic, can dimerize to form a transmembrane pore that is permeable to monovalent ions (Figure 5.29).

Compounds such as valinomycin and crown ethers that can affect metal ion transport are called *ionophores*, or ion carriers.

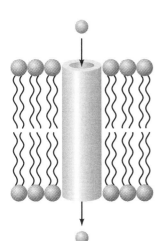

Figure 5.29
Schematic representation of an antibiotic with the ability to form a channel containing an aqueous pore that allows the transport of ions.

APPENDIX 5.1

Notes on Electrostatics

An electric charge (q_A) is said to produce an *electric field* (E) in the space around itself. This field exerts a force on any charge (q_B) within that space. According to Coulomb's law, the potential energy (V) between these two charges separated by distance r in a vacuum is given by

$$V = \frac{q_A q_B}{4\pi\varepsilon_0 r} \tag{1}$$

and the electrostatic force, F, between the charges is

$$F = \frac{q_A q_B}{4\pi\varepsilon_0 r^2} \tag{2}$$

where ε_0 is the permittivity of the vacuum (see p. 156). The electric field is the electrostatic force on a unit positive charge. Thus, the electric field at q_B due to q_A is F divided by q_B, or

$$E = \frac{q_A q_B}{4\pi\varepsilon_0 r^2 q_B} = \frac{q_A}{4\pi\varepsilon_0 r^2} \tag{3}$$

Note that E is a vector and is directed away from q_A toward q_B. Its units are V m^{-1} or V cm^{-1}.

Another important property of the electric field is its *electric potential*, ϕ, which is the potential energy of a unit positive charge in the electric field. Its units are J/C or V. A unit positive charge in an electric field E experiences a force equal in magnitude to E. When the charge is moved through a certain distance, dr, the potential energy change is equal to $qE\,dr$ or $E\,dr$ because $q = 1$ C. Because the repulsive potential energy increases as the unit positive charge comes closer to the positive charge q_A that generates the electric field, the change in potential energy, $d\phi$, is $-E\,dr$. (The negative sign ensures that as dr decreases, $-E\,dr$ is a positive quantity, signifying the increase in repulsion between the two positive charges.) The electric potential at a certain point at a distance r from the charge q_A is the potential energy change that occurs in bringing the unit positive charge from infinity to distance r from the charge:

$$\phi = -\int_{r=\infty}^{r=r} E\,dr = -\int_{r=\infty}^{r=r} \frac{q_A}{4\pi\varepsilon_0 r^2}\,dr = \frac{q_A}{4\pi\varepsilon_0 r} \tag{4}$$

Note that $\phi = 0$ at $r = \infty$. From Equation 4, we can define the electric potential difference between points 2 and 1 in an electric field as the work done in bringing a unit

charge from 1 to 2; that is,

$$\Delta\phi = \phi_2 - \phi_1 \tag{5}$$

This difference is commonly referred to as the voltage between points 1 and 2.

Dielectric Constant (ε) and Capacitance (C)

When a nonconducting substance (called a *dielectric*) is placed between two flat, parallel metal plates with opposite charges that are equal in magnitude (called a *capacitor*), the substance becomes polarized. The reason is that the electric field of the plates either orients the permanent dipoles of the dielectric or induces dipole moments, as shown in Figure 5.30. The *dielectric constant* of the substance is defined as

$$\varepsilon = \frac{E_0}{E} \tag{6}$$

where E_0 and E are the electric fields in the space between the plates of the capacitor in the absence of a dielectric (a vacuum) and presence of a dielectric, respectively. Keep in mind that the orientation of the dipoles (or induced dipoles) reduces the electric field between the capacitor plates so that $E < E_0$ and $\varepsilon > 1$. For ions in aqueous solution, this decrease in electric field reduces the attraction between the cation and the anion (see Figure 5.15).

The *capacitance* (C) of a capacitor measures its ability to hold charges for a given electric potential difference between the plates; that is, it is given by the ratio of charge to potential difference. The capacitances of a capacitor when the space is filled with a dielectric (C) and when there is a vacuum (C_0) between the plates are given by

$$C = \frac{Q}{\Delta\phi} \tag{7}$$

and

$$C_0 = \frac{Q}{\Delta\phi_0} \tag{8}$$

Figure 5.30
(a) The charge separation of a capacitor. The electric field E is directed from the positive plate to the negative plate, separated by distance d. With a vacuum between the plates, the dielectric constant is ε_0. (b) Orientation of the dipoles of a dielectric in a capacitor. The degree of orientation is exaggerated. The dielectric material decreases the electric field between the capacitor plates.

Because $\Delta\phi = Ed$, where d is the distance between the plates, Equation 6 can also be written as

$$\varepsilon = \frac{(\Delta\phi_0/d)}{(\Delta\phi/d)} = \frac{(Q/C_0)}{(Q/C)} = \frac{C}{C_0} \tag{9}$$

Capacitance is an experimentally measurable quantity so the dielectric constant of a substance can be determined. It has the unit farad (F), where $1\ F = 1\ C/1\ V$. Note that the ratio C/C_0 in Equation 9 makes ε a dimensionless quantity.

Membrane Capacitance

As stated in the chapter, ions can cross the membrane lipid bilayer only through ion channels. The unequal concentrations of ions in the cytoplasm and extracellular fluid produce a charge displacement, as shown in Figure 5.31. Because ions cannot cross

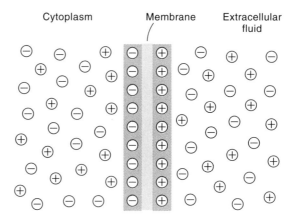

Figure 5.31
The cell membrane acts as a capacitor. The unequal concentrations of K^+ and Na^+ ions cause a charge separation as shown. At equilibrium, the electrostatic interaction holds the charges in a narrow region immediately adjacent to the two surfaces of the cell membrane. Note that only a very small fraction of the ions in solution take part in establishing the charge separation.

the lipid bilayer due to the bilayer's hydrophobic nature, they accumulate on the two surfaces of the membrane. The membrane thus stores electric charges in the same way charges are stored by a capacitor in an electric circuit.

Based on a lipid-layer thickness of about 5 nm and a dielectric constant of 3, which is roughly that of an 18-carbon fatty acid, the membrane capacitance has been calculated to be about 1 microfarad (1 μF) per square centimeter. This result is in good agreement with the experimentally determined value.

Suggestions for Further Reading

BOOKS

Aidley, D. J. and P. R. Stanfield, *Ion Channel*, Cambridge University Press, New York, 1996.

Gennis, R. B. *Biomembranes: Molecular Structure and Function*, Springer–Verlag, New York, 1989.

Harrison, R. and G. G. Lunt, *Biological Membranes*, Halsted Press, New York, 1976.

Hunt, J. P. *Metal Ions in Aqueous Solution*, W. A. Benjamin, Menlo Park, CA, 1963.

Jain, M. K. and R. C. Wagner, *Introduction to Biological Membranes*, Wiley–Interscience, New York, 1980.

Pass, G. *Ions in Solution*, Clarendon Press, Oxford, 1973.

Schultz, S. G. *Basic Principles of Membrane Transport*, Cambridge University Press, New York, 1980.

Tombs, M. P. and A. R. Peacocke, *The Osmotic Pressure of Biological Macromolecules*, Clarendon Press, New York, 1975.

ARTICLES

Nonelectrolyte and Electrolyte Solutions

"Ideal Solutions," W. A. Oates, *J. Chem. Educ.* **46**, 501 (1969).

"Ion Pairs and Complexes: Free Energies, Enthalpies, and Entropies," J. E. Prue, *J. Chem. Educ.* **46**, 12 (1969).

"Electrolyte Theory and SI Units," R. I. Holliday, *J. Chem. Educ.* **53**, 21 (1976).

"The Motion of Ions in Solution Under the Influence of an Electric Field," C. A. Vincent, *J. Chem. Educ.* **53**, 490 (1976).

"Thermodynamics of Ion Solvation and Its Significance in Various Systems," C. M. Criss and M. Salomon, *J. Chem. Educ.* **53**, 763 (1976).

"Ionic Hydration Enthalpies," D. W. Smith, *J. Chem. Educ.* **54**, 540 (1977).

"Standard States of Real Solutions," A. Lainez and G. Tardajos, *J. Chem. Educ.* **62**, 678 (1985).

"Thermodynamics of Mixing of Ideal Gases: A Persistent Pitfall," E. F. Meyer, *J. Chem. Educ.* **64**, 676 (1987).

"Paradox of the Activity Coefficient $\gamma_{\pm}$," E.-I. Ochiai, *J. Chem. Educ.* **67**, 489 (1990).

"Standard Enthalpies of Formation of Ions in Solution," T. Solomon, *J. Chem. Educ.* **68**, 41 (1991).

"Henry's Law: A Historical View," J. J. Carroll, *J. Chem. Educ.* **70**, 91 (1993).

"Understanding Chemical Potential," M. P. Tarazona and E. Saiz, *J. Chem. Educ.* **72**, 882 (1995).

"Determination of the Thermodynamic Solubility Product, K_{sp}°, of PbI$_2$ Assuming Nonideal Behavior," D. B. Green, G. Rechtsteiner, and A. Honodel, *J. Chem. Educ.* **73**, 789 (1996).

"The Definition and Unit of Ionic Strength," T. Solomon, *J. Chem. Educ.* **78**, 1691 (2001).

Phase Equilibria

"The Direct Relation Between Altitude and Boiling Point," L. Earl, *J. Chem. Educ.* **67**, 45 (1990).

"Phase Diagrams for Aqueous Systems," R. E. Treptow, *J. Chem. Educ.* **70**, 616 (1993).

"Journey Around a Phase Diagram," N. K. Kildahl, *J. Chem. Educ.* **71**, 1052 (1994).

Colligative Properties

"The Kidney," H. W. Smith, *Sci. Am.* January 1953.

"The Mechanism of Vapor Pressure Lowering," K. J. Mysels, *J. Chem. Educ.* **32**, 179 (1955).

"Osmotic Pressure," D. W. Kupke, *Adv. Protein Chem.* **15**, 57 (1960).

"Desalting Water by Freezing," A. E. Snyder, *Sci. Am.* December 1962.

"Deviations from Raoult's Law," M. L. McGlashan, *J. Chem. Educ.* **40**, 516 (1963).

"Desalination of Water by Reverse Osmosis," C. E. Hecht, *J. Chem. Educ.* **44**, 53 (1967).

"Demonstrating Osmotic and Hydrostatic Pressures in Blood Capillaries," J. W. Ledbetter, Jr., and H. D. Jones, *J. Chem. Educ.* **44**, 362 (1967).

"Reverse Osmosis," M. J. Suess, *J. Chem. Educ.* **48**, 190 (1971).

"Desalination," R. F. Probstein, *Am. Sci.* **61**, 280 (1973).

"Colligative Properties," F. Rioux, *J. Chem. Educ.* **50**, 490 (1973).

"Osmotic Pressure in the Physics Course for Students of the Life Sciences," R. K. Hobbie, *Am. J. Phys.* **42**, 188 (1974).

"Colligative Properties of a Solution," H. T. Hammel, *Science* **192**, 748 (1976).

"Reverse Osmosis," K. W. Boddeker, *Angew. Chem. Int. Ed.* **16**, 607 (1977).

"The Donnan Equilibrium and Osmotic Pressure," R. Chang and L. J. Kaplan, *J. Chem. Educ.* **54**, 218 (1977).

"Removal of an Assumption in Deriving the Phase Change Formula $\Delta T = Km$," F. E. Schubert, *J. Chem. Educ.* **56**, 259 (1979).

"Reappearing Phases," J. S. Walker and C. A. Vause, *Sci. Am.* May 1987.

"The Freezing Point Depression Law in Physical Chemistry," H. F. Franzen, *J. Chem. Educ.* **65**, 1077 (1988).

"Osmosis," B. Freeman in *Encyclopedia of Applied Physics*, G. L. Trigg, Ed., VCH Publishers, New York, 1995, Vol. 13, p. 59.

"Regulating Cell Volume," F. Lang and S. Waldegger, *Am. Sci.* **85**, 456 (1997).

"Transporting Water in Plants," M. J. Canny, *Am. Sci.* **86**, 152 (1998).

Biological Membranes

"The Structure of Cell Membranes," C. F. Fox, *Sci. Am.* February 1972.

"The Fluid Mosaic Model of the Structure of Cell Membranes," S. J. Singer and G. L. Nicolson, *Science* **175**, 720 (1972).

"Membrane Structure: Some General Principles," M. S. Bretscher, *Science* **181**, 622 (1973).

"A Dynamic Model of Cell Membranes," R. A. Capaldi, *Sci. Am.* March 1974.

"Crown Ethers," A. C. Knipe, *J. Chem. Educ.* **53**, 618 (1976).

"Ion Fluxes Through Membranes," M. E. Starzak, *J. Chem. Educ.* **54**, 200 (1977).

"The Assembly of Cell Membranes," H. F. Lodish and J. E. Rothman, *Sci. Am.* January 1979.

"Ion Channels in the Nerve Cell Membrane," R. D. Keynes, *Sci. Am.* March 1979.

"Lithium and Mania," D. C. Tosterson, *Sci. Am.* April 1981.

"The Structure of Proteins in Biological Membranes," N. Unwin and R. Henderson, *Sci. Am.* February 1984.

"The Molecules of the Cell Membrane," M. S. Bretscher, *Sci. Am.* October 1985.

"The Blood-Brain Barrier," G. W. Goldstein and A. L. Betz, *Sci. Am.* September 1986.

"Consequences of the Lipid Bilayer to Membrane-Associated Reactions," M. O. Eze, *J. Chem. Educ.* **67**, 17 (1990).

"Growth and Form of the Bacterial Cell Wall," A. L. Koch, *Am. Sci.* **78**, 326 (1990).

"Biological Membranes," S. D. Kohlwein, *J. Chem. Educ.* **69**, 3 (1992).

"How Cells Absorb Glucose," G. E. Lienhard, J. W. Slot, D. E. James, and M. M. Mueckler, *Sci. Am.* January 1992.

"The Patch Clamp Technique," E. Neher and B. Sakmann, *Sci. Am.* March 1992.

"Regulating Cell Volume," F. Lang and S. Waldegger, *Am. Sci.* **85**, 456 (1997).

"Building Doors Into Cells," H. Bayley, *Sci. Am.* September 1997.

"How Water Molecules Pass Through Aquaporins," T. Zeuthen, *Trends Biochem. Sci.* **26**, 77 (2001).

General

"The State of Water in Red Cells," A. K. Solomon, *Sci. Am.* February 1971.

"A Biological Antifreeze," R. E. Feeney, *Am. Sci.* **62**, 172 (1974).

"Antarctic Fishes," J. T. Eastman and A. C. DeVries, *Sci. Am.* November 1986.

"Cell Wounding and Healing," P. L. McNeil, *Am. Sci.* **79**, 222 (1991).

"G Proteins," M. E. Linder and A. G. Gilman, *Sci. Am.* July 1992.

"The Thermodynamics of Home-Made Ice Cream," D. L. Gibbon, K. Kennedy, N. Reading, and M. Quierox, *J. Chem. Educ.* **69**, 658 (1992).

Problems

Concentration Units

5.1 How many grams of water must be added to 20.0 g of urea to prepare a 5.00% aqueous urea solution by weight?

5.2 What is the molarity of a 2.12 mol kg^{-1} aqueous sulfuric acid solution? The density of this solution is 1.30 g cm^{-3}.

5.3 Calculate the molality of a 1.50 M aqueous ethanol solution. The density of the solution is 0.980 g cm^{-3}.

5.4 The concentrated sulfuric acid we use in the laboratory is 98.0% sulfuric acid by weight. Calculate the molality and molarity of concentrated sulfuric acid if the density of the solution is 1.83 g cm^{-3}.

5.5 Convert a 0.25 mol kg^{-1} sucrose solution into percent by weight. The density of the solution is 1.2 g cm^{-3}.

5.6 For dilute aqueous solutions in which the density of the solution is roughly equal to that of the pure solvent, the molarity of the solution is equal to its molality. Show that this statement is correct for a 0.010 M aqueous urea $[(NH_2)_2CO]$ solution.

5.7 The blood sugar (glucose) level of a diabetic patient is approximately 0.140 g of glucose/100 mL of blood. Every time the patient ingests 40 g of glucose, her blood glucose level rises to approximately 0.240 g/100 mL of blood. Calculate the number of moles of glucose per milliliter of blood and the total number of moles and grams of glucose in the blood before and after consumption of glucose. (Assume that the total volume of blood in her body is 5.0 L.)

5.8 The strength of alcoholic beverages is usually described in terms of "proof," which is defined as twice the percentage by volume of ethanol. Calculate the number of grams of alcohol in 2 quarts of 75-proof gin. What is the molality of the gin? (The density of ethanol is 0.80 g cm^{-3}; 1 quart = 0.946 L.)

Thermodynamics of Mixing

5.9 Liquids A and B form a nonideal solution. Provide a molecular interpretation for each of the following situations: $\Delta_{mix}H > 0$, $\Delta_{mix}H < 0$, $\Delta_{mix}V > 0$, $\Delta_{mix}V < 0$.

5.10 Calculate the changes in entropy for the following processes: **(a)** mixing of 1 mole of nitrogen and 1 mole of oxygen, and **(b)** mixing of 2 moles of argon, 1 mole of helium, and 3 moles of hydrogen. Both **(a)** and **(b)** are carried out under conditions of constant temperature (298 K) and constant pressure. Assume ideal behavior.

5.11 At 25°C and 1 atm pressure, the absolute third-law entropies of methane and ethane are 186.19 J K^{-1} mol^{-1} and 229.49 J K^{-1} mol^{-1}, respectively, in the gas phase. Calculate the absolute third-law entropy of a "solution" containing 1 mole of each gas. Assume ideal behavior.

Chemical Potential

5.12 Which of the following has a higher chemical potential? If neither, answer "same." **(a)** $H_2O(s)$ or $H_2O(l)$ at water's normal melting point, **(b)** $H_2O(s)$ at −5°C and 1 bar or $H_2O(l)$ at −5°C and 1 bar, **(c)** benzene at 25°C and 1 bar or benzene in a 0.1 M toluene solution in benzene at 25°C and 1 bar.

5.13 A solution of ethanol and n-propanol behaves ideally. Calculate the chemical potential of ethanol in solution relative to that of pure ethanol when its mole fraction is 0.40 at its boiling point (78.3°C).

5.14 Write the phase equilibrium conditions for a liquid solution of methanol and water in equilibrium with its vapor.

Henry's Law

5.15 Prove the statement that an alternative way to express Henry's law of gas solubility is to say that the volume of gas that dissolves in a fixed volume of solution is independent of pressure at a given temperature.

5.16 A miner working 900 ft below the surface had a soft drink beverage during the lunch break. To his surprise, the drink seemed very flat (that is, not much effervescence was observed upon removing the cap). Shortly after lunch, he took the elevator up to the surface. During the trip up, he felt a great urge to belch. Explain.

5.17 The Henry's law constant of oxygen in water at 25°C is 773 atm mol^{-1} kg of water. Calculate the molality of oxygen in water under a partial pressure of 0.20 atm. Assuming that the solubility of oxygen in blood at 37°C is roughly the same as that in water at 25°C, comment on the prospect for our survival without hemoglobin molecules. (The total volume of blood in the human body is about 5 L.)

5.18 The solubility of N_2 in blood at 37°C and a partial pressure of 0.80 atm is 5.6×10^{-4} mol L^{-1}. A deep-sea diver breathes compressed air with a partial pressure of N_2 equal to 4.0 atm. Assume that the total volume of blood in the body is 5.0 L. Calculate the amount of N_2 gas released (in liters) when the diver returns to the surface of water, where the partial pressure of N_2 is 0.80 atm.

Colligative Properties

5.19 List the important assumptions in the derivation of Equation 5.35.

5.20 Liquids A (bp = T_A°) and B (bp = T_B°) form an ideal solution. Predict the range of boiling points of solutions formed by mixing different amounts of A and B.

5.21 A mixture of ethanol and n-propanol behaves ideally at 36.4°C. **(a)** Determine graphically the mole fraction of n-propanol in a mixture of ethanol and n-propanol that boils at 36.4°C and 72 mmHg. **(b)** What is the total vapor pressure over the mixture at 36.4°C when the mole fraction of n-propanol is 0.60? **(c)** Calculate the composition of the vapor in **(b)**. (The equilibrium vapor pressures of ethanol and n-propanol at 36.4°C are 108 mmHg and 40.0 mmHg, respectively.)

5.22 Two beakers, 1 and 2, containing 50 mL of 0.10 M urea and 50 mL of 0.20 M urea, respectively, are placed under a tightly sealed bell jar at 298 K. Calculate the mole fraction of urea in the solutions at equilibrium. Assume ideal behavior. (*Hint:* Use Raoult's law and note that at equilibrium, the mole fraction of urea is the same in both solutions.)

5.23 At 298 K, the vapor pressure of pure water is 23.76 mmHg and that of seawater is 22.98 mmHg. Assuming that seawater contains only NaCl, estimate its concentration. (*Hint:* Sodium chloride is a strong electrolyte.)

5.24 Trees in cold climates may be subjected to temperatures as low as −60°C. Estimate the concentration of an aqueous solution in the body of the tree that would remain unfrozen at this temperature. Is this a reasonable concentration? Comment on your result.

5.25 Explain why jams can be stored under atmospheric conditions for long periods of time without spoilage.

5.26 Provide a molecular interpretation for the positive and negative deviations in the boiling-point curves.

5.27 The freezing-point-depression measurement of benzoic acid in acetone yields a molar mass of 122 g; the same measurement in benzene gives a value of 242 g. Account for this discrepancy. (*Hint:* Consider solvent–solute and solute–solute interactions.)

5.28 A common antifreeze for car radiators is ethylene glycol, $CH_2(OH)CH_2(OH)$. How many milliliters of this substance would you add to 6.5 L of water in the radiator if the coldest day in winter is −20°C? Would you keep this substance in the radiator in the summer to prevent the water from boiling? (The density and boiling point of ethylene glycol are 1.11 g cm^{-3} and 470 K, respectively.)

5.29 For intravenous injections, great care is taken to ensure that the concentration of solutions to be injected is comparable to that of blood plasma. Why?

5.30 The tallest trees known are the redwoods in California. Assuming the height of a redwood to be 105 m (about 350 ft), estimate the osmotic pressure required to push water up from the roots to the treetop.

5.31 A mixture of liquids A and B exhibits ideal behavior. At 84°C, the total vapor pressure of a solution containing 1.2 moles of A and 2.3 moles of B is 331 mmHg. Upon the addition of another mole of B to the solution, the vapor pressure increases to 347 mmHg. Calculate the vapor pressures of pure A and B at 84°C.

5.32 Fish breathe the dissolved air in water through their gills. Assuming the partial pressures of oxygen and nitrogen in air to be 0.20 atm and 0.80 atm, respectively, calculate the mole fractions of oxygen and nitrogen in water at 298 K. Comment on your results.

5.33 Liquids A (molar mass 100 g mol^{-1}) and B (molar mass 110 g mol^{-1}) form an ideal solution. At 55°C, A has a vapor pressure of 95 mmHg and B a vapor pressure of 42 mmHg. A solution is prepared by mixing equal weights of A and B. **(a)** Calculate the mole fraction of each component in the solution. **(b)** Calculate the partial pressures of A and B over the solution at 55°C. **(c)** Suppose that some of the vapor described in **(b)** is condensed to a liquid. Calculate the mole fraction of each component in this liquid and the vapor pressure of each component above this liquid at 55°C.

5.34 Lysozyme extracted from chicken egg white has a molar mass of 13,930 g mol^{-1}. Exactly 0.1 g of this protein is dissolved in 50 g of water at 298 K. Calculate the vapor pressure lowering, the depression in freezing point, the elevation of boiling point, and the osmotic pressure of this solution. The vapor pressure of pure water at 298 K is 23.76 mmHg.

5.35 The following argument is frequently used to explain the fact that the vapor pressure of the solvent is lower over a solution than over the pure solvent and that lowering is proportional to the concentration. A dynamic equilibrium exists in both cases, so that the rate at which molecules of solvent evaporate from the liquid is always equal to that at which they condense. The rate of condensation is proportional to the partial pressure of the vapor, whereas that of evaporation is unimpaired in the pure solvent but is impaired by solute molecules in the surface of the solution. Hence the rate of escape is reduced in proportion to the concentration of the solute, and maintenance of equilibrium requires a corresponding lowering of the rate of condensation and therefore of the partial pressure of the vapor phase. Explain why this argument is incorrect. [*Source:* K. J. Mysels, *J. Chem. Educ.* **32**, 179 (1955).]

5.36 A compound weighing 0.458 g is dissolved in 30.0 g of acetic acid. The freezing point of the solution is found to be 1.50 K below that of the pure solvent. Calculate the molar mass of the compound.

5.37 Two aqueous urea solutions have osmotic pressures of 2.4 atm and 4.6 atm, respectively, at a certain temperature. What is the osmotic pressure of a solution prepared by mixing equal volumes of these two solutions at the same temperature?

5.38 A forensic chemist is given a white powder for analysis. She dissolves 0.50 g of the substance in 8.0 g of benzene. The solution freezes at 3.9°C. Can the chemist conclude that the compound is cocaine ($C_{17}H_{21}NO_4$)? What assumptions are made in the analysis? The freezing point of benzene is 5.5°C.

5.39 "Time-release" drugs have the advantage of releasing the drug to the body at a constant rate so that the drug concentration at any time is not high enough to have harmful side effects or so low as to be ineffective. A schematic diagram of a pill that works on this basis is shown below. Explain how it works.

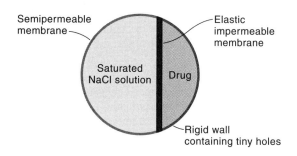

5.40 A nonvolatile organic compound, Z, was used to make up two solutions. Solution A contains 5.00 g of Z dissolved in 100 g of water, and solution B contains 2.31 g of Z dissolved in 100 g of benzene. Solution A has a vapor pressure of 754.5 mmHg at the normal boiling point of water, and solution B has the same vapor pressure at the normal boiling point of benzene. Calculate the molar mass of Z in solutions A and B, and account for the difference.

5.41 Acetic acid is a polar molecule that can form hydrogen bonds with water molecules. Therefore, it has a high solubility in water. Yet acetic acid is also soluble in benzene (C_6H_6), a nonpolar solvent that lacks the ability to form hydrogen bonds. A solution of 3.8 g of CH_3COOH in 80 g C_6H_6 has a freezing point of 3.5°C. Calculate the molar mass of the solute, and suggest what its structure might be. (*Hint:* Acetic acid molecules can form hydrogen bonds among themselves.)

5.42 At 85°C, the vapor pressure of A is 566 torr and that of B is 250 torr. Calculate the composition of a mixture of A and B that boils at 85°C when the pressure is 0.60 atm. Also, calculate the composition of the vapor mixture. Assume ideal behavior.

5.43 Comment on whether each of the following statements is true or false, and briefly explain your answer: **(a)** If one component of a solution obeys Raoult's law, then the other component must also obey the same law. **(b)** Intermolecular forces are small in ideal solutions. **(c)** When 15.0 mL of an aqueous 3.0 *M* ethanol solution is mixed with 55.0 mL of an aqueous 3.0 M ethanol solution, the total volume is 70.0 mL.

5.44 Liquids A and B form an ideal solution at a certain temperature. The vapor pressures of pure A and B are 450 torr and 732 torr, respectively, at this temperature. **(a)** A sample of the solution's vapor is condensed. Given that the original solution contains 3.3 moles of A and 8.7 moles of B, calculate the composition of the condensate in mole fractions. **(b)** Suggest a method for measuring the partial pressures of A and B at equilibrium.

5.45 Nonideal solutions are the result of unequal intermolecular forces between components. Based on this knowledge, comment on whether a racemic mixture of a liquid compound would behave as an ideal solution.

5.46 Calculate the molal boiling-point elevation constant (K_b) for water. The molar enthalpy of vaporization of water is 40.79 kJ mol^{-1} at 100°C.

5.47 Explain the following phenomena. **(a)** A cucumber placed in concentrated brine (saltwater) shrivels into a pickle. **(b)** A carrot placed in fresh water swells in volume.

5.48 The following data give the pressures for carbon disulfide–acetone solutions at 35.2°C. Calculate the activity coefficients of both components based on deviations from Raoult's law and Henry's law. (*Hint:* First determine Henry's law constants graphically.)

x_{CS_2}	0	0.20	0.45	0.67	0.83	1.00
P_{CS_2}/torr	0	272	390	438	465	512
$P_{C_3H_6O}$/torr	344	291	250	217	180	0

5.49 A solution is made up by dissolving 73 g of glucose ($C_6H_{12}O_6$; molar mass 180.2 g) in 966 g of water. Calculate the activity coefficient of glucose in this solution if the solution freezes at −0.66°C.

5.50 A certain dilute solution has an osmotic pressure of 12.2 atm at 20°C. Calculate the difference between the chemical potential of the solvent in the solution and that of pure water. Assume that the density is the same as that of water. (*Hint:* Express the chemical potential in terms of mole fraction, x_1, and rewrite the osmotic pressure equation as $\pi V = n_2 RT$, where n_2 is the number of moles of the solute and $V = 1$ L.)

5.51 At 45°C, the vapor pressure of water for a glucose solution in which the mole fraction of glucose is 0.080 is 65.76 mmHg. Calculate the activity and activity coefficient of the water in the solution. The vapor pressure of pure water at 45°C is 71.88 mmHg.

5.52 Consider a binary liquid mixture A and B, where A is volatile and B is nonvolatile. The composition of the solution in terms of mole fraction is $x_A = 0.045$ and $x_B = 0.955$. The vapor pressure of A from the mixture is 5.60 mmHg, and that of pure A is 196.4 mmHg at the same temperature. Calculate the activity coefficient of A at this concentration.

Ionic Activity and Debye–Hückel Limiting Law

5.53 Express the mean activity, mean activity coefficient, and mean molality in terms of the individual ionic quantities ($a_+, a_-, \gamma_+, \gamma_-, m_+,$ and m_-) for the following electrolytes: KI, SrSO$_4$, CaCl$_2$, Li$_2$CO$_3$, K$_3$Fe(CN)$_6$, and K$_4$Fe(CN)$_6$.

5.54 Calculate the ionic strength and the mean activity coefficient for the following solutions at 298 K: **(a)** 0.10 m NaCl, **(b)** 0.010 m MgCl$_2$, and **(c)** 0.10 m K$_4$Fe(CN)$_6$.

5.55 The mean activity coefficient of a 0.010 m H$_2$SO$_4$ solution is 0.544. What is its mean ionic activity?

5.56 A 0.20 m Mg(NO$_3$)$_2$ solution has a mean ionic activity coefficient of 0.13 at 25°C. Calculate the mean molality, the mean ionic activity, and the activity of the compound.

5.57 The Debye–Hückel limiting law is more reliable for 1:1 electrolytes than for 2:2 electrolytes. Explain.

5.58 In theory, the size of the ionic atmosphere is $1/\kappa$, called the Debye radius, and κ is given by

$$\kappa = \left(\frac{e^2 N_A}{\varepsilon_0 \varepsilon k_B T} \right)^{1/2} \sqrt{I}$$

where e is the electronic charge, N_A Avogadro's constant, ε_0 the permittivity of vacuum (8.854 × 10^{-12} C^2 N^{-1} m^{-2}), ε the dielectric constant of the solvent, k_B the Boltzmann constant, T the absolute temperature, and I the ionic strength (see the physical chemistry texts listed in Chapter 1). Calculate the Debye radius in a 0.010 m aqueous Na$_2$SO$_4$ solution at 25°C.

5.59 Explain why it is preferable to take the geometric mean rather than the arithmetic mean when defining mean activity, mean molality, and mean activity coefficient.

5.60 Calculate the ionic strength of a 0.0020 m aqueous solution of MgCl$_2$ at 298 K. Use the Debye–Hückel limiting law to estimate **(a)** the activity coefficients of the Mg^{2+} and Cl$^-$ ions in this solution and **(b)** the mean ionic activity coefficients of these ions.

Additional Problems

5.61 Calculate the change in the Gibbs energy at 37°C when the human kidneys secrete 0.275 mole of urea per kilogram of water from blood plasma to urine if the concentrations of urea in blood plasma and urine are 0.005 mol kg^{-1} and 0.326 mol kg^{-1}, respectively.

5.62 **(a)** Which of the following expressions is incorrect as a representation of the partial molar volume of component A in a two-component solution? Why? How would you correct it?

$$\left(\frac{\partial V_m}{\partial n_A} \right)_{T,P,n_B} \qquad \left(\frac{\partial V_m}{\partial x_A} \right)_{T,P,x_B}$$

(b) Given that the molar volume of this mixture (V_m) is given by

$$V_m = 0.34 + 3.6x_A x_B + 0.4x_B(1 - x_A) \text{ L mol}^{-1}$$

derive an expression for the partial molar volume for A at $x_A = 0.20$.

5.63 The partial molar volumes for a benzene–carbon tetrachloride solution at 25°C at a mole fraction of 0.5 are $\bar{V}_b = 0.106$ L mol^{-1} and $\bar{V}_c = 0.100$ L mol^{-1}, respectively, where the subscripts b and c denote C_6H_6 and CCl_4. **(a)** What is the volume of a solution made up of one mole of each? **(b)** Given that the molar volumes are $C_6H_6 = 0.089$ L mol^{-1} and $CCl_4 = 0.097$ L mol^{-1}, what is the change in volume on mixing 1 mole each of C_6H_6 and CCl_4? **(c)** What can you deduce about the nature of intermolecular forces between C_6H_6 and CCl_4?

5.64 The osmotic pressure of poly(methyl methacrylate) in toluene has been measured at a series of concentrations at 298 K. Determine graphically the molar mass of the polymer.

π/atm	8.40×10^{-4}	1.72×10^{-3}	2.52×10^{-3}
c/g · L^{-1}	8.10	12.31	15.00

π/atm	3.23×10^{-3}	7.75×10^{-3}
c/g · L^{-1}	18.17	28.05

5.65 Benzene and toluene form an ideal solution. Prove that to achieve the maximum entropy of mixing, the mole fraction of each component must be 0.5.

5.66 Suppose 2.6 moles of He at 0.80 atm and 25°C are mixed with 4.1 moles of Ne at 2.7 atm and 25°C. Calculate the Gibbs energy change for the process. Assume ideal behavior.

5.67 Two beakers are placed in a closed container. Beaker A initially contains 0.15 mole of naphthalene $(C_{10}H_8)$ in 100 g of benzene (C_6H_6) and beaker B initially contains 31 g of an unknown compound dissolved in 100 g of benzene. At equilibrium, beaker A is found to have lost 7.0 g. Assuming ideal behavior, calculate the molar mass of the unknown compound. State any assumptions made.

5.68 From the following data, calculate the heat of solution for KI:

	NaCl	NaI	KCl	KI
Lattice energy/kJ · mol^{-1}	787	700	716	643
Heat of solution/kJ · mol^{-1}	3.8	−5.1	17.1	?

5.69 The concentrations of K^+ and Na^+ ions in the intracellular fluid of a nerve cell are approximately 400 mM and 50 mM, respectively, but in the extracellular fluid the K^+ and Na^+ concentrations are 20 mM and

440 mM, respectively. Given that the electric potential inside the cell is −70 mV relative to the outside, calculate the Gibbs energy change for the transfer of 1 mole of each type of ion against the concentration gradient at 37°C.

5.70 In this chapter (see Figures 5.12, 5.17, and 5.19), we extrapolated concentration-dependent values to zero solute concentration. Explain what these extrapolated values mean physically and why they differ from the value obtained for the pure solvent.

5.71 **(a)** The root cells of plants contain a solution that is hypertonic in relation to water in the soil. Thus, water can move into the roots by osmosis. Explain why salts (NaCl and CaCl$_2$) spread on roads to melt ice can be harmful to nearby trees. **(b)** Just before urine leaves the human body, the collecting ducts in the kidney (which contain the urine) pass through a fluid whose salt concentration is considerably greater than is found in the blood and tissues. Explain how this action helps conserve water in the body.

5.72 A very long pipe is capped at one end with a semi-permeable membrane. How deep (in meters) must the pipe be immersed into the sea for fresh water to begin passing through the membrane? Assume seawater is at 20°C and treat it as a 0.70 M NaCl solution. The density of seawater is 1.03 g cm^{-3}.

5.73 **(a)** Using the Debye–Hückel limiting law, calculate the value of $\gamma_\pm$ for a 2.0×10^{-3} m Na$_3$PO$_4$ solution at 25°C. **(b)** Calculate the values of γ_+ and γ_- for the Na$_3$PO$_4$ solution, and show that they give the same value for $\gamma_\pm$ as that obtained in **(a)**.

5.74 Calculate the solubility of BaSO$_4$ (in g L^{-1}) in **(a)** water and **(b)** a 6.5×10^{-5} M MgSO$_4$ solution. The solubility product of BaSO$_4$ is 1.1×10^{-10}. Assume ideal behavior.

5.75 The thermodynamic solubility product of AgCl is 1.6×10^{-10}. What is [Ag$^+$] in **(a)** a 0.020 M KNO$_3$ solution and **(b)** a 0.020 M KCl solution?

5.76 Oxalic acid, (COOH)$_2$, is a poisonous compound present in many plants and vegetables, including spinach. Calcium oxalate is only slightly soluble in water $(K_{sp} = 3.0 \times 10^{-9}$ at 25°C) and its ingestion can result in kidney stones. Calculate **(a)** the apparent and thermodynamic solubility of calcium oxalate in water, and **(b)** the concentrations of calcium and oxalate ions in a 0.010 M Ca(NO$_3$)$_2$ solution. Assume ideal behavior in **(b)**.

5.77 The freezing-point depression of a 0.010 m acetic acid solution is 0.0193 K. Calculate the degree of dissociation for acetic acid at this concentration.

5.78 A 0.010 m aqueous solution of the ionic compound Co(NH$_3$)$_5$Cl$_3$ has a freezing-point depression of 0.0558 K. What can you conclude about its structure? Assume the compound is a strong electrolyte.

5.79 The osmotic pressure of blood plasma is approximately 7.5 atm at 37°C. Estimate the total concentration of dissolved species and the freezing point of blood plasma.

5.80 Referring to Figure 5.22, calculate the osmotic pressure for the following cases at 298 K: **(a)** The left compartment contains 200 g of hemoglobin in 1 liter of solution; the right compartment contains pure water. **(b)** The left compartment contains the same hemoglobin solution as in part **(a)**; the right compartment initially contains 6.0 g of NaCl in 1 liter of solution. Assume that the pH of the solution is such that the hemoglobin molecules are in the Na^+Hb^- form. (The molar mass of hemoglobin is 65,000 g mol^{-1}.)

5.81 The antibiotic Gramicidin A can transport Na^+ ions into a certain cell at the rate of 5.0×10^7 Na^+ ions $channel^{-1}$ s^{-1}. Calculate the time in seconds to transport enough Na^+ ions to increase its concentration by 8.0×10^{-3} M in a cell whose intracellular volume is 2.0×10^{-10} mL.

5.82 The concentration of glucose inside a cell is 0.12 mM and that outside a cell is 12.3 mM. Calculate the Gibbs energy change for the transport of 3 moles of glucose into the cell at 37°C.

5.83 Referring to Figure 5.22, suppose we have a protein solution ($Na_{30}P$) in the left compartment at 0.0010 M concentration. Calculate the osmotic pressure of the solution if the concentration of the NaCl solution in the right compartment is at 0 M, 0.10 M, 2.0 M, and 10 M. The temperature is at 298 K.

5.84 Use Equations 5.34 and 5.37 to calculate K_b and K_f for water shown in Table 5.2.

5.85 Describe how you would experimentally distinguish between two types of cells containing and lacking aquaporins for transporting water.

CHAPTER 6

Chemical Equilibrium

Turning from our discussion of nonelectrolyte and electrolyte solutions and physical equilibria in Chapter 5, we shall focus in this chapter on chemical equilibrium in gaseous and condensed phases. Equilibrium is a state in which there are no observable changes over time; at equilibrium, the concentrations of reactants and products in a chemical reaction remain constant. Much activity occurs at the molecular level, however, because reactant molecules continue to form product molecules while product molecules react to yield reactant molecules. This process is an example of dynamic equilibrium. The laws of thermodynamics help us predict equilibrium composition under different reaction conditions. We shall also apply thermodynamic principles to biological processes.

6.1 Chemical Equilibrium in Gaseous Systems

In this section, we shall derive an expression relating the Gibbs energy change for a reaction in the gas phase to the concentrations of the reacting species and temperature. We first consider the case in which all gases exhibit ideal behavior.

Ideal Gases

Examples of the simplest type of chemical equilibrium,

$$A(g) \rightleftharpoons B(g)$$

are *cis–trans* isomerization, racemization, and the cyclopropane-ring-opening reaction to form propene. The progress of the reaction can be monitored by the quantity ξ (Greek letter xi), called the *extent of reaction*. When an infinitesimal amount of A is converted to B, the change in A is $dn_A = -d\xi$ and that in B is $dn_B = +d\xi$, where dn denotes the change in number of moles. The change in Gibbs energy for this transformation at constant T and P is given by

$$dG = \mu_A \, dn_A + \mu_B \, dn_B$$
$$= -\mu_A \, d\xi + \mu_B \, d\xi$$
$$= (\mu_B - \mu_A)d\xi \tag{6.1}$$

where μ_B and μ_A are the chemical potentials of A and B, respectively. Equation 6.1

193

can be rearranged to give

$$\left(\frac{\partial G}{\partial \xi}\right)_{T,P} = \mu_B - \mu_A \tag{6.2}$$

For simplicity, we use the units kJ mol^{-1} rather than kJ (mol reaction)$^{-1}$.

The quantity $(\partial G/\partial \xi)_{T,P}$ is represented by $\Delta_r G$, which is the change in Gibbs energy per mole of reaction; it has the units kJ mol^{-1}.

During the reaction, the chemical potentials vary with composition. The reaction proceeds in the direction of decreasing G; that is, $(\partial G/\partial \xi)_{T,P} < 0$. Therefore, the forward reaction (A $\rightarrow$ B) is spontaneous when $\mu_A > \mu_B$, whereas the reverse reaction (B $\rightarrow$ A) is spontaneous when $\mu_B > \mu_A$. At equilibrium, $\mu_A = \mu_B$ so that

$$\left(\frac{\partial G}{\partial \xi}\right)_{T,P} = 0$$

Figure 6.1 shows a plot of Gibbs energy versus extent of reaction. At constant T and P, we have

$$\Delta_r G < 0 \quad \text{Forward reaction is spontaneous}$$

$$\Delta_r G > 0 \quad \text{Reverse reaction is spontaneous}$$

$$\Delta_r G = 0 \quad \text{Reacting system is at equilibrium}$$

Let us now consider a more complicated case:

$$a\mathrm{A}(g) \rightleftharpoons b\mathrm{B}(g)$$

where a and b are stoichiometric coefficients. According to Equation 5.10, the chemical potential of the ith component in a mixture, assuming ideal behavior, is given by

$$\mu_i = \mu_i^{\circ} + RT \ln \frac{P_i}{P^{\circ}}$$

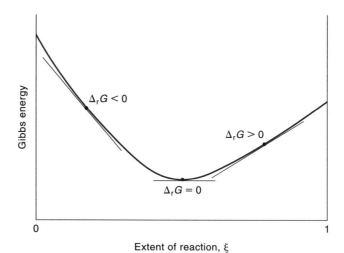

Figure 6.1
A plot of Gibbs energy versus extent of reaction. For a reacting system at equilibrium, the slope of the curve is zero.

where P_i is the partial pressure of component i in the mixture, μ_i° is the standard chemical potential of component i, and $P^\circ = 1$ bar. Therefore, we can write

$$\mu_A = \mu_A^\circ + RT \ln \frac{P_A}{P^\circ} \tag{6.3a}$$

$$\mu_B = \mu_B^\circ + RT \ln \frac{P_B}{P^\circ} \tag{6.3b}$$

The Gibbs energy change for the reaction, $\Delta_r G$, can be expressed as

$$\Delta_r G = b\mu_B - a\mu_A \tag{6.4}$$

Substituting the expressions in Equation 6.3 into Equation 6.4, we get

$$\Delta_r G = b\mu_B^\circ - a\mu_A^\circ + bRT \ln \frac{P_B}{P^\circ} - aRT \ln \frac{P_A}{P^\circ} \tag{6.5}$$

The standard Gibbs energy change of the reaction, $\Delta_r G^\circ$, is just the difference between the standard Gibbs energies of products and reactants; that is,

$$\Delta_r G^\circ = b\mu_B^\circ - a\mu_A^\circ$$

Therefore, we can write Equation 6.5 as

$$\Delta_r G = \Delta_r G^\circ + RT \ln \frac{(P_B/P^\circ)^b}{(P_A/P^\circ)^a} \tag{6.6}$$

By definition, $\Delta_r G = 0$ at equilibrium, so Equation 6.6 becomes

$$0 = \Delta_r G^\circ + RT \ln \frac{(P_B/P^\circ)^b}{(P_A/P^\circ)^a}$$

$$0 = \Delta_r G^\circ + RT \ln K_P$$

or

$$\Delta_r G^\circ = -RT \ln K_P \tag{6.7}$$

K_P, the equilibrium constant (where the subscript P denotes that concentrations are expressed in partial pressures), is given by

$$K_P = \frac{(P_B/P^\circ)^b}{(P_A/P^\circ)^a} = \frac{P_B^b}{P_A^a}(P^\circ)^{a-b} \tag{6.8}$$

Equation 6.7 is one of the most important and useful equations in chemical thermodynamics. It relates the equilibrium constant, K_P, and the standard Gibbs energy change of the reaction, $\Delta_r G^\circ$, in a remarkably simple fashion. Keep in mind that at a given temperature, $\Delta_r G^\circ$ is a constant whose value depends only on the nature of the reactants and products. In our example, $\Delta_r G^\circ$ is the standard Gibbs energy change when reactant A at 1 bar pressure and temperature T is converted to product B at 1 bar pressure and the same temperature per mole of reaction, as shown above. Figure 6.2 shows the Gibbs energy versus the extent of reaction for $\Delta_r G^\circ < 0$. When there is no mixing of reactants with products, the Gibbs energy decreases linearly as the

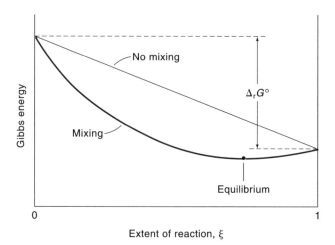

Figure 6.2
Total Gibbs energy versus extent of reaction for the $aA(g) \rightleftharpoons bB(g)$ reaction assuming $\Delta_r G° < 0$. At equilibrium, product is favored over reactant. Note that the equilibrium point, which is at the minimum of Gibbs energy, is a compromise between $\Delta_r G°$ and the Gibbs energy of mixing.

reaction progresses, and eventually the reactants will be completely converted to products. As Equation 5.11 shows, however, $\Delta_{mix} G$ is a negative quantity; therefore, the Gibbs energy for the actual path will be lower than that for the nonmixing case. Consequently, the equilibrium point, which is at the *minimum* of the Gibbs energy, is a compromise between these two opposing tendencies; that is, between the conversion of reactants to products and the mixing of products with reactants.

Equation 6.7 tells us that if we know the value of $\Delta_r G°$, we can calculate the equilibrium constant K_P, and vice versa. The standard Gibbs energy of a reaction is just the difference between the standard Gibbs energies of formation ($\Delta_f G°$) of the products and the reactants, discussed in Section 4.5. Thus, once we have defined a reaction, we can usually calculate the equilibrium constant from the $\Delta_f G°$ values listed in Appendix 2 and Equation 6.7. Note that these values all refer to 298 K. Later in this chapter (Section 6.4), we shall learn how to calculate the value of K_P at another temperature if its value at 298 K is known.

Finally, note that the equilibrium constant is a function of temperature alone (because $\mu°$ depends only on temperature) and is dimensionless. This follows from the fact that in the expression for K_P, each pressure term is divided by its standard-state value of 1 bar, which cancels the pressure unit but does not alter P numerically.

Example 6.1

From the thermodynamic data listed in Appendix 2, calculate the equilibrium constant for this reaction at 298 K:

$$N_2(g) + 3H_2(g) \rightleftharpoons 2NH_3(g)$$

ANSWER

The equilibrium constant for the equation is given by

$$K_P = \frac{(P_{NH_3}/P°)^2}{(P_{N_2}/P°)(P_{H_2}/P°)^3}$$

To calculate the value of K_P, we need Equation 6.7 and the value of $\Delta_r G^\circ$. From Equation 4.28 and Appendix 2, we have

$$\Delta_r G^\circ = 2\Delta_f \bar{G}^\circ(\text{NH}_3) - \Delta_f \bar{G}^\circ(\text{N}_2) - 3\Delta_f \bar{G}^\circ(\text{H}_2)$$
$$= (2)(-16.6 \text{ kJ mol}^{-1}) - (0) - (3)(0)$$
$$= -33.2 \text{ kJ mol}^{-1}$$

From Equation 6.7,

$$-33{,}200 \text{ J mol}^{-1} = -(8.314 \text{ J K}^{-1} \text{ mol}^{-1})(298 \text{ K}) \ln K_P$$
$$\ln K_P = 13.4$$

or

$$K_P = 6.6 \times 10^5$$

COMMENT

Note that if the reaction were written as

$$\tfrac{1}{2}\text{N}_2(g) + \tfrac{3}{2}\text{H}_2(g) \rightleftharpoons \text{NH}_3(g)$$

the value of $\Delta_r G^\circ$ would be $-16.6 \text{ kJ mol}^{-1}$, and the equilibrium constant would be calculated as follows:

$$K_P = \frac{(P_{\text{NH}_3}/P^\circ)}{(P_{\text{N}_2}/P^\circ)^{1/2}(P_{\text{H}_2}/P^\circ)^{3/2}} = 8.1 \times 10^2$$

Thus, whenever we multiply a balanced equation throughout by a factor n, we change the equilibrium constant K_P to K_P^n. Here $n = \tfrac{1}{2}$, so we have changed K_P to $K_P^{1/2}$.

A Closer Look at Equation 6.7

The change in standard Gibbs energy, $\Delta_r G^\circ$, is generally not equal to zero. According to Equation 6.7, if $\Delta_r G^\circ$ is negative, the equilibrium constant must be greater than unity; in fact, the more negative $\Delta_r G^\circ$ is at a given temperature, the larger K_P is. The reverse holds true if $\Delta_r G^\circ$ is a positive number. Here the equilibrium constant is less than unity. Of course, just because $\Delta_r G^\circ$ is positive does not mean that no reaction will take place. For example, if $\Delta_r G^\circ = 10 \text{ kJ mol}^{-1}$ and $T = 298 \text{ K}$, then $K_P = 0.018$. While 0.018 is a small number compared to unity, an appreciable amount of products can still be obtained at equilibrium if we use large quantities of reactants for the reaction. There is also the special case in which $\Delta_r G^\circ = 0$, which corresponds to a K_P of unity, meaning that the products and reactants are equally favored at equilibrium.

It is instructive to look at the factors that affect $\Delta_r G^\circ$ and hence K_P. From Equation 4.21 we have

$$\Delta_r G^\circ = \Delta_r H^\circ - T\Delta_r S^\circ$$

so the equilibrium constant at temperature T is governed by two terms: the change in enthalpy and temperature times the change in entropy. For many exothermic reactions ($\Delta_r H^\circ < 0$) at room temperature and below, the first term on the right side of

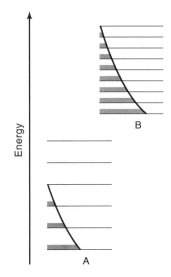

Energy

B

A

Figure 6.3
The occupancy of energy levels in B is greater than that in A. Consequently, there are more microstates in B and the entropy of B is greater than that of A. At equilibrium, B dominates even though the A → B reaction is endothermic.

The K_P for this reaction is practically zero.

the above equation dominates. This means that K_P is greater than one, so products are favored over reactants. For an endothermic reaction ($\Delta_r H° > 0$), the equilibrium composition will favor products only if $\Delta_r S > 0$ and the reaction is run at a high temperature. Consider the following reaction in which the A → B step is endothermic

$$A \rightleftharpoons B$$

As Figure 6.3 shows, the energy levels of A are below those of B, so the conversion of A to B is energetically unfavorable. This is, in fact, the nature of all endothermic reactions. But because the energy levels of B are more closely spaced together, the Boltzmann distribution law (see Equation 3.22) tells us that the population spread of the B molecules over the energy levels is greater than that of the A molecules. Consequently $\Delta_r S° > 0$ because the entropy of B is greater than that of A. At a sufficiently high temperature, then, the $T\Delta_r S°$ term will outweigh the $\Delta_r H°$ term in magnitude and $\Delta_r G°$ will be a negative quantity.

As an illustration of the relative importance of $\Delta_r H°$ versus $T\Delta_r S°$, let's consider the thermal decomposition of limestone or chalk ($CaCO_3$):

$$CaCO_3(s) \rightleftharpoons CaO(s) + CO_2(g) \qquad \Delta_r H° = 177.8 \text{ kJ mol}^{-1}$$

Using the data in Appendix 2 we can show that $\Delta_r S° = 160.5 \text{ J K}^{-1} \text{ mol}^{-1}$. At 298 K, we have

$$\Delta_r G° = 177.8 \text{ kJ mol}^{-1} - (298 \text{ K})(160.5 \text{ J K}^{-1} \text{ mol}^{-1})(1 \text{ kJ}/1000 \text{ J})$$
$$= 130.0 \text{ kJ mol}^{-1}$$

Because $\Delta G°$ is a large positive quantity, we conclude that the reaction is not favored for product formation at 298 K. Indeed, the pressure of CO_2 is so low at room temperature that it cannot be measured. In order to make $\Delta_r G°$ negative, we first have to find the temperature at which $\Delta_r G°$ is zero; that is,

$$0 = \Delta_r H° - T\Delta_r S°$$

or

$$T = \frac{\Delta_r H°}{\Delta_r S°}$$
$$= \frac{(177.8 \text{ kJ mol}^{-1})(1000 \text{ J}/1 \text{ kJ})}{160.5 \text{ J K}^{-1} \text{ mol}^{-1}}$$
$$= 1108 \text{ K or } 835°C$$

At a temperature higher than 835°C, $\Delta_r G°$ becomes negative, indicating that the reaction now favors the formation of CaO and CO_2. For example, at 840°C, or 1113 K,

$$\Delta_r G° = \Delta_r H° - T\Delta_r S°$$
$$= 177.8 \text{ kJ mol}^{-1} - (1113 \text{ K})(160.5 \text{ J K}^{-1} \text{ mol}^{-1})\left(\frac{1 \text{ kJ}}{1000 \text{ J}}\right)$$
$$= -0.8 \text{ kJ mol}^{-1}$$

Two points are worth making about such a calculation. First, we used the $\Delta_r H°$ and $\Delta_r S°$ values at 25°C to calculate changes that occur at a much higher temperature. Because both $\Delta_r H°$ and $\Delta_r S°$ change with temperature, this approach will not give us an accurate value of $\Delta_r G°$, but it is good enough for "ballpark" estimates. Second, we should not be misled into thinking that nothing happens below 835°C and that at 835°C $CaCO_3$ suddenly begins to decompose. Far from it. The fact that $\Delta_r G°$ is a positive value at some temperature below 835°C does not mean that no CO_2 is produced, but rather that the pressure of the CO_2 gas formed at that temperature will be below 1 bar (its standard-state value). The significance of 835°C is that this is the temperature at which the equilibrium pressure of CO_2 reaches 1 bar. Above 835°C, the equilibrium pressure of CO_2 exceeds 1 bar.

A Comparison of $\Delta_r G°$ with $\Delta_r G$

Suppose we start a gaseous reaction with all the reactants in their standard states (that is, all at 1 bar). As soon as the reaction starts, the standard-state condition no longer exists for the reactants or the products because their pressures are different from 1 bar. Under conditions that are not standard state, we must use $\Delta_r G$ rather than $\Delta_r G°$ to predict the direction of a reaction.

We can use the change in standard Gibbs energy ($\Delta_r G°$) and Equation 6.6 to find $\Delta_r G$. The value of $\Delta_r G$ is determined by two terms: $\Delta_r G°$ and a concentration dependent term. At a given temperature the value of $\Delta_r G°$ is fixed, but we can change the value of $\Delta_r G$ by adjusting the partial pressures of the gases. Although the quotient composed of the pressures of reactants and products has the form of an equilibrium constant, it is *not* equal to the equilibrium constant unless P_A and P_B are the partial pressures at equilibrium. In general, we can rewrite Equation 6.6 as

The sign of $\Delta_r G$ and not that of $\Delta_r G°$ determines the direction of reaction spontaneity.

$$\Delta_r G = \Delta_r G° + RT \ln Q \qquad (6.9)$$

where Q is the *reaction quotient* and $Q \neq K_P$ unless $\Delta_r G = 0$. The usefulness of Equation 6.6 or Equation 6.9 is that it tells us the direction of a spontaneous change if the concentrations of the reacting species are known. If $\Delta_r G°$ is a large positive or a large negative number (say, 50 kJ mol^{-1} or more), then the direction of the reaction (or the sign of $\Delta_r G$) is primarily determined by $\Delta_r G°$ alone, unless either the reactants or the products are present in a much larger amount so that the $RT \ln Q$ term in Equation 6.9 is comparable to $\Delta_r G°$ in magnitude but opposite in sign. If $\Delta_r G°$ is a small number, either positive or negative (say, 10 kJ mol^{-1} or less), then the reaction can go either way.*

---

*Alternatively, we can determine the direction of a reaction by comparing Q with K_P. From Equations 6.7 and 6.9 we can show that $\Delta_r G = RT \ln(Q/K_P)$. Therefore if $Q < K_P$, $\Delta_r G$ is negative and the reaction will proceed in the forward direction (left to right). If $Q > K_P$, $\Delta_r G$ is positive. Here, the reaction will proceed in the reverse direction (right to left).

Example 6.2

The equilibrium constant (K_P) for the reaction

$$N_2O_4(g) \rightleftharpoons 2NO_2(g)$$

is 0.113 at 298 K, which corresponds to a standard Gibbs energy change of 5.40 kJ mol^{-1}. In a certain experiment, the initial pressures are $P_{NO_2} = 0.122$ bar and $P_{N_2O_4} = 0.453$ bar. Calculate $\Delta_r G$ for the reaction at these pressures and predict the direction of the net reaction.

ANSWER

To determine the direction of the net reaction, we need to calculate the Gibbs energy change under non-standard-state conditions ($\Delta_r G$) using Equation 6.9 and the given $\Delta_r G°$ value. Note that the partial pressures are expressed as dimensionless quantities in the reaction quotient Q because each pressure is divided by its standard-state value of 1 bar.

$$\Delta G = \Delta G° + RT \ln Q$$

$$= \Delta G° + RT \ln \frac{P_{NO_2}^2}{P_{N_2O_4}}$$

$$= 5.40 \times 10^3 \text{ J mol}^{-1} + (8.314 \text{ J K}^{-1} \text{ mol}^{-1})(298 \text{ K}) \times \ln \frac{(0.122)^2}{0.453}$$

$$= 5.40 \times 10^3 \text{ J mol}^{-1} - 8.46 \times 10^3 \text{ J mol}^{-1}$$

$$= -3.06 \times 10^3 \text{ J mol}^{-1} = -3.06 \text{ kJ mol}^{-1}$$

Because $\Delta_r G < 0$, the net reaction proceeds from left to right to reach equilibrium.

COMMENT

Note that although $\Delta_r G° > 0$, the reaction can be made initially to favor product formation by having a small concentration (pressure) of the product compared to that of the reactant.

Real Gases

What form does the equilibrium constant take for real gases? As we saw in Chapter 2, the behavior of real gases cannot be described by the ideal gas equation and requires a more accurate equation of state, such as the van der Waals equation or the virial equation. However, if we tried to calculate P using the van der Waals equation for every gas, say, and substituted this quantity in the equilibrium constant expression, the final form would be very unwieldy. Instead, we adopt a simpler procedure that is analogous to the use of activity for concentration in Chapter 5. For real gases, we define a new variable called *fugacity* (f) to replace partial pressure. Equation 6.3 applies only to an ideal gas. For a real gas we must write

$$\mu = \mu° + RT \ln \frac{f}{P°} \tag{6.10}$$

Fugacity has the same dimensions as pressure. At low pressures, the gas behaves ideally and fugacity equals pressure, but deviation occurs when the pressure increases. The standard state for the fugacity of a real gas is defined as the hypothetical state in which the gas is at 1 bar pressure and behaving ideally. In general, we have

$$\underset{P \to 0}{\text{Lim}} f = P$$

$\gamma < 1$ indicates that the attractive intermolecular forces are dominant. Conversely, $\gamma > 1$ means repulsive intermolecular forces are dominant.

The fugacity coefficient γ is given by

$$\gamma = \frac{f}{P} \tag{6.11}$$

and

$$\lim_{P \to 0} \gamma = 1$$

It is important to understand that although pressure is directly measurable, fugacity can only be calculated using P–V–T data of gases and the appropriate equations of state.*

Starting with Equation 6.10, we can derive an equilibrium constant K_f (where the subscript f denotes fugacity) for the hypothetical reaction discussed earlier ($a\text{A} \rightleftharpoons b\text{B}$):

$$K_f = \frac{(f_\text{B}/1 \text{ bar})^b}{(f_\text{A}/1 \text{ bar})^a} \tag{6.12}$$

Because $f = \gamma P$, Equation 6.12 can be rewritten as

$$K_f = \frac{\gamma_\text{B}^b}{\gamma_\text{A}^a} \frac{(P_\text{B}/1 \text{ bar})^b}{(P_\text{A}/1 \text{ bar})^a} = K_\gamma K_P \tag{6.13}$$

where K_γ is given by $\gamma_\text{B}^b/\gamma_\text{A}^a$ and K_P is given by $(P_\text{B}^b/P_\text{A}^a)(1 \text{ bar})^{a-b}$. K_f, as defined by Equation 6.12 or Equation 6.13, is called the *thermodynamic equilibrium constant* and gives the exact result (if the fugacities are accurately known). K_P, on the other hand, is called the *apparent equilibrium constant* because its value is not constant, but depends on pressure at a given temperature. For an ideal gas reaction, K_f is equal to K_P. The use of fugacities is essential for calculating equilibrium constants of many industrial processes; for example, the synthesis of ammonia from hydrogen and nitrogen is carried out at hundreds of bars of the reactant gases. As Table 6.1 shows, K_P can differ appreciably from K_f at such high pressures. Because biological processes usually occur under atmospheric conditions where gases behave nearly ideally, reliable estimates of equilibrium constants can be obtained by using pressures only. For this reason, it is unnecessary to use fugacities in studying equilibrium processes in biological systems.

Table 6.1
Equilibrium Constants for the Reaction
$\frac{1}{2}\text{N}_2(g) + \frac{3}{2}\text{H}_2(g) \rightleftharpoons \text{NH}_3(g)$ at 450°C[a]

Total Pressure/bar	K_P	K_f
10.2	0.0064	0.0064
30.3	0.0066	0.0064
50.6	0.0068	0.0065
101.0	0.0072	0.0064
302.8	0.0088	0.0062
606	0.0130	0.0065

[a] Data from A. J. Larson, *J. Am. Chem. Soc.* **46**, 367 (1924).

6.2 Reactions in Solutions

The treatment of chemical equilibrium in solution is analogous to that in the gas phase, although we normally express concentrations of reacting species in solu-

---

* See the physical chemistry texts listed on p. 6 for a discussion of how fugacity is determined.

tion in molality or molarity. We again start with a hypothetical reaction, this time in solution:

$$aA \rightleftharpoons bB$$

where A and B are the nonelectrolyte solutes. Assuming ideal behavior and expressing the solute concentrations in molalities, we have, from Equation 5.27,

$$\mu_A = \mu_A^\circ + RT \ln \frac{m_A}{m^\circ}$$

where m° represents 1 mol (kg solvent)$^{-1}$. Following the same procedure as that employed for ideal gases in Section 6.1, we arrive at the standard Gibbs energy change:

$$\Delta_r G^\circ = -RT \ln K_m \tag{6.14}$$

where

$$K_m = \frac{(m_B/m^\circ)^b}{(m_A/m^\circ)^a}$$

If we express the solute concentrations in molarities, the equilibrium constant takes the form

$$K_c = \frac{([B]/1 \ M)^b}{([A]/1 \ M)^a}$$

where the square brackets signify mol L^{-1}. Again, both K_m and K_c are dimensionless quantities, because each concentration term is divided by its standard-state value (1 m or 1 M). For nonequilibrium reactions, the Gibbs energy change is given by

$$\Delta_r G = \Delta_r G^\circ + RT \ln Q$$

where Q is the reaction quotient.

For nonideal solutions, we must replace concentrations with activities. From Equation 5.25, we write the chemical potential of the ith component as

$$\mu_i = \mu_i^\circ + RT \ln a_i$$

The substitution of activity for concentration is analogous to the substitution of fugacity for pressure. Starting with the above chemical potential expression, we obtain the thermodynamic equilibrium constant, K_a:

$$K_a = \frac{a_B^b}{a_A^a} \tag{6.15}$$

Because $a = \gamma m$, Equation 6.15 can be written as

$$K_a = \frac{\gamma_B^b}{\gamma_A^a} \times \frac{(m_B/m^\circ)^b}{(m_A/m^\circ)^a}$$

$$= K_\gamma K_m \tag{6.16}$$

where K_γ is given by (γ_B^b/γ_A^a) and K_m, the apparent equilibrium constant for a non-ideal solution reaction, is given by $(m_B^b/m_A^a)(m^\circ)^{a-b}$.

6.3 Heterogeneous Equilibria

So far, we have concentrated on homogeneous equilibria—that is, reactions that occur in a single phase. Here, we shall discuss heterogeneous equilibria in which the reactants and products are present in more than one phase.

Consider the thermal decomposition of calcium carbonate in a closed system:

$$CaCO_3(s) \rightleftharpoons CaO(s) + CO_2(g)$$

The two solids and one gas constitute three separate phases. We might write the equilibrium constant of this reaction as

$$K_c' = \frac{[CaO][CO_2]}{[CaCO_3]}$$

By common practice, however, we do not include the concentrations of solids in the equilibrium-constant expression. The concentration of any pure solid is the ratio of the total number of moles present in the solid divided by the volume of the solid. If part of the solid is removed, the number of moles of the solid will decrease, but so will its volume. The same holds true for the addition of more solid substance—increasing the number of moles of the solid results in an increase in volume. For this reason, the ratio of moles to volume always remains unchanged. Thus, the relative amounts of CO_2 and CaO produced are always the same, regardless of the amount of $CaCO_3$ employed in the beginning, as long as some of the solid is present at equilibrium. The equilibrium-constant expression given above can now be arranged as follows:

$$\frac{[CaCO_3]}{[CaO]} K_c' = [CO_2]$$

Because both $[CaCO_3]$ and $[CaO]$ are constants, every term on the left side is a constant, and we write

$$K_c = [CO_2]$$

where K_c, the "new" equilibrium constant, is given by $[CaCO_3]K_c'/[CaO]$. More conveniently, we can measure the pressure of CO_2 and obtain

$$K_P = P_{CO_2}$$

Both K_c and K_P are dimensionless because $[CO_2]$ is divided by its standard state value of 1 M, and P_{CO_2} is divided by 1 bar. The equilibrium constants K_c and K_P are related to each other in a simple manner (see Problem 6.1).

Heterogeneous equilibria are more easily handled if we write the thermodynamic equilibrium constant instead of the apparent equilibrium constant. Replacing concentrations with activities, we have

$$K_a = \frac{a_{CaO}a_{CO_2}}{a_{CaCO_3}}$$

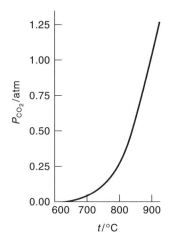

Figure 6.4
A plot of the equilibrium pressure (P) of CO_2 over CaO and $CaCO_3$ as a function of temperature (t).

By convention, the activities of pure solids (and pure liquids) in their standard states (i.e., at 1 bar) are equal to unity; that is, $a_{CaO} = 1$ and $a_{CaCO_3} = 1$. For reactions carried out under moderate pressure conditions, we can assume that their values do not change appreciably from unity and write the equilibrium constant in terms of the fugacity of the CO_2 gas:

$$K_a = \frac{f_{CO_2}}{1 \text{ bar}}$$

or, assuming ideal behavior,

$$K_P = \frac{P_{CO_2}}{1 \text{ bar}}$$

where the fugacity and pressure are in bars. Figure 6.4 shows the equilibrium pressure of CO_2 for the decomposition of $CaCO_3$ as a function of temperature.

Example 6.3

Calculate the equilibrium constant for the following reaction at 298 K, using the data listed in Appendix 2:

$$2H_2(g) + O_2(g) \rightleftharpoons 2H_2O(l)$$

ANSWER

The thermodynamic equilibrium constant is given by

$$K_a = \frac{a_{H_2O}^2}{(f_{H_2}/1 \text{ bar})^2 (f_{O_2}/1 \text{ bar})} = \frac{1}{f_{H_2}^2 f_{O_2}} (1 \text{ bar})^3$$

Assuming ideal behavior, we write

$$K_P = \frac{1}{P_{H_2}^2 P_{O_2}} (1 \text{ bar})^3$$

The standard Gibbs energy change for the reaction is given by

$$\begin{aligned}
\Delta_r G^\circ &= 2\Delta_f \bar{G}^\circ(H_2O) - 2\Delta_f \bar{G}^\circ(H_2) - \Delta_f \bar{G}^\circ(O_2) \\
&= (2)(-237.2 \text{ kJ mol}^{-1}) \\
&= -474.4 \text{ kJ mol}^{-1}
\end{aligned}$$

Finally, from Equation 6.7,

$$-474.4 \times 10^3 \text{ J mol}^{-1} = -(8.314 \text{ J K}^{-1} \text{ mol}^{-1})(298 \text{ K}) \ln K_P$$
$$K_P = 1.4 \times 10^{83}$$

The very large K_P value indicates that the reaction virtually goes to completion.

6.4 The Influence of Temperature, Pressure, and Catalysts on the Equilibrium Constant

In this section we shall consider the influence of temperature, pressure, and the use of a catalyst on the equilibrium constant.

The Effect of Temperature

Although Equation 6.7 relates the change in the standard Gibbs energy to the equilibrium constant at *any* temperature, calculating the K value at 298 K is usually the most convenient because of the availability of thermodynamic data. In practice, however, a reaction may be run at temperatures other than 298 K, and so we either have to know the $\Delta_r G^\circ$ values at that particular temperature or find some other way to calculate the K value. The question we ask is: If we know the equilibrium constant of a reaction K_1 at temperature T_1, can we calculate the equilibrium constant of the same reaction K_2 at temperature T_2? The answer is yes.

A very useful equation relating the equilibrium constant to temperature can be derived as follows. Substituting Equation 6.7 into the Gibbs–Helmholtz equation (Equation 4.31) written for the changes in the standard state, we obtain

$$\left[\frac{\partial\left(\dfrac{\Delta_r G^\circ}{T}\right)}{\partial T}\right]_P = -\frac{\Delta_r H^\circ}{T^2}$$

$$\left[\frac{\partial\left(\dfrac{-RT\ln K}{T}\right)}{\partial T}\right]_P = -\frac{\Delta_r H^\circ}{T^2}$$

$$\left(\frac{\partial \ln K}{\partial T}\right)_P = \frac{\Delta_r H^\circ}{RT^2} \tag{6.17}$$

Equation 6.17 is known as the *van't Hoff equation*. Assuming $\Delta_r H^\circ$ to be temperature independent, this equation can be integrated to give

$$\ln\frac{K_2}{K_1} = \frac{\Delta_r H^\circ}{R}\left(\frac{1}{T_1} - \frac{1}{T_2}\right)$$

$$\ln\frac{K_2}{K_1} = \frac{\Delta_r H^\circ}{R}\left(\frac{T_2 - T_1}{T_1 T_2}\right) \tag{6.18}$$

Because

$$\ln K = -\frac{\Delta_r G^\circ}{RT}$$

and

$$\Delta_r G^\circ = \Delta_r H^\circ - T\Delta_r S^\circ$$

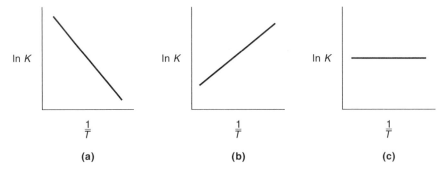

Figure 6.5
Graph of $\ln K$ versus $1/T$ for the van't Hoff equation. The slope is equal to $-\Delta_r H^\circ / R$ and the intercept on the ordinate ($\ln K$) is equal to $\Delta_r S^\circ / R$. (a) $\Delta_r H^\circ > 0$, (b) $\Delta_r H^\circ < 0$, (c) $\Delta_r H^\circ = 0$. In all cases, $\Delta_r H^\circ$ is assumed to be temperature independent.

it follows that

$$\ln K = -\frac{\Delta_r H^\circ}{RT} + \frac{\Delta_r S^\circ}{R} \qquad (6.19)$$

Equation 6.19 is another form of the van't Hoff equation in which $\Delta_r S^\circ$ is the standard entropy change of the reaction.* Thus, by plotting $\ln K$ versus $1/T$, we obtain a straight line with a slope equal to $-\Delta_r H^\circ / R$, and the intercept on the ordinate gives $\Delta_r S^\circ / R$ (Figure 6.5). This is a convenient way to determine the values of $\Delta_r H^\circ$ and $\Delta_r S^\circ$ for a reaction. Note that this graph will yield a straight line only if both $\Delta_r H^\circ$ and $\Delta_r S^\circ$ are independent of temperature. For relatively small temperature ranges (say 50 K or less), this method provides a fairly good approximation.

Some interesting observations can be made from Equation 6.18. If the reaction is endothermic in the forward direction (that is, $\Delta_r H^\circ$ is positive) and if we assume that $T_2 > T_1$, the quantity on the left side of Equation 6.18 is positive, which means that $K_2 > K_1$. On the other hand, if the reaction is exothermic in the forward direction (that is, $\Delta_r H^\circ$ is negative), then the quantity on the left side of Equation 6.18 is negative, or $K_2 < K_1$. Thus, we can conclude that an increase in temperature shifts the equilibrium from left to right (favors the formation of products) in an endothermic reaction and shifts the equilibrium from right to left (favors the formation of reactants) in an exothermic reaction. These results are consistent with *Le Chatelier's principle* (after the French chemist Henri Louis Le Chatelier, 1850–1936), which states that if an external stress is applied to a system at equilibrium, the system will adjust itself in such a way as to partially offset the stress as it tries to re-establish equilibrium. In this case the "stress" is the change in temperature.

---

*If $\Delta_r H^\circ$ is independent of temperature, then the change in heat capacities between the products and reactants is zero, which means that $\Delta_r S^\circ$ is also independent of temperature. The term $\Delta_r S^\circ / R$ is the integration constant for the indefinite integral of Equation 6.17.

Example 6.4

The equilibrium constants for the gas-phase dissociation of molecular iodine,

$$I_2(g) \rightleftharpoons 2I(g)$$

have been measured at the following temperatures:

T/K	872	973	1073	1173
K_P	1.8×10^{-4}	1.8×10^{-3}	1.08×10^{-2}	0.0480

Determine graphically the values of $\Delta_r H^\circ$ and $\Delta_r S^\circ$ for the reaction.

ANSWER

We need Equation 6.19. First, we construct the table,

K/T	1.15×10^{-3}	1.03×10^{-3}	9.32×10^{-4}	8.53×10^{-4}
$\ln K_P$	-8.62	-6.32	-4.53	-3.04

Next, we plot $\ln K_P$ versus $1/T$ (Figure 6.6). The points fall on a straight line whose

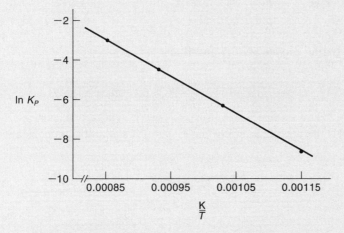

Figure 6.6
The van't Hoff plot for the dissociation of iodine vapor.

equation is $\ln K_P = -1.875 \times 10^4 \ K/T + 12.954$. Assuming that $\Delta_r H^\circ$ is temperature independent, we have

$$-\frac{\Delta_r H^\circ}{R} = -1.875 \times 10^4 \ K$$

or

$$\Delta_r H^\circ = 1.56 \times 10^2 \ kJ \ mol^{-1}$$

The intercept on the ordinate is equal to $\Delta_r S^\circ / R$, so

$$\Delta_r S^\circ = 12.954(8.314 \ J \ K^{-1} \ mol^{-1})$$
$$= 108 \ J \ K^{-1} \ mol^{-1}$$

Effect of Pressure

As we saw earlier (p. 201), the thermodynamic equilibrium constant (K_f), expressed in fugacities, does not depend on pressure but the apparent equilibrium constant (K_P) does because of deviation from ideal behavior. For reactions involving ideal gases, $K_f = K_P$, so at constant temperature T we write

$$\left(\frac{\partial K_P}{\partial P}\right)_T = 0$$

The fact that K_P is not affected by pressure does not mean that the amounts of various gases at equilibrium do not change with pressure. To illustrate this point, let us consider the ideal gas-phase reaction at equilibrium:

$$A(g) \rightleftharpoons 2B(g)$$
$$n(1-\alpha) \qquad 2n\alpha$$

where n is the number of moles of A originally present and α is the fraction of the A molecules dissociated. The total number of moles of molecules present at equilibrium is $n(1+\alpha)$, and the mole fractions of A and B are

$$x_A = \frac{n(1-\alpha)}{n(1+\alpha)} = \frac{(1-\alpha)}{(1+\alpha)} \quad \text{and} \quad x_B = \frac{2n\alpha}{n(1+\alpha)} = \frac{2\alpha}{1+\alpha}$$

and the partial pressures of A and B are

$$P_A = \frac{(1-\alpha)}{(1+\alpha)} P \quad \text{and} \quad P_B = \frac{2\alpha}{1+\alpha} P$$

where P is the total pressure of the system. The equilibrium constant is given by

For simplicity, we omit the $P°$ terms in K_P.

$$K_P = \frac{P_B^2}{P_A} = \left(\frac{2\alpha}{1+\alpha} P\right)^2 \Big/ \frac{(1-\alpha)}{(1+\alpha)} P$$

$$= \left(\frac{4\alpha^2}{1-\alpha^2}\right) P$$

Rearranging the last equation, we obtain

$$\alpha = \sqrt{\frac{K_P}{K_P + 4P}}$$

Pressure has no effect on the equilibrium position for condensed phases or if there is no change in the number of moles of gases from reactants to products.

Because K_P is a constant, the value of α depends only on P. If P is large, α is small; if P is small, α is large. These predictions again remind us of Le Chatelier's principle. If the stress applied to the system is an increase in pressure, then the equilibrium will shift to the side that produces fewer molecules, which is right to left in our example, and hence α decreases. The reverse holds true for a decrease in pressure.

The Effect of a Catalyst

By definition, a catalyst is a substance that can speed up the rate of a reaction without itself being used up. Will the addition of a catalyst to a reacting system at equilibrium shift the equilibrium in a particular direction? To answer this question,

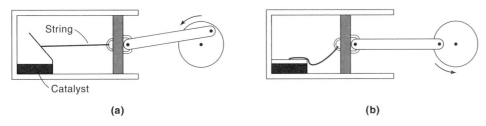

Figure 6.7
A perpetual-motion machine operating with a hypothetical catalyst capable of shifting the equilibrium position of a gaseous reaction in only one direction.

let us again consider the gas-phase equilibrium,

$$A(g) \rightleftharpoons 2B(g)$$

as a thought experiment. Suppose a catalyst exists that favors the reverse reaction $(2B \rightarrow A)$ but not the forward reaction $(A \rightarrow 2B)$. We can then construct an apparatus such as that shown in Figure 6.7. A small box is placed inside a cylinder fitted with a movable piston. The lid of the box is connected to the piston by a piece of string, so that the cover can be closed and opened by the movement of the piston. We start with the equilibrium mixture of gases A and B in the cylinder and then add the catalyst to the box (Figure 6.7a). Immediately, the equilibrium shifts toward the formation of A. Because two molecules of B are consumed to produce one molecule of A, this step decreases the total number of molecules present in the cylinder and hence the internal gas pressure. Consequently, the piston will be pushed inward by the external pressure until the internal and external pressures are again balanced. At this stage, the box lid drops (Figure 6.7b). Without the catalyst, the gases gradually return to their original concentrations, and the increase in the number of molecules as a result of the formation of B pushes the piston from left to right until the cover is lifted. The whole process will then repeat itself.

You may have noticed something strange about the whole arrangement. The piston can be made to perform work even though there is no energy input or net consumption of chemicals. Such a device is known as a perpetual-motion machine. Construction of a perpetual-motion machine is impossible, however, because it violates the laws of thermodynamics. The reason such a machine can never be realized is that no catalysts can speed up the rate of a reaction in one direction without affecting the rate of the reverse reaction in a similar manner. The important conclusion from this simple illustration is that a catalyst cannot shift the position of an equilibrium.* For a reaction mixture that is not at equilibrium, a catalyst will increase both the forward and reverse rate processes so that the reaction will reach equilibrium sooner, but the same equilibrium state will be attained eventually even if no catalyst is present.

6.5 Binding of Ligands and Metal Ions to Macromolecules

The interaction of small molecules (ligands) with proteins and with specific receptor sites on membrane surfaces is one of the most extensively studied biochemical phenomena. Examples of these reversible interactions include the binding and release of

---

*The fact that a catalyst must enhance both the forward and reverse rates of a reaction is predicted by the principle of microscopic reversibility (see Chapter 9).

protons by the acidic and basic groups in proteins, the association of cations such as Mg^{2+} and Ca^{2+} with proteins and nucleic acids, antibody–antigen reactions, and the reversible binding of oxygen by myoglobin and hemoglobin. These processes are closely related to the combination of enzymes with substrates and inhibitors, a subject we shall study in Chapter 10.

In this section, we shall apply the equilibrium treatment to the study of the binding of ligands and metal ions to macromolecules in solution. We concentrate on two cases: one in which the macromolecule possesses one binding site per molecule, and another in which there are *n equivalent* and *independent* binding sites per molecule. Because our approach here is strictly thermodynamic, we need not discuss the structure of macromolecules and the nature of covalent and other intermolecular forces responsible for the binding.

One Binding Site per Macromolecule

This is the simplest case, in which one site of a macromolecule, P, binds one molecule (or ion) of a ligand, L. The reaction can be represented as

$$P + L \rightleftharpoons PL$$

The equilibrium constant for this association reaction, K_a, is

We assume ideal behavior and use concentrations instead of activities.

$$K_a = \frac{[PL]}{[P][L]}$$

Frequently, working with the dissociation constant, K_d, is more convenient:

$$K_d = \frac{[P][L]}{[PL]} \qquad (6.20)$$

The smaller the value of K_d, the "tighter" the complex PL is. These equilibrium constants are related by the simple relation $K_a K_d = 1$.

To determine the value of K_d, we first define a quantity Y, called the *fractional saturation of sites*, such that

$$Y = \frac{\text{concentration of L bound to P}}{\text{total concentration of all forms of P}}$$

$$= \frac{[PL]}{[P] + [PL]} \qquad (6.21)$$

The value of Y ranges from zero, when $[PL] = 0$, to 1, when $[P] = 0$. For example, when $Y = 0.5$, half of the P molecules are complexed with L and half of the P molecules are in the free form, so that $[P] = [PL]$ and $[L] = K_d$. To determine the value of K_d, we first rearrange Equation 6.20 to give

$$[PL] = \frac{[P][L]}{K_d}$$

Substituting the expression for [PL] in Equation 6.21, we obtain

$$Y = \frac{[P][L]/K_d}{[P] + [P][L]/K_d}$$

$$Y = \frac{[L]}{[L] + K_d} \tag{6.22}$$

Keep in mind that [L] is the concentration of free ligands at equilibrium. The value of Y (see Equation 6.21) can be determined in the following way. Initially a known concentration of L, $[L]_0$, is added to a known concentration of P, $[P]_0$. At equilibrium, we can measure either [PL] or [L] because $[L]_0$ is known at the outset, and mass balance requires that $[L] + [PL] = [L]_0$. The value of [P] need not be measured because $[P] + [PL] = [P]_0$, so that $[P] = [P]_0 - [PL]$. We shall soon discuss an experimental procedure for determining [L] and [PL].

By taking the reciprocal of Equation 6.22, we get

$$\frac{1}{Y} = 1 + \frac{K_d}{[L]} \tag{6.23}$$

A plot of $1/Y$ versus $1/[L]$ gives a straight line of slope K_d. Alternatively, Equation 6.22 can be rearranged to give

$$\frac{Y}{[L]} = \frac{1}{K_d} - \frac{Y}{K_d} \tag{6.24}$$

In this case, a plot of $Y/[L]$ versus Y gives a straight line whose slope is $-1/K_d$ (Figure 6.8).

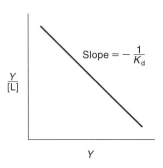

Figure 6.8
A plot of $Y/[L]$ versus Y according to Equation 6.24.

n Equivalent Binding Sites per Macromolecule

Now consider the case in which a macromolecule has n equivalent sites; that is, each binding site has the same K_d value, regardless of whether other sites on the same molecule are occupied. We shall take $n = 2$ first and then generalize the result to cases for which $n > 2$.

If a macromolecule has two equivalent binding sites, there will be two binding equilibria:

$$P + L \rightleftharpoons PL \qquad K_1 = \frac{[P][L]}{[PL]}$$

$$PL + L \rightleftharpoons PL_2 \qquad K_2 = \frac{[PL][L]}{[PL_2]}$$

where K_1 and K_2 are the dissociation constants (Figure 6.9). This time, we define Y such that

$$Y = \frac{\text{concentration of L bound to P}}{\text{total concentration of all forms of P}}$$

$$= \frac{[PL] + 2[PL_2]}{[P] + [PL] + [PL_2]} \tag{6.25}$$

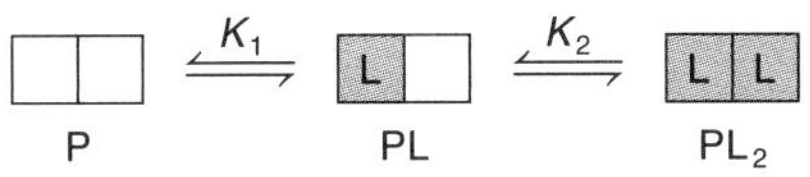

Figure 6.9
The successive bindings of ligand L to a macromolecule that has two equivalent binding sites.

Note that because PL_2 has two molecules of L bound to one molecule of P, its concentration must be multiplied by 2. Substituting the relations

$$[PL] = \frac{[P][L]}{K_1} \quad \text{and} \quad [PL_2] = \frac{[PL][L]}{K_2}$$

into Equation 6.25, we write

$$
\begin{aligned}
Y &= \frac{[P][L]/K_1 + 2[P][L]^2/K_1 K_2}{[P] + [P][L]/K_1 + [P][L]^2/K_1 K_2} \\
&= \frac{[L]/K_1 + 2[L]^2/K_1 K_2}{1 + [L]/K_1 + [L]^2/K_1 K_2}
\end{aligned}
\tag{6.26}
$$

At first, K_1 might appear to be equal to K_2 because both sites are independent and have the same dissociation constant. But this is not the case because of certain statistical factors. Let K be the dissociation constant at a *given* site (in this case, K is called the *intrinsic dissociation constant*). Then $2K_1 = K$ because there are two ways for L to be attached to P and one way for it to detach from PL. (K_1 would be equal to K if there were only one site present.) After an L is bound, there is one way for L to be attached to PL and two ways for it to detach from PL_2, so $K_2 = 2K$. The general relationship between the ith dissociation constant K_i and K is given by

$$K_i = \left(\frac{i}{n-i+1}\right) K \tag{6.27}$$

For our example $i = 1, 2$ and $n = 2$, so from Equation 6.27,

$$K_1 = \frac{K}{2} \quad \text{and} \quad K_2 = 2K$$

Thus, purely on the basis of statistical analysis, we find that the second dissociation constant is four times as large as the first dissociation constant; that is, $K_2 = 4K_1$. Note that K is the geometric mean of the individual dissociation constants; that is,

$$K = \sqrt{K_1 K_2} \tag{6.28}$$

Equation 6.26 can now be written as

$$
\begin{aligned}
Y &= \frac{2[L]/K + 2[L]^2/K^2}{1 + 2[L]/K + [L]^2/K^2} \\
&= \frac{(2[L]/K)(1 + [L]/K)}{(1 + [L]/K)^2} = \frac{2[L]}{[L] + K}
\end{aligned}
\tag{6.29}
$$

Equation 6.29 is the result obtained for two equivalent sites.

In general, for n equivalent sites, we have*

$$Y = \frac{n[L]}{[L] + K} \tag{6.30}$$

---

* For n equivalent sites, we have $K = (K_1 K_2 K_3 \cdots K_n)^{1/n}$.

Equation 6.30 can be rearranged into several forms suitable for graphing. We shall look at three of the most common procedures.

1. The Direct Plot. Figure 6.10 shows a graph of Y versus the ligand concentration. A direct plot of this type yields a hyperbolic curve, which is characteristic of simple binding (that is, the binding sites are all equivalent and noninteracting). When $[L] = K$, we have $Y = n/2$, and at very high ligand concentrations, we can assume that $[L] \gg K$, so that $Y = n$. The direct plot is generally not very useful in determining the values of n and K, however, because determining n at very high ligand concentrations is often difficult and we need to know the value of n to determine that of K.

2. The Double Reciprocal Plot. By taking the reciprocal of each side of Equation 6.30, we obtain

$$\frac{1}{Y} = \frac{1}{n} + \frac{K}{n[L]} \tag{6.31}$$

Thus, a plot of $1/Y$ versus $1/[L]$ gives a straight line of slope K/n and an intercept on the ordinate of $1/n$ (Figure 6.11). This plot is known as the Hughes–Klotz plot.

3. The Scatchard Plot. Starting with Equation 6.30, we have

$$Y[L] + KY = n[L]$$

$$\frac{Y[L]}{K} + Y = \frac{n[L]}{K}$$

$$Y = \frac{n[L]}{K} - \frac{Y[L]}{K}$$

or

$$\frac{Y}{[L]} = \frac{n}{K} - \frac{Y}{K} \tag{6.32}$$

Equation 6.32 is known as the Scatchard equation (after the American chemist George Scatchard, 1892–1973). Thus, plotting $Y/[L]$ versus Y gives a straight line of slope $-1/K$ and an intercept on the abscissa of n (Figure 6.12).

Experimental Studies of Binding Equilibria

Having discussed the theoretical aspects of the binding of ligands to macromolecules, we now look at two experimental methods for determining n and K—*equilibrium dialysis and isothermal titration calorimetry.*

Equilibrium Dialysis. Dialysis is the process of exchanging small ions and other solute molecules from a protein solution through a semipermeable membrane. Suppose that in an experiment we have precipitated hemoglobin from a solution by using the salting-out technique with ammonium sulfate (see Section 5.5). The protein can be freed of the $(NH_4)_2SO_4$ salt as follows. The precipitate is dissolved in water or, more typically, in a buffer solution. The protein solution is then placed in a cellophane bag, which, in turn, is immersed in a beaker containing the same buffer (Figure 6.13). Because both the NH_4^+ and SO_4^{2-} ions are small enough to diffuse through the membrane but the protein molecules are not, the ions in the bag will begin to

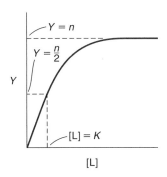

Figure 6.10
A plot of fractional saturation (Y) versus ligand concentration ($[L]$).

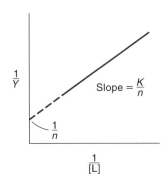

Figure 6.11
A plot of $1/Y$ versus $1/[L]$. This graph is also known as the Hughes–Klotz plot.

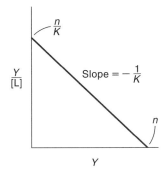

Figure 6.12
A plot of $Y/[L]$ versus Y. This graph is also known as the Scatchard plot.

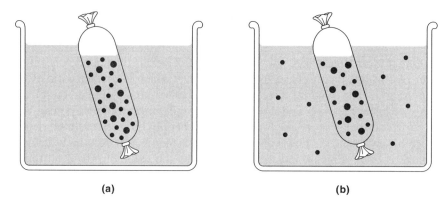

Figure 6.13
A dialysis experiment. The small dots denote ions and the larger dots proteins. (a) At the start of dialysis. (b) At equilibrium, some of the small ions have diffused out of the cellophane bag. By repeatedly replacing the buffer solution in the beaker, it is possible to remove all of the small ions not bound to the protein.

enter into the outside solution, where the chemical potentials are lower:

$$\left(\mu_{NH_4^+}\right)_{inside} > \left(\mu_{NH_4^+}\right)_{outside}$$

$$\left(\mu_{SO_4^{2-}}\right)_{inside} > \left(\mu_{SO_4^{2-}}\right)_{outside}$$

The flow of ions out of the bag continues until the chemical potentials of the cation and anion outside of the bag equal those on the inside, and an equilibrium is established. If desired, all the $(NH_4)_2SO_4$ can be removed by continually changing the buffer solution in the beaker.

The procedure described above may be reversed to study the binding of ions or small ligands to proteins. In this case we begin by placing the protein without ligands in a buffer solution (called phase 1) inside the cellophane bag, which is then immersed in a similar buffer solution (called phase 2) that contains the ligand (L) of a known concentration. At equilibrium, the chemical potentials of the free (unbound) ligands (μ_L) in both phases must be the same (Figure 6.14), so

$$\left(\mu_L\right)^1_{unbound} = \left(\mu_L\right)^2_{unbound} \tag{6.33}$$

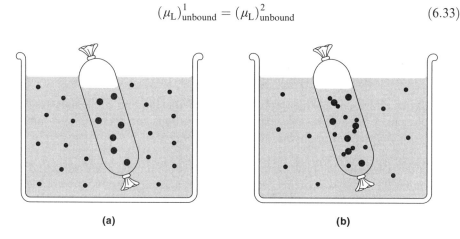

Figure 6.14
Equilibrium dialysis. The small dots denote ligands and the large dots proteins. (a) Initially, a cellophane bag containing a protein solution is immersed in a buffer solution containing the ligand molecules. (b) At equilibrium, some of the ligand molecules have diffused into the bag and are bound to the protein molecules.

or

$$(\mu^\circ + RT \ln a_L)^1_{\text{unbound}} = (\mu^\circ + RT \ln a_L)^2_{\text{unbound}} \tag{6.34}$$

Because the standard chemical potential μ° is the same on both sides, we have

$$(a_L)^1_{\text{unbound}} = (a_L)^2_{\text{unbound}} \tag{6.35}$$

If the solutions are dilute, we can replace activities with concentrations so that

$$[L]^1_{\text{unbound}} = [L]^2_{\text{unbound}} \tag{6.36}$$

However, the total concentration of the ligands inside the bag is given by

$$[L]^1_{\text{total}} = [L]^1_{\text{bound}} + [L]^1_{\text{unbound}} \tag{6.37}$$

Therefore, the concentration of ligands bound to protein molecules is

$$[L]^1_{\text{bound}} = [L]^1_{\text{total}} - [L]^1_{\text{unbound}} \tag{6.38}$$

The first quantity on the right-hand side of Equation 6.38 can be determined by analyzing the solution in the bag *after* its removal from the beaker; the second quantity is obtained by measuring the concentration of the ligands in the solution remaining in the beaker. (Remember that the concentrations of the unbound ligands are equal in phases 1 and 2.)

We can now see how both the intrinsic dissociation constant K and the number of binding sites n can be obtained by using the equilibrium dialysis technique. Suppose that we start with a protein solution of known concentration in phase 1 and a ligand solution of known concentration in phase 2. At equilibrium, the quantity Y is given by (see definition of Y in Equation 6.25)

$$Y = \frac{[L]^1_{\text{bound}}}{[P]_{\text{total}}} \tag{6.39}$$

where $[P]_{\text{total}}$ is the concentration of the original protein solution. The experiment can be repeated by using different concentrations for the protein and ligand solutions and the values of K and n can be determined from either the Hughes–Klotz plot or the Scatchard plot. Keep in mind that the quantity $[L]$ in Equations 6.31 and 6.32 refers to the concentration of unbound ligands at equilibrium, which is $[L]^1_{\text{unbound}}$ in our case. It is useful to keep in mind that we have implicitly assumed that the ligand is a nonelectrolyte so that the concentrations of unbound ligands are equal in phases 1 and 2 at equilibrium. If the ligand were an electrolyte, we would have to apply the Donnan effect in treating the dialysis data.

For many years equilibrium dialysis had been successfully employed to bind drugs, hormones, and other small molecules to proteins and nucleic acids. It is a slow and tedious procedure, however, and does not readily provide all the relevant thermodynamic information (for example, the $\Delta_r H^\circ$ and $\Delta_r S^\circ$ values). Although this technique is seldom used now to study binding equilibria, the dialysis phenomenon itself is still useful for purifying macromolecules.

Isothermal Titration Calorimetry. Isothermal titration calorimetry (ITC) is based on heat measurements for monitoring any chemical reaction initiated by the addition of a binding component. Most isothermal titration calorimeters measure the differential heat effects between a sample cell containing the macromolecule in a buffer solution

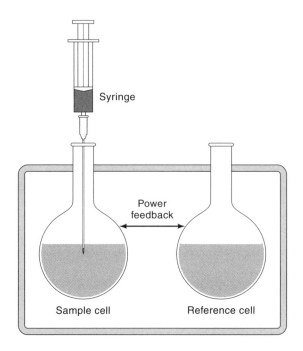

Figure 6.15
Schematic diagram of an isothermal titration calorimeter. As ligand solution is added to the sample cell, there will be a temperature change due to binding interaction. A power feedback mechanism then adjusts to keep the sample cell at nearly the same temperature as the reference cell. Thus, the reaction is carried out under constant-temperature conditions. The ITC technique is highly sensitive. Relevant thermodynamic quantities can be obtained from very dilute solutions (10^{-6} M or less).

and a reference cell containing only the buffer solution (Figure 6.15). Electronic circuitry (power feedback) is used to keep the two cells at nearly the same temperature (hence the term *isothermal*). In a typical experiment, aliquots of ligand solution are added to the sample solution (as in a titration) through a syringe at regular time intervals. The sign of temperature change from the binding interaction depends on whether the reaction is exothermic or endothermic. When the ligand binds to a protein molecule, say, heat is evolved. This interaction leads to a peak downward in the cell feedback power because heat is no longer required to maintain the temperature of the sample cell (Figure 6.16). Because the cell feedback has units of power, the time integral of the peak yields a measurement of $\Delta_r H^\circ$ for the binding interaction at each stage of the addition.

For each injection of the ligand solution, the heat change (q) is given by

$$q = V \Delta_r H^\circ \Delta[L]_{bound}$$

where V is the reaction volume, $\Delta_r H^\circ$ the enthalpy change, and $\Delta[L]_{bound}$ the change in the concentration of the ligand bound to the macromolecule. Consider the following binding equilibrium involving n binding sites:

$$P + nL \rightleftharpoons PL_n$$

From the cumulative heat change (Q) (that is, the total heat change when all the sites are saturated), it is possible to obtain the values of the association constant K_a, n,

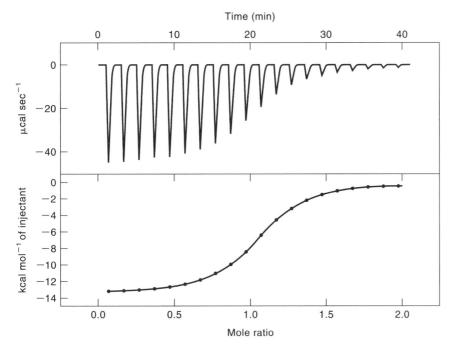

Figure 6.16
Top: Experimental output of power (energy released per unit time) versus time for an isothermal reaction involving the ligand binding to a protein. Each downward peak corresponds to an injection of the ligand solution. The small peaks on the extreme right are due to heat of dilution and mixing. They are subtracted from the data before analysis. Bottom: A binding isotherm for the reaction. This curve is created by plotting the areas under the peaks in the top diagram against the mole ratio ($[L]_{total}/[P]_{total}$). By fitting Q to the curve (see text), we can obtain K_a, n, and $\Delta_r H^\circ$ for the reaction. Note that the energy unit is in calories in these plots. (Courtesy of MicroCal, Inc.)

and $\Delta_r H^\circ$ by a curve fitting procedure all in a single experiment.* Furthermore, from the relationships $\Delta_r G^\circ = -RT \ln K_a$ and $\Delta_r G^\circ = \Delta_r H^\circ - T \Delta_r S^\circ$, we can also obtain $\Delta_r G^\circ$ and $\Delta_r S^\circ$ for the same reaction.

A typical ITC experiment requires only about an hour to perform and can provide a complete thermodynamic profile of the ligand–macromolecular interaction. It has become the method of choice for a whole range of studies of binding equilibria.

6.6 Bioenergetics

Bioenergetics is the study of energy transformation in living systems. As scientists begin to understand many biological and biochemical phenomena at the molecular level, they are learning to apply thermodynamics to the study of living systems. For this discussion of the energetics of biochemical reactions, we shall use glycolysis as an example. In particular, we shall look at the role of adenosine 5′-triphosphate (ATP) in these processes.

---

*The relation between Q and the thermodynamic parameters is rather complicated. Interested readers should see E. Freire, O. L. Mayorga, and M. Straume, *Analytical Chemistry* **62** (18), 950A (1990).

The Standard State in Biochemistry

Physical chemists and biochemists have different definitions for the *standard states*. In physical chemistry, the standard state for an ideal solution is one in which all the reactants and products are at unit molar (or molal) concentration. In biochemistry, we define the hydrogen ion concentration for the standard state to be $10^{-7} \, M$, because the physiological pH is about 7. Consequently, the change in the standard Gibbs energy according to these two conventions will be different for reactions involving uptake or liberation of hydrogen ions, depending on which convention is used. We shall therefore replace $\Delta_r G^\circ$ with $\Delta_r G^{\circ\prime}$ when discussing biochemical processes. Consider the reaction

$$A + B \rightarrow C + xH^+$$

The Gibbs energy change ($\Delta_r G$) for the process is given by

$$\Delta_r G = \Delta_r G^\circ + RT \ln \frac{([C]/1 \, M)([H^+]/1 \, M)^x}{([A]/1 \, M)([B]/1 \, M)}$$

where $1 \, M$ represents the physical chemists' standard state for solute in solution. Because the biochemists' standard state for H^+ ions is $1 \times 10^{-7} \, M$, the Gibbs energy change for the same process is given by

$$\Delta_r G = \Delta_r G^{\circ\prime} + RT \ln \frac{([C]/1 \, M)([H^+]/1 \times 10^{-7} \, M)^x}{([A]/1 \, M)([B]/1 \, M)}$$

Note that regardless of the convention for the standard state, $\Delta_r G$ does not change. From the previous two equations we obtain

$$\Delta_r G^\circ = \Delta_r G^{\circ\prime} + xRT \ln \frac{1}{1 \times 10^{-7}} \tag{6.40}$$

If $x = 1$ and $T = 298$ K, then

$$\Delta_r G^\circ = \Delta_r G^{\circ\prime} + 39.93 \text{ kJ mol}^{-1}$$

This means that for reactions producing H^+ ions, the value of $\Delta_r G^\circ$ is greater than that of $\Delta_r G^{\circ\prime}$ by 39.93 kJ per mole of H^+ ions released. Hence, under standard-state conditions the reaction is more spontaneous at pH 7 than at pH 0. On the other hand, if H^+ ion appears as the reactant,

$$C + xH^+ \rightarrow A + B$$

then

$$\Delta_r G^\circ = \Delta_r G^{\circ\prime} - xRT \ln \frac{1}{1 \times 10^{-7}} \tag{6.41}$$

or, for $x = 1$

$$\Delta_r G^\circ = \Delta_r G^{\circ\prime} - 39.93 \text{ kJ mol}^{-1}$$

This reaction will be more spontaneous at pH 0 than at pH 7. For reactions that do not involve H^+ ions, $\Delta_r G°$ is equal to $\Delta_r G°'$.

Example 6.5

NAD^+ and NADH are the oxidized and reduced forms of nicotinamide adenine dinucleotide. For the oxidation of NADH,

$$NADH + H^+ \rightarrow NAD^+ + H_2$$

$\Delta_r G°$ is -21.8 kJ mol^{-1} at 298 K. Calculate the values of $\Delta_r G°'$, K, and K' for the reaction, where $\Delta_r G°' = -RT \ln K'$. Also calculate the Gibbs energy change ($\Delta_r G$) using both the physical chemical standard state and biochemical standard state for the reaction when $[NADH] = 1.5 \times 10^{-2}$ M, $[H^+] = 3.0 \times 10^{-5}$ M, $[NAD^+] = 4.6 \times 10^{-3}$ M, and $P_{H_2} = 0.010$ bar.

ANSWER

Because H^+ ion appears as the reactant, we use Equation 6.41:

$$\Delta_r G° = \Delta_r G°' - 39.93 \text{ kJ mol}^{-1}$$

From the relation $\Delta_r G° = -RT \ln K$,

$$-21,800 \text{ J mol}^{-1} = -(8.314 \text{ J K}^{-1} \text{ mol}^{-1})(298 \text{ K}) \ln K$$

Hence,

$$\ln K = 8.80$$
$$K = 6.6 \times 10^3$$

Next, we calculate the value of $\Delta_r G°'$

$$\Delta_r G°' = \Delta_r G° + 39.93 \text{ kJ mol}^{-1}$$
$$= -21.8 \text{ kJ mol}^{-1} + 39.93 \text{ kJ mol}^{-1}$$
$$= 18.13 \text{ kJ mol}^{-1}$$

Now, from $\Delta_r G°' = -RT \ln K'$,

$$18,130 \text{ J mol}^{-1} = -(8.314 \text{ J K}^{-1} \text{ mol}^{-1})(298 \text{ K}) \ln K'$$
$$\ln K' = -7.32$$
$$K' = 6.6 \times 10^{-4}$$

Thus,

$$\frac{K}{K'} = 10^7$$

This ratio corresponds to the difference in the two standard states for $[H^+]$. As mentioned earlier, the value of $\Delta_r G$ for the reaction should be the same regardless of what standard state we employ.

Physical Chemists' Standard State

$$\Delta_r G = \Delta_r G^\circ + RT \ln \frac{([NAD^+]/1\ M)(p_{H_2}/1\ \text{bar})}{([NADH]/1\ M)([H^+]/1\ M)}$$

$$= -21{,}800\ \text{J mol}^{-1} + (8.314\ \text{J K}^{-1}\ \text{mol}^{-1})(298\ \text{K})$$

$$\times \ln \frac{(4.6 \times 10^{-3})(0.010)}{(1.5 \times 10^{-2})(3.0 \times 10^{-5})}$$

$$= -10.3\ \text{kJ mol}^{-1}$$

Biochemists' Standard State

$$\Delta_r G = \Delta_r G^{\circ\prime} + RT \ln \frac{([NAD^+]/1\ M)(p_{H_2}/1\ \text{bar})}{([NADH]/1\ M)([H^+]/10^{-7}\ M)}$$

$$= 18{,}130\ \text{J mol}^{-1} + (8.314\ \text{J K}^{-1}\ \text{mol}^{-1})(298\ \text{K})$$

$$\times \ln \frac{(4.6 \times 10^{-3})(0.010)}{(1.5 \times 10^{-2})(3.0 \times 10^{-5}/10^{-7})}$$

$$= -10.3\ \text{kJ mol}^{-1}$$

ATP—The Currency of Energy

Adenosine 5′-triphosphate (Figure 6.17) is the primary energy source for numerous biological reactions, ranging from protein synthesis and ion transport to muscle contraction and electrical activities in nerve cells. The energy required to carry out these processes is derived from the hydrolysis reaction at pH 7:

$$ATP^{4-} + H_2O \rightarrow ADP^{3-} + H^+ + HPO_4^{2-}$$

This reaction is accompanied by a decrease in the standard Gibbs energy of as much as 25 to 40 kJ per mole of ATP hydrolyzed. The exact value depends on the pH, temperature, and the metal ions present. One of the most accurately studied systems is the Mg–ATP complex,* for which $\Delta_r G^{\circ\prime} = -30.5\ \text{kJ mol}^{-1}$ at pH = 7 and $T = 310$

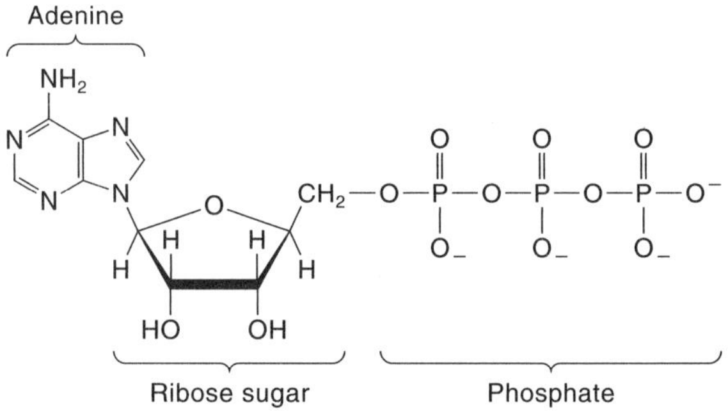

Figure 6.17
Structure of adenosine 5′-triphosphate (ATP). Upon hydrolysis, ATP loses the end phosphate group to form adenosine 5′-diphosphate (ADP). ADP may undergo further hydrolysis to form adenosine 5′-monophosphate (AMP).

*R. A. Alberty, *J. Biol. Chem.* **243**, 1337 (1968). This work is presented at a lower level by the same author in *J. Chem. Educ.* **46**, 713 (1969).

K. Note that biochemists prefer a different way of representing the hydrolysis of ATP:

$$ATP + H_2O \rightarrow ADP + P_i$$

In this equation, the H atoms and charges are not balanced, and symbols like ATP and P_i represent the sum of all forms in which ATP and inorganic phosphate exist. Thus, P_i includes the species PO_4^{3-}, HPO_4^{2-}, and $H_2PO_4^-$; ATP includes ATP^{4-}, $HATP^{3-}$, H_2ATP^{2-}, and so on.

The large decrease in the Gibbs energy that results from the hydrolysis of ATP and ADP prompted some biochemists to use the term *high-energy phosphate bond* in describing these compounds. Unfortunately, this term wrongly implies that the P–O bond in these molecules is somehow different from the normal covalent bond, which is untrue. Then why does $\Delta_r G^{\circ\prime}$ have such a large negative value? To answer this question, we must examine the structures of ATP and its hydrolysis products, ADP and HPO_4^{2-}, because $\Delta G^{\circ\prime}$ depends on the *difference* between the standard Gibbs energies of the products and reactants. At least two factors must be taken into consideration: electrostatic repulsion and resonance stabilization. At pH 7, the triphosphate unit of ATP carries four negative charges:

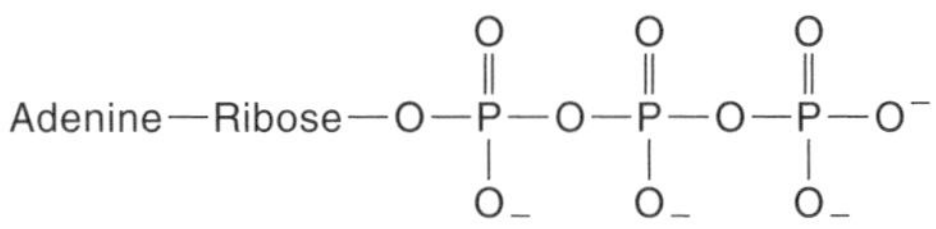

Due to the proximity of the charges, there is considerable electrostatic repulsion. This repulsion is reduced when ATP is hydrolyzed to ADP and HPO_4^{2-} because there are only three negative charges on ADP. The other factor contributing to the large $\Delta_r G^{\circ\prime}$ value is that ADP and HPO_4^{2-} possess more resonance structures than ATP. For example, HPO_4^{2-} has several significant resonance structures:

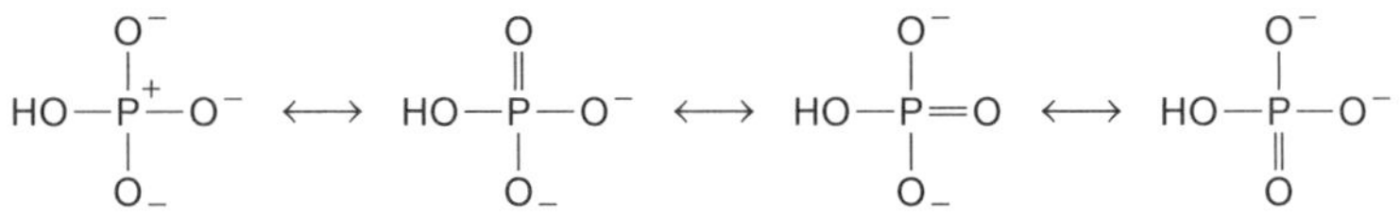

The terminal portion of ATP has fewer significant resonance structures per phosphate group. The following resonance structure for ATP is improbable because one of the oxygen atoms has three bonds, and there is a positive charge on the oxygen atom adjacent to a positively charged phosphorus atom:

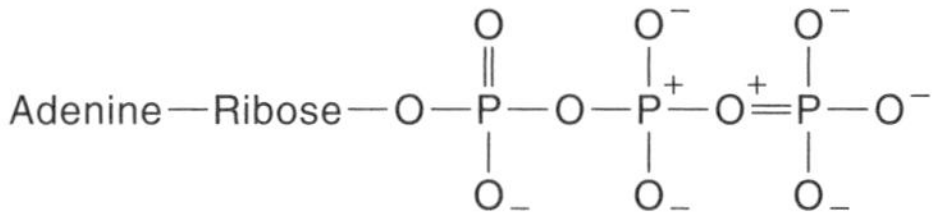

Finally, to a lesser extent, release of the steric crowding among oxygen atoms on adjacent phosphate groups in ATP upon hydrolysis also contributes to the large decrease in the standard Gibbs energy.

As Table 6.2 shows, the hydrolysis of ATP is by no means the most exergonic (see p. 102) hydrolysis reaction. The significance of its relative position among other phosphates will be explained shortly. But first we shall examine the role of coupled reactions in biological processes.

Table 6.2
Standard Gibbs Energies of Hydrolysis of Some
Phosphate Compounds at pH 7[a]

Phosphates	$\Delta_r G^{\circ\prime}/\text{kJ} \cdot \text{mol}^{-1}$
Phosphoenolpyruvate	−61.9
Acetyl phosphate	−43.1
Creatine phosphate	−43.1
Pyrophosphate	−33.5
Adenosine 5′-triphosphate	−30.5
Glucose-1-phosphate	−20.9
Glucose-6-phosphate	−13.8
Glycerol-1-phosphate	−9.2

[a] From H. A. Sober, Ed., *Handbook of Biochemistry*, © The Chemical Rubber Co., 1968. Used by permission of The Chemical Rubber Co.

Principles of Coupled Reactions

Many chemical and biological reactions are endergonic and therefore are not spontaneous under standard-state conditions. However, in some cases, these reactions can be carried out to an appreciable extent by coupling them with an exergonic reaction. Let us consider a chemical process for the extraction of copper from its ore, Cu_2S. As the following equation shows, heating the ore alone will not yield much copper because $\Delta_r G^\circ$ for the reaction is a large positive quantity:

$$Cu_2S(s) \rightarrow 2Cu(s) + S(s) \qquad \Delta_r G^\circ = 86.2 \text{ kJ mol}^{-1}$$

If we couple the thermal decomposition of Cu_2S with the oxidation of sulfur to sulfur dioxide, however, the outcome changes dramatically:

The price we pay for this coupled reaction is acid rain due to the formation of SO_2.

$$Cu_2S(s) \rightarrow 2Cu(s) + S(s) \qquad \Delta_r G^\circ = 86.2 \text{ kJ mol}^{-1}$$
$$S(s) + O_2(g) \rightarrow SO_2(g) \qquad \Delta_r G^\circ = -300.1 \text{ kJ mol}^{-1}$$
$$\text{Overall:} \quad Cu_2S(s) + O_2(g) \rightarrow 2Cu(s) + SO_2(g) \qquad \Delta_r G^\circ = -213.9 \text{ kJ mol}^{-1}$$

The Gibbs energy change for the overall reaction is the sum of the Gibbs energy changes of the two reactions. Because the negative Gibbs energy changes for the oxidation of sulfur is considerably larger than the positive Gibbs energy change for the decomposition of Cu_2S, the overall reaction has a large negative Gibbs energy change, and therefore it favors the formation of Cu. Figure 6.18 shows a mechanical analog of a coupled reaction.

Coupled biological reactions have the following characteristics: (1) an endergonic

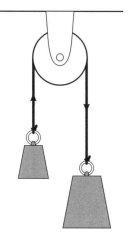

Figure 6.18
A mechanical analog for coupled reactions. Normally weights fall downward under the influence of gravity (a spontaneous process). However, it is possible to make the smaller weight move upwards (a nonspontaneous process) by coupling it with the falling of a larger weight. Overall the process is still spontaneous. Similarly, a reaction with a large negative $\Delta_r G^\circ$ can cause another reaction with a smaller positive $\Delta_r G^\circ$ to proceed in its nonspontaneous direction.

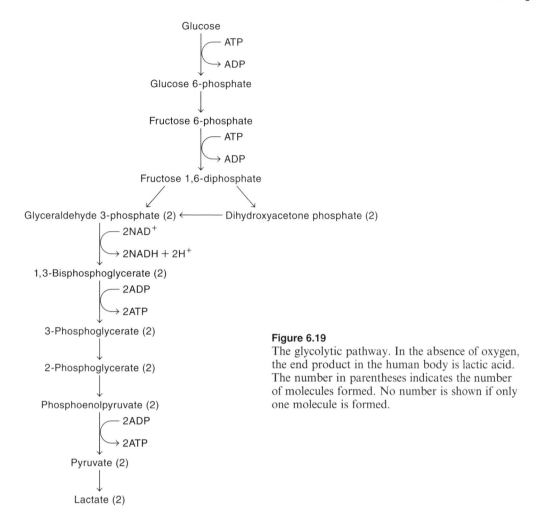

Glucose

ATP

ADP

Glucose 6-phosphate

Fructose 6-phosphate

ATP

ADP

Fructose 1,6-diphosphate

Glyceraldehyde 3-phosphate (2) ⟵ Dihydroxyacetone phosphate (2)

2NAD$^+$

2NADH + 2H$^+$

1,3-Bisphosphoglycerate (2)

2ADP

2ATP

3-Phosphoglycerate (2)

2-Phosphoglycerate (2)

Phosphoenolpyruvate (2)

2ADP

2ATP

Pyruvate (2)

Lactate (2)

Figure 6.19
The glycolytic pathway. In the absence of oxygen, the end product in the human body is lactic acid. The number in parentheses indicates the number of molecules formed. No number is shown if only one molecule is formed.

reaction is coupled with an exergonic reaction so that the combined, coupled reaction is exergonic overall, (2) the exergonic reaction is usually the hydrolysis of ATP to ADP and HPO_4^{2-}, and (3) the coupled reaction is invariably catalyzed by an enzyme. We shall now see examples of coupled reactions in glycolysis.

Glycolysis

All living organisms require energy for growth and function. Plants derive their energy from the sun through the process of photosynthesis. They are called *auto-trophs* because they contain self-feeding cells. We, as *heterotrophs*, survive by feeding on other organisms. The food we eat consists mostly of carbohydrates, proteins, and fats. Through stepwise oxidation, we obtain the necessary energy stored in these molecules. Carbohydrates, for example, serve two purposes: they provide the building blocks for biosynthesis, and they produce energy through oxidation. We shall focus on energy production.

Figure 6.19 shows the metabolic pathway from glucose to pyruvate. A total of nine steps are involved in breaking a six-carbon molecule, glucose, into two three-carbon molecules, pyruvate.* Each of the steps is enzymatically catalyzed. ATP is

---

*For a detailed discussion of glycolysis, refer to the standard biochemistry texts listed in Chapter 1.

utilized in some steps and synthesized in others. The process, however, results in a net gain of 2 moles of ATP per mole of glucose metabolized to pyruvate.

The first step along the glycolytic pathway involves the conversion of glucose to glucose 6-phosphate:

Glucose   Glucose 6-phosphate

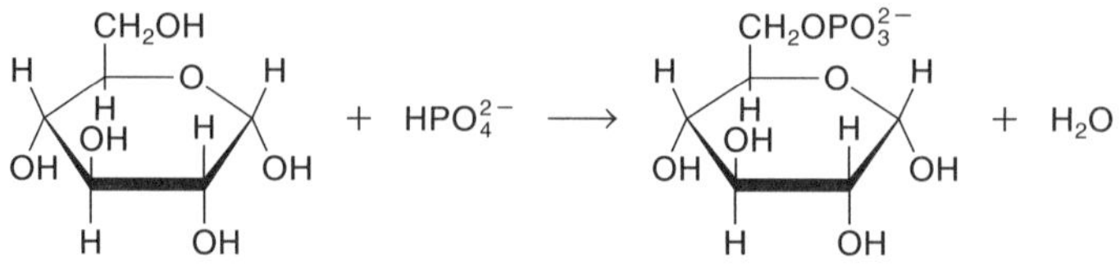

Because this is an endergonic process, the reaction will not favor the formation of products. This reaction can be driven, however, by coupling it with an exergonic reaction, that is, the hydrolysis of ATP. We can write the *coupled* reaction as follows:

$$\text{glucose} + \text{ATP} \rightarrow \text{glucose-6-phosphate} + \text{ADP} \qquad \Delta_r G^{\circ\prime} = -17.2 \text{ kJ mol}^{-1}$$

This reaction will take place spontaneously because $\Delta_r G^{\circ\prime}$ is a large negative quantity.

The coupling of glucose with ATP illustrates the importance of coupled reactions in biological systems. Essentially, we have a reaction that is energetically unfavorable but necessary for metabolism. It can be made to proceed by connecting it with a reaction that releases a large amount of Gibbs energy. The presence of the enzyme glucokinase (or hexokinase) is what makes the coupling reaction possible.

Another coupled reaction converts fructose 6-phosphate to fructose 1,6-diphosphate. Again, 1 mole of ATP is hydrolyzed for each mole of fructose 6-phosphate that is phosphorylated.

The investment of ATP in these reactions is more than compensated for in subsequent steps, which result in a total synthesis of 4 moles of ATP; hence, there is a net gain of 2 moles of ATP. At pH 7.5 and 298 K, the synthesis of ATP from ADP and HPO_4^{2-} is accompanied by an increase in Gibbs energy, $\Delta_r G^{\circ\prime} = 31.4$ kJ mol^{-1}. Thus, this reaction must be linked to an exergonic reaction, which in this case is the oxidation of glyceraldehyde 3-phosphate to 1,3-bisphosphoglycerate (formerly known as 1,3-diphosphoglycerate):

Glyceraldehyde 3-phosphate   1,3-Bisphosphoglycerate

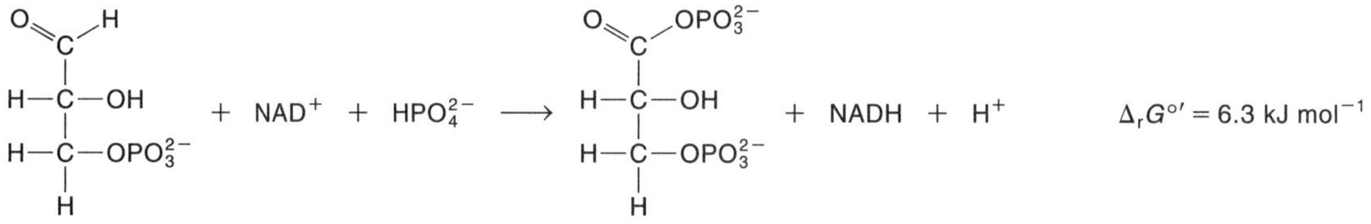

where NAD$^+$ and NADH are the oxidized and reduced forms, respectively, of nicotinamide adenine dinucleotide (Figure 6.20). NAD$^+$, another important biological molecule, functions as an electron carrier. The preceding reaction is slightly endergonic and therefore cannot be used to drive any other energetically unfavorable processes. What is important, however, is that the hydrolysis of the product, 1,3-bisphosphoglycerate, is highly exergonic, so that in the presence of the enzyme phos-

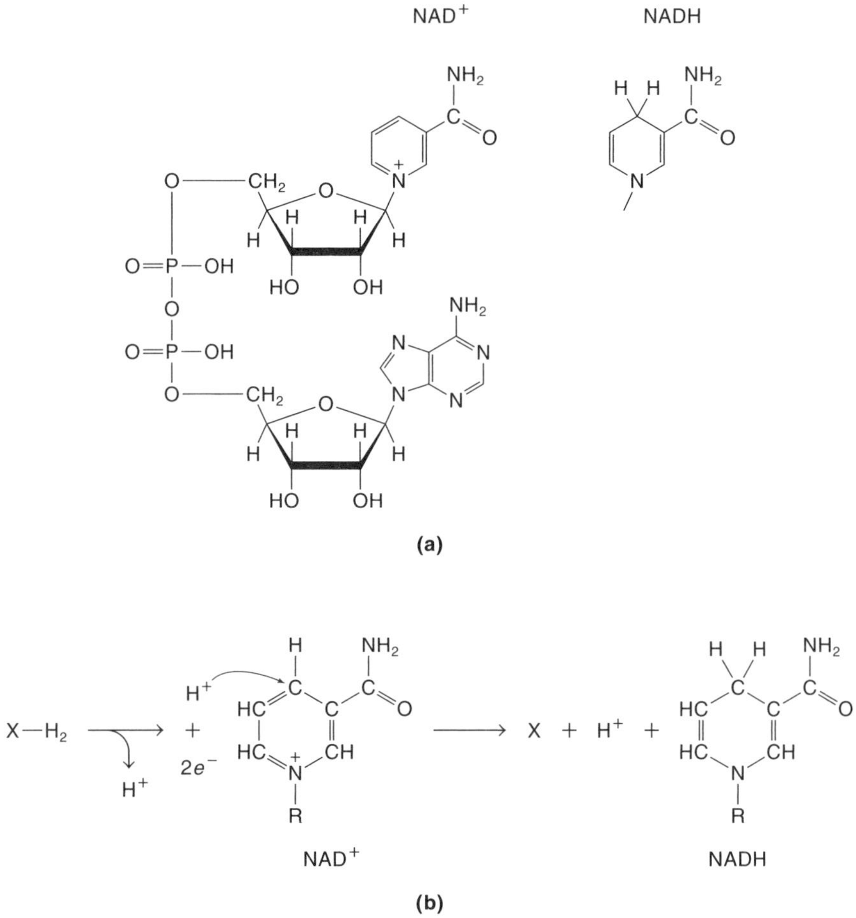

Figure 6.20
(a) Structure of NAD$^+$ and NADH. (b) The reduction of NAD$^+$ to NADH. R represents the rest of the molecule shown in (a), and XH$_2$ is the substrate molecule being oxidized as NAD$^+$ is being reduced. The substrate, XH$_2$, loses two hydrogen ions and two electrons. One of the hydrogen ions is released into the solution. The other hydrogen ion, accompanied by two electrons, combines with the NAD$^+$ to form NADH. The reaction can be written as NAD$^+$ + 2H$^+$ + 2e^- → NADH + H$^+$ or simply as NAD$^+$ + H$_2$ → NADH + H$^+$.

phoglycerate kinase, we have

$$1,3\text{-bisphosphoglycerate} + \text{ADP} \rightarrow 3\text{-phosphoglycerate} + \text{ATP}$$

$$\Delta_r G^{\circ\prime} = -18.8 \text{ kJ mol}^{-1}$$

The large decrease in the standard Gibbs energy for this process ensures the direction of the reaction from left to right. Stoichiometrically, for every mole of glucose decomposed, 2 moles of 1,3-bisphosphoglycerate are produced and hence 2 moles of ATP are formed. The conversion of phosphoenopyruvate to pyruvate also takes place via a coupling process, resulting in an additional 2 moles of ATP.

Having considered the energetics for some individual steps in glycolysis, next we would like to know how the equilibrium constant can be applied to such a process.

For a sequence of reactions, the important quantity is not the equilibrium constant for an isolated reaction; rather, we are concerned with the equilibrium constant for the overall process. Consider the following reactions:

$$A + B \rightleftharpoons C + D \quad K_1 = \frac{[C][D]}{[A][B]}$$

$$C + D \rightleftharpoons E + F \quad K_2 = \frac{[E][F]}{[C][D]}$$

$$E + F \rightleftharpoons G + H \quad K_3 = \frac{[G][H]}{[E][F]}$$

The equilibrium constant for the overall process

$$A + B \rightleftharpoons G + H$$

is given by

$$K_4 = \frac{[G][H]}{[A][B]} = K_1 K_2 K_3$$

The key point here is that even if one of the intermediate steps has a small equilibrium constant, an appreciable amount of the products can still be formed if the other steps are favored from left to right. For example, if $K_1 = 10^5$, $K_2 = 10^{-4}$, and $K_3 = 10^2$, we still have $K_4 = 10^3$. Furthermore, since

$$RT \ln K_4 = RT \ln K_1 + RT \ln K_2 + RT \ln K_3$$

it follows that

$$\Delta_r G_4^{\circ\prime} = \Delta_r G_1^{\circ\prime} + \Delta_r G_2^{\circ\prime} + \Delta_r G_3^{\circ\prime}$$

In reality, glycolysis is never at equilibrium. This analysis merely shows that $\Delta_r G^{\circ\prime}$ for the overall process can be part of the driving force even though some of the steps are not thermodynamically favorable.

The glycolytic pathway discussed so far is an *anaerobic* process; that is, the reactions are carried out in the absence of molecular oxygen. If oxygen is lacking, the reaction does not stop at pyruvate but proceeds through one more step:

$$\underset{\text{pyruvate}}{CH_3COCOO^-} + NADH + H^+ \rightarrow \underset{\text{lactate}}{CH_3CH(OH)COO^-} + NAD^+$$

For anaerobic cells, this is the end of the line—the lactate formed is finally discharged from the cell.* This process is not very efficient because much of the Gibbs energy could still be extracted from a molecule as complex as lactate. The conversion of glucose to lactate via the glycolytic pathway probably arose very early in evolutionary development when little or no molecular oxygen was present on Earth. More recently, perhaps a couple of billion years ago, as a result of the change in Earth's atmosphere, *aerobic* cells developed. In these cells, molecular oxygen is used for the

---

* In humans, the lactic acid causes muscle cramps, or a "charley horse." This painful condition occurs when muscles are exerted suddenly with an inadequate supply of oxygen. In other organisms, such as yeast, ethanol is formed instead of lactic acid.

further degradation of pyruvic acid to carbon dioxide and water via the citric acid cycle and the terminal respiratory chain (see Chapter 7). Consequently, an additional 36 moles of ATP are synthesized, so that the complete degradation of 1 mole of glucose yields a total of 38 moles of ATP:

$$C_6H_{12}O_6 + 38H^+ + 38ADP^{3-} + 38HPO_4^{2-} + 6O_2 \rightarrow 38ATP^{4-} + 6CO_2 + 44H_2O$$

Estimating the efficiency of a biological process such as the one discussed here is interesting. As we saw in Section 4.5, the complete combustion of glucose in air gives

$$C_6H_{12}O_6(s) + 6O_2(g) \rightarrow 6CO_2(g) + 6H_2O(l) \qquad \Delta_r G^\circ = -2879 \text{ kJ mol}^{-1}$$

On the other hand, the synthesis of ATP from ADP and HPO_4^{2-} is as follows:

$$ADP^{3-} + H^+ + HPO_4^{2-} \rightarrow ATP^{4-} + H_2O \qquad \Delta_r G^{\circ\prime} = 31.4 \text{ kJ mol}^{-1}$$

Thus, the efficiency for the degradation of glucose to carbon dioxide and water via glycolysis is given by

$$\text{efficiency} = \frac{\text{Gibbs energy stored in ATP molecules}}{\text{total Gibbs energy released}}$$

$$= \frac{38 \times 31.4 \text{ kJ}}{2879 \text{ kJ}} \times 100\% = 41\%$$

This result is a lower limit value. When the physiological concentrations of various components are taken into account, the efficiency is believed to be more than 50%. What would the efficiency be if the human body behaved as an internal combustion engine? From Equation 4.7 and using 37°C (310 K) as the body temperature and 25°C (298 K) as the temperature of the surroundings, we write

$$\text{efficiency} = \left(\frac{310 \text{ K} - 298 \text{ K}}{310 \text{ K}}\right) = 0.039, \text{ or } 3.9\%$$

which is too inefficient to sustain normal bodily functions.

The impressively high biological efficiency is the result of trial and error over several billion years. In our earlier discussion of gas expansion (see Section 3.1), we saw that a reversible process can perform much more work than an irreversible one. The combustion of glucose in air is a highly irreversible reaction; consequently, the energy stored in the parent molecule is released in a much less useful form (that is, heat). Once the reaction is broken down into several steps with the aid of enzymes, much of the energy can be stored via the synthesis of ATP.

Although high efficiencies are preferred in general, we must also consider the *rate* of a process. As mentioned earlier, maximum efficiency is obtained only for a truly reversible process, but the process would take an infinite amount of time to complete. Thus, if glycolysis were lengthened by 10 more steps, the efficiency would undoubtedly go up because the process would become more reversible. The rate would then be slower, however, perhaps dangerously so for survival. Most likely, a compromise between efficiency and rate will finally be reached through evolution.

One reason ATP is unique in its participation in so many biological reactions is undoubtedly its intermediate value of $\Delta_r G^{\circ\prime}$ for hydrolysis (see Table 6.2). A more negative $\Delta_r G^{\circ\prime}$ value would also mean that more energy would be required for its synthesis, an undesirable situation. On the other hand, a less negative $\Delta_r G^{\circ\prime}$ value for hydrolysis would make ATP much less useful in coupled reactions.

Some Limitations of Thermodynamics in Biology

Our discussion of how some of the thermodynamic concepts can be applied to help us understand the nature of biochemical processes has been rather general and qualitative. In this section, we consider a more fundamental question: Under what conditions is thermodynamics applicable to biology? The answer is not trivial, and considerable debate centers on the subject.

The limitation of standard free energy, $\Delta_r G^{\circ\prime}$, is that it is valid only for reactions in which reactants in their standard states are converted to products in their standard states. Such reactions are *never* the case for biological processes *in vivo*. Therefore, the direction of a reaction of the type

$$A + B \rightleftharpoons C + D$$

is given by $\Delta_r G$, where

$$\Delta_r G = \Delta_r G^{\circ\prime} + RT \ln \frac{[C][D]}{[A][B]}$$

Thus, the fact that the hydrolysis of ATP at 310 K and pH 7 results in a decrease in the standard Gibbs energy equal to -30.5 kJ mol^{-1} does not mean that this number is also the value for $\Delta_r G$. In living cells, the physiological temperature, pH, concentrations of reactants and products, and metal ions can vary from system to system, and these factors must affect the value of $\Delta_r G$. Accurate measurements of the concentrations of various species in solution are difficult, and in some cases impossible to carry out. Nevertheless, with improved instrumentation, we can make reasonable estimates of the concentrations in some cases and hence of the sign of $\Delta_r G$.

Thermodynamics also deals only with closed systems at equilibrium. Living systems are open systems, and they are maintained at a steady state rather than at equilibrium.* In fact, a cell at true equilibrium is a dead cell. Thus, the rates of the biochemical reactions may be more relevant than the values of the equilibrium constants.

Finally, reactions in a living cell, like all other reactions, can be classified into two categories—those that are thermodynamically controlled and those that are kinetically controlled. An example of the former is the synthesis of a dipeptide from two amino acids:

$$\text{alanine} + \text{glycine} \rightarrow \text{alanylglycine} + H_2O \qquad \Delta_r G^{\circ\prime} = 17.2 \text{ kJ mol}^{-1}$$

The corresponding equilibrium constant for this reaction is approximately 1×10^{-3} at 298 K. Clearly, such a process will not proceed to an appreciable extent by itself. If this process could be coupled to the hydrolysis of ATP by the action of an enzyme, however, the reaction would proceed from left to right (Figure 6.21). Such a reaction is said to be thermodynamically controlled because it is not spontaneous, and energy must be supplied from an outside source.

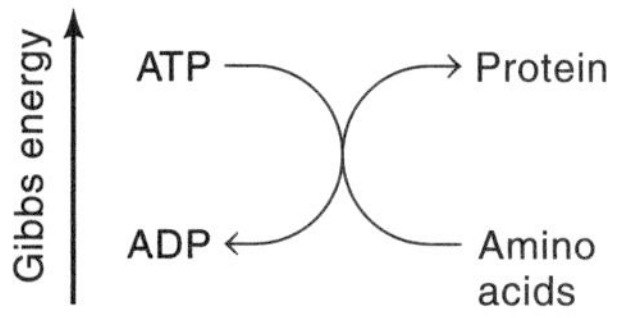

A kinetically controlled reaction is one in which the overall $\Delta_r G$ value is negative (and hence it is thermodynamically favorable), but the rate is negligibly small in the absence of appropriate enzyme catalysis. The phosphorylation of glucose to glucose

Figure 6.21
Schematic representation of the Gibbs energy changes that occur in protein synthesis.

---

*A steady state bears some superficial resemblance to equilibrium in that the concentrations of species remain unchanged with time. To achieve this state, however, we need to constantly supply and remove materials from the system. Furthermore, a system at equilibrium is a homogeneous system, whereas concentration gradients exist in a system maintained at a steady state. The treatment of open systems requires nonequilibrium thermodynamics, which is beyond the scope of this book.

6-phosphate (coupled to ATP hydrolysis) is certainly an exergonic process, but the reaction occurs at a immeasurably slow rate in the absence of the enzyme hexokinase.

Suggestions for Further Reading

BOOKS

Cramer, W. A. and D. B. Knaff, *Energy Transduction in Biological Membranes*, Springer-Verlag, New York, 1991.

Edsall, J. T. and H. Gutfreund, *Biothermodynamics*, John Wiley & Sons, New York, 1983.

Harold, F. M. *The Vital Force: A Study of Bioenergetics*, W. H. Freeman, New York, 1986.

Harris, D. A. *Bioenergetics at a Glance*, Blackwell Science, Oxford, 1995.

Klotz, I. M. *Energy Changes in Biochemical Reactions*, Academic Press, New York, 1967.

Klotz, I. M. *Ligand–Receptor Energetics*, John Wiley & Sons, New York, 1997.

Lehninger, A. L. *Bioenergetics*, W. A. Benjamin, New York, 1965.

Nicholls, D. G. *Bioenergetics*, Academic Press, New York, 1982.

Nicholls, D. G. and S. J. Ferguson, *Bioenergetics*, Academic Press, New York, 1992.

Wyman, J. and S. J. Gill, *Binding and Linkage: Functional Chemistry of Biological Macromolecules*, University Science Books, Sansalito, CA, 1990.

ARTICLES

General

"Some Observations Concerning the van't Hoff Equation," J. T. MacQueen, *J. Chem. Educ.* **44**, 755 (1967).

"Effect of Ionic Strength on Equilibrium Constants," M. D. Seymour and Q. Fernando, *J. Chem. Educ.* **54**, 225 (1977).

"Clarifying the Concept of Equilibrium in Chemically Reacting Systems," W. F. Harris, *J. Chem. Educ.* **59**, 1034 (1982).

"On the Dynamic Nature of Chemical Equilibrium," M. L. Hernandez and J. M. Alvarino, *J. Chem. Educ.* **60**, 930 (1983).

"Le Chatelier's Principle—a Redundant Principle?" R. T. Allsop and N. H. George, *Educ. Chem.* **21**, 82 (1984).

"Le Chatelier's Principle and the Law of van't Hoff," J. Gold and V. Gold, *Educ. Chem.* **22**, 82 (1985).

"A Better Way of Dealing with Chemical Equilibrium," R. J. Tykodi, *J. Chem. Educ.* **63**, 582 (1986).

"The Effect of Temperature and Pressure on Equilibria: A Derivation of the van't Hoff Rules," H. R. Kemp, *J. Chem. Educ.* **64**, 482 (1987).

"Entropy of Mixing and Homogeneous Equilibrium," S. R. Logan, *Educ. Chem.* **25**, 44 (1988).

"Equilibrium, Free Energy, and Entropy: Rates and Differences," J. J. MacDonald, *J. Chem. Educ.* **67**, 380 (1990).

"Equilibria and $\Delta G°$," J. J. MacDonald, *J. Chem. Educ.* **67**, 745 (1990).

"Chemical Equilibrium," A. A. Gordus, *J. Chem. Educ.* **68**, 138, 215, 291, 397, 566, 656, 759, 927 (1991).

"Reaction Thermodynamics: A Flawed Derivation," F. M. Horuack, *J. Chem. Educ.* **69**, 112 (1992).

"The Conversion of Chemical Energy," D. J. Wink, *J. Chem. Educ.* **69**, 264 (1992).

"Practical Calculation of the Equilibrium Constant and the Enthalpy of Reaction at Different Temperatures," K. Anderson, *J. Chem. Educ.* **71**, 474 (1994).

"Teaching Chemical Equilibrium and Thermodynamics in Undergraduate General Chemistry Classes," A. C. Banerjee, *J. Chem. Educ.* **72**, 879 (1995). Also see *J. Chem. Educ.* **73**, A261 (1996).

"Thermodynamics and Spontaneity," R. S. Ochs, *J. Chem. Educ.* **73**, 952 (1996).

"Free Energy Versus Extent of Reaction," R. S. Treptow, *J. Chem. Educ.* **73**, 51 (1996). Also see *J. Chem. Educ.* **74**, 22 (1997).

"The Iron Blast Furnace: A Study in Chemical Thermodynamics," R. S. Treptow and L. Jean, *J. Chem. Educ.* **75**, 43 (1998).

"A Mechanical Analogue for Chemical Potential, Extent of Reaction, and the Gibbs Energy," S. V. Glass and R. L. DeKock, *J. Chem. Educ.* **75**, 190 (1998).

"The Temperature Dependence of $\Delta G°$ and the Equilibrium Constant, K_{eq}; Is There a Paradox?" F. H. Chapple, *J. Chem. Educ.* **75**, 342 (1998).

"How Thermodynamic Data and Equilibrium Constants Changed When the Standard-State Pressure Became 1 Bar," R. S. Treptow, *J. Chem. Educ.* **76**, 212 (1999).

Binding Equilibria

"Equilibrium Dialysis," S. A. Katz, C. Parfitt, and R. Purdy, *J. Chem. Educ.* **47**, 721 (1970).

"Spectroscopic Determination of Protein-Ligand Binding Constants," A. Orstan and J. F. Wojcik, *J. Chem. Educ.* **64**, 814 (1987).

"Dialysis," R. H. Barth in *Encyclopedia of Applied Physics*, G. L. Trigg, Ed., VCH Publishers, New York, 1992, Vol. 4, p. 533.

"Analysis of Receptor-Ligand Interactions," A. D. Atlie and R. T. Raines, *J. Chem. Educ.* **72**, 119 (1995).

"A Thermodynamic Study of Azide Binding to Myoglobin," A. T. Marcoline and T. E. Elgren, *J. Chem. Educ.* **75**, 1622 (1998).

Biochemical Thermodynamics
"Thermodynamics and Biology," B. E. C. Banks, *Chem. Brit.* **5**, 514 (1969).
"Structure of High-Energy Molecules," L. Pauling, *Chem. Brit.* **6**, 468 (1970).
"Thermodynamics and Biology," D. Wilkie, *Chem. Brit.* **6**, 473 (1970).

"Energetics of Muscle," A. F. Huxley, *Chem. Brit.* **6**, 477 (1970).
"How Cells Make ATP," P. C. Hinkle and R. E. McCarthy, *Sci. Am.* March 1972.
"Composite Formulated Biochemical Equilibria," R. R. Richards, *J. Chem. Educ.* **56**, 514 (1979).
"Energetics of Biological Processes," I. H. Segel and L. D. Segel in *Encyclopedia of Applied Physics*, G. L. Trigg, Ed., VCH Publishers, New York, 1993, Vol. 6, p. 207.
"Biochemical Thermodynamics," R. A. Alberty, *Biochem. Biophys. Acta* **1207**, 1 (1994).

Problems

Chemical Equilibrium

6.1 Equilibrium constants of gaseous reactions can be expressed in terms of pressures only (K_P), concentrations only (K_c), or mole fractions only (K_x). For the hypothetical reaction

$$a\text{A}(g) \rightleftharpoons b\text{B}(g)$$

derive the following relationships: **(a)** $K_P = K_c(RT)^{\Delta n}(P^\circ)^{-\Delta n}$ and **(b)** $K_P = K_x P^{\Delta n}(P^\circ)^{-\Delta n}$, where Δn is the difference in the number of moles of products and reactants, and P is the total pressure of the system. Assume ideal-gas behavior.

6.2 At 1024°C, the pressure of oxygen gas from the decomposition of copper(II) oxide (CuO) is 0.49 bar:

$$4\text{CuO}(s) \rightleftharpoons 2\text{Cu}_2\text{O}(s) + \text{O}_2(g)$$

(a) What is the value of K_P for the reaction? **(b)** Calculate the fraction of CuO that will decompose if 0.16 mole of it is placed in a 2.0-L flask at 1024°C. **(c)** What would the fraction be if a 1.0-mole sample of CuO were used? **(d)** What is the smallest amount of CuO (in moles) that would establish the equilibrium?

6.3 Gaseous nitrogen dioxide is actually a mixture of nitrogen dioxide (NO_2) and dinitrogen tetroxide (N_2O_4). If the density of such a mixture is 2.3 g L^{-1} at 74°C and 1.3 atm, calculate the partial pressures of the gases and the value of K_P for the dissociation of N_2O_4.

6.4 About 75% of the hydrogen produced for industrial use is produced by the *steam-reforming* process. This process is carried out in two stages called primary and secondary reforming. In the primary stage, a mixture of steam and methane at about 30 atm is heated over a nickel catalyst at 800°C to give hydrogen and carbon monoxide:

$$\text{CH}_4(g) + \text{H}_2\text{O}(g) \rightleftharpoons \text{CO}(g) + 3\text{H}_2(g)$$
$$\Delta_r H^\circ = 206 \text{ kJ mol}^{-1}$$

The secondary stage is carried out at about 1000°C, in the presence of air, to convert the remaining methane to hydrogen:

$$\text{CH}_4(g) + \tfrac{1}{2}\text{O}_2(g) \rightleftharpoons \text{CO}(g) + 2\text{H}_2(g)$$
$$\Delta_r H^\circ = 35.7 \text{ kJ mol}^{-1}$$

(a) What conditions of temperature and pressure would favor the formation of products in both the primary and secondary stages? **(b)** The equilibrium constant, K_c, for the primary stage is 18 at 800°C. **(i)** Calculate the value of K_P for the reaction. **(ii)** If the partial pressures of methane and steam were both 15 atm at the start, what would the pressures of all the gases be at equilibrium?

6.5 Consider the reaction

$$\text{PCl}_5(g) \rightleftharpoons \text{PCl}_3(g) + \text{Cl}_2(g)$$

for which $K_P = 1.05$ at 250°C. A quantity of 2.50 g of PCl_5 is placed in an evacuated flask of volume 0.500 L and heated to 250°C. **(a)** Calculate the pressure of PCl_5 if it did not dissociate. **(b)** Calculate the partial pressure of PCl_5 at equilibrium. **(c)** What is the total pressure at equilibrium? **(d)** What is the degree of dissociation of PCl_5? (The degree of dissociation is given by the fraction of PCl_5 that has undergone dissociation.)

6.6 The vapor pressure of mercury is 0.002 mmHg at 26°C. **(a)** Calculate the values of K_c and K_P for the process $\text{Hg}(l) \rightleftharpoons \text{Hg}(g)$. **(b)** A chemist breaks a thermometer and spills mercury onto the floor of a laboratory measuring 6.1 m long, 5.3 m wide, and 3.1 m high. Calculate the mass of mercury (in grams) vaporized at equilibrium and the concentration of mercury vapor in mg m^{-3}. Does this concentration exceed the safety limit of 0.05 mg m^{-3}? (Ignore the volume of furniture and other objects in the laboratory.)

6.7 A quantity of 0.20 mole of carbon dioxide was heated to a certain temperature with an excess of graphite in a closed container until the following equilibrium was reached:

$$C(s) + CO_2(g) \rightleftharpoons 2CO(g)$$

Under these conditions, the average molar mass of the gases was 35 g mol^{-1}. **(a)** Calculate the mole fractions of CO and CO$_2$. **(b)** What is the value of K_P if the total pressure is 11 atm? (*Hint:* The average molar mass is the sum of the products of the mole fraction of each gas times its molar mass.)

van't Hoff Equation

6.8 Consider the thermal decomposition of CaCO$_3$:

$$CaCO_3(s) \rightleftharpoons CaO(s) + CO_2(g)$$

The equilibrium vapor pressures of CO$_2$ are 22.6 mmHg at 700°C and 1829 mmHg at 950°C. Calculate the standard enthalpy of the reaction.

6.9 Consider the following reaction:

$$CO_2(g) + H_2(g) \rightleftharpoons CO(g) + H_2O(g)$$

The equilibrium constant is 0.534 at 960 K and 1.571 at 1260 K. What is the enthalpy of the reaction?

6.10 The vapor pressure of dry ice (solid CO$_2$) is 672.2 torr at −80°C and 1486 torr at −70°C. Calculate the molar heat of sublimation of CO$_2$.

6.11 Nitric oxide from car exhaust is a primary air pollutant. Calculate the equilibrium constant for the reaction

$$N_2(g) + O_2(g) \rightleftharpoons 2NO(g)$$

at 25°C using the data listed in Appendix 2. Assume that both $\Delta_r H°$ and $\Delta_r S°$ are temperature independent. Calculate the equilibrium constant at 1500°C, which is the typical temperature inside the cylinders of a car's engine after it has been running for some time.

$\Delta_r G°$ and K

6.12 Calculate the value of $\Delta_r G°$ for each of the following equilibrium constants: 1.0×10^{-4}, 1.0×10^{-2}, 1.0, 1.0×10^2, and 1.0×10^4 at 298 K.

6.13 Use the data listed in Appendix 2 to calculate the equilibrium constant, K_P, for the synthesis of HCl at 298 K:

$$H_2(g) + Cl_2(g) \rightleftharpoons 2HCl(g)$$

What is the value of K_P if the equilibrium is expressed as

$$\tfrac{1}{2}H_2(g) + \tfrac{1}{2}Cl_2(g) \rightleftharpoons HCl(g)$$

6.14 The dissociation of N$_2$O$_4$ into NO$_2$ is 16.7% complete at 298 K and 1 atm:

$$N_2O_4(g) \rightleftharpoons 2NO_2(g)$$

Calculate the equilibrium constant and the standard Gibbs energy change for the reaction. [*Hint:* Let α be the degree of dissociation and show that $K_P = 4\alpha^2 P/(1 - \alpha^2)$, where P is the total pressure.]

6.15 The standard Gibbs energies of formation of gaseous *cis*- and *trans*-2-butene are 67.15 kJ mol^{-1} and 64.10 kJ mol^{-1}, respectively. Calculate the ratio of equilibrium pressures of the gaseous isomers at 298 K.

6.16 Consider the decomposition of magnesium carbonate:

$$MgCO_3(s) \rightleftharpoons MgO(s) + CO_2(g)$$

Calculate the temperature at which the decomposition begins to favor products. Assume that $\Delta_r H°$ and $\Delta_r S°$ are temperature independent. Use the data in Appendix 2 for your calculation.

6.17 Use the data in Appendix 2 to calculate the equilibrium constant (K_P) for the following reaction at 25°C:

$$2SO_2(g) + O_2(g) \rightleftharpoons 2SO_3(g)$$

Calculate K_P for the reaction at 60°C **(a)** using the van't Hoff equation (that is, Equation 6.18); **(b)** using the Gibbs–Helmholtz equation (that is, Equation 4.31) to find $\Delta_r G°$ at 60°C and hence K_P at the same temperature; and **(c)** using $\Delta_r G° = \Delta_r H° - T\Delta_r S°$ to find $\Delta_r G°$ at 60°C and hence K_P at the same temperature. State the approximations employed in each case and compare your results. (*Hint:* From Equation 4.31, you can derive the relationship

$$\frac{\Delta_r G_2}{T_2} - \frac{\Delta_r G_1}{T_1} = \Delta_r H\left(\frac{1}{T_2} - \frac{1}{T_1}\right)$$

Le Chatelier's Principle

6.18 Consider the reaction

$$2NO_2(g) \rightleftharpoons N_2O_4(g) \qquad \Delta_r H° = -58.04 \text{ kJ mol}^{-1}$$

Predict what happens to the system at equilibrium if **(a)** the temperature is raised, **(b)** the pressure on the system is increased, **(c)** an inert gas is added to the system at constant pressure, **(d)** an inert gas is added to the system at constant volume, and **(e)** a catalyst is added to the system.

6.19 Referring to Problem 6.14, calculate the degree of dissociation of N$_2$O$_4$ if the total pressure is 10 atm. Comment on your result.

6.20 At a certain temperature, the equilibrium pressures of NO_2 and N_2O_4 are 1.6 bar and 0.58 bar, respectively. If the volume of the container is doubled at constant temperature, what would be the partial pressures of the gases when equilibrium is re-established?

6.21 Eggshells are composed mostly of calcium carbonate ($CaCO_3$) formed by the reaction

$$Ca^{2+}(aq) + CO_3^{2-}(aq) \rightleftharpoons CaCO_3(s)$$

The carbonate ions are supplied by carbon dioxide produced during metabolism. Explain why eggshells are thinner in the summer when the rate of panting by chickens is greater. Suggest a remedy for this situation.

6.22 Photosynthesis can be represented by

$$6CO_2(g) + 6H_2O(l) \rightleftharpoons C_6H_{12}O_6(s) + 6O_2(g)$$
$$\Delta_r H° = 2801 \text{ kJ mol}^{-1}$$

Explain how the equilibrium would be affected by the following changes: **(a)** the partial pressure of CO_2 is increased, **(b)** O_2 is removed from the mixture, **(c)** $C_6H_{12}O_6$ (glucose) is removed from the mixture, **(d)** more water is added, **(e)** a catalyst is added, **(f)** the temperature is decreased, and **(g)** more sunlight shines on the plants.

6.23 When a gas was heated at atmospheric pressure and 25°C, its color deepened. Heating above 150°C caused the color to fade, and at 550°C the color was barely detectable. At 550°C, however, the color was partially restored by increasing the pressure of the system. Which of the following scenarios best fits the above description? Justify your choice. **(a)** A mixture of hydrogen and bromine, **(b)** pure bromine, or **(c)** a mixture of nitrogen dioxide and dinitrogen tetroxide. (*Hint:* Bromine is reddish, and nitrogen dioxide is brown. The other gases are colorless.)

6.24 Industrially, sodium metal is obtained by electrolyzing molten sodium chloride. The reaction at the cathode is $Na^+ + e^- \rightarrow Na$. We might expect that potassium metal could also be prepared by electrolyzing molten potassium chloride. Potassium metal is soluble in molten potassium chloride, however, and is therefore hard to recover. Furthermore, potassium vaporizes readily at the operating temperature, creating hazardous conditions. Instead, potassium is prepared by the distillation of molten potassium chloride in the presence of sodium vapor at 892°C:

$$Na(g) + KCl(l) \rightleftharpoons NaCl(l) + K(g)$$

Considering that potassium is a stronger reducing agent than sodium, explain why this approach works. (The boiling points of sodium and potassium are 892°C and 770°C, respectively.)

6.25 People living at high altitudes have higher hemoglobin content in their red blood cells than those living near sea level. Explain.

Binding Equilibria

6.26 Derive Equation 6.23 from 6.21.

6.27 The calcium ion binds to a certain protein to form a 1:1 complex. The following data were obtained in an experiment:

Total Ca^{2+}/μM	60	120	180	240	480
Ca^{2+} bound to Protein/μM	31.2	51.2	63.4	70.8	83.4

Determine graphically the dissociation constant of the Ca^{2+}–protein complex. The protein concentration was kept at 96 μM for each run. (1 $\mu M = 1 \times 10^{-6}$ M.)

6.28 An equilibrium dialysis experiment showed that the concentrations of the free ligand, bound ligand, and protein are 1.2×10^{-5} M, 5.4×10^{-6} M, and 4.9×10^{-6} M, respectively. Calculate the dissociation constant for the reaction $PL \rightleftharpoons P + L$. Assume there is one binding site per protein molecule.

Bioenergetics

6.29 The reaction

L-glutamate + pyruvate → α-ketoglutarate + L-alanine

is catalyzed by the enzyme L-glutamate-pyruvate aminotransferase. At 300 K, the equilibrium constant for the reaction is 1.11. Predict whether the forward reaction (left to right) will occur spontaneously if the concentrations of the reactants and products are [L-glutamate] $= 3.0 \times 10^{-5}$ M, [pyruvate] $= 3.3 \times 10^{-4}$ M, [α-ketoglutarate] $= 1.6 \times 10^{-2}$ M, and [L-alanine] $= 6.25 \times 10^{-3}$ M.

6.30 As mentioned in the chapter, the standard Gibbs energy for the hydrolysis of ATP to ADP at 310 K is approximately -30.5 kJ mol^{-1}. Calculate the value of $\Delta_r G°'$ for the reaction in the muscle of a polar sea fish at -1.5°C. (*Hint:* $\Delta_r H°' = -20.1$ kJ mol^{-1}.)

6.31 Under standard-state conditions, one of the steps in glycolysis does not occur spontaneously:

$$\text{glucose} + HPO_4^{2-} \rightarrow \text{glucose-6-phosphate} + H_2O$$
$$\Delta_r G°' = 13.4 \text{ kJ mol}^{-1}$$

Can the reaction take place in the cytoplasm of a cell where the concentrations are [glucose] $= 4.5 \times 10^{-2}$ M, [HPO_4^{2-}] $= 2.7 \times 10^{-3}$ M, and [glucose-6-phosphate] $= 1.6 \times 10^{-4}$ M and the temperature is 310 K?

6.32 The formation of a dipeptide is the first step toward the synthesis of a protein molecule. Consider the following reaction:

$$glycine + glycine \rightarrow glycylglycine + H_2O$$

Use the data in Appendix 2 to calculate the value of $\Delta_r G^{\circ\prime}$ and the equilibrium constant at 298 K, keeping in mind that the reaction is carried out in an aqueous buffer solution. Assume that the value of $\Delta_r G^{\circ\prime}$ is essentially the same at 310 K. What conclusion can you draw about your result?

6.33 From the following reactions at 25°C:

$$fumarate^{2-} + NH_4^+ \rightarrow aspartate^-$$
$$\Delta_r G^{\circ\prime} = -36.7 \text{ kJ mol}^{-1}$$
$$fumarate^{2-} + H_2O \rightarrow malate^{2-}$$
$$\Delta_r G^{\circ\prime} = -2.9 \text{ kJ mol}^{-1}$$

calculate the standard Gibbs energy change and the equilibrium constant for the following reaction:

$$malate^{2-} + NH_4^+ \rightarrow aspartate^- + H_2O$$

6.34 A polypeptide can exist in either the helical or random coil forms. The equilibrium constant for the equilibrium reaction of the helix to the random coil transition is 0.86 at 40°C and 0.35 at 60°C. Calculate the values of $\Delta_r H^\circ$ and $\Delta_r S^\circ$ for the reaction.

Additional Problems

6.35 List two important differences between a steady state and an equilibrium state.

6.36 At a certain temperature, the equilibrium partial pressures are $P_{NH_3} = 321.6$ atm, $P_{N_2} = 69.6$ atm, and $P_{H_2} = 208.8$ atm, respectively. **(a)** Calculate the value of K_P for the reaction described in Example 6.1. **(b)** Calculate the thermodynamic equilibrium constant if $\gamma_{NH_3} = 0.782$, $\gamma_{N_2} = 1.266$, and $\gamma_{H_2} = 1.243$.

6.37 Based on the material covered so far in the text, describe as many ways as you can for calculating the $\Delta_r G^\circ$ value of a process.

6.38 The solubility of *n*-heptane in water is 0.050 g per liter of solution at 25°C. What is the Gibbs energy change for the hypothetical process of dissolving *n*-heptane in water at a concentration of 2.0 g L^{-1} at the same temperature? (*Hint:* First calculate the value of $\Delta_r G^\circ$ from the equilibrium process and then the $\Delta_r G$ value using Equation 6.6.)

6.39 In this chapter, we introduced the quantity $\Delta_r G^{\circ\prime}$, which is the standard Gibbs energy change for a reaction in which the reactants and products are in their biochemical standard states. The discussion focused on the uptake or liberation of H^+ ions. The $\Delta_r G^{\circ\prime}$ can also be applied to reactions involving the uptake and liberation of gases such as O_2 and CO_2. In these cases, the

biochemical standard states are $P_{O_2} = 0.2$ bar and $P_{CO_2} = 0.0003$ bar, where 0.2 bar and 0.0003 bar are the partial pressures of O_2 and CO_2 in air, respectively. Consider the reaction

$$A(aq) + B(aq) \rightarrow C(aq) + CO_2(g)$$

where A, B, and C are molecular species. Derive a relation between $\Delta_r G^\circ$ and $\Delta_r G^{\circ\prime}$ for this reaction at 310 K.

6.40 The binding of oxygen to hemoglobin (Hb) is quite complex, but for our purpose we can represent the reaction as

$$Hb(aq) + O_2(g) \rightarrow HbO_2(aq)$$

If the value of $\Delta_r G^\circ$ for the reaction is -11.2 kJ mol^{-1} at 20°C, calculate the value of $\Delta_r G^{\circ\prime}$ for the reaction. (*Hint:* Refer to the result in Problem 6.39.)

6.41 The K_{sp} value of AgCl is 1.6×10^{-10} at 25°C. What is its value at 60°C?

6.42 Many hydrocarbons exist as structural isomers, which are compounds that have the same molecular formula but different structures. For example, both butane and isobutane have the same molecular formula: C_4H_{10}. Calculate the mole percent of these molecules in an equilibrium mixture at 25°C, given that the standard Gibbs energy of formation of butane is -15.9 kJ mol^{-1} and that of isobutane is -18.0 kJ mol^{-1}. Does your result support the notion that straight-chain hydrocarbons (that is, hydrocarbons in which the C atoms are joined in a line) are less stable than branch-chain hydrocarbons?

6.43 Consider the equilibrium system $3A \rightleftharpoons B$. Sketch the change in the concentrations of A and B with time for the following situations: **(a)** initially only A is present; **(b)** initially only B is present; and **(c)** initially both A and B are present (with A in higher concentration). In each case, assume that the concentration of B is higher than that of A at equilibrium.

6.44 Comment on the validity of using concentrations instead of activities when discussing reactions in biological cells.

6.45 The dissociation constant (K_d) of a 1:1 ligand (L)–protein (P) complex is 2.0×10^{-6}. Calculate the fractional saturation (Y) at an initial protein concentration of 1.3×10^{-6} M and initial ligand concentration of **(a)** 1.8×10^{-6} M and **(b)** 6.4×10^{-5} M.

6.46 The following data show the binding of Mg^{2+} ions with a protein containing n equivalent sites:

$[Mg^{2+}]_{total}/\mu M$	108	180	288	501	752
$[Mg^{2+}]_{free}/\mu M$	35	65	115	248	446

Apply the Scatchard plot to determine n and K_d. The protein concentration is 98 μM.

Electrochemistry

Electrochemical reactions reverse the action of electrolysis. While electrolysis converts electrical energy to chemical energy, electrochemical reactions convert chemical energy directly to electrical energy. There is a convenient difference between electrochemical reactions and chemical reactions: The Gibbs energy change for an electrochemical reaction is equivalent to the maximum electrical work done, which can be readily measured.

In this chapter we shall discuss the basic principles of electrochemistry and their applications to chemical and biological systems, including membrane potentials.

7.1 Electrochemical Cells

When a piece of zinc metal is placed in a $CuSO_4$ solution, two things happen. Some of the zinc metal enters the solution as Zn^{2+} ions and, more obviously, some of the Cu^{2+} ions are converted to metallic copper at the electrode. This spontaneous redox reaction is represented by

$$Zn(s) + Cu^{2+}(aq) \rightarrow Zn^{2+}(aq) + Cu(s)$$

In time, the blue of the $CuSO_4$ solution fades. Similarly, if a piece of copper wire is placed in a $AgNO_3$ solution, silver metal is deposited on the copper wire, and the solution gradually turns blue due to the presence of the hydrated Cu^{2+} ions. In each case, nothing will happen if we exchange the roles of the metals involved.

Now, suppose zinc and copper metals are placed in two separate compartments containing $ZnSO_4$ and $CuSO_4$ solutions, respectively, as shown in Figure 7.1. These

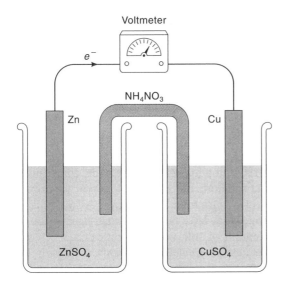

Figure 7.1
Schematic diagram of a galvanic cell. Electrons flow externally from the zinc electrode to the copper electrode. In solution, the anions (SO_4^{2-} and NO_3^-) move toward the zinc anode while the cations (Zn^{2+}, Cu^{2+}, and NH_4^+) move toward the copper cathode.

235

Agar-agar is a polysaccharide.

solutions are connected by a *salt bridge*, a tube that contains an inert electrolyte solution, such as NH_4NO_3 or KCl. This solution is kept from flowing into the compartments by either a sintered disc on each end of the tube or a gelatinous material, such as agar-agar, mixed with the electrolyte solution. When the two electrodes are connected by a piece of metal wire, electrons will flow from the zinc electrode to the copper electrode through the external wire. At the same time, zinc will enter the solution in the left compartment as Zn^{2+} ions, and Cu^{2+} ions will be converted to metallic copper at the copper electrode. The purpose of the salt bridge is to complete the electrical circuit between the two solutions and to facilitate the movement of ions from one compartment to the other.

The setup described above is known as the *Daniell cell*, a type of *galvanic* or *voltaic cell*. The operation of galvanic cells is based on oxidation–reduction, or *redox*, reactions. For the zinc–copper cell, the redox reactions can be expressed in terms of two *half-cell reactions* at the electrodes:

$$\text{Anode:} \quad Zn(s) \rightarrow Zn^{2+}(aq) + 2e^-$$
$$\text{Cathode:} \quad Cu^{2+}(aq) + 2e^- \rightarrow Cu(s)$$

The zinc electrode is called the *anode*, where oxidation (loss of electrons) takes place; the copper electrode is called the *cathode*, where reduction (gain of electrons) takes place. The *cell diagram* for the Daniell cell is given by

$$Zn(s)|ZnSO_4(1.00\ M)||CuSO_4(1.00\ M)|Cu(s)$$

The single vertical line represents a phase boundary. The double vertical lines denote the salt bridge. By convention, the anode is written first, to the left of the double lines, and the other components appear in the order in which we would encounter them in moving from the anode to the cathode. The concentrations of the solutions are usually indicated in the cell diagram.

The fact that electrons flow from the anode to the cathode means that there is a potential difference between the electrodes, which we call the *electromotive force*, or emf (E), of the cell. For the Daniell cell, $E = 1.104$ V at 298 K and equal molar concentrations of $CuSO_4$ and $ZnSO_4$.

7.2 Single Electrode Potentials

Just as monitoring the activity of a single ion is impossible, so is measuring the potential of a single electrode. Any complete circuit must contain two electrodes. By convention, we measure the potential of all electrodes in reference to the *standard hydrogen electrode*, or SHE (Figure 7.2). The potential of the SHE at 298 K, 1 bar H_2 pressure, and 1 M H^+ concentration (more correctly, unit activity) is arbitrarily set to be zero; that is,

$$H^+(1\ M) + e^- \rightleftharpoons \tfrac{1}{2}H_2(1\ \text{bar}) \qquad E° = 0\ V$$

The double arrows mean that the SHE can act as either a cathode or an anode. The measured emf, then, is the potential of the other electrode. Note that we do not need to employ the standard hydrogen electrode for all measurements. As we shall see, it is more convenient to use other electrodes that have been calibrated against the SHE to measure the standard reduction potentials of still other electrodes.

Table 7.1 lists the *standard reduction potentials* for some common half-cell reactions. The more positive the reduction potential, the greater the strength of the

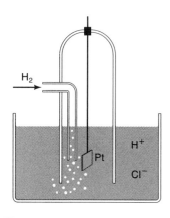

Figure 7.2
Schematic diagram of a hydrogen gas electrode. Hydrogen gas is bubbled into a solution containing H^+ ions. The half-cell redox reaction takes place on the platinum metal immersed in the solution.

Table 7.1
Standard Reduction Potentials, $E°$, for Half-Cells at 298 K (pH = 0)^a

Electrode	Electrode Reaction	$E°$/V
$Pt\|F_2\|F^-$	$F_2(g) + 2e^- \rightarrow 2F^-$	+2.87
$Pt\|Co^{3+}, Co^{2+}$	$Co^{3+} + e^- \rightarrow Co^{2+}$	+1.92
$Pt\|Ce^{4+}, Ce^{3+}$	$Ce^{4+} + e^- \rightarrow Ce^{3+}$	+1.72
$Pt\|MnO_4^-, Mn^{2+}$	$MnO_4^- + 8H^+ + 5e^- \rightarrow Mn^{2+} + 4H_2O$	+1.507
$Pt\|Mn^{3+}, Mn^{2+}$	$Mn^{3+} + e^- \rightarrow Mn^{2+}$	+1.54
$Au\|Au^{3+}$	$Au^{3+} + 3e^- \rightarrow Au$	+1.498
$Pt\|Cl_2\|Cl^-$	$Cl_2(g) + 2e^- \rightarrow 2Cl^-$	+1.36
$Pt\|Cr_2O_7^{2-}, Cr^{3+}$	$Cr_2O_7^{2-} + 14H^+ + 6e^- \rightarrow 2Cr^{3+} + 7H_2O$	+1.23
$Pt\|Tl^{3+}, Tl^+$	$Tl^{3+} + 2e^- \rightarrow Tl^+$	+1.252
$Pt\|O_2, H_2O$	$O_2(g) + 4H^+ + 4e^- \rightarrow 2H_2O$	+1.229
$Pt\|Br_2, Br^-$	$Br_2 + 2e^- \rightarrow 2Br^-$	+1.087
$Pt\|Hg^{2+}, Hg_2^{2+}$	$2Hg^{2+} + 2e^- \rightarrow Hg_2^{2+}$	+0.92
$Hg\|Hg^{2+}$	$Hg^{2+} + 2e^- \rightarrow Hg$	+0.851
$Ag\|Ag^+$	$Ag^+ + e^- \rightarrow Ag$	+0.800
$Pt\|Fe^{3+}, Fe^{2+}$	$Fe^{3+} + e^- \rightarrow Fe^{2+}$	+0.771
$Pt\|I_2, I^-$	$I_2 + 2e^- \rightarrow 2I^-$	+0.536
$Pt\|O_2, OH^-$	$O_2(g) + 2H_2O + 4e^- \rightarrow 4OH^-$	+0.401
$Pt\|Fe(CN)_6^{3-}, Fe(CN)_6^{4-}$	$Fe(CN)_6^{3-} + e^- \rightarrow Fe(CN)_6^{4-}$	+0.36
$Cu\|Cu^{2+}$	$Cu^{2+} + 2e^- \rightarrow Cu$	+0.342
$Pt\|Hg\|Hg_2Cl_2\|Cl^-$	$Hg_2Cl_2 + 2e^- \rightarrow 2Hg + 2Cl^-$	+0.268
$Ag\|AgCl\|Cl^-$	$AgCl + e^- \rightarrow Ag + Cl^-$	+0.222
$Pt\|Sn^{4+}, Sn^{2+}$	$Sn^{4+} + 2e^- \rightarrow Sn^{2+}$	+0.151
$Pt\|Cu^{2+}, Cu^+$	$Cu^{2+} + e^- \rightarrow Cu^+$	+0.153
$Ag\|AgBr\|Br^-$	$AgBr + e^- \rightarrow Ag + Br^-$	+0.0713
$Pt\|H_2\|H^+$	$2H^+ + 2e^- \rightarrow H_2(g)$	0.0
$Pb\|Pb^{2+}$	$Pb^{2+} + 2e^- \rightarrow Pb$	−0.126
$Sn\|Sn^{2+}$	$Sn^{2+} + 2e^- \rightarrow Sn$	−0.138
$Co\|Co^{2+}$	$Co^{2+} + 2e^- \rightarrow Co$	−0.277
$Tl\|Tl^+$	$Tl^+ + e^- \rightarrow Tl$	−0.336
$Pb\|PbSO_4\|SO_4^{2-}$	$PbSO_4 + 2e^- \rightarrow Pb + SO_4^{2-}$	−0.359
$Cd\|Cd^{2+}$	$Cd^{2+} + 2e^- \rightarrow Cd$	−0.403
$Pt\|Cr^{3+}, Cr^{2+}$	$Cr^{3+} + e^- \rightarrow Cr^{2+}$	−0.41
$Fe\|Fe^{2+}$	$Fe^{2+} + 2e^- \rightarrow Fe$	−0.447
$Zn\|Zn^{2+}$	$Zn^{2+} + 2e^- \rightarrow Zn$	−0.762
$Pt\|H_2O\|H_2, OH^-$	$2H_2O + 2e^- \rightarrow H_2(g) + 2OH^-$	−0.828
$Mn\|Mn^{2+}$	$Mn^{2+} + 2e^- \rightarrow Mn$	−1.180
$Al\|Al^{3+}$	$Al^{3+} + 3e^- \rightarrow Al$	−1.662
$Mg\|Mg^{2+}$	$Mg^{2+} + 2e^- \rightarrow Mg$	−2.372
$Na\|Na^+$	$Na^+ + e^- \rightarrow Na$	−2.714
$Ca\|Ca^{2+}$	$Ca^{2+} + 2e^- \rightarrow Ca$	−2.868
$Sr\|Sr^{2+}$	$Sr^{2+} + 2e^- \rightarrow Sr$	−2.899
$Ba\|Ba^{2+}$	$Ba^{2+} + 2e^- \rightarrow Ba$	−2.905
$K\|K^+$	$K^+ + e^- \rightarrow K$	−2.931
$Li\|Li^+$	$Li^+ + e^- \rightarrow Li$	−3.05

^{a} Data mostly from *CRC Handbook of Chemistry and Physics*, 74th ed., CRC Press, Boca Raton, FL, 1993.

oxidizing agent. Thus, F_2 is the strongest oxidizing agent because it has the greatest tendency to pick up electrons, and F^- is the weakest reducing agent. The weakest oxidizing agent is Li^+, which makes lithium metal the most powerful reducing agent. The standard reduction potentials are measured at 298 K for aqueous solutions in which the concentration of each dissolved species is at 1 M and the gas is at 1 bar.

Electrode potential is an intensive property, the value of which depends only on the type of substance, concentration, and temperature, and not on the size of the electrode or on the amount of solution present. Furthermore, the half-cell reactions are reversible. Depending on the conditions, any electrode can act either as an anode or as a cathode. When we reverse a half-cell reaction, the numerical value of $E°$ remains the same, but its sign changes. For example,

$$Sr^{2+}(aq) + 2e^- \rightarrow Sr(s) \qquad\qquad E° = -2.899 \text{ V}$$

and

$$Sr(s) \rightarrow Sr^{2+}(aq) + 2e^- \qquad\qquad E° = 2.899 \text{ V}$$

The standard electrode potential for any electrochemical cell can be readily obtained from the data in Table 7.1. By convention, the emf of a galvanic cell ($E°$) is given by

$$E° = E°_{cathode} - E°_{anode} \qquad\qquad (7.1)$$

where both $E°_{cathode}$ and $E°_{anode}$ refer to the standard reduction potential. Using the Daniell cell as an example, we write

Anode: $Zn(s) \rightarrow Zn^{2+}(aq) + 2e^-$

Cathode: $Cu^{2+}(aq) + 2e^- \rightarrow Cu(s)$

Overall: $Zn(s) + Cu^{2+}(aq) \rightarrow Zn^{2+}(aq) + Cu(s)$

Thus, we calculate the emf of the cell as

$$E° = 0.342 \text{ V} - (-0.762 \text{ V})$$
$$= 1.104 \text{ V}$$

Finally, note that as the electrode reactions progress, the concentrations in both the anode and the cathode compartments change, and the solutes are no longer in their standard-state concentrations. Therefore, the emf of a cell refers only to the initial measured value.

7.3 Thermodynamics of Electrochemical Cells

For us to relate the electrochemical energy associated with a cell to $\Delta_r G$, the cell must behave reversibly in the following manner. If we apply an external potential that is exactly equal but opposite to that of the cell, no reaction occurs within the cell. An infinitesimal decrease or increase in the external potential would lead to either the normal or the reverse cell reaction. Any cell that satisfies these conditions is called a *reversible cell*. The situation described here is analogous to the reversible expansion of a gas discussed in Section 3.1. Under normal conditions, though, cells never operate reversibly; if they did, no current would ever flow through them. By measuring the emf of a cell, however, we learn what it would take to reverse the reaction.

Consider an electrochemical cell reaction in which electrons are transferred from one electrode to the other (that is, from the anode to the cathode). The quantity of charge per mole of reaction, Q (in coulombs), is given by

$$Q = vF$$

where v is the stoichiometric coefficient and F is the faraday constant, after the English chemist and physicist Michael Faraday (1791–1867), which is the charge carried by 1 mole of electrons; that is,

$$\text{faraday constant} = \text{charge of electron} \times \text{number of electrons per mole}$$
$$= 1.6022 \times 10^{-19} \text{ C} \times 6.022 \times 10^{23} \text{ mol}^{-1}$$
$$= 96{,}485 \text{ C mol}^{-1}$$

Except for very accurate work, we round the value of the faraday constant to 96,500 C mol^{-1}. In principle, the total electric current generated by an electrochemical cell can be used for work. The amount of electrical work done is given by the product of potential and charge, $-vFE$, where the potential is the emf E (in volts), charge (in coulombs) is given by vF, and units are related by $1 \text{ J} = 1 \text{ V} \times 1 \text{ C}$. The negative sign indicates that work is done by the cell on the surroundings, in keeping with the convention established in Section 3.1.

For a reversible cell at a given temperature and pressure, $-vFE$ is the maximum work done, which is equal to the decrease in the Gibbs energy of the system (see Equation 4.27):

$$\Delta_r G = -vFE \qquad (7.2)$$

where $\Delta_r G$ is the difference between the Gibbs energy of the products and that of the reactants. Equation 7.2 may be rewritten as

Electrochemical measurements provide the most direct determination of $\Delta_r G$ (or $\Delta_r G^\circ$) for a process.

$$E = \frac{-\Delta_r G}{vF} \qquad (7.3)$$

Recall that at constant temperature and pressure, the criterion for a reaction to proceed spontaneously in the direction written is that $\Delta_r G$ must be a negative value. According to Equation 7.3, this condition corresponds to a positive emf E. Thus, when E is positive for an electrochemical cell, the cell is galvanic, and the cell reaction proceeds as written. When E for an electrochemical cell reaction as written is negative, the forward reaction is electrolytic. An external voltage larger than E must be supplied to the cell to sustain the nonspontaneous process, which is electrolysis.

A special case of Equation 7.3 is a cell in which all reactants and products are in their standard states. In this case, the emf is called the standard emf E°, which is related to the change in the standard Gibbs energy as follows:

$$E^\circ = \frac{-\Delta_r G^\circ}{vF} \qquad (7.4)$$

Because $\Delta_r G^\circ$ is related to the equilibrium constant by Equation 6.7, we have

$$E^\circ = \frac{RT \ln K}{vF} \qquad (7.5)$$

or

$$K = e^{\nu FE^\circ / RT} \qquad (7.6)$$

Thus, a knowledge of E° enables us to calculate the equilibrium constant of the redox cell reaction.

Example 7.1

Calculate the equilibrium constant for the following reaction at 25°C:

$$Sn(s) + 2Ag^+(aq) \rightleftharpoons Sn^{2+}(aq) + 2Ag(s)$$

ANSWER

The two half-reactions for the overall process are

Oxidation: $Sn(s) \rightarrow Sn^{2+}(aq) + 2e^-$

Reduction: $2[Ag^+(aq) + e^- \rightarrow Ag(s)]$

Overall: $\overline{Sn(s) + 2Ag^+(aq) \rightarrow Sn^{2+}(aq) + 2Ag(s)}$

In Table 7.1, we find the standard reduction potentials for these half-reactions and use Equation 7.1 to calculate the E° value for the reaction

$$E^\circ = 0.800 \text{ V} - (-0.138 \text{ V})$$
$$= 0.938 \text{ V}$$

Because E° is a positive quantity, the reaction favors the formation of products at equilibrium. Because $\nu = 2$ (two electrons are transferred in the overall reaction), from Equation 7.6, we write

$$K = \exp\left[\frac{(2)(96,500 \text{ C mol}^{-1})(0.938 \text{ V})}{(8.314 \text{ J K}^{-1} \text{ mol}^{-1})(298 \text{ K})}\right]$$
$$= 5.4 \times 10^{31}$$

Example 7.2

Based on the following electrode potentials,

$$Fe^{2+}(aq) + 2e^- \rightarrow Fe(s) \quad (1) \qquad\qquad E_1^\circ = -0.447 \text{ V}$$

and

$$Fe^{3+}(aq) + e^- \rightarrow Fe^{2+}(aq) \quad (2) \qquad\qquad E_2^\circ = 0.771 \text{ V}$$

calculate the standard reduction potential for the half-reaction

$$Fe^{3+}(aq) + 3e^- \rightarrow Fe(s) \quad (3) \qquad\qquad E_3^\circ = ?$$

ANSWER

It might appear that because the sum of the first two half-reactions gives Equation 3, E_3° is given by $E_1^\circ + E_2^\circ$, or 0.324 V. This is not the case, however. The reason is that emf is not an extensive property, so we cannot set $E_3^\circ = E_1^\circ + E_2^\circ$. On the other hand, the Gibbs energy is an extensive property, so we can add the separate Gibbs energy changes to get the overall Gibbs energy change; that is,

$$\Delta_r G_3^\circ = \Delta_r G_1^\circ + \Delta_r G_2^\circ$$

Substituting the relationship $\Delta_r G^\circ = -\nu F E^\circ$, we obtain

$$\nu_3 F E_3^\circ = \nu_1 F E_1^\circ + \nu_2 F E_2^\circ$$

or

$$E_3^\circ = \frac{\nu_1 E_1^\circ + \nu_2 E_2^\circ}{\nu_3} \quad (\nu_1 = 2, \nu_2 = 1, \nu_3 = 3)$$

$$= \frac{(2)(-0.447\ \text{V}) + (0.771\ \text{V})}{3}$$

$$= -0.041\ \text{V}$$

Equation 7.4 enables us to determine the E° values of the alkali metals and certain alkaline earth metals that react with water. Suppose we want to determine the standard reduction potential of lithium:

$$\text{Li}^+(aq) + e^- \rightarrow \text{Li}(s) \qquad\qquad E^\circ = ?$$

We cannot place a lithium electrode in water, because lithium reacts with water to form hydrogen gas and lithium hydroxide, but we can imagine the following electrochemical process:

$$\text{Li}^+(aq) + \tfrac{1}{2}\text{H}_2(g) \rightarrow \text{Li}(s) + \text{H}^+(aq)$$

This is not a spontaneous reaction.

in which lithium ions are reduced at the lithium electrode, and H_2 molecules are oxidized at the hydrogen electrode. From data in Appendix 2, we can calculate the $\Delta_r H^\circ$ and $\Delta_r S^\circ$ values as follows.

$$\Delta_r H^\circ = \Delta_f \overline{H}^\circ[\text{Li}(s)] + \Delta_f \overline{H}^\circ[\text{H}^+(aq)] - \Delta_f \overline{H}^\circ[\text{Li}^+(aq)] - \left(\tfrac{1}{2}\right)\Delta_f \overline{H}^\circ[\text{H}_2(g)]$$

$$= -(-278.5\ \text{kJ mol}^{-1})$$

$$= 278.5\ \text{kJ mol}^{-1}$$

$$\Delta_r S^\circ = \overline{S}^\circ[\text{Li}(s)] + \overline{S}^\circ[\text{H}^+(aq)] - \overline{S}^\circ[\text{Li}^+(aq)] - \left(\tfrac{1}{2}\right)\overline{S}^\circ[\text{H}_2(g)]$$

$$= 28.03\ \text{J K}^{-1}\ \text{mol}^{-1} - (14.23\ \text{J K}^{-1}\ \text{mol}^{-1}) - \left(\tfrac{1}{2}\right)(130.6\ \text{J K}^{-1}\ \text{mol}^{-1})$$

$$= -51.5\ \text{J K}^{-1}\ \text{mol}^{-1}$$

Thus, at 298 K,

$$\Delta_r G^\circ = \Delta_r H^\circ - T\Delta_r S^\circ$$

$$= 278.5 \text{ kJ mol}^{-1} - (298 \text{ K})\left(\frac{-51.5 \text{ J K}^{-1}}{1000 \text{ J/kJ}}\right) \text{mol}^{-1}$$

$$= 293.8 \text{ kJ mol}^{-1}$$

Finally, from Equation 7.4,

$$E^\circ = \frac{-\Delta_r G^\circ}{\nu F}$$

$$= \frac{-293.8 \times 1000 \text{ J mol}^{-1}}{96,485 \text{ C mol}^{-1}}$$

$$= -3.05 \text{ V}$$

By the same procedure, we can calculate the E° values of other reactive metals and fluorine (F_2), which also reacts with water (see Problem 7.43).

The Nernst Equation

We can now derive an equation relating the emf of a cell to variables such as temperature and the concentrations of reacting species. The Gibbs energy change for the cell reaction

$$a\text{A} + b\text{B} \rightarrow c\text{C} + d\text{D}$$

is given by (see Equation 6.6)

$$\Delta_r G = \Delta_r G^\circ + RT \ln \frac{a_C^c a_D^d}{a_A^a a_B^b}$$

where a denotes activity. Dividing this equation throughout by $-\nu F$ and using both Equations 7.3 and 7.4, we obtain

$$E = E^\circ - \frac{RT}{\nu F} \ln \frac{a_C^c a_D^d}{a_A^a a_B^b} \qquad (7.7)$$

Equation (7.7) is known as the *Nernst equation*, after the German chemist Walter Hermann Nernst (1864–1941). Here, E is the observed emf of the cell and E° is the standard emf of the cell (that is, the emf when all the reactants and products are in their unit activity standard states). At equilibrium, $E = 0$, so

$$E^\circ = \frac{RT}{\nu F} \ln K = \frac{-\Delta_r G^\circ}{\nu F}$$

Because most electrochemical cells operate at or near room temperature, we can evaluate the quantity RT/F by taking $R = 8.314 \text{ J K}^{-1} \text{ mol}^{-1}$, $T = 298 \text{ K}$, and $F = 96,500 \text{ C mol}^{-1}$. Therefore,

$$\frac{(8.314 \text{ J K}^{-1} \text{ mol}^{-1})(298 \text{ K})}{96,500 \text{ C mol}^{-1}} = 0.0257 \text{ J C}^{-1}$$

$$= 0.0257 \text{ V}$$

Finally, Equation 7.7 can be expressed as

$$E = E^\circ - \frac{0.0257 \text{ V}}{v} \ln \frac{a_C^c a_D^d}{a_A^a a_B^b} \tag{7.8}$$

Example 7.3

Predict whether the following reaction would proceed spontaneously as written at 298 K:

$$Cd(s) + Fe^{2+}(aq) \rightarrow Cd^{2+}(aq) + Fe(s)$$

given that $[Cd^{2+}] = 0.15 \ M$ and $[Fe^{2+}] = 0.68 \ M$.

ANSWER

The half-cell reactions are

$$\text{Anode:} \quad Cd(s) \rightarrow Cd^{2+}(aq) + 2e^-$$
$$\text{Cathode:} \quad Fe^{2+}(aq) + 2e^- \rightarrow Fe(s)$$

From Equation 7.1 and Table 7.1, we write

$$E^\circ = -0.447 \text{ V} - (-0.403 \text{ V})$$
$$= -0.044 \text{ V}$$

Assuming ideal behavior and noting that the activities of the solids are unity, the Nernst equation for this reaction is

$$E = -0.044 \text{ V} - \frac{0.0257 \text{ V}}{2} \ln \frac{[Cd^{2+}]}{[Fe^{2+}]}$$

$$= -0.044 \text{ V} - \frac{0.0257 \text{ V}}{2} \ln \frac{0.15 \ M}{0.68 \ M}$$

$$= -0.025 \text{ V}$$

Because E is negative, the reaction is not spontaneous as written, and the reaction must be

$$Cd^{2+}(aq) + Fe(s) \rightarrow Cd(s) + Fe^{2+}(aq)$$

At what ratio of $[Cd^{2+}]$ to $[Fe^{2+}]$ does the reaction in Example 7.3 become spontaneous as written? To find out, we first set E equal to zero, which corresponds to the equilibrium situation, and we write the Nernst equation as

$$0 = -0.044 \text{ V} - \frac{0.0257 \text{ V}}{2} \ln \frac{[Cd^{2+}]}{[Fe^{2+}]}$$

or

$$\frac{[Cd^{2+}]}{[Fe^{2+}]} = 0.033 = K$$

Thus, the reaction will be spontaneous if the ratio $[Cd^{2+}]/[Fe^{2+}]$ is smaller than 0.033, so that E is positive.

Temperature Dependence of EMF

Thermodynamic values for a cell reaction can be obtained from the temperature dependence of the emf. Starting with

$$\Delta_r G° = -\nu F E°$$

and differentiating $\Delta_r G°$ with respect to temperature at constant pressure, we get

$$\left(\frac{\partial \Delta_r G°}{\partial T}\right)_P = -\nu F \left(\frac{\partial E°}{\partial T}\right)_P$$

Equation 4.29, when expressed in terms of changes in G and S, is given by

$$\left(\frac{\partial \Delta_r G°}{\partial T}\right)_P = -\Delta_r S°$$

so that

$$\Delta_r S° = \nu F \left(\frac{\partial E°}{\partial T}\right)_P \qquad (7.9)$$

Thus, from the variation of $E°$ with temperature,* we can measure the standard entropy change of a cell reaction. Suppose that we want to determine the value of $(\partial E°/\partial T)_P$ for the Daniell cell. The simplest way is to set $[Zn^{2+}] = 1.00\ M$ and $[Cu^{2+}] = 1.00\ M$ (the standard states) and then measure the emf of the cell at several different temperatures. Once $\Delta_r S°$ and $\Delta_r G°$ are known at a certain temperature T, we can calculate the value of $\Delta_r H°$ as follows:

$$\Delta_r G° = \Delta_r H° - T \Delta_r S°$$

or

$$\Delta_r H° = \Delta_r G° + T \Delta_r S°$$

$$\Delta_r H° = -\nu F E° + \nu F T \left(\frac{\partial E°}{\partial T}\right)_P \qquad (7.10)$$

In general, $\Delta_r H°$ and $\Delta_r S°$ are roughly temperature independent (for a range of 50 K or smaller), but $\Delta_r G°$ varies with temperature. Note that Equation 7.10 provides us with a noncalorimetric way of determining the enthalpy change of a reaction.

---

*The temperature dependence of the emf of most car batteries is generally quite small, of the order of 5×10^{-4} V K^{-1}, which is insufficient to explain why cars will not start on a cold morning. For an interesting explanation of the real cause, see L. K. Nash, *J. Chem. Educ.* **47**, 382 (1970).

7.4 Types of Electrochemical Cells

The galvanic cell discussed earlier is one of several types of electrochemical cells in use. Two examples we shall consider here are concentration cells and fuel cells.

Concentration Cells

A concentration cell contains electrodes made of the same metal and solutions containing the same ions but at different concentrations. An example is the $ZnSO_4$ concentration cell:

$$Zn(s)|ZnSO_4(0.10\ M)||ZnSO_4(1.0\ M)|Zn(s)$$

The electrode reactions are given by

Anode:	$Zn(s) \rightarrow Zn^{2+}(0.10\ M) + 2e^-$
Cathode:	$Zn^{2+}(1.0\ M) + 2e^- \rightarrow Zn(s)$
Overall:	$Zn^{2+}(1.0\ M) \rightarrow Zn^{2+}(0.10\ M)$

Note that the cathode compartment in a concentration cell always contains the more concentrated solution because it has a greater tendency to accept electrons.

This is a dilution process, so as the electrode reactions progress, the anode compartment becomes more concentrated in Zn^{2+} ions while the cathode department becomes more diluted in Zn^{2+} ions. Eventually, when the concentrations in the two compartments are the same, the cell ceases to function. The initial emf of the cell at 298 K is

$$E = E^\circ - \frac{RT}{vF} \ln \frac{[Zn^{2+}]_{dil}}{[Zn^{2+}]_{conc}}$$

$$= -\frac{0.0257\ V}{2} \ln \frac{0.10\ M}{1.0\ M}$$

$$= 0.030\ V$$

Because the same electrode is used in the cell, E° is zero in the Nernst equation. The concentration cell generally produces small emfs and has no practical use. The concept of its operation is important in the study of membrane potentials, however, as we shall see later.

Fuel Cells

Fossil fuel is a major source of energy at present. Unfortunately, the combustion of fossil fuel is a highly irreversible process, and so its thermodynamic efficiency is low. Fuel cells, however, can make combustion largely reversible by converting a greater amount of chemical energy to useful work. Furthermore, they do not operate like a heat engine and therefore are not subject to the same kind of thermodynamic limitations in energy conversion (see Equation 4.7).

Consider the hydrogen–oxygen fuel cell, the simplest example. Such a cell consists of an electrolyte solution, such as sulfuric acid or sodium hydroxide, and two inert electrodes. Hydrogen and oxygen gases are bubbled through the anode and cathode compartments, where the following reactions take place:

Anode:	$H_2(g) + 2OH^-(aq) \rightarrow 2H_2O(l) + 2e^-$
Cathode:	$\frac{1}{2}O_2(g) + H_2O(l) + 2e^- \rightarrow 2OH^-(aq)$
Overall:	$H_2(g) + \frac{1}{2}O_2(g) \rightarrow H_2O(l)$

E° of the cell is 1.229 V at 298 K.

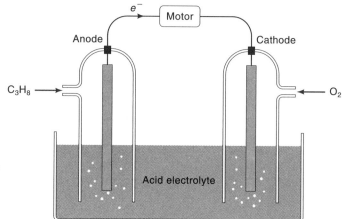

Figure 7.3
Schematic diagram of a
propane–oxygen fuel cell.

The overall reaction is the same as the combustion of hydrogen gas in air. A potential difference is established between the two electrodes, and electrons flow from the anode to the cathode through the wire connecting the two electrodes.

The function of the electrodes is twofold. First, the anode acts as a source of electrons, and the cathode as a sink. Second, the electrodes provide the necessary surface for the initial decomposition of the molecules into atomic species. They are *electrocatalysts*. Metals such as platinum, iridium, and rhodium are good electrocatalysts.

Another type of fuel cell is the propane–oxygen fuel cell shown in Figure 7.3. The half-cell reactions are

Anode: $C_3H_8(g) + 6H_2O(l) \rightarrow 3CO_2(g) + 20H^+(aq) + 20e^-$

Cathode: $5O_2(g) + 20H^+(aq) + 20e^- \rightarrow 10H_2O(l)$

Overall: $C_3H_8(g) + 5O_2(g) \rightarrow 3CO_2(g) + 4H_2O(l)$

The overall reaction is identical to burning propane in oxygen. With proper design, the efficiency of a propane–oxygen fuel cell may be as high as 70%, about twice that of an internal combustion engine. In addition, fuel cells generate electricity without the noise, vibration, heat transfer, and other problems normally associated with conventional power plants. The advantages are so attractive that fuel cells are most likely to become operational on a large scale in the 21st century. At present, much effort is being spent in search of suitable electrocatalysts for various gases.

7.5 Applications of EMF Measurements

We shall now discuss two important applications of emf measurements.

Determination of Activity Coefficients

Emf measurements provide one of the most convenient and accurate methods for determining the activity coefficient of ions. As an example, let us consider the following cell arrangement:

$$Pt|H_2(1 \text{ bar})|HCl(m)|AgCl(s)|Ag$$

The overall reaction for the cell is

$$\tfrac{1}{2}H_2(g) + AgCl(s) \rightarrow Ag(s) + H^+(aq) + Cl^-(aq)$$

and the emf of the cell at 298 K is given by

$$E = E^\circ - 0.0257 \text{ V} \ln \frac{a_{H^+} a_{Cl^-} a_{Ag}}{f_{H_2}^{1/2} a_{AgCl}}$$

Because both Ag and AgCl are solids, their activities are unity; the fugacity of hydrogen gas at 1 bar is also approximately unity, so the preceding equation reduces to

$$E = E^\circ - 0.0257 \text{ V} \ln a_{H^+} a_{Cl^-}$$

From Equation 5.55, we find that

$$a_{H^+} a_{Cl^-} = \gamma_\pm^2 m_\pm^2$$
$$= \gamma_\pm^2 m^2$$

For a 1:1 electrolyte such as HCl, $m_\pm = m$. Thus, the emf of the cell can be expressed as

$$E = E^\circ - 0.0257 \text{ V} \ln(\gamma_\pm m)^2$$
$$= E^\circ - 0.0514 \text{ V} \ln m - 0.0514 \text{ V} \ln \gamma_\pm$$

The above equation can be rearranged to give

$$E + 0.0514 \text{ V} \ln m = E^\circ - 0.0514 \text{ V} \ln \gamma_\pm$$

By measuring E over a range of molalities of HCl, the quantity $(E + 0.0514 \text{ V} \ln m)$ can be calculated at various molalities. If we plot $(E + 0.0514 \text{ V} \ln m)$ against m and extrapolate to zero m, we determine the value of E°, because at zero m the mean activity coefficient is unity so $\ln \gamma_\pm = 0$. Once we know the value of E°, $\gamma_\pm$ can be found at a particular value of m (see Problem 7.45).

Determination of pH

Determining pH from emf measurements is a standard analytical technique. Because using the hydrogen electrode itself for this purpose is impractical, the actual setup involves the use of a *glass electrode*, which responds specifically to H^+ ions and a reference electrode, called the *calomel electrode*.* The cell arrangement is as follows:

$$\underbrace{Ag(s)|AgCl(s)|HCl(aq), NaCl(aq)}_{\text{glass electrode}} \underbrace{|HCl(aq)|}_{\substack{\text{solution of}\\ \text{unknown pH}}} \underbrace{KCl(sat'd)|Hg_2Cl_2(s)|Hg(l)}_{\text{calomel electrode}}$$

The overall emf E for this arrangement at 298 K is given by

$$E = E_{ref} - 0.0591 \text{ V} \log a_{H^+}$$
$$= E_{ref} + 0.0591 \text{ V pH}$$

We change from ln to log because of the definition of pH.

---

*See the physical chemistry texts listed on p. 6 for a description of the glass electrode and the calomel electrode.

where $pH = -\log a_{H^+}$ and E_{ref} is the standard electrode potential difference between the glass electrode and the calomel electrode. In practice, we can replace a_{H^+} with $[H^+]$, except in very precise work. Rearranging the above equation, we get

$$pH = \frac{E - E_{ref}}{0.0591 \text{ V}}$$

We can determine the value of E_{ref} by measuring E for a number of solutions of accurately known pH. Once E_{ref} is known, the combination of the glass electrode and the calomel electrode can be used to find the pH of other solutions from the values of E. The practical arrangement of this combination is called a pH meter.

7.6 Biological Oxidations

In Chapter 6, we saw that glycolysis is an inefficient process. If the system lacks oxygen, the end product of glycolysis, pyruvate, is reduced to lactate with NADH as the reducing agent. In an aerobic process, the degradation of glucose proceeds through two more steps, the *citric acid cycle* and terminal *respiratory chain* (Figure 7.4), and the end products are carbon dioxide and water.* The NADH and $FADH_2$†

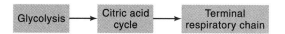

Figure 7.4
The three main stages in the breakdown of glucose to carbon dioxide and water.

formed in glycolysis and the citric acid cycle can reduce molecular oxygen, releasing a large amount of Gibbs energy that can be used to synthesize ATP. *Oxidative phosphorylation* is the process in which ATP is formed as a result of the transfer of electrons from NADH or $FADH_2$ to O_2 by a series of electron carriers. It is the primary source of ATP in aerobic organisms.

In some biological oxidation–reduction reactions, or simply *biological oxidations*, hydrogen ions are transferred along with electrons. Thus, we may have a reaction of the form

$$AH_2 + B \rightarrow [A + 2H^+ + 2e^- + B] \rightarrow A + BH_2$$

In other instances, the substance being oxidized may lose hydrogen ions, while transferring only its electrons to the substance being reduced:

$$AH_2 + B \rightarrow [A + 2H^+ + 2e^- + B] \rightarrow A + B^{2-} + 2H^+$$

The third type of biological oxidation involves only the transfer of electrons:

$$A^{2-} + B \rightarrow [A + 2e^- + B] \rightarrow A + B^{2-}$$

---

*Organic compounds are thermodynamically unstable with respect to CO_2 and H_2O, which are products of their reactions with O_2. Fortunately for us, O_2 possesses a triplet ground state (two unpaired electrons), whereas almost everything else possesses singlet ground states (paired electrons only). Because of this difference in ground electronic states, the reactions are spin forbidden and there is a large activation energy barrier. This kinetic barrier is what makes the existence of life possible.

†FAD and $FADH_2$ are the oxidized and reduced forms of flavin adenine dinucleotide, an oxidation–reduction carrier molecule of the same general type as NAD^+ and NADH.

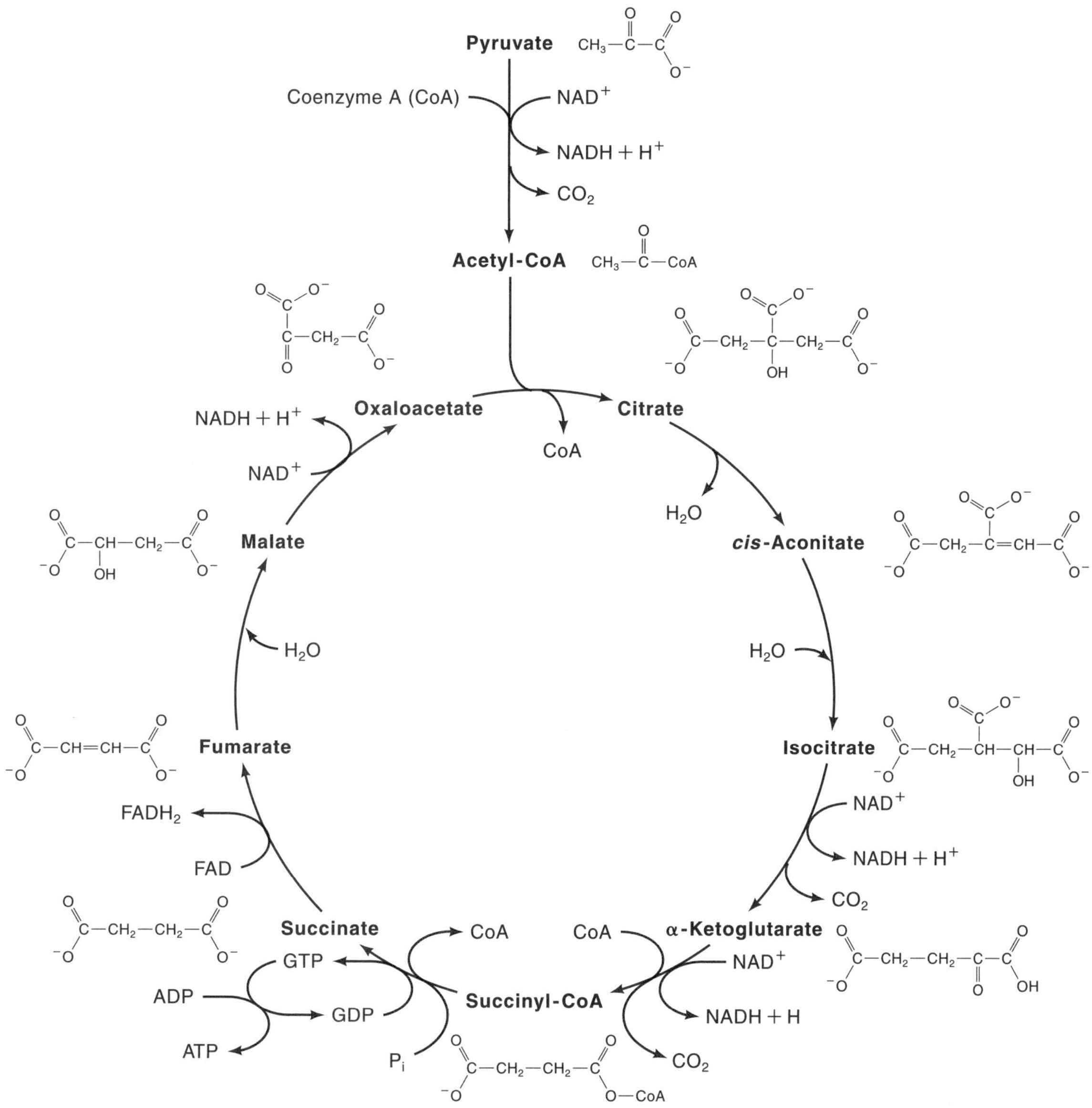

Figure 7.5
The citric acid cycle (also known as the Krebs cycle).

Biological oxidation, seldom, if ever, take places in a simple, direct manner. Generally, the mechanism is quite complex and involves several enzymes. In this section, we briefly describe how knowledge of redox potentials can be applied to the study of some biological processes.

The "primer" step that precedes the citric acid cycle is the combination of pyruvate with a molecule called reduced coenzyme A (CoA) to form acetyl coenzyme A:

$$CH_3COCOO^- + NAD^+ + CoA \rightarrow \quad CH_3COCoA \quad + CO_2 + NADH$$
$$\text{pyruvate} \qquad\qquad\qquad\qquad \text{acetyl coenzyme A}$$

Because one mole of glucose yields two moles of pyruvate in glycolysis (see Figure 6.19), two complete trips around the citric acid cycle, shown in Figure 7.5, are necessary for every mole of glucose degraded. The citirc acid cycle thus produces two

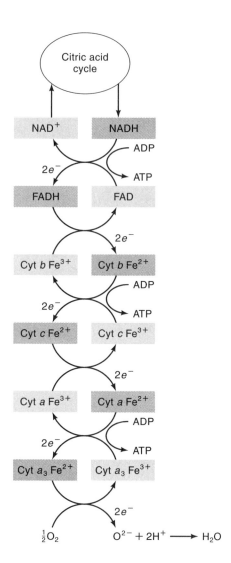

Figure 7.6
The terminal respiratory chain. This chain consists of two separate sets of reactions: electron transport and phosphorylation. Electrons derived from reactions of the citric acid cycle are passed in sequence from one carrier to another. Each carrier alternates between the reduced state (red) and oxidized state (gray). The final electron acceptor is molecular oxygen.

moles of ATP for every mole of glucose degraded, and also generates reduced forms of the carrier molecules NADH and FADH$_2$, which are used in the additional synthesis of ATP in the terminal respiratory chain (Figure 7.6), the last stage in the metabolic process.

In the terminal respiratory chain, the electrons donated by glucose and carried by NADH are transferred from one electron carrier molecule to another. Eventually, they are passed on to a molecule of oxygen, which thereby becomes converted to water. The cytochromes are electron-carrying proteins containing the heme group. The letters b, c, a, and a_3 denote different forms of cytochrome. The iron atom of each cytochrome molecule can exist in the oxidized (Fe^{3+}) or reduced (Fe^{2+}) form. Cytochromes a and a_3 together are called *cytochrome oxidase*, or the *respiratory enzyme*, which can transfer its electron directly to molecular oxygen. The sequence of the electron carriers in the respiratory chain is determined by their relative redox potentials. Table 7.2 lists the standard redox potentials for several important biological systems.

Notice that the standard reduction potentials ($E^{\circ\prime}$) are based on the hydrogen electrode scale at pH 7, rather than pH 0, the reference point for the values listed in Table 7.1. A similar situation involving $\Delta_r G^\circ$ and $\Delta_r G^{\circ\prime}$ was discussed in Section 6.6.

Table 7.2
Standard Reduction Potentials $E^{\circ\prime}$ for Some Biological Half-Reactions at 298 K (pH 7)^a

System	Half-Cell Reaction	$E^{\circ\prime}$/V
O_2/H_2O	$O_2(g) + 4H^+ + 4e^- \rightarrow 2H_2O$	+0.816
Cu^{2+}/Cu^+ hemocyanin	$Cu^{2+} + e^- \rightarrow Cu^+$	+0.540
Cyt f^{3+}/Cyt f^{2+}	$Fe^{3+} + e^- \rightarrow Fe^{2+}$	+0.365
Cyt a^{3+}/Cyt a^{2+}	$Fe^{3+} + e^- \rightarrow Fe^{2+}$	+0.29
Cyt c^{3+}/Cyt c^{2+}	$Fe^{3+} + e^- \rightarrow Fe^{2+}$	+0.254
Fe^{3+}/Fe^{2+} hemoglobin	$Fe^{3+} + e^- \rightarrow Fe^{2+}$	+0.17
Fe^{3+}/Fe^{2+} myoglobin	$Fe^{3+} + e^- \rightarrow Fe^{2+}$	+0.046
Fumarate/succinate	$^-OOCCH{=}CHCOO^- + 2H^+ + 2e^- \rightarrow {}^-OOCCH_2CH_2COO^-$	+0.031
MB/MBH$_2$^{b}	$MB + 2H^+ + 2e^- \rightarrow MBH_2$	+0.011
Oxaloacetate/malate	$^-OOC{-}COCH_2COO^- + 2H^+ + 2e^- \rightarrow {}^-OOCCHOHCH_2COO^-$	−0.166
Pyruvate/lactate	$CH_3COCOO^- + 2H^+ + 2e^- \rightarrow CH_3CHOHCOO^-$	−0.185
Acetaldehyde/ethanol	$CH_3CHO + 2H^+ + 2e^- \rightarrow CH_3CH_2OH$	−0.197
FAD/FADH$_2$	$FAD + 2H^+ + 2e^- \rightarrow FADH_2$	−0.219
NAD$^+$/NADH	$NAD^+ + H^+ + 2e^- \rightarrow NADH$	−0.320
NADP$^+$/NADPH	$NADP^+ + H^+ + 2e^- \rightarrow NADPH$	−0.324
CO_2/formate	$CO_2(g) + H^+ + 2e^- \rightarrow HCOO^-$	−0.414
H^+/H_2	$2H^+ + 2e^- \rightarrow H_2(g)$	−0.421
Fe^{3+}/Fe^{2+} ferredoxin	$Fe^{3+} + e^- \rightarrow Fe^{2+}$	−0.432
Acetic acid/acetaldehyde	$CH_3COOH + 2H^+ + 2e^- \rightarrow CH_3CHO + H_2O$	−0.581
Acetate/pyruvate	$CH_3COOH + CO_2(g) + 2H^+ + 2e^- \rightarrow CH_3COCOOH + H_2O$	−0.70

^{a} From *Handbook of Biochemistry*, Sober, H. A., Ed. © The Chemical Rubber Co., 1968. Used by permission of The Chemical Rubber Co.
^{b} The symbols MB and MBH$_2$ represent the oxidized and reduced forms of methylene blue, respectively, which is used as a redox indicator.

Again, we consider the reaction that produces H^+ ions (see p. 218):

$$A + B \rightarrow C + xH^+$$

for which we write (at $T = 298$ K and $x = 1$)

$$\Delta_r G^\circ = \Delta_r G^{\circ\prime} + 39.93 \text{ kJ mol}^{-1}$$

Dividing the last equation by $-vF$, we obtain

$$E^\circ = E^{\circ\prime} - \frac{39{,}930 \text{ J mol}^{-1}}{v(96{,}500 \text{ C mol}^{-1})}$$

$$E^\circ = E^{\circ\prime} - \frac{0.414}{v} \text{ V}$$

This result means that for reactions producing H^+ ions, the value of E° is less than that of $E^{\circ\prime}$ by $0.414/v$ volt per mole of H^+ ions produced. Hence, the reaction is more spontaneous at the biochemists' standard state with pH 7 than at the physical chemists' standard state with pH 0. On the other hand, if H^+ ions appear as a reactant,

$$C + xH^+ \rightarrow A + B$$

the following reaction can be readily derived:

$$E^\circ = E^{\circ\prime} + \frac{0.414}{\nu} \text{ V}$$

Here, the reaction is more spontaneous at pH 0 than at pH 7. For reactions not involving hydrogen ions, $E^\circ = E^{\circ\prime}$.

We can calculate the Gibbs energy change as a pair of electrons move through the terminal respiratory chain. The conversion of NADH to NAD^+ releases two electrons (the first step), which are used to reduce molecular oxygen to water. From Table 7.2, the half-reactions are

$$NAD^+ + H^+ + 2e^- \rightarrow NADH \qquad\qquad E^{\circ\prime} = -0.32 \text{ V}$$

$$\tfrac{1}{2}O_2 + 2H^+ + 2e^- \rightarrow H_2O \qquad\qquad E^{\circ\prime} = 0.816 \text{ V}$$

The overall reaction is

$$NADH + H^+ + \tfrac{1}{2}O_2 \rightarrow NAD^+ + H_2O \qquad\qquad E^{\circ\prime} = 1.136 \text{ V}$$

Because $\Delta_r G^{\circ\prime} = -\nu F E^{\circ\prime}$, we write

$$\Delta_r G^{\circ\prime} = -(2)(96{,}500 \text{ C mol}^{-1})(1.136 \text{ V})$$
$$= -219 \text{ kJ mol}^{-1}$$

The Gibbs energy released here is considerably larger than that which can be stored in the synthesis of ATP molecules ($\Delta_r G^{\circ\prime} = 31.4 \text{ kJ mol}^{-1}$). The solution to this dilemma is a series of smaller steps that release Gibbs energy along the chain shown in Figure 7.6. There are actually three sites of ATP synthesis: one between NADH and FAD, another between cytochrome b and cytochrome c, and a third between cytochrome a and cytochrome a_3.

Finally, let us calculate the total number of moles of ATP synthesized when glucose is completely oxidized to H_2O and CO_2. The number of moles of ATP formed during glycolysis (2) and the citric acid cycle (2) are unequivocally known. An estimated 34 more are formed by oxidative phosphorylation.* Thus, a total of 38 ATP molecules are synthesized for every glucose molecule degraded to H_2O and CO_2, and most of the released Gibbs energy is stored in the terminal respiratory chain (Table 7.3).

The Chemiosmotic Theory of Oxidative Phosphorylation

In the series of metabolic reactions that we have considered, reduced coenzymes, formed in the citric acid cycle, pass their electrons to an assembly of electron acceptors called the electron transport chain. As the electrons travel down this chain, much of the energy they release is coupled to the phosphorylation of ADP to ATP. How is the transfer of electrons through the series of carrier molecules coupled to the synthesis of ATP? The answer is provided by the chemiosmotic theory proposed by the English biochemist Peter Mitchell (1920–1992) in 1961.

Oxidative phosphorylation in eukaryotic cells (that is, cells with nuclei) occurs in the mitochondria. Mitochondria are subcellular organelles that are responsible

---

*There is no agreement among biochemistry texts about the total number of ATP molecules synthesized by the complete oxidation of glucose to carbon dioxide and water. The number ranges from 30 to 38.

Table 7.3
Products from the Biological Degradation of 1 Mole of Glucose to Carbon Dioxide and Water

	Products (mol)		
Process	NADH	FADH	ATP
Glycolysis			
ATP needed for coupled reactions			−2
ATP produced			+4
NADH produced	+2		
Citric acid cycle			
ATP produced			+2
NADH produced	+8		
FADH produced		+2	
Terminal respiratory chain			
ATP produced from 2 mol of NADH from glycolysis (2 × 3)			+6
ATP produced from 8 mol of NADH from the citric acid cycle (8 × 3)			+24
ATP produced from 2 mol of FADH from the citric acid cycle (2 × 2)			+4
ATP from glycolysis			2 (5%)
ATP from the citric acid cycle			2 (5%)
ATP from the terminal respiratory chain			34 (90%)

for the reactions of respiratory metabolism, among other things. Figure 7.7a shows a schematic diagram of a mitochondrion, which has two membrane systems: an outer one and an inner one. Embedded in the inner membrane are the enzymes and other components of the respiratory chain, where oxidative phosphorylation takes place. According to Mitchell, the energy released during the transport of the electrons along

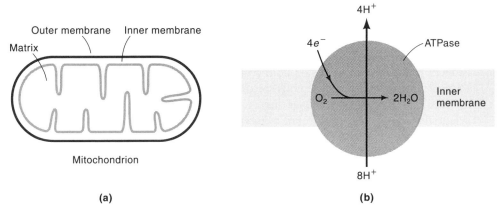

(a)   **(b)**

Figure 7.7
(a) A mitochondrion has two membrane systems: an outer membrane and a folded inner membrane. The two compartments in mitochondria are the intermembrane space between the outer and inner membranes, and the matrix, which is bounded by the inner membrane.
(b) Oxidative phosphorylation takes place in the inner membrane while the citric acid cycle occurs in the matrix. It takes four electrons to reduce one oxygen molecule. In the process, eight protons are picked up from the matrix side of the membrane. Four of these protons are used to provide the hydrogens for water. The remaining four are pumped to intermembrane space.

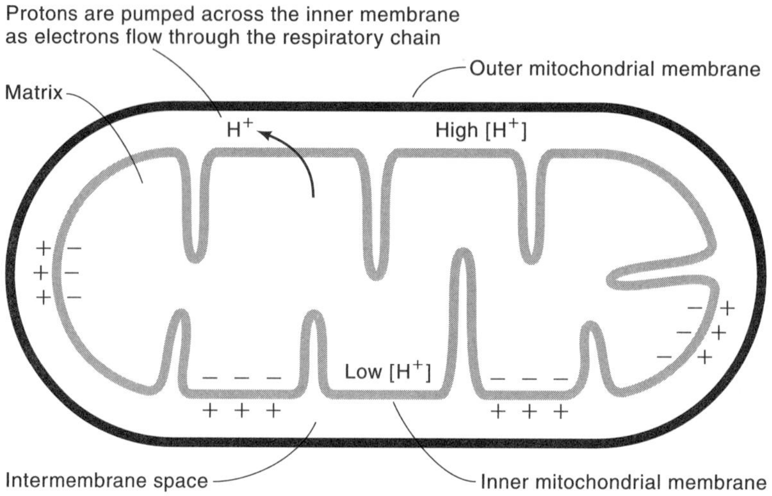

Figure 7.8
According to the chemiosmotic theory, the flow of electrons through the electron-transport network pumps H^+ ions across the inner membrane from the matrix to the intermembrane space. The back flow of H^+ ions into the matrix drives the formation of ATP from ADP and P_i.

the carrier chain is conserved in a hydrogen ion gradient and an electrical gradient, which then drive the oxidative phosphorylation. As electrons flows down the electron transport chain, hydrogen ions are expelled from the matrix to the intermembrane space (Figure 7.7b). This expulsion results in a rise in pH on the inside and a fall in pH on the outside of the inner membrane, and a pH gradient is maintained (Figure 7.8). The electrical potential also rises across the membrane because more positive ions (H^+) are on the outside than on the inside. Protons at the outer surface will seek to move back to the inside, down the potential gradient; this proton gradient, which is analogous to the electric current produced by a battery, can be drawn upon to do work. The flow of protons to the synthesis of ATP involves the enzyme ATP-synthase (ATPase), located in the inner membrane. Current estimates are that the transport of four H^+ ions from the external medium through the ATPase complex into the mitochondrial matrix provides the necessary energy for the formation of one molecule of ATP from ADP and P_i. The term *chemiosmotic theory* is based on the fact that chemical reactions can drive, or be driven by, the movement of molecules or ions between osmotically distinct regions separated by membranes.

The first convincing evidence for the chemiosmotic theory was obtained with thylakoids (see Section 15.2). The principles by which ATP synthesis takes place in chloroplasts are nearly identical to those for oxidative phosphorylation in mitochondria.

Finally, how much Gibbs energy is released by moving protons back into the matrix? The Gibbs energy change depends both on the ratio of the proton concentrations on the two sides of the membrane and on the difference between the electrical potentials on the two sides. We start by writing the chemical potential of the H^+ ion:

$$\mu_{H^+} = \mu_{H^+}^\circ + RT \ln[H^+] + zF\psi \qquad (7.11)$$

The term $zF\psi$ represents the electrochemical potential, where z is the charge of the ion, F the faraday constant, and ψ the electrical potential. Because $pH = -\log[H^+]$

and $z = 1$, Equation 7.11 becomes

$$\mu_{H^+} = \mu_{H^+}^\circ - 2.3RT\,\text{pH} + F\psi \qquad (7.12)$$

The Gibbs energy change when one mole of H^+ ions move across the membrane into the matrix is given by

$$\Delta_r G = \mu_{H^+(in)} - \mu_{H^+(out)} = -2.3RT\Delta\text{pH} + F\Delta\psi \qquad (7.13)$$

where subscripts *in* and *out* denote the inner and outer regions of the inner membrane, respectively, $\Delta\text{pH} = \text{pH}_{in} - \text{pH}_{out}$, and $\Delta\psi = \psi_{in} - \psi_{out}$. Setting $\Delta\text{pH} = 0.5$, $\Delta\psi = -0.15$ V, and $T = 310$ K, we obtain

$$\Delta_r G = -2.3(8.314 \text{ J K}^{-1} \text{ mol}^{-1})(310 \text{ K})(0.5) + (96,500 \text{ C mol}^{-1})(-0.15 \text{ V})$$

$$= -3.0 \text{ kJ mol}^{-1} - 14.5 \text{ kJ mol}^{-1}$$

$$= -17.5 \text{ kJ mol}^{-1}$$

If four protons move across the membrane for each ATP molecule synthesized, the $\Delta_r G$ associated with the proton movement is $4 \times (-17.5 \text{ kJ mol}^{-1})$, or -70 kJ mol^{-1}, which is more than twice the value of $\Delta_r G^{\circ\prime}$ for ATP synthesis (31.4 kJ mol^{-1}). Thus, the Gibbs energy decrease is large enough to account for the [ATP] to [ADP][P$_i$] ratio that mitochondria generate under physiological conditions.

7.7 Membrane Potential

Electrical potentials exist across the membranes of various kinds of cells. Some cells, such as nerve cells and muscle cells, are said to be excitable because they are capable of transmitting a change of potential along their membranes. In this section, we shall briefly discuss the nature of membrane potentials.

A human nerve cell consists of a cell body and a single long fiber extension approximately 10^{-5} to 10^{-3} cm in diameter, called the *axon*, which transmits impulses from the cell body to the adjacent nerve cell (Figure 7.9). Table 7.4 shows the ion distribution of a typical nerve cell. The membrane of the axon is similar in structure

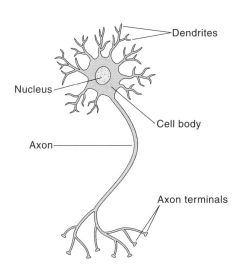

Figure 7.9
Schematic diagram of a neuron (nerve cell), made up of a cell body, an axon, and dendrites. The dendrites carry unidirectional nerve impulses from other neurons toward the cell body. The axon transmits impulses to the adjacent neurons.

Table 7.4
Distribution of Major Ions on Opposite Sides of the Membrane of a Typical Nerve Cell

Ion	Concentration/mM	
	Intracellular	Extracellular
Na^+	15	150
K^+	150	5
Cl^-	10	110

to other cell membranes (see Section 5.6) and similar in composition to the fluid in the cell body. The electrical potential established by the difference in ionic concentrations across the membrane is known as the *membrane potential.*

To understand how membrane potentials arise, let us consider the simple chemical systems shown in Figure 7.10. Figure 7.10a depicts two KCl solutions, both at a concentration of 0.01 *M*, separated by a membrane permeable to K^+ but not to Cl^- ions, so K^+ diffuses across the membrane without its counterion (Cl^-). Because the concentration in the two compartments is the same, the net transport of K^+ ions in either direction is zero and so is the electrical potential across the membrane. The arrangement in Figure 7.10b shows that the concentration in the left compartment is 10 times that in the right compartment. In this case, more K^+ will diffuse from left to right, producing an increase in positive charge on the right and establishing a potential difference across the membrane. The movement of K^+ ions ceases when the extra positive charges in the right compartment begin repelling additional positive charges, while the electrostatic attraction of the excess negative charges in the left compartment holds K^+ ions there. The potential difference due to charge separation across

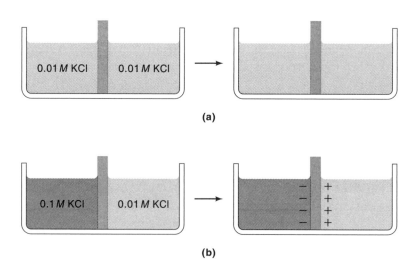

(a)

(b)

Figure 7.10
Two compartments are separated by a membrane permeable only to K^+ ions. (a) Because the concentrations in the two compartments are equal, there is no net flow of ions across the membrane and no electrical potential. (b) A difference in concentration causes K^+ ions to move from the left compartment to the right one. At equilibrium, an electrical potential is established across the membrane due to an accumulation of negative charges on the left side and positive charges on the right. Only a small fraction of the K^+ ions take part in establishing the membrane potential.

the membrane at equilibrium is the equilibrium membrane potential, or simply the membrane potential, of K^+ ions.*

We can calculate the membrane potential for K^+ ions as follows. The Nernst equation for a single type of ion at 298 K is

$$E_{K^+} = E_{K^+}^\circ - \frac{0.0257\text{ V}}{\nu}\ln[K^+]$$

By convention, the electrical potential inside a nerve cell (or other living cells), E_{in}, is expressed relative to the potential outside the cell, E_{ex}. That is, the membrane potential is given as $E_{in} - E_{ex}$. Because $\nu = 1$, we write the membrane potential of the K^+ ions, ΔE_{K^+}, as

$$\Delta E_{K^+} = E_{K^+,in} - E_{K^+,ex} = 0.0257\text{ V}\ln\frac{[K^+]_{ex}}{[K^+]_{in}}$$

From Table 7.4, we obtain

$$\Delta E_{K^+} = 0.0257\text{ V}\ln\frac{5}{150} = -8.7\times10^{-2}\text{ V} = -87\text{ mV}$$

Using an arrangement like the one shown in Figure 7.11, we find that the membrane potential of a nerve cell is only about -70 mV. The reason for the discrepancy is that there is also a membrane potential due to the presence of Na^+ ions. Because the concentration of Na^+ ions is much higher outside the cell, the movement of Na^+ ions into the cell makes the inside more positive. Referring again to Table 7.4, we write

$$\Delta E_{Na^+} = 0.0257\text{ V}\ln\frac{[Na^+]_{ex}}{[Na^+]_{in}}$$

$$= 0.0257\text{ V}\ln\frac{150}{15} = 5.9\times10^{-2}\text{ V} = 59\text{ mV}$$

Because the membrane is much more permeable to K^+ ions than to Na^+ ions, however, the measured potential is closer to the membrane potential of K^+ ions.

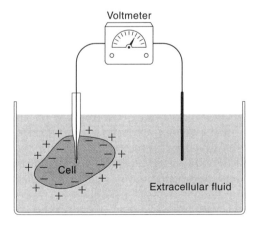

Figure 7.11
Arrangement for measuring the membrane potential of a cell.

---

*If the membrane is impermeable to certain ions (Cl^- in this case), then their presence will have no effect on the membrane potential.

Recall that the experimentally measured membrane potential does not equal the K^+ membrane potential because some Na^+ ions continually move into the cell, canceling the effect of K^+ ions that are simultaneously moving out. If such net ion movements occur, why does the concentration of intracellular Na^+ not progressively increase and that of intracellular K^+ progressively decrease? The reason is that there is a specific membrane protein called Na^+–K^+ ATPase that transports Na^+ out of the cell and K^+ into it using energy from ATP hydrolysis (see Figure 5.27).

The Goldman Equation

The Nernst equation enables us to calculate membrane potential for only one ionic species at a time; it is not applicable for several types of ions that are distributed in unequal concentrations across the same membrane. To calculate membrane potential in such cases, we must employ the Goldman equation (after the American biophysicist David Eliot Goldman, 1910–1998), a generalization of the Nernst equation that has been extended to include the relative permeability of each ionic species. Applied to a nerve cell at 298 K, the Goldman equation takes the form

$$E = 0.0257 \text{ V} \ln \frac{[K^+]_{ex} P_{K^+} + [Na^+]_{ex} P_{Na^+} + [Cl^-]_{ex} P_{Cl^-}}{[K^+]_{in} P_{K^+} + [Na^+]_{in} P_{Na^+} + [Cl^-]_{in} P_{Cl^-}} \tag{7.14}$$

where P is the permeability of the membrane to an ion. Nerve cell membranes in the resting (unperturbed) state are about 100 times more permeable to K^+ ions than to Na^+ ions and are nearly impermeable to Cl^- ions (that is, $P_{Cl^-} \approx 0$). Under these conditions, Equation 7.14 becomes

$$E = 0.0257 \text{ V} \ln \frac{[K^+]_{ex} P_{K^+} + [Na^+]_{ex} P_{Na^+}}{[K^+]_{in} P_{K^+} + [Na^+]_{in} P_{Na^+}}$$

$$= 0.0257 \text{ V} \ln \frac{[K^+]_{ex} P_{K^+}/P_{Na^+} + [Na^+]_{ex}}{[K^+]_{in} P_{K^+}/P_{Na^+} + [Na^+]_{in}} \tag{7.15}$$

Because $P_{K^+}/P_{Na^+} \approx 100$, we write

$$E = 0.0257 \text{ V} \ln \frac{5 \times 100 + 150}{150 \times 100 + 15}$$

$$= -81 \text{ mV}$$

This value is closer to the experimentally determined membrane potential.

The Action Potential

If a nerve cell is stimulated electrically, chemically, or mechanically, the cell membrane becomes much more permeable to Na^+ ions than to K^+ ions so $P_{K^+}/P_{Na^+} \approx 0.17$. (The permeability of the membrane to K^+ ions does not change very much at first but there is a 600-fold increase in the permeability to Na^+ ions.) Nerve-cell stimulation causes a small fraction of the Na^+ ions to rush into the cell, thus changing the membrane potential (the membrane is said to be *depolarized*). From Equation 7.15, we write

$$E = 0.0257 \text{ V} \ln \frac{5 \times 0.17 + 150}{150 \times 0.17 + 15}$$

$$= 34 \text{ mV}$$

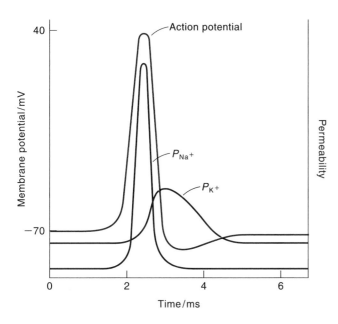

Figure 7.12
The rise and fall of an action potential and the changes in membrane permeability to Na^+ and K^+ ions during this event.

During a very short time (less than 1 ms), the membrane potential changes from −70 mV to about 35 mV (inside positive) and then rapidly returns to its original value (Figure 7.12). The sudden spike in the membrane potential is called the *action potential*.

What causes the membrane potential to return so quickly to its resting value? Two factors are involved. First, the increased Na^+ permeability is rapidly turned off after the initial influx of Na^+ ions into the cell. Second, the membrane's permeability to K^+ ions increases relative to its resting value over a short time (about 1 ms). For this reason, the membrane potential actually dips below −70 mV initially before it returns to its normal value (see Figure 7.12), at which point the cell is ready to "fire" again when it receives another signal. The small number of excess Na^+ ions present in the cell are eventually pumped out of the cell.

The events that give rise to an action potential occur in and around a small area of the nerve cell membrane. The action potential is then propagated along the axon of a neuron. Looking at Figure 7.9, you may be tempted to assume that axons act like electrical cables. After all, the axon does have a cablelike structure. It has a core of electrolytic solution, and it is surrounded by a membrane that acts as an electrical insulator. The resistance of axoplasm (the cytoplasm within the axon), however, is some 100 million times greater than that of copper of the same size. Therefore, the axon would be a relatively poor electrical conductor. Yet we know that when an action potential is generated at a particular site on a neuron, it moves rapidly and without any decrease in magnitude along the axon. Figure 7.13 shows the mechanism of action-potential propagation. The depolarizing influx of Na^+ ions at the immediate site of the action potential causes the membrane potential in adjacent regions to depolarize slowly. When this slow depolarization has pushed the potential of the adjacent membrane beyond a certain value, called the *threshold potential*,* the membrane's permeability to Na^+ ions increases drastically and there is an influx of Na^+ ions into the cell that is greater than the efflux of K^+ ions. Consequently, the potential becomes more positive, and an action potential is generated at this site. This

---

*The threshold potential is the potential at which gradual depolarization is replaced by explosive depolarization. It is about 20 to 40 mV more positive than the resting membrane potential; that is, it is somewhere between −30 mV and −50 mV.

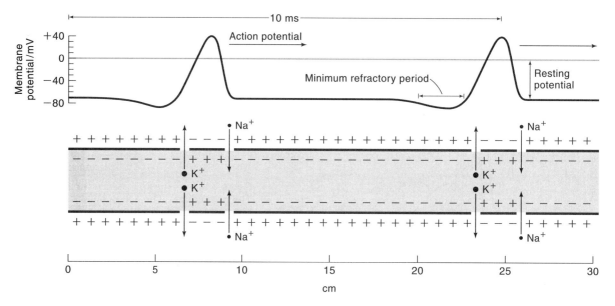

Figure 7.13
Propagation of nerve impulses along the axon coincides with a localized influx of Na^+ ions followed by an outflux of K^+ ions through channels that are "gated," or controlled, by voltage changes across the axon membrane. The electrical event that sends a nerve impulse traveling down the axon normally originates in the cell body. The impulse begins with a slight depolarization, or reduction in the negative potential, across the membrane of the axon where it leaves the cell body. This slight voltage shift induces the opening of the Na^+ channels, shifting the voltage still further. The influx of Na^+ ions accelerates until the inner surface of the membrane is locally positive. The voltage reversal closes the Na^+ channels and opens the K^+ channels. The outflux of K^+ ions quickly restores the negative potential. The voltage reversal, known as the action potential, propagates itself along the axon. After a brief refractory period, a second impulse can follow. (A refractory period is the period during and following an action potential in which an excitable membrane cannot be reexcited.) The impulse-propagation speed is that measured in the giant axon of the squid.

event, in turn, causes a slow depolarization at an adjacent site farther down from the original site, and so on. In this manner, the action potential moves down the neuron without any decrease in magnitude. The speed at which the fastest action potentials in human nerves move along their axon is about 30 m s^{-1}.

The action potential travels along the axon until it reaches either a *synaptic junction* (the connection between nerve cells) or a *neuromuscular junction* (the connection between a nerve cell and a muscle cell). The arrival of an action potential at a synapse triggers the release of a *neurotransmitter*, which is a small, diffusible molecule such as acetylcholine present in the synaptic vesicles. The acetylcholine molecules then diffuse to the postsynaptic membranes, where they produce a large change in the permeability of the membranes. The conductance of both Na^+ and K^+ ions increases markedly, resulting in a large inward current of Na^+ ions and a small outward current of K^+ ions. The inward flow of Na^+ ions again depolarizes the postsynaptic membrane and triggers an action potential in the adjacent axon. Finally, acetylcholine is hydrolyzed to acetate and choline by the enzyme acetylcholinesterase as follows:

$$CH_3-\overset{\overset{\displaystyle O}{\|}}{C}-O-CH_2-CH_2-\overset{+}{N}(CH_3)_3 \ + \ H_2O \ \longrightarrow \ HO-CH_2-CH_2-\overset{+}{N}(CH_3)_3 \ + \ CH_3COO^- \ + \ H^+$$

Acetylcholine   Choline

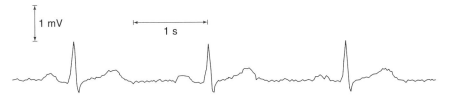

Figure 7.14
A printout of the author's EKG. The appearance is more complicated than the action potential shown in Figure 7.13 due to the depolarization and repolarization in the atria and ventricles. Because measurements are made on the skin, the magnitude of the action potential is much smaller than that at the heart.

In a similar manner, an action potential generated in a nerve cell can be transmitted to a muscle cell. In the muscle cell of the heart, a large action potential is generated during each heartbeat. This potential produces enough current to be detected by placing electrodes on the chest. After amplification, the signals can be recorded either on a moving chart or displayed on an oscilloscope. The record, called an *electrocardiogram* (ECG, also known as EKG, where K is from the German *kardio*, heart), is of great value in diagnosing heart diseases (Figure 7.14).

Suggestions for Further Reading

BOOKS
Brett, C. M. A. and A. M. Oliveira Brett, *Electrochemistry: Principles, Methods, and Applications*, Oxford University Press, New York, 1993.
Rieger, P. H., *Electrochemistry*, Prentice Hall, Englewood Cliffs, NJ, 1987.
Sawyer, D. T., A. Sobkowiak, and J. L. Roberts, Jr., *Electrochemistry for Chemists*, John Wiley & Sons, New York, 1995.

ARTICLES
General
"Fuel Cells—Electrochemical Converts of Chemical to Electrical Energy," J. Weissbart, *J. Chem. Educ.* **38**, 267 (1961).
"Mechanisms of Oxidation-Reduction Reactions," H. Taube, *J. Chem. Educ.* **45**, 452 (1968).
"Fuel Cells—Present and Future," D. P. Gregory, *Chem. Brit.* **5**, 308 (1969).
"Thermodynamic Parameters from an Electrochemical Cell," C. A. Vincent, *J. Chem. Educ.* **47**, 365 (1970).
"Electrochemical Principles Involved in a Fuel Cell," A. K. Vijh, *J. Chem. Educ.* **47**, 680 (1970).
"Ion-Selective Electrodes in Science, Medicine and Technology," R. A. Durst, *Am. Sci.* **59**, 353 (1971).
"Electrochemical Cells for Space Power," R. M. Lawrence and W. H. Bowman, *J. Chem. Educ.* **48**, 359 (1971).

"On the Relationship Between Cell Potential and Half-Cell Reactions," D. N. Bailey, A. Moe, Jr., and J. N. Spencer, *J. Chem. Educ.* **53**, 77 (1976).
"Dental Filling Discomforts Illustrates the Electrochemical Potential of Metals," R. E. Treptow, *J. Chem. Educ.* **55**, 189 (1978).
"Ion and Bio-Selective Membrane Electrodes," G. A. Rechnitz, *J. Chem. Educ.* **60**, 282 (1983).
"Cathodes, Terminals, and Signs," J. J. MacDonald, *Educ. Chem.* **25**, 52 (1988).
"Alleviating the Common Confusion Caused by Polarity in Electrochemistry," P. J. Morgan, *J. Chem. Educ.* **66**, 912 (1989).
"Electrochemistry," R. R. Adzic and E. B. Yeager in *Encyclopedia of Applied Physics*, G. L. Trigg, Ed., VCH Publishers, New York, 1993, Vol. 5, p. 223.
"Energetics of Biological Processes," I. H. Segel and L. D. Segel in *Encyclopedia of Applied Physics*, G. L. Trigg, Ed., VCH Publishers, New York, 1993, Vol. 6, p. 207.
"The Nernst Equation," A. S. Feiner and A. J. McEvoy, *J. Chem. Educ.* **71**, 493 (1994).
"Tendency of Reaction, Electrochemistry, and Units," R. L. DeKock, *J. Chem. Educ.* **73**, 955 (1996).
"Electric Potential Distribution in an Electrochemical Cell," P. Millet, *J. Chem. Educ.* **73**, 956 (1996).

"Students' Misconceptions in Electrochemistry," M. J. Sanger and T. J. Greenbowe, *J. Chem. Educ.* **74**, 819 (1997).

"The Future of Fuel Cells," (Three articles), *Sci. Am.* July (1999).

Bioelectrochemistry

"The Nerve Axon," P. F. Baker, *Sci. Am.* March 1966.

"Biogalvanic Cells," W. D. Hobey, *J. Chem. Educ.* **49**, 413 (1972).

"Electron Transfer in Chemical and Biological Systems," N. Sutin, *Chem. Brit.* **8**, 148 (1972).

"Enzyme Electrodes," D. A. Gough and J. D. Andrade, *Science* **180**, 380 (1973).

"Neurotransmitters," I. Axelrod, *Sci. Am.* June 1974.

"Electrochemistry in Organism," T. P. Chirpith. *J. Chem. Educ.* **52**, 99 (1975).

"Membrane Electrode Probes for Biological Systems," G. A. Rechnitz, *Science* **190**, 234 (1975).

"Chemistry and Nerve Conduction," K. A. Rubinson, *J. Chem. Educ.* **54**, 345 (1977).

"How Cells Make ATP," P. C. Hinkle and R. E. McCarty, *Sci. Am.* March 1978.

"Ion Channels in the Nerve-Cell Membrane," R. D. Keynes, *Sci. Am.* March 1979.

"The Neuron," C. F. Stevens, *Sci. Am.* September 1979.

"Keilin's Respiratory Chain Concept and Its Chemiosmotic Consequences," P. Mitchell, *Science* **206**, 1148 (1979).

"The Transport of Substances in Nerve Cells," J. H. Schwartz, *Sci. Am.* April 1980.

"Davy's Electrochemistry: Nature's Protochemistry," P. Mitchell, *Chem. Brit.* **17**, 14 (1981).

"Lithium and Mania," D. C. Tosteson, *Sci. Am.* April 1981.

"The Release of Acetylcholine," Y. Dunant and M. Israel, *Sci. Am.* April 1985.

"Bio-Electrochemistry," H. A. O. Hill, *Pure Appl. Chem.* **59**, 743 (1987).

"Direct and Indirect Electron Transfer Between Electrodes and Redox Proteins," J. E. Frew and H. A. O. Hill, *Eur. J. Biochem.* **172**, 261 (1988).

"Classical Neurotransmitters and Their Significance within the Nervous System," A. Veca and J. H. Dreisbach, *J. Chem. Educ.* **65**, 108 (1988).

"Experimenting with Biosensors," B. R. Eggins and G. McAteer, *Educ. Chem.* **34**, 20 (1997).

Problems

Emf of Electrochemical Cells

7.1 Calculate the standard emf for the following reaction at 298 K:

$$Fe(s) + Tl^{3+} \rightarrow Fe^{2+} + Tl^{+}$$

7.2 Calculate the emf of the Daniell cell at 298 K when the concentrations of $CuSO_4$ and $ZnSO_4$ are 0.50 M and 0.10 M, respectively. What would the emf be if activities were used instead of concentrations? (The $\gamma_{\pm}$ values for $CuSO_4$ and $ZnSO_4$ at their respective concentrations are 0.068 and 0.15, respectively.)

7.3 The half-reaction at an electrode is

$$Al^{3+}(aq) + 3e^{-} \rightarrow Al(s)$$

Calculate the number of grams of aluminum that can be produced by passing 1.00 faraday through the electrode.

7.4 Consider a Daniell cell operating under non-standard-state conditions. Suppose that the cell's reaction is multiplied by 2. What effect does this have on each of the following quantities in the Nernst equation? **(a)** E, **(b)** E°, **(c)** Q, **(d)** $\ln Q$, and **(e)** ν.

7.5 A student is given two beakers in the laboratory. One beaker contains a solution that is 0.15 M in Fe^{3+} and 0.45 M in Fe^{2+}, and the other beaker contains a solution that is 0.27 M in I^{-} and 0.050 M in I_2. A piece of platinum wire is dipped into each solution. **(a)** Calculate the potential of each electrode relative to a standard hydrogen electrode at 25°C. **(b)** Predict the chemical reaction that will occur when these two electrodes are connected and a salt bridge is used to join the two solutions together.

7.6 From the standard reduction potentials listed in Table 7.1 for $Cu^{2+}|Cu$ and $Pt|Cu^{2+}, Cu^{+}$, calculate the standard reduction potential for $Cu^{+}|Cu$.

Thermodynamics of Electrochemical Cells and the Nernst Equation

7.7 Complete the following table, indicating in the third column whether the cell reaction is spontaneous:

E	$\Delta_r G$	Cell Reaction
+		
	+	
0		

7.8 Calculate the values of E°, $\Delta_r G^{\circ}$, and K for the following reactions at 25°C:

(a) $Zn + Sn^{4+} \rightleftharpoons Zn^{2+} + Sn^{2+}$

(b) $Cl_2 + 2I^{-} \rightleftharpoons 2Cl^{-} + I_2$

(c) $5Fe^{2+} + MnO_4^{-} + 8H^{+} \rightleftharpoons Mn^{2+} + 4H_2O + 5Fe^{3+}$

7.9 The equilibrium constant for the reaction

$$Sr + Mg^{2+} \rightleftharpoons Sr^{2+} + Mg$$

is 6.56×10^{17} at 25°C. Calculate the value of $E°$ for a cell made up of the $Sr|Sr^{2+}$ and $Mg|Mg^{2+}$ half-cells.

7.10 Consider a concentration cell consisting of two hydrogen electrodes. At 25°C, the cell emf is found to be 0.0267 V. If the pressure of hydrogen gas at the anode is 4.0 bar, what is the pressure of hydrogen gas at the cathode?

7.11 An electrochemical cell consists of a half-cell in which a piece of platinum wire is dipped into a solution that is 2.0 M in KBr and 0.050 M in Br_2. The other half-cell consists of magnesium metal immersed in a 0.38 M Mg^{2+} solution. **(a)** Which electrode is the anode and which is the cathode? **(b)** What is the emf of the cell? **(c)** What is the spontaneous cell reaction? **(d)** What is the equilibrium constant of the cell reaction? Assume that the temperature is 25°C.

7.12 From the standard reduction potentials listed in Table 7.1 for $Sn^{2+}|Sn$ and $Pb^{2+}|Pb$, calculate the ratio of $[Sn^{2+}]$ to $[Pb^{2+}]$ at equilibrium at 25°C and the $\Delta_r G°$ value for the reaction.

7.13 Consider the following cell:

$$Ag(s)|AgCl(s)|NaCl(aq)|Hg_2Cl_2(s)|Hg(l)$$

(a) Write the half-cell reactions. **(b)** The standard emfs of the cell at several temperatures are as follows:

T/K	291	298	303	311
$E°/mV$	43.0	45.4	47.1	50.1

Calculate the values of $\Delta_r G°, \Delta_r S°$, and $\Delta_r H°$ for the reaction at 298 K.

7.14 Calculate the emf of the following concentration cell at 298 K:

$$Mg(s)|Mg^{2+}(0.24\ M)||Mg^{2+}(0.53\ M)|Mg(s)$$

7.15 An electrochemical cell consists of a silver electrode in contact with 346 mL of 0.100 M $AgNO_3$ solution and a magnesium electrode in contact with 288 mL of 0.100 M $Mg(NO_3)_2$ solution. **(a)** Calculate the value of E for the cell at 25°C. **(b)** A current is drawn from the cell until 1.20 g of silver have been deposited at the silver electrode. Calculate the value of E for the cell at this stage of operation.

Bioelectrochemistry

7.16 For the reaction

$$NAD^+ + H^+ + 2e^- \rightarrow NADH$$

$E°'$ is −0.320 V at 25°C. Calculate the value of E' at

pH = 1. Assume that both NAD^+ and NADH are at unimolar concentration.

7.17 From the $E°'$ value for the following reaction in Table 7.2,

$$CH_3CHO + 2H^+ + 2e^- \rightarrow C_2H_5OH$$

calculate the value of E' at pH 5.0 and 298 K, given that $[C_2H_5OH] = 5.0 \times 10^{-6}\ M$, and $[CH_3CHO] = 2.4 \times 10^{-4}\ M$.

7.18 Look up the $E°'$ values in Table 7.2 for the reactions

$$CH_3CHO + 2H^+ + 2e^- \rightarrow C_2H_5OH$$
$$NAD^+ + H^+ + 2e^- \rightarrow NADH$$

Calculate the equilibrium constant for the following reaction at 298 K.

$$CH_3CHO + NADH + H^+ \rightleftharpoons C_2H_5OH + NAD^+$$

7.19 The following reaction, which takes place just before the citric acid cycle, is catalyzed by the enzyme lactate dehydrogenase:

$$CH_3COCOO^- + NADH + H^+$$
$$\text{pyruvate}$$
$$\rightleftharpoons CH_3CH(OH)COO^- + NAD^+$$
$$\text{lactate}$$

From the data listed in Table 7.2, calculate the value of $\Delta_r G°'$ and the equilibrium constant for the reaction at 298 K.

7.20 Calculate the number of moles of cytochrome c^{3+} formed from cytochrome c^{2+} with the Gibbs energy derived from the oxidation of 1 mole of glucose. ($\Delta_r G° = -2879$ kJ for the degradation of 1 mole of glucose to CO_2 and H_2O.)

7.21 The terminal respiratory chain involves the redox couples $NAD^+|NADH$ and $FAD|FADH_2$. Calculate the $\Delta_r G°'$ value for the following reaction at 298 K:

$$NADH + FAD + H^+ \rightarrow NAD^+ + FADH_2$$

Is this Gibbs energy change sufficient to synthesize ATP from ADP and inorganic phosphate? Draw a diagram showing the experimental arrangement for measuring the emf of a cell consisting of these two couples.

7.22 The oxidation of malate to oxaloacetate is a key reaction in the citric acid cycle:

$$\text{malate} + NAD^+ \rightarrow \text{oxaloacetate} + NADH + H^+$$

Calculate the value of $\Delta_r G°'$ and the equilibrium constant for the reaction at pH 7 and 298 K.

7.23 Calculate the value of $\Delta_r G^{\circ\prime}$ for the oxidation of succinate to fumarate by cytochrome c at 298 K.

7.24 Flavin adenine dinucleotide (FAD) participates in several biological redox reactions according to the half-reaction

$$FAD + 2H^+ + 2e^- \rightarrow FADH_2$$

If the value of $E^{\circ\prime}$ of this couple is -0.219 V at 298 K and pH 7, calculate its reduction potential at this temperature and pH when the solution contains **(a)** 85% of the oxidized form and **(b)** 15% of the oxidized form.

7.25 According to the chemiosmotic theory, the synthesis of 1 mole of ATP is coupled to the movement of 4 moles of H^+ ions from the low-pH side of the membrane to the high-pH side. **(a)** Derive an expression for ΔG for the movement of $4H^+$. **(b)** Calculate the change in pH across the membrane that is required at 25°C to synthesize one mole of ATP from ADP and P_i under standard-state conditions. $\Delta_r G^{\circ\prime} = 31.4$ kJ for the synthesis of 1 mole of ATP.

7.26 The nitrite in soil is oxidized to nitrate by the bacteria *nitrobacter agilis* in the presence of oxygen. The half-reduction reactions are

$$NO_3^- + 2H^+ + 2e^- \rightarrow NO_2^- + H_2O \qquad E^{\circ\prime} = 0.42 \text{ V}$$
$$\tfrac{1}{2}O_2 + 2H^+ + 2e^- \rightarrow H_2O \qquad E^{\circ\prime} = 0.82 \text{ V}$$

Calculate the yield of ATP synthesis per mole of nitrite oxidized, assuming an efficiency of 55%. (The $\Delta_r G^{\circ\prime}$ value for ATP synthesis from ADP and P_i is 31.4 kJ mol^{-1}.)

Membrane Potentials

7.27 Describe an experiment that would show that the nerve cell membrane is much more permeable to K^+ than to Na^+.

7.28 A membrane permeable only to K^+ ions is used to separate the following two solutions:

α	$[KCl] = 0.10$ M	$[NaCl] = 0.050$ M
β	$[KCl] = 0.050$ M	$[NaCl] = 0.10$ M

Calculate the membrane potential at 25°C, and determine which solution has the more negative potential.

7.29 Referring to Figure 7.14b, carry out the following operations: **(a)** Calculate the membrane potential due to K^+ ions at 25°C. **(b)** Given that biological membranes typically have a capacitance of approximately 1 μF cm^{-2}, calculate the charge in coulombs on a unit area (1 cm^2) of the membrane. (See Appendix 5.1 for the units of capacitance.) **(c)** Convert the charge in **(b)** to number of K^+ ions. **(d)** Compare the result in **(c)** with the number of K^+ ions in 1 cm^3 of the solution in the left compartment. What can you conclude about the

relative number of K^+ ions needed to establish the membrane potential?

Additional Problems

7.30 Look up the values of E° for the following half-cell reactions:

$$Ag^+ + e^- \rightarrow Ag$$
$$AgBr + e^- \rightarrow Ag + Br^-$$

Describe how you would use these values to determine the solubility product (K_{sp}) of AgBr at 25°C.

7.31 A well-known organic redox system is the quinone–hydroquinone couple. In an aqueous solution at a pH below 8, we have

This system can be prepared by dissolving quinhydrone, QH (a complex consisting of equimolar amounts of Q and HQ), in water. A quinhydrone electrode can be constructed by immersing a piece of platinum wire in a quinhydrone solution. **(a)** Derive an expression for the electrode potential of this couple in terms of E° and the hydrogen ion concentration. **(b)** When the quinone–hydroquinone couple is joined to a saturated calomel electrode, the emf of the cell is found to be 0.18 V. In this arrangement, the saturated calomel electrode acts as the anode. Calculate the pH of the quinhydrone solution. Assume the temperature is 25°C.

7.32 One way to prevent a buried iron pipe from rusting is to connect it with a piece of wire to a magnesium or zinc rod. What is the electrochemical principle for this action?

7.33 Aluminum has a more negative standard reduction potential than iron. Yet aluminum does not form rust or corrode as easily as iron. Explain.

7.34 Given that the $\Delta_r S^\circ$ value for the Daniell cell is -21.7 J K^{-1} mol^{-1}, calculate the temperature coefficient $(\partial E^\circ / \partial T)_P$ of the cell and the emf of the cell at 80°C.

7.35 For years it was not clear whether mercury(I) ions existed in solution as Hg^+ or as Hg_2^{2+}. To distinguish between these two possibilities, we could set up the following system:

$$Hg(l)|\text{soln A}||\text{soln B}|Hg(l)$$

where solution A contained 0.263 g mercury(I) nitrate

per liter and solution B contained 2.63 g mercury(I) nitrate per liter. If the measured emf of such a cell is 0.0289 V at 18°C, what can you deduce about the nature of the mercury ions?

7.36 Given the following standard reduction potentials, calculate the ion-product K_w value ($[H^+][OH^-]$) at 25°C:

$$2H^+(aq) + 2e^- \rightarrow H_2(g) \qquad E° = 0.00 \text{ V}$$
$$2H_2O(l) + 2e^- \rightarrow H_2(g) + 2OH^-(aq)$$
$$E° = -0.828 \text{ V}$$

7.37 Given that

$$2Hg^{2+}(aq) + 2e^- \rightarrow Hg_2^{2+}(aq) \qquad E° = 0.920 \text{ V}$$
$$Hg_2^{2+}(aq) + 2e^- \rightarrow 2Hg(l) \qquad E° = 0.797 \text{ V}$$

calculate the values of $\Delta_r G°$ and K for the following process at 25°C:

$$Hg_2^{2+}(aq) \rightarrow Hg^{2+}(aq) + Hg(l)$$

(The above reaction is an example of a *disproportionation reaction*, in which an element in one oxidation state is both oxidized and reduced.)

7.38 The magnitudes of the standard electrode potentials of two metals, X and Y, are

$$X^{2+} + 2e^- \rightarrow X \qquad |E°| = 0.25 \text{ V}$$
$$Y^{2+} + 2e^- \rightarrow Y \qquad |E°| = 0.34 \text{ V}$$

where the | | notation denotes that only the magnitude (but *not* the sign) of the $E°$ value is shown. When the half-cells of X and Y are connected, electrons flow from X to Y. When X is connected to a SHE, electrons flow from X to SHE. **(a)** Which value of $E°$ is positive and which is negative? **(b)** What is the standard emf of a cell made up of X and Y?

7.39 An electrochemical cell is constructed as follows. One half-cell consists of a platinum wire immersed in a solution containing 1.0 M Sn^{2+} and 1.0 M Sn^{4+}, and the other half-cell has a thallium rod immersed in a solution of 1.0 M Tl^+. **(a)** Write the half-cell reactions and the overall reaction. **(b)** What is the equilibrium constant at 25°C? **(c)** What is the cell voltage if the Tl^+ concentration is increased tenfold?

7.40 Given the standard reduction potential for Au^{3+} in Table 7.1 and

$$Au^+(aq) + e^- \rightarrow Au(s) \qquad E° = 1.69 \text{ V}$$

answer the following questions. **(a)** Why does gold not

tarnish in air? **(b)** Will the following disproportionation occur spontaneously?

$$3Au^+(aq) \rightarrow Au^{3+}(aq) + 2Au(s)$$

(c) Predict the reaction between gold and fluorine gas.

7.41 Consider the Daniell cell shown in Figure 7.1. In the diagram, the anode appears to be negative and the cathode positive (electrons are flowing from the anode to the cathode). Yet the anions in solution are moving toward the anode, which must therefore seem positive to the anions. Because the anode cannot simultaneously be negative and positive, explain this apparently contradictory situation.

7.42 Calculate the pressure of H_2 (in bar) required to maintain equilibrium with respect to the following reaction at 25°C:

$$Pb(s) + 2H^+(aq) \rightleftharpoons Pb^{2+}(aq) + H_2(g)$$

given that $[Pb^{2+}] = 0.035 \ M$ and the solution is buffered at pH 1.60.

7.43 Use the data in Appendix 2 and the convention that $\Delta_f \bar{G}°[H^+(aq)] = 0$ to determine the standard reduction potentials for sodium and fluorine. (Like sodium, fluorine also reacts violently with water.)

7.44 Use the data in Table 7.1 to determine the value of $\Delta_f \bar{G}°$ for $Fe^{2+}(aq)$.

7.45 Consider the following cell:

$$Pt|H_2(1 \text{ bar})|HCl(m)|AgCl(s)|Ag$$

At 25°C, the emf values at various molalities are

$m/(\text{mol kg}^{-1})$	0.124	0.0539	0.0256	0.0134	0.00914
E/V	0.342	0.382	0.418	0.450	0.469

$m/(\text{mol kg}^{-1})$	0.00562	0.00322
E/V	0.493	0.521

(a) Determine the value of $E°$ graphically. Compare your value of $E°$ with that listed in Table 7.1. **(b)** Calculate the mean activity coefficient ($\gamma_\pm$) for HCl at 0.124 m.

7.46 Calculate the $E°$ for the propane fuel cell discussed on p. 246. The $\Delta_f \bar{G}°$ for C_3H_8 is -23.5 kJ mol^{-1}.

7.47 Compare the $E°$ value for the half reaction $Fe^{3+} + e^- \rightarrow Fe^{2+}$ in aqueous solution (see Table 7.1) with the half reactions occurring in the cytochromes (see Table 7.2). Comment on the biological significance of the differences.

CHAPTER 8

Acids and Bases

Acids and bases form a particularly important class of electrolytes. No chemical equilibria are as widespread as those involving acids and bases. The precise balance of their concentrations or pH in our bodies is necessary for the proper function of enzymes, maintenance of osmotic pressure, and so on. A deviation from the normal pH value by as small an amount as one-tenth of a unit can lead to disease or even death.

A proper understanding of acid–base balance in chemical and biological systems requires a clear understanding of the behavior of weak acids and weak bases, and of the hydrogen ion. In this chapter we discuss general acid–base reactions, buffers, and amino acids.

8.1 Definitions of Acid and Base

In 1923 the Danish chemist Johannes Nicolaus Brønsted (1879–1947) advanced the view that an acid is a substance that can donate protons and a base is a substance that can accept protons. According to Brønsted's definition, HCl is an acid because it can donate a proton to a water molecule as follows:

$$\underset{\text{acid}}{HCl(aq)} + H_2O(l) \rightarrow H_3O^+(aq) + \underset{\text{conjugate base}}{Cl^-(aq)}$$

The chloride ion is a base because of its ability to accept protons. HCl and Cl^- are said to be *conjugate* or paired to each other. A strong acid such as HCl, which is assumed to be largely or totally dissociated, has a weak conjugate base, Cl^-, while a weak acid has a strong conjugate base. For example, consider the dissociation of acetic acid:

$$\underset{\text{acid}}{CH_3COOH(aq)} + H_2O(l) \rightleftharpoons \underset{\text{conjugate base}}{CH_3COO^-(aq)} + H_3O^+(aq)$$

Because acetic acid is a weak acid, its conjugate base, the acetate ion, is a much stronger base than the Cl^- ion.

Ammonia, for example, is a Brønsted base because it can accept a proton,

$$\underset{\text{base}}{NH_3(aq)} + H_2O(l) \rightleftharpoons \underset{\text{conjugate acid}}{NH_4^+(aq)} + OH^-(aq)$$

and the ammonium ion is the conjugate acid. On the other hand, metallic hydroxides such as NaOH or KOH are not themselves bases in Brønsted's definition because they cannot accept a proton. For all practical purposes, however, these compounds completely dissociate into Na^+, K^+, and OH^- ions in water. The hydroxide ion itself is a base because it can accept a proton ($OH^- + H^+ \rightarrow H_2O$).

Brønsted's definitions of acids and bases are not limited to substances that exchange protons in water. In pure liquid form, NH_3 also qualifies as a Brønsted acid and a Brønsted base because of the reaction

$$NH_3 + NH_3 \rightleftharpoons NH_4^+ + NH_2^-$$

We shall shortly see that this is a property also shared by water.

A broader definition of an acid and a base was provided by Gilbert Lewis in 1923. According to Lewis, an acid is any substance that can accept an electron pair and a base is any substance that can donate an electron pair. This definition not only covers the reactions discussed above, but it also describes acid–base reactions that do not involve protons. Examples are

$$\underset{\text{acid}}{BF_3} + \underset{\text{base}}{F^-} \rightarrow BF_4^-$$

$$\underset{\text{acid}}{BCl_3} + \underset{\text{base}}{(CH_3)_3N} \rightarrow (CH_3)_3NBCl_3$$

$$\underset{\text{acid}}{Ag^+} + \underset{\text{base}}{2CN^-} \rightarrow Ag(CN)_2^-$$

Lewis's definition applies to a greater range of substances and is more fundamental from a theoretical viewpoint. In this chapter, however, we shall be concerned mostly with the Brønsted concept because our main focus is on acid–base reactions in aqueous solutions.

8.2 The Acid–Base Properties of Water

Water is a unique solvent. As we shall see, one of its special properties is its ability to act either as an acid or as a base. Water is a very weak electrolyte and therefore a poor conductor of electricity, but it does undergo dissociation to a small extent:

$$H_2O(l) + H_2O(l) \rightleftharpoons H_3O^+(aq) + OH^-(aq)$$

Water behaves both as a Brønsted acid and as a Brønsted base.

This reaction is sometimes called the autoionization of water. In aqueous medium the proton exists in the hydrated form and the H_3O^+ species is called the hydronium ion. Evidence suggests that besides H_3O^+, more complex structures such as $H_9O_4^+$ are also present.

However, the thermodynamic treatment of acid–base equilibria is the same regardless of which species we employ. For simplicity, we shall use the H^+ notation in most cases. Thus, the autoionization of water can be represented as

$$H_2O(l) \rightleftharpoons H^+(aq) + OH^-(aq)$$

and the dissociation constant, which is called the *ion product of water*, K_w, is

$$K_w = \frac{a_{H^+} a_{OH^-}}{a_{H_2O}}$$

Setting $a_{H_2O} = 1$, we write

$$K_w = a_{H^+} a_{OH^-} \tag{8.1}$$

or, in terms of concentrations,

$$K_w = [H^+][OH^-] \tag{8.2}$$

Like most equilibrium constants, K_w is a function of temperature:

T/K	273	298	313	373
K_w	0.12×10^{-14}	1.0×10^{-14}	2.9×10^{-14}	5.4×10^{-13}

As the values of K_w indicate, water is an extremely weak acid as well as an extremely weak base. In a neutral solution, $[H^+] = [OH^-] = 1.0 \times 10^{-7}$ M at 298 K. If the solution is acidic, $[H^+] > [OH^-]$. In a basic solution, $[OH^-] > [H^+]$.

pH—A Measure of Acidity

The H^+ ion plays a central role in many chemical and biological processes, and its concentration can vary over a wide range. It is therefore useful to set up a pH scale such that

$$pH = -\log a_{H^+} \tag{8.3}$$

where a_{H^+} is the activity of the H^+ ions in solution, given by $\gamma_{H^+}[H^+]$. As we saw in Chapter 5, only mean ionic activity coefficients can be determined experimentally, and so we can only estimate the value of γ_{H^+}. For example, for a 0.050 M HCl solution, we write the ionic activity for the H^+ ion as (see Equation 5.61)

$$\log \gamma_{H^+} = -0.509z^2\sqrt{I}$$
$$= -0.509(1^2)\sqrt{0.050}$$
$$= -0.114$$

or

$$\gamma_{H^+} = 0.77$$

Finally,

$$pH = -\log(0.77)(0.050)$$
$$= 1.4$$

Generally, for relatively dilute solutions at low ionic strengths ($[H^+] \leq 0.1$ M, $I \leq 0.1$), we can use the following approximate equation:*

$$pH = -\log[H^+] \tag{8.4}$$

Bear in mind that the *measured* pH of a solution is seldom equal to the *calculated* pH based on Equation 8.4 due to nonideal behavior.[†]
We can also define a pOH scale as follows:

$$pOH = -\log[OH^-] \tag{8.5}$$

Taking the negative logarithm of Equation 8.2 we get

$$-\log K_w = -\log[H^+] - \log[OH^-]$$
$$pK_w = pH + pOH \tag{8.6}$$

where $pK_w = -\log K_w$. At 25°C, Equation 8.6 becomes

$$pH + pOH = 14.00 \tag{8.7}$$

In terms of concentrations, we can express the acidity of a solution as follows:

Acidic solutions: $[H^+] > 1 \times 10^{-7}$ M, pH < 7

Basic solutions: $[H^+] < 1 \times 10^{-7}$ M, pH > 7

Neutral solutions: $[H^+] = 1 \times 10^{-7}$ M, pH = 7

Although the practical pH range is between 1 and 14, negative pH values exist, as do pH values greater than 14. For example, a 2.0 M HCl solution has a negative pH value, whereas a 2.0 M NaOH solution has a pH greater than 14.

8.3 Dissociation of Acids and Bases

As an approximation, strong acids such as HCl and HNO_3 are assumed to be completely dissociated in solution:

$$HCl(aq) + H_2O(l) \rightarrow H_3O^+(aq) + Cl^-(aq)$$
$$HNO_3(aq) + H_2O(l) \rightarrow H_3O^+(aq) + NO_3^-(aq)$$

and it is a simple procedure to calculate the concentrations of the species (H_3O^+ and conjugate base) in solution. The dissociation of a weak acid, HA, on the other hand, is incomplete

$$HA(aq) + H_2O(l) \rightleftharpoons H_3O^+(aq) + A^-(aq)$$

The thermodynamic equilibrium constant for this process is

---

* Strictly speaking, $[H^+]$ should be divided by its standard-state value of 1 M. This step is usually omitted for simplicity, however.
† See T. P. Dirkse, *J. Chem. Educ.* **38**, 261 (1961); S. J. Hawkes, *ibid*, **71**, 747 (1994).

$$K_a = \frac{a_{H_3O^+} a_{A^-}}{a_{HA} a_{H_2O}} \tag{8.8}$$

where a represents the activities of the species. Because the concentration of water in most aqueous solutions is close to that of pure water [1 liter of pure water contains 1000 g/(18.02 g mol^{-1}), or 55.5 mol], it is essentially unchanged by the dissociation process. Therefore, H_2O can be considered to be in its standard state (that is, pure liquid), for which its activity is unity. If we replace $a_{H_3O^+}$ with a_{H^+} for simplicity, Equation 8.8 becomes

$$K_a = \frac{a_{H^+} a_{A^-}}{a_{HA}}$$

$$= \frac{[H^+]\gamma_+ [A^-]\gamma_-}{[HA]\gamma_{HA}} \tag{8.9}$$

Because HA is an uncharged species, we can set $\gamma_{HA} \approx 1$ for dilute solutions, and from Equation 5.56 we have $\gamma_\pm^2 = \gamma_+ \gamma_-$, so

$$K_a = \frac{[H^+][A^-]\gamma_\pm^2}{[HA]} \tag{8.10}$$

If the acid is sufficiently weak, the concentrations of the ionic species are low ($\leq 0.050\ M$), and so we can, as a good approximation, ignore $\gamma_\pm^2$ and write

$$K_a = \frac{[H^+][A^-]}{[HA]} \tag{8.11}$$

The strength of the acid is indicated by the magnitude of K_a; that is, the larger the K_a value, the stronger the acid. Another way to measure the strength of an acid is to calculate its percent dissociation, defined by

$$\text{percent dissociation} = \frac{[H^+]_{eq}}{[HA]_0} \times 100\% \tag{8.12}$$

where $[H^+]_{eq}$ is the hydrogen ion concentration at equilibrium and $[HA]_0$ is the initial concentration of the acid. Strong acids are assumed to be 100 percent dissociated. Table 8.1 lists the dissociation constants of a number of common acids, which we can use to calculate the concentrations of dissociated ions and undissociated acid in solution at equilibrium.

Example 8.1

Calculate the concentrations of the undissociated acid, the H^+ ions, and the CN^- ions of a 0.050 M HCN solution and the percent dissociation at 25°C.

ANSWER

Let x be the concentrations of H^+ and CN^- at equilibrium. Thus, we have

$$\begin{array}{cccc} HCN(aq) & \rightleftharpoons & H^+(aq) + & CN^-(aq) \\ (0.050 - x)\ M & & x\ M & x\ M \end{array}$$

272   Chapter 8: Acids and Bases

Using the value of K_a from Table 8.1, we write

$$\frac{x^2}{0.050 - x} = 4.9 \times 10^{-10}$$

We can either solve the quadratic equation for x or apply an approximation. Because K_a values for weak acids are generally known to an accuracy of only $\pm 5\%$, it is reasonable to require x to be less than 5% of 0.050, the number from which it is subtracted. Assuming that this approximation holds ($0.050 - x \approx 0.050$), we have

$$x^2 = 2.5 \times 10^{-11}$$

or

$$x = 5.0 \times 10^{-6} \ M$$

Therefore, at equilibrium,

$$[H^+] = 5.0 \times 10^{-6} \ M$$
$$[CN^-] = 5.0 \times 10^{-6} \ M$$
$$[HCN] = 0.050 \ M - 5.0 \times 10^{-6} \ M \approx 0.050 \ M$$

Finally, the percent dissociation is given by

$$\frac{5.0 \times 10^{-6} \ M}{0.050 \ M} \times 100\% = 1.0 \times 10^{-2}\%$$

Note that the percent dissociation calculation also shows that our "5%" approximation is valid in this case.

COMMENT

Note that in this calculation we have neglected the contribution to the hydrogen ion concentration from water molecules. This assumption always holds unless we are dealing with very dilute solutions. A more rigorous treatment of acid dissociation is given in Appendix 8.1 on p. 298. (See also Problem 8.11.)

If we substitute hydrofluoric acid (HF) for HCN in Example 8.1, we have

$$\frac{x^2}{0.050 - x} = 7.1 \times 10^{-4}$$

and the "5%" approximation does not hold. In this case, we can either solve the quadratic equation to get $x = 5.6 \times 10^{-3} \ M$ or apply the *method of successive approximations* as follows. We first solve for x by assuming $0.050 - x \approx 0.050$. This approximation gives a value of $6.0 \times 10^{-3} \ M$. Next, we use the approximate value of x ($6.0 \times 10^{-3} \ M$) to find a more exact value of the concentration for HF at equilibrium:

$$[HF] = 0.050 \ M - 6.0 \times 10^{-3} \ M = 0.044 \ M$$

Table 8.1
Dissociation Constants of Common Weak Acids at 298 K

Acid	K_a	pK_a
HF	7.1×10^{-4}	3.15
HCN	4.9×10^{-10}	9.31
HNO_2	4.5×10^{-4}	3.35
H_2S	5.7×10^{-8} (K_a')	7.24 (pK_a')
HS^-	1.2×10^{-15} (K_a'')	14.92 (pK_a'')
H_2CO_3	4.2×10^{-7} (K_a')	6.38 (pK_a')
HCO_3^-	4.8×10^{-11} (K_a'')	10.32 (pK_a'')
H_2SO_4	Very high (K_a')	—
HSO_4^-	1.3×10^{-2} (K_a'')	1.89
H_3BO_3 [a]	7.3×10^{-10}	9.14
H_3PO_4	7.5×10^{-3} (K_a')	2.13 (pK_a')
$H_2PO_4^-$	6.2×10^{-8} (K_a'')	7.21 (pK_a'')
HPO_4^{2-}	4.8×10^{-13} (K_a''')	12.32 (pK_a''')
CH_3COOH	1.75×10^{-5}	4.76
C_6H_5COOH	6.30×10^{-5}	4.20
HCOOH	1.77×10^{-4}	3.75
$ClCH_2COOH$	1.36×10^{-3}	2.87
C_6H_5OH	1.30×10^{-10}	9.89
HOOCCOOH (oxalic acid)	6.5×10^{-2} (K_a')	1.19 (pK_a')
	6.1×10^{-5} (K_a'')	4.21 (pK_a'')
$(CH_2COOH)_2$ (succinic acid)	6.4×10^{-5} (K_a')	4.19 (pK_a')
	2.7×10^{-6} (K_a'')	5.57 (pK_a'')
$C_6H_8O_6$ (ascorbic acid)	8×10^{-5} (K_a')	4.1 (pK_a')
	1.6×10^{-12} (K_a'')	11.79 (pK_a'')
$(CH_2COOH)_2$ (fumaric acid)	9.3×10^{-4} (K_a')	3.03 (pK_a')
	3.4×10^{-5} (K_a'')	4.47 (pK_a'')
$CH_3CH(OH)COOH$ (lactic acid)	1.39×10^{-4}	3.86
$HOOCCH(OH)CH_2COOH$ (malic acid)	4×10^{-4} (K_a')	3.40 (pK_a')
	9×10^{-6} (K_a'')	5.50 (pK_a'')
$HOOCCH_2C(OH)COOHCH_2COOH$ (citric acid)	8.7×10^{-4} (K_a')	3.06 (pK_a')
	1.8×10^{-5} (K_a'')	4.74 (pK_a'')
	4.0×10^{-6} (K_a''')	5.40 (pK_a''')

(acetylsalicylic acid, or aspirin) 3×10^{-4} 3.5

[a] Boric acid does not ionize in water to produce an H^+ ion. Its reaction with water is $B(OH)_3(aq) + H_2O(l) \rightleftharpoons B(OH)_4^-(aq) + H^+(aq)$.

Substituting this in the expression for K_a, we write

$$\frac{x^2}{0.044} = 7.1 \times 10^{-4}$$

$$x = 5.6 \times 10^{-3}\ M$$

which is the same as the result obtained using the quadratic equation. In general, we apply the method of successive approximation until the value of x obtained for the last step is identical to the value found in the previous step. In most cases, applying this method twice will produce the correct answer.

The treatment of the dissociation of bases is the same as that for acids. For example, when ammonia dissolves in water, it reacts as follows:

Ammonium hydroxide, or NH_4OH, does not exist.

$$NH_3 + H_2O \rightleftharpoons NH_4^+ + OH^-$$

By analogy with the acid-dissociation constant, we can write the base-dissociation constant, K_b, as

$$K_b = \frac{a_{NH_4^+}\,a_{OH^-}}{a_{NH_3}\,a_{H_2O}}$$

Setting $a_{H_2O} = 1$ and replacing activities with concentrations, we obtain

$$K_b = \frac{[NH_4^+][OH^-]}{[NH_3]}$$

Table 8.2 lists the dissociation constants of some common bases.

Table 8.2
Dissociation Constants of Weak Bases at 298 K

Base	K_b	pK_b
Ammonia	1.8×10^{-5}	4.75
Aniline	3.80×10^{-10}	9.42
Caffeine	4.1×10^{-4}	3.39
Cocaine	2.57×10^{-6}	5.59
Creatine	1.92×10^{-11}	10.72
Ethylamine	5.6×10^{-4}	3.25
Methylamine	4.38×10^{-4}	3.36
Morphine	7.4×10^{-7}	6.13
Nicotine	7×10^{-7}	6.2
Novocaine	7×10^{-6}	5.2
Pyridine	1.71×10^{-9}	8.77
Quinine	$1.1 \times 10^{-6}\ (K_b')$	5.96 (pK_b')
	$1.35 \times 10^{-10}\ (K_b'')$	9.87 (pK_b'')
Strychnine	$1 \times 10^{-6}\ (K_b')$	6.0 (pK_b')
	$2 \times 10^{-12}\ (K_b'')$	11.7 (pK_b'')
Urea	1.5×10^{-14}	13.82

The Relationship Between the Dissociation Constant of an Acid and That of Its Conjugate Base

An important relationship between the acid-dissociation constant and the dissociation constant of its conjugate base can be derived as follows, using acetic acid as an example:

$$CH_3COOH(aq) \rightleftharpoons H^+(aq) + CH_3COO^-(aq)$$

$$K_a = \frac{[H^+][CH_3COO^-]}{[CH_3COOH]}$$

The conjugate base, CH_3COO^-, supplied by a sodium acetate (CH_3COONa) solution, reacts with water according to the equation

$$CH_3COO^-(aq) + H_2O(l) \rightleftharpoons CH_3COOH(aq) + OH^-(aq)$$

We can write the base-dissociation constant as

$$K_b = \frac{[CH_3COOH][OH^-]}{[CH_3COO^-]}$$

The product of K_a and K_b is given by

$$K_a K_b = \frac{[H^+][CH_3COO^-]}{[CH_3COOH]} \times \frac{[CH_3COOH][OH^-]}{[CH_3COO^-]}$$

$$= [H^+][OH^-]$$

$$K_a K_b = K_w \qquad (8.13)$$

This equation enables us to draw an important conclusion: The stronger the acid (the larger the K_a), the weaker its conjugate base (the smaller the K_b), and vice versa.

Finally, we note that the pK values listed in Tables 8.1 and 8.2 are defined as the negative logarithm of the dissociation constant:

$$pK = -\log K \qquad (8.14)$$

Remember that the larger the pK value, the weaker the acid or base.

Salt Hydrolysis

Dissolving NaCl or K_2SO_4 in water creates an essentially neutral solution. A solution of sodium acetate or ammonium chloride, on the other hand, is anything but neutral. Depending on the concentration, the pH of a sodium acetate solution can be appreciably higher than 7, whereas the pH of an ammonium chloride solution is appreciably lower than 7. The departures from pH 7 result from *salt hydrolysis*, which is the reaction of an anion or a cation of a salt, or both, with water.

Consider the reaction between sodium acetate and water:

$$CH_3COONa(s) \xrightarrow{H_2O} CH_3COO^-(aq) + Na^+(aq)$$

$$CH_3COO^-(aq) + H_2O(l) \rightleftharpoons CH_3COOH(aq) + OH^-(aq)$$

Sodium acetate, a strong electrolyte, dissociates completely. Acetic acid is a weak acid, so the equilibrium for the second step shifts to the right, producing a surplus of hydroxide ions and a basic solution. The equilibrium constant is the same as the base-dissociation constant:

$$K_b = \frac{[CH_3COOH][OH^-]}{[CH_3COO^-]}$$

From Equation 8.13 and Table 8.1,

$$K_b = \frac{1.0 \times 10^{-14}}{1.75 \times 10^{-5}} = 5.7 \times 10^{-10}$$

We assume that the Na^+ ion does not react with water. When ammonium chloride dissolves in water, the reactions are

$$NH_4Cl(s) \xrightarrow{H_2O} NH_4^+(aq) + Cl^-(aq)$$
$$NH_4^+(aq) \rightleftharpoons NH_3(aq) + H^+(aq)$$

The dissociation of the ammonium ion causes the solution to become acidic. Note that the chloride ion is an extremely weak conjugate base of the HCl acid and does not undergo hydrolysis.

Small, highly charged cations, such as Be^{2+}, Al^{3+}, and Bi^{4+} also hydrolyze in water. For example, an $AlCl_3$ solution is acidic because of the following reaction:

$$AlCl_3(s) \xrightarrow{H_2O} Al^{3+}(aq) + 3Cl^-(aq)$$

The Al^{3+} ion polarizes the O–H bonds in the H_2O molecules in the hydration sphere, resulting in the loss of a proton to the bulk water:

$$Al(H_2O)_6^{3+} + H_2O \rightleftharpoons Al(H_2O)_5(OH)^{2+} + H_3O^+$$

The $Al(H_2O)_5(OH)^{2+}$ ion can dissociate further and so on. Strictly speaking, *all* metal cations, including the alkali and alkaline earth metal cations, hydrolyze to a certain extent so that their presence can also influence (lower) the pH of the solution.

Finally, we note that the assumption that salts completely dissociate in aqueous solution is at best an approximation. For example, depending on concentration, a $CaCl_2$ solution actually contains Ca^{2+} and Cl^- ions as well as the charged and neutral ion pairs $Ca^{2+}Cl^-$ and $Cl^-Ca^{2+}Cl^-$. An accurate study of solution properties must take all these species into account.*

8.4 Diprotic and Polyprotic Acids

Our discussion so far has focused on monoprotic acids. The treatment of acid–base equilibria is more complicated for acids that have two or more dissociable protons. In this section, we shall consider two acids of considerable importance in living systems—carbonic acid and phosphoric acid.

Carbon dioxide readily dissolves in water, but only a small percentage (about 0.25%) of the dissolved CO_2 is converted to what we might call the hydrated form,

---

*See S. J. Hawkes, *J. Chem. Educ.* **73**, 421 (1996).

H_2CO_3. The equilibrium constant for the reaction

$$CO_2(aq) + H_2O(l) \rightleftharpoons H_2CO_3(aq)$$

is only 0.00258. Because we cannot distinguish experimentally between dissolved CO_2 gas and H_2CO_3, we can write the first dissociation of "carbonic acid" as

$$CO_2(aq) + H_2O(l) \rightleftharpoons H^+(aq) + HCO_3^-(aq)$$

or more conveniently as

$$H_2CO_3(aq) \rightleftharpoons H^+(aq) + HCO_3^-(aq)$$

The form of the equilibrium constant (that is, the first acid-dissociation constant), would be the same as long as we used the total concentration of CO_2 in water as the acid concentration. The common practice is to use the last equation in describing the dissociation and to assume that all the dissolved CO_2 is in the form of H_2CO_3. On this basis, the first acid-dissociation constant has the value (see Table 8.1)

$$K_a' = \frac{[H^+][HCO_3^-]}{[H_2CO_3]} = 4.2 \times 10^{-7}$$

The conjugate base for the first dissociation HCO_3^- becomes the acid in the second dissociation step:

$$HCO_3^-(aq) \rightleftharpoons H^+(aq) + CO_3^{2-}(aq)$$

and we have

$$K_a'' = \frac{[H^+][CO_3^{2-}]}{[HCO_3^-]} = 4.8 \times 10^{-11}$$

Thus, K_a' is greater than K_a'' by some four orders of magnitude. (It is harder to remove an H^+ ion from an anion than from a neutral species.)

An order of magnitude is a factor of ten.

Example 8.2

Example 5.3 shows that the solubility of carbon dioxide in equilibrium with water at 298 K and a partial pressure of 3.3×10^{-4} atm is 1.1×10^{-5} mol CO_2 (kg $H_2O)^{-1}$. What are the concentrations of all the species in this solution?

ANSWER

Because the solution is dilute, we can equate molality to molarity and assume ideal behavior. Thus, the initial concentration of H_2CO_3 is 1.1×10^{-5} M. In this solution, there are three equilibria to consider:

$$H_2CO_3 \rightleftharpoons H^+ + HCO_3^-$$
$$HCO_3^- \rightleftharpoons H^+ + CO_3^{2-}$$
$$H_2O \rightleftharpoons H^+ + OH^-$$

There are altogether five unknowns: $[H^+]$, $[OH^-]$, $[H_2CO_3]$, $[HCO_3^-]$, and $[CO_3^{2-}]$. From the mass balance of species containing the carbonate group,

$$1.1 \times 10^{-5} \, M = [\text{H}_2\text{CO}_3] + [\text{HCO}_3^-] + [\text{CO}_3^{2-}] \tag{1}$$

Moreover, electrical neutrality requires that

$$[\text{H}^+] = [\text{HCO}_3^-] + 2[\text{CO}_3^{2-}] + [\text{OH}^-] \tag{2}$$

Because each carbonate ion carries two negative charges, we need to multiply its concentration by a factor of 2. These five unknowns can be determined from five independent equations: 1, 2, K_a', K_a'', and K_w. Certain assumptions will simplify the procedure. Because $K_a' \gg K_a''$ and K_a' itself is a small number, we have

$$[\text{H}_2\text{CO}_3] \gg [\text{HCO}_3^-] \gg [\text{CO}_3^{2-}]$$

From the first stage of dissociation,

$$\begin{array}{ccccc}
\text{H}_2\text{CO}_3 & \rightleftharpoons & \text{H}^+ & + & \text{HCO}_3^- \\
(1.1 \times 10^{-5} - x) \, M & & x \, M & & x \, M
\end{array}$$

we write

$$4.2 \times 10^{-7} = \frac{x^2}{1.1 \times 10^{-5} - x}$$

Solving the quadratic equation gives

$$x = 1.9 \times 10^{-6} \, M = [\text{H}^+] = [\text{HCO}_3^-]$$

The second-stage dissociation is given by

$$\begin{array}{ccccc}
\text{HCO}_3^- & \rightleftharpoons & \text{H}^+ & + & \text{CO}_3^{2-} \\
(1.9 \times 10^{-6} - y) \, M & & (1.9 \times 10^{-6} + y) \, M & & y \, M
\end{array}$$

Here, we have

$$4.8 \times 10^{-11} = \frac{(1.9 \times 10^{-6} + y)y}{(1.9 \times 10^{-6} - y)}$$

Because $1.9 \times 10^{-6} \gg y$, the above equation becomes

$$y = 4.8 \times 10^{-11} \, M$$

Finally,

$$[\text{OH}^-] = \frac{1.0 \times 10^{-14}}{1.9 \times 10^{-6}} = 5.3 \times 10^{-9} \, M$$

At equilibrium, the concentrations are

$$[\text{H}^+] = 1.9 \times 10^{-6} \, M$$
$$[\text{OH}^-] = 5.3 \times 10^{-9} \, M$$
$$[\text{H}_2\text{CO}_3] = 1.1 \times 10^{-5} \, M$$
$$[\text{HCO}_3^-] = 1.9 \times 10^{-6} \, M$$
$$[\text{CO}_3^{2-}] = 4.8 \times 10^{-11} \, M$$

COMMENT

This calculation shows that water in equilibrium with atmospheric carbon dioxide becomes acidic ($pH = 5.7$). Also, within two significant figures, the equilibrium concentration of H_2CO_3 is the same as the initial concentration.

The results from Example 8.2 can be generalized as follows. Because $K_a' \gg K_a''$ for most diprotic acids, we can assume that the concentration of H^+ ions is produced mainly by the first-stage dissociation and that the concentration of the conjugate base for the second-stage dissociation is *numerically* equal to K_a''. The large difference between K_a' and K_a'' also ensures that no more than two of the successive species associated with the dissociation of a diprotic acid are present in significant concentrations at a particular pH. For carbonic acid, then, at any pH the predominant species in solution will be H_2CO_3 or HCO_3^- or CO_3^{2-}. Or, it will be the H_2CO_3/HCO_3^- couple or the HCO_3^-/CO_3^{2-} couple. Figure 8.1 shows the *distribution diagram* for carbonic acid, from which we can deduce the relative amounts of the species present as a function of pH.

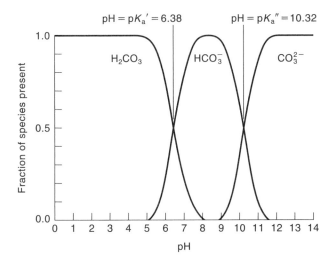

Figure 8.1
Distribution diagram for the carbonic acid system as a function of pH. Note that no more than two species predominate at any given pH.

A particularly important triprotic acid is phosphoric acid, H_3PO_4. The three stages of dissociation are

$$H_3PO_4 \rightleftharpoons H^+ + H_2PO_4^-$$
$$H_2PO_4^- \rightleftharpoons H^+ + HPO_4^{2-}$$
$$HPO_4^{2-} \rightleftharpoons H^+ + PO_4^{3-}$$

Using the same procedure as in Example 8.2, we can calculate the concentrations of all the species present at equilibrium. Figure 8.2 shows the distribution diagram for H_3PO_4.

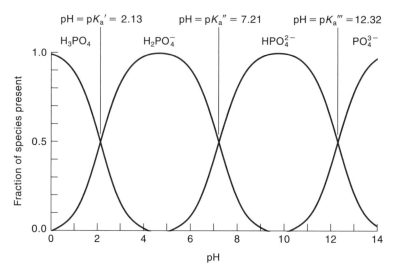

Figure 8.2
Distribution diagram for the phosphoric acid system as a function of pH.
Note that no more than two species predominate at any given pH.

8.5 Buffer Solutions

One of the most important applications of acid–base solutions is buffering. A buffer solution is a solution of (1) a weak acid or a weak base and (2) its salt; both components must be present. The solution has the ability to resist changes in pH upon the addition of small amounts of either an acid or a base. Buffer solutions, or simply buffers, play a crucial role in chemical and biological systems. The pH of human body fluids varies greatly, depending on the location. For example, the pH of blood plasma is about 7.4, whereas that of gastric juice, a fluid produced by glands in the mucosa membrane lining the stomach, is about 1.2. Maintained by buffers, these pHs are essential for the proper action of enzymes, the balance of osmotic pressure, and so on. Table 8.3 lists the pHs of a number of fluids.

Consider the dissociation of a weak acid HA:

$$HA \rightleftharpoons H^+ + A^-$$

$$K_a = \frac{[H^+][A^-]}{[HA]}$$

Rearranging this equation gives

$$[H^+] = K_a \frac{[HA]}{[A^-]} \tag{8.15}$$

Taking the negative logarithm of Equation 8.15, we get

$$-\log[H^+] = -\log K_a - \log \frac{[HA]}{[A^-]}$$

$$pH = pK_a + \log \frac{[A^-]}{[HA]} \tag{8.16}$$

Table 8.3
The pHs of Some Common Fluids

Sample	pH Value
Gastric juice in the stomach	1.0–2.0
Lemon juice	2.4
Vinegar	3.0
Grapefruit juice	3.2
Orange juice	3.5
Urine	4.8–7.5
Water exposed to air[a]	5.5
Saliva	6.4–6.9
Milk	6.5
Pure water	7.0
Blood	7.35–7.45
Tears	7.4
Milk of magnesia	10.6
Household ammonia	11.5

[a] Water exposed to air for a long period of time absorbs atmospheric CO_2 to form carbonic acid, H_2CO_3.

Equation 8.16 is known as the Henderson–Hasselbalch equation, which is generally expressed as

$$pH = pK_a + \log \frac{[\text{conjugate base}]}{[\text{acid}]}$$ (8.17)

Note the similarity between Equations 8.17 and 7.7. Acid–base and redox reactions are analogous in many respects.

For a given acid (that is, for a given K_a), Equation 8.15 and 8.17 show that the hydrogen ion concentration or the pH of the solution depends on the relative amounts of the acid and the conjugate base present at equilibrium. An acid solution alone cannot act as a buffer because the conjugate base concentration is too low. But we can increase $[A^-]$ by adding a sodium salt (NaA), say, to the solution so that at equilibrium we have $[HA] \approx [A^-]$. Let us consider a simple buffer solution prepared by adding equal molar amounts of acetic acid (CH_3COOH) and its salt, sodium acetate (CH_3COONa), to water. In solution, acetic acid dissociates as

Equation 8.17 shows that when $[A^-] = [HA]$, pH = pK_a.

$$CH_3COOH \rightleftharpoons CH_3COO^- + H^+$$

and the acetate ion (from CH_3COONa) hydrolyzes:

$$CH_3COO^- + H_2O \rightleftharpoons CH_3COOH + OH^-$$

Because CH_3COOH is a weak acid, however, it dissociates only to a small degree. The extent of the acetate ion hydrolysis is also negligible (see Problem 8.25). Furthermore, the presence of the acid suppresses the hydrolysis of the conjugate base and the presence of the conjugate base suppresses the dissociation of the acid. In essence, the reaction between the acid and its conjugate base does not produce any new species:

$$CH_3COOH + CH_3COO^- \rightleftharpoons CH_3COO^- + CH_3COOH$$

For these reasons, as a good approximation, we can treat the concentrations of the acid and its conjugate base at equilibrium to be the same as the initial concentrations. This solution acts as a buffer because CH_3COOH will neutralize any added base,

$$CH_3COOH + OH^- \rightarrow CH_3COO^- + H_2O$$

and the presence of CH_3COO^- can combine with any added acid:

$$CH_3COO^- + H^+ \rightarrow CH_3COOH$$

Example 8.3

Calculate the pH of a buffer system containing 0.40 M CH_3COOH and 0.55 M CH_3COONa. What is the pH of the buffer after the addition of 0.10 mole of HCl to 1.0 L of the solution? Assume no change in the volume of the solution.

ANSWER

Before the addition of HCl and from Equation 8.17 and Table 8.1,

$$pH = 4.76 + \log\frac{0.55\ M}{0.40\ M}$$

$$= 4.90$$

The addition of 0.10 mole of HCl produces 0.10 mole of H^+ ions. Originally, 0.40 mole of CH_3COOH and 0.55 mole of CH_3COO^- were present. After neutralization of the HCl acid by CH_3COO^-, which we write

$$\begin{array}{cccc} CH_3COO^- + & H^+ & \rightarrow CH_3COOH \\ 0.10\ \text{mol} & 0.10\ \text{mol} & 0.10\ \text{mol} \end{array}$$

so that the pH of the buffer becomes

$$pH = 4.76 + \log\frac{(0.55 - 0.10)\ \text{mol}}{(0.40 + 0.10)\ \text{mol}}$$

$$= 4.71$$

COMMENT

Because the volume of the solution is the same for both species, we replaced the ratio of their molar concentrations with the ratio of the number of moles present. As an exercise, you can show that the addition of 0.10 mole of NaOH to the above buffer solution will raise the pH to 5.10.

In the buffer solution examined in Example 8.3, there is a decrease in pH as a result of added HCl. We can also compare the changes in $[H^+]$ as follows:

Before addition of HCl: $[H^+] = 10^{-4.90} = 1.3 \times 10^{-5}\ M$

After addition of HCl: $[H^+] = 10^{-4.71} = 1.9 \times 10^{-5}\ M$

Table 8.4
Common Buffer Solutions

Buffer	pH Range[a]
Na acetate/acetic acid	3.8–5.8
Na borate/boric acid	8.1–10.1
Na citrate/citric acid	2.1–4.1
KH phthalate/phthalic acid	2.1–4.1
KNa phthalate/KH phthalate	4.4–6.4
Na_2CO_3/$NaHCO_3$	9.3–11.3
Na_2HPO_4/KH_2PO_4	4.4–6.4
Na_3PO_4/Na_2HPO_4	11.3–13.3
HEPES[b]	6.6–8.6
Tris/HCl[c]	7.1–9.1

[a] Defined as $pK_a = pH \pm 1$.
[b] HEPES is N-2-hydroxyethylpiperazine-N'-2-ethanesulfonic acid.
[c] Tris is tris(hydroxymethyl)aminomethane.

Thus, the H^+ ion concentration increases by a factor of

$$\frac{1.9 \times 10^{-5}\ M}{1.3 \times 10^{-5}\ M} = 1.5$$

To appreciate the effectiveness of the buffer, let us find out what would happen if 0.10 mole of HCl were added to 1 L of water and compare the increase in H^+ concentration.

Before addition of HCl: $\quad [H^+] = 1.0 \times 10^{-7}\ M$

After addition of HCl: $\quad [H^+] = 0.10\ M$

As a result of the addition of HCl, the H^+ ion concentration increases by a factor

$$\frac{0.10\ M}{1.0 \times 10^{-7}\ M} = 1.0 \times 10^6$$

amounting to a millionfold increase!

Our discussion thus far applies equally well to the buffer system of a weak base (B) such as NH_3 and its conjugate acid (BH^+), NH_4^+, for which you should be able to derive the following Henderson–Hasselbalch equation:

$$pH = pK_a + \log\frac{[B]}{[BH^+]}$$

Table 8.4 lists the pH ranges of common buffer systems.

The Effect of Ionic Strength and Temperature on Buffer Solutions

Equation 8.17 was derived assuming ideal behavior. A more accurate treatment begins with the replacement of concentrations with activities. Starting with the weak monoprotic acid HA, we write

$$K_a = \frac{a_{H^+} a_{A^-}}{a_{HA}}$$

For the neutral HA species, we can replace a_{HA} with [HA]. Rearranging the above equation, we get

$$a_{H^+} = K_a \frac{[HA]}{\gamma_{A^-}[A^-]}$$

where $a_{A^-} = \gamma_{A^-}[A^-]$. Taking the negative logarithm of the equation, we get

$$pH = pK_a + \log \frac{[A^-]}{[HA]} + \log \gamma_{A^-}$$

From Equation 5.61,

$$\log \gamma_{A^-} = -0.509(-1)^2 \sqrt{I} = -0.509\sqrt{I}$$

Finally, we arrive at the modified Henderson–Hasselbalch equation as

$$pH = pK_a + \log \frac{[A^-]}{[HA]} - 0.509\sqrt{I} \qquad (8.18)$$

Note that in Example 8.3, the ionic strength of the buffer solution before the addition of HCl, ignoring the dissociation of CH_3COOH, is due to both the Na^+ and the CH_3COO^- ions (see Equation 5.59):

$$I = \tfrac{1}{2}(0.55)(-1)^2 + \tfrac{1}{2}(0.55)(1)^2 = 0.55 \ M$$

We assume molarity is equal to molality.

so the pH of the buffer is

$$pH = 4.76 + \log \frac{0.55}{0.40} - 0.509\sqrt{0.55} = 4.52$$

which is quite different from 4.90.

Another correction that should be made for accurate work involving buffer solutions takes into account temperature variation. Many buffer solutions are characterized at 25°C, but biochemical reactions commonly occur at temperatures between 30°C and 40°C, and so it is important to convert K_a values to higher temperatures. Looking at Equations 8.17 or 8.18, we see that both the concentration and ionic strength terms are roughly temperature independent. Therefore, the pH of a buffer solution will change by the *same* amount as the pK_a. From knowledge of the K_a value at 25°C and the enthalpy of dissociation ($\Delta H°$) and using Equation 6.18, we can calculate the value of K_a at 37°C, say, and then the pH at that temperature (see Problem 8.80).

Preparing a Buffer Solution with a Specific pH

Suppose we wish to prepare a buffer solution with a specific pH. How do we go about it? Equation 8.17 indicates that if the molar concentrations of the acid and its conjugate base are about equal, then

$$\log \frac{[\text{conjugate base}]}{[\text{acid}]} \approx 0$$

or

$$\text{pH} \approx \text{p}K_\text{a}$$

Thus, to prepare a buffer solution, we work backwards. First, we choose a weak acid whose $\text{p}K_\text{a}$ value is close to the desired pH. Next, we substitute the pH and $\text{p}K_\text{a}$ values in Equation 8.17 to obtain the ratio [conjugate base]/[acid]. This ratio can then be converted to molar quantities for the preparation of the buffer solution.

Example 8.4

Phosphate buffers (that is, buffers containing the phosphate group) are present in biological systems such as the blood plasma. Describe how you would prepare a phosphate buffer with a pH of 7.40, assuming ideal behavior.

ANSWER

From Table 8.1, we have

$$\text{H}_3\text{PO}_4 \rightleftharpoons \text{H}^+ + \text{H}_2\text{PO}_4^- \quad K_\text{a}' = 7.5 \times 10^{-3} \quad \text{p}K_\text{a}' = 2.13$$
$$\text{H}_2\text{PO}_4^- \rightleftharpoons \text{H}^+ + \text{HPO}_4^{2-} \quad K_\text{a}'' = 6.2 \times 10^{-8} \quad \text{p}K_\text{a}'' = 7.21$$
$$\text{HPO}_4^{2-} \rightleftharpoons \text{H}^+ + \text{PO}_4^{3-} \quad K_\text{a}''' = 4.8 \times 10^{-13} \quad \text{p}K_\text{a}''' = 12.32$$

The most suitable of the three buffer systems is the $\text{HPO}_4^{2-}/\text{H}_2\text{PO}_4^-$ couple, because the $\text{p}K_\text{a}''$ value of the acid H_2PO_4^- is closest to the desired pH. From Equation 8.17, we write

$$7.40 = 7.21 + \log \frac{[\text{HPO}_4^{2-}]}{[\text{H}_2\text{PO}_4^-]}$$

or

$$\frac{[\text{HPO}_4^{2-}]}{[\text{H}_2\text{PO}_4^-]} = 1.5$$

Thus, one way to prepare the buffer is to dissolve disodium hydrogen phosphate (Na_2HPO_4) and sodium dihydrogen phosphate (NaH_2PO_4) in a mole ratio of 1.5:1.0 in water to make up a buffer solution of desired volume and pH.

In addition to the method described in Example 8.4, we can prepare a buffer solution (1) by partially neutralizing a weak acid with NaOH until equal concentrations of the acid and its conjugate base have formed, or (2) by adding HCl acid to a solution of the sodium salt of the weak acid until the concentration of the weak acid and conjugate base are equal (see Problem 8.47).

Buffer Capacity

We measure the effectiveness of a buffer in terms of its *buffer capacity* (β), which is the amount of acid or base that must be added to the buffer to produce a unit change of pH. Hence,

$$\beta = \frac{d[\text{B}]}{d\text{pH}} \tag{8.19}$$

where $d[B]$ is the increase (in mol L^{-1}) in concentration of base B, and dpH is the corresponding increase in pH. If an acid is added to a buffer solution, the concentration of the base will decrease and so will the pH of the solution. In this case, the buffer capacity is given by

$$\beta = \frac{-d[B]}{-d\text{pH}}$$

where the negative signs denote the decreases. Thus, $d[B]$ and dpH always have the same sign, and the buffer capacity always has a positive value. The buffer capacity depends not only on the nature of the buffer and the concentrations of the acid and conjugate base, but also on the pH.

Figure 8.3 shows a plot of buffer capacity versus pH for the CH$_3$COONa/CH$_3$COOH system. As we can see, this buffer functions best around its pK_a value of 4.76. This result is not surprising because at this pH we have pH = pK_a so that [CH$_3$COOH] = [CH$_3$COO$^-$], and there are equal amounts of acid and conjugate base to react with the added base or acid. In general, a buffer will maintain a nearly constant pH as long as the [conjugate base]/[acid] ratio stays between 0.1 and 10 because of the slow variation of the logarithmic term over this range of ratios. Thus, it is useful to speak of the *buffer range*, which is the pH range over which the buffer is most effective; that is

$$\text{pH} = \text{p}K_a \pm 1 \tag{8.20}$$

Therefore, the buffer range for the CH$_3$COONa/CH$_3$COOH couple is 4.76 ± 1, or between 3.76 and 5.76.

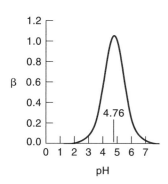

Figure 8.3
Buffer capacity of the 1 M CH$_3$COONa/1 M CH$_3$COOH buffer system. The maximum of the peak occurs at pH = 4.76, which is also equal to pK_a.

8.6 Acid–Base Titrations

This section deals with one of the most common and important techniques in the arsenal of analytical chemistry. Acid–base titration is a simple experimental procedure. A base is added from a buret to an acid solution until the equivalence point is reached. If the concentration of one of the solutions is known, the concentration of the other solution can be easily calculated from a knowledge of the volumes of the acid and the base. A pH meter is the most convenient instrument for monitoring the titration, but many other devices are employed as well.

A strong acid and a strong base provide the most clear-cut results in a titration experiment. A good example is the titration of HCl versus NaOH:

$$\text{HCl} + \text{NaOH} \rightarrow \text{NaCl} + \text{H}_2\text{O}$$

Initially, the pH of the solution is determined largely by the acid present and increases only slowly with the addition of the base. Near the equivalence point, the pH rises steeply upon the addition of a small amount of the base. Beyond the equivalence point, the pH of the solution is determined by the excess base present. Because Na$^+$ and Cl$^-$ ions do not hydrolyze appreciably, the solution at the equivalence point should be nearly neutral, with a pH of 7. Now consider the titration of a weak acid such as acetic acid with NaOH:

$$\text{CH}_3\text{COOH} + \text{NaOH} \rightarrow \text{CH}_3\text{COONa} + \text{H}_2\text{O}$$

As mentioned earlier, the acetate ion (from CH$_3$COONa) does hydrolyze to a certain extent to produce OH$^-$ ions. Consequently, at the point at which the acetic acid is

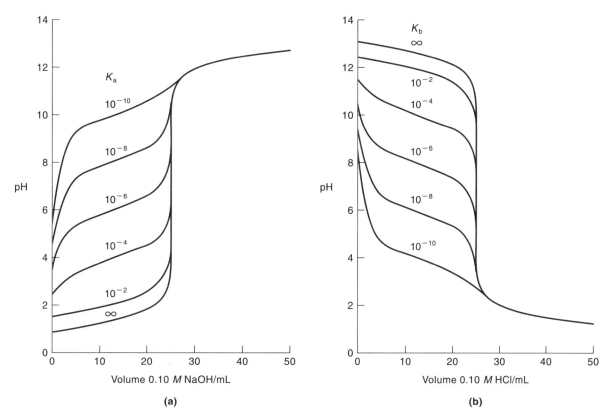

Figure 8.4
(a) Titration curves of 0.10 M NaOH with a series of acids. The ∞ sign denotes a strong acid. (b) Titration curves of 0.10 M HCl with a series of bases. The ∞ sign denotes a strong base.

completely neutralized by NaOH, the solution is basic instead of neutral. Figure 8.4a shows the titration curves of a 0.10 M NaOH solution versus 25.0 mL of a 0.10 M acid solution. Note that as the strength of the acid decreases, the steep portion of the curve becomes progressively shorter.

An analogous situation is the titration of a strong acid against a weak base:

$$HCl + NH_3 \rightarrow NH_4^+ + Cl^-$$

As a result of the hydrolysis of the NH_4^+ ion, the pH at the equivalence point is below 7. Figure 8.4b shows the titration curves of weak bases being added to a strong acid. Again, the steep portion becomes shorter as the base strength decreases. Finally, we note that because of salt hydrolysis involving both the cation and the anion, it is generally difficult to study the titration of a weak acid against a weak base.

Acid–Base Indicators

Another way to monitor titration is to use an indicator whose color depends on the pH of the solution. An indicator is itself an acid or a base. Let us imagine a compound whose acid (HIn) form is distinctly different in color than its conjugate base (In^-) form. In solution, the following equilibrium is established:

$$HIn \rightleftharpoons H^+ + In^-$$

Table 8.5
Some Common Acid–Base Indicators

Indicator	Color Acid	Color Base	pK_{In}	pH Range[a]
Thymol blue	Red	Yellow	1.51	1.2–2.8
Bromophenol blue	Yellow	Blue	3.98	3.0–4.6
Chlorophenol blue	Yellow	Red	5.98	4.8–6.4
Bromothymol blue	Yellow	Blue	7.0	6.0–7.6
Cresol red	Yellow	Red	8.3	7.2–8.8
Methyl orange	Orange	Yellow	3.7	3.1–4.4
Methyl red	Red	Yellow	5.1	4.2–6.3
Phenolphthalein	Colorless	Pink	9.4	8.3–10.0

[a] These values are determined experimentally, rather than calculated using Equation 8.21.

for which we write

$$K_{In} = \frac{[H^+][In^-]}{[HIn]}$$

Applying Equation 8.17, we write

$$pH = pK_{In} + \log \frac{[In^-]}{[HIn]}$$

The following ratios apply to most indicators:

$$\frac{[In^-]}{[HIn]} \leq 0.1 \quad \text{acid color} \qquad \frac{[In^-]}{[HIn]} \geq 10 \quad \text{base color}$$

Therefore, the pH range over which the indicator changes color is

$$pH = pK_{In} \pm 1 \tag{8.21}$$

To choose an indicator for a particular titration, we must be sure that the indicator's range lies on the steep portion of the titration curve.

Table 8.5 lists a number of commonly employed acid–base indicators and their pH ranges. Keep in mind that the pH at which the color of the indicator changes drastically is called the *end point*, which is not the same as the equivalence point. However, if the pH range of the indicator lies on the steep portion of the titration curve, then the end point is close enough to be taken as the equivalence point.

8.7 Amino Acids

Amino acids are the building blocks of proteins. To understand the structure and function of proteins, we must first investigate the properties of individual amino acids. By definition, an amino acid contains at least one amino group ($-NH_2$) and at least one carboxyl group ($-COOH$). All the proteins in the human body are made from the 20 amino acids shown in Table 8.6. All except glycine are chiral and have the L-configuration.

Table 8.6
Amino Acids Isolated from Proteins $\overset{\overset{+}{N}H_3}{R-CH-COOH}$

Name	Side chain (R)	$pK_a{}'$ —COOH	$pK_a{}''$ $-\overset{+}{N}H_3$	$pK_a{}'''$ R	p*I*	Abbreviation[a]	
Alanine	CH_3-	2.35	9.69		6.02	Ala	A
Arginine	$\overset{\overset{+}{N}H_2}{H_2N-\overset{\|}{C}-NH(CH_2)_3-}$	2.17	9.04	12.48	10.76	Arg	R
Asparagine	$\overset{O}{H_2N-\overset{\|}{C}-CH_2-}$	2.02	8.80		5.41	Asn	N
Aspartic acid	$HOOC-CH_2-$	2.09	9.82	3.86	2.98	Asp	D
Cysteine	$HS-CH_2-$	1.71	8.90	8.50	5.02	Cys	C
Glutamic acid	$HOOC-CH_2-CH_2-$	2.19	9.67	4.25	3.22	Glu	E
Glutamine	$\overset{O}{H_2N-\overset{\|}{C}-CH_2-CH_2-}$	2.17	9.13		5.70	Gln	Z
Glycine	$H-$	2.34	9.60		5.97	Gly	G
Histidine	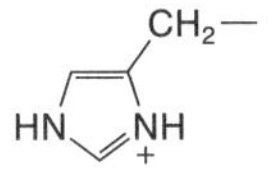 (imidazole) CH_2-	1.82	9.17	6.00	7.59	His	H
Isoleucine	$\overset{CH_3-CH_2}{\underset{CH_3}{\diagdown}CH-}$	2.36	9.68		6.02	Ile	I
Leucine	$\overset{CH_3}{\underset{CH_3}{\diagup}}CH-CH_2-$	2.36	9.60		5.98	Leu	L
Lysine	$H_3\overset{+}{N}(CH_2)_3CH_2-$	2.18	8.95	10.53	9.74	Lys	K
Methionine	$CH_3S-CH_2-CH_2-$	2.28	9.21		5.75	Met	M
Phenylalanine	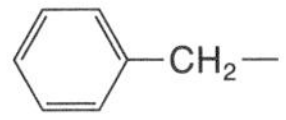 (phenyl)$-CH_2-$	1.83	9.13		5.48	Phe	F
Proline[b]	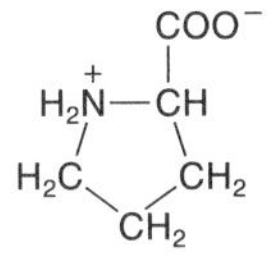 $\overset{COO^-}{\underset{}{H_2\overset{+}{N}-CH}}$ (pyrrolidine ring)	1.95	10.65		6.30	Pro	P
Serine	$HO-CH_2-$	2.21	9.15		5.68	Ser	S
Threonine	$\overset{OH}{CH_3-\overset{\|}{CH}-}$	2.09	9.10		5.60	Thr	T
Tryptophan	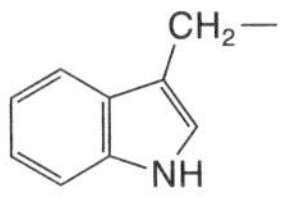 (indole) CH_2-	2.38	9.39		5.88	Trp	W
Tyrosine	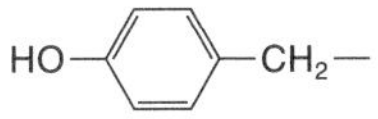 $HO-$(phenyl)$-CH_2-$	2.20	9.11	10.07	5.67	Tyr	Y
Valine	$\overset{CH_3}{\underset{CH_3}{\diagup}}CH-$	2.32	9.62		5.97	Val	V

[a]Amino acids are identified by three-letter and one-letter abbreviations.
[b]Proline differs from the other common amino acids in that its side chain is bonded to the backbone nitrogen atom as well as to the C_α atom. The structure shown here is the entire molecule, not just the R group.

Dissociation of Amino Acids

Amino acids, like water, are ampholytes; that is, they behave both as acids and as bases. For many years, it was uncertain whether these substances exist in solution as $NH_2CHRCOOH$ or as $^+NH_3CHRCOO^-$, which is called a *dipolar ion*, or a *zwitterion*. Now, much evidence suggests that zwitterions are predominant in solution. Among the characteristics supporting this conclusion are their high dipole moments and high solubilities in polar solvents.

Zwitter is the German word for hybrid.

Let us start with the simplest amino acid, glycine. In solution, glycine exists as the zwitterion, $^+NH_3CH_2COO^-$. It behaves as a base when titrated with hydrochloric acid:

$$^+NH_3CH_2COO^- + HCl \rightarrow {}^+NH_3CH_2COOH + Cl^-$$

and behaves as an acid when titrated with sodium hydroxide:

$$^+NH_3CH_2COO^- + NaOH \rightarrow NH_2CH_2COO^- + Na^+ + H_2O$$

Figure 8.5 shows the titration curve of glycine versus HCl and NaOH. The pH at the first half-equivalence point is equal to pK_a', and the pH at the second half-equivalence point is equal to pK_a'' (see Appendix 8.1 for details). At the equivalence point, the pH is given by

$$pH = \frac{pK_a' + pK_a''}{2} = \frac{2.34 + 9.60}{2}$$
$$= 5.97$$

At this pH, the zwitterion predominates. The dissociation of glycine can be sum-

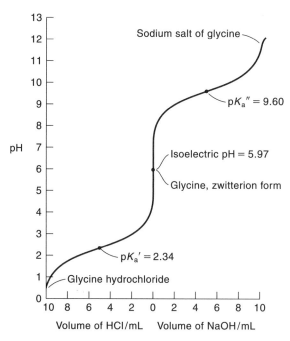

Figure 8.5
Titration curve for 0.10 M glycine with equivalent amounts of HCl and NaOH.

marized as follows:

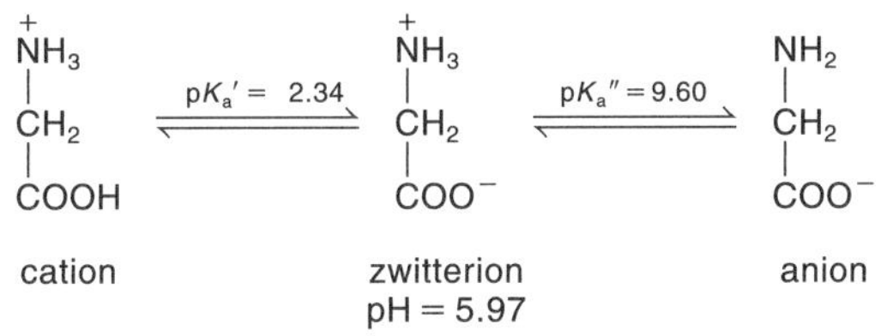

cation   zwitterion   anion

pH = 5.97

The Isoelectric Point (p*I*)

When the net charge of a molecule is zero, as it is for the zwitterion of glycine, for example, the species is electrically neutral. In this condition, the molecule is said to be *isoelectric*. The pH at which the zwitterion does not migrate in an electric field is called the *isoelectric point*, or p*I*. As we saw earlier, the isoelectric point of glycine is equal to 5.97.

For acids containing more than two dissociable protons, the situation is more complex. Consider aspartic acid, which dissociates as follows:

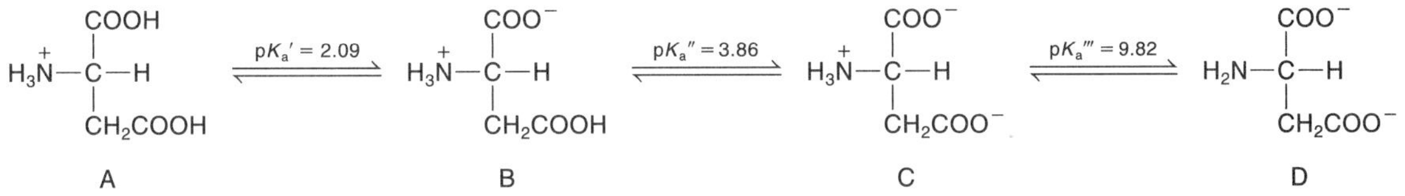

Because only B has the same number of positive and negative charges, the isoelectric point is the mean of the pK_a' and pK_a'' values; that is,

$$pI = \frac{2.09 + 3.86}{2} = 2.98$$

Another polyprotic amino acid is histidine. This amino acid is important because the dissociation of a proton from the imidazole ring is primarily responsible for the buffering action of proteins—particularly hemoglobin—in blood (see Section 8.7). Figure 8.6 shows the stepwise dissociation of a fully protonated histidine molecule.

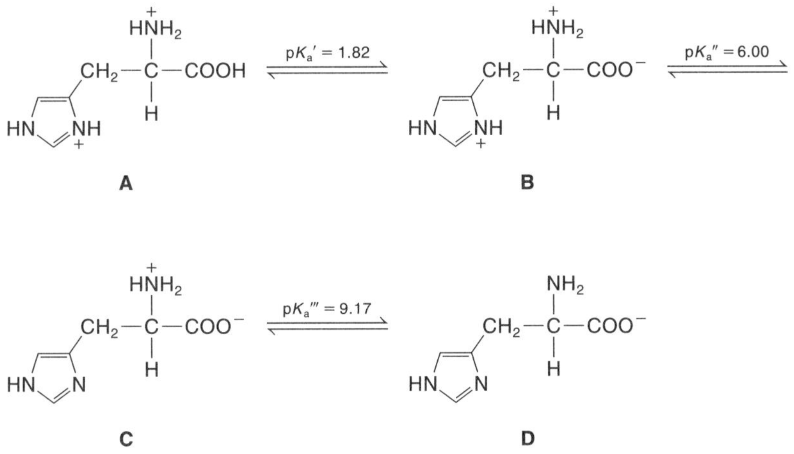

Figure 8.6
Stepwise dissociation of a fully protonated histidine molecule. The dissociated H^+ ions are not shown.

Because only C has the same number of positive and negative charges, the isoelectric point is given by the mean of pK_a'' and pK_a''' values; that is,

$$pI = \frac{6.00 + 9.17}{2} = 7.59$$

Table 8.6 lists the pI values of all 20 amino acids.

In Chapter 16 we shall see how the different values of pI are used to separate a mixture of proteins in a technique called *isoelectric focusing*.

Titration of Proteins

Having discussed the acid–base properties of individual amino acids, we shall see how these properties are affected when the amino acids are incorporated into a protein molecule. One way to do this is by a titration experiment. It may seem that proteins are far too complex to handle by titration. Actually, the situation is more manageable than we might at first expect. Although a protein molecule has numerous dissociable protons, most of them can be characterized according to the pK values using Table 8.6 as a reference. Because of the spread in pK's, however, the titration curve is broad and much care is needed for accurate assignments.

The primary reason for titrating proteins is to count the number of dissociable protons, identify them, and compare the result with amino acid analysis. The technique can also yield additional information regarding the conformation of the macromolecule in solution. Figure 8.7 shows the titration curve of the enzyme ribonuclease. The shape of the curve depends somewhat on the ionic strength of the solution. This curve can be divided into three regions: between pH 1 and 5, eleven protons are dissociated; between pH 5 and 8, five protons are dissociated; between pH 8 and 12, seventeen protons are dissociated. The assignments of the protons

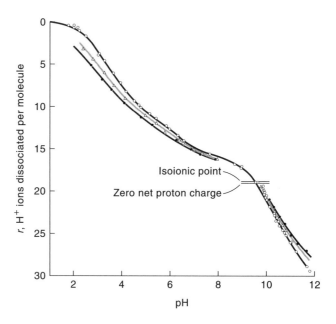

Figure 8.7
Titration curve of ribonuclease. The three curves correspond to three different ionic strengths of the solution. [Reprinted with permission from C. Tanford and J. D. Hauenstein, *J. Am. Chem. Soc.* **78**, 5288 (1956). Copyright by the American Chemical Society.]

Table 8.7
Titration of Ribonuclease[a]

Acid Group	Number of Protons		pK	
	Titration	Amino Acid Analysis	Observed	Normal
α-COOH	1	1	3.75	3.75
β,γ-COOH	10	10		4.6
Imidazole	4	4	6.5	6.5–7
α-$\overset{+}{N}H_3$	1	1	7.8	7.8
ε-$\overset{+}{N}H_3$	10	10	10.2	10.1–10.6
Phenolic	3	6	9.5	9.6
Guanidyl	4	4	$\geq$12	>12

[a] Reprinted with permission from C. Tanford and J. D. Hauenstein, *J. Am. Chem. Soc.* **78**, 5290 (1956). Copyright by the American Chemical Society.

are shown in Table 8.7. With the exception of the phenolic groups, agreement is remarkably good. This exception is easily explained if we assume that the three nontitratable groups are located in the interior region of the protein molecule, inaccessible to acid–base reaction. Similar instances are also found in other systems. For example, only 6 of the 12 imidazole groups in myoglobin are titratable. In each case, the buried groups can be brought to the surface of the molecule by denaturation, and can then be accounted for by titration.

8.8 Maintaining the pH of Blood

The pH of most intracellular fluids ranges between 6.8 and 7.8. Among the numerous buffer systems that help maintain the proper pH are HCO_3^-/H_2CO_3 and $HPO_4^{2-}/H_2PO_4^-$. They react with acids and bases as follows:

$$HA + HCO_3^- \rightleftharpoons A^- + H_2CO_3$$
$$B + H_2CO_3 \rightleftharpoons BH^+ + HCO_3^-$$

and

$$HA + HPO_4^{2-} \rightleftharpoons A^- + H_2PO_4^-$$
$$B + H_2PO_4^- \rightleftharpoons BH^+ + HPO_4^{2-}$$

In an adult weighing 70 kg, approximately 0.1 mole of H^+ ions and 15 moles of CO_2 are produced everyday as a result of metabolism. The body has two mechanisms for handling the acid produced by metabolism to prevent a drop in pH: buffering and excreting H^+. We shall discuss the buffering action of blood first.

Blood consists essentially of two components: blood plasma—a complex solution containing many vital biochemical compounds (carbohydrates, amino acids, proteins, enzymes, hormones, vitamins, and inorganic ions)—and erythrocytes (red blood cells). The blood of the average adult contains about 5 million erythrocytes per milliliter of blood. Hemoglobin, the "respiratory" protein, is present inside the erythrocyte. There are about 2×10^5 hemoglobin molecules per erythrocyte. Blood plasma is maintained at pH 7.4 largely by the HCO_3^-/H_2CO_3 and $HPO_4^{2-}/H_2PO_4^-$ buffers and various proteins. Proteins are polymers of amino acids, which can act as buffers themselves. As Table 8.6 shows, most carboxyl groups and amino groups

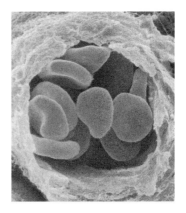

Electron micrograph of red blood cells in the capillary of an artery. (Professor P. P. Botta and S. Correr. Science Picture Library/Photo Researchers, Inc. Reprinted with permission.)

have pK_a values that are far removed from the proper pH range. The only exception is histidine, whose imidazole group has a pK_a value of 6.0, which is best suited for buffering $[H^+]$ (see Figure 8.6).

We can estimate the relative concentrations of H_2CO_3 and HCO_3^- in the blood as follows. At pH 7.4 (the pH of blood plasma), Equation 8.17 becomes*

$$7.4 = 6.1 + \log \frac{[HCO_3^-]}{[H_2CO_3]}$$

so that $[HCO_3^-]/[H_2CO_3] = 20$. Normally, the concentrations of CO_2 and HCO_3^- are approximately 1.2×10^{-3} M and 0.024 M, respectively, giving us a ratio of $0.024/1.2 \times 10^{-3}$, or 20, as we calculated above. In the erythrocytes, the buffers are HCO_3^-/H_2CO_3 and the histidine groups in hemoglobin. The pH is about 7.25. Again from Equation 8.17, the $[HCO_3^-]/[H_2CO_3]$ is found to be about 14. The membrane of the red blood cell, unlike those of most other cells, is more permeable to anions such as HCO_3^-, OH^-, and Cl^- than to K^+ and Na^+ cations.

Oxyhemoglobin, formed by the combination of oxygen with hemoglobin in the lungs, is carried in the arterial blood to the tissues, where the oxygen is unloaded to myoglobin. Both hemoglobin and oxyhemoglobin are weak acids, although the latter is considerably stronger than the former:

$$HHb \rightleftharpoons H^+ + Hb^- \qquad pK_a = 8.2$$

$$HHbO_2 \rightleftharpoons H^+ + HbO_2^- \qquad pK_a = 6.95$$

where HHb and $HHbO_2$ represent "monoprotic" hemoglobin and oxyhemoglobin, respectively. Thus, at pH = 7.25, about 65% of $HHbO_2$ is in the dissociated form, whereas only 10% of HHb is dissociated. The release of oxygen by $HHbO_2$ is strongly influenced by the presence of carbon dioxide. In metabolizing tissues, the partial pressure of CO_2, P_{CO_2}, is higher in the interstitial fluid (that is, fluid within the tissue space) than in the plasma. Consequently, CO_2 diffuses into the blood vessels and then into the erythrocytes. Here, most of the CO_2 is converted to H_2CO_3 by the enzyme carbonic anhydrase:

$$CO_2 + H_2O \rightleftharpoons H_2CO_3$$

The presence of H_2CO_3 lowers the pH, which has a direct effect on the release of oxygen. Oxygen may be released from either $HHbO_2$ or HbO_2^- as shown:

$$HHbO_2 \rightleftharpoons HHb + O_2$$

$$HHbO_2 \rightleftharpoons H^+ + HbO_2^-$$

$$HbO_2^- \rightleftharpoons Hb^- + O_2$$

Because $HHbO_2$ releases oxygen more readily than HbO_2^-, a drop in pH increases the concentration of $HHbO_2$ and promotes the first step. The conjugate base Hb^- of the weaker acid HHb has a greater tendency to react with H_2CO_3 as follows:

$$Hb^- + H_2CO_3 \rightleftharpoons HHb + HCO_3^-$$

The bicarbonate ion formed passes through the cell membrane and is carried away in

The nature of oxygen–hemoglobin binding is discussed in Chapter 9.

---

*In Table 8.1, we find that $pK_a' = 6.38$ for HCO_3^-/H_2CO_3, which is the dissociation constant at 25°C. However, at the ionic strength of blood and the physiological temperature of 37°C, $pK_a' = 6.1$.

the plasma. This is the primary mechanism for the elimination of CO_2.* When the venous blood circulates back to the lungs, where P_{CO_2} is low and P_{O_2} is high, hemoglobin recombines with oxygen to form oxyhemoglobin:

$$HHb + O_2 \rightleftharpoons HHbO_2$$

The bicarbonate ions in blood plasma now diffuse into the erythrocyte to raise the pH, and we have

$$HHbO_2 + HCO_3^- \rightleftharpoons HbO_2^- + H_2CO_3$$

The H_2CO_3 is then converted to CO_2, catalyzed by carbonic anhydrase:

$$H_2CO_3 \rightleftharpoons CO_2 + H_2O$$

Because of the lower P_{CO_2} in the lungs, the CO_2 formed diffuses out of the erythrocyte and is then exhaled into the atmosphere.

What causes the bicarbonate ions formed in red blood cells to preferentially diffuse into plasma? The Donnan effect, discussed in Chapter 5, provides the answer to this question. The concentrations of Hb^- and HbO_2^- are quite high in the erythrocytes; consequently, there is an unequal distribution of the diffusible anions in the erythrocytes and in the plasma. Now, according to the Donnan effect, the concentrations of the HCO_3^-, Cl^-, and OH^- ions are greater in the plasma than in the erythrocyte (assuming the proteins are in the anion form). Further, for any given salt MX, we can write

$$(\mu_{MX})^c = (\mu_{MX})^p$$

where c and p denote red blood cell and blood plasma, respectively. Following the procedure used for the Donnan effect (see p. 172), we arrive at the result

$$[M^+]_c[X^-]_c = [M^+]_p[X^-]_p$$

or

$$\frac{[M^+]_p}{[M^+]_c} = \frac{[X^-]_c}{[X^-]_p}$$

Thus, for a given cation and varying anions, we can show that

$$\frac{[HCO_3^-]_c}{[HCO_3^-]_p} = \frac{[Cl^-]_c}{[Cl^-]_p}$$

At the tissues, the CO_2 diffuses through the capillaries into the red blood cell and is converted to H_2CO_3. The carbonic acid reacts with Hb^- and HbO_2^- to form bicarbonate ion, causing the ratio $[HCO_3^-]_c/[HCO_3^-]_p$ to increase. To balance ionic concentrations, the bicarbonate ions diffuse into the plasma, while the chloride ions

---

* Another mechanism that accounts for a small amount of CO_2 transported involves the reaction of CO_2 with the amino groups of hemoglobin to form carbaminohemoglobin: $CO_2 + RNH_2 \rightarrow RNHCOO^- + H^+$. Carbaminohemoglobin has a lower affinity for O_2 than hemoglobin does. Thus, when the concentration of CO_2 in blood is high (at the metabolizing tissues), the affinity of hemoglobin for oxygen is decreased. The reverse is true in the lungs.

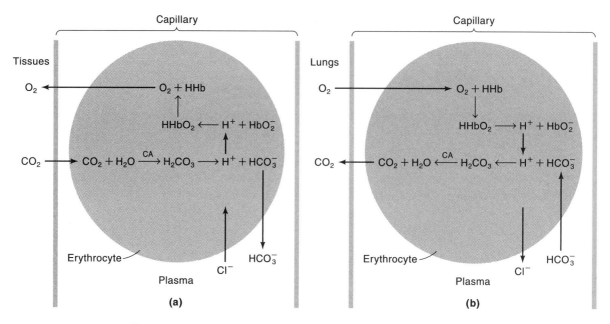

Figure 8.8
Oxygen–carbon dioxide transport and release by blood. (a) In metabolizing tissues, the partial pressure of CO_2 is higher in the interstitial fluid (fluid in the tissues) than in the plasma. Thus, CO_2 diffuses into the blood capillaries and then into erythrocytes. There it is converted to carbonic acid by the enzyme carbonic anhydrase (CA). The protons provided by the carbonic acid then combine with the oxyhemoglobin anions to form $HHbO_2$, which eventually dissociates into HHb and O_2. Because the partial pressure of O_2 is higher in the erythrocytes than in the interstitial fluid, oxygen molecules diffuse out of the erythrocytes and then into the tissues. The bicarbonate ions also diffuse out of the erythrocytes and are carried by the plasma to the lungs. A small portion of the CO_2 also binds to hemoglobin to form carbaminohemoglobin. (b) In the lungs, the processes are reversed.

diffuse into the cell, maintaining the overall electrical neutrality until the above equalities are restored. There is a corresponding decrease in pH in the erythrocyte caused by the departure of HCO_3^- ions, but this decrease is balanced by the flow of OH^- ions in the reverse direction. Because

$$[H^+]_c[OH^-]_c = [H^+]_p[OH^-]_p$$

or

$$\frac{[OH^-]_c}{[OH^-]_p} = \frac{[H^+]_p}{[H^+]_c}$$

we see that the different pH levels are always maintained in the red blood cells and in the plasma. Figure 8.8 summarizes this discussion.

The phenomenon described above is sometimes called the *bicarbonate–chloride shift*. In the lungs, the process is exactly reversed. There, the bicarbonate ion reacts with oxygenated hemoglobin, causing the ratio $[HCO_3^-]_c/[HCO_3^-]_p$ to decrease. The HCO_3^- ions diffuse from the plasma into the erythrocyte, whereas the Cl^- and OH^- ions diffuse in the opposite direction, until all the concentration ratios are again equal.

The delivery of oxygen to tissues and the removal of carbon dioxide illustrate the efficient and fascinating ways our bodies make use of buffers. But buffering alone is not enough to sustain physiological processes. The excretion of constantly forming H^+ ions also plays an important role in maintaining a normal blood pH. The kidneys excrete H^+ ions and return HCO_3^- ions back to the blood, where they combine with more H^+ ions. Normally, the pH of urine lies between 4.8 and 7.5. But from blood that has a pH of 7.4, the kidneys can produce a urine with a pH as low as 4.5! Two other mechanisms through which the kidneys can excrete H^+ ions are worth noting. The first is the excretion of anions of weak acids, particularly phosphoric acid. At pH 7.4, $H_2PO_4^-$ makes up roughly one-third of the phosphate present, and HPO_4^{2-} accounts for the other two-thirds. In an acidic urine, however, most of the phosphate excreted is $H_2PO_4^-$. The second mechanism involves the formation of NH_4^+ ions. Amino acids are degraded by the kidneys to form ammonia, which combines with H^+ ions as follows:

$$NH_3 + H^+ \rightarrow NH_4^+$$

The ammonium ions are then excreted in the urine.

A More Exact Treatment of Acid–Base Equilibria

In this appendix, we shall carry out a more exact derivation of the equilibrium equations for weak acids and their salts and examine acid–base titrations more closely.

1. Weak Acid Dissociation

For simplicity, we shall use concentrations rather than activities.

Suppose the initial concentration of a weak acid HA is $[HA]_0$ (mol L^{-1}). At equilibrium, there are four unknown concentrations: $[H^+]$, $[HA]$, $[A^-]$, and $[OH^-]$. In addition, there are four equations relating these unknowns:

$$K_a = \frac{[H^+][A^-]}{[HA]} \tag{1}$$

$$K_w = [H^+][OH^-] \tag{2}$$

Mass balance for the A^- anion:

$$[HA]_0 = [HA] + [A^-] \tag{3}$$

Charge balance:

$$[H^+] = [A^-] + [OH^-] \tag{4}$$

Equation (4) can now be written as

$$[A^-] = [H^+] - [OH^-]$$

$$= [H^+] - \frac{K_w}{[H^+]} \tag{5}$$

From Equation (3), we have

$$[HA] = [HA]_0 - [A^-]$$

$$= [HA]_0 - [H^+] + \frac{K_w}{[H^+]} \tag{6}$$

Substituting Equations (5) and (6) into Equation (1), we obtain

$$K_a = \frac{[H^+]([H^+] - K_w/[H^+])}{[HA]_0 - [H^+] + K_w/[H^+]} \tag{7}$$

This is a cubic equation in $[H^+]$, and in general it is quite tedious to solve. In the majority of cases, however, we can show that $K_w/[H^+] \ll [H^+]$ so that Equation (7) is reduced to

$$K_a = \frac{[H^+]^2}{[HA]_0 - [H^+]}$$

This equation is only quadratic in $[H^+]$ and can be readily solved. If the acid is very weak, then $[H^+] \ll [HA]_0$ and we have

$$K_a = \frac{[H^+]^2}{[HA]_0}$$

This simple equation was employed in Example 8.1.

2. Weak Acids and Their Salts

Here we consider the case of a solution initially containing $[HA]_0$ of a weak acid HA and $[NaA]_0$ of its salt NaA. We shall first derive a general expression for the acid dissociation constant and then look at some special situations.

In addition to Equations (1) and (2), we have the following relations:

Mass balance for the A^- anion:

$$[HA]_0 + [NaA]_0 = [HA] + [A^-] \tag{8}$$

Mass balance for the Na^+ cation:

$$[NaA]_0 = [Na^+] \tag{9}$$

Charge balance:

$$[Na^+] + [H^+] = [A^-] + [OH^-] \tag{10}$$

Keep in mind that all the quantities in square brackets in Equation (10) refer to equilibrium concentrations. From Equation (10), we write

$$[A^-] = [Na^+] + [H^+] - [OH^-]$$
$$= [NaA]_0 + [H^+] - [OH^-]$$

and from Equation (8),

$$[HA] = [HA]_0 + [NaA]_0 - [A^-]$$
$$= [HA]_0 + [NaA]_0 - [NaA]_0 - [H^+] + [OH^-]$$
$$= [HA]_0 - [H^+] + [OH^-]$$

Substituting the expressions for $[A^-]$ and $[HA]$ into Equation (1), we obtain

$$K_a = \frac{[H^+]([NaA]_0 + [H^+] - [OH^-])}{[HA]_0 - [H^+] + [OH^-]}$$
$$= \frac{[H^+]([NaA]_0 + [H^+] - K_w/[H^+])}{[HA]_0 - [H^+] + K_w/[H^+]} \tag{11}$$

Equation (11) is in a sense the "master" equation. Let us now consider two special cases.

Case 1. If no salt is present (that is, if we are dealing only with an acid), then $[NaA]_0 = 0$ and Equation (11) becomes

$$K_a = \frac{[H^+]([H^+] - K_w/[H^+])}{[HA]_0 - [H^+] + K_w/[H^+]}$$

This equation is the same as Equation (7).

Case 2. In anionic salt hydrolysis, the initial concentration of the acid is zero (that is, $[HA]_0 = 0$). Furthermore, the hydrogen ion concentration is usually negligible compared to the hydroxide ion concentration; that is, $[H^+] \ll K_w/[H^+]$. Thus, Equation (11) now takes the form

$$K_a = \frac{[H^+]([NaA]_0 - K_w/[H^+])}{K_w/[H^+]}$$

$$= \frac{(K_w/[OH^-])([NaA]_0 - [OH^-])}{[OH^-]}$$

Rearrangement of the equation gives

$$\frac{K_w}{K_a} = K_b = \frac{[OH^-]^2}{[NaA]_0 - [OH^-]}$$

3. Titration of a Weak Monoprotic Acid with a Strong Base

Consider the titration of acetic acid (in an Erlenmeyer flask) with sodium hydroxide (added from a buret):

$$CH_3COOH + OH^- \rightarrow CH_3COO^- + H_2O$$

We can check the pH of the mixture at every stage of the titration using Equation (11). Initially, before any base has been added, $[NaA]_0 = 0$ and $[H^+] \gg [OH^-]$, so that Equation (11) can be written as

$$K_a = \frac{[H^+][A^-]}{[HA]}$$

which is equivalent to Equation (1). After some base has been added, $[NaA]_0 > 0$, but because we still have $[H^+] \gg [OH^-]$, Equation (11) takes the form

$$K_a = \frac{[H^+]([NaA]_0 + [H^+])}{[HA]_0 - [H^+]}$$

Solving for $[H^+]$ gives us the hydrogen ion concentration and hence the pH. Note that because of dilution, $[HA]_0$ and $[NaA]_0$ must be calculated at each stage by taking the increase in volume into account. As the titration progresses, the concentration of the OH^- ions begins to build up, so that $[H^+] \approx [OH^-]$, and we need to use Equation (11) or (7) to solve for $[H^+]$. Because of anionic salt hydrolysis, we predict that the pH at the equivalence point will be greater than 7, so that at and beyond the equivalence point we have $[OH^-] \gg [H^+]$, or $K_w/[H^+] \gg [H^+]$, and Equation (11) becomes

$$K_a = \frac{[H^+]([NaA]_0 - K_w/[H^+])}{[HA]_0 + K_w/[H^+]}$$

$$= \frac{[H^+]^2[NaA]_0 - [H^+]K_w}{[H^+][HA]_0 + K_w}$$

4. Titration of a Weak Diprotic Acid with a Strong Base

The rigorous treatment of the titration of a diprotic acid H_2A with a strong base, such as NaOH, is quite complex and will not be presented here. Instead, we shall look at some qualitative features of the titration curve and see how to obtain the pK_a values.

For a weak diprotic acid H_2A, the dissociations are

$$H_2A \rightleftharpoons H^+ + HA^- \qquad\qquad K_a' = \frac{[H^+][HA^-]}{[H_2A]}$$

$$HA^- \rightleftharpoons H^+ + A^{2-} \qquad\qquad K_a'' = \frac{[H^+][A^{2-}]}{[HA^-]}$$

The additional equations are:

$$K_w = [H^+][OH^-]$$

Mass balance for the A^{2-} anion:

$$[H_2A]_0 = [H_2A] + [HA^-] + [A^{2-}]$$

Charge balance:

$$[H^+] + [Na^+] = [HA^-] + 2[A^{2-}] + [OH^-]$$

Note that the coefficient "2" is needed to account for the two negative charges on each A^{2-} anion. Basically, these six equations tell us all we need to know about such a titration. But instead of deriving the necessary mathematical equations, let us divide the titration into five stages and examine some of the salient features.

(i) At the Beginning of the Titration. Here we are concerned only with the first stage of dissociation:

$$H_2A \rightleftharpoons H^+ + HA^-$$

$$K_a' = \frac{[H^+][HA^-]}{[H_2A]} \approx \frac{[H^+]^2}{[H_2A]_0}$$

or

$$[H^+] = \sqrt{K_a'[H_2A]_0}$$

(ii) Halfway to the First Equivalence Point. At this point, half of H_2A has been converted to HA^- so that

$$[HA^-] = [H_2A]$$

giving us

$$K_a' = [H^+]$$

or

$$pK_a' = pH$$

We use this relationship to determine the first dissociation constant of the acid.

These are fairly good approximations for most diprotic acids.

(iii) At the First Equivalence Point. At this stage, the first dissociable protons have been completely neutralized by the base. If we ignore the ionization of HA^- ($HA^- \rightleftharpoons H^+ + A^{2-}$) and the hydrolysis of HA^- ($HA^- + H_2O \rightleftharpoons H_2A + OH^-$), then we can set $[H_2A] = [A^{2-}]$ because they are both derived from the disproportionation reaction (also called autoprotolysis): $2HA^- \rightleftharpoons H_2A + A^{2-}$. The product of the acid dissociation constants gives

$$K_a' K_a'' = \frac{[H^+][HA^-]}{[H_2A]} \times \frac{[H^+][A^{2-}]}{[HA^-]}$$

$$= [H^+]^2$$

$$[H^+] = \sqrt{K_a' K_a''}$$

or

$$pH = \frac{pK_a' + pK_a''}{2}$$

(iv) Halfway Between the First and Second Equivalence Points. This condition corresponds to the halfway point in the neutralization of the acid HA^-. Thus, we have $[HA^-] = [A^{2-}]$. Because

$$K_a'' = \frac{[H^+][A^{2-}]}{[HA^-]}$$

at this point, we can write

$$K_a'' = [H^+]$$

or

$$pK_a'' = pH$$

In this way we determine the second dissociation constant of the acid.

(v) At the Second Equivalence Point.   At this point, we have a solution of Na_2A for which we can write the hydrolysis equilibrium:

$$A^{2-} + H_2O \rightleftharpoons HA^- + OH^-$$

and

$$K_b'' = \frac{[HA^-][OH^-]}{[A^{2-}]} = \frac{K_w}{K_a''}$$

Also, we may assume that $[HA^-] = [OH^-]$ and $[A^{2-}] = [H_2A]_0$. Thus,

$$[OH^-]^2 = \frac{[H_2A]_0 K_w}{K_a''}$$

Because $[OH^-] = K_w/[H^+]$, we have

$$[OH^-]^2 = \frac{K_w^2}{[H^+]^2} = \frac{[H_2A]_0 K_w}{K_a''}$$

or

$$[H^+] = \sqrt{\frac{K_a'' K_w}{[H_2A]_0}}$$

Figure 8.9 shows the titration curve for the H_2A acid and the relationships between pK_a', pK_a'', and pH discussed above. In arriving at Figure 8.9, we have made the implicit assumption that $K_a' \gg K_a''$. Depending on the nature of the diprotic acid, this assumption may or may not be borne out in practice. In theory, unless the K_a'

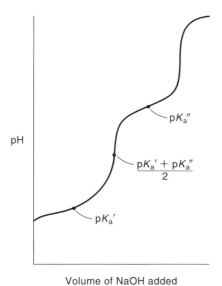

Figure 8.9
Titration of a diprotic acid with sodium hydroxide of equivalent strength.

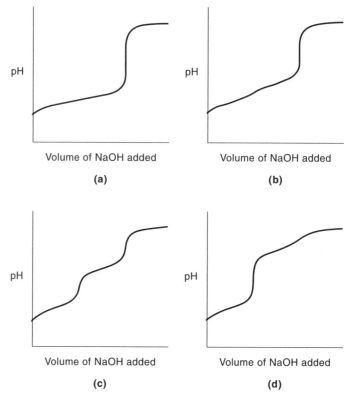

Figure 8.10
Titration of a diprotic acid with sodium hydroxide of equivalent strength. (a) $pK_a' = 4.0$ and $pK_a'' = 4.60$, (b) $pK_a' = 4.0$ and $pK_a'' = 6.0$, (c) $pK_a' = 4.0$ and $pK_a'' = 8.0$, and (d) $pK_a' = 4.0$ and $pK_a'' = 10.0$. As can be seen, unless $K_a' \gg K_a''$ (by roughly four orders of magnitude), it is difficult to locate the first equivalence point.

and K_a'' values differ by about a factor of 10^4 or greater, the first equivalence point will be difficult if not impossible to detect (Figure 8.10).

Suggestions for Further Reading

BOOKS

Bell, R. P., *Acids and Bases*, Methuen & Co., London (Barnes & Noble, New York), 1969.

Christensen, H. N., *Body Fluids and the Acid–Base Balance*, W. B. Saunders, Philadelphia, 1964.

King, E. J., *Acid–Base Equilibria*, Pergamon Press, Inc., Elmsford, NY, 1965.

Lowenstein, J., *Acids and Bases*, Oxford University Press, New York, 1993.

Masoro, E. J. and P. D. Siegel, *Acid–Base Regulation: Its Physiology and Pathophysiology*, W. B. Saunders, Philadelphia, 1971.

ARTICLES

General

"Development of the pH Concept," F. Szabadvary, *J. Chem. Educ.* **41**, 105 (1964).

"The Hydration of Carbon Dioxide," P. Jones, M. L. Haggett, and J. L. Longridge, *J. Chem. Educ.* **41**, 610 (1964).

"Thermodynamics of the Ionization of Acetic and Chloroacetic Acids," H. A. Neidig and R. T. Yingling, *J. Chem. Educ.* **42**, 484 (1965).

"Acid–Base Titration and Distribution Curves," J. Waser, *J. Chem. Educ.* **44**, 274 (1967).

"pK Values for D_2O and H_2O," R. G. Bates, R. A. Robinson, and A. K. Covington, *J. Chem. Educ.* **44**, 635 (1967).

"The pK_a of a Weak Acid as a Function of Temperature and Ionic Strength," J. L. Bada, *J. Chem. Educ.* **46**, 689 (1969).

"Actual Effects Controlling the Acidity of Carboxylic Acids," G. V. Calder and T. J. Barton, *J. Chem. Educ.* **48**, 338 (1971).

"Measurement of pH of Distilled Water," H. L. Youmans, *J. Chem. Educ.* **49**, 429 (1972).

"Conjugate Acid–Base and Redox Theory," R. A. Pacer, *J. Chem. Educ.* **50**, 178 (1973).

"Effect of Ionic Strength on Equilibrium Constants," M. D. Seymour and Q. Fernando, *J. Chem. Educ.* **54**, 225 (1977).

"The Great Fallacy of the H^+ Ion," P. A. Giguere, *J. Chem. Educ.* **56**, 571 (1979).

"Assigning the pK_a's of Polyprotic Acids," G. M. Bodner, *J. Chem. Educ.* **63**, 246 (1986).

"Arrhenius Confuses Students," S. J. Hawkes, *J. Chem. Educ.* **69**, 542 (1992).

"The Determination of 'Apparent' pK_a's," J. J. Cawley, *J. Chem. Educ.* **70**, 596 (1993); **72**, 88 (1995).

"There is No Such Thing as H_2SO_3," M. Laing, *Educ. Chem.* **30**, 140 (1993).

"The Uncertainty of pH," G. Schmitz, *J. Chem. Educ.* **71**, 117 (1994).

"Teaching the Truth About pH," S. J. Hawkes, *J. Chem. Educ.* **71**, 747 (1994).

"Calculating $[H^+]$," L. Cardellini, *Educ. Chem.* **33**, 161 (1996).

"pK_w Is Almost Never 14.0," S. J. Hawkes, *J. Chem. Educ.* **72**, 799 (1995).

"Salts Are Mostly NOT Ionized," S. J. Hawkes, *J. Chem. Educ.* **73**, 421 (1996).

"All Positive Ions Give Acid Solutions in Water," S. J. Hawkes, *J. Chem. Educ.* **73**, 516 (1996).

Acid–Base Equilibria of Biological Systems

"Hydrogen Ion Buffers for Biological Research," N. E. Good et al., *Biochemistry*, **5**, 467 (1966).

"Determination of the Microscopic Ionization Constants of Cysteine," G. E. Clement and T. P. Hartz, *J. Chem. Educ.* **48**, 395 (1971).

"Why the Stomach Does Not Digest Itself," H. W. Davenport, *Sci. Am.* January 1972.

"Stomach Upset Caused by Aspirin," W. D. Hobey, *J. Chem. Educ.* **50**, 212 (1973).

"Cystinuria: The Relationship of pH to the Origin and Treatment of a Disease," C. Minnier, *J. Chem. Educ.* **50**, 427 (1973).

"Physiochemical Properties of Antacids," S. L. Hein, *J. Chem. Educ.* **52**, 383 (1975).

"Imidazole—Versatile Today, Prominent Tomorrow," C. A. Matuszak and A. J. Matuszak, *J. Chem. Educ.* **53**, 280 (1976).

"Pepsin and Antacid Therapy: A Dilemma," W. B. Batson and P. H. Laswick, *J. Chem. Educ.* **56**, 484 (1979).

"What is the Energy Difference Between H_2NCH_2COOH and $^+H_3NCH_2CO_2^-$?" P. Haberfield, *J. Chem. Educ.* **57**, 346 (1980).

"Carbon Dioxide Flooding: A Classroom Case Study Derived from Surgical Practice," R. C. Kerber, *J. Chem. Educ.* **80**, 1437 (2003).

Problems

Acids, Bases, Dissociation Constants, and pH

8.1 Classify each of the following species as a Brønsted acid or base, or both: (a) H_2O, (b) OH^-, (c) H_3O^+, (d) NH_3, (e) NH_4^+, (f) NH_2^-, (g) NO_3^-, (h) CO_3^{2-}, (i) HBr, (j) HCN, and (k) HCO_3^-.

8.2 Write the formulas for the conjugate bases of the following acids: (a) HI, (b) H_2SO_4, (c) H_2S, (d) HCN, and (e) HCOOH (formic acid).

8.3 Classify each of the following species as a weak or strong acid: (a) HNO_3, (b) HF, (c) H_2SO_4, (d) HSO_4^-, (e) H_2CO_3, (f) HCO_3^-, (g) HCl, (h) HCN, and (i) HNO_2.

8.4 Classify each of the following species as a weak or strong base: (a) LiOH, (b) CN^-, (c) H_2O, (d) ClO_4^-, and (e) NH_2^-.

8.5 Calculate the pH of the following solutions: (a) 1.0 *M* HCl, (b) 0.10 *M* HCl, (c) 1.0×10^{-2} *M* HCl, (d) 1.0×10^{-2} *M* NaOH, and (e) 1.0×10^{-2} *M* $Ba(OH)_2$. Assume ideal behavior.

8.6 A 0.040 *M* solution of a monoprotic acid is 13.5% dissociated. What is the dissociation constant of the acid?

8.7 Write the equation relating K_a for a weak acid and K_b for its conjugate base. Use NH_3 and its conjugate acid NH_4^+ to derive the relationship between K_a and K_b.

8.8 The dissociation constant of a monoprotic acid at 298 K is 1.47×10^{-3}. Calculate the degree of dissociation by (a) assuming ideal behavior and (b) using a mean activity coefficient $\gamma_\pm = 0.93$. The concentration of the acid is 0.010 *M*.

8.9 The ion product of D_2O is 1.35×10^{-15} at 25°C. **(a)** Calculate the value of pD for pure D_2O where p$D = -\log[D^+]$. **(b)** For what values of pD will a solution be acidic in D_2O? **(c)** Derive a relation between pD and pOD.

8.10 HF is a weak acid, but its strength increases with concentration. Explain. (*Hint:* F^- reacts with HF to form HF_2^-. The equilibrium constant for this reaction is 5.2 at 25°C.)

8.11 When the concentration of a strong acid is not substantially higher than 1.0×10^{-7} M, the ionization of water must be taken into account in the calculation of the solution's pH. **(a)** Derive an expression for the pH of a strong acid solution, including the contribution to $[H^+]$ from H_2O. **(b)** Calculate the pH of a 1.0×10^{-7} M HCl solution.

8.12 What are the concentrations of HSO_4^-, SO_4^{2-}, and H^+ in a 0.20 M $KHSO_4$ solution? (*Hint:* H_2SO_4 is a strong acid; K_a for $HSO_4^- = 1.3 \times 10^{-2}$.)

8.13 Calculate the concentrations of H^+, HCO_3^-, and CO_3^{2-} in a 0.025 M H_2CO_3 solution.

8.14 To which of the following would the addition of an equal volume of 0.60 M NaOH lead to a solution having a lower pH? **(a)** Water, **(b)** 0.30 M HCl, **(c)** 0.70 M KOH, and **(d)** 0.40 M NaNO$_3$.

8.15 A solution contains a weak monoprotic acid, HA, and its sodium salt, NaA, both at 0.1 M concentration. Show that $[OH^-] = K_w/K_a$.

8.16 A solution of methylamine (CH_3NH_2) has a pH of 10.64. How many grams of methylamine are in 100.0 mL of the solution?

8.17 Hydrocyanic acid (HCN) is a weak acid and a deadly poisonous compound that is used in gas chambers in the gaseous form (hydrogen cyanide). Why is it dangerous to treat sodium cyanide with acids (such as HCl) without proper ventilation?

8.18 Novocaine, used as a local anesthetic by dentists, is a weak base ($K_b = 8.91 \times 10^{-6}$). What is the ratio of the concentration of the base to that of its acid in the blood plasma (pH = 7.40) of a patient?

8.19 Calculate the percent dissociation of HF at the following concentrations: **(a)** 0.50 M and **(b)** 0.050 M. Comment on your results.

8.20 Explain why phenol is a stronger acid than methanol:

Phenol   Methanol

8.21 Calculate the concentrations of all species in a 0.100 M H_3PO_4 solution.

8.22 The disagreeable odor of fish is mainly due to organic compounds (RNH_2) containing an amino group, $-NH_2$, where R is the rest of the molecule. Amines are bases just like ammonia. Explain why putting some lemon juice on fish can greatly reduce the odor.

Salt Hydrolysis

8.23 Specify which of the following salts will undergo hydrolysis: KF, $NaNO_3$, NH_4NO_2, $MgSO_4$, KCN, C_6H_5COONa, RbI, Na_2CO_3, $CaCl_2$, and HCOOK.

8.24 Calculate the pH of a 0.10 M NH_4Cl solution.

8.25 Calculate the pH and percent hydrolysis of a 0.36 M CH_3COONa solution.

Acid–Base Titration

8.26 A student added NaOH solution from a buret to an Erlenmeyer flask containing HCl solution and used phenolphthalein as indicator. At the equivalence point of the titration, she observed a faint reddish-pink color. However, after a few minutes, the solution gradually turned colorless. What do you suppose happened?

8.27 The ionization constant, K_a, of an indicator, HIn, is 1.0×10^{-6}. The color of the nonionized form is red and that of the ionized form is yellow. What is the color of this indicator in a solution whose pH is 4.00?

8.28 The K_a of a certain indicator is 2.0×10^{-6}. The color of HIn is green, and that of In$^-$ is red. A few drops of the indicator are added to a HCl solution, which is then titrated against a NaOH solution. At what pH will the indicator change color?

8.29 The pK_a of the indicator methyl orange is 3.46. Over what pH range does this indicator change from 90% HIn to 90% In$^-$?

8.30 A 200-mL volume of NaOH solution was added to 400 mL of a 2.00 M HNO_2 solution. The pH of the mixed solution was 1.50 units greater than that of the original acid solution. Calculate the molarity of the NaOH solution.

8.31 A volume of 25.0 mL of 0.100 M HCl is titrated with a 0.100 M CH_3NH_2 solution. Calculate the pH values of the solution **(a)** after 10.0 mL of CH_3NH_2 solution have been added, **(b)** after 25.0 mL of CH_3NH_2 solution have been added, and **(c)** after 35.0 mL of CH_3NH_2 solution have been added.

8.32 Phenolphthalein is the common indicator for the titration of a strong acid with a strong base. **(a)** If the pK_a of phenolphthalein is 9.10, what is the ratio of the nonionized form of the indicator (colorless) to the ionized form (reddish pink) at pH 8.00? **(b)** If 2 drops of 0.060 M phenolphthalein are used in a titration involving a 50.0-mL volume, what is the concentration of the ionized form at pH 8.00? (Assume that 1 drop = 0.050 mL.)

8.33 Shown below is a titration curve for carbonic acid versus sodium hydroxide. Fill in the missing species and the pH and pK_a values.

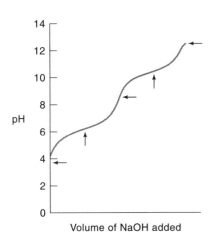

Buffer Solutions

8.34 Specify which of the following systems can be classified as a buffer system: **(a)** KCl/HCl, **(b)** NH_3/NH_4NO_3, **(c)** Na_2HPO_4/NaH_2PO_4, **(d)** KNO_2/HNO_2, **(e)** $KHSO_4/H_2SO_4$, and **(f)** $HCOOK/HCOOH$.

8.35 Derive the Henderson–Hasselbalch equation for the buffer system NH_4^+/NH_3.

8.36 Calculate the pH of the 0.20 M NH_3/0.20 M NH_4Cl buffer. What is the pH of the buffer after the addition of 10.0 mL of 0.10 M HCl to 65.0 mL of the buffer?

8.37 Calculate the pH of 1.00 L of the buffer 1.00 M CH_3COONa/1.00 M CH_3COOH before and after the addition of **(a)** 0.080 mol NaOH and **(b)** 0.12 mol HCl. (Assume that there is no change in volume.)

8.38 A quantity of 26.4 mL of a 0.45 M acetic acid solution is added to 31.9 mL of a 0.37 M sodium hydroxide solution. What is the pH of the final solution?

8.39 What is the pH of the buffer 0.10 M Na_2HPO_4/0.10 M KH_2PO_4? Calculate the concentration of all the species in solution.

8.40 A phosphate buffer has a pH equal to 7.30. **(a)** What is the predominant conjugate pair present in this buffer? **(b)** If the concentration of this buffer is 0.10 M, what is the new pH after the addition of 5.0 mL of 0.10 M HCl to 20.0 mL of this buffer solution?

8.41 Tris [tris(hydroxymethyl)aminomethane] is a common buffer for studying biological systems:

(a) Calculate the pH of the tris buffer after mixing 15.0 mL of 0.10 M HCl solution with 25.0 mL of 0.10 M tris. **(b)** This buffer was used to study an enzyme-catalyzed reaction. As a result of the reaction, 0.00015 mole of H^+ was consumed. What is the pH of the buffer at the end of the reaction? **(c)** What would be the final pH if no buffer were present?

8.42 Describe the number of different ways to prepare 1 liter of a 0.050 M phosphate buffer with a pH of 7.8.

8.43 Calculate the concentration of all the species present in a solution that is 0.12 M in HCN and 0.34 M in NaCN. What is the pH of the solution? Does the solution possess buffer capacity?

8.44 The pH of blood plasma is 7.40. Assuming the principal buffer system is HCO_3^-/H_2CO_3, calculate the ratio $[HCO_3^-]/[H_2CO_3]$. Is this buffer more effective against an added acid or an added base?

8.45 A student is asked to prepare a buffer solution with pH = 8.60, using one of the following weak acids: HA ($K_a = 2.7 \times 10^{-3}$), HB ($K_a = 4.4 \times 10^{-6}$), HC ($K_a = 2.6 \times 10^{-9}$). Which acid should she choose?

8.46 The buffer range is defined by the equation pH = $pK_a \pm 1$. Calculate the range of the ratio [conjugate base]/[acid] that corresponds to this equation.

8.47 Describe how you would prepare 1 L of 0.20 M CH_3COONa/0.20 M CH_3COOH buffer system by **(a)** mixing a solution of CH_3COOH with a solution of CH_3COONa, **(b)** reacting a solution of CH_3COOH with a solution of NaOH, and **(c)** reacting a solution of CH_3COONa with a solution of HCl.

8.48 How many milliliters of 1.0 M NaOH must be added to 200 mL of 0.10 M NaH_2PO_4 to make a buffer solution with a pH of 7.50?

8.49 Suggest two chemical tests that would allow you to distinguish an acid solution and a buffer solution both at pH = 3.5.

8.50 How would you prepare a CH_3COONa/CH_3COOH buffer with a pH of 4.40 and an ionic strength of 0.050 m? Treat molarity the same as molality.

8.51 The pH of a phosphate buffer is 7.10 at 25°C. What is the pH of the buffer at 37°C? The $\Delta_r H^\circ$ for the relevant dissociation step is 3.75 kJ mol^{-1}.

Amino Acids

8.52 Which of the amino acids listed in Table 8.4 have a buffer capacity in the physiological region of pH 7?

8.53 Calculate the ionic strength of a 0.035 M serine buffer at pH 9.15.

8.54 From the pK_a values listed in Table 8.4, calculate the pI value for amino acids lysine and valine.

8.55 Sketch the titration curve for 100 mL of 0.1 M aspartic acid hydrogen chloride titrated with sodium hydroxide.

8.56 At neutral pH, amino acids exist as dipolar ions. Using glycine as an example, and given that the pK_a of the carboxyl group is 2.3 and that of the ammonium group is 9.6, predict the predominant form of the molecule at pHs of 1, 7, and 12. Justify your answers using Equation 8.16.

Additional Problems

8.57 Describe a procedure that would allow you to compare the strength of Lewis acids.

8.58 From the dependence of K_w on temperature (see p. 269), calculate the enthalpy of dissociation for water.

8.59 Freshly distilled, deionized water has a pH of 7. Left standing in air, however, the water gradually becomes acidic. Calculate the pH of the "solution" at equilibrium. (*Hint:* First calculate the solubility of CO_2 in water according to Example 5.3. Assume the partial pressure of CO_2 is 0.00030 atm.)

8.60 Show that the acid dissociation constant, K_a, of a weak monoprotic acid in water is related to its concentration, c (mol L^{-1}), and its degree of dissociation, α, by $K_a = \alpha^2 c/(1 - \alpha)$ if the self-dissociation of water is ignored. If the latter is taken into account, show that $K_a = \frac{1}{2}\alpha^2 c[1 + (1 + 4K_w\alpha^{-2}c^{-2})^{1/2}]/(1 - \alpha)$.

8.61 To correct for the effect of ionic strength, we can write the dissociation constant of an acid as

$$pK_a' = pK_a - \frac{0.509\sqrt{I}}{1 + \sqrt{I}}$$

where K_a is the acid dissociation at zero ionic strength and K_a' the corresponding value at ionic strength, I. Calculate the dissociation constant of acetic acid in a 0.15 m KCl solution at 298 K. You may neglect the ionic strength contribution due to the dissociation of the acid itself.

8.62 Depending on the pH of the solution, ferric ions (Fe^{3+}) may exist in the free-ion form or form the insoluble precipitate $Fe(OH)_3$ ($K_{sp} = 1.0 \times 10^{-36}$). Calculate the pH at which 90% of the Fe^{3+} ions in a 4.5×10^{-5} M Fe^{3+} solution would be precipitated. What conclusion can you draw about the Fe^{3+} ion concentration in blood plasma whose pH is 7.40?

8.63 A 0.020 M aqueous solution of benzoic acid has a freezing point of $-0.0392°C$. Calculate the dissociation constant of benzoic acid. Assume ideal behavior, and assume that molarity is equal to molality at this low concentration.

8.64 The pH of gastric juice is about 1.00 and blood plasma is 7.40. Calculate the Gibbs energy required to secrete a mole of H^+ ions from blood plasma to the stomach at 37°C. Assume ideal behavior.

8.65 Chemical analysis shows that 20.0 mL of a certain sample of blood yields 12.5 mL of CO_2 gas (measured at 25°C and 1 atm) when treated with an acid. Calculate **(a)** the number of moles of CO_2 originally present in the blood, **(b)** the concentration of CO_2 and HCO_3^- at equilibrium, and **(c)** the partial pressure of CO_2 over the blood solution at equilibrium. Assume ideal behavior. The pH of blood is 7.40, and the Henry's law constant for CO_2 in blood is 29.3 atm mol^{-1} (kg H_2O).

8.66 Calcium oxalate is a major component of kidney stones. From the dissociation constants listed in Table 8.1 and given that the solubility product of CaC_2O_4 is 3.0×10^{-9}, predict whether the formation of kidney stones can be minimized by increasing or decreasing the pH of the fluid present in the kidneys. The pH of normal kidney fluid is about 8.2.

8.67 What is the pH of a 0.050 M glycine solution at 298 K?

8.68 From the dissociation constant of formic acid listed in Table 8.1, calculate the Gibbs energy and the standard Gibbs energy for the dissociation of formic acid at 298 K.

8.69 **(a)** Calculate the percent ionization of a 0.20 M solution of the monoprotic acetylsalicylic acid (aspirin, $C_9H_8O_4$), for which $K_a = 3.0 \times 10^{-4}$. **(b)** The pH of gastric juice in the stomach of a certain individual is 1.00. After a few aspirin tablets have been swallowed, the concentration of acetylsalicylic acid in the stomach is 0.20 M. Calculate the percent ionization of the acid under these conditions. What effect does the nonionized acid have on the membranes lining the stomach?

8.70 A 0.400 M formic acid (HCOOH) solution freezes at $-0.758°C$. Calculate the value of K_a at that temperature. (*Hint:* Assume that molarity is equal to molality.)

8.71 Explain the action of smelling salts, which is ammonium nitrate [$(NH_4)_2CO_3$]. (*Hint:* The thin film of aqueous solution that lines the nasal passage is slightly basic.)

8.72 Acid–base reactions usually go to completion. Confirm this statement by calculating the equilibrium constant for each of the following cases: **(a)** a strong acid reacting with a strong base, **(b)** a strong acid reacting with a weak base (NH_3), **(c)** a weak acid (CH_3COOH) reacting with a strong base, and **(d)** a weak acid (CH_3COOH) reacting with a weak base (NH_3). (*Hint:* Strong acids exist as H^+ ions, and strong bases exist as OH^- ions in solution. You need to look up K_a, K_b, and K_w values.)

8.73 When lemon juice is squirted into tea, the color becomes lighter. In part, the color change is due to dilution, but the main reason for the change is an acid–base reaction. What is the reaction? (*Hint:* Tea contains "polyphenols," which are weak acids, and lemon juice contains citric acid.)

8.74 One of the most common antibiotics is penicillin G (benzylpenicillinic acid), which has the following structure:

It is a weak monoprotic acid:

$$HP \rightleftharpoons H^+ + P^- \quad K_a = 1.64 \times 10^{-3}$$

where HP denotes the parent acid and P^- the conjugate base. Penicillin G is produced by growing molds in fermentation tanks at 25°C and a pH range of 4.5 to 5.0. The crude form of this antibiotic is obtained by extracting the fermentation broth with an organic solvent in which the acid is soluble. **(a)** Identify the acidic hydrogen atom. **(b)** In one stage of purification, the organic extract of the crude penicillin G is treated with a buffer solution at pH = 6.50. What is the ratio of the conjugate base of penicillin G to the acid at this pH? Would you expect the conjugate base to be more soluble in water than the acid? **(c)** Penicillin G is not suitable for oral administration, but the sodium salt (NaP) is because it is soluble. Calculate the pH of a 0.12 *M* NaP solution formed when a tablet containing the salt is dissolved in a glass of water.

8.75 Derive an equation (using the Debye–Hückel theory) showing the influence of ionic strength on the first dissociation constant (pK_a) of carbonic acid.

8.76 Referring to the buffer system listed in Problem 8.36, calculate the pH of the buffer after it has been diluted by a factor of 100. What would be the change in pH if the base component of the buffer were diluted by the same factor?

8.77 (a) Derive a relationship between $\Delta_r G^{\circ\prime}$ and pK_a. **(b)** Calculate $\Delta_r G^{\circ\prime}$ for the dissociation of benzoic acid, which has a pK_a of 4.20. (Assume $T = 298$ K.)

8.78 Tris [tris(hydroxymethyl)aminomethane] is a widely used buffer by biochemists. It has a pK_a of 8.30 at 20°C. Calculate the buffer capacity of a 0.10 *M* Tris buffer (containing both Tris and its conjugate acid TrisH$^+$) at **(a)** pH = 8.30 and **(b)** pH = 10.30. In each case 0.020 mol H$^+$ ions is added to one liter of the buffer solution.

8.79 The pK_as listed in Table 8.6 all refer to the free acids in aqueous solution. Their values actually depend on the environment; that is, the pK_a of an amino acid can change appreciably when it is part of a protein molecule. Consider glutamic acid in this respect. How would its side chain pK_a be affected (increase, decrease, or no change) when **(a)** the terminal $-COO^-$ group is brought into close proximity, **(b)** the terminal $-NH_3^+$ group is brought into close proximity, **(c)** it is exposed on the surface of the protein, and **(d)** it is buried in the interior of the protein.

8.80 The pK_a of the Tris buffer is 8.30 at 20°C. What is its value at 37°C? The molar enthalpy of dissociation for Tris is 48.0 kJ mol^{-1}.

8.81 (a) Use Equation 8.17 to derive an expression showing the fraction of an acid (f) in terms of its pK_a and the pH of the solution. **(b)** Aspartame is the artificial sweetener containing the dipeptide aspartyl phenylalanine,

where the numbers denote the pK_a values of the acidic groups. Calculate the net charge on the dipeptide at pH = 3.00 and 7.00. Assume the pK_a's are unaffected by other groups present. Also, calculate the isoelectric point of the dipeptide.

8.82 The volume of a 1 *m* glycine solution depends on the pH. At what pH would the volume be a minimum? Why? Assume temperature is kept constant.

CHAPTER 9

Chemical Kinetics

The aims of studying chemical kinetics are to determine experimentally the rate of a reaction and its dependence on parameters such as concentration, temperature, and catalysts, and to understand the mechanism of a reaction—that is, the number of steps involved and the nature of intermediates formed.

The subject of chemical kinetics is conceptually easier to understand than some other topics in physical chemistry, such as thermodynamics and quantum mechanics, although rigorous theoretical treatment of the energetics involved is possible only for very simple systems in the gas phase. Nevertheless, the macroscopic, empirical approach to the subject can provide much useful information.

In this chapter we discuss general topics in chemical kinetics and consider some important examples, including fast reactions. Enzyme kinetics will be treated in Chapter 10.

9.1 Reaction Rates

The rate of a reaction is expressed as the change in reactant concentration with time. Consider the stoichiometrically simple reaction

$$R \rightarrow P$$

Let the concentrations (in mol L^{-1}) of R at times t_1 and t_2 ($t_2 > t_1$) be $[R]_1$ and $[R]_2$. The rate of the reaction over the time interval $(t_2 - t_1)$ is given by

$$\frac{[R]_2 - [R]_1}{t_2 - t_1} = \frac{\Delta[R]}{\Delta t}$$

Because $[R]_2 < [R]_1$, we introduce a minus sign so that the rate will be a positive quantity:

$$\text{rate} = -\frac{\Delta[R]}{\Delta t}$$

The rate can be expressed also in terms of the appearance of a product

$$\text{rate} = \frac{[P]_2 - [P]_1}{t_2 - t_1} = \frac{\Delta[P]}{\Delta t}$$

In this case, we have $[P]_2 > [P]_1$. In practice, we find that the quantity of interest is not the rate over a certain time interval (because this is only an average quantity whose value depends on the particular value of Δt); rather, we are interested in the instantaneous rate. In the language of calculus, as Δt becomes smaller and eventually approaches zero, the rate of the foregoing reaction at a specific time t is given by

$$\text{rate} = -\frac{d[R]}{dt} = \frac{d[P]}{dt}$$

The units of reaction rates are usually $M\ s^{-1}$ or $M\ min^{-1}$.

For stoichiometrically more complicated reactions, the rate must be expressed in an unambiguous manner. Suppose that the reaction of interest is

$$2R \rightarrow P$$

The ratios $-d[R]/dt$ and $d[P]/dt$ still express the rate of change of the reactant and the product, respectively, but they are no longer equal to each other because the reactant is disappearing twice as fast as the product is appearing. For this reason, we write the rate of this reaction as

$$\text{rate} = -\frac{1}{2}\frac{d[R]}{dt} = \frac{d[P]}{dt}$$

In general, for the reaction

$$a\text{A} + b\text{B} \rightarrow c\text{C} + d\text{D}$$

the rate is given by

$$\text{rate} = -\frac{1}{a}\frac{d[A]}{dt} = -\frac{1}{b}\frac{d[B]}{dt} = \frac{1}{c}\frac{d[C]}{dt} = \frac{1}{d}\frac{d[D]}{dt} \tag{9.1}$$

where the expressions in brackets refer to the concentrations of the reactants and products at time t after the start of the reaction.

9.2 Reaction Order

The relationship between the rate of a chemical reaction and the concentrations of the reactants is a complex one that must be determined experimentally. Referring to the general equation above, however, we find that usually (but by no means always) the reaction rate can be expressed as

$$\text{rate} \propto [A]^x[B]^y$$
$$= k[A]^x[B]^y \tag{9.2}$$

This equation, known as the *rate law*, tells us that the rate of a reaction is not constant; its value at any time, t, is proportional to the concentrations of A and B raised to some powers. The proportionality constant, k, is called the *rate constant*. The rate law is defined in terms of the reactant concentrations, but the rate constant for a given reaction does not depend on the concentrations of the reactants. The rate constant is affected only by temperature, as we shall see later.

Expressing the rate of a reaction as shown in Equation 9.2 enables us to define the *order of a reaction*. We say that the reaction is x order with respect to A and y order with respect to B. Thus, the reaction has an overall order of $(x + y)$. It is important to understand that, in general, there is no connection between the order of a reactant in the rate expression and its stoichiometric coefficient in the balanced chemical equation. For example, the rate of the reaction

$$2N_2O_5(g) \rightarrow 4NO_2(g) + O_2(g)$$

is given by

$$\text{rate} = k[N_2O_5]$$

The reaction is first order in N_2O_5—not second order as we might have inferred from the balanced equation.

The order of a reaction specifies the empirical dependence of the rate on concentrations. It may be zero, an integer, or even a noninteger. We can use the rate law to determine the concentrations of reactants at any time during the course of a reaction. To do so, we need to integrate the rate law expressions. For simplicity, we shall focus only on reactions that have integral orders.

This must be the case because a chemical equation can be balanced in many different ways.

Zero-Order Reactions

The rate law for a zero-order reaction of the type

$$A \rightarrow \text{product}$$

is given by

$$\text{rate} = -\frac{d[A]}{dt} = k[A]^0 = k \tag{9.3}$$

The quantity k ($M\ s^{-1}$) is the zero-order rate constant. As you can see, the rate of the reaction is independent of the reactant concentration (Figure 9.1). Rearranging Equation 9.3, we obtain

$$d[A] = -k\,dt$$

Integration between $t = 0$ and $t = t$ at concentrations $[A]_0$ and $[A]$ gives

$$\int_{[A]_0}^{[A]} d[A] = [A] - [A]_0 = -\int_0^t k\,dt = -kt$$

or

$$[A] = [A]_0 - kt \tag{9.4}$$

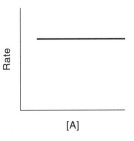

Figure 9.1
Plot of rate versus concentration for a zero-order reaction.

Note that Equation 9.4 gives the time dependence of $[A]$, but cannot be the full description of the factors affecting the rate. To illustrate this point, consider the decomposition of gaseous ammonia on a tungsten surface:

$$NH_3(g) \rightarrow \tfrac{1}{2}N_2(g) + \tfrac{3}{2}H_2(g)$$

Under certain conditions, this reaction obeys a zero-order rate law. Such a zero-order reaction can occur if the rate is limited, for example, by the concentration of a catalyst. The rate of the reaction is given by

$$\text{rate} = k'\theta A$$

where k' is a constant, θ is the fraction of metal surface covered by the adsorbed ammonia molecules, and A is the total catalyst surface area. If the pressure of ammonia is large enough, $\theta = 1$, and the reaction is zero order in ammonia. At sufficiently low pressures, however, θ is proportional to $[NH_3]$ in the gas phase and the reaction becomes first order in ammonia. Note that the rate will also depend on the amount of catalyst (that is, on the area A).

First-Order Reactions

A first-order reaction is one in which the rate of the reaction depends only on the concentration of the reactant raised to the first power:

$$\text{rate} = -\frac{d[A]}{dt} = k[A] \tag{9.5}$$

where k (s^{-1}) is the first-order rate constant. Rearranging Equation 9.5, we get

$$-\frac{d[A]}{[A]} = k\,dt$$

Integrating between $t = 0$ and $t = t$ at concentrations $[A]_0$ and $[A]$, we obtain

$$\int_{[A]_0}^{[A]} \frac{d[A]}{[A]} = -\int_0^t k\,dt$$

$$\ln\frac{[A]}{[A]_0} = -kt \tag{9.6}$$

or

$$[A] = [A]_0 e^{-kt} \tag{9.7}$$

A plot of $\ln([A]/[A]_0)$ versus t gives a straight line whose slope, which is negative, is given by $-k$ (Figure 9.2a). Equation 9.7 shows that in first-order reactions, the decrease in reactant concentration with time is exponential (Figure 9.2b).

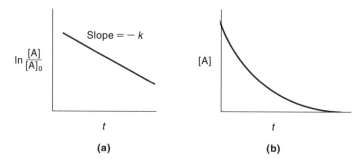

Figure 9.2
First-order reaction. (a) Plot based on Equation 9.6 with a slope of $-k$. (b) Exponential decay of $[A]$ with time according to Equation 9.7.

Radioactive decays fit first-order kinetics. An example is

$$^{222}_{86}Rn \rightarrow {}^{218}_{84}Po + \alpha$$

where α represents the helium nucleus (He^{2+}). The thermal decomposition of N_2O_5 mentioned earlier is first order in N_2O_5. Another example is the rearrangement of methyl isonitrile to acetonitrile:

$$CH_3NC(g) \rightarrow CH_3CN(g)$$

Half-Life of a Reaction. A measure of considerable practical importance in kinetic studies is the *half-life* ($t_{1/2}$) of a reaction. The half-life of a reaction is defined as the time it takes for the concentration of the reactant to decrease by half of its original value. For example, in a first-order reaction, as $[A] = [A]_0/2$, $t = t_{1/2}$ and Equation 9.6 becomes

$$\ln \frac{[A]_0/2}{[A]_0} = -kt_{1/2}$$

or

$$t_{1/2} = \frac{\ln 2}{k} = \frac{0.693}{k} \tag{9.8}$$

Thus, the half-life of a first-order reaction is *independent* of the initial concentration (Figure 9.3). For A to decrease from 1 M to 0.5 M takes just as much time as it does for A to decrease from 0.1 M to 0.05 M. Table 9.1 lists the half-lives of radioactive isotopes that are used extensively in biochemical research and medicine.

In contrast to first-order reactions, the half-lives of other types of reaction all depend on the initial concentration. In general, we can show the dependence of half-life on the initial concentration as follows:

$$t_{1/2} \propto \frac{1}{[A]_0^{n-1}} \tag{9.9}$$

where n is the order of the reaction.

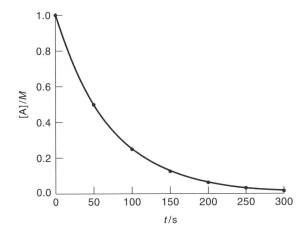

Figure 9.3
The half-lives of a first-order reaction (A $\rightarrow$ product). The initial concentration is arbitrarily set at 1 M and A reacts with a constant half-life of 50 s.

Table 9.1
Half-lives of Common Radioisotopes

Isotope	Decay Process	$t_{1/2}$
$^{3}_{1}\text{H}$	$^{3}_{1}\text{H} \rightarrow {}^{3}_{2}\text{He} + {}^{0}_{-1}\beta$	12.3 yr
$^{14}_{6}\text{C}$	$^{14}_{6}\text{C} \rightarrow {}^{14}_{7}\text{N} + {}^{0}_{-1}\beta$	5.73×10^3 yr
$^{24}_{11}\text{Na}$	$^{24}_{11}\text{Na} \rightarrow {}^{24}_{12}\text{Mg} + {}^{0}_{-1}\beta$	15 h
$^{32}_{15}\text{P}$	$^{32}_{15}\text{P} \rightarrow {}^{32}_{16}\text{S} + {}^{0}_{-1}\beta$	14.3 d
$^{35}_{16}\text{S}$	$^{35}_{16}\text{S} \rightarrow {}^{35}_{17}\text{Cl} + {}^{0}_{-1}\beta$	88 d
$^{60}_{27}\text{Co}$	Emission of γ rays	5.26 yr
$^{99m}_{43}\text{Tc}^a$	Emission of γ rays	6 h
$^{131}_{53}\text{I}$	$^{131}_{53}\text{I} \rightarrow {}^{131}_{54}\text{Xe} + {}^{0}_{-1}\beta$	8.05 d

^{a} The superscript m denotes the excited nuclear energy state.

Example 9.1

The thermal decomposition of 2,2′-azobisisobutyronitrile (AIBN)

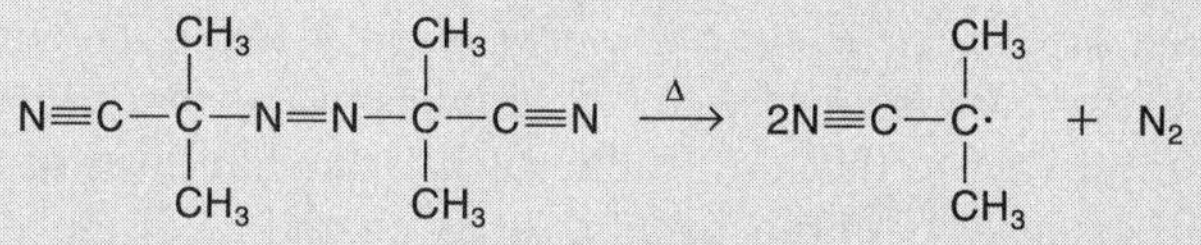

has been studied in an inert organic solvent at room temperature. The progress of the reaction can be monitored by the optical absorption of AIBN at 350 nm. The following data are obtained:

t/s	A
0	1.50
2,000	1.26
4,000	1.07
6,000	0.92
8,000	0.81
10,000	0.72
12,000	0.65
∞	0.40

where A is the absorbance. Assume that the reaction is first order in AIBN, and calculate the rate constant.

ANSWER

From Equation 9.6, we have

$$\ln \frac{[\text{AIBN}]}{[\text{AIBN}]_0} = -kt$$

The difference in absorbance at $t = 0$ and at $t = \infty$, $(A_0 - A_\infty)$, is proportional to the initial concentration of AIBN in the solution. Similarly, the difference $(A_t - A_\infty)$, where

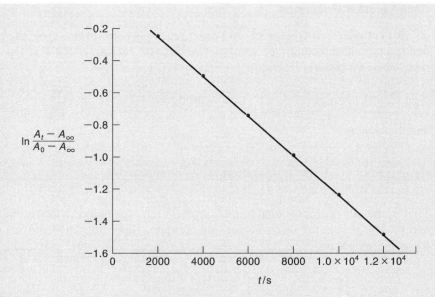

Figure 9.4
The fit of the linear equation is $y = -0.000124x - 0.00167$. Therefore, the first-order rate constant, which is equal to the negative slope, is 1.24×10^{-4} s^{-1}.

A_t is the absorbance of AIBN at time t, is proportional to the instantaneous concentration [AIBN].* The rate equation can now be expressed as

$$\ln \frac{A_t - A_\infty}{A_0 - A_\infty} = -kt$$

Because $A_0 = 1.50$ and $A_\infty = 0.40$, we have

t/s	$\ln \dfrac{A_t - A_\infty}{A_0 - A_\infty}$
2,000	−0.246
4,000	−0.496
6,000	−0.749
8,000	−0.987
10,000	−1.240
12,000	−1.482

The first-order rate constant can be obtained by plotting the natural logarithmic term versus t, as shown in Figure 9.4. It is given by the slope of the straight line, which is 1.24×10^{-4} s^{-1}.

---

*These statements hold true if little or no AIBN remains unreacted as t approaches infinity and the absorbance of products does not interfere with that of AIBN at 350 nm.

Second-Order Reactions

We consider two types of second-order reactions here. In one type, there is just one reactant. The second type involves two different reactants. The first type is represented by the general reaction

$$A \rightarrow products$$

and that rate is

$$rate = -\frac{d[A]}{dt} = k[A]^2 \tag{9.10}$$

That is, the rate is proportional to the concentration of A raised to the second power, and k $(M^{-1} s^{-1})$ is the second-order rate constant. Separating the variables and integrating, we obtain

$$\int_{[A]_0}^{[A]} \frac{d[A]}{[A]^2} = -\int_0^t k \, dt$$

$$\frac{1}{[A]} - \frac{1}{[A]_0} = kt$$

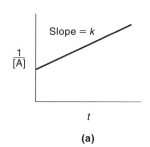

Slope = k

$\frac{1}{[A]}$

t

(a)

or

$$\boxed{\frac{1}{[A]} = kt + \frac{1}{[A]_0}} \tag{9.11}$$

where $[A]_0$ is the initial concentration. Thus, a plot of $1/[A]$ versus t gives a straight line with a slope equal to k (Figure 9.5a). To derive the half-life of a second-order equation, we set $[A] = [A]_0/2$ in Equation 9.11 and write

$$\frac{1}{[A]_0/2} = kt_{1/2} + \frac{1}{[A]_0}$$

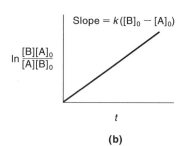

Slope = $k([B]_0 - [A]_0)$

$\ln \frac{[B][A]_0}{[A][B]_0}$

t

(b)

Figure 9.5
Second-order reaction.
(a) Plot based on Equation 9.11.
(b) Plot based on Equation 9.14.
The slope gives the rate constant.

or

$$t_{1/2} = \frac{1}{k[A]_0} \tag{9.12}$$

As mentioned earlier, except for first-order reactions, the half-lives of all other reactions are concentration dependent.

The second type of second-order reactions is represented by

$$A + B \rightarrow products$$

and

$$rate = -\frac{d[A]}{dt} = -\frac{d[B]}{dt} = k[A][B] \tag{9.13}$$

This reaction is first order in A, first order in B, and second order overall. Let

$$[A] = [A]_0 - x$$

$$[B] = [B]_0 - x$$

where x (in mol L^{-1}) is the amount of A and B consumed in time t. From Equation 9.13,

$$-\frac{d[A]}{dt} = -\frac{d([A]_0 - x)}{dt} = \frac{dx}{dt} = k[A][B]$$

$$= k([A]_0 - x)([B]_0 - x)$$

Rearranging, we obtain

$$\frac{dx}{([A]_0 - x)([B]_0 - x)} = k\,dt$$

By the somewhat tedious but straightforward method of integration by partial functions, we can obtain the final result:

$$\frac{1}{[B]_0 - [A]_0} \ln \frac{([B]_0 - x)[A]_0}{([A]_0 - x)[B]_0} = kt$$

or

$$\frac{1}{[B]_0 - [A]_0} \ln \frac{[B][A]_0}{[A][B]_0} = kt \tag{9.14}$$

Equation 9.14 was derived by assuming that $[A]_0 < [B]_0$. If $[A]_0 = [B]_0$, the solution is the same as that in Equation 9.11. (Note that Equation 9.11 cannot be obtained from Equation 9.14 by setting $[A]_0 = [B]_0$.) A plot of Equation 9.14 is shown in Figure 9.5b.

Below are a few examples of second-order reactions:

$$Cl(g) + H_2(g) \rightarrow HCl(g) + H(g)$$

$$2NO_2(g) \rightarrow 2NO(g) + O_2(g)$$

$$C_2H_5Br(aq) + OH^-(aq) \rightarrow C_2H_5OH(aq) + Br^-(aq)$$

An interesting special case of second-order reactions occurs when one of the reactants is present in great excess. An example is the hydrolysis of acetyl chloride:

$$CH_3COCl(aq) + H_2O(l) \rightarrow CH_3COOH(aq) + HCl(aq)$$

Because the concentration of water in the acetyl chloride solution is quite high (about 55.5 M, the concentration of pure water) and the concentration of acetyl chloride is of the order of 1 M or less, the amount of water consumed is negligible compared with the amount of water originally present. Thus, we can express the rate as

In one liter of water, there are 1000 g/(18.02 g mol^{-1}) or 55.5 moles of water.

$$-\frac{d[CH_3COCl]}{dt} = k'[CH_3COCl][H_2O]$$

$$= k[CH_3COCl]$$

Table 9.2
Summary of Rate Equations for A → Products

Order	Differential Form	Integrated Form	Half-Life	Units of the Rate Constant
0	$-\dfrac{d[A]}{dt} = k$	$[A]_0 - [A] = kt$	$\dfrac{[A]_0}{2k}$	$M\,s^{-1}$
1	$-\dfrac{d[A]}{dt} = k[A]$	$[A] = [A]_0 e^{-kt}$	$\dfrac{\ln 2}{k}$	s^{-1}
2	$-\dfrac{d[A]}{dt} = k[A]^2$	$\dfrac{1}{[A]} - \dfrac{1}{[A]_0} = kt$	$\dfrac{1}{[A]_0 k}$	$M^{-1}\,s^{-1}$
2^a	$-\dfrac{d[A]}{dt} = k[A][B]$	$\dfrac{1}{[B]_0 - [A]_0} \ln \dfrac{[B][A]_0}{[A][B]_0} = kt$	—	$M^{-1}\,s^{-1}$

^{a} For A + B → products.

where $k = k'[H_2O]$. The reaction therefore appears to follow first-order kinetics and is called a *pseudo*-first-order reaction.

Table 9.2 summarizes the rate laws and half-life expressions for zero-, first-, and second-order reactions. Third-order reactions are known but are uncommon, and so we shall not discuss them.

Renaturation of DNA—A Case Study. A well-known example of a second-order reaction is the renaturation of DNA in solution. Kinetic studies of this process provide information regarding the sequence of the DNA molecule. In a typical experiment, a large piece of DNA is broken down into smaller fragments of about the same size by sonication (agitation by ultrasonic vibrations). Heating the solution briefly to 90°C denatures the fragments into single strands. As they cool, the single strands recombine (renature) to form double-stranded fragments (Figure 9.6). The rate of renaturation depends on the makeup of the DNA molecule. If the molecule has a unique base-pair sequence, then the probability of a single fragment strand meeting its complementary strand in solution is small, and the rate of renaturation is slow. On the other hand, if a DNA molecule has many repeat, or "redundant," sequences, the concentration of similar strands will be high and so will the rate of renaturation. In the extreme case of synthetic DNA containing, say, only adenine (A) and thymine (T) complementary base pairs, the renaturation rate would be faster than the rates of the other two examples.

The kinetic analysis of DNA renaturation begins with the combination of two complementary strands, A and B, to form a double helix:

$$A + B \rightarrow AB$$

Because $[A] = [B]$, the rate of this second-order reaction is given by

$$\text{rate} = k[A][B]$$
$$= k[A]^2$$

where k is the second-order rate constant. From Equation 9.11 we write

$$[A] = \frac{[A]_0}{1 + [A]_0 kt} \tag{9.15}$$

See Chapter 16 for the structure and composition of DNA.

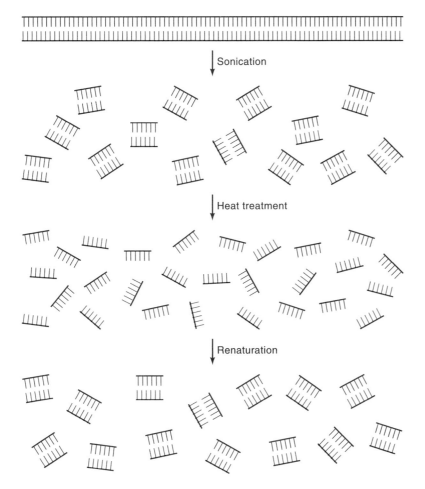

Figure 9.6
DNA renaturation experiment. A native DNA molecule is broken into double-stranded fragments of about the same size by sonication. The small fragments are denatured by heat treatment. Renaturation of the complementary single strands follows second-order kinetics. If the original DNA molecule has little or no repeat sequence, then the concentration of the single-strand fragments is very small compared to the total concentration of all the fragments. Consequently, the rate of renaturation will be slow. In contrast, for a synthetic DNA containing the same base pair (say A and T), the concentration of the single-strand fragments will be high and the rate of renaturation will be much faster.

or

$$\frac{[A]}{[A]_0} = f = \frac{1}{1 + [A]_0 kt} \tag{9.16}$$

where f is the fraction of strands that dissociate. Note that $[A]_0$ is the initial concentration (in moles of nucleotide per liter) of fragment A that is complementary to fragment B (before denaturation) and $[A]$ is the concentration of the single strands A at time t, also in moles of nucleotide per liter. Let C_0 be the total concentration (in moles of nucleotide per liter) of all the single strands before renaturation. Therefore, $[A]_0$ is related to C_0 by the relation

$$[A]_0 = \frac{C_0}{2N} \tag{9.17}$$

where N is the smallest repeating sequence, called the *complexity* of the nucleotide pairs (the larger N is, the more complex the sequence). For example, for the synthetic poly A · poly T DNA, we have $N = 1$, because the repeating sequence is one (every A–T base pair is the same as the next), and $[A]_0 = C_0/2$. On the other hand, if a DNA extracted from some organism has no repeating sequence, then N is the same as the total number of base pairs present, which may be on the order of 1×10^6 or greater. In this case, $[A]_0$ is a very small number because each fragment has a unique sequence.

Equation 9.16 can now be written as

$$f = \frac{1}{1 + C_0 tk/2N} \tag{9.18}$$

At $f = \frac{1}{2}$, the fraction dissociated is equal to the fraction reassociated.

The half-life of the reaction is the amount of time it takes for half of the single strands to renature, so $f = \frac{1}{2}$ and

$$\frac{C_0 t_{1/2} k}{2N} = 1$$

or

$$C_0 t_{1/2} = \frac{N}{k'} \tag{9.19}$$

where $k' = k/2$. Figure 9.7 shows plots of f versus $C_0 t$ for different DNA samples.

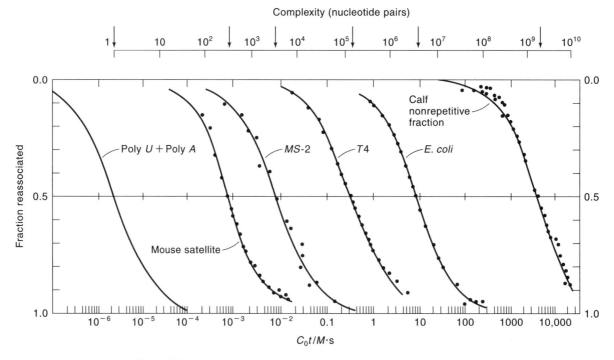

Figure 9.7
Plot of fraction reassociated versus $C_0 t$ (logarithmic scale) for DNA from different sources. The top scale gives the complexity of the base pairs (N). The smallest value of N is unity, which corresponds to the synthetic poly U · poly A. From the $C_0 t$ values at $f = \frac{1}{2}$ and the second-order rate constant, we can calculate N for other DNA. [From R. J. Britten and D. E. Kohne, *Science* **161**, 529 (1968). Copyright 1968 by the American Association for the Advancement of Science.]

For poly U · poly A (a synthetic double-stranded RNA molecule that behaves similarly to DNA), the $C_0 t_{1/2}$ value at $f = \frac{1}{2}$ is 2×10^{-6} M s. Because $N = 1$, from Equation 9.19 we write

$$k' = \frac{1}{2 \times 10^{-6}\ M\ \text{s}}$$
$$= 5 \times 10^5\ M^{-1}\ \text{s}^{-1}$$

We can now use the second-order rate constant to determine the value of N for DNA from other sources. For example, at $f = \frac{1}{2}$ for calf DNA, $C_0 t_{1/2}$ is 3×10^3 M s. Therefore,

We assume that the second-order rate constant is the same for different single strands.

$$N = k' C_0 t_{1/2}$$
$$= (5 \times 10^5\ M^{-1}\ \text{s}^{-1})(3 \times 10^3\ M\ \text{s})$$
$$= 2 \times 10^9$$

We can conclude that this DNA sample has a very high order of complexity.

Determination of Reaction Order

In the study of chemical kinetics, one of the first tasks is to determine the order of the reaction. Several methods are available for determining the order of a reaction, and we shall briefly discuss four common approaches.

1. Integration Method. An obvious procedure is to measure the concentration of the reactant(s) at various time intervals of a reaction and to substitute the data into the equations listed in Table 9.2. The equation giving the most constant value of the rate constant for a series of time intervals is the one that corresponds best to the correct order of the reaction. In practice, this method is not precise enough to do more than to distinguish between, say, first- and second-order reactions.

2. Differential Method. This method was developed by van't Hoff in 1884. Because the rate of an nth-order reaction (v) is proportional to the nth power of the concentration of the reactant, we write

$$v = k[\text{A}]^n$$

The methods described above apply only in ideal cases. In practice, determining reaction order can be very difficult because of uncertainty in concentration measurements (for example, when there are small concentration changes in initial rate determinations) as well as the complexity of the reactions (for example, when reactions are reversible and reactions occur between reactants and products). To a certain extent, the procedure is one of trial and error. The use of computers has significantly facilitated the analysis of kinetic data.

Once the order of the reaction has been determined, the rate constant at a particular temperature can be calculated from the ratio of the rate and the concentrations of the reactants, each raised to the power of its order. Knowledge of the order and rate constant then enables us to write the rate law for the reaction. Taking common logarithms of both sides, we obtain

$$\log v = n \log[\text{A}] + \log k$$

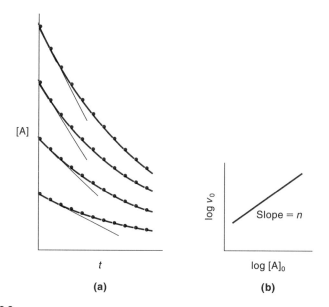

Figure 9.8
(a) Measurement of the initial rates, v_0, of a reaction at different concentrations.
(b) Plot of $\log v_0$ versus $\log[A]_0$.

Thus, by measuring v at several different concentrations of A, we can obtain the value of n from a plot of $\log v$ versus $\log[A]$. A satisfactory procedure is to measure the initial rates (v_0) of the reaction for several different starting concentrations of A, as shown in Figure 9.8. The advantages of using initial rates are (1) that it avoids possible complications due to the presence of products that might affect the order of the reaction and (2) that the reactant concentrations are known most accurately at this time.

3. Half-Life Method. Another simple method of determining reaction order is to find the dependence of the half-life of a reaction on the initial concentration, again using the equations in Table 9.2 or Equation 9.9. Thus, measuring the half-life of a reaction will help us determine the order of the reaction. This procedure is particularly useful for first-order reactions because their half-lives are independent of concentration.

4. Isolation Method. If a reaction involves more than one type of reactant, we can keep the concentrations of all but one reactant constant and measure the rate as a function of its concentration. Any change in rate must be due to that reactant alone. Once we have determined the order with respect to this reactant, we repeat the procedure for the second reactant, and so on. In this way, we can obtain the overall order of the reaction.

9.3 Molecularity of a Reaction

Knowledge of the order of a reaction is but one step toward a detailed understanding of how a reaction occurs. A reaction seldom takes place in the manner suggested by a balanced chemical equation. Typically the overall reaction is the sum of several steps; the sequence of steps by which a reaction occurs is called the *mechanism* of the reaction. To know the mechanism of a reaction is to know how molecules approach one another during a collision, and how chemical bonds are broken and formed, charges transferred, and so on when the reactant molecules are in close proximity. The mechanism proposed for a given reaction must account for the overall stoichiometry, the rate law, and other known facts. Consider the reaction between nitrogen dioxide and carbon monoxide:

$$NO_2(g) + CO(g) \rightarrow NO(g) + CO_2(g)$$

Experimentally, the rate law is found to be

$$\text{rate} = k[NO_2]^2$$

The reaction is more complicated than the balanced equation shows because one of the reactants, CO, does not appear in the rate law expression.

Whereas the word *order* reflects the overall change in going from the reactants to products, the *molecularity* of a reaction refers to a single, definite kinetic process that may be only one step in the overall reaction. Evidence shows that the reaction takes place in two steps, as follows:

$$\text{Step (1)} \quad NO_2(g) + NO_2(g) \xrightarrow{k_1} NO_3(g) + NO(g)$$

$$\text{Step (2)} \quad NO_3(g) + CO(g) \xrightarrow{k_2} NO_2(g) + CO_2(g)$$

Each of these so-called *elementary steps* describes what actually happens at the molecular level. How do we account for the observed rate dependence in terms of these two steps? We simply assume that the rate for the first step is much slower than that for the second step (that is, $k_1 \ll k_2$). The overall rate of reaction then is completely controlled by the rate of the first step, which is aptly called the *rate-determining step*, so rate $= k_1[NO_2]^2$, where $k_1 = k$. Note that the sum of steps (1) and (2) gives us the overall reaction, because the species NO_3 cancels out. Such a species is called an *intermediate* because it appears in the mechanism of the reaction but is not in the overall balanced equation. Keep in mind that an intermediate is always formed in an early elementary step and consumed in a later elementary step.

The preceding discussion shows that our insight into a reaction comes from an understanding of molecularity, not of order. Once we know the mechanism and the rate-determining step, we can write the rate law for the reaction, which must agree with the experimentally determined rate law. Although most reactions are kinetically complex, the mechanisms for a number of them are sufficiently well understood to be discussed in molecular terms. In general, however, proving the uniqueness of a mechanism is very difficult, or impossible, especially for a complex reaction.

As in a court of law, we ask only for proof beyond a reasonable doubt.

We shall now examine three different types of molecularity. Unlike the order of a reaction, molecularity cannot be zero or a noninteger.

Unimolecular Reactions

Reactions such as *cis–trans* isomerization, thermal decomposition, ring opening, and racemization are usually *unimolecular*—that is, they involve only one reactant molecule in the elementary step. For example, the following gas-phase elementary steps are unimolecular:

$$N_2O_4(g) \longrightarrow 2NO_2(g)$$

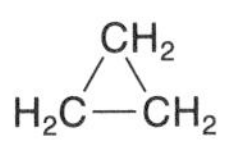

$$\begin{array}{c} CH_2 \\ / \quad \backslash \\ H_2C - CH_2 \end{array} \longrightarrow CH_3CH{=}CH_2$$

Cyclopropane   Propene

Unimolecular reactions often follow a first-order rate law. Because these reactions presumably occur as the result of a binary collision through which the reactant molecules acquire the necessary energy to change forms, we would expect them to be bimolecular processes and hence second-order reactions. How do we account for the discrepancy between the predicted and observed rate laws? To answer this question, let us consider the treatment put forward by the British chemist Frederick Alexander Lindemann (1886–1957) in 1922.[†] Every now and then a reactant molecule, A, collides with another A molecule, and one becomes energetically excited at the expense of the other:

$$A + A \xrightarrow{k_1} A + A^*$$

where the asterisk denotes the activated molecule. The activated molecule can form the desired product according to the elementary step

$$A^* \xrightarrow{k_2} product$$

Another process that may also be going on is the deactivation of the A^* molecule:

$$A^* + A \xrightarrow{k_{-1}} A + A$$

The rate of product formation is given by

$$\frac{d[\text{product}]}{dt} = k_2[A^*]$$

All that remains for us to do is to derive an expression for $[A^*]$. Because A^* is an energetically excited species, it has little stability and a short lifetime. Its concentration in the gas phase is not only low but probably fairly constant as well. Using this assumption, we can apply the steady-state approximation as follows. The rate of change of $[A^*]$ is given by the steps leading to the formation of A^* minus the steps leading to the removal of A^*. According to the steady-state approximation, however, this rate of change can be treated as zero. Mathematically, we have[‡]

$$\frac{d[A^*]}{dt} = 0 = k_1[A]^2 - k_{-1}[A][A^*] - k_2[A^*]$$

Solving for $[A^*]$, we obtain

$$[A^*] = \frac{k_1[A]^2}{k_2 + k_{-1}[A]}$$

The rate of product formation is now given by

$$\frac{d[\text{product}]}{dt} = k_2[A^*] = \frac{k_1 k_2 [A]^2}{k_2 + k_{-1}[A]}$$

Two important limiting cases may be applied to the above equation. At higher

---

[†] A similar treatment was proposed independently and almost simultaneously by the Danish chemist Jens Anton Christiansen (1888–1969).

[‡] Note that the steady-state approximation does not always apply to intermediates. Its use must be justified by either experimental evidence or theoretical considerations.

pressures (≥ 1 atm), most A^* molecules will be deactivated instead of forming product, and we have

$$k_{-1}[A][A^*] \gg k_2[A^*]$$

or

$$k_{-1}[A] \gg k_2$$

The rate in this case is given by

$$\frac{d[\text{product}]}{dt} = \frac{k_1 k_2}{k_{-1}}[A]$$

and the reaction is first order in A. On the other hand, if the reaction is run at low pressures (≤ 0.5 atm) so that most A^* molecules form the product instead of being deactivated, the following inequality will hold:

$$k_{-1}[A][A^*] \ll k_2[A^*]$$

or

$$k_{-1}[A] \ll k_2$$

The rate now becomes

$$\frac{d[\text{product}]}{dt} = k_1[A]^2$$

which is second order in A.

Lindemann's theory has been tested for a number of systems and is found to be essentially correct. The analysis for the intermediate case (that is, $k_{-1}[A][A^*] \approx k_2[A^*]$) is more complex and will not be discussed here.

Bimolecular Reactions

Any elementary step that involves two reactant molecules is a *bimolecular reaction*. Some of the examples in the gas phase are

$$H + H_2 \rightarrow H_2 + H$$
$$NO_2 + CO \rightarrow NO + CO_2$$
$$2NOCl \rightarrow 2NO + Cl_2$$

In the solution phase we have

$$2CH_3COOH \rightarrow (CH_3COOH)_2 \quad \text{(in nonpolar solvents)}$$
$$Fe^{2+} + Fe^{3+} \rightarrow Fe^{3+} + Fe^{2+}$$

Termolecular Reactions

Finally, we note that an elementary step that involves the simultaneous encounter of three reactant molecules is called a *termolecular reaction*. The probability of a three-body collision is usually quite small and only a few such reactions are known. Interestingly, they all involve nitric oxide as one of the reactants:

$$2NO(g) + X_2(g) \rightarrow 2NOX(g)$$

where X = Cl, Br, or I. Another type of termolecular "reaction" involves atomic recombinations in the gas phase; for example,

$$H + H + M \rightarrow H_2 + M$$

$$I + I + M \rightarrow I_2 + M$$

where M is usually some inert gas such as N_2 or Ar. When atoms combine to form diatomic molecules, they possess an excess of kinetic energy, which is converted to vibrational motion, resulting in bond dissociation. Through three-body collisions, the M species can take away some of this excess energy to prevent the breakup of the diatomic molecules.

No elementary steps with a molecularity greater than three are known.

9.4 More Complex Reactions

All the reactions discussed so far are simple in the sense that only one reaction is taking place in each case. Unfortunately, this condition is often not satisfied in actual practice. Three examples of more complex reactions will now be discussed.

Reversible Reactions

Most reactions are reversible to a certain extent, and we must consider both the forward and reverse rates. For the reversible reaction that proceeds by two elementary steps:

$$A \underset{k_{-1}}{\overset{k_1}{\rightleftharpoons}} B$$

we represent the net rate of change in [A] as

$$\frac{d[A]}{dt} = -k_1[A] + k_{-1}[B]$$

At equilibrium, there is no net change in the concentration of A with time; that is, $d[A]/dt = 0$, so that

$$k_1[A] = k_{-1}[B]$$

This expression leads to

$$\frac{[B]}{[A]} = \frac{k_1}{k_{-1}} = K$$

where K is the equilibrium constant.

The discussion of the relationship between reaction rates and equilibria is rooted in a principle of great importance in chemical kinetics. The *principle of microscopic reversibility* states that at equilibrium, the rates of the forward and reverse processes are equal for every elementary reaction occurring.* It means that the process $A \rightarrow B$ is exactly balanced by $B \rightarrow A$ so that equilibrium cannot be maintained by a cyclic process in which the forward reaction is $A \rightarrow B$ and the reverse reaction is $B \rightarrow C \rightarrow A$:

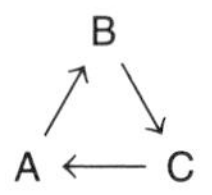

$$
\begin{array}{ccc}
 & B & \\
\nearrow & & \searrow \\
A & \longleftarrow & C
\end{array}
$$

Instead, for every elementary reaction we must write a reverse reaction as follows:

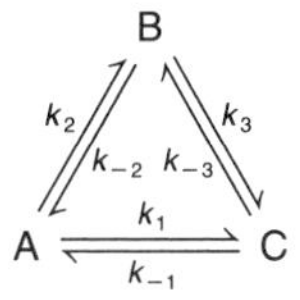

$$
\begin{array}{ccc}
 & B & \\
{}^{k_2}\!\!\nearrow\!\!\nwarrow^{} & & {}^{}\!\!\nearrow\!\!\searrow^{k_3} \\
{}^{k_{-2}} & {}^{k_{-3}} & \\
A & \underset{k_{-1}}{\overset{k_1}{\rightleftharpoons}} & C
\end{array}
$$

such that

$$k_2[A] = k_{-2}[B]$$

$$k_3[B] = k_{-3}[C]$$

$$k_1[A] = k_{-1}[C]$$

These rate constants are not all independent. By simple algebraic manipulation, we can show that $k_{-1}k_2k_3 = k_1k_{-2}k_{-3}$ (see Problem 9.56). The usefulness of the principle of microscopic reversibility is that it tells us that the reaction pathway for the reverse of a reaction at equilibrium is the exact opposite of the pathway for the forward reaction. Therefore, the transition states[†] for the forward and reverse reactions are identical. Consider the base-catalyzed esterification between acetic acid and ethanol:

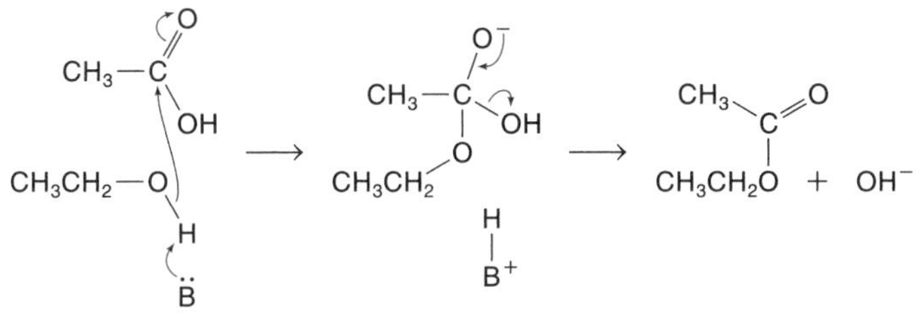

where B is a base (for example, OH^-). The species formed in the first step is a tetrahedral intermediate. Now, according to the principle of microscopic reversibility, the reverse reaction (that is, the hydrolysis of ethyl acetate) must involve the acid-catalyzed expulsion of ethoxide ion from the same tetrahedral intermediate:

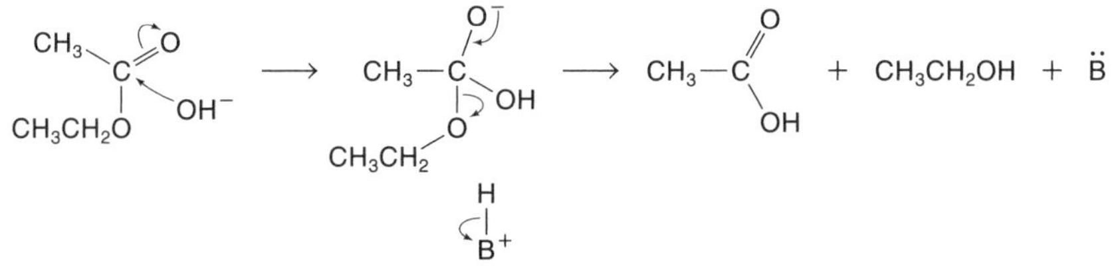

---

*The principle of microscopic reversibility is a consequence of the fact that the fundamental equations for the macroscopic or microscopic dynamics of a system (that is, Newton's laws or the Schrödinger equation) have the same form when time t is replaced by $-t$ and when the signs of all velocities are also reversed. See B. H. Mahan, *J. Chem. Educ.* **52**, 299 (1975).

[†]The transition state of a reaction is the complex formed between the reactants and products along the reaction coordinate (discussed further in Section 9.7).

Thus, when the likelihood of a certain mechanism is being considered, we can always turn to the principle for guidance. If the reverse mechanism looks implausible, then chances are that the proposed mechanism is wrong and we must search for another mechanism.

Consecutive Reactions

A consecutive reaction is one in which the product from the first step becomes the reactant for the second step, and so on. The thermal decomposition of acetone in the gas phase is an example:

$$CH_3COCH_3 \rightarrow CH_2{=}CO + CH_4$$

$$CH_2{=}CO \rightarrow CO + \tfrac{1}{2}C_2H_4$$

Many nuclear decays are also consecutive reactions. For example, upon the capture of a neutron, a uranium-238 isotope is converted to a uranium-239 isotope, which then decays as follows:

$$^{239}_{92}U \rightarrow ^{239}_{93}Np + ^{0}_{-1}\beta$$

$$^{239}_{93}Np \rightarrow ^{239}_{94}Pu + ^{0}_{-1}\beta$$

For a two-step consecutive reaction, we have

$$A \xrightarrow{k_1} B \xrightarrow{k_2} C$$

Because each step is first order, the rate law equations are

$$\frac{d[A]}{dt} = -k_1[A] \tag{9.20}$$

$$\frac{d[B]}{dt} = k_1[A] - k_2[B] \tag{9.21}$$

$$\frac{d[C]}{dt} = k_2[B] \tag{9.22}$$

We assume that initially only A is present and its concentration is $[A]_0$ so that

$$[A] = [A]_0 e^{-k_1 t} \tag{9.23}$$

The rate equation for the intermediate B is quite complex and will not be fully discussed here. The treatment can be simplified, however, by applying the steady-state approximation to B, that is, by assuming that the concentration of B remains constant over a certain time period so that we can write

$$\frac{d[B]}{dt} = 0 = k_1[A] - k_2[B] \tag{9.24}$$

or

$$[B] = \frac{k_1}{k_2}[A] = \frac{k_1}{k_2}[A]_0 e^{-k_1 t} \tag{9.25}$$

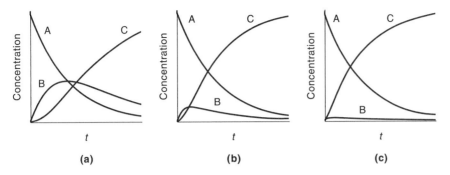

Figure 9.9
Variation in the concentrations of A, B, C with time for a consecutive reaction A → B → C.
(a) $k_2 = k_1$; (b) $k_2 = 5k_1$; (c) $k_2 = 25k_1$.

Equation (9.25) holds if $k_2 \gg k_1$. Under this condition, B molecules are converted to C as soon as they are formed so [B] is kept constant and low compared to [A].

To get an expression for [C], we note that at any instant we have $[A]_0 = [A] + [B] + [C]$. Therefore, from Equations 9.23 and 9.25 we obtain

$$[C] = [A]_0 - [A] - [B]$$

$$= [A]_0 \left(1 - e^{-k_1 t} - \frac{k_1}{k_2} e^{-k_1 t} \right)$$

$$= [A]_0 (1 - e^{-k_1 t}) \tag{9.26}$$

The $(k_1/k_2)\exp(-k_1 t)$ term is eliminated because it is much smaller than 1.

Figure 9.9 shows plots of [A], [B], and [C] with time for different rate constant ratios. In all cases, [A] falls steadily from $[A]_0$ to zero while [C] rises from zero to $[A]_0$. The concentration of B rises from zero to a maximum and then falls to zero. Note that as k_2 becomes much larger than k_1, the steady-state approximation becomes valid over the time period when [B] remains constant (Figure 9.9c).

A more complicated but common consecutive reaction is shown below:

$$A + B \underset{k_{-1}}{\overset{k_1}{\rightleftharpoons}} C \overset{k_2}{\rightarrow} P$$

where P denotes product. This scheme involves *pre-equilibrium*, in which an intermediate is in equilibrium with the reactants. A pre-equilibrium arises when the rates of formation of the intermediate and of its decay back into reactants are much faster than its rate of formation of products—that is, when $k_{-1} \gg k_2$. Because A, B, and C are assumed to be in equilibrium, we can write

$$K = \frac{[C]}{[A][B]} = \frac{k_1}{k_{-1}}$$

and the rate of formation of P is given by

$$\frac{d[P]}{dt} = k_2[C]$$

$$= k_2 K[A][B]$$

In Chapter 10 we shall see applications of both the steady-state approximation and the pre-equilibrium treatment to enzyme kinetics.

Chain Reactions

One of the best-known gas-phase chain reactions involves the formation of hydrogen bromide from molecular hydrogen and bromine between 230°C and 300°C:

$$H_2(g) + Br_2(g) \rightarrow 2HBr(g)$$

The complexity of this reaction is indicated by the rate equation

$$\frac{d[HBr]}{dt} = \frac{\alpha[H_2][Br_2]^{1/2}}{1 + \beta[HBr]/[Br_2]} \tag{9.27}$$

where α and β are some constants. Thus, the reaction does not have an integral reaction order. It has taken many experiments and considerable chemical intuition to come up with Equation 9.27. We assume that a chain of reactions proceeds as follows:

$$Br_2 \xrightarrow{k_1} 2Br \qquad \text{chain initiation}$$

$$Br + H_2 \xrightarrow{k_2} HBr + H \quad \text{chain propagation}$$

$$H + Br_2 \xrightarrow{k_3} HBr + Br \quad \text{chain propagation}$$

$$H + HBr \xrightarrow{k_4} H_2 + Br \quad \text{chain inhibition}$$

$$Br + Br \xrightarrow{k_5} Br_2 \qquad \text{chain termination}$$

The following reactions play only a minor role in determining the rate:

$$H_2 \rightarrow 2H \qquad \text{chain initiation}$$

$$Br + HBr \rightarrow Br_2 + H \quad \text{chain inhibition}$$

$$H + H \rightarrow H_2 \qquad \text{chain termination}$$

$$H + Br \rightarrow HBr \qquad \text{chain termination}$$

For this reason, they are not included in the kinetic analysis. By applying the steady-state approximation to the intermediates H and Br, we can derive Equation 9.27 using the first five elementary steps (see Problem 9.20).

9.5 The Effect of Temperature on Reaction Rate

Figure 9.10 shows four types of temperature dependence for reaction rate constants. Type (a) represents normal reactions whose rates increase with increasing temperature. Type (b) shows a rate that initially increases with temperature, reaches a maximum, and finally decreases with further temperature rise. Type (c) shows a steady decrease of rate with temperature. The behavior outlined in (b) and (c) may be surprising, because we might expect the rate of a reaction to depend on two quantities: the number of collisions per second and the fraction of collisions that activate molecules for the reaction. Both quantities should increase with increasing temperature. The complex nature of the reaction mechanism accounts for this ostensibly unusual behavior. For example, in an enzyme-catalyzed reaction, the enzyme molecule must be in a specific conformation to react with the substrate molecule. When the enzyme

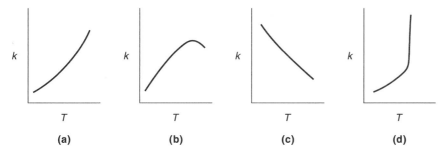

Figure 9.10
Four types of temperature dependence for rate constants. See text.

is in the native state, the reaction rate does increase with temperature. At higher temperatures, the molecule may undergo denaturation, thereby losing its effectiveness as a catalyst. Consequently, the rate will decrease with increasing temperature.

The behavior shown in Figure 9.10c is known for only a few systems. Consider the formation of nitrogen dioxide from nitric oxide and oxygen:

$$2NO(g) + O_2(g) \rightleftharpoons 2NO_2(g)$$

The rate law is

$$\text{rate} = k[NO]^2[O_2]$$

The mechanism is believed to involve two bimolecular steps:

$$\text{Rapid:}\quad 2NO \rightleftharpoons (NO)_2 \quad K = \frac{[(NO)_2]}{[NO]^2}$$

$$\text{Slow, rate determining:}\quad (NO)_2 + O_2 \xrightarrow{k'} 2NO_2$$

This is an example of pre-equilibrium, discussed earlier.

Thus, the overall rate is

$$\text{rate} = k'[(NO)_2][O_2] = k'K[NO]^2[O_2]$$
$$= k[NO]^2[O_2]$$

where $k = k'K$. Furthermore, the equilibrium between 2NO and $(NO)_2$ is exothermic from left to right. Because the decrease in K with temperature outweighs the increase in k' with temperature, the overall rate decreases with increasing temperature over a certain range of temperature.

Finally, we note that the behavior shown in Figure 9.10d corresponds to a chain reaction. At first, the rate rises gradually with temperature. At a particular temperature, the chain propagation reactions become significant, and the reaction is literally explosive.

The Arrhenius Equation

In 1889, the Swedish chemist Svante Arrhenius (1859–1927) discovered that the temperature dependence of many reactions could be described by the following equation:

$$k = Ae^{-E_a/RT} \tag{9.28}$$

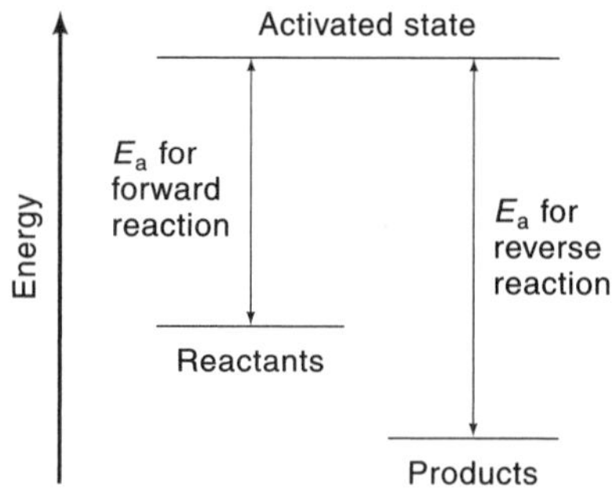

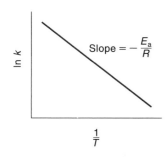

Figure 9.11
Schematic diagram of activation energy
for an exothermic reaction.

where k is the rate constant, A is called the frequency factor or pre-exponential factor, E_a is the activation energy (kJ mol^{-1}), R is the gas constant, and T is the absolute temperature. The *activation energy* is the minimum amount of energy required to initiate a chemical reaction. The frequency factor, A, represents the frequency of collisions between reactant molecules. The factor $\exp(-E_a/RT)$ resembles the Boltzmann distribution law (see Equation 3.22); it represents the fraction of molecular collisions that have energy equal to or greater than the activation energy, E_a (Figure 9.11). Because the exponential term is a number, the units of A are the same as the units of the rate constant (s^{-1} for first-order rate constants; M^{-1} s^{-1} for second-order rate constants, and so on).

As we shall see later, because the frequency factor A is related to molecular collisions, it is temperature dependent. For a limited temperature range (≤ 50 K), however, the predominant temperature variation is embraced by the exponential term. Taking the natural logarithm of Equation 9.28, we obtain

$$\ln k = \ln A - \frac{E_a}{RT} \tag{9.29}$$

Thus, a plot of $\ln k$ versus $1/T$ yields a straight line whose slope, which is negative, is equal to $-E_a/R$ (Figure 9.12). Note that in Equation 9.29 k and A are treated as dimensionless quantities.

Alternatively, if we know the rate constants k_1 and k_2 at T_1 and T_2, we have, from Equation 9.29,

$$\ln k_1 = \ln A - \frac{E_a}{RT_1}$$

$$\ln k_2 = \ln A - \frac{E_a}{RT_2}$$

Taking the difference between these two equations, we obtain

$$\ln \frac{k_2}{k_1} = -\frac{E_a}{R}\left(\frac{1}{T_2} - \frac{1}{T_1}\right) \tag{9.30}$$

Equation 9.30 enables us to calculate the rate constant at a different temperature if E_a is known.

From the standpoint of Arrhenius's rate equation, a complete understanding of the factors determining the rate constant of a reaction requires that we be able to calculate the values of both A and E_a. Considerable effort has been devoted to this problem, as we shall see below.

Figure 9.12
Plot of $\ln k$ versus $1/T$. The slope of the straight line is equal to $-E_a/R$.

9.6 Potential-Energy Surfaces

To discuss activation energy in more detail, we need to learn something about the energetics of a reaction. One of the simplest reactions is the combination of two atoms to form a diatomic molecule, such as $H + H \rightarrow H_2$. Basically, we would like to describe more complex reactions in terms of a potential-energy curve such as that shown in Figure 3.16. However, potential-energy diagrams are prohibitively complex for all but the simplest systems. One of the simplest and most studied systems is the exchange reaction between the hydrogen atom and the hydrogen molecule:

$$H + H_2 \rightarrow H_2 + H$$

Even for a three-atom system such as this, we need a four-dimensional plot, describing three bond lengths, or two bond lengths and a bond angle, versus energy. The problem is greatly simplified by assuming that the minimum energy configuration is linear so that only two bond lengths need to be specified. Consequently, only a three-dimensional plot is required (Figure 9.13). Labeling the atoms A, B, and C, we can represent the reaction as

$$H_A + H_B\text{–}H_C \rightarrow [H_A \cdots H_B \cdots H_C] \rightarrow H_A\text{–}H_B + H_C$$
$$\text{activated complex}$$

The plot, called the *potential-energy surface*, is a contour map of potential energies corresponding to different values of r_{AB} and r_{BC}, which are separations between the atoms. Although the reaction can proceed along any path, the one that requires the minimum amount of energy is shown by the red curve. The system travels along this path through the first valley and over the saddle point, which is the location of the activated complex, and then moves down the second valley. We represent this path in a plot of the potential energy versus the reaction coordinate, which describes the

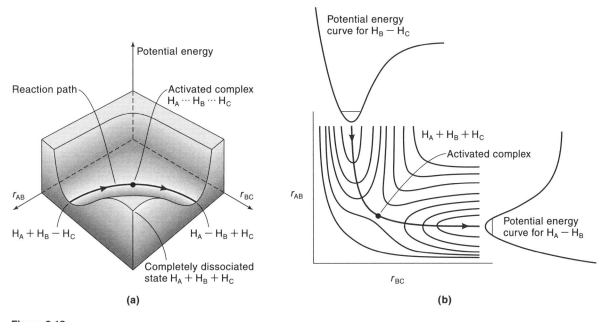

Figure 9.13
The $H + H_2 \rightarrow H_2 + H$ reaction. (a) Potential-energy surface. (b) Contour diagram of the potential-energy surface.

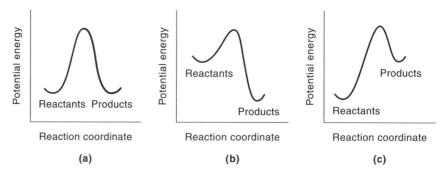

Figure 9.14
Potential-energy profile along the minimum energy path for (a) the $H + H_2 \rightarrow H_2 + H$ reaction, (b) an exothermic reaction, and (c) an endothermic reaction.

positions of the atoms during the course of a reaction. Figure 9.14a shows the plot for the $H + H_2$ reaction. The plots shown in Figures 9.14b and 9.14c are customarily employed for reactions in general, where the products differ from the reactants. You need to understand that these plots provide only a qualitative description for the reaction path because of the complexities involved for large molecules.

Much effort has gone into the calculation of the activation energy for the $H + H_2$ reaction. The close correspondence between the calculated and measured values of E_a (36.8 kJ mol^{-1}) lends support to the validity of the model (that is, a linear activated complex).* An interesting side note is that if the reaction took the path involving the dissociation of the H_2 molecule followed by recombination, an activation energy of 432 kJ mol^{-1} would be required.

9.7 Theories of Reaction Rates

At this point, we are ready to consider two important theories of reaction rates: the collision theory and the transition-state theory. These theories provide us with greater insight into the energetic and mechanistic aspects of reactions.

Collision Theory

The collision theory of reaction rates is based on the kinetic theory of gases, discussed in Chapter 2. In its simplest form, it applies only to bimolecular reactions in the gas phase. Consider the bimolecular elementary reaction

$$A + A \rightarrow product$$

From Equation 2.31, the number of binary collisions per cubic meter per second between "hard-sphere" A molecules is given by

We set $V = 1$ m^3 so N_A is now the number of molecules in 1 m^3.

$$Z_{AA} = \frac{\sqrt{2}}{2} \pi d^2 \bar{c} \left(\frac{N_A}{V}\right)^2$$

$$= \frac{\sqrt{2}}{2} \pi d^2 \bar{c} N_A^2$$

---

*A theoretical analysis shows that the activation energy for this reaction is 40.21 kJ mol^{-1}. The calculation for this simple reaction required 80 days of computer time! See D. D. Diedrich and J. B. Anderson, *Science* **258**, 786 (1992).

According to Equation 2.28,

$$\bar{c} = \sqrt{\frac{8k_B T}{\pi m}}$$

so that

$$Z_{AA} = 2N_A^2 d^2 \sqrt{\frac{\pi k_B T}{m_A}} \tag{9.31}$$

For a bimolecular reaction of the type

$$A + B \rightarrow product$$

the binary collision number is

$$Z_{AB} = N_A N_B d_{AB}^2 \sqrt{\frac{8\pi k_B T}{\mu}} \tag{9.32}$$

where d_{AB} is the collision diameter between A and B, and μ, the *reduced mass*, is given by

$$\mu = \frac{m_A m_B}{m_A + m_B} \tag{9.33}$$

(Equation 9.33 enables us to treat a two-body system as if it contained only a single particle of mass μ and hence the term "reduced mass.") Now, if the collisions were 100% effective—that is, a product formed as a result of every binary collision—the rate of the reaction would be equal to either Z_{AA} or Z_{AB}. But this is not the case. In a gas at a pressure of 1 atm, the collision number is about 10^{31} L^{-1} s^{-1} at 298 K. If every collision led to the formation of a product, all gas-phase reactions would be complete in about 10^{-9} s, which is contrary to our experience. The additional factor needed for Equations 9.31 and 9.32 is the term that contains the activation energy. For the A + B $\rightarrow$ product reaction, we write

$$rate = Z_{AB} e^{-E_a/RT}$$

$$= N_A N_B d_{AB}^2 \sqrt{\frac{8\pi k_B T}{\mu}} e^{-E_a/RT} \tag{9.34}$$

Division of the rate by $N_A N_B$ gives a rate constant in molecular units (SI unit: m^3 molecule^{-1} s^{-1})*

$$k = \frac{rate}{N_A N_B} = z_{AB} e^{-E_a/RT} \tag{9.35}$$

where

$$z_{AB} = d_{AB}^2 \sqrt{\frac{8\pi k_B T}{\mu}}$$

---

*Equation 9.35 can be expressed in molar units (M^{-1} s^{-1}) by multiplication by the factor $[6.022 \times 10^{23}$ molecules mol$^{-1}/(10^{-3}$ m^3 L$^{-1})]$.

Comparing Equation 9.29 with Equation 9.35, we get

$$A = z_{AB} = d_{AB}^2 \sqrt{\frac{8\pi k_B T}{\mu}}$$

Thus, the frequency factor, A, is temperature dependent. In practice, we usually treat it as a temperature-independent quantity in the calculation of E_a values. Doing so does not introduce any serious error, however, because the exponential term [that is, $\exp(-E_a/RT)$] depends much more on temperature than the square-root term does on temperature.

The collision theory (Equation 9.35) predicts the value of the rate constant fairly accurately for reactions that involve atomic species or simple molecules if the activation energy is known. Significant deviations are found, however, for reactions involving complex molecules. For these reactions, the rate constants tend to be smaller than that predicted by Equation 9.35, sometimes by a factor of 10^6 or more. The reason is that the simple kinetic theory counts every sufficiently energetic collision as an effective one. In reality, the molecules may not approach each other in the right way for a reaction to occur, even if plenty of energy is available. To correct for the discrepancies observed, we modify Equation 9.35 as follows:

We use z to represent binary collision number in general.

$$k = Pze^{-E_a/RT} \tag{9.36}$$

where P, called the *probability*, or *steric*, *factor*, takes into account the fact that in a collision complex, molecules must be properly oriented to undergo the reaction. This modification is an improvement, but the evaluation of P is rather difficult. The comparison of Equation 9.36 with Equation 9.28 shows that $A = Pz$.

Transition-State Theory

Although the collision theory is intuitively appealing and does not involve complicated mathematics, it does suffer from some serious drawbacks. Because it is based on the kinetic theory of gases, it assumes the reacting species are hard spheres and totally ignores the structures of molecules. For this reason, it cannot satisfactorily account for the probability factor at the molecular level. Furthermore, without quantum mechanics, we cannot use collision theory to calculate activation energy. A different approach, called the transition-state theory (also known as the activated complex theory), was developed by the American chemist Henry Eyring (1901–1981) and others in the 1930s, to provide greater insight into the details of a reaction on the molecular scale. It also enables us to calculate the rate constant with considerable accuracy.

The starting point of transition-state theory is similar to collision theory. In a bimolecular collision, an activated complex (also called a transition-state complex) of relatively high energy is formed. Consider the elementary reaction

$$A + B \rightleftharpoons X^{\ddagger} \xrightarrow{k} C + D$$

where A and B are reactants and $X^{\ddagger}$ represents an activated complex. A fundamental assumption of transition-state theory (and one that differentiates it from collision theory) is that the reactants are always in equilibrium with $X^{\ddagger}$. The activated complex should not be thought of as a stable, isolatable intermediate because it is assumed to be always in the process of decomposing. It is, in fact, neither stable nor isolatable.* Thus, the equilibrium between the reactants and the activated complex is not the conventional type. Nevertheless, we can write the equilibrium constant as

$$K^{\ddagger} = \frac{[X^{\ddagger}]}{[A][B]} \qquad (9.37)$$

The rate of the reaction is equal to the concentration of the activated complex at the top of the energy barrier, multiplied by the frequency, v, of crossing the barrier. Hence,

rate = number of activated complexes decomposing to form products

$$= v[X^{\ddagger}]$$

$$= v[A][B]K^{\ddagger}$$

Because the rate can also be written as

$$\text{rate} = k[A][B]$$

where k is the rate constant, it follows that

$$k = vK^{\ddagger}$$

where v is the frequency of vibration of the activated complex leading to the formation of products; it has the unit s^{-1}. Calculation of the value of k now depends on our ability to evaluate both v and $K^{\ddagger}$. Using statistical thermodynamics, we can show that $v = k_B T/h$,§ where h is Planck's constant, so that

$$k = \frac{k_B T}{h} K^{\ddagger}(M^{1-m}) \qquad (9.38)$$

Note that to make the units equal on both sides of Equation 9.38, we added the term (M^{1-m}), where M is molarity and m is the molecularity of the reaction. For a unimolecular reaction, $m = 1$ and $(M^{1-1}) = 1$, so the first-order rate constant, k, has the same unit as $k_B T/h$. (At 298 K, $k_B T/h = 6.21 \times 10^{12}$ s^{-1}.) For bimolecular reactions, $m = 2$, and the units on the right side are M^{-1} s^{-1}, which are consistent with those of a second-order rate constant. The equilibrium constant $K^{\ddagger}$ can also be calculated from fundamental physical properties, such as bond length, atomic masses, and vibrational frequencies of the reactants. This approach has also been called the absolute rate theory because it enables us to calculate the value of k from absolute, or fundamental, molecular properties.

---

* This statement is not universally true. Using fast lasers, chemists in recent years have obtained spectroscopic evidence of an activated complex. See A. H. Zewail, *Science* **242**, 1645 (1988) and *Sci. Am.* December 1990.

§ When the thermal energy $(k_B T)$ is comparable to the vibrational energy (hv), the activated complex dissociates into products. See the physical chemistry texts listed in Chapter 1 (p. 6).

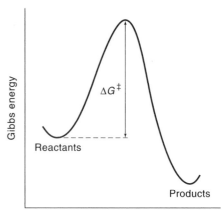

Figure 9.15
Definition of $\Delta G^{\ddagger}$ for a reaction.

Thermodynamic Formulation of Transition-State Theory

The rate constant expressed in Equation 9.38 can be related to the thermodynamic properties of a reaction. We write

$$\Delta G^{\circ\ddagger} = -RT \ln K^{\ddagger}$$

Hence,

$$K^{\ddagger} = e^{-\Delta G^{\circ\ddagger}/RT} \tag{9.39}$$

where $\Delta G^{\circ\ddagger}$, the standard molar Gibbs energy of activation (Figure 9.15) is given by

$$\Delta G^{\circ\ddagger} = G^{\circ}(\text{activated complex}) - G^{\circ}(\text{reactants})$$

The rate constant can be written as

$$k = \frac{k_B T}{h} e^{-\Delta G^{\circ\ddagger}/RT} (M^{1-m}) \tag{9.40}$$

Because $k_B T/h$ is independent of the nature of A and B, the rate of any reaction at a given temperature is determined by $\Delta G^{\circ\ddagger}$. Furthermore,

$$\Delta G^{\circ\ddagger} = \Delta H^{\circ\ddagger} - T\Delta S^{\circ\ddagger}$$

so Equation 9.40 becomes

$$k = \frac{k_B T}{h} e^{\Delta S^{\circ\ddagger}/R} e^{-\Delta H^{\circ\ddagger}/RT} (M^{1-m}) \tag{9.41}$$

where $\Delta S^{\circ\ddagger}$ and $\Delta H^{\circ\ddagger}$ are the standard molar entropy and standard molar enthalpy of activation, respectively. Equation 9.41 is the thermodynamic formulation of transition-state theory. A more rigorous approach includes a factor known as the transmission coefficient on the right side of Equation 9.41, but this factor is generally close to unity and may be ignored.

It is useful to compare the three expressions for rate constants discussed so far:

$$k = Ae^{-E_a/RT}$$

$$k = Pze^{-E_a/RT}$$

$$k = \frac{k_\mathrm{B} T}{h} e^{\Delta S^{\circ\ddagger}/R} e^{-\Delta H^{\circ\ddagger}/RT} (M^{1-m})$$

The first equation (Equation 9.28) is an empirical one; both A and E_a must be determined experimentally. The second equation (Equation 9.36) is based in part on collision theory; the value of z can be calculated from the kinetic theory of gases. On the other hand, it is very difficult in general to estimate the magnitude of P accurately. The last equation (Equation 9.41) is based on transition-state theory. This equation provides us with the thermodynamic formulation of the reaction rate constant and is the most reliable of the three approaches. It is also the most difficult of the three equations to apply.

What is the significance of $\Delta S^{\circ\ddagger}$ and $\Delta H^{\circ\ddagger}$? Comparing Equation 9.41 with Equation 9.36 and assuming $\Delta H^{\circ\ddagger} = E_\mathrm{a}$, we obtain

$$A = Pz = \frac{k_\mathrm{B} T}{h} e^{\Delta S^{\circ\ddagger}/R} \tag{9.42}$$

This equation enables us to interpret the probability factor in terms of the standard molar entropy of activation. If reactants are atoms or simple molecules, then relatively little energy is redistributed among the various types of motion in the activated complex. Consequently, $\Delta S^{\circ\ddagger}$ will be either a small positive or a small negative number, so that $\exp(\Delta S^{\circ\ddagger}/R)$—or P—is close to unity. But if complex molecules are involved in a reaction, $\Delta S^{\circ\ddagger}$ will be either a large positive or a large negative number. In the former case, the reaction will proceed much faster than predicted by collision theory; in the latter case, a much slower rate will be observed.

The standard molar enthalpy of activation, $\Delta H^{\circ\ddagger}$, is closely related to the ease of bond breaking and making in the generation of the activated complex. The lower the $\Delta H^{\circ\ddagger}$ value, the faster the rate. If we compare the coefficients of $1/T$ in Equations 9.41 and 9.36, we obtain $E_\mathrm{a} = \Delta H^{\circ\ddagger}$. However, a more rigorous treatment shows that*

$$E_\mathrm{a} = \Delta U^{\circ\ddagger} + RT \tag{9.43}$$

where $\Delta U^{\circ\ddagger}$ is the standard molar internal energy of activation. At constant pressure, we have

$$\Delta H^{\circ\ddagger} = \Delta U^{\circ\ddagger} + P \Delta V^{\circ\ddagger}$$

The quantity $\Delta V^{\circ\ddagger}$ is known as the standard molar volume of activation. Equation 9.43 can now be written as

$$E_\mathrm{a} = \Delta H^{\circ\ddagger} - P \Delta V^{\circ\ddagger} + RT \tag{9.44}$$

For reactions occurring in solution, the term $P \Delta V^{\circ\ddagger}$ is quite small compared with $\Delta H^{\circ\ddagger}$ and can usually be neglected so that $\Delta H^{\circ\ddagger} \approx E_\mathrm{a} - RT$, and Equation 9.41 can be written as

$$k = \frac{k_\mathrm{B} T}{h} e^{\Delta S^{\circ\ddagger}/R} e^{-(E_\mathrm{a} - RT)/RT} (M^{1-m})$$

$$= e \frac{k_\mathrm{B} T}{h} e^{\Delta S^{\circ\ddagger}/R} e^{-E_\mathrm{a}/RT} (M^{1-m}) \quad \text{in solution} \tag{9.45}$$

For reactions in the gas phase, we use the relationship $P \Delta V^{\circ\ddagger} = \Delta n^{\ddagger} RT$ in Equation

---

*For the derivation of Equation 9.43, see Physical Chemistry reference 3 (p. 496) on p. 6 of this text.

9.44 to get

$$E_a = \Delta H^{\circ\ddagger} - \Delta n^{\ddagger} RT + RT \qquad (9.46)$$

For a unimolecular reaction, $\Delta n^{\ddagger} = 0$, and Equation 9.41 becomes

$$k = e \frac{k_B T}{h} e^{\Delta S^{\circ\ddagger}/R} e^{-E_a/RT} (M^{1-m}) \quad \text{unimolecular, gas phase} \qquad (9.47)$$

which is the same as Equation 9.45. For a bimolecular reaction, $\Delta n^{\ddagger} = -1$, so $E_a = \Delta H^{\circ\ddagger} + 2RT$ and Equation 9.41 takes the form

$$k = e^2 \frac{k_B T}{h} e^{\Delta S^{\circ\ddagger}/R} e^{-E_a/RT} (M^{1-m}) \quad \text{bimolecular, gas phase} \qquad (9.48)$$

Example 9.2

The pre-exponential factor and activation energy for the unimolecular reaction

$$CH_3NC(g) \rightarrow CH_3CN(g)$$

are $4.0 \times 10^{13} \text{ s}^{-1}$ and 272 kJ mol^{-1}, respectively. Calculate the values of $\Delta H^{\circ\ddagger}$, $\Delta S^{\circ\ddagger}$, and $\Delta G^{\circ\ddagger}$ at 300 K.

ANSWER

Equating the pre-exponential factor in Equation 9.47 to the experimental data, we have

$$e \frac{k_B T}{h} e^{\Delta S^{\circ\ddagger}/R} = 4.0 \times 10^{13} \text{ s}^{-1}$$

so

$$e^{\Delta S^{\circ\ddagger}/R} = \frac{(4.0 \times 10^{13} \text{ s}^{-1})h}{e k_B T}$$

$$= \frac{(4.0 \times 10^{13} \text{ s}^{-1})(6.626 \times 10^{-34} \text{ J s})}{(2.718)(1.381 \times 10^{-23} \text{ J K}^{-1})(300 \text{ K})} = 2.354$$

$$\Delta S^{\circ\ddagger} = 7.12 \text{ J K}^{-1} \text{ mol}^{-1}$$

From Equation 9.46, noting that $\Delta n^{\ddagger} = 0$, we write

$$\Delta H^{\circ\ddagger} = E_a - RT$$

$$= 272 \text{ kJ mol}^{-1} - \left[\frac{8.314}{1000} \text{ kJ K}^{-1} \text{ mol}^{-1} (300 \text{ K}) \right]$$

$$= 270 \text{ kJ mol}^{-1}$$

Finally,

$$\Delta G^{\circ\ddagger} = \Delta H^{\circ\ddagger} - T\Delta S^{\circ\ddagger}$$

$$= 270 \text{ kJ mol}^{-1} - (300 \text{ K}) \left(\frac{7.12}{1000} \text{ kJ K}^{-1} \text{ mol}^{-1} \right)$$

$$= 268 \text{ kJ mol}^{-1}$$

COMMENT

For unimolecular reactions, $\Delta S^{\circ\ddagger}$ is a small positive or negative quantity, so this is largely an enthalpy-driven process. ($\Delta S^{\circ\ddagger}$ will be a negative quantity for bimolecular gas-phase reactions because two molecules combine to form a single entity—the activated complex.) In general, regardless of the molecularity of the reactions, $\Delta H^{\circ\ddagger}$ is approximately equal to E_a.

9.8 Isotope Effects in Chemical Reactions

When an atom in a reactant molecule is replaced by one of its isotopes, both the equilibrium constant of the reaction and the rate constants may change. The term *equilibrium isotope effects* refers to changes in equilibrium constants that result from isotope substitution. Rate variations caused by this exchange are known as *kinetic isotope effects*. The study of isotope effects provides information on reaction mechanisms that has applications in many branches of chemistry. The underlying theory is complex, requiring both quantum mechanics and statistical mechanics; therefore, only a qualitative description will be given here.

Isotopic replacement in a molecule does not result in a change in the electronic structure of the molecule or in the potential-energy surface for any reaction the molecule might undergo, yet the rate of the reaction can be profoundly affected by the substitution. To see why, let us consider the H_2, HD, and D_2 molecules, whose zero-point energies—that is, the ground-state vibrational energies—are 26.5 kJ mol^{-1}, 21.6 kJ mol^{-1}, and 17.9 kJ mol^{-1}, respectively.* Because D_2 has the lowest zero-point energy (due to the fact that it has the largest reduced mass), more energy is required to dissociate this molecule than to break apart H_2 or HD (Figure 9.16a).

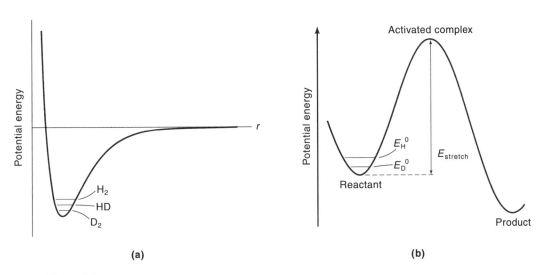

Figure 9.16
(a) Zero-point energy levels for H_2, HD, and D_2. The internuclear distance is r.
(b) Activation energies for bond rupture in H_2 and D_2.

---

* The zero-point energy is given by $E_{vib} = \frac{1}{2}h\nu$, where ν is the fundamental frequency of vibration. This frequency is given by $\nu = (1/2\pi)\sqrt{k/\mu}$, where k is the force constant of the bond and μ the reduced mass, given by $m_1 m_2/(m_1 + m_2)$. Because D_2 has the largest value of μ, it possesses the smallest frequency of vibration and hence of E_{vib}. The reverse holds true for H_2. More will be said about molecular vibration in Chapter 14.

Consequently, the reaction rate for $D_2 \rightarrow 2D$ will be the slowest compared with the other two corresponding dissociations. As a rough estimate, we can calculate the ratio of the rate constants for the dissociation of H_2 and D_2, k_H/k_D, as follows. According to Figure 9.16b, the activation energies for these two processes, E_H and E_D, are given by

$$E_H = E_{\text{stretch}} - E_H^0$$

$$E_D = E_{\text{stretch}} - E_D^0$$

where E_H^0 and E_D^0 are the zero-point energies and E_{stretch} is the difference between the lowest possible potential energy and the potential energy of the activated complex. Using the Arrhenius expression (Equation 9.28), we write

$$\frac{k_H}{k_D} = \frac{Ae^{-(E_{\text{stretch}} - E_H^0)/RT}}{Ae^{-(E_{\text{stretch}} - E_D^0)/RT}}$$

$$= e^{(E_H^0 - E_D^0)/RT}$$

$$= e^{(26.5 - 17.9) \times 1000 \text{ J mol}^{-1}/(8.314 \text{ J K}^{-1} \text{ mol}^{-1})(300 \text{ K})}$$

$$\approx 31$$

which is quite a large number.

The example above dramatizes the difference between the rate constants of D_2 and H_2. In practice, however, we are concerned more with the breaking of a bond between hydrogen and some other atom, such as carbon. Consider, for example, the following reactions:

where B is some group that can take up a hydrogen atom. Again, we would predict a kinetic isotope effect, because the fundamental frequencies of vibration are different for the C–H and C–D bonds. But the difference is not as great as that between H_2 and D_2 because the reduced masses are closer to each other in this case (see Problem 9.39). Still, the ratio k_{C-H}/k_{C-D} might be appreciable, of the order of 5 or so.

Isotope effects that reflect isotope substitution for an atom involved in a bond-breaking process are called primary kinetic isotope effects. Such effects are most pronounced for light elements such as H, D, and T. Reactions involving isotopes of mercury (^{199}Hg and ^{201}Hg), for example, would show hardly any detectable difference in rates. A secondary kinetic isotope effect occurs when the isotope is not directly involved in the bond rupture. We would expect a small change in the reaction rate in this case. Indeed, this prediction has been confirmed experimentally.

Kinetic isotope effects are important in unraveling the secrets of organic and biological processes. Generally, the primary kinetic isotope effects for light elements (up to about chlorine) are sufficiently large so that many rates can be conveniently and accurately measured. One application of the primary kinetic isotope effect is in the elucidation of enzyme mechanisms. Alcohol dehydrogenase is an enzyme that oxidizes a wide range of aliphatic and aromatic alcohols to their corresponding aldehydes and ketones using NAD^+ as a coenzyme. The enzyme is found in a variety

We assume that this bond-breaking process is also the rate-determining step.

of sources, including yeast, horses, and humans. The reaction is represented by

$$CH_3CH_2OH + NAD^+ \rightarrow CH_3CHO + NADH + H^+$$

The question chemists asked was: Does the oxidation of ethanol catalyzed by alcohol dehydrogenase proceed with the direct transfer of hydrogen from ethanol to NAD^+? The following experiment provided the answer. The oxidation of CH_3CD_2OH was carried out in H_2O, resulting in reduced NAD^+, NADH:

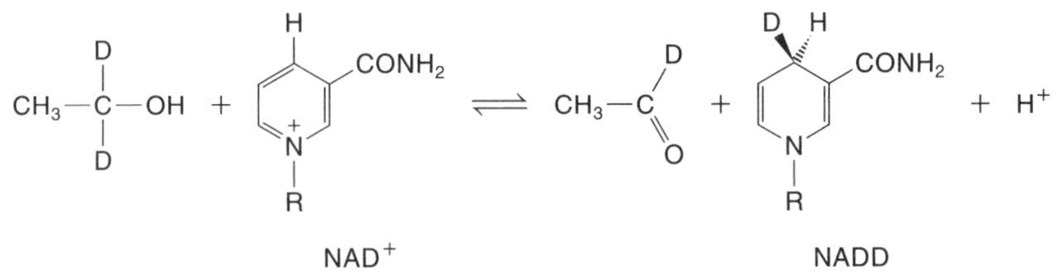

The NADH formed was found to contain one deuterium atom per molecule. (Note that NADH has one D atom derived from the deuterated ethanol in addition to the H atom originally present in the C-4 position in NAD^+. Therefore, it is represented as NADD.) When this reaction was carried out with undeuterated ethanol in D_2O, no deuterium turned up in NADH. These experiments showed that the oxidation of ethanol by NAD^+ proceeds with "direct" hydrogen transfer; that is, there is no hydrogen exchange with the solvent. When the product, NADD, was reacted with the enzyme and unlabeled aldehyde, CH_3CHO, all the deuterium was transferred back to aldehyde. Thus, we may conclude that the deuterium or hydrogen is transferred stereospecifically to one face of the NAD^+ (in the oxidation of ethanol) and transferred back from the same face (in the reduction of aldehyde). A nonspecific transfer to both faces would lead to a transfer of 50% of the deuterium back to acetaldehyde (or considerably less than 50% because of the kinetic isotope effect). The measured kinetic isotope effect (k_H/k_D) of about 5 shows that the proton transfer is indeed the rate-determining step.

How does the isotope effect change an equilibrium process? Although the forward and reverse directions of an equilibrium process must trace the same reaction pathway, the isotope effect on the two rate constants need not be the same. Consequently, there can be an isotope effect on the equilibrium constant. As a simple example, let us consider the dissociation of a monoprotic acid, such as acetic acid, in H_2O and D_2O. The dissociations are

$$CH_3COOH \rightleftharpoons CH_3COO^- + H^+ \quad K_H = \frac{[H^+][CH_3COO^-]}{[CH_3COOH]}$$

$$CH_3COOD \rightleftharpoons CH_3COO^- + D^+ \quad K_D = \frac{[D^+][CH_3COO^-]}{[CH_3COOD]}$$

(In D_2O, all the ionizable protons are replaced by deuterons.) Experimentally, we find that $K_H/K_D = 3.3$. The greater acid strength of CH_3COOH over CH_3COOD can be explained by noting that the undeuterated molecule has a higher zero-point vibrational level (for the O–H bond), and less energy is required to dissociate the hydrogen than to dissociate the deuterium in CH_3COOD. A useful general rule with regard to the isotope effect on equilibria is that substitution with a heavier isotope will favor the formation of a stronger bond. As for acetic acid, when we replace H with D, the O–D bond becomes stronger, and the resulting molecule has less of a tendency to dissociate.

9.9 Reactions in Solution

The major difference between gas-phase reactions and reactions in solution lies in the role of the solvent. In many cases, the solvent plays a minor role, and rates do not differ much in the two phases. In terms of simple kinetic theory, the frequency of collisions between reacting molecules depends only on the concentrations of the reactants; it is not affected by solvent molecules. There is a difference, however, in the outcome of an encounter between reactant molecules in solution compared with the collision of molecules in the gas phase. If two molecules collide in the gas phase and do not react, they will normally move away from each other. There is very little likelihood that this same pair will collide again. In contrast, when two solute molecules diffuse together in a solution, they cannot move apart again quickly after the initial encounter because they are surrounded closely by solvent molecules. In this case, the reactants are temporarily trapped in a "cage" of solvent (Figure 9.17). To be sure, the cage is not rigid, as the solvent molecules are constantly in motion and changing positions. Nevertheless, the cage effect causes the reactant molecules to remain together for a longer time than they would in the gas phase, and they may collide with each other hundreds of times before they drift apart.* For reactions that have relatively low activation energies, the cage effect virtually ensures reaction during each encounter; the steric factor no longer plays an important role, because the reacting molecules would sooner or later become properly oriented for reaction during the collisions. Under these conditions, the rate of the reaction is limited only by how fast the reactants can diffuse together. We shall return to this type of reaction in the next section.

The situation is quite different if the reactants are charged species. The solvation of ions can be an appreciable factor in determining the sign and magnitude of $\Delta S^{\ddagger}$. In cases involving charged species, the value of $\Delta S^{\ddagger}$ depends on the relative net charge of the activated complex. If the activated complex has a greater charge than the reactants, we would expect $\Delta S^{\ddagger}$ to be negative because of the increase in solvation around the complex:

$$(C_2H_5)_3N + C_2H_5I \rightarrow (C_2H_5)_4N^+I^- \qquad \Delta S^{\ddagger} = -172 \text{ J K}^{-1} \text{ mol}^{-1}$$

On the other hand, if an activated complex carries a lower net charge than the reactants, we would predict a positive $\Delta S^{\ddagger}$ value:

$$Co(NH_3)_5Br^{2+} + OH^- \rightarrow Co(NH_3)_4Br(OH)^+ + NH_3 \quad \Delta S^{\ddagger} = 83.7 \text{ J K}^{-1} \text{ mol}^{-1}$$

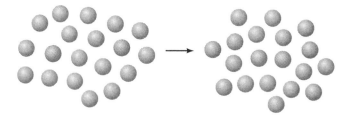

Figure 9.17
Solute molecules (red spheres) diffuse into a solvent "cage" and encounter each other. Hundreds of collisions between solute molecules occur before the cage is destroyed.

---

*We speak of molecular collisions in the gas phase and molecular encounters in solution. The difference is that after each encounter in solution, the molecules might collide many times before they move away from each other.

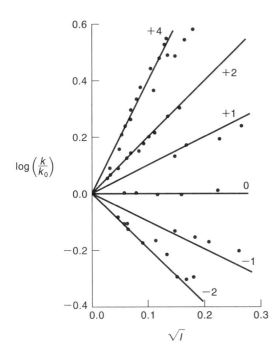

Figure 9.18
Effect of ionic strength on the rate of reaction between two ions. The reactions are:

$$+4: \quad Co(NH_3)_5Br^{2+} + Hg^{2+};$$
$$+2: \quad S_2O_8^{2-} + I^-;$$
$$+1: \quad [NO_2NCO_2C_2H_5]^- + OH^-;$$
$$0: \quad CH_3CO_2C_2H_5 + OH^-;$$
$$-1: \quad H_2O_2 + H^+ + Br^-;$$
$$-2: \quad Co(NH_3)_5Br^{2+} + OH^-.$$

The slopes are determined by $z_A z_B$. [From V. K. LaMer, *Chem. Rev.* **10**, 179 (1932). Used with permission of Williams & Wilkins, Baltimore.]

We should keep in mind that steric and other factors also contribute to the large entropy changes.

As we would expect, the rate of a reaction involving ions depends strongly on the ionic strength of the solution. This dependence is known as the *kinetic salt effect*. The effect of ionic strength on the rate constant is given by*

$$\log \frac{k}{k_0} = z_A z_B B \sqrt{I} \tag{9.49}$$

where B is a constant that depends only on the temperature and the nature of the solvent; k and k_0 are the rate constants at ionic strength I (of an inert salt) and at infinitely dilute concentration ($I = 0$), respectively; and z_A and z_B are the ionic charges of the reactants A and B. Equation 9.49 predicts that (1) if A and B have the same charges, $z_A z_B$ is positive and the rate constant k increases with $\sqrt{I}$; (2) if A and B have opposite signs, $z_A z_B$ is negative and k decreases with $\sqrt{I}$; and (3) if either A or B is uncharged, $z_A z_B = 0$ and k is independent of the ionic strength of the solution. Figure 9.18 confirms these predictions.

9.10 Fast Reactions in Solution

Roughly speaking, first- and second-order reactions that have rate constants between 10 and 10^9 can be described as fast reactions. Examples of fast reactions include recombination of reactive species in the gas phase and in solution, acid–base neutralization, and electron- and proton-exchange reactions. These reactions have engendered considerable interest because of their importance to chemistry and biology and because of a widespread desire to design experiments to measure processes whose half-lives are seconds or shorter.

---

*Equation 9.49 is derived from the Debye–Hückel limiting law and therefore applies only to dilute solutions.

How fast can a reaction occur in solution? The limit is set by the rate of approach by reacting molecules, which in turn is governed by the rate of diffusion. Thus, the fastest reaction is a *diffusion-controlled reaction* in which a reaction occurs with every encounter of reactant molecules. Suppose that we have a solution of two uncharged reactant molecules, B and C, with radii r_B and r_C. The Polish-American physicist Roman Smoluckowski (1910–) showed that the rate constant k_D of the elementary diffusion-controlled reaction $B + C \rightarrow$ product is given by

$$k_D = 4\pi N_A (r_B + r_C)(D_B + D_C)$$

where N_A is the Avogardro constant and D_B and D_C are the diffusion coefficients. If we assume that $D_B = D_C = D$, $r_B = r_C = r$, and use the expression $D = k_B T/6\pi\eta r$, where η is the viscosity of the solution, then

$$k_D = 4\pi N_A (2D)(2r)$$

$$k_D = \frac{16\pi N_A k_B T r}{6\pi\eta r} = \frac{8}{3}\frac{RT}{\eta} \tag{9.50}$$

A truly diffusion-controlled reaction has two characteristics. First, such a reaction has a zero activation energy [note the absence of the $\exp(-E_a/RT)$ term in Equation 9.50]. Second, the rate is inversely proportional to the viscosity of the medium. The dependence on viscosity is interesting in that η itself depends on temperature as follows:

$$\eta = Be^{E_a/RT}$$

where E_a is the "activation energy" of viscosity (note that η decreases with increasing temperature) and B is a constant characteristic of the solvent. Thus, Equation 9.50 can now be written as

$$k_D = \frac{8RT}{3B}e^{-E_a/RT} \tag{9.51}$$

Equation 9.51 has the form of an Arrhenius equation.

Example 9.3

Estimate the rate constant for a diffusion-controlled reaction in water at 298 K, given that the viscosity of water is 8.9×10^{-4} N s m^{-2}.

ANSWER

Because 1 J = 1 N m, the units of viscosity can also be expressed as J s m^{-3}. From Equation 9.50,

$$k_D = \frac{8(8.314 \text{ J K}^{-1} \text{ mol}^{-1})(298 \text{ K})}{3(8.9 \times 10^{-4} \text{ J s m}^{-3})}$$

$$= 7.4 \times 10^6 \text{ m}^3 \text{ mol}^{-1} \text{ s}^{-1}$$

$$= 7.4 \times 10^9 \text{ } M^{-1} \text{ s}^{-1}$$

COMMENT

According to Table 9.2, the half-life for a diffusion-controlled process, assuming that the starting reactants are identical and their concentrations equal 1 M, is

$$t_{1/2} = \frac{1}{1 \ M \times 7.4 \times 10^9 \ M^{-1} \ s^{-1}} = 1.4 \times 10^{-10} \ s$$

which is a very small number.

Many ingenious methods have been devised to study fast reactions. Two examples will be briefly discussed.

The Flow Method

There are two types of flow apparatus. In the continuous-flow apparatus, two reactant solutions are first brought together in a mixing chamber, and the mixed solution is then passed along an observation tube. By monitoring the concentration of either the reactant or product at different points along the tube spectrophotometrically (that is, by measuring the absorption of light by either the reactant or the product), we can plot the extent of reaction versus time (Figure 9.19). The limiting

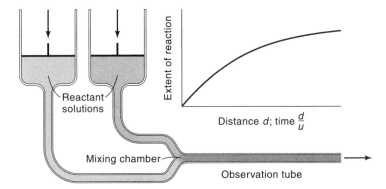

Figure 9.19
Schematic diagram for a continuous-flow experiment. The velocity of the mixed solution is u. Measurements are carried out along the length of the tube.

factor here is the time required for mixing, which can be as short as 0.001 s. This technique uses large quantities of solutions for every run, a major disadvantage.

Figure 9.20 shows a stopped-flow apparatus. The advantage of the stopped-flow technique is that only small samples of reactants are needed; therefore, it is particularly suited for biochemical processes such as enzyme-catalyzed reactions.

The Relaxation Method

A system initially at equilibrium is subjected to an external perturbation, such as a temperature or pressure change. If the change is applied suddenly, there will be a time lag while the system approaches (or relaxes toward) a new equilibrium. This time lag, called the *relaxation time*, can be related to the forward and reverse rate

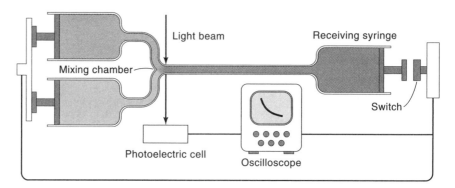

Figure 9.20
Schematic diagram for a stopped-flow experiment. As in the continuous-flow apparatus, two solutions containing different reactants are injected into the mixing region, usually with mechanically coupled hypodermic syringes. Another syringe on the right receives the effluent solution and is arranged so that when its plunger strikes a barrier, the flow is abruptly stopped. At the same instant, an oscilloscope is triggered to establish the time origin. The time scale is provided by the sweep frequency of the oscilloscope, which displays a plot of the transmitted light intensity against time. In this arrangement, the distance between the mixing region and the observation point (that is, the point at which absorption of light is monitored) is small.

constants. Depending on the systems, reactions with half-lives between 1 s and 10^{-10} s can be studied with the relaxation technique.

Any property (X) that varies linearly with the extent of the reaction (for example, electrical conductance or spectroscopic absorption) is measured as a function of time following the disruption:

$$X_t = X_0 e^{-t/\tau}$$

where X_t and X_0 are the values of the property at time $t = t$ and $t = 0$, respectively, and τ is the relaxation time. When $\tau = t$,

$$X_t = \frac{X_0}{e} = \frac{X_0}{2.718}$$

Thus, by measuring the time it takes for X_0 to decrease to $X_0/2.718$, we can determine the relaxation time (Figure 9.21).

We can derive a relationship between τ and rate constant using an equilibrium chemical system in solution at some temperature:

$$A + B \underset{k_r'}{\overset{k_f'}{\rightleftharpoons}} C$$

At equilibrium, we have

$$\frac{d[C]}{dt} = k_f'[A][B] - k_r'[C] = 0$$

where k_f' and k_r' are the rate constants for the forward and reverse steps. In a temperature-jump experiment, the temperature of the solution can be increased by 5 K or

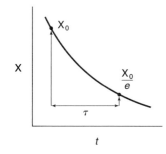

Figure 9.21
Definition of relaxation time τ.

so in a time as short as 10^{-6} s, either by discharging a capacitor through the solution or by applying a short, powerful laser pulse. After the temperature jump has occurred, the concentrations of A, B, and C will change as the system "relaxes" toward the new equilibrium at the elevated temperature, with altered rate constants k_f and k_r. Let x be the time-dependent reaction progress variable expressed as a concentration. According to the reaction stoichiometry, at any time t following the temperature jump we have

$$[A] = [A]_{eq} + x$$
$$[B] = [B]_{eq} + x$$
$$[C] = [C]_{eq} - x$$

where the subscript eq denotes the new equilibrium concentration. Note that x could be positive or negative, depending on the direction in which the equilibrium is shifted. The rate of change of [C] is now given by

$$\frac{d[C]}{dt} = \frac{d\{[C]_{eq} - x\}}{dt} = -\frac{dx}{dt} = k_f([A]_{eq} + x)([B]_{eq} + x) - k_r([C]_{eq} - x)$$

$$= \underbrace{k_f[A]_{eq}[B]_{eq} - k_r[C]_{eq}}_{\text{first term}} + \underbrace{\{k_f[A]_{eq} + k_f[B]_{eq} + k_r\}x}_{\text{second term}} + \underbrace{k_f x^2}_{\text{third term}}$$

The first term is equal to zero because at equilibrium the forward and reverse rates are equal; the third term can be neglected because x is a small number (due to a small rise in temperature) so that $k_f x^2 \ll k_f[A]_{eq}x$ (or $k_f[B]_{eq}x$). Therefore

$$\frac{dx}{dt} = -\{k_f[A]_{eq} + k_f[B]_{eq} + k_r\}x = -\frac{x}{\tau}$$

where τ is given by

$$\tau = \frac{1}{k_f([A]_{eq} + [B]_{eq}) + k_r} \qquad (9.52)$$

Because $[A]_{eq}$ and $[B]_{eq}$ can be determined in separate experiments, τ can be measured by monitoring the change in concentrations during the time subsequent to the temperature jump (see Figure 9.21). Coupled with the equilibrium constant $(K = k_f/k_r)$ of the reaction at the elevated temperature, these values enable us to determine the values of both k_f and k_r (two equations for two unknowns).

Example 9.4

A sample of pure water was subjected to a temperature jump. The relaxation time for the system (water) to reach the new equilibrium at 25°C was found to be 36 μs (36×10^{-6} s). Calculate the values of k_f and k_r for the following reactions:

$$\mathrm{H^+ + OH^- \underset{k_r}{\overset{k_f}{\rightleftharpoons}} H_2O}$$

ANSWER

From Equation 9.52,

$$\tau = \frac{1}{k_f([H^+]_{eq} + [OH^-]_{eq}) + k_r}$$

The equilibrium condition is

$$k_f[H^+]_{eq}[OH^-]_{eq} = k_r[H_2O]_{eq}$$

Substituting $k_r = k_f[H^+]_{eq}[OH^-]_{eq}/[H_2O]_{eq}$ and using $[H^+]_{eq} = [OH^-]_{eq}$, we get

$$\tau = \frac{1}{k_f(2[H^+]_{eq}) + k_f[H^+]_{eq}^2/[H_2O]_{eq}}$$

$$= \frac{1}{k_f(2[H^+]_{eq} + [H^+]_{eq}^2/[H_2O]_{eq})}$$

Using $[H^+]_{eq} = 1.0 \times 10^{-7}$ M and $[H_2O]_{eq} = 55.5$ M, we write

$$36 \times 10^{-6} \text{ s} = \frac{1}{k_f[2(1.0 \times 10^{-7} \text{ } M) + (1.0 \times 10^{-7} \text{ } M)^2/55.5 \text{ } M]}$$

or

$$k_f = 1.4 \times 10^{11} \text{ } M^{-1} \text{ s}^{-1}$$

To calculate the value of k_r, we need first evaluate the equilibrium constant K, given by

$$K = \frac{k_r}{k_f} = \frac{[H^+]_{eq}[OH^-]_{eq}}{[H_2O]_{eq}} = \frac{(1.0 \times 10^{-7} \text{ } M)(1.0 \times 10^{-7} \text{ } M)}{(55.5 \text{ } M)} = 1.8 \times 10^{-16} \text{ } M$$

Therefore,

$$k_r = Kk_f = (1.8 \times 10^{-16} \text{ } M)(1.4 \times 10^{11} \text{ } M^{-1} \text{ s}^{-1})$$

$$k_r = 2.5 \times 10^{-5} \text{ s}^{-1}$$

COMMENTS

(1) Note that the equilibrium constant, K, is related to the ion-product (K_w) by $K = K_w/[H_2O]_{eq}$. Watch out for the units: k_f is a second-order rate constant (M^{-1} s^{-1}), whereas k_r is a first-order rate constant (s^{-1}). The very large value of k_f shows that the combination of H$^+$ and OH$^-$ ions in solution is diffusion controlled. (2) K normally is treated as a dimensionless quantity, but for calculating rate constants, it is assigned the unit M in this case. See K. J. Laidler, *J. Chem. Educ.* **67**, 88 (1990).

Reactions are seldom as simple as the one described above. Often, a reaction may have several relaxation times and the analysis can be very complex. Nevertheless, relaxation methods are among the most useful and versatile techniques in the study of fast chemical and biochemical processes.

9.11 Oscillating Solutions

Chemical reactions normally proceed until the reactants have been exhausted or until an equilibrium state is reached. For certain complicated reactions, however, the concentrations of intermediates can oscillate. Although such *oscillating reactions* have been known since the late 19th century, for a long time they were dismissed by most chemists as being nonreproducible phenomena or artifacts due to impurities. According to the second law of thermodynamics, at constant temperature and pressure in a closed system, the Gibbs energy, G, of a reacting mixture must continually decrease as the reaction approaches equilibrium. An oscillating reaction would seem to violate the second law.

The oscillating reaction that finally convinced the skeptics was discovered by the Russian chemist B. P. Belousov in 1958 and later studied in detail by the Russian chemist A. M. Zhabotinsky. The Belousov–Zhabotinsky reaction, or the BZ reaction as it is now commonly called, occurs when malonic acid $[CH_2(COOH)_2]$ and sulfuric acid are dissolved in water with potassium bromate $(KBrO_3)$ and a cerium salt (containing the ceric ion, Ce^{4+}). The overall reaction is represented by

$$2H^+ + 2BrO_3^- + 3CH_2(COOH)_2 \rightarrow 2BrCH(COOH)_2 + 3CO_2 + 4H_2O$$

The mechanism of this reaction has been investigated extensively over the past 30 years and is believed to involve 18 elementary steps and 20 different chemical species! During the reaction, the color of the solution changes periodically from pale yellow (Ce^{4+}) to colorless (Ce^{3+}). Figure 9.22 shows the oscillations in the log of $[Br^-]$ and $[Ce^{4+}]/[Ce^{3+}]$.

The thermodynamic explanation of oscillating reactions comes from the Belgian chemist Ilya Prigogine (1917–2004). Reactions in a closed system cannot oscillate about their equilibrium state because such behavior is prohibited by the principle of microscopic reversibility (see p. 329). According to Prigogine, however, if a system is far from equilibrium, periodic oscillations in the concentrations of intermediate species in a chemical reaction can occur. Eventually, these oscillations die off as the system nears its equilibrium state. The initial reactants and final products cannot

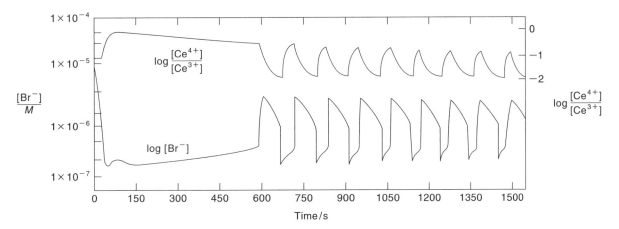

Figure 9.22
Oscillation in the log of $[Br^-]$ and $[Ce^{4+}]/[Ce^{3+}]$ in the BZ reaction. Note that Br^-, Ce^{4+}, and Ce^{3+} are not reactants or products in the overall reaction. [Reprinted with permission from R. J. Field, E. Körös, and R. M. Noyes, *J. Am. Chem. Soc.* **94**, 8649 (1972). Copyright 1972 American Chemical Society.]

participate in oscillations because they are not intermediates. However, in open systems, where exchanges of both energy and mass with the surroundings are allowed, a steady state exists rather than an equilibrium state, so oscillations may occur and last indefinitely.

The study of oscillating reactions has added a new dimension to chemical kinetics and is one of the most rapidly growing branches of chemistry. It has yielded useful insight into chemical dynamics and reaction mechanisms. Such reactions may have considerable importance in biological systems. The periodic beating of the heart is one example. Oscillatory behavior has also been detected in glycolysis.* The atmosphere is another open system that shows periodic oscillations in the concentrations of various gas constituents.

---

* A. Gosh and B. Chance, *Biochem. Biophys. Res. Commun.* **16**, 174 (1964).

Suggestions for Further Reading

BOOKS

Brouard, M., *Reaction Dynamics*, Oxford University Press, New York, 1998.

Espenson, J. H., *Chemical Kinetics and Reaction Mechanism*, 2nd edition, WCB/McGraw–Hill, New York, 1995.

Hague, D. N., *Fast Reactions*, Wiley–Interscience, New York, 1971.

Hammes, G. G., *Principles of Chemical Kinetics*, Academic Press, New York, 1978.

Houston, P. L., *Chemical Kinetics and Reaction Dynamics*, McGraw–Hill, Dubuque, IA, 2001.

Laidler, K. J., *Chemical Kinetics*, 3rd ed., Harper & Row, New York, 1987.

Moore, J. W. and R. G. Pearson, *Kinetics and Mechanisms*, John Wiley & Sons, New York, 1981.

Nicholas, J., *Chemical Kinetics*, John Wiley & Sons, New York, 1976.

Pilling, M. J. and P. W. Seakins, *Reaction Kinetics*, Oxford University Press, New York, 1995.

Scott, S. K., *Oscillations, Waves, and Chaos in Chemical Kinetics*, Oxford University Press, New York, 1994.

Steinfeld, J. I., J. S. Francisco, and W. L. Hase, *Chemical Kinetics and Dynamics*, 2nd ed., Prentice Hall, Englewood Cliffs, NJ, 1999.

ARTICLES
General

"Method for Determining Order of a Reaction," H. K. Zimmerman, *J. Chem. Educ.* **40**, 356 (1963).

"Concept of Time in Chemistry," O. T. Benfey, *J. Chem. Educ.* **40**, 574 (1963).

"Unimolecular Gas Reactions at Low Pressures," B. Perlmutter-Hayman, *J. Chem. Educ.* **44**, 605 (1967).

"Tables of Conversion Factors for Reaction Rate Constants," D. D. Drysdale and A. C. Lloyd, *J. Chem. Educ.* **46**, 54 (1969).

"Mechanistic Ambiguities of Rate Laws," J. P. Birk, *J. Chem. Educ.* **47**, 805 (1970).

"Along the Reaction Coordinate," W. F. Sheehan, *J. Chem. Educ.* **47**, 853 (1970).

"Unconventional Applications of Arrhenius's Law," K. J. Laidler, *J. Chem. Educ.* **49**, 343 (1972).

"Drinking Too Fast Can Cause Sudden Death," C. D. Eskelson, *J. Chem. Educ.* **50**, 365 (1973).

"The Influence of Solvents on Chemical Reactivity," M. R. J. Dack, *J. Chem. Educ.* **51**, 231 (1974).

"Steady State and Equilibrium Approximations in Reaction Kinetics," L. Volk, W. Richardson, K. H. Lau, M. Hall, and S. H. Lin, *J. Chem. Educ.* **54**, 95 (1977).

"Some Common Oversimplifications in Teaching Chemical Kinetics," R. K. Boyd, *J. Chem. Educ.* **55**, 84 (1978).

"Chemical Reactions Without Solvation," R. T. McIver, Jr., *Sci. Am.* November 1980.

"What Is the Rate-Determining Step of a Multistep Reaction?" J. R. Murdoch, *J. Chem. Educ.* **58**, 32 (1981).

"The Origin and Status of the Arrhenius Equation," S. R. Logan, *J. Chem. Educ.* **59**, 279 (1982).

"The Steady State and Equilibrium Assumptions in Chemical Kinetics," D. C. Tardy and D. C. Cater, *J. Chem. Educ.* **60**, 109 (1983).

"The Extent of Reaction and Chemical Kinetics," H. Maskill, *Educ. Chem.* **21**, 122 (1984).

"Experimental Approaches to Studying Biological Electron Transfer," R. A. Scott, A. G. Mauk, and H. B. Gray, *J. Chem. Educ.* **62**, 932 (1985).

"Quantum Chemical Reactions in the Deep Cold," V. I. Goldanskii, *Sci. Am.* February 1986.

"The Meaning and Significance of 'the Activation Energy' of a Chemical Reaction," R. Logan, *Educ. Chem.* **23**, 148 (1986).

"The Transition State," (no author's name given) *J. Chem. Educ.* **64**, 41 (1987).

"Rate-Controlling Step: A Necessary and Useful Concept?" K. J. Laidler, *J. Chem. Educ.* **65**, 250 (1988).

"Just What Is a Transition State?" K. J. Laidler, *J. Chem. Educ.* **65**, 540 (1988).

"An Intuitive Approach to Steady-State Kinetics," R. T. Raines and D. E. Hansen, *J. Chem. Educ.* **65**, 757 (1988).

"Electron Transfer in Biology," R. J. P. Williams, *Molec. Phys.* **68**, 1 (1989).

"The Birth of Molecules," A. H. Zewail, *Sci. Am.* December 1990.

"The Arrhenius Equation," H. Maskill, *Educ. Chem.* **27**, 111 (1990).

"Some Provocative Opinions on the Terminology of Chemical Kinetics," C. Reeve, *J. Chem. Educ.* **68**, 728 (1991).

"Chemical Kinetics," R. W. Carr in *Encyclopedia of Applied Physics*, G. L. Trigg, Ed., VCH Publishers, New York, 1992, Vol. 3, p. 345.

"Applying the Principles of Chemical Kinetics to Population Growth Problems," G. F. Swiegers, *J. Chem. Educ.* **70**, 364 (1993).

"A Simplified Integration Technique for Reaction Rate Laws of Integral Order in Several Substances," G. Eberhardt and E. Levin, *J. Chem. Educ.* **72**, 193 (1995).

"Reaction Dynamics in Organic Chemistry," B. Carpenter, *Am. Sci.* **85**, 138 (1997).

"Anatomy of Elementary Chemical Reactions," A. J. Alexander and R. N. Zare, *J. Chem. Educ.* **75**, 1105 (1998).

"Relevance of Chemical Kinetics for Medicine: The Case of Nitric Oxide," A. T. Balabau and W. Seitz, *J. Chem. Educ.* **80**, 662 (2003).

Kinetic Isotope Effect

"The Exposition of Isotope Effects on Rates and Equilibria," M. M. Kreevoy, *J. Chem. Educ.* **41**, 636 (1964).

"Application of Isotope Effects," V. Gold, *Chem. Brit.* **6**, 292 (1970).

"The World of Isotope Effects," A. M. Rouhi, *Chem. Eng. News* December 22, 1997.

"Primary Kinetic Isotope Effect—A Lecture Demonstration," R. Chang, *Chem. Educator*, 1997, 2(3): S1430–4171 (97) 03121-X. Avail. URL: http://journals.springer-ny.com/chedr.

Catalysis

"Catalysis: New Reaction Pathways, Not Just a Lowering of Activation Energy," A. Haim, *J. Chem. Educ.* **66**, 935 (1989).

"Catalysis," G. L. Haller in *Encyclopedia of Applied Physics*, G. L. Trigg, Ed., VCH Publishers, New York, 1991, Vol. 3, p. 67.

"Catalysis on Surfaces," C. M. Friend, *Sci. Am.* April 1993.

Relaxation Kinetics

"Relaxation Kinetics," J. H. Swinehart, *J. Chem. Educ.* **44**, 524 (1967).

"The Temperature-Jump Method for the Study of Fast Reactions," J. E. Finholt, *J. Chem. Educ.* **45**, 394 (1968).

"Relaxation Methods in Chemistry," L. Faller, *Sci. Am.* May 1969.

"Temperature-Jump Techniques," E. Caldin, *Chem. Brit.* **11**, 4 (1975).

Oscillating Reactions

"Rotating Chemical Reactions," A. T. Winfree, *Sci. Am.* June 1974.

"Oscillating Chemical Reactions," I. R. Epstein, K. Kustin, P. De Kepper, and M. Orbán, *Sci. Am.* March 1983.

"Self-Organizing Structures," B. F. Madone and W. L. Freedman, *Am. Sci.* **75**, 252 (1987).

"Some Models of Chemical Oscillators," R. M. Noyes, *J. Chem. Educ.* **66**, 190 (1989).

"Oscillating Chemical Reactions and Nonlinear Dynamics," R. J. Field and F. W. Schneider, *J. Chem. Educ.* **66**, 195 (1989).

"Recipes for Belousov-Zhabotinsky Reagents," W. Jahnke and A. T. Winfree, *J. Chem. Educ.* **68**, 320 (1991).

"The Kinetics of Oscillating Reactions," R. F. Melka, G. Olsen, L. Beavers, and J. A. Draeger, *J. Chem. Educ.* **69**, 596 (1992).

"An Oscillating Reaction as a Demonstration of Principles Applied in Chemistry and Chemical Engineering," J. J. Weimer, *J. Chem. Educ.* **71**, 325 (1994).

"The BZ Reaction: Experimental and Model Studies in the Physical Chemistry Laboratory," O. Benini, R. Cervellat, and P. Fetto, *J. Chem. Educ.* **73**, 865 (1996).

Problems

Reaction Order, Rate Law

9.1 Write the rates for the following reactions in terms of the disappearance of reactants and appearance of products:

(a) $3O_2 \rightarrow 2O_3$

(b) $C_2H_6 \rightarrow C_2H_4 + H_2$

(c) $ClO^- + Br^- \rightarrow BrO^- + Cl^-$

(d) $(CH_3)_3CCl + H_2O \rightarrow (CH_3)_3COH + H^+ + Cl^-$

(e) $2AsH_3 \rightarrow 2As + 3H_2$

9.2 The rate law for the reaction

$$NH_4^+(aq) + NO_2^-(aq) \rightarrow N_2(g) + 2H_2O(l)$$

is given by rate $= k[NH_4^+][NO_2^-]$. At 25°C, the rate constant is 3.0×10^{-4} M^{-1} s^{-1}. Calculate the rate of the reaction at this temperature if $[NH_4^+] = 0.26$ M and $[NO_2^-] = 0.080$ M.

9.3 What are the units of the rate constant for a third-order reaction?

9.4 The following reaction is found to be first order in A:

$$A \rightarrow B + C$$

If half of the starting quantity of A is used up after 56 s, calculate the fraction that will be used up after 6.0 min.

9.5 A certain first-order reaction is 34.5% complete in 49 min at 298 K. What is its rate constant?

9.6 (a) The half-life of the first-order decay of radioactive ^{14}C is about 5720 years. Calculate the rate constant for the reaction. (b) The natural abundance of ^{14}C isotope is 1.1×10^{-13} mol % in living matter. Radiochemical analysis of an object obtained in an archeological excavation shows that the ^{14}C isotope content is 0.89×10^{-14} mol %. Calculate the age of the object. State any assumptions.

9.7 The first-order rate constant for the gas-phase decomposition of dimethyl ether,

$$(CH_3)_2O \rightarrow CH_4 + H_2 + CO$$

is 3.2×10^{-4} s^{-1} at 450°C. The reaction is carried out in a constant-volume container. Initially, only dimethyl ether is present, and the pressure is 0.350 atm. What is the pressure of the system after 8.0 min? Assume ideal-gas behavior.

9.8 When the concentration of A in the reaction $A \rightarrow B$ was changed from 1.20 M to 0.60 M, the half-life increased from 2.0 min to 4.0 min at 25°C. Calculate the order of the reaction and the rate constant.

9.9 The progress of a reaction in the aqueous phase was monitored by the absorbance of a reactant at various times:

Time/s	0	54	171	390	720	1010	1190
Absorbance	1.67	1.51	1.24	0.847	0.478	0.301	0.216

Determine the order of the reaction and the rate constant.

9.10 Cyclobutane decomposes to ethylene according to the equation

$$C_4H_8(g) \rightarrow 2C_2H_4(g)$$

Determine the order of the reaction and the rate constant based on the following pressures, which were recorded when the reaction was carried out at 430°C in a constant-volume vessel:

Time/s	$P_{C_4H_8}$/mmHg
0	400
2000	316
4000	248
6000	196
8000	155
10000	122

9.11 What is the half-life of a compound if 75% of a given sample of the compound decomposes in 60 min? Assume first-order kinetics.

9.12 The rate constant for the second-order reaction

$$2NO_2(g) \rightarrow 2NO(g) + O_2(g)$$

is 0.54 M^{-1} s^{-1} at 300°C. How long (in seconds) would it take for the concentration of NO_2 to decrease from 0.62 M to 0.28 M?

9.13 The decomposition of N_2O to N_2 and O_2 is a first-order reaction. At 730°C, the half-life of the reaction is 3.58×10^3 min. If the initial pressure of N_2O is 2.10 atm at 730°C, calculate the total gas pressure after one half-life. Assume that the volume remains constant.

9.14 The integrated rate law for the zero-order reaction $A \rightarrow B$ is $[A] = [A]_0 - kt$. (a) Sketch the following plots: (i) rate versus [A] and (ii) [A] versus t. (b) Derive an expression for the half-life of the reaction. (c) Calculate the time in half-lives when the integrated rate law is no longer valid (that is, when [A] = 0).

9.15 In the nuclear industry, workers use a rule of thumb that the radioactivity from any sample will be relatively harmless after 10 half-lives. Calculate the fraction of a radioactive sample that remains after this time period. (*Hint:* Radioactive decays obey first-order kinetics.)

9.16 Many reactions involving heterogeneous catalysis are zero order; that is, rate $= k$. An example is the decomposition of phosphine (PH_3) over tungsten (W):

$$4PH_3(g) \rightarrow P_4(g) + 6H_2(g)$$

The rate for this reaction is independent of $[PH_3]$ as long as phosphine's pressure is sufficiently high (≥ 1 atm). Explain.

9.17 If the first half-life of a zero-order reaction is 200 s, what will be the duration of the next half-life?

9.18 Consider the following nuclear decay

$$^{64}Cu \rightarrow {}^{64}Zn + {}_{-1}^{0}\beta \quad t_{1/2} = 12.8 \text{ h}$$

Starting with one mole of ^{64}Cu, calculate the number of grams of ^{64}Zn formed after 25.6 hours.

Reaction Mechanisms

9.19 The reaction

$$S_2O_8^{2-} + 2I^- \rightarrow 2SO_4^{2-} + I_2$$

proceeds slowly in aqueous solution, but it can be catalyzed by the Fe^{3+} ion. Given that Fe^{3+} can oxidize I^- and Fe^{2+} can reduce $S_2O_8^{2-}$, write a plausible two-step mechanism for this reaction. Explain why the uncatalyzed reaction is slow.

9.20 Derive Equation 9.27 using the steady-state approximation for both the H and Br atoms.

9.21 An excited ozone molecule, O_3^*, in the atmosphere can undergo one of the following reactions:

$$O_3^* \xrightarrow{k_1} O_3 \qquad (1) \text{ fluorescence}$$
$$O_3^* \xrightarrow{k_2} O + O_2 \qquad (2) \text{ decomposition}$$
$$O_3^* + M \xrightarrow{k_3} O_3 + M \quad (3) \text{ deactivation}$$

where M is an inert molecule. Calculate the fraction of ozone molecules undergoing decomposition in terms of the rate constants.

9.22 The following data were collected for the reaction between hydrogen and nitric oxide at 700°C:

$$2H_2(g) + 2NO(g) \rightarrow 2H_2O(g) + N_2(g)$$

Experiment	$[H_2]/M$	$[NO]/M$	Initial rate/$M \cdot s^{-1}$
1	0.010	0.025	2.4×10^{-6}
2	0.0050	0.025	1.2×10^{-6}
3	0.010	0.0125	0.60×10^{-6}

(a) What is the rate law for the reaction? **(b)** Calculate the rate constant for the reaction. **(c)** Suggest a plausible reaction mechanism that is consistent with the rate

law. (*Hint:* Assume that the oxygen atom is the intermediate.) **(d)** More careful studies of the reaction show that the rate law over a wide range of concentrations of reactants should be

$$\text{rate} = \frac{k_1[NO]^2[H_2]}{1 + k_2[H_2]}$$

What happens to the rate law at very high and very low hydrogen concentrations?

9.23 The rate law for the decomposition of ozone to molecular oxygen ($2O_3 \rightarrow 3O_2$) is

$$\text{rate} = k\frac{[O_3]^2}{[O_2]}$$

The mechanism proposed for this process is

$$O_3 \underset{k_{-1}}{\overset{k_1}{\rightleftharpoons}} O + O_2$$

$$O + O_3 \xrightarrow{k_2} 2O_2$$

Derive the rate law from these elementary steps. Clearly state the assumptions you use in the derivation. Explain why the rate decreases with increasing O_2 concentration.

9.24 The gas-phase reaction between H_2 and I_2 to form HI involves a two-step mechanism:

$$I_2 \rightleftharpoons 2I$$
$$H_2 + 2I \rightarrow 2HI$$

The rate of formation of HI increases with the intensity of visible light. **(a)** Explain why this fact supports the two-step mechanism given. (*Hint:* The color of I_2 vapor is purple.) **(b)** Explain why the visible light has no effect on the formation of H atoms.

9.25 In recent years, ozone in the stratosphere has been depleted at an alarmingly fast rate by chlorofluorocarbons (CFCs). A CFC molecule such as $CFCl_3$ is first decomposed by UV radiation:

$$CFCl_3 \rightarrow CFCl_2 + Cl$$

The chlorine radical then reacts with ozone as follows:

$$Cl + O_3 \rightarrow ClO + O_2$$
$$ClO + O \rightarrow Cl + O_2$$

(a) Write the overall reaction for the last two steps. **(b)** What are the roles of Cl and ClO? **(c)** Why is the fluorine radical not important in this mechanism? **(d)** One suggestion for reducing the concentration of chlorine radicals is to add hydrocarbons such as ethane (C_2H_6) to the stratosphere. How will this approach work? **(e)** Draw potential energy versus reaction progress diagrams for the uncatalyzed and catalyzed (by Cl) destruction of ozone: $O_3 + O \rightarrow 2O_2$. Use the thermodynamic data in Appendix 2 to determine whether the reaction is exothermic or endothermic.

Activation Energy

9.26 Use Equation 9.23 to calculate the rate constant at 300 K for $E_a = 0, 2$, and 50 kJ mol^{-1}. Assume that $A = 10^{11}$ s^{-1} in each case.

9.27 Many reactions double their rates with every $10°$ rise in temperature. Assume that such a reaction takes place at 305 K and 315 K. What must its activation energy be for this statement to hold?

9.28 Over a range of about $\pm 3°$C from normal body temperature the metabolic rate, M_T, is given by $M_T = M_{37}(1.1)^{\Delta T}$, where M_{37} is the normal rate and ΔT is the change in T. Discuss this equation in terms of a possible molecular interpretation. [*Source:* "Eco-Chem," J. A. Campbell, *J. Chem. Educ.* **52**, 327 (1975).]

9.29 The rate of bacterial hydrolysis of fish muscle is twice as great at 2.2°C as at $-1.1°$C. Estimate an E_a value for this reaction. Is there any relation to the problem of storing fish for food? [*Source:* "Eco-Chem," J. A. Campbell, *J. Chem. Educ.* **52**, 390 (1975).]

9.30 The rate constants for the first-order decomposition of an organic compound in solution are measured at several temperatures:

k/s^{-1}	4.92×10^{-3}	0.0216	0.0950	0.326	1.15
t/°C	5.0	15	25	35	45

Determine graphically the pre-exponential factor and the energy of activation for the reaction.

9.31 The energy of activation for the reaction $2HI \rightarrow H_2 + I_2$ is 180 kJ mol^{-1} at 556 K. Calculate the rate constant using Equation 9.23. The collision diameter for HI is 3.5×10^{-8} cm. Assume that the pressure is 1 atm.

9.32 The rate constant of a first-order reaction is 4.60×10^{-4} s^{-1} at 350°C. If the activation energy is 104 kJ mol^{-1}, calculate the temperature at which its rate constant is 8.80×10^{-4} s^{-1}.

9.33 The rate at which tree crickets chirp is 2.0×10^2 per minute at 27°C but only 39.6 per minute at 5°C. From these data, calculate the "activation energy" for the chirping process. (*Hint:* The ratio of rates is equal to the ratio of rate constants.) Find the chirping rate at 15°C.

9.34 Consider the following parallel reactions

$$A \quad \substack{\xrightarrow{k_1} B \\ \xrightarrow{k_2} C}$$

The activation energies are 45.3 kJ mol^{-1} for k_1 and 69.8 kJ mol^{-1} for k_2. If the rate constants are equal at 320 K, at what temperature will $k_1/k_2 = 2.00$?

Thermodynamic Formulation of Transition-State Theory

9.35 The thermal isomerization of cyclopropane to propene in the gas phase has a rate constant of 5.95×10^{-4} s^{-1} at 500°C. Calculate the value of $\Delta G^{\circ\ddagger}$ for the reaction.

9.36 The rate of the electron-exchange reaction between naphthalene ($C_{10}H_8$) and its anion radical ($C_{10}H_8^-$) is diffusion controlled:

$$C_{10}H_8^- + C_{10}H_8 \rightleftharpoons C_{10}H_8 + C_{10}H_8^-$$

The reaction is bimolecular and second order. The rate constants are

T/K	307	299	289	273
$k/10^9$ $M^{-1} \cdot$ s^{-1}	2.71	2.40	1.96	1.43

Calculate the values of $E_a, \Delta H^{\circ\ddagger}, \Delta S^{\circ\ddagger}$ and $\Delta G^{\circ\ddagger}$ at 307 K for the reaction. [*Hint:* Rearrange Equation 9.41 and plot $\ln(k/T)$ versus $1/T$.]

9.37 (a) The pre-exponential factor and activation energy for the hydrolysis of *t*-butyl chloride are 2.1×10^{16} s^{-1} and 102 kJ mol^{-1}, respectively. Calculate the values of $\Delta S^{\circ\ddagger}$ and $\Delta H^{\circ\ddagger}$ at 286 K for the reaction. **(b)** The pre-exponential factor and activation energy for the gas-phase cycloaddition of maleic anhydride and cyclopentadiene are 5.9×10^7 M^{-1} s^{-1} and 51 kJ mol^{-1}, respectively. Calculate the values of $\Delta S^{\circ\ddagger}$ and $\Delta H^{\circ\ddagger}$ at 293 K for the reaction.

Kinetic Isotope Effect

9.38 A person may die after drinking D_2O instead of H_2O for a prolonged period (on the order of days). Explain. Because D_2O has practically the same properties as H_2O, how would you test the presence of large quantities of the former in a victim's body?

9.39 The rate-determining step of the bromination of acetone involves breaking a carbon–hydrogen bond. Estimate the ratio of the rate constants k_{C-H}/k_{C-D} for the reaction at 300 K. The frequencies of vibration for the particular bonds are $\tilde{v}_{C-H} \approx 3000$ cm^{-1} and $\tilde{v}_{C-D} \approx 2100$ cm^{-1}. The wavenumber ($\tilde{v}$) is given by v/c, where v is the frequency and c is the velocity of light.

9.40 Lubricating oils for watches or other mechanical objects are made of long-chain hydrocarbons. Over long periods of time they undergo auto-oxidation to form solid polymers. The initial step in this process involves hydrogen abstraction. Suggest a chemical means for prolonging the life of these oils.

Additional Problems

9.41 A flask contains a mixture of compounds A and B. Both compounds decompose by first-order kinetics. The half-lives are 50.0 min for A and 18.0 min for B. If the concentrations of A and B are equal initially, how long will it take for the concentration of A to be four times that of B?

9.42 The term *reversible* is used in both thermodynamics (see Chapter 3) and in this chapter. Does it convey the same meaning in these two instances?

9.43 The recombination of iodine atoms in an organic solvent, such as carbon tetrachloride, is a diffusion-controlled process:

$$I + I \rightarrow I_2$$

Given that the viscosity of CCl_4 is 9.69×10^{-4} N s m^{-2} at 20°C, calculate the rate of recombination at this temperature.

9.44 The equilibrium between dissolved CO_2 and carbonic acid can be represented by

$$H^+ + HCO_3^- \underset{k_{21}}{\overset{k_{12}}{\rightleftharpoons}} H_2CO_3$$

$$k_{13} \Big\Updownarrow k_{31} \qquad\qquad k_{23} \Big\Updownarrow k_{32}$$

$$CO_2 \quad + \quad H_2O$$

Show that

$$-\frac{d[CO_2]}{dt} = (k_{31} + k_{32})[CO_2] - \left(k_{13} + \frac{k_{23}}{K}\right)[H^+][HCO_3^-]$$

where $K = [H^+][HCO_3^-]/[H_2CO_3]$.

9.45 Polyethylene is used in many items, including water pipes, bottles, electrical insulation, toys, and mailing envelopes. It is a *polymer*, a molecule with a very high molar mass made by joining many ethylene molecules (the basic unit is called a *monomer*) together. The initiation step is

$$R_2 \overset{k_i}{\rightarrow} 2R\cdot \quad \text{initiation}$$

The R· species (called a radical) reacts with an ethylene molecule (M) to generate another radical

$$R\cdot + M \rightarrow M_1\cdot$$

Reaction of $M_1\cdot$ with another monomer leads to the growth or propagation of the polymer chain:

$$M_1\cdot + M \overset{k_p}{\rightarrow} M_2\cdot \quad \text{propagation}$$

This step can be repeated with hundreds of monomer units. The propagation terminates when two radicals combine

$$M'\cdot + M''\cdot \overset{k_t}{\rightarrow} M'-M'' \quad \text{termination}$$

The initiator in the polymerization of ethylene commonly is benzoyl peroxide [$(C_6H_5COO)_2$]:

$$[(C_6H_5COO)_2] \rightarrow 2C_6H_5COO\cdot$$

This is a first-order reaction. The half-life of benzoyl peroxide at 100°C is 19.8 min. **(a)** Calculate the rate constant (in min^{-1}) of the reaction. **(b)** If the half-life of benzoyl peroxide is 7.30 h, or 438 min, at 70°C, what is the activation energy (in kJ/mol) for the decomposition of benzoyl peroxide? **(c)** Write the rate laws for the elementary steps in the above polymerization process and identify the reactant, product, and intermediates. **(d)** What condition would favor the growth of long high-molar-mass polyethylenes?

9.46 In a certain industrial process involving a heterogeneous catalyst, the volume of the catalyst (in the shape of a sphere) is 10.0 cm^3. **(a)** Calculate the surface area of the catalyst. **(b)** If the sphere is broken down into eight spheres, each of which has a volume of 1.25 cm^3, what is the total surface area of the spheres? **(c)** Which of the two geometric configurations is the more effective catalyst? (*Hint:* The surface area of a sphere is $4\pi r^2$, where r is the radius of the sphere.)

9.47 Explain why grain dust in elevators can be explosive.

9.48 At a certain elevated temperature, ammonia decomposes on the surface of tungsten metal as follows:

$$NH_3 \rightarrow \tfrac{1}{2}N_2 + \tfrac{3}{2}H_2$$

The kinetic data are expressed as the variation of the half-life with the initial pressure of NH_3:

P/torr	264	130	59	16
$t_{1/2}$/s	456	228	102	60

(a) Determine the order of the reaction. **(b)** How does the order depend on the initial pressure? **(c)** How does the mechanism of the reaction vary with pressure?

9.49 The *activity* of a radioactive sample is the number of nuclear disintegrations per second, which is equal to the first-order rate constant times the number of radioactive nuclei present. The fundamental unit of radioactivity is the *curie* (Ci), where 1 Ci corresponds to exactly 3.70×10^{10} disintegrations per second. This decay rate is equivalent to that of 1 g of radium-226. Calculate the rate constant and half-life for the radium decay. Starting with 1.0 g of the radium sample, what is the activity after 500 years? The molar mass of Ra-226 is 226.03 g mol^{-1}.

9.50 The reaction X → Y has a reaction enthalpy of -64 kJ mol^{-1} and an activation energy of 22 kJ mol^{-1}. What is the activation energy for the Y → X reaction?

9.51 Consider the following parallel first-order reactions:

$$A \underset{k_2}{\overset{k_1}{\nearrow\searrow}} \begin{matrix} B \\ C \end{matrix}$$

(a) Write the expression for $d[B]/dt$ at time t, given that $[A]_0$ is the concentration of A at $t = 0$. **(b)** What is the ratio of $[B]/[C]$ upon completion of the reactions?

9.52 As a result of being exposed to the radiation released during the Chernobyl nuclear accident, a person had a level of iodine-131 in his body equal to 7.4 mCi (1 mCi $= 1 \times 10^{-3}$ Ci). Calculate the number of atoms of I-131 to which this radioactivity corresponds. Why were people who lived close to the nuclear reactor site urged to take large amounts of potassium iodide after the accident?

9.53 A certain protein molecule, P, of molar mass $\mathcal{M}$ dimerizes when it is allowed to stand in solution at room temperature. A plausible mechanism is that the protein molecule is first denatured before it dimerizes:

$$P \xrightarrow{k} P^*(\text{denatured}) \quad \text{slow}$$
$$2P^* \rightarrow P_2 \quad \quad \quad \text{fast}$$

The progress of this reaction can be followed by making viscosity measurements of the average molar mass, $\mathcal{M}$. Derive an expression for $\mathcal{M}$ in terms of the initial concentration, $[P]_0$, and the concentration at time t, $[P]$, and $\mathcal{M}$. Write a rate equation consistent with this scheme.

9.54 The bromination of acetone is acid catalyzed:

$$CH_3COCH_3 + Br_2 \xrightarrow{H^+} CH_3COCH_2Br + H^+ + Br^-$$

The rate of disappearance of bromine was measured for several different concentrations of acetone, bromine, and H^+ ions at a certain temperature:

	$[CH_3COCH_3]/M$	$[Br_2]/M$	$[H^+]/M$	Rate of Disappearance of $Br_2/M \cdot s^{-1}$
(1)	0.30	0.050	0.050	5.7×10^{-5}
(2)	0.30	0.10	0.050	5.7×10^{-5}
(3)	0.30	0.050	0.10	1.2×10^{-4}
(4)	0.40	0.050	0.20	3.1×10^{-4}
(5)	0.40	0.050	0.050	7.6×10^{-5}

(a) What is the rate law for the reaction? **(b)** Determine the rate constant. **(c)** The following mechanism has been proposed for the reaction:

Show that the rate law deduced from the mechanism is consistent with that shown in **(a)**.

9.55 The rate law for the reaction $2NO_2(g) \rightarrow N_2O_4(g)$ is rate $= k[NO_2]^2$. Which of the following changes will alter the value of k? **(a)** The pressure of NO_2 is doubled. **(b)** The reaction is run in an organic solvent. **(c)** The volume of the container is doubled. **(d)** The temperature is decreased. **(e)** A catalyst is added to the container.

9.56 For the cyclic reactions shown on p. 329, show that $k_{-1}k_2k_3 = k_1k_{-2}k_{-3}$.

9.57 Oxygen for metabolism is taken up by hemoglobin (Hb) to form oxyhemoglobin (HbO_2) according to the simplified equation

$$Hb(aq) + O_2(aq) \xrightarrow{k} HbO_2(aq)$$

where the second-order rate constant is 2.1×10^6 M^{-1} s^{-1} at 37°C. For an average adult, the concentrations of Hb and O_2 in the blood in the lungs are 8.0×10^{-6} M and 1.5×10^{-6} M, respectively. **(a)** Calculate the rate of formation of HbO_2. **(b)** Calculate the rate of consumption of O_2. **(c)** The rate of formation of HbO_2 increases to 1.4×10^{-4} M s^{-1} during exercise to meet the demand of an increased metabolic rate. Assuming the Hb concentration remains the same, what oxygen concentration is necessary to sustain this rate of HbO_2 formation?

9.58 Sucrose ($C_{12}H_{22}O_{11}$), commonly called table sugar, undergoes hydrolysis (reaction with water) to produce fructose ($C_6H_{12}O_6$) and glucose ($C_6H_{12}O_6$):

$$C_{12}H_{22}O_{11} + H_2O \rightarrow \underset{\text{fructose}}{C_6H_{12}O_6} + \underset{\text{glucose}}{C_6H_{12}O_6}$$

This reaction has particular significance in the candy industry. First, fructose is sweeter than sucrose. Second, a mixture of fructose and glucose, called *invert* sugar, does not crystallize, so candy made with this combination is chewier and not brittle as crystalline sucrose is. Sucrose is dextrorotatory $(+)$, whereas the mixture of glucose and fructose resulting from inversion is levorotatory $(-)$. Thus, a decrease in the concentration of sucrose will be accompanied by a proportional decrease in the optical rotation. **(a)** From the following kinetic data, show that the reaction is first order, and determine the rate constant:

time/min	0	7.20	18.0	27.0	∞
optical rotation (α)	$+24.08°$	$+21.40°$	$+17.73°$	$+15.01°$	$-10.73°$

(b) Explain why the rate law does not include $[H_2O]$ even though water is a reactant.

9.59 Thallium(I) is oxidized by cerium(IV) in solution as follows:

$$Tl^+ + 2Ce^{4+} \rightarrow Tl^{3+} + 2Ce^{3+}$$

The elementary steps, in the presence of Mn(II), are as follows:

$$Ce^{4+} + Mn^{2+} \rightarrow Ce^{3+} + Mn^{3+}$$

$$Ce^{4+} + Mn^{3+} \rightarrow Ce^{3+} + Mn^{4+}$$

$$Tl^+ + Mn^{4+} \rightarrow Tl^{3+} + Mn^{2+}$$

(a) Identify the catalyst, intermediates, and the rate-determining step if the rate law is rate $= k[Ce^{4+}][Mn^{2+}]$. (b) Explain why the reaction is slow without the catalyst. (c) Classify the type of catalysis (homogeneous or heterogeneous).

9.60 Under certain conditions the gas-phase decomposition of ozone is found to be second order in O_3 and inhibited by molecular oxygen. Apply the steady-state approximation to the following mechanism to show that the rate law is consistent with the experimental observation:

$$O_3 \underset{k_{-1}}{\overset{k_1}{\rightleftharpoons}} O_2 + O$$

$$O + O_3 \overset{k_2}{\rightarrow} 2O_2$$

State any assumption made in the derivation.

9.61 The rate constants for the reaction

$$CH_2{=}CH{-}CH{=}CH_2 \;+\; CH_2{=}CH{-}CHO \;\longrightarrow$$

have been measured at several temperatures:

$10^3 k/M^{-1}\cdot s^{-1}$	0.138	1.63	7.2	36.8	81
$t/°C$	155.3	208.3	246.5	295.8	330.8

Calculate the values of the pre-exponential factor, $E_a, \Delta S^{\circ\ddagger}$, and $\Delta H^{\circ\ddagger}$ for the reaction. Use 516 K as the mean temperature for your calculation. [Data taken from G. B. Kistiakowsky and J. R. Lacher, *J. Am. Chem. Soc.* **58**, 123 (1936).]

9.62 A reaction $X + Y \rightarrow Z$ proceeds by two different mechanisms, one of which is pH dependent. The rate law for the reaction is

$$\frac{d[Z]}{dt} = k_1[X] + k_2[Y][H^+]$$

At pH $= 3.4$ the rates of the two reactions are equal. What is the ratio k_1/k_2? Assume $[X] = [Y]$.

9.63 To prevent brain damage, a standard procedure is to lower the body temperature of someone who has been resuscitated after suffering cardiac arrest. What is the physiochemical basis for this procedure?

9.64 The Polish-American physicist Roman Smoluchowski showed that the rate constant for a diffusion-controlled reaction between molecules A and B, k_D, is given by

$$k_D = 4\pi N_A (D_A + D_B) r_{AB}$$

where r_{AB} is the distance (in cm) between the molecules and D_A and D_B are their diffusion coefficients. Calculate k_D, given that $D_A = 6.0 \times 10^{-5}$ cm^2 s^{-1}, $D_B = 2.5 \times 10^{-5}$ cm^2 s^{-1} at 20°C, and $r_{AB} = 1.0 \times 10^{-8}$ cm.

9.65 The diameter of the methyl radical ($\cdot CH_3$) is 3.80 Å. Calculate the rate constant for the second-order gas phase reaction

$$2 \cdot CH_3 \rightarrow C_2H_6$$

at 50°C. Is this the maximum possible rate constant? Explain.

CHAPTER 10

Enzyme Kinetics

One of the most fascinating areas of study in chemical kinetics is enzyme catalysis. The phenomenon of enzyme catalysis usually results in a very large increase in reaction rate (on the order of 10^6 to 10^{18}) and high specificity. By specificity, we mean that an enzyme molecule is capable of selectively catalyzing certain reactants, called *substrates*, while discriminating against other molecules.

This chapter presents the basic mathematical treatment of enzyme kinetics and discusses the topics of enzyme inhibition, allosterism, and the effect of pH on enzyme kinetics.

10.1 General Principles of Catalysis

A *catalyst* is a substance that increases the rate of a reaction without itself being consumed by the process. A reaction in which a catalyst is involved is called a *catalyzed reaction*, and the process is called *catalysis*. In studying catalysis, keep in mind the following characteristics:

1. A catalyst lowers the Gibbs energy of activation by providing a different mechanism for the reaction (Figure 10.1). This mechanism enhances the rate and it applies to *both* the forward and the reverse directions of the reaction.

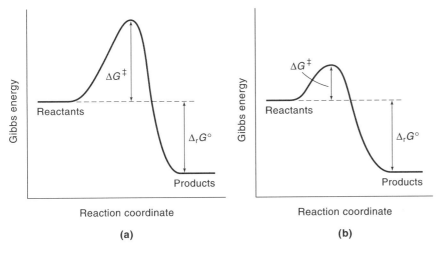

Figure 10.1
Gibbs energy change for (a) an uncatalyzed reaction and (b) a catalyzed reaction. A catalyzed reaction must involve the formation of at least one intermediate (between the reactant and the catalyst). The $\Delta_r G^\circ$ is the same in both cases.

2. A catalyst forms an intermediate with the reactant(s) in the initial step of the mechanism and is released in the product-forming step. The catalyst does not appear in the overall reaction.

3. Regardless of the mechanism and the energetics of a reaction, a catalyst cannot affect the enthalpies or Gibbs energies of the reactants and products. Thus, catalysts increase the rate of approach to equilibrium, but cannot alter the thermodynamic equilibrium constant.

Humans have used catalysts for thousands of years in food preparation and wine making. Industrially, hundreds of billions of dollars worth of chemicals are produced annually with the aid of catalysts. There are three types of catalysis: heterogeneous, homogeneous, and enzymatic. In a heterogeneously catalyzed reaction, the reactants and the catalyst are in different phases (usually gas/solid or liquid/solid). Well-known examples are the Haber synthesis of ammonia and the Ostwald manufacture of nitric acid. The bromination of acetone, catalyzed by acids,

$$CH_3COCH_3 + Br_2 \xrightarrow{H^+} CH_2BrCOCH_3 + HBr$$

is an example of homogeneous catalysis because the reactants and the catalyst (H^+) are all present in the aqueous medium. Enzyme catalysis is also mostly homogeneous in nature. However, because it is of biological origin and is the most complex of the three types of catalysis, enzyme catalysis is treated as a separate category. Whether or not their mechanisms are well understood, enzymes have been used widely in food and beverage production, as well as in the manufacture of drugs and other chemicals.

Enzyme Catalysis

Since 1926, when the American biochemist James Sumner (1887–1955) crystallized urease (an enzyme that catalyzes the cleavage of urea to ammonia and carbon dioxide), it has come to be known that most enzymes are proteins.* An enzyme usually contains one or more *active sites*, where reactions with substrates take place. An active site may comprise only a few amino acid residues; the rest of the protein is required for maintaining the three-dimensional integrity of the network. The specificity of enzymes for substrates varies from molecule to molecule. Many enzymes exhibit stereochemical specificity in that they catalyze the reactions of one conformation but not the other (Figure 10.2). For example, proteolytic enzymes catalyze only

Figure 10.2
Diagram showing how two enantiomers bind differently to an enzyme. Because the geometry of an enzyme's active site is normally fixed (that is, it can have only one of the above two arrangements), a reaction occurs for only one of the two enantiomers. Specificity requires a minimum of three contact points between the substrate and the enzyme.

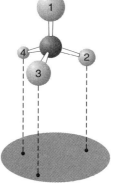
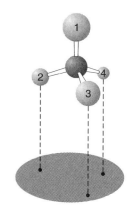
Enzyme binding site   Enzyme binding site

*In the early 1980s, chemists discovered that certain RNA molecules, called ribozymes, also possess catalytic properties.

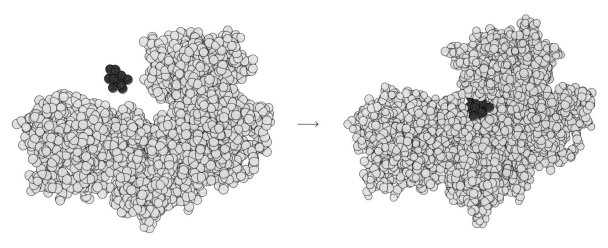

Figure 10.3
The conformational change that occurs when glucose binds to hexokinase, which is an enzyme in the metabolic pathway. [From W. S. Bennet and T. A. Steitz, *J. Mol. Biol.* **140**, 211 (1980).]

the hydrolysis of peptides made up of L-amino acids. Some enzymes are catalytically inactive in the absence of certain metal ions.

In the 1890s the German chemist Emil Fischer (1852–1919) proposed a lock-and-key theory of enzyme specificity. According to Fischer, the active site can be assumed to have a rigid structure, similar to a lock. A substrate molecule then has a complementary structure and functions as a key. Although appealing in some respects, this theory has been modified to take into account the flexibility of proteins in solution. We now know that the binding of the substrate to the enzyme results in a distortion of the substrate into the conformation of the transition state. At the same time, the enzyme itself also undergoes a change in conformation to fit the substrate (Figure 10.3). The flexibility of the protein also explains the phenomenon of cooperativity. *Cooperativity* means the binding of a substrate to an enzyme with multiple binding sites can alter the substrate's affinity for enzyme binding at its other sites.

Enzymes, like other catalysts, increase the rate of a reaction. An understanding of the efficiency of enzymes can be gained by examining Equation 9.41:

$$k = \frac{k_\mathrm{B} T}{h} e^{-\Delta G^{\circ\ddagger}/RT} (M^{1-m})$$

$$= \frac{k_\mathrm{B} T}{h} e^{\Delta S^{\circ\ddagger}/R} e^{-\Delta H^{\circ\ddagger}/RT} (M^{1-m})$$

There are two contributions to the rate constant: $\Delta H^{\circ\ddagger}$ and $\Delta S^{\circ\ddagger}$. The enthalpy of activation is approximately equal to the energy of activation (E_a) in the Arrhenius equation (see Equation 9.28). Certainly a reduction in E_a by the action of a catalyst would enhance the rate constant. In fact, this is usually the explanation of how a catalyst works, but it is not always true for enzyme catalysis. Entropy of activation, $\Delta S^{\circ\ddagger}$, may also be an important factor in determining the efficiency of enzyme catalysis.

Consider the bimolecular reaction

$$\mathrm{A} + \mathrm{B} \rightarrow \mathrm{AB}^{\ddagger} \rightarrow \text{product}$$

where A and B are both nonlinear molecules. Before the formation of the activated complex, each A or B molecule has three translational, three rotational, and three vibrational degrees of freedom. These motions all contribute to the entropy of the molecule. At 25°C, the greatest contribution comes from translational motion (about 120 J K^{-1} mol^{-1}), followed by rotational motion (about 80 J K^{-1} mol^{-1}). Vibrational motion makes the smallest contribution (about 15 J K^{-1} mol^{-1}). The translational and rotational entropies of the activated complex are only slightly larger than those of an individual A or B molecule (these entropies increase slowly with size); therefore, there is a net loss in entropy of about 200 J K^{-1} mol^{-1} when the activated complex is formed. This loss in entropy is compensated for to a small extent by new modes of internal rotation and vibration in the activated complex. For unimolecular reactions, such as the *cis–trans* isomerization of an alkene, however, there is very little entropy change because the activated complex is formed from a single molecular species. A theoretical comparison of a unimolecular reaction with a bimolecular one shows a difference of as much as 3×10^{10} in the $e^{\Delta S^{\circ \ddagger}/R}$ term, favoring the unimolecular reaction.

Consider a simple enzyme-catalyzed reaction in which one substrate (S) is transformed into one product (P). The reaction proceeds as follows:

$$E + S \rightleftharpoons ES \rightleftharpoons ES^{\ddagger} \rightleftharpoons EP \rightleftharpoons E + P$$

In this scheme, the enzyme and the substrate must first encounter each other in solution to form the enzyme–substrate intermediate, ES. This is a reversible reaction but when [S] is high, the formation of ES is favored. When the substrate is bound, forces within the active site can align the substrate and enzyme reactive groups into proper orientation, leading to the activated complex. The reaction takes place in the single entity enzyme–substrate intermediate to form the enzyme–substrate activated complex ($ES^{\ddagger}$), as in a unimolecular reaction, so the loss in entropy will be much less. In other words, the loss of the translational and rotational entropies occurred during the formation of ES, and not during the $ES \rightarrow ES^{\ddagger}$ step. (This loss of entropy is largely compensated for by the substrate binding energy.) Once formed, $ES^{\ddagger}$ proceeds energetically downhill to the enzyme–product intermediate and finally to the product with the regeneration of the enzyme. Figure 10.4 summarizes the steps on a diagram of Gibbs energy versus reaction coordinate.

Figure 10.4
Plot of Gibbs energy versus reaction coordinate for an enzyme-catalyzed reaction.

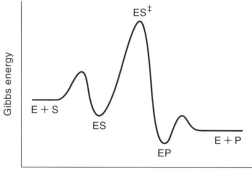

10.2   The Equations of Enzyme Kinetics

In enzyme kinetics, it is customary to measure the *initial rate* (v_0) of a reaction to minimize reversible reactions and the inhibition of enzymes by products. Furthermore, the initial rate corresponds to a known fixed substrate concentration. As time proceeds, the substrate concentration will drop.

Figure 10.5 shows the variation of the initial rate (v_0) of an enzyme-catalyzed reaction with substrate concentration [S], where the subscript zero denotes the initial value. The rate increases rapidly and linearly with [S] at low substrate concentrations, but it gradually levels off toward a limiting value at high concentrations of the substrate. In this region, all the enzyme molecules are bound to the substrate molecules, and the rate becomes zero order in substrate concentration. Mathematical analysis shows that the relationship between v_0 and [S] can be represented by an equation of a rectangular hyperbola:

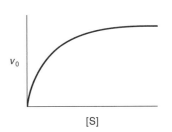

Figure 10.5
Plot of the initial rate of an enzyme-catalyzed reaction versus substrate concentration.

$$v_0 = \frac{a[\text{S}]}{b + [\text{S}]} \tag{10.1}$$

where a and b are constants. Our next step is to develop the necessary equations to account for the experimental data.

Michaelis–Menten Kinetics

In 1913, the German biochemist Leonor Michaelis (1875–1949) and the Canadian biochemist Maud L. Menten (1879–1960), building on the work of the French chemist Victor Henri (1872–1940), proposed a mechanism to explain the dependence of the initial rate of enzyme-catalyzed reactions on concentration. They considered the following scheme, in which ES is the enzyme–substrate complex:

$$\text{E} + \text{S} \underset{k_{-1}}{\overset{k_1}{\rightleftharpoons}} \text{ES} \overset{k_2}{\rightarrow} \text{P} + \text{E}$$

The initial rate of product formation, v_0, is given by

$$v_0 = \left(\frac{d[\text{P}]}{dt}\right)_0 = k_2[\text{ES}] \tag{10.2}$$

To derive an expression for the rate in terms of the more easily measurable substrate concentration, Michaelis and Menten assumed that $k_{-1} \gg k_2$ so that the first step (formation of ES) can be treated as a rapid equilibrium process. The dissociation constant, K_S, is given by

This step corresponds to the pre-equilibrium case discussed on p. 331.

$$K_\text{S} = \frac{k_{-1}}{k_1} = \frac{[\text{E}][\text{S}]}{[\text{ES}]}$$

The total concentration of the enzyme at a time shortly after the start of the reaction is

$$[\text{E}]_0 = [\text{E}] + [\text{ES}]$$

so that

$$K_\text{S} = \frac{([\text{E}]_0 - [\text{ES}])[\text{S}]}{[\text{ES}]} \tag{10.3}$$

Solving for [ES], we obtain

$$[\text{ES}] = \frac{[\text{E}]_0[\text{S}]}{K_\text{S} + [\text{S}]} \tag{10.4}$$

Substituting Equation 10.4 into Equation 10.2 yields

$$v_0 = \left(\frac{d[P]}{dt}\right)_0 = \frac{k_2[E]_0[S]}{K_S + [S]} \qquad (10.5)$$

Thus, the rate is always proportional to the total concentration of the enzyme.

Equation 10.5 has the same form as Equation 10.1, where $a = k_2[E]_0$ and $b = K_S$. At low substrate concentrations $[S] \ll K_S$, so Equation 10.5 becomes $v_0 = (k_2/K_S)[E]_0[S]$; that is, it is a second-order reaction (first order in $[E]_0$ and first order in $[S]$). This rate law corresponds to the initial linear portion of the plot in Figure 10.5. At high substrate concentrations, $[S] \gg K_S$, so Equation 10.5 can be written

$$v_0 = \left(\frac{d[P]}{dt}\right)_0 = k_2[E]_0$$

Under these conditions, all the enzyme molecules are in the enzyme–substrate complex form; that is, the reacting system is saturated with S. Consequently, the initial rate is zero order in $[S]$. This rate law corresponds to the horizontal portion of the plot. The curved portion in Figure 10.5 represents the transition from low to high substrate concentrations.

When all the enzyme molecules are complexed with the substrate as ES, the measured initial rate must be at its maximum value (V_{max}), so that

$$V_{max} = k_2[E]_0 \qquad (10.6)$$

where V_{max} is called the *maximum rate*. Now consider what happens when $[S] = K_S$. From Equation 10.5 we find that this condition gives $v_0 = V_{max}/2$, so K_S equals the concentration of S when the initial rate is half its maximum value.

Steady-State Kinetics

The British biologists George Briggs (1893–1978) and John Haldane (1892–1964) showed in 1925 that it is unnecessary to assume that enzyme and substrate are in thermodynamic equilibrium with the enzyme–substrate complex to derive Equation 10.5. They postulated that soon after enzyme and substrate are mixed, the concentration of the enzyme–substrate complex will reach a constant value so that we can apply the steady-state approximation as follows (Figure 10.6):*

$$\frac{d[ES]}{dt} = 0 = k_1[E][S] - k_{-1}[ES] - k_2[ES]$$
$$= k_1([E]_0 - [ES])[S] - (k_{-1} + k_2)[ES]$$

Solving for [ES], we get

$$[ES] = \frac{k_1[E]_0[S]}{k_1[S] + k_{-1} + k_2} \qquad (10.7)$$

---

*Chemists are also interested in *pre-steady-state kinetics*—that is, the period before steady state is reached. Pre-steady-state kinetics is more difficult to study but provides useful information regarding the mechanism of enzyme catalysis. But steady-state kinetics is more important for the understanding of metabolism, because it measures the rates of enzyme-catalyzed reactions in the steady-state conditions that exist in the cell.

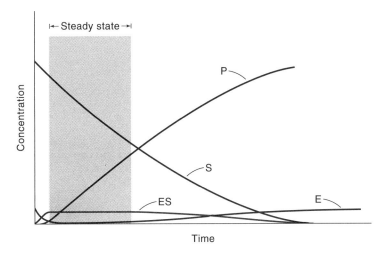

Figure 10.6
Plot of the concentrations of the various species in an enzyme-catalyzed reaction $E + S \rightleftharpoons ES \rightarrow P + E$ versus time. We assume that the initial substrate concentration is much larger than the enzyme concentration and that the rate constants k_1, k_{-1}, and k_2 (see text) are of comparable magnitudes.

Substituting Equation 10.7 into 10.2 gives

$$v_0 = \left(\frac{d[P]}{dt}\right)_0 = k_2[ES] = \frac{k_1 k_2 [E]_0 [S]}{k_1[S] + k_{-1} + k_2}$$

$$= \frac{k_2[E]_0[S]}{[(k_{-1} + k_2)/k_1] + [S]}$$

$$= \frac{k_2[E]_0[S]}{K_M + [S]} \qquad (10.8)$$

where K_M, the *Michaelis constant*, is defined as

$$K_M = \frac{k_{-1} + k_2}{k_1} \qquad (10.9)$$

Comparing Equation 10.8 with Equation 10.5, we see that they have a similar dependence on substrate concentration; however, $K_M \neq K_S$ in general unless $k_{-1} \gg k_2$.

The Briggs–Haldane treatment defines the maximum rate exactly as Equation 10.6 does. Because $[E]_0 = V_{max}/k_2$, Equation 10.8 can also be written as

$$v_0 = \frac{V_{max}[S]}{K_M + [S]} \qquad (10.10)$$

Equation 10.10 is a fundamental equation of enzyme kinetics, and we shall frequently refer to it. When the initial rate is equal to half the maximum rate, Equation 10.10 becomes

$$\frac{V_{max}}{2} = \frac{V_{max}[S]}{K_M + [S]}$$

or

$$K_M = [S]$$

Note that the larger the K_M (the weaker the binding), the larger the [S] needed to reach the half maximum rate.

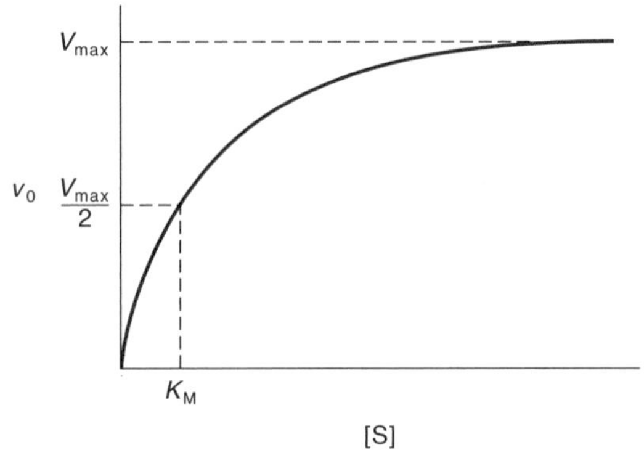

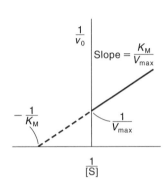

Figure 10.7
Graphical determination
of $V_{\max}$ and K_M.

Thus, both $V_{\max}$ and K_M can be determined, at least in principle, from a plot such as the one in Figure 10.7. In practice, however, we find that the plot of v_0 versus [S] is not very useful in determining the value of $V_{\max}$ because locating the asymptotic value $V_{\max}$ at very high substrate concentrations is often difficult. A more satisfactory approach, suggested by the American chemists H. Lineweaver (1907–) and Dean Burk (1904–1988), is to employ the double-reciprocal plot of $1/v_0$ versus $1/[S]$. From Equation 10.10, we write

$$\frac{1}{v_0} = \frac{K_M}{V_{\max}[S]} + \frac{1}{V_{\max}} \qquad (10.11)$$

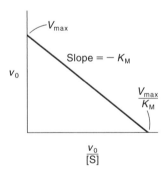

Figure 10.8
Lineweaver–Burk plot for an enzyme-catalyzed reaction obeying Michaelis–Menten kinetics.

As Figure 10.8 shows, both K_M and $V_{\max}$ can be obtained from the slope and intercepts of the straight line.

Although useful and widely employed in enzyme kinetic studies, the Lineweaver–Burk plot has the disadvantage of compressing the data points at high substrate concentrations into a small region and emphasizing the points at lower substrate concentrations, which are often the least accurate. Of the several other ways of plotting the kinetic data, we shall mention the Eadie–Hofstee plot. Multiplying both sides of Equation 10.11 by $v_0 V_{\max}$, we obtain

$$V_{\max} = v_0 + \frac{v_0 K_M}{[S]}$$

Rearrangement gives

$$v_0 = V_{\max} - \frac{v_0 K_M}{[S]} \qquad (10.12)$$

Figure 10.9
Eadie–Hofstee plot for the reaction graphed in Figure 10.7.

This equation shows that a plot of v_0 versus $v_0/[S]$, the so-called Eadie–Hofstee plot, gives a straight line with slope equal to $-K_M$ and intercepts $V_{\max}$ on the v_0 axis and $V_{\max}/K_M$ on the $v_0/[S]$ axis (Figure 10.9).

The Significance of K_M and $V_{\max}$

The Michaelis constant, K_M, varies considerably from one enzyme to another, and also with different substrates for the same enzyme. By definition, it is equal to the substrate concentration at half the maximum rate. Put another way, K_M represents the substrate concentration at which half the enzyme active sites are filled by substrate molecules. The value of K_M is sometimes equated with the dissociation constant of the enzyme–substrate complex, ES (the larger the K_M, the weaker the binding). As can be seen from Equation 10.9, however, this is true only when $k_2 \ll k_{-1}$ so that $K_M = k_{-1}/k_1$. In general, K_M must be expressed in terms of three rate constants. Nevertheless, K_M (in units of molarity) is customarily reported together with other kinetic parameters for enzyme-catalyzed reactions. To begin with, it is a quantity that can be measured easily and directly. Furthermore, K_M depends on temperature, the nature of the substrate, pH, ionic strength, and other reaction conditions; therefore, its value serves to characterize a particular enzyme–substrate system under specific conditions. Any variation in K_M (for the same enzyme and substrate) is often an indication of the presence of an inhibitor or activator. Useful information about evolution can also be obtained by comparing the K_M values of a similar enzyme from different species. For the majority of enzymes, K_M lies between 10^{-1} M and 10^{-7} M.

The maximum rate, V_{max}, has a well-defined meaning, both theoretically and empirically. It represents the maximum rate attainable; that is, it is the rate at which the total enzyme concentration is present as the enzyme–substrate complex. According to Equation 10.6, if $[E]_0$ is known, the value of k_2 can be determined from the value of V_{max} measured by one of the plots mentioned earlier. Note that k_2 is a first-order rate constant and has the unit of per unit time (s^{-1} or min^{-1}). It is called the *turnover number* (also referred to as k_{cat}, the *catalytic constant*). The turnover number of an enzyme is the number of substrate molecules (or moles of substrate) that are converted to product per unit time, when the enzyme is fully saturated with the substrate. For most enzymes, the turnover number varies between 1 and 10^5 s^{-1} under physiological conditions. Carbonic anhydrase, an enzyme that catalyzes the hydration of carbon dioxide and the dehydration of carbonic acid,

$$CO_2 + H_2O \rightleftharpoons H_2CO_3$$

has one of the largest turnover numbers known ($k_2 = 1 \times 10^6$ s^{-1}) at 25°C. Thus, a 1×10^{-6} M solution of the enzyme can catalyze the formation of 1 M H_2CO_3 from CO_2 (produced by metabolism) and H_2O per second; that is,

$$V_{max} = (1 \times 10^6 \ s^{-1})(1 \times 10^{-6} \ M)$$
$$= 1 \ M \ s^{-1}$$

Without the enzyme, the pseudo first-order rate constant is only about 0.03 s^{-1}. [Note that if the purity of the enzyme or the number of active sites per molecule is unknown, we cannot calculate the turnover number. In that case, the activity of the enzyme may be given as *units of activity per milligram of protein* (called the *specific activity*). One *international unit* is the amount of enzyme that produces one micromole (1 μmol) of product per minute.]

As stated, we can determine the turnover number by measuring the rate under saturating substrate conditions; that is, when $[S] \gg K_M$ (see Equation 10.8). Under physiological conditions, the ratio $[S]/K_M$ is seldom greater than one; in fact, it is frequently much smaller than one. When $[S] \ll K_M$, Equation 10.8 becomes

$$v_0 = \frac{k_2}{K_M}[E]_0[S]$$
$$= \frac{k_{cat}}{K_M}[E]_0[S] \tag{10.13}$$

Note that Equation 10.13 expresses the rate law of a second-order reaction. It is interesting that the ratio k_{cat}/K_M (which has the units $M^{-1} s^{-1}$) is a measure of the

Table 10.1
Values of K_M, k_{cat}, and k_{cat}/K_M for Some Enzymes and Substrates

Enzyme	Substrate	K_M/M	k_{cat}/s^{-1}	$(k_{cat}/K_M)/M^{-1} \cdot s^{-1}$
Acetylcholin-esterase	Acetylcholine	9.5×10^{-5}	1.4×10^4	1.5×10^8
Catalase	H_2O_2	2.5×10^{-2}	1.0×10^7	4.0×10^8
Carbonic anhydrase	CO_2	0.012	1.0×10^6	8.3×10^7
Chymotrypsin	N-acetylglycine ethyl ester	0.44	5.1×10^{-2}	0.12
Fumarase	Fumarate	5.0×10^{-6}	8.0×10^2	1.6×10^8
Urease	Urea	2.5×10^{-2}	1.0×10^4	4.0×10^5

catalytic efficiency of an enzyme. A large ratio favors the formation of product. The reverse holds true for a small ratio.

Finally we ask the question: What is the upper limit of the catalytic efficiency of an enzyme? From Equation 10.9, we find

$$\frac{k_{cat}}{K_M} = \frac{k_2}{K_M} = \frac{k_1 k_2}{k_{-1} + k_2} \tag{10.14}$$

This ratio is a maximum when $k_2 \gg k_{-1}$; that is, k_1 is rate-determining and the enzyme turns over a product as soon as an ES complex is formed. However, k_1 can be no greater than the frequency of encounter between the enzyme and the substrate molecule, which is controlled by the rate of diffusion in solution.* The rate constant of a diffusion-controlled reaction is on the order of 10^8 M^{-1} s^{-1}. Therefore, enzymes with such k_{cat}/K_M values must catalyze a reaction almost every time they collide with a substrate molecule. Table 10.1 shows that acetylcholinesterase, catalase, fumarase, and perhaps carbonic anhydrase, have achieved this state of catalytic perfection.

10.3 Chymotrypsin: A Case Study

Having developed the basic equations of enzyme kinetics, we shall now consider some reactions catalyzed by chymotrypsin, a digestive enzyme. Aside from its important role in digestion, chymotrypsin catalysis is significant for being the system whose study provided the first evidence for the general existence of covalent enzyme–substrate complexes.

Chymotrypsin is one of the serine proteases, a family of protein-cutting enzymes that includes trypsin, elastase, and subtilisin. It has a molar mass of 24,800 daltons, 246 amino acid residues, and one active site (containing the serine residue) per molecule. Chymotrypsin is produced in the mammalian pancreas, where it takes the form of an inactive precursor, chymotrypsinogen. Once this precursor has entered the intestine, it is activated by another enzyme, trypsin, to become chymotrypsin. In this way, it avoids self-destruction before it can digest food. The enzyme can be prepared in highly purified form by crystallization.

---

*The rates of some enzyme-catalyzed reactions actually exceed the diffusion-controlled limit. When enzymes are associated with organized assemblies (for example, in cellular membranes), the product of one enzyme is channeled to the next enzyme, much as in an assembly line. In such cases, the rate of catalysis is not limited by the rate of diffusion in solution.

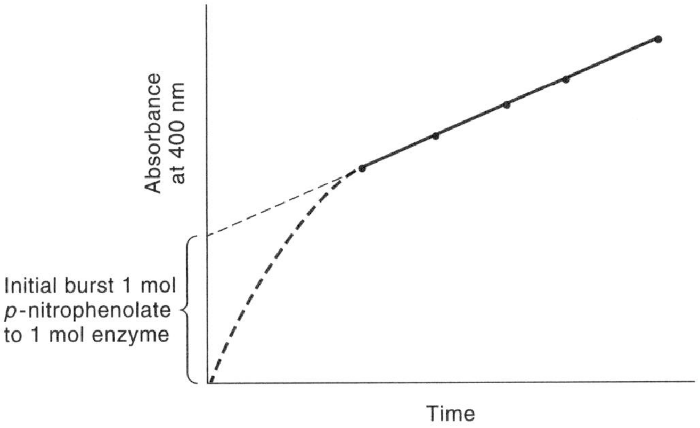

Figure 10.10
Chymotrypsin-catalyzed hydrolysis of *p*-nitrophenyl acetate. The reaction shows an initial burst of *p*-nitrophenolate. Extrapolation of the absorbance to zero time shows 1 : 1 stoichiometry between the *p*-nitrophenolate produced and the amount of enzyme used.

In 1953, the British chemists B. S. Hartley and B. A. Kilby studied the hydrolysis of *p*-nitrophenyl acetate (PNPA), catalyzed by chymotrypsin to yield *p*-nitrophenolate ion and acetate ion:

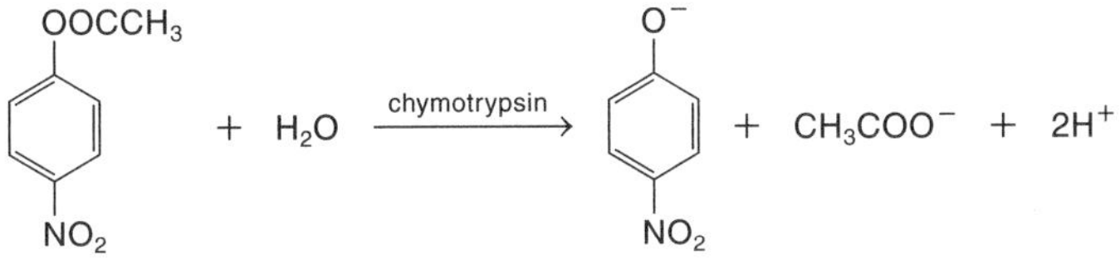

This reaction can be monitored spectrophotometrically because *p*-nitrophenyl acetate is colorless, whereas *p*-nitrophenolate is bright yellow, with a maximum absorbance at 400 nm. Hartley and Kilby found that in the presence of a large excess of *p*-nitrophenyl acetate,* the release of *p*-nitrophenolate was linear with time. When they extrapolated the absorbance at 400 nm back to zero time, however, they found that it did not converge to zero absorbance (Figure 10.10). Kinetic measurements showed that the reaction proceeds with an initial burst of *p*-nitrophenolate release, followed by the usual zero-order release of *p*-nitrophenolate from turnover of the enzyme when it reaches the steady-state limit. The burst corresponds to one mole of *p*-nitrophenolate for each mole of enzyme, suggesting that the burst is the result of a chemical reaction between *p*-nitrophenyl acetate and chymotrypsin.

The chymotrypsin study clearly demonstrated that the reaction is *biphasic* (proceeds in two phases): the rapid reaction of the substrate with the enzyme, which yields a stoichiometric amount of *p*-nitrophenolate followed by a slower, steady-state reaction that produces the acetate ion. The following kinetic scheme is consistent with Hartley and Kilby's observations:

$$\mathrm{E} + \mathrm{S} \underset{k_{-1}}{\overset{k_1}{\rightleftharpoons}} \mathrm{ES} \xrightarrow{k_2} \mathrm{ES}' + \mathrm{P}_1 \xrightarrow{k_3} \mathrm{E} + \mathrm{P}_2$$

where P_1 is *p*-nitrophenolate and P_2 is acetate. Furthermore, k_3 is the rate-

* A large excess of *p*-nitrophenyl acetate was used in the study because the enzyme has a very high K_M value.

determining step in the hydrolysis reaction. The reaction mechanism is

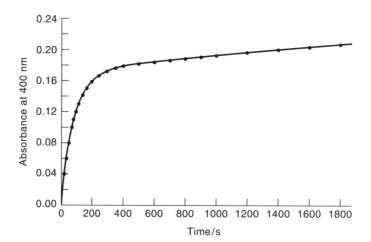

where X represents a nucleophilic group on the enzyme (En), which is the hydroxyl group of the serine residue at the active site. The first step is the rapid acylation of X by *p*-nitrophenol actetate, with the release of one equivalent mole of *p*-nitrophenolate in the burst.* Next is the slow hydrolysis of this acyl–enzyme intermediate (ES′), followed by the fast reacylation of the free enzyme by *p*-nitrophenol actetate, which accounts for the slow turnover of *p*-nitrophenolate production.

The chymotrypsin-catalyzed hydrolysis of *p*-nitrophenol acetate and related compounds is an example of *covalent hydrolysis*, a pathway in which part of the substrate forms a covalent bond with the enzyme to give an intermediate chemical species. In a second step, the intermediate undergoes another reaction to form the product and regenerate the free enzyme. The initial phase of the catalyzed reaction with *p*-nitrophenol acetate is so rapid that a stopped-flow apparatus must be employed to measure the progress of the reaction. However, the chymotrypsin-catalyzed hydrolysis of *p*-nitrophenyl trimethylacetate to *p*-nitrophenolate and trimethylacetate has the same characteristics as *p*-nitrophenyl acetate hydrolysis but proceeds much more slowly because the methyl groups constitute a steric barrier. Consequently, this reaction can be studied conveniently by means by a conventional spectrometer. Figure 10.11 shows a plot of the absorbance of *p*-nitrophenolate versus time with *p*-nitrophenyl trimethylacetate as the substrate.

Figure 10.11
The α-chymotrypsin-catalyzed hydrolysis of *p*-nitrotrimethylacetate at 298 K. [From M. L. Bender, F. J. Kézdy, and F. C. Wedler, *J. Chem. Educ.* **44**, 84 (1967).]

---

*In the formation of the ES complex, the proton from the hydroxyl group is transferred to a nearby histidine residue on chymotrypsin.

The kinetic analysis of this reaction—that is, the theoretical fit for the curve in Figure 10.10—starts with the following equations:

$$[E]_0 = [E] + [ES] + [ES']$$

$$\frac{d[P_1]}{dt} = k_2[ES]$$

$$\frac{d[P_2]}{dt} = k_3[ES']$$

$$\frac{d[ES']}{dt} = k_2[ES] - k_3[ES']$$

Because there are five unknowns ($k_2, k_3, [E]_0$, and two of the following three quantities: [E], [ES], and [ES']) and only four equations, we need one more equation. For this equation, we assume that the first step is a rapid equilibrium; that is,

$$E + S \underset{k_{-1}}{\overset{k_1}{\rightleftharpoons}} ES$$

and we write

$$K_S = \frac{k_{-1}}{k_1} = \frac{[E][S]}{[ES]}$$

From these equations, we can fit the curve shown in Figure 10.10 and solve for the pertinent kinetic constants.* Table 10.2 shows the results. For this mechanism, the quantity k_{cat} (catalytic rate constant) is defined by

$$k_{cat} = \frac{k_2 k_3}{k_2 + k_3} \tag{10.15}$$

For ester hydrolysis, $k_2 \gg k_3$, so k_{cat} is essentially equal to k_3.

Table 10.2
Kinetic Constants of the α-Chymotrypsin-Catalyzed Hydrolysis of p-Nitrophenyl Trimethylacetate at pH 8.2[a,b]

k_2	0.37 ± 0.11 s^{-1}
k_3	$(1.3 \pm 0.03) \times 10^{-4}$ s^{-1}
K_s	$(1.6 \pm 0.5) \times 10^{-3}$ M
k_{cat}	1.3×10^{-4} s^{-1}
K_M	5.6×10^{-7} M^{-1}

[a] From M. L. Bender, F. J. Kézdy, and F. C. Wedler, *J. Chem. Educ.* **44**, 84 (1967).
[b] 0.01 M tris–HCl buffer, ionic strength 0.06, $25.6 \pm 0.1°$C, 1.8% (v/v) acetonitrile–water.

10.4 Multisubstrate Systems

So far, we have considered enzyme catalysis involving only a single substrate, but in many cases, the process involves two or more substrates. For example, the reaction

$$C_2H_5OH + NAD^+ \rightleftharpoons CH_3CHO + NADH + H^+$$

is catalyzed by the enzyme alcohol dehydrogenase, which binds both NAD^+ and the substrate that is to be oxidized. Many of the principles developed for a single-substrate system may be extended to multisubstrate systems. Ignoring mathematical details, we shall briefly examine the different types of bisubstrate reactions—that is, reactions involving two substrates.

The overall picture of a bisubstrate reaction can be represented by

$$A + B \rightleftharpoons P + Q$$

where A and B are the substrates and P and Q the products. In most cases, these

* For the derivation, see Reference 3 on p. 6 of this text.

reactions involve the transfer of a specific functional group from one substrate (A) to the other (B). The binding of A and B to the enzyme can take place in different ways, which can be categorized as sequential or nonsequential mechanisms.

The Sequential Mechanism

In some reactions, the binding of both substrates must take place before the release of products. A sequential process can be further classified as follows.

Ordered Sequential Mechanism. In this mechanism, one substrate must bind before a second substrate can bind.

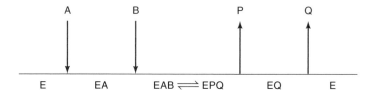

The enzyme and enzyme–substrate complexes are represented by horizontal lines, and successive additions of substrates and release of products are denoted by vertical arrows. Each vertical arrow actually represents the forward and reverse reaction. This mechanism is often observed in the oxidation of substrates by NAD^+.

Random Sequential Mechanism. The case in which the binding of substrates and the release of products do not follow a definite obligatory order is known as a random sequential mechanism. The general pathway is as follows:

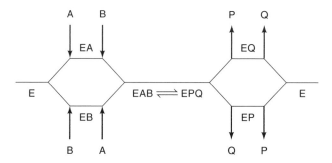

The phosphorylation of glucose by ATP to form glucose-6-phosphate, in which hexokinase is the enzyme in the first step of glycolysis, appears to follow such a mechanism.

The Nonsequential or "Ping-Pong" Mechanism

In this mechanism, one substrate binds, and one product is released. Then, a second substrate binds, and a second product is released.

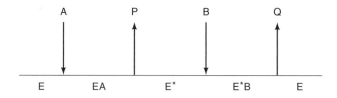

This process is called the "Ping-Pong mechanism" to emphasize the bouncing of the enzyme between the two states E and E^*, where E^* is a modified state of E, which often carries a fragment of A. An example of the Ping-Pong mechanism is the action by chymotrypsin (discussed on p. 372).

10.5 Enzyme Inhibition

Inhibitors are compounds that decrease the rate of an enzyme-catalyzed reaction. The study of enzyme inhibition has enhanced our knowledge of specificity and the nature of functional groups at the active site. The activity of certain enzymes is regulated by a feedback mechanism such that an end product inhibits the enzyme's function in an initial stage of a sequence of reactions (Figure 10.12). The glycolytic pathway is

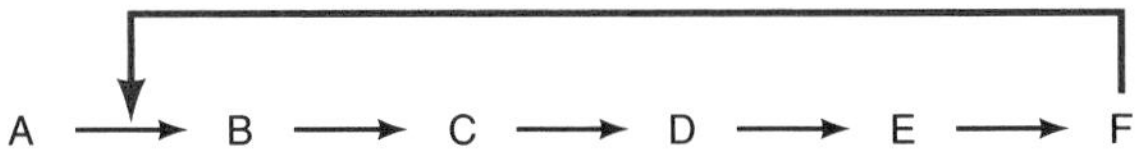

$$A \longrightarrow B \longrightarrow C \longrightarrow D \longrightarrow E \longrightarrow F$$

Figure 10.12
Control of regulatory enzymes frequently involves feedback mechanisms. In this sequence of reactions catalyzed by enzymes, the first enzyme in the series is inhibited by product F. At the early stages of the reaction, the concentration of F is low and its inhibitory effect is minimal. As the concentration of F reaches a certain level, it can lead to total inhibition of the first enzyme and hence turns off its own source of production. This action is analogous to a thermostat turning off heat supply when the ambient temperature reaches a preset level.

an example of this feedback mechanism. In effect, enzyme inhibition controls the amount of products formed.

The action of an inhibitor on an enzyme can be described as either reversible or irreversible. In *reversible inhibition*, an equilibrium exists between the enzyme and the inhibitor. In *irreversible inhibitions*, inhibition progressively increases with time. Complete inhibition results if the concentration of the irreversible inhibitor exceeds that of the enzyme.

Reversible Inhibition

There are three important types of reversible inhibition: competitive inhibition, noncompetitive inhibition, and uncompetitive inhibition. We shall discuss each type in turn.

Competitive Inhibition. In this case, both the substrate S and the inhibitor I compete for the same active site (Figure 10.13a). The reactions are

$$E + S \underset{k_{-1}}{\overset{k_1}{\rightleftharpoons}} ES \overset{k_2}{\longrightarrow} P + E$$

$$+$$

$$I$$

$$K_I \Big\updownarrow$$

$$EI$$

where

$$K_I = \frac{[E][I]}{[EI]} \qquad (10.16)$$

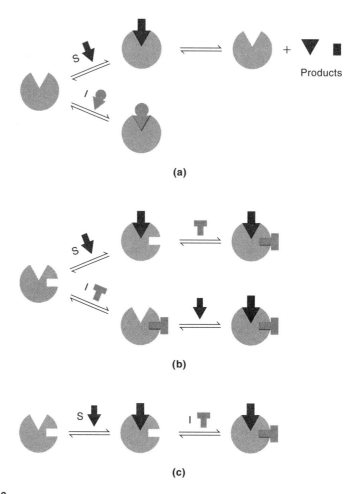

Figure 10.13
Three types of reversible inhibition. (a) Competitive inhibition. Both the substrate and the inhibitor compete for the same active site. Only the ES complex leads to production formation. (b) Noncompetitive inhibition. The inhibitor binds to a site other than the active site. The ESI complex does not lead to product formation. (c) Uncompetitive inhibition. The inhibitor binds only to the ES complex. The ESI complex does not lead to product formation.

Note that the complex EI does not react with S to form products. Applying the steady-state approximation for ES, we obtain*

$$v_0 = \frac{V_{max}[S]}{K_M\left(1 + \frac{[I]}{K_I}\right) + [S]} \tag{10.17}$$

Equation 10.17 has the same form as Equation 10.10, except that the K_M term has been modified by $(1 + [I]/K_I)$, thereby reducing v_0. The Lineweaver–Burk equation is given by

$$\frac{1}{v_0} = \frac{K_M}{V_{max}}\left(1 + \frac{[I]}{K_I}\right)\frac{1}{[S]} + \frac{1}{V_{max}} \tag{10.18}$$

---

*For derivations of Equations 10.17 and 10.19, see Reference 3 on p. 6 of this text.

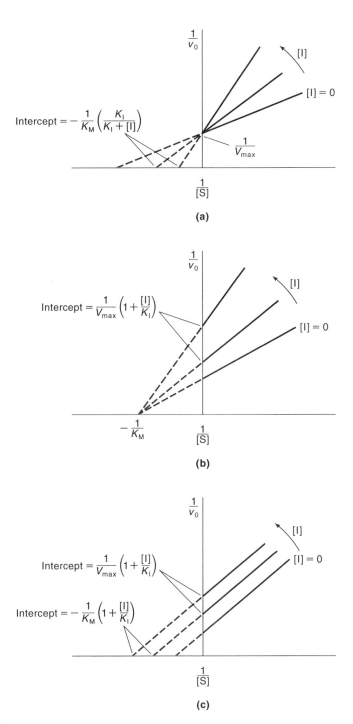

Figure 10.14
Lineweaver–Burk plots: (a) competitive inhibition, (b) noncompetitive inhibition, and (c) uncompetitive inhibition.

Thus, a straight line results when $1/v_0$ is plotted versus $1/[S]$ at constant [I] (Figure 10.14a). The difference between Equations 10.18 and 10.11 is that in the former, the slope is enhanced by the factor $(1 + [I]/K_I)$. The intercept on the $1/v_0$ axis is the same for Figures 10.13a and 10.8 because V_{max} does not change.

A well-known example of a competitive inhibitor is malonic acid, $CH_2(COOH)_2$, which competes with succinic acid in the dehydrogenation reaction catalyzed by succinic dehydrogenase:

$$
\begin{array}{c}
\text{COOH} \\
| \\
\text{CH}_2 \\
| \\
\text{CH}_2 \\
| \\
\text{COOH}
\end{array}
\quad
\overset{\text{Succinic}}{\underset{\text{dehydrogenase}}{\rightleftharpoons}}
\quad
\begin{array}{c}
\text{H}-\text{C}-\text{COOH} \\
\parallel \\
\text{HOOC}-\text{C}-\text{H}
\end{array}
$$

Succinic acid   Fumaric acid

Because malonic acid resembles succinic acid in structure, it can combine with the enzyme, although no product is formed in this reaction.

Dividing Equation 10.10 by Equation 10.17, we obtain

$$
\frac{v_0}{(v_0)_{\text{inhibition}}} = 1 + \frac{K_M[I]}{K_M K_I + [S]K_I}
$$

To overcome competitive inhibition, we need to increase the substrate concentration relative to that of the inhibitor; that is, at high substrate concentrations, $[S]K_I \gg K_M K_I$, so that

$$
\frac{v_0}{(v_0)_{\text{inhibition}}} \approx 1 + \frac{K_M[I]}{[S]K_I} \approx 1
$$

Noncompetitive Inhibition. A noncompetitive inhibitor binds to the enzyme at a site that is distinct from the substrate binding site; therefore, it can bind to both the free enzyme and the enzyme–substrate complex (see Figure 10.13b). The binding of the inhibitor has no effect on the substrate binding, and vice versa. The reactions are

$$
\begin{array}{ccccccc}
\text{E} & + & \text{S} & \underset{k_{-1}}{\overset{k_1}{\rightleftharpoons}} & \text{ES} & \overset{k_2}{\longrightarrow} & \text{E} + \text{P} \\
+ & & & & + & & \\
\text{I} & & & & \text{I} & & \\
K_I \updownarrow & & & & K_I \updownarrow & & \\
\text{EI} & + & \text{S} & \rightleftharpoons & \text{ESI} & &
\end{array}
$$

Neither EI nor ESI forms products. Because I does not interfere with the formation of ES, noncompetitive inhibition cannot be reversed by increasing the substrate concentration. The initial rate is given by

$$
v_0 = \frac{\dfrac{V_{\max}}{\left(1 + \dfrac{[I]}{K_I}\right)}[S]}{K_M + [S]} \tag{10.19}
$$

Comparing Equation 10.19 with Equation 10.10, we see that $V_{\max}$ has been reduced by the factor $(1 + [I]/K_I)$ but K_M is unchanged. The Lineweaver–Burk equation becomes

$$
\frac{1}{v_0} = \frac{K_M}{V_{\max}}\left(1 + \frac{[I]}{K_I}\right)\frac{1}{[S]} + \frac{1}{V_{\max}}\left(1 + \frac{[I]}{K_I}\right) \tag{10.20}
$$

From Figure 10.14b we see that a plot of $1/v_0$ versus $1/[S]$ gives a straight line with an increase in slope and intercept on the $1/v_0$ axis compared with that in Figure 10.8. Dividing Equation 10.10 by Equation 10.19, we get

$$\frac{v_0}{(v_0)_{\text{inhibition}}} = 1 + \frac{[I]}{K_I}$$

This result confirms our earlier statement that the extent of noncompetitive inhibition is independent of $[S]$ and depends only on $[I]$ and K_I.

Noncompetitive inhibition is very common with multisubstrate enzymes. Other examples are the reversible reactions between the sulfhydryl groups of cysteine residues on enzymes with heavy metal ions:

$$2\text{–SH} + \text{Hg}^{2+} \rightleftharpoons \text{–S–Hg–S–} + 2\text{H}^+$$

$$\text{–SH} + \text{Ag}^+ \rightleftharpoons \text{–S–Ag} + \text{H}^+$$

Uncompetitive Inhibition. An uncompetitive inhibitor does not bind to the free enzyme; instead, it binds reversibly to the enzyme–substrate complex to yield an inactive ESI complex (see Figure 10.13c). The reactions are

$$\text{E} + \text{S} \underset{k_{-1}}{\overset{k_1}{\rightleftharpoons}} \text{ES} \xrightarrow{k_2} \text{E} + \text{P}$$
$$+$$
$$\text{I}$$
$$K_I \updownarrow$$
$$\text{ESI}$$

where

$$K_I = \frac{[\text{ES}][\text{I}]}{[\text{ESI}]} \tag{10.21}$$

The ESI complex does not form a product. Again, because I does not interfere with the formation of ES, uncompetitive inhibition cannot be reversed by increasing the substrate concentration. The initial rate is given by (see Problem 10.16)

$$v_0 = \frac{\dfrac{V_{\text{max}}}{\left(1 + \dfrac{[I]}{K_I}\right)}[S]}{\dfrac{K_M}{\left(1 + \dfrac{[I]}{K_I}\right)} + [S]} \tag{10.22}$$

Comparison of Equation 10.22 with Equation 10.10 shows that both V_{max} and K_M have been reduced by the factor $(1 + [I]/K_I)$. The Lineweaver–Burk equation is given by

$$\frac{1}{v_0} = \frac{K_M}{V_{\text{max}}} \frac{1}{[S]} + \frac{1}{V_{\text{max}}} \left(1 + \frac{[I]}{K_I}\right) \tag{10.23}$$

Thus, a straight line is obtained by plotting $1/v_0$ versus $1/[S]$ at constant $[I]$ (see Figure 10.14c). The difference between Equation 10.23 and 10.11 is that the intercept on the $1/v_0$ axis is altered by the factor $(1 + [I]/K_I)$, but the slope remains the same. Dividing Equation 10.10 by Equation 10.22, we get

$$\frac{v_0}{(v_0)_{\text{inhibition}}} = \frac{K_M + [S](1 + [I]/K_I)}{K_M + [S]}$$

If conditions are such that $[S] \gg K_M$, then the equation above becomes

$$\frac{v_0}{(v_0)_{\text{inhibition}}} = \frac{[S] + [S][I]/K_I}{[S]} = 1 + \frac{[I]}{K_I}$$

Again we see that increasing the substrate concentration cannot overcome the effect of I in uncompetitive inhibition, just as in the case of noncompetitive inhibition.

Uncompetitive inhibition is rarely observed in one-substrate systems. Multisubstrate enzymes, however, often give parallel line plots with inhibitors.

Example 10.1

A chemist measured the initial rate of an enzyme-catalyzed reaction in the absence and presence of inhibitor A and, in a separate procedure, inhibitor B. In each case, the inhibitor's concentration was 8.0 mM $(8.0 \times 10^{-3}\ M)$. The following data were obtained:

$[S]/M$	$v_0/M \cdot \text{s}^{-1}$ No Inhibitor	$v_0/M \cdot \text{s}^{-1}$ Inhibitor A	$v_0/M \cdot \text{s}^{-1}$ Inhibitor B
5.0×10^{-4}	1.25×10^{-6}	5.8×10^{-7}	3.8×10^{-7}
1.0×10^{-3}	2.0×10^{-6}	1.04×10^{-6}	6.3×10^{-7}
2.5×10^{-3}	3.13×10^{-6}	2.00×10^{-6}	1.00×10^{-6}
5.0×10^{-3}	3.85×10^{-6}	2.78×10^{-6}	1.25×10^{-6}
1.0×10^{-2}	4.55×10^{-6}	3.57×10^{-6}	1.43×10^{-6}

(a) Determine the values of K_M and $V_{\max}$ of the enzyme. (b) Determine the type of inhibition imposed by inhibitors A and B, and calculate the value of K_I in each case.

ANSWER

Our first step is to convert the data to $1/[S]$ and $1/v_0$:

$(1/[S])/M^{-1}$	$(1/v_0)/M^{-1} \cdot \text{s}$ No Inhibitor	$(1/v_0)/M^{-1} \cdot \text{s}$ Inhibitor A	$(1/v_0)/M^{-1} \cdot \text{s}$ Inhibitor B
2.0×10^3	8.0×10^5	1.72×10^6	2.63×10^6
1.0×10^3	5.0×10^5	9.6×10^5	1.60×10^6
4.0×10^2	3.2×10^5	5.0×10^5	1.00×10^5
2.0×10^2	2.6×10^5	3.6×10^5	8.0×10^5
1.0×10^2	2.2×10^5	2.8×10^5	7.0×10^5

Next, we draw the Lineweaver–Burk plots for these three sets of kinetic data, as shown in Figure 10.15. Comparing Figure 10.15 with Figures 10.14a, 10.14b, and 10.14c shows that A is a competitive inhibitor and B is a noncompetitive inhibitor.

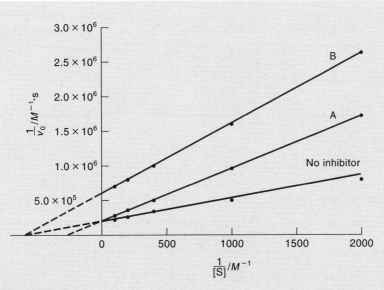

Figure 10.15
Lineweaver–Burk plots to determine the kinetic parameters and types of inhibition for Example 10.1.

(a) The computer linear fit for no inhibition is

$$\frac{1}{v_0} = 302.6 \frac{1}{[S]} + 1.96 \times 10^5$$

From Equation 10.11 we find that

$$\frac{1}{V_{\text{max}}} = 1.96 \times 10^5 \ M^{-1} \ s$$

or

$$V_{\text{max}} = 5.1 \times 10^{-6} \ M \ s^{-1}$$

From the slope of the line,

$$302.6 \ s = \frac{K_{\text{M}}}{V_{\text{max}}}$$

so

$$K_{\text{M}} = (302.6 \ s)(5.1 \times 10^{-6} \ M \ s^{-1})$$
$$= 1.5 \times 10^{-3} \ M$$

(b) The computer linear fit for inhibitor A is

$$\frac{1}{v_0} = 757.8 \frac{1}{[S]} + 2.03 \times 10^5$$

[Note that the slight difference in the $1/V_{\text{max}}$ value (2.03×10^5) compared with 1.96×10^5 for the no-inhibition plot is due to experimental uncertainty.] From Equation

10.18 the slope is equated as follows:

$$757.8 \text{ s} = \frac{K_M}{V_{max}}\left(1 + \frac{[I]}{K_I}\right)$$
$$= \frac{1.5 \times 10^{-3} \ M}{5.1 \times 10^{-6} \ M \ s^{-1}}\left(1 + \frac{[I]}{K_I}\right)$$

Because $[I] = 8.0 \times 10^{-3} \ M$,

$$K_I = 5.1 \times 10^{-3} \ M$$

The computer linear fit for inhibitor **B** is

$$\frac{1}{v_0} = 1015.3 \frac{1}{[S]} + 5.95 \times 10^5$$

From Equation 10.20, we express the slope as

$$1015.3 \text{ s} = \frac{K_M}{V_{max}}\left(1 + \frac{[I]}{K_I}\right)$$
$$= \frac{1.5 \times 10^{-3} \ M}{5.1 \times 10^{-6} \ M \ s^{-1}}\left(1 + \frac{[I]}{K_I}\right)$$

Because $[I] = 8.0 \times 10^{-3} \ M$,

$$K_I = 3.3 \times 10^{-3} \ M$$

Irreversible Inhibition

Michaelis–Menten kinetics cannot be applied to irreversible inhibition. The inhibitor forms a covalent linkage with the enzyme molecule and cannot be removed. The effectiveness of an irreversible inhibitor is determined not by the equilibrium constant but by the rate at which the binding takes place. Iodoacetamides and maleimides act as irreversible inhibitors to the sulfhydryl groups:

$$-SH + ICH_2CONH_2 \rightarrow -S-CH_2CONH_2 + HI$$

Another example is the action of diisopropyl phosphofluoridate (a nerve gas) on the enzyme acetylcholinesterase. When a nerve makes a muscle cell contract, it gives the cell a tiny squirt of acetylcholine molecules. Acetylcholine is called a neurotransmitter because it acts as a messenger between the nerve and the final destination (in this case, the muscle cell). Once they have performed the proper function, the acetylcholine molecules must be destroyed; otherwise, the resulting excess of this substance will hyperstimulate glands and muscle, producing convulsions, choking, and other distressing symptoms. Many victims of exposure to this nerve gas suffer paralysis or even death. The effective removal of excess acetylcholine is by means of a hydrolysis reaction (see Section 7.5):

$$CH_3COOCH_2CH_2-\overset{+}{N}(CH_3)_3 + H_2O \rightarrow HOCH_2CH_2-\overset{+}{N}(CH_3)_3 + CH_3COOH$$
$$\text{Acetylcholine} \qquad\qquad\qquad\qquad \text{Choline}$$

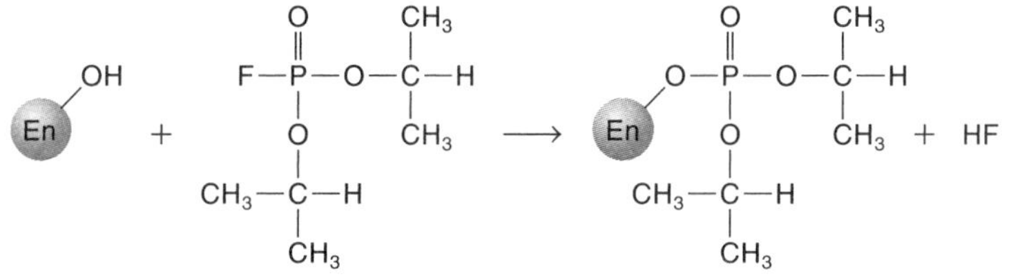

Diisopropyl phosphorofluoridate

Figure 10.16
An example of irreversible inhibition. The nerve gas diisopropyl phosphorofluoridate forms a strong covalent bond with the hydroxyl group of the serine residue at the active site of acetylcholinesterase.

The catalyst for this reaction is acetylcholinesterase. The irreversible inhibition of this enzyme takes place via the formation of a covalent bond between the phosphorus atom and the hydroxyl oxygen of the serine residue in the enzyme (Figure 10.16). The complex formed is so stable that for practical purposes the restoration of normal nerve function must await the formation of new enzyme molecules by the exposed person's body.

10.6 Allosteric Interactions

One class of enzymes has kinetics that do not obey the Michaelis–Menten description. Instead of the usual hyperbolic curve (see Figure 10.5), the rate equations of these enzymes produce a sigmoidal, or S-shaped, curve. This behavior is typically exhibited by enzymes that possess multiple binding sites and whose activity is regulated by the binding of inhibitors or activators. Sigmoidal curves are characteristic of positive cooperativity, which means that the binding of the ligand at one site increases the enzyme's affinity for another ligand at a different site. Enzymes that show cooperativity are called *allosteric* (from the Greek words *allos*, meaning different, and *steros*, meaning space or solid, which means conformation in our discussion). The term *effector* describes the ligand that can affect the binding at a different site on the enzyme. There are four types of allosteric interactions, depending on whether the ligands are of the same type (*homotropic effect*) or different type (*heterotropic effect*): positive or negative homotropic effect and positive or negative heterotropic effect. The words *positive* and *negative* here describe the enzyme's affinity for other ligands as a result of the binding to the effector.

Oxygen Binding to Myoglobin and Hemoglobin

The phenomenon of cooperativity was first observed for the oxygen–hemoglobin system. Although hemoglobin is not an enzyme, its mode of binding with oxygen is analogous to binding by allosteric enzymes. Figure 10.17 shows the percent saturation curves for hemoglobin and myoglobin. A hemoglobin molecule is made up of four polypeptide chains, two α chains of 141 amino acid residues each and two β chains of 146 amino acid residues each. Each chain contains a heme group. The iron atom in the heme group has octahedral geometry; it is bonded to the four nitrogen atoms of the heme group and the nitrogen atom of the histidine residue, leaving a sixth coordination site open for ligand binding (water or molecular oxygen). The four chains fold to form similar three-dimensional structures. In an intact hemoglobin molecule, these four chains, or *subunits*, are joined together to form a tetramer. The

Hemoglobin is sometimes referred to as the honorary enzyme.

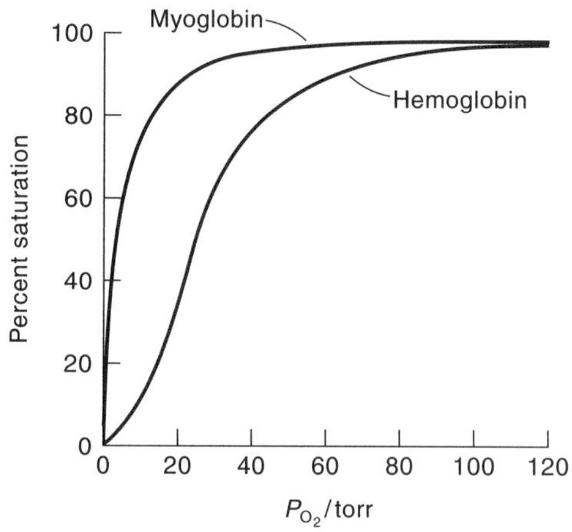

Figure 10.17
Oxygen saturation curves for myoglobin and hemoglobin.

less complex myoglobin molecule possesses only one polypeptide of 153 amino acids. It contains one heme group and is structurally similar to the β chain of hemoglobin. As we can see in Figure 10.17, the curve for myoglobin is hyperbolic, indicating that it binds noncooperatively with oxygen. This observation is consistent with the fact that there is only one heme group and hence only one binding site. On the other hand, the curve for hemoglobin is sigmoidal, indicating that its affinity for oxygen increases with the binding of oxygen.

Because of the great physiological significance of the binding of oxygen to hemoglobin, we need to look at the process in more detail. The oxygen affinity for hemoglobin depends on the concentration of several species in the red blood cell: protons, carbon dioxide, chloride ions, and 2,3-bisphosphoglycerate (BPG) [once known as 2,3-diphosphoglycerate (DPG)],

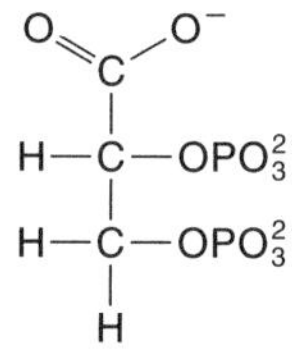

An increase in the concentration of any of these species shifts the oxygen binding curve (see Figure 10.17) to the right, which indicates a decrease in the oxygen affinity. Thus, all of these ligands act as negative heterotropic effectors. In the tissues, where the partial pressure of carbon dioxide and the concentration of H^+ ions are high, the oxyhemoglobin molecules have a greater tendency to dissociate into hemoglobin and oxygen, and the latter is taken up by myoglobin for metabolic processes. About two protons are taken up by the hemoglobin molecule for every four oxygen molecules released. The reverse effect occurs in the alveolar capillaries of the lungs. The high concentration of oxygen in the lungs drives off protons and carbon dioxide bound to deoxyhemoglobin. This reciprocal action, known as the *Bohr effect*, was first reported by the Danish physiologist Christian Bohr (1855–1911) in 1904. Figure 10.18a shows the effect of pH on the oxygen affinity of hemoglobin.

The effect of BPG on the oxygen affinity for hemoglobin was discovered by the American biochemists Reinhold Benesch (1919–1986) and Ruth Benesch (1925–) in 1967. They found that BPG binds only to deoxyhemoglobin and not to oxyhemoglobin, and that BPG reduces the oxygen affinity by a factor of about 25 (Figure 10.18b). The number of BPG molecules in the red blood cell is roughly the same as the number of hemoglobin molecules (280 million), but a shortage of oxygen triggers an increase in BPG, which promotes the release of oxygen. Interestingly, when

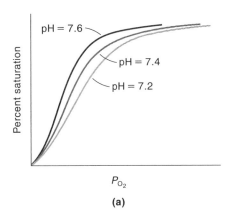
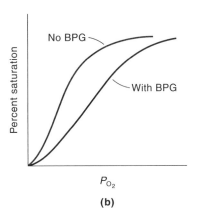

Figure 10.18
(a) The Bohr effect. A decrease in pH leads to a lowering of oxygen affinity for hemoglobin.
(b) The presence of BPG decreases the oxygen affinity for hemoglobin.

a person travels quickly from sea level to a high-altitude region, where the partial pressure of oxygen is low, the level of BPG in his or her red blood cells increases. This increase lowers the oxygen affinity of hemoglobin and helps maintain a higher concentration of free oxygen. The human fetus has its own kind of hemoglobin, called hemoglobin F, which consists of two α chains and two γ chains. This hemoglobin differs from adult hemoglobin A, which consists of two α chains and two β chains. Under normal physiological conditions, hemoglobin F has a higher oxygen affinity than hemoglobin A. This difference in affinity promotes the transfer of oxygen from the maternal to the fetal circulation. The higher oxygen affinity for hemoglobin F is due to the fact that this molecule binds BPG less strongly than hemoglobin A. The comparison of these systems also shows that BPG binds only to the β chains in hemoglobin A and only to the γ chains in hemoglobin F.

Finally, the binding of oxygen to myoglobin is not affected by any of these factors. It does vary, however, with temperature. The oxygen affinity of both myoglobin and hemoglobin decreases with increasing temperature.

The Hill Equation

We now present a phenomenological description for the binding of oxygen to myoglobin and hemoglobin. Consider, first, the binding of oxygen to myoglobin (Mb), because it is a simpler system. The reaction is

$$Mb + O_2 \rightleftharpoons MbO_2$$

The dissociation constant is given by

$$K_d = \frac{[Mb][O_2]}{[MbO_2]} \tag{10.24}$$

We define a quantity Y, the fractional saturation, as follows (see p. 210):

$$Y = \frac{[MbO_2]}{[MbO_2] + [Mb]} \tag{10.25}$$

From Equation 10.24 and 10.25,

$$Y = \frac{[O_2]}{[O_2] + K_d} \qquad (10.26)$$

Because O_2 is a gas, expressing its concentration in terms of its partial pressure is more convenient. Furthermore, if we represent the oxygen affinity for myoglobin as P_{50}, which is the partial pressure of oxygen when half, or 50%, of the binding sites are filled (that is, when $[Mb] = [MbO_2]$), it follows that

$$K_d = \frac{[Mb]P_{O_2}}{[MbO_2]} = P_{O_2} = P_{50}$$

and Equation 10.26 becomes

$$Y = \frac{P_{O_2}}{P_{O_2} + P_{50}} \qquad (10.27)$$

Rearranging, we have

$$\frac{Y}{1 - Y} = \frac{P_{O_2}}{P_{50}}$$

Taking the logarithm of both sides of the above equation, we obtain

$$\log \frac{Y}{1 - Y} = \log P_{O_2} - \log P_{50} \qquad (10.28)$$

Thus, a plot of $\log(Y/1 - Y)$ versus $\log P_{O_2}$ gives a straight line with a slope of unity (Figure 10.19).

Equation 10.28 describes the binding of myoglobin with oxygen quite well, but it does not hold for hemoglobin. Instead, it must be modified as follows:

$$\log \frac{Y}{1 - Y} = n \log P_{O_2} - n \log P_{50} \qquad (10.29)$$

A similar plot in this case yields a straight line with a slope of 2.8 (that is, $n = 2.8$), also shown in Figure 10.19. The fact that the slope is greater than unity indicates that the binding of hemoglobin with oxygen is cooperative. Note that we cannot explain

Figure 10.19
Plots of $\log(Y/1 - Y)$ versus $\log P_{O_2}$ for hemoglobin and myoglobin.

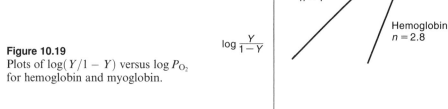

the binding phenomenon in this case by assuming that it is a higher-order reaction, because n is not an integer and it is not identical to the number of sites. (If all four sites were equivalent and independent of one another, we would analyze the binding curve using Equation 6.30.) Furthermore, the fact that the slope of the line at very low and very high partial pressures of oxygen tends to unity is inconsistent with a high-order mechanism, which predicts a constant slope at all partial pressures of oxygen. Equation 10.29 is often referred to as the *Hill equation* (after the British biochemist Archibald Vivian Hill, 1886–1977), and n is known as the *Hill coefficient*. The Hill coefficient is a measure of cooperativity—the higher n is, the higher the cooperativity. If $n = 1$, there is no cooperativity; if $n < 1$, there is negative cooperativity. The upper limit of n is the number of binding sites, which is 4 for hemoglobin.

What is the significance of cooperativity? In essence, it enables hemoglobin to be a more efficient oxygen transporter than myoglobin. The partial pressure of oxygen is about 100 torr in the lungs, compared with about 20 torr in the capillaries of muscle. Furthermore, the partial pressure for 50% saturation of hemoglobin is about 26 torr (see Figure 10.17). From Equation 10.29,

$$\frac{Y}{1-Y} = \left(\frac{P_{O_2}}{P_{50}}\right)^n$$

In the lungs,

$$\frac{Y_{\text{lung}}}{1 - Y_{\text{lung}}} = \left(\frac{100}{26}\right)^{2.8}$$

or

$$Y_{\text{lung}} = 0.98$$

In the muscles,

$$\frac{Y_{\text{muscle}}}{1 - Y_{\text{muscle}}} = \left(\frac{20}{26}\right)^{2.8}$$

$$Y_{\text{muscle}} = 0.32$$

The amount of oxygen delivered is proportional to ΔY, given by

$$\Delta Y = Y_{\text{lung}} - Y_{\text{muscle}} = 0.66$$

What would happen if the binding between hemoglobin and oxygen were not cooperative? In this case, $n = 1$ and we have, from Equation 10.27,

$$Y_{\text{lung}} = \frac{100}{100 + 26} = 0.79$$

In the muscles,

$$Y_{\text{muscle}} = \frac{20}{20 + 26} = 0.43$$

so that $\Delta Y = 0.36$. Thus, almost twice as much oxygen is delivered to the tissues when the binding of hemoglobin with oxygen is cooperative.

Equation 10.29 is an empirical approach to cooperativity; it says nothing about the mechanism involved. Over the past 60 years, several theories have been proposed to explain cooperativity. Next, we shall briefly discuss two theories that have played important roles in our understanding of allosteric interactions.

The Concerted Model

In 1965, Monod, Wyman, and Changeux proposed a theory, called the *concerted model*, to explain cooperativity.* Their theory makes the following assumptions: (1) Proteins are oligomers; that is, they contain two or more subunits. (2) Each protein molecule can exist in either of two states, called T (tense) and R (relaxed), which are in equilibrium. (3) In the absence of substrate molecules, the T state is favored. When substrate molecules are bound to the enzyme, the equilibrium gradually shifts to the R state, which has a higher affinity for the ligand. (4) All binding sites in each state are equivalent and have an identical dissociation constant for the binding of ligands (K_T for the T state and K_R for the R state). Figure 10.20 shows the concerted model for the binding of oxygen with hemoglobin.

The equilibrium constant, L_0, for the two states in the absence of oxygen (denoted by the subscript 0) is given by

$$L_0 = \frac{[T_0]}{[R_0]} \tag{10.30}$$

Because the T state is favored in the absence of O_2, L_0 is large, and only negligible amounts of the R state are present. When oxygen is present, the equilibrium shifts gradually to the R state, which has a higher affinity for oxygen. When four molecules of ligand are bound, virtually all of the hemoglobin molecules will be in the R state, which corresponds to the conformation of oxyhemoglobin. We define c as the ratio of the dissociation constants:

$$c = \frac{K_R}{K_T} \tag{10.31}$$

Because the R state has the higher affinity for O_2, c must be smaller than 1. The affinity of a subunit for the ligand depends solely on whether it is in the T or R state and not on whether the sites on neighboring units are occupied; thus, K_R and K_T are the same for all stages of saturation. When one ligand is bound, the [T]/[R] ratio changes by the factor c; when two ligands are bound, the [T]/[R] ratio changes by c^2, and so on. We can represent the successive ligand binding depicted in Figure 10.20 with the following equations (see Problem 10.28):

$$K_1 = cL_0$$
$$K_2 = cK_1 = c^2L_0$$
$$K_3 = cK_2 = c^3L_0$$
$$K_4 = cK_3 = c^4L_0$$

We see that the equilibrium between T and R shifts to the R form as more O_2 molecules are bound.

Note that if hemoglobin were always entirely in the T state, its binding of oxygen, although weak, would be completely noncooperative and characterized only by

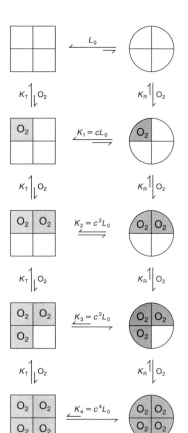

Figure 10.20
The concerted model for binding of oxygen with hemoglobin. The squares represent the tense state; the quarter circles represent the relaxed state.

*J. Monod, J. Wyman, and P. P. Changeaux, *J. Mol. Biol.* **12**, 88 (1965).

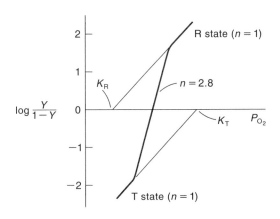

Figure 10.21
Plot of $\log(Y/1 - Y)$ versus P_{O_2} for hemoglobin.

K_T. Conversely, if hemoglobin were always entirely in the R state, its binding of oxygen, although strong, would also be completely noncooperative and characterized only by K_R. This noncooperativity is attributable to the fact that in any given hemoglobin molecule, all four subunits must either be in the R state (R_4) or the T state (T_4). Mixed forms, such as R_3T or R_2T_2, are considered nonexistent. For this reason, the model is called the "concerted," "all-or-none," or "symmetry-conserved" model. By fitting the oxygen saturation curve (Figure 10.21), we find that $L_0 = 9,000$ and $c = 0.014$. Thus, in the absence of oxygen, equilibrium greatly favors the T state by a factor of 9,000. On the other hand, the value of c shows that the binding of oxygen to a site in the R state is $1/0.014$, or 71 times stronger than to one in the T state. As the above equations show, the progressive binding of oxygen changes the [T]/[R] ratio from 9,000 (no O_2 bound) to 126 (one O_2 bound), 1.76 (two O_2 bound), 0.25 (3 O_2 bound), and 0.00035 (4 O_2 bound).

The concerted model cannot account for negative homotropic cooperativity. (Negative cooperativity means that as a result of the first ligand binding, the second ligand would bind *less* readily. Glyceraldehyde-3-phosphate dehydrogenase, an important enzyme in glycolysis, exhibits this behavior.) Nevertheless, it is remarkable that the allosteric behavior of hemoglobin (and enzymes) can be described by just three equilibrium constants (L_0, K_R, and K_T).

The Sequential Model

An alternative model of cooperativity suggested by Koshland, Némethy, and Filmer[*] assumes that the affinity of vacant sites for a particular ligand changes progressively as sites are taken up. Referring to the binding of oxygen to hemoglobin, this means that when an oxygen molecule binds to a vacant site on one of the four subunits, the interaction causes the site to change its conformation, which in turn affects the binding constants of the three sites that are still vacant (Figure 10.22). For this reason, this model is called the *sequential model*. Unlike the concerted model, the sequential model can have tetramers that consist of both R- and T-state subunits such as R_2T_2 or R_3T. This approach, too, predicts a sigmoidal curve. The affinity for O_2 molecules increases from left to right in Figure 10.22.

At present, the concerted and sequential models are both employed by biochemists in the study of enzymes. For hemoglobin, the actual mechanism seems more complex, and both models probably should be treated as limiting cases. In some cases, the sequential model has an advantage over the concerted model in that it can also account for negative homotropic cooperativity. Overall, these two models have

---

[*] D. E. Koshland, Jr., G. Némethy, and D. Filmer, *Biochemistry* **5**, 365 (1966).

Figure 10.22
The sequential model for oxygen binding with hemoglobin. The squares represent the tense state; the quarter circles represent the relaxed state. The binding of a ligand to a subunit changes the conformation of that subunit (from T to R). This transition increases the affinity of the remaining subunits for the ligand. The dissociation constants decrease from K_1 to K_4.

provided biochemists with deeper insight into the structure and function of many enzymes.

Conformational Changes in Hemoglobin Induced by Oxygen Binding

Finally, we ask the question: If the four heme groups are well separated from one another in the hemoglobin molecule (the closest distance between any two Fe atoms is approximately 25 Å), then how are they able to transmit information regarding binding of oxygen? We can reasonably assume that the communication among the heme groups, called "heme–heme interaction," takes place by means of some kind of conformational change in the molecule. Deoxyhemoglobin and oxyhemoglobin are known to form different crystals. X-ray crystallographic studies show that there are indeed structural differences between the completely oxygenated and completely deoxygenated hemoglobin molecules. At present, an intense research effort is underway to understand how the binding of O_2 to one heme group can trigger such extensive structural changes from one subunit to another. Nature has apparently devised a most ingenious mechanism for cooperativity in hemoglobin. The Fe^{2+} ion in deoxyhemoglobin is in the high-spin state ($3d^6$, with four unpaired electrons),* and it is too large to fit into the plane of the porphyrin ring of the heme group (Figure 10.23). Consequently, the iron atom lies about 0.4 Å above a slightly domed porphyrin. Upon binding to O_2, the Fe^{2+} ion becomes low spin and shrinks sufficiently

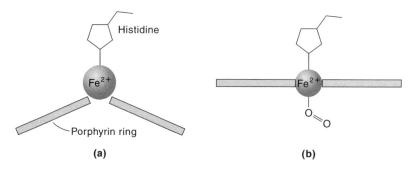

(a)   (b)

Figure 10.23
Schematic diagram showing the changes that occur when the heme group in hemoglobin binds an oxygen molecule. (a) The heme group in deoxyhemoglobin. The radius of the high-spin Fe^{2+} ion is too large to fit into the porphyrin ring. (b) When O_2 binds to Fe^{2+}, however, the ion shrinks somewhat so that it can fit in the plane of the ring. This movement pulls the histidine residue toward the ring and sets off a sequence of structural changes, thereby signaling the presence of an oxygen molecule at that heme group. The structural changes also drastically affect the affinity of the remaining heme groups for oxygen molecules.

---

*The electronic structure of the iron atom in the heme group is discussed in Chapter 12.

so that it can fit into the plane of the porphyrin ring, an arrangement that is energetically more favorable. (We can understand the change in size by recognizing that in the high-spin state, the $3d$ electrons are prohibited from coming too close to one another by the Pauli exclusion principle. Hence, the high-spin-state ion is larger than the low-spin-state ion.) When the Fe^{2+} ion moves into the porphyrin ring, it pulls the histidine ligand with it and sets off the chain of events that eventually lead to conformational changes in other parts of the molecule. This sequence is the means by which the binding of oxygen at one heme site is communicated to the other sites. To test this idea, chemists have replaced the histidine side chain with an imidazole ligand that resembles histidine but is detached from the polypeptide chain of the subunit:*

Imidazole

The movement of imidazole when oxygen is bound to the iron will have no effect on the conformation of the protein molecule. Indeed, results show that in this modified system, cooperativity is much attenuated but not totally eliminated. Apparently, other structural changes that are not yet fully understood also contribute to cooperativity.

10.7 pH Effects on Enzyme Kinetics

A useful way to understand the enzyme mechanism is to study the rate of an enzyme-catalyzed reaction as a function of pH. The activities of many enzymes vary with pH in a manner that can often be explained in terms of the dissociation of acids and bases. This is not too surprising because most active sites function as general acids and general bases in catalysis. Figure 10.24 shows a plot of the initial rate versus pH for the reaction catalyzed by the enzyme fumarase. As can be seen, the plot gives a bell-shaped curve. The pH at the maximum of the curve is called the pH *optimum*, which corresponds to the maximum activity of the enzyme; above or below this pH, the activity declines. Most enzymes that are active within cells have a pH optimum fairly close to the range of pH within which cells normally function. For example, the pH optima of two digestive enzymes, pepsin and trypsin, occur at about pH 2 and pH 8, respectively. The reasons are not hard to understand. Pepsin is secreted into the lumen of the stomach, where the pH is around 2. On the other hand, trypsin is secreted into and functions in the alkaline environment of the intestine, where the pH

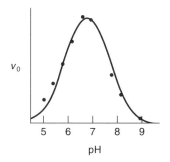

Figure 10.24
The effect of pH on the initial rate of the reaction catalyzed by the enzyme fumarase. [After C. Tanford, *Physical Chemistry of Macromolecules.* Copyright 1967 by John Wiley & Sons. Reprinted by permission of Charles Tanford.]

---

* See D. Barrick, N. T. Ho, V. Simplaceanu, F. W. Dahlquist, and C. Ho, *Nat. Struct. Biol.* **4**, 78 (1997).

is about 8. For general assays of enzyme activity, then, the solution should be buffered at the pH optimum for catalysis. Finally, when studying the influence of pH on enzyme activity, we should be careful to avoid gross structural changes brought about by the large changes in pH, such as protein denaturation.

The initial rate versus pH plot shown in Figure 10.24 yields much useful kinetic and mechanistic information about enzyme catalysis. In the simplest case, let us assume that an enzyme has two dissociable protons (say, from the –COOH and –NH$_3^+$ groups) with the zwitterion as the active form:

$$
\underset{\text{EnH}_2^+}{\overset{\text{HOOC} \quad \overset{+}{\text{NH}_3}}{\text{En}}} \quad \overset{pK_a}{\rightleftharpoons} \quad \underset{\text{EnH active}}{\overset{\text{H}^+ \\ + \\ {}^-\text{OOC} \quad \overset{+}{\text{NH}_3}}{\text{En}}} \quad \overset{pK_a'}{\rightleftharpoons} \quad \underset{\text{En}^-}{\overset{\text{H}^+ \\ + \\ {}^-\text{OOC} \quad \text{NH}_2}{\text{En}}}
$$

The concentration of the EnH form goes through a maximum as the pH is varied, so that the rate also passes through a maximum. The enzyme–substrate complex also may exist in three states of dissociation (as in the case of the free enzyme), with only the intermediate form capable of giving rise to products. Figure 10.25a shows the kinetic scheme for this reaction. At low substrate concentrations, the enzyme exists

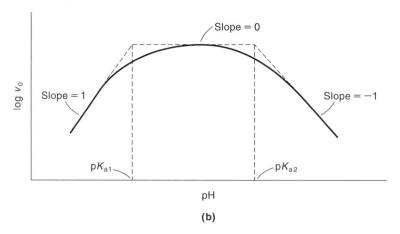

Figure 10.25
Effect of pH on enzyme kinetics. (a) Reaction scheme for the enzyme-catalyzed reaction. (b) Plot of $\log v_0$ versus pH according to Equations 10.34, 10.35, and 10.36. At the intercept of the lines with slopes 1 and 0, pH = pK_{a1}. Similarly, at the intercept of the lines with slopes −1 and 0, pH = pK_{a2}.

mostly in the free form; therefore, the pH is controlled by the dissociation of the free enzyme. Thus, analysis of the experimental pH dependence of the initial rate at low substrate concentrations provides information about pK_{a1} and pK_{a2} of the free enzyme. On the other hand, at high substrate concentrations, when the enzyme is saturated with substrate, analysis of pH dependence allows the determination of pK'_{a1} and pK'_{a2}, which relate to the dissociation of the enzyme–substrate complex.

In Figure 10.25a the rate is given by*

$$v_0 = \frac{k_2[E]_0[S]}{K_S\left(1 + \dfrac{K_{a2}}{[H^+]} + \dfrac{[H^+]}{K_{a1}}\right) + [S]\left(1 + \dfrac{K'_{a2}}{[H^+]} + \dfrac{[H^+]}{K'_{a1}}\right)} \tag{10.32}$$

At low (and constant) substrate concentrations, we can ignore the second term in the denominator in Equation 10.32, so that

$$v_0 = \frac{k_2[E]_0[S]}{K_S\left(1 + \dfrac{K_{a2}}{[H^+]} + \dfrac{[H^+]}{K_{a1}}\right)} \tag{10.33}$$

Consider the following three cases:

CASE 1. At low pH or high $[H^+]$, the term $[H^+]/K_{a1}$ predominates in the denominator of Equation 10.33, and we write

$$v_0 = \frac{k_2[E]_0[S]K_{a1}}{K_S[H^+]}$$

or

$$\log v_0 = \log\frac{k_2[E]_0[S]K_{a1}}{K_S} - \log[H^+]$$

$$= \text{constant} + \text{pH} \tag{10.34}$$

Thus, a plot of $\log v_0$ versus pH yields a straight line with a slope of $+1$ at low pH values.

CASE 2. At high pH or low $[H^+]$, the $K_{a2}/[H^+]$ term predominates in the denominator of Equation 10.33, and we have

$$v_0 = \frac{k_2[E]_0[S][H^+]}{K_S K_{a2}}$$

or

$$\log v_0 = \log\frac{k_2[E]_0[S]}{K_S K_{a2}} + \log[H^+]$$

$$= \text{constant} - \text{pH} \tag{10.35}$$

In this case, a plot of $\log v_0$ versus pH gives a straight line with a slope of -1.

---

* For the derivation, see Reference 3 on p. 6.

Table 10.3
pK_a Values of Amino Acids

Side Chain	Free State	Active Site	Enzyme
Glu	3.9	6.5	Lysozyme
His	6.0	5.2	Ribonuclease
Cys	8.3	4.0	Papain
Lys	10.8	5.9	Acetoacetate decarboxylase

CASE 3. At intermediate pH values, the first term (that is, 1) is the predominant term in the denominator of Equation 10.33. Therefore,

$$v_0 = \frac{k_2[E]_0[S]}{K_S}$$

or

$$\log v_0 = \log \frac{k_2[E]_0[S]}{K_S} \tag{10.36}$$

Because the term on the right is a constant, $\log v_0$ is *independent* of pH. The plot in Figure 10.25b shows these three situations and the determination of pK_{a1} and pK_{a2}.

Two points are worth noting. First, the above treatment is based on Michaelis–Menten kinetics. In reality, there may be more intermediates with additional dissociation constants, even for a one-substrate reaction. Second, as Table 10.3 shows, the pK_a values for the amino acid residues at the active site can be quite different from those of the corresponding free amino acids in solution (see Table 8.6). This deviation in pK_a values is the result of hydrogen bonding, electrostatic, and other types of interactions at the active site. Thus, as a rule, we do not rely solely on pK_a values to identify amino acids in enzyme catalysis; often pH dependence measurements are used in conjunction with spectroscopic and X-ray diffraction studies to construct a three-dimensional picture of the active site.

Suggestions for Further Reading

BOOKS

Bender, M. L. and L. J. Braubacher, *Catalysis and Enzyme Action*, McGraw-Hill, New York, 1973.

Boyer, P. D., Ed. *The Enzymes*, Academic Press, New York, 1970.

Copeland, R. A., *Enzymes: A Practical Introduction to Structure, Mechanism, and Data Analysis*, VCH Publishers, New York, 1996.

Dixon, M. and E. C. Webb, *Enzymes*, Academic Press, New York, 1964.

Fersht, A., *Structure and Mechanism in Protein Science*, W. H. Freeman, New York, 1999.

Gutfreund, H., *Enzymes: Physical Principles*, John Wiley & Sons, New York, 1975.

Klotz, I. M., *Ligand–Receptor Energetics*, John Wiley & Sons, New York, 1997.

Perutz, M., *Mechanisms of Cooperativity and Allosteric Regulation in Proteins*, Cambridge University Press, New York, 1990.

Segal, I. H., *Enzyme Kinetics*, John Wiley & Sons, New York, 1975.

Walsh, C., *Enzymatic Reaction Mechanisms*, W. H. Freeman, San Francisco, 1979.

Wyman, J. and S. J. Gill, *Binding and Linkage: Functional Chemistry of Biological Macromolecules*, University Science Books, Sausalito, CA, 1990.

ARTICLES

General

"α-Chymotrypsin: Enzyme Concentration and Kinetics," M. L. Bender, F. J. Kézedy, and F. C. Wedler, *J. Chem. Educ.* **44**, 84 (1967).

"Interactions of Enzymes and Inhibitors," B. R. Baker, *J. Chem. Educ.* **44**, 610 (1967).

"Enzyme Catalysis and Transition-State Theory," G. E. Linehard, *Science* **180**, 149 (1973).

"An Introduction to Enzyme Kinetics," A. Ault, *J. Chem. Educ.* **51**, 381 (1974).

"What Limits the Rate of an Enzyme-Catalyzed Reaction?" W. W. Cleland, *Acc. Chem. Res.* **8**, 145 (1975).

"Collision and Transition State Theory Approaches to Acid-Base Catalysis," H. B. Dunford, *J. Chem. Educ.* **52**, 578 (1975).

"Mechanisms of Action of Naturally Occurring Irreversible Enzyme Inhibitors," R. R. Rando, *Acc. Chem. Res.* **8**, 281 (1975).

"The Study of Enzymes," G. K. Radda and R. J. P. Williams, *Chem. Brit.* **12**, 124 (1976).

"Free Energy Diagrams and Concentration Profiles for Enzyme-Catalyzed Reactions," I. M. Klotz, *J. Chem. Educ.* **53**, 159 (1976).

"A Kinetic Investigation of an Enzyme-Catalyzed Reaction," W. G. Nigh, *J. Chem. Educ.* **53**, 668 (1976).

"Determination of the Kinetic Constants in a Two-Substrate Enzymatic Reaction," W. T. Yap, B. F. Howell, and R. Schaffer, *J. Chem. Educ.* **54**, 254 (1977).

"Entropy, Binding Energy, and Enzyme Catalysis," M. I. Page, *Angew. Chem. Int. Ed.* **16**, 449 (1977).

"K_m as an Apparent Dissociation Constant," J. A. Cohlberg, *J. Chem. Educ.* **56**, 512 (1979).

"RNA as an Enzyme," T. R. Cech, *Sci. Am.* November 1986.

"RNA's as Catalysts," G. M. McCorkle and S. Altman, *J. Chem. Educ.* **64**, 221 (1987).

"Enzyme Kinetics," O. Moe and R. Cornelius, *J. Chem. Educ.* **65**, 137 (1988).

"Homogeneous, Heterogeneous, and Enzymatic Catalysis," S. T. Oyama and G. A. Somorjai, *J. Chem. Educ.* **65**, 765 (1988).

"How Do Enzymes Work?" J. Krant, *Science* **242**, 533 (1988).

"A Kinetic Study of Yeast Alcohol Dehydrogenase," R. E. Utecht, *J. Chem. Educ.* **71**, 436 (1994).

"Chemical Oscillations in Enzyme Kinetics," K. L. Queeney, E. P. Marin, C. M. Campbell, and E. Peacock-López, **1996**, 1(3): S1430–4171 (96) 03035-X. Avali. URL: http://journals.springer-ny.com/chedr.

"Proteins and Enzymes," H. Bisswanger in *Encyclopedia of Applied Physics*, G. L. Trigg, Ed., VCH Publishers, New York, 1996, Vol. 15, p. 185.

"On the Meaning of K_m and V/K in Enzyme Kinetics," D. B. Northrop, *J. Chem. Educ.* **75**, 1153 (1998).

"Understanding Enzyme Inhibition," R. S. Ochs, *J. Chem. Educ.* **77**, 1453 (2000).

"The Temperature Optima of Enzymes: A New Perspective on an Old Phenomenon," R. M. Daniel, M. J. Danson, and R. Eisenthal, *Trends Biochem. Sci.* **26**, 223 (2001).

"A Perspective on Enzyme Catalysis," S. J. Benkovic and S. Hammes-Schiffer, *Science* **301**, 1196 (2003).

Allosteric Interactions

"The Control of Biochemical Reactions," J. P. Changeux, *Sci. Am.* April 1965.

"Demonstration of Allosteric Behavior," W. H. Sawyer, *J. Chem. Educ.* **49**, 777 (1972).

"Protein Shape and Biological Control," D. E. Koshland, Jr., *Sci. Am.* October 1973.

"Probe-Dependent Cooperativity in Hill-Plots," L. D. Byers, *J. Chem. Educ.* **54**, 352 (1977).

"Hemoglobin Structure and Respiratory Transport," M. F. Perutz, *Sci. Am.* December 1978.

"Ligand Binding to Macromolecules: Allosteric and Sequential Models of Cooperativity," V. L. Hess and A. Szabo, *J. Chem. Educ.* **56**, 289 (1979).

"A Structural Model for the Kinetic Behavior of Hemoglobin," K. Moffat, J. F. Deatherage, and D. W. Seybert, *Science* **206**, 1035 (1979).

"The Synthetic Analogs of O_2-Binding Heme Proteins," K. S. Suslick and T. J. Reinert, *J. Chem. Educ.* **62**, 974 (1985).

"An Alternative Analogy for the Dissociation of Oxyhemoglobin," T. S. Rao, R. B. Dabke, and D. B. Patil, *J. Chem. Educ.* **69**, 793 (1992).

"Hemoglobin as a Remarkable Molecular Pump," N. M. Senozan and E. Burton, *J. Chem. Educ.* **71**, 282 (1994).

"The Oxygen Dissociation Curve of Hemoglobin: Bridging the Gap between Biochemistry and Physiology," J. Gonez-Cambronero, *J. Chem. Educ.* **78**, 757 (2001).

"Cooperative Hemoglobin: Conserved Fold, Diverse Quaternary Assemblies and Allosteric Mechanisms," W. E. Royer, Jr., J. E. Knapp, K. Strand, and H. A. Heaslet, *Trends Biochem. Sci.* **26**, 297 (2001).

Problems

Michaelis–Menten Kinetics

10.1 Explain why a catalyst must affect the rate of a reaction in both directions.

10.2 Measurements of a certain enzyme-catalyzed reaction give $k_1 = 8 \times 10^6 \ M^{-1} \ s^{-1}$, $k_{-1} = 7 \times 10^4 \ s^{-1}$, and $k_2 = 3 \times 10^3 \ s^{-1}$. Does the enzyme–substrate binding follow the equilibrium or steady-state scheme?

10.3 The hydrolysis of acetylcholine is catalyzed by the enzyme acetylcholinesterase, which has a turnover rate of 25,000 s^{-1}. Calculate how long it takes for the enzyme to cleave one acetylcholine molecule.

10.4 Derive the following equation from Equation 10.10,

$$\frac{v_0}{[S]} = \frac{V_{max}}{K_M} - \frac{v_0}{K_M}$$

and show how you would obtain values of K_M and V_{max} graphically from this equation.

10.5 An enzyme that has a K_M value of $3.9 \times 10^{-5} \ M$ is studied at an initial substrate concentration of 0.035 M. After 1 min, it is found that 6.2 μM of product has been produced. Calculate the value of V_{max} and the amount of product formed after 4.5 min.

10.6 The hydrolysis of N-glutaryl-L-phenylalanine-p-nitroanilide (GPNA) to p-nitroaniline and N-glutaryl-L-phenylalanine is catalyzed by α-chymotrypsin. The following data are obtained:

$[S]/10^{-4} \ M$	2.5	5.0	10.0	15.0
$v_0/10^{-6} \ M \cdot min^{-1}$	2.2	3.8	5.9	7.1

where [S] = GPNA. Assuming Michaelis–Menten kinetics, calculate the values of V_{max}, K_M, and k_2 using the Lineweaver–Burk plot. Another way to treat the data is to plot v_0 versus $v_0/[S]$, which is the Eadie–Hofstee plot. Calculate the values of V_{max}, K_M, and k_2 from the Eadie–Hofstee treatment, given that $[E]_0 = 4.0 \times 10^{-6} \ M$. [*Source*: J. A. Hurlbut, T. N. Ball, H. C. Pound, and J. L. Graves, *J. Chem. Educ.* **50**, 149 (1973).]

10.7 The K_M value of lysozyme is $6.0 \times 10^{-6} \ M$ with hexa-N-acetylglucosamine as a substrate. It is assayed at the following substrate concentrations: **(a)** $1.5 \times 10^{-7} \ M$, **(b)** $6.8 \times 10^{-5} \ M$, **(c)** $2.4 \times 10^{-4} \ M$, **(d)** $1.9 \times 10^{-3} \ M$, and **(e)** 0.061 M. The initial rate measured at 0.061 M was 3.2 $\mu M \ min^{-1}$. Calculate the initial rates at the other substrate concentrations.

10.8 The hydrolysis of urea,

$$(NH_2)_2CO + H_2O \rightarrow 2NH_3 + CO_2$$

has been studied by many researchers. At 100°C, the (pseudo) first-order rate constant is $4.2 \times 10^{-5} \ s^{-1}$. The reaction is catalyzed by the enzyme urease, which at 21°C has a rate constant of $3 \times 10^4 \ s^{-1}$. If the enthalpies of activation for the uncatalyzed and catalyzed reactions are 134 kJ mol^{-1} and 43.9 kJ mol^{-1}, respectively, **(a)** calculate the temperature at which the non-enzymatic hydrolysis of urea would proceed at the same rate as the enzymatic hydrolysis at 21°C; **(b)** calculate the lowering of $\Delta G^{\ddagger}$ due to urease; and **(c)** comment on the sign of $\Delta S^{\ddagger}$. Assume that $\Delta H^{\ddagger} = E_a$ and that $\Delta H^{\ddagger}$ and $\Delta S^{\ddagger}$ are independent of temperature.

10.9 An enzyme is inactivated by the addition of a substance to a solution containing the enzyme. Suggest three ways to find out whether the substance is a reversible or an irreversible inhibitor.

10.10 Silver ions are known to react with the sulfhydryl groups of proteins and therefore can inhibit the action of certain enzymes. In one reaction, 0.0075 g of $AgNO_3$ is needed to completely inactivate a 5-mL enzyme solution. Estimate the molar mass of the enzyme. Explain why the molar mass obtained represents the minimum value. The concentration of the enzyme solution is such that 1 mL of the solution contains 75 mg of the enzyme.

10.11 The initial rates at various substrate concentrations for an enzyme-catalyzed reaction are as follows:

$[S]/M$	$v_0/10^{-6} \ M \cdot min^{-1}$
2.5×10^{-5}	38.0
4.00×10^{-5}	53.4
6.00×10^{-5}	68.6
8.00×10^{-5}	80.0
16.0×10^{-5}	106.8
20.0×10^{-5}	114.0

(a) Does this reaction follow Michaelis–Menten kinetics? **(b)** Calculate the value of V_{max} of the reaction. **(c)** Calculate the K_M value of the reaction. **(d)** Calculate the initial rates at $[S] = 5.00 \times 10^{-5} \ M$ and $[S] = 3.00 \times 10^{-1} \ M$. **(e)** What is the total amount of product formed during the first 3 min at $[S] = 7.2 \times 10^{-5} \ M$? **(f)** How would an increase in the enzyme concentration by a factor of 2 affect each of the following quantities: K_M, V_{max}, and v_0 (at $[S] = 5.00 \times 10^{-5} \ M$)?

10.12 An enzyme has a K_M value of $2.8 \times 10^{-5} \ M$ and a V_{max} value of 53 $\mu M \ min^{-1}$. Calculate the value of v_0 if $[S] = 3.7 \times 10^{-4} \ M$ and $[I] = 4.8 \times 10^{-4} \ M$ for **(a)** a competitive inhibitor, **(b)** a noncompetitive inhibitor, and **(c)** an uncompetitive inhibitor. ($K_I = 1.7 \times 10^{-5} \ M$ for all three cases.)

10.13 The degree of inhibition i is given by $i\% = (1 - \alpha)100\%$, where $\alpha = (v_0)_{\text{inhibition}}/v_0$. Calculate the percent inhibition for each of the three cases in Problem 10.12.

10.14 An enzyme-catalyzed reaction ($K_M = 2.7 \times 10^{-3} M$) is inhibited by a competitive inhibitor I ($K_I = 3.1 \times 10^{-5} M$). Suppose that the substrate concentration is $3.6 \times 10^{-4} M$. How much of the inhibitor is needed for 65% inhibition? How much does the substrate concentration have to be increased to reduce the inhibition to 25%?

10.15 Calculate the concentration of a noncompetitive inhibitor ($K_I = 2.9 \times 10^{-4} M$) needed to yield 90% inhibition of an enzyme-catalyzed reaction.

10.16 Derive Equation 10.22.

10.17 The metabolism of ethanol in our bodies is catalyzed by liver alcohol dehydrogenase (LADH) to acetaldehyde and finally to acetate. In contrast, methanol is converted to formaldehyde (also catalyzed by LADH), which can cause blindness or even death. An antidote for methanol is ethanol, which acts as a competitive inhibitor for LADH. The excess methanol can then be safely discharged from the body. How much absolute (100%) ethanol would a person have to consume after ingesting 50 mL of methanol (a lethal dosage) to reduce the activity of LADH to 3% of the original value? Assume that the total fluid volume in the person's body is 38 liters and that the densities of ethanol and methanol are 0.789 g mL^{-1} and 0.791 g mL^{-1}, respectively. The K_M value for methanol is $1.0 \times 10^{-2} M$, and the K_I value for ethanol is $1.0 \times 10^{-3} M$. State any assumptions.

Allosteric Interactions

10.18 **(a)** What is the physiological significance of cooperative O_2 binding by hemoglobin? Why is O_2 binding by myoglobin not cooperative? **(b)** Compare the concerted model with the sequential model for the binding of oxygen with hemoglobin.

10.19 Fatality usually results when more than 50% of a human being's hemoglobin is complexed with carbon monoxide. Yet a person whose hemoglobin content is diminished by anemia to half its original content can often function normally. Explain.

10.20 Competitive inhibitors, when present in small amounts, often act as activators to allosteric enzymes. Why?

10.21 What is the advantage of having the heme group in a hydrophobic region in the myoglobin and hemoglobin molecule?

10.22 What is the effect of each of the following actions on oxygen affinity of adult hemoglobin (Hb A) *in vitro*? **(a)** Increase pH, **(b)** increase partial pressure of CO_2, **(c)** decrease [BPG], **(d)** dissociate the tetramer into monomers, and **(e)** oxidize Fe(II) to Fe(III).

10.23 Although it is possible to carry out X-ray diffraction studies of fully deoxygenated hemoglobin and fully oxygenated hemoglobin, it is much more difficult, if not impossible, to obtain crystals in which each hemoglobin molecule is bound to only one, two, or three oxygen molecules. Explain.

10.24 When deoxyhemoglobin crystals are exposed to oxygen, they shatter. On the other hand, deoxymyoglobin crystals are unaffected by oxygen. Explain.

Additional Problems

10.25 An enzyme contains a single dissociable group at its active site. For catalysis to occur, this group must be in the dissociated (that is, negative) form. The substrate bears a net positive charge. The reaction scheme can be represented by

$$EH \rightleftharpoons H^+ + E^-$$
$$E^- + S^+ \rightleftharpoons ES \rightarrow E + P$$

(a) What kind of inhibitor is H^+? **(b)** Write an expression for the initial rate of the reaction in the presence of the inhibitor.

10.26 The discovery in the 1980s that certain RNA molecules (the ribozymes) can act as enzymes was a surprise to many chemists. Why?

10.27 The activation energy for the decomposition of hydrogen peroxide,

$$2H_2O_2(aq) \rightarrow 2H_2O(l) + O_2(g)$$

is 42 kJ mol^{-1}, whereas when the reaction is catalyzed by the enzyme catalase, it is 7.0 kJ mol^{-1}. Calculate the temperature that would cause the nonenzymatic catalysis to proceed as rapidly as the enzyme-catalyzed decomposition at 20°C. Assume the pre-exponential factor to be the same in both cases.

10.28 Referring to the concerted model discussed on p. 390, show that $K_1 = cL_0$.

10.29 The following data were obtained for the variation of $V_{\max}$ with pH for a reaction catalyzed by α-amylase at 24°C. What can you conclude about the pK_a values of the ionizing groups at the active site?

pH	3.0	3.5	4.0	4.5	5.0	5.5
$V_{\max}$ (arbitrary units)	200	501	1584	1778	3300	5248

pH	6.0	6.5	7.0	7.5	8.0	8.5	9.0
$V_{\max}$ (arbitrary units)	5250	5251	2818	2510	1585	398	158

10.30 **(a)** Comment on the following data obtained for an enzyme-catalyzed reaction (no calculations are needed):

$t/°C$	10	15	20	25	30	35	40	45
$V_{\max}$ (arbitrary units)	1.0	1.7	2.3	2.6	3.2	4.0	2.6	0.2

(b) Referring to Equation 10.8, under what conditions will an Arrhenius plot (that is, $\ln k$ versus $1/T$) yield a straight line?

10.31 Crocodiles can be submerged in water for a prolonged period of time (up to an hour), while drowning their preys. It is known that BPG does not bind to the crocodile deoxyhemoglobin but the bicarbonate ion does. Explain how this action enables crocodiles to utilize practically all of the oxygen bound to hemoglobin.

10.32 Give an explanation for the Lineweaver–Burk plot for a certain enzyme-catalyzed reaction shown below.

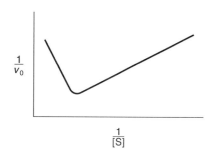

10.33 The following Arrhenius plot has been obtained for a certain enzyme. Account for the shape of the plot.

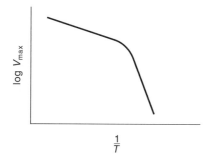

10.34 In Lewis Carroll's tale "Through the Looking Glass," Alice wonders whether looking-glass milk on the other side of the mirror would be fit to drink. What do you think?

10.35 Referring to Problem 9.64, calculate k_D for an enzyme–substrate reaction such as that between an enzyme ($D \approx 4 \times 10^{-7}$ cm^2 s^{-1}) and a substrate ($D \approx 5 \times 10^{-6}$ cm^2 s^{-1}) at 20°C. The distance between the enzyme and the substrate may be taken as 5×10^{-8} cm. Compare your result with the k_{cat}/K_M values listed in Table 10.1.

10.36 When fruits such as apples and pears are cut, the exposed areas begin to turn brown. This is the result of an oxidation reaction. Often the browning action can be prevented or slowed by adding a few drops of lemon juice. What is the chemical basis for this treatment?

10.37 "Dark meat" (the legs) and "white meat" (the breast) are one's choices when eating a turkey. Explain what causes the meat to assume different colors.

10.38 Despite what you may have read in science fiction novels or seen in horror movies, it is extremely unlikely that insects can ever grow to human size. Why?

10.39 The first-order rate constant for the dehydration of carbonic acid,

$$H_2CO_3(aq) \rightleftharpoons CO_2(g) + H_2O(l)$$

is about 1×10^2 s^{-1}. In view of this rather high rate constant, explain why it is necessary to have the enzyme carbonic anhydrase to enhance the rate of dehydration in the lungs.

10.40 Referring to Equation 10.12, sketch the Eadie–Hofstee plots for **(a)** a competitive inhibitor and **(b)** a noncompetitive inhibitor.

CHAPTER 11

Quantum Mechanics and Atomic Structure

Thus far, we have focused mainly on the bulk properties of matter. Thermodynamics and chemical kinetics provide important information regarding chemical processes, but they do not explain what takes place at the molecular level during these processes. Now we shall take a close look at the properties of atoms and molecules. To do so, we need to become familiar with quantum mechanics. In this chapter, we briefly describe the development of quantum theory, proposed by Max Planck in 1900. To understand Planck's quantum theory, we must first have some idea about the nature of radiation. Because radiation involves the emission and transmission of energy in the form of waves through space, we shall start with a discussion of the properties of waves and the wave theory of light.

11.1 The Wave Theory of Light

The first quantitative investigation of the nature of light was carried out by Newton in the 17th century. Using a glass prism, he showed that sunlight is composed of seven different colors and that these colors can be recombined with the aid of a second prism, turned opposite to the first, to produce white light. The work of physicists in the 18th and 19th centuries firmly established the fact that light has wave properties.

Figure 11.1 shows the propagation of a sinusoidal wave along the x direction. The velocity of the wave, v, is given by

$$v = \lambda v \tag{11.1}$$

where λ is the wavelength (in cm or m) and v is the frequency of the wave (in s^{-1} or hertz, Hz, after the German physicist Heinrich Rudolf Hertz, 1857–1894).

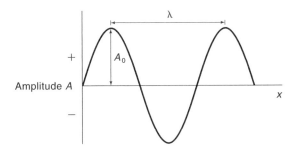

Figure 11.1
Sinusoidal wave of the form $A = A_0 \sin x$, where A_0 is the amplitude of the wave.

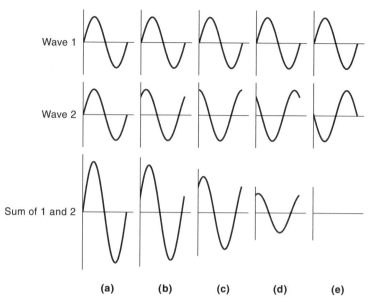

Figure 11.2
Constructive and destructive interference between two waves of equal wavelength and amplitude: (a) two waves completely in phase; (b)–(d) two waves partially out of phase; and (e) two waves exactly out of phase.

The interference phenomenon is a convincing demonstration of the wave theory of light. Consider the interaction of two waves in space, as shown in Figure 11.2. Depending on the relative displacement or the *phase difference* (that is, whether the maxima and minima of the waves occur at the same points in space), the interaction can lead to constructive or destructive interference. Experimentally, this phenomenon can be observed by using the arrangement shown in Figure 11.3. A light source is directed toward a filter, which selects light of approximately one wavelength. Slits S_1 and S_2 are small openings (compared with the distance between them) that can act as two separate light sources. Interference occurs between these two waves, and constructive and destructive patterns are observed on the screen as alternate bright and dark regions.

Maxwell showed in 1873 that light is just one form of electromagnetic radiation. Others are microwaves, infrared, ultraviolet (UV), X rays, and so on. An elec-

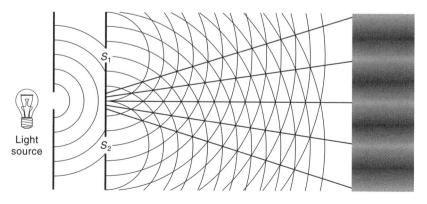

Figure 11.3
Two-slit experiment demonstrating the interference phenomenon. The pattern on the screen consists of alternating bright and dark bands.

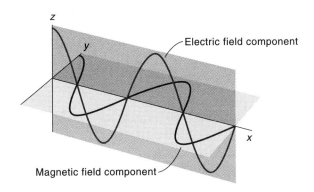

Figure 11.4
Electric-field component and magnetic-field component of an electromagnetic wave.
The wave travels along the x direction.

tromagnetic wave consists of an electric-field component and a mutually perpendicular magnetic-field component oscillating in space with a frequency v. The direction of oscillation is perpendicular to the direction of wave propagation (Figure 11.4). For ordinary, unpolarized light, the electric and magnetic components can and do rotate about the x axis (the direction of propagation), although they are always perpendicular to each other. In polarized light, these two components can oscillate only within the two fixed planes (the xy and xz planes). We shall discuss this point further in Chapter 14.

Figure 11.5 shows the regions of the electromagnetic spectrum, along with their wavelengths and frequencies. The speed of light depends on the medium through which it travels, but for most purposes (in air or in a vacuum) it can be taken as 3.00×10^8 m s^{-1}.

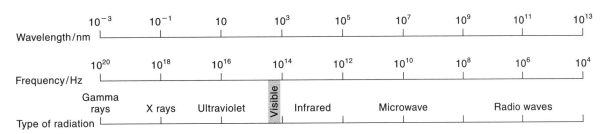

Figure 11.5
Types of electromagnetic radiation. The range of visible light extends from a wavelength of 400 nm (violet) to 700 nm (red).

11.2 Planck's Quantum Theory

Toward the end of the 19th century, physics was in a secure state. The wave theory of light was well established, and Newtonian mechanics, since its formulation in the 17th century, successfully described the motion of systems ranging from billiard balls to planets. The science of thermodynamics had become a powerful tool for solving chemical and physical problems. This comfortable state of affairs was not a lasting one, however. In 1899, the German physicists Otto R. Lummer (1860–1925) and

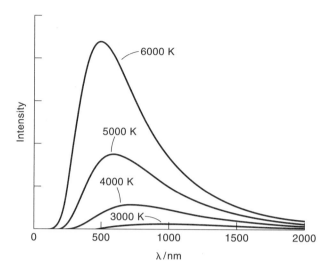

Figure 11.6
Blackbody radiation curves at various temperatures.

Ernst Pringsheim (1859–1917), among others, studied the emission of radiation by solids as a function of temperature and obtained a series of curves that could not be explained by the wave theory or by the laws of thermodynamics. The search for a proper explanation quickly led to a new and exciting era in physics.

All bodies at a temperature above absolute zero emit radiation over a range of wavelengths. The red glow of an electric heater and the bright white light of an electric bulb are familiar examples. If we measured the intensity of radiation versus the wavelength emitted at different temperatures, we would obtain a series of curves similar to the ones shown in Figure 11.6. These plots are commonly referred to as blackbody radiation curves. A *blackbody* is defined as a perfect absorber because it absorbs all the radiation that strikes it. Because it is in thermal equilibrium with its surroundings, a blackbody is also a perfect emitter of radiation.

In 1900, the German physicist Max Planck (1858–1947) solved the mystery of the blackbody radiation curves with an assumption that departed drastically from classical physics. In classical physics, it was assumed that the radiant energy emitted by a collection of oscillators (atoms or molecules) in a solid could have any energy value within a continuous range. This approach, however, predicts a radiation profile that has no maximum and goes to infinite intensity at a very short wavelength (an effect called the *ultraviolet catastrophe*). What Planck proposed was that radiant energy could not have any arbitrary value; instead, the energy could be emitted only in small, discrete amounts that he called *quanta*. The energy of the emitted radiation, E, is proportional to the frequency v of the oscillator:

$$E \propto v$$

$$E = hv \tag{11.2}$$

where h is Planck's constant, equal to 6.626×10^{-34} J s. According to Planck's quantum theory, energy is always emitted in multiples of hv; for example, $hv, 2hv, 3hv, \ldots$, but never $1.68hv$, or $3.52hv$. The difference between the classical model and Planck's model for energy variation is illustrated in Figure 11.7.

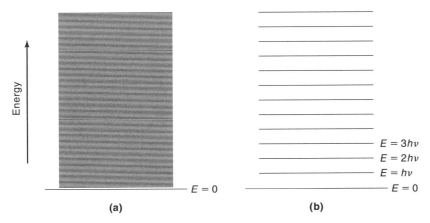

Figure 11.7
Energy variation for an oscillator: (a) the classical model and (b) Planck's model. The spacing between successive levels in (a) is so small that energy can be considered to vary continuously.

11.3 The Photoelectric Effect

In science, a single major discovery or the formulation of an important theory can trigger an avalanche of activity. This was the case with the quantum theory. Within a few years, Planck's hypothesis helped explain many previously puzzling observations. One of these conundrums was the photoelectric effect.

When light of a certain frequency shines on a clean metal surface, electrons are ejected from the metal. Experimentally, it is found that (1) the number of electrons ejected is proportional to the intensity of light; (2) the kinetic energy of the ejected electrons is proportional to the frequency of incident light; and (3) no electrons can be ejected if the frequency of the light is lower than a certain value, called the *threshold frequency* (v_0). Figure 11.8 shows an apparatus for studying the photoelectric effect.

According to the wave theory of light, the energy of radiation is proportional to the square of the amplitude. Thus, the energy is related to the intensity, not the frequency, of radiation. This seems to contradict point 2 above. In 1905, the German–American physicist Albert Einstein (1879–1955) resolved this difficulty in the following manner. He assumed that light consisted of particles called light quanta, or *photons*, of energy hv, where v is the frequency of light.* We can think of light striking on the metal as a collision between photons and electrons. According to the law of conservation of energy, we have energy input equal to energy output. If v is above the threshold frequency, then the Einstein photoelectric equation is

$$hv = \Phi + \tfrac{1}{2}m_e v^2 \tag{11.3}$$

where Φ—called the work function—represents the energy that the photon must possess to remove an electron from the metal, and $\tfrac{1}{2}m_e v^2$ is the kinetic energy of the ejected electron. The work function, Φ, is a measure of how strongly the electrons are

---

*The energy of the photon has the same expression as Equation 11.2 because electromagnetic radiation is emitted and absorbed in the form of photons.

Figure 11.8
An apparatus for studying the photoelectric effect. Light of a certain frequency falls on a clean metal surface. Ejected electrons are attracted toward the positive electrode. The flow of electrons is registered by a detecting meter. Not shown is the grid applying the retarding potential that allows the kinetic energy of the ejected electrons to be measured.

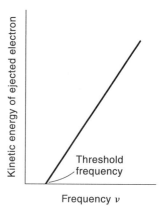

Figure 11.9
Plot of the kinetic energy of ejected electrons versus the frequency of incident radiation.

held in the metal. A plot of the kinetic energy of the ejected electrons versus the frequency of light is shown in Figure 11.9.

Equation 11.3 enables us to explain experimental observations. Because the number of photons increases with the intensity of light, more electrons are ejected at higher intensities. Furthermore, the energy of the photons increases with the frequency of light so that electrons ejected at higher frequencies will also possess higher kinetic energies.

Example 11.1

The color of chlorophyll is a consequence of the absorption by this molecule of blue light at about 435 nm and red light at about 680 nm, so that mostly green light is transmitted. Calculate the energy per mole of photons at these wavelengths.

ANSWER

From Equation 11.1 and $E = h\nu$ for photons, we calculate the energy of the photons with a wavelength of 435 nm as

$$E = \frac{hc}{\lambda} = \frac{(6.626 \times 10^{-34} \text{ J s})(3.00 \times 10^8 \text{ m s}^{-1})}{435 \text{ nm } (1 \times 10^{-9} \text{ m}/1 \text{ nm})}$$
$$= 4.57 \times 10^{-19} \text{ J}$$

This is the energy of one photon at this wavelength. For one mole of photons, we have

$$E = (4.57 \times 10^{-19} \text{ J})(6.022 \times 10^{23} \text{ mol}^{-1})$$
$$= 2.75 \times 10^5 \text{ J mol}^{-1}$$
$$= 275 \text{ kJ mol}^{-1}$$

Similarly, for the photons at 680 nm,

$$E = 176 \text{ kJ mol}^{-1}$$

By answering one question about light, Equation 11.3 poses another: What is the nature of light? On the one hand, the wave properties of light have been proved beyond doubt. On the other hand, the photoelectric effect can be explained only in terms of particulate photons. Can light be both wavelike and particlelike? This idea was strange and unfamiliar at the time the quantum theory was postulated, but scientists were beginning to realize that submicroscopic particles behave very differently from macroscopic objects.

11.4 Bohr's Theory of the Hydrogen Emission Spectrum

Einstein's work paved the way for the solution of yet another 19th-century mystery in physics: the emission spectra of atoms.

It had long been known that when atoms are subjected to high temperatures or an electrical discharge, they emit electromagnetic radiation that has characteristic frequencies. Figure 11.10 shows the arrangement for studying the emission spectrum of atomic hydrogen, which consists of a series of sharp, well-defined lines. Different atoms give rise to different sets of frequencies. Although the origin of these lines was not understood, this phenomenon was used to identify elements in unknown samples or in distant stars by matching their spectra with those of known elements.

Based on experimental data, the Swedish physicist Johannes Rydberg (1854–1919) formulated the following equation, which fits all the observed lines in the hydrogen emission spectra:

$$\tilde{v} = \frac{1}{\lambda} = R_H \left(\frac{1}{n_f^2} - \frac{1}{n_i^2} \right) \tag{11.4}$$

Equation 11.4 is known as the Rydberg formula, where $\tilde{v}$ is the wavenumber (number of waves per centimeter or per meter; it is a common unit in spectroscopy), R_H is the *Rydberg constant*, given by 109,737 cm^{-1}, and n_f and n_i are integers ($n_i > n_f$). The emission lines could be grouped according to particular values of n_f. Table 11.1 lists five series in the emission spectrum of hydrogen, which are named for their discoverers.

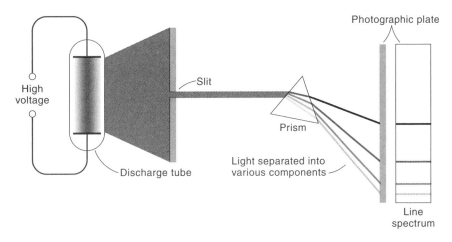

Figure 11.10
Experimental arrangement for studying the emission spectra of atoms and molecules. The gas (hydrogen) under study is placed in a discharge tube containing two electrodes. As electrons flow from the cathode to the anode, they collide with the H$_2$ molecules, which then dissociate into atoms. The H atoms are formed in an excited state and quickly decay to the ground state with emission of light. The emitted light is spread into various components by a prism. Each component color is focused at a definite position according to its wavelength and forms an image of the slit on the screen (or photographic plate). The color images of the slit are called spectral lines.

Table 11.1
Series in the Emission Spectrum of Atomic Hydrogen

Series	n_f	n_i	Region
Lyman	1	$2, 3, \ldots$	UV
Balmer	2	$3, 4, \ldots$	Visible, UV
Paschen	3	$4, 5, \ldots$	IR
Brackett	4	$5, 6, \ldots$	IR
Pfund	5	$6, 7, \ldots$	IR

While useful, the Rydberg formula does not explain the origin of the spectral lines. In 1913 the Danish physicist Niels Bohr (1885–1962) presented a theory that accounts for the hydrogen emission spectrum. Bohr's model was that the electron in the hydrogen atom moves around the nucleus only in certain allowed circular orbits. This means that its energies are quantized and we can speak of the electron being in a particular energy level corresponding to a specific orbit. Using classical physics and Planck's quantum theory, Bohr showed that the energies an electron can possess in the hydrogen atom are given by

$$E_n = -hcR_H\left(\frac{1}{n^2}\right) \tag{11.5}$$

where E_n is the energy of the electron in the nth level ($n = 1, 2, 3 \ldots$), h the Planck constant, and c the speed of light. The negative sign in this equation means that the allowed energies of the electron are *less* than the case in which the electron and the proton are infinitely separated, which is arbitrarily assigned to be zero. The more negative E_n is, the stronger the attraction between the electron and the proton. Thus, the most stable state is the one given by $n = 1$, which is called the *ground state*. An *excited state* is one for which $n \geq 2$.

Equation 11.5 provides a basis for analyzing the emission spectra of atomic hydrogen. Consider a certain electronic transition from a higher energy level (E_i) to a lower one (E_f). The condition for this transition is

This is sometimes referred to as the resonance condition.

$$\Delta E = E_f - E_i = h\nu \tag{11.6}$$

where $h\nu$ is the energy of the emitted photon and $E_i > E_f$ (Figure 11.11a). From Equations 11.5 and 11.6 we write

$$\Delta E = h\nu = hcR_H\left(\frac{1}{n_i^2} - \frac{1}{n_f^2}\right) \tag{11.7}$$

Figure 11.11
Interaction of electromagnetic radiation with atoms and molecules. (a) Absorption, and (b) emission. In each case, the energy of the photon ($h\nu$) is equal to ΔE, the energy difference between the two levels.

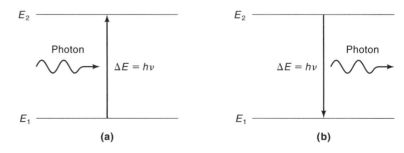

Hence

$$\frac{v}{c} = \frac{1}{\lambda} = \tilde{v} = R_\mathrm{H}\left(\frac{1}{n_\mathrm{i}^2} - \frac{1}{n_\mathrm{f}^2}\right) \tag{11.8}$$

A word about the signs for ΔE and v in Equations 11.7 and 11.8 is in order. In absorption (see Figure 11.11b), $n_\mathrm{f} > n_\mathrm{i}$, so both ΔE and $\tilde{v}$ are positive. In emission, $n_\mathrm{f} < n_\mathrm{i}$, so ΔE is a negative value, which is consistent with the fact that energy is given off by the system to the surroundings. However, $\tilde{v}$ also becomes a negative value, which is not physically meaningful. We can ensure that the calculated value of $\tilde{v}$ will always be positive, regardless of whether the transition is an absorption or emission, by taking the *absolute value* (that is, the magnitude but not the sign) of $[(1/n_\mathrm{i}^2) - (1/n_\mathrm{f}^2)]$.

Figure 11.12 shows the energy-level diagram of the hydrogen atom and the various emissions that give rise to the spectral series listed in Table 11.1.

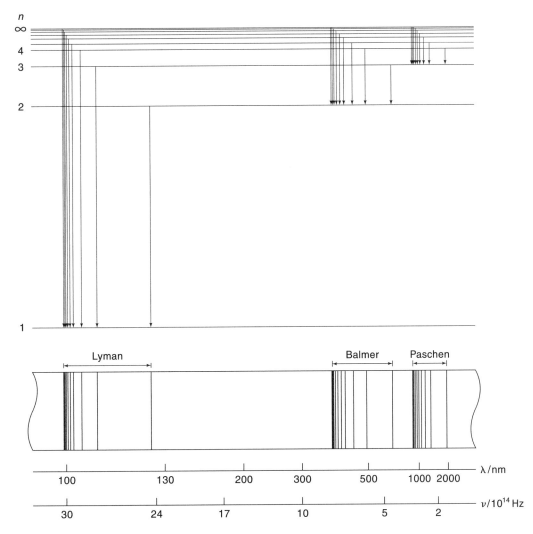

Figure 11.12
Energy levels and some of the hydrogen emission spectra series. (Adapted from D. A. McQuarrie and J. D. Simon, *Physical Chemistry*, University Science Books, Sausalito, CA, 1997.)

Example 11.2

Calculate the wavelength in nanometers of the $n = 4 \rightarrow 2$ transition in the hydrogen atom.

ANSWER

This is an emission process. Because $n_f = 2$, this line belongs to the Balmer series. We calculate the absolute value of $\tilde{v}$ from Equation 11.8

$$\tilde{v} = (109{,}737 \text{ cm}^{-1}) \left| \left(\frac{1}{4^2} - \frac{1}{2^2} \right) \right|$$

$$= 2.058 \times 10^4 \text{ cm}^{-1}$$

Therefore,

$$\lambda = \frac{1}{\tilde{v}} = \frac{1}{2.058 \times 10^4 \text{ cm}^{-1}}$$

$$= 4.86 \times 10^{-5} \text{ cm}$$

$$= 486 \text{ nm}$$

COMMENT

Four spectral lines in the Balmer series are in the visible region, including this case.

11.5 de Broglie's Postulate

Physicists were both mystified and intrigued by Bohr's theory. They questioned why the energies of the hydrogen electron would be quantized. Or, phrasing the question more concretely, why is the electron in a Bohr atom restricted to orbiting the nucleus at certain fixed distances? For a decade, no one—not even Bohr himself—had a logical explanation. In 1924, the French physicist Louis de Broglie (1892–1977) provided the answer.

de Broglie deduced the connection between particle and wave properties from the Einstein–Planck expression for the energy of an electromagnetic wave and the classical result for the momentum of such a wave. The two expressions are

$$E = hv$$

$$p = \frac{E}{c} \tag{11.9}$$

where p is the momentum, and c is the velocity of light. If we replace E by $hv = hc/\lambda$, we arrive at the de Broglie relation:

$$p = \frac{h}{\lambda}$$

or

$$\lambda = \frac{h}{p} = \frac{h}{mv} \tag{11.10}$$

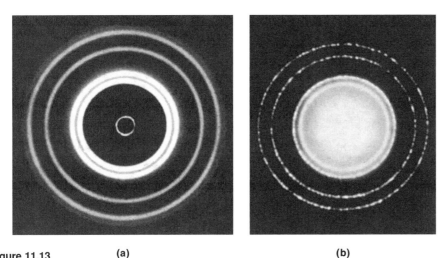

Figure 11.13   **(a)**   **(b)**
(a) X-ray diffraction pattern of aluminum foil. (b) Electron diffraction pattern of aluminum foil. The similarity of these two patterns shows that electrons can behave like X rays and display wavelike properties. (Reprinted with permission from *Education Development Center, Newton, MA.*)

Equation 11.10 says that any particle of mass m moving with velocity v will possess wavelike properties characterized by wavelength λ.

The experimental confirmation of Equation 11.10 was provided by the American physicists Clinton Davisson (1881–1958) and Lester Germer (1896–1972) in 1927, and the British physicist G. P. Thomson (1892–1975) in 1928. When Thomson bombarded a thin sheet of gold foil with electrons, the resulting pattern of concentric rings produced on a screen resembled the pattern made by X rays, which were known to be waves. Figure 11.13 shows the diffraction patterns of electron waves and X rays arising from aluminum foil.

Example 11.3

The fastest serves in tennis are about 150 mph, or 66 m s^{-1}. Calculate the wavelength associated with a 6.0×10^{-2} kg tennis ball traveling at this speed. Repeat the calculation for an electron traveling at the same speed.

ANSWER

Using Equation 11.10, we write

$$\lambda = \frac{6.626 \times 10^{-34} \text{ J s}}{(6.0 \times 10^{-2} \text{ kg})(66 \text{ m s}^{-1})}$$

The conversion factor is 1 J = 1 kg m^2 s^{-2}. Therefore,

$$\lambda = 1.7 \times 10^{-34} \text{ m}$$

This is an exceedingly small wavelength, because the size of an atom itself is on the order of 1×10^{-10} m. For this reason, the wave properties of such a tennis ball cannot be detected by any existing measuring device.

For the electron, we have

$$\lambda = \frac{6.626 \times 10^{-34} \text{ J s}}{(9.10939 \times 10^{-31} \text{ kg})(66 \text{ m s}^{-1})}$$

$$= 1.1 \times 10^{-5} \text{ m}$$

$$= 1.1 \times 10^4 \text{ nm}$$

which is in the infrared region.

COMMENT

This example shows that the de Broglie equation is important only for submicroscopic objects such as electrons, atoms, and molecules.

According to de Broglie, an electron bound to the nucleus behaves like a *standing wave*. Standing waves can be generated by plucking, say, a guitar string. The waves are described as standing or stationary, because they do not travel along the string (Figure 11.14). Some points on the string, called *nodes*, do not move at all; that is, the

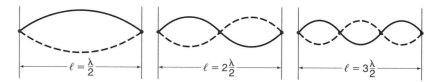

Figure 11.14
The standing waves generated by plucking a guitar string. The length of the string, *l*, must be equal to a whole number times one-half the wavelength ($\lambda/2$).

amplitude of the wave at these points is zero. The greater the frequency of vibration, the shorter the wavelength of the standing wave and the greater the number of nodes. As Figure 11.14 shows, there can be only certain wavelengths in any of the allowed motions of the string. de Broglie argued that if an electron does behave like a standing wave in the hydrogen atom, the length of the wave must fit the circumference of the orbit exactly (Figure 11.15). Otherwise the wave would partially cancel itself on

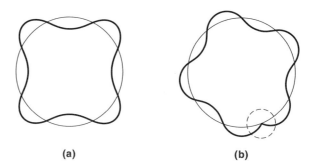

(a)   (b)

Figure 11.15
(a) The circumference of the orbit is equal to an integral number of wavelengths. This is an allowed orbit. (b) The circumference of the orbit is not equal to an integral number of wavelengths. As a result, the electron wave does not close on itself evenly. This is a disallowed orbit.

each successive orbit. Eventually, the amplitude of the wave would be reduced to zero, and the wave would not exist. Thus, the wave properties of the electron restrict the orbits it can occupy, and hence the energies it can possess.

One practical application of the wavelike behavior of electrons is in the use of the electron microscope. Human eyes are sensitive to light of wavelengths from about 400 nm to 700 nm. The ability to see details of small structures is limited by the resolving power, or resolution, of our optical systems. Resolution refers to the minimum distance at which objects can be distinguished as separate entities. Any two objects separated by less than that distance will blur together into a single object. The lower limit of resolution of the unaided human eye is approximately 0.2 mm, below which we cannot see individual objects. On the other hand, the lower limit of resolution of a light microscope is approximately 200 nm, or 0.2 μm. This means that with the aid of a light microscope, we can see objects that are the size of about one-half the wavelength of violet light (400 nm) but not smaller. Greater resolution is possible with an electron microscope because a beam of electrons has properties that correspond to wavelengths 100,000 times shorter than visible light. When a beam of electrons is directed through an accelerating electrostatic field (two parallel plates with a potential difference of V volts), the potential energy gained by each electron, eV, can be equated to its kinetic energy as follows:

$$eV = \tfrac{1}{2}m_e v^2$$

or

$$v = \sqrt{\frac{2eV}{m_e}}$$

where e is the electronic charge. Using the above expression for velocity in Equation 11.10, we obtain

$$\lambda = \frac{h}{\sqrt{2m_e eV}} \qquad (11.11)$$

Example 11.4

What is the wavelength of an electron when it is accelerated by 1.00×10^3 V?

ANSWER

From Equation 11.11, we write

$$\lambda = \frac{6.626 \times 10^{-34} \text{ J s}}{\sqrt{2(9.109 \times 10^{-31} \text{ kg})(1.602 \times 10^{-19} \text{ C})(1000 \text{ V})}}$$

Using the conversion factor $1 \text{ J} = 1 \text{ C} \times 1 \text{ V}$, we find that

$$\lambda = 3.88 \times 10^{-11} \text{ m}$$
$$= 0.0388 \text{ nm}$$

Obtaining voltage in the kilo- or even mega-volt range is relatively easy, so that very small wavelengths can be achieved. Thus, an electron microscope differs from a light microscope in that visible light is replaced by a beam of electrons. The much shorter wavelength produces better resolution. This technique allows us to "see" large molecules as well as heavy atoms. The major advantage of electron microscopy over X-ray diffraction is that electrons are charged particles, so they can be focused easily and thus imaged by electric and magnetic fields, which act as lenses. X rays are uncharged so they cannot be focused this way; no condensing lenses are known for X rays.

11.6 The Heisenberg Uncertainty Principle

In 1927, the German physicist Werner Heisenberg (1901–1976) proposed a principle that has utmost importance in the philosophical groundwork for quantum mechanics. He deduced that when the uncertainties in the *simultaneous* measurements of momentum and position for a particle are multiplied together, the product is approximately equal to Planck's constant divided by 4π. Mathematically, this can be expressed as

$$\Delta x \Delta p \geq \frac{h}{4\pi} \tag{11.12}$$

where Δ means "uncertainty of." Thus, Δx is the uncertainty of position and Δp is the uncertainty of momentum. If the measured uncertainties of position and momentum are large, their product can be substantially greater than $h/4\pi$. The significance of Equation 11.12, which is the mathematical statement of the *Heisenberg uncertainty principle*, is that even in the most favorable conditions for measuring position and momentum, the lower limit of uncertainty is always given by $h/4\pi$.

Conceptually, we can see why the uncertainty principle should exist. Any measurement of a system must, by necessity, result in some disturbance on the system. Suppose that we want to determine the position of a quantum mechanical object, say an electron. To locate the electron within a distance Δx, we might employ light with a wavelength on the order of $\lambda \approx \Delta x$. During the interaction (collision) between the photon and the electron, part of the photon's momentum ($p = h/\lambda$) will be transferred to the electron. Thus, the very act of trying to "see" the electron has changed its momentum. If we want to locate the electron more accurately, then we must use light of a smaller wavelength. Consequently, the photons of the light will possess greater momentum, resulting in a correspondingly larger change in the momentum of the electron. In essence, to make Δx as small as possible, the uncertainty in the momentum (Δp) will become correspondingly large at the same time. Similarly, if we design an experiment to determine the momentum of an electron as accurately as we can, then the uncertainty in its position will simultaneously become large. Keep in mind that this uncertainty is *not* the result of poor measurements or experimental techniques—it is a fundamental property of the act of measurement itself.

What about macroscopic objects? Because of their sheer size compared with that of quantum mechanical systems, the inaccuracies due to the interactions of observation in measuring the position and momentum of a baseball, for example, are completely negligible. Thus, we can accurately determine the position and momentum of a macroscopic object simultaneously. Planck's constant is such a small number that it becomes important only when we are dealing with particles on the atomic scale.

Example 11.5

(a) Bohr's theory also enables us to calculate the radius of an orbit in the hydrogen atom. For the ground state ($n = 1$), the radius is 0.529 Å or 5.29×10^{-11} m, which is called the *Bohr radius*. Assuming that we know the position of an electron in this orbit to an accuracy of 1% of the radius, calculate the uncertainty in the velocity of the electron. (b) A baseball (0.15 kg) thrown at 100 mph has a momentum of 6.7 kg m s^{-1}. If the uncertainty in measuring this momentum is 1.0×10^{-7} of the momentum, calculate the uncertainty in the baseball's position.

ANSWER

(a) The uncertainty in the electron's position is

$$\Delta x = 0.01 \times 5.29 \times 10^{-11} \text{ m}$$
$$= 5.29 \times 10^{-13} \text{ m}$$

From Equation 11.12,

$$\Delta p = \frac{h}{4\pi \Delta x}$$
$$= \frac{6.626 \times 10^{-34} \text{ J s}}{4\pi (5.29 \times 10^{-13} \text{ m})}$$
$$= 9.97 \times 10^{-23} \text{ kg m s}^{-1}$$

Because $\Delta p = m_e \Delta v$, the uncertainty in velocity is given by

$$\Delta v = \frac{9.97 \times 10^{-23} \text{ kg m s}^{-1}}{9.1095 \times 10^{-31} \text{ kg}}$$
$$= 1.1 \times 10^8 \text{ m s}^{-1}$$

We see that the uncertainty in the electron's velocity is of the same magnitude as the speed of light (3×10^8 m s^{-1}). At this level of uncertainty, we have virtually no idea of the velocity of the electron.

(b) The uncertainty in the position of the baseball is

$$\Delta x = \frac{h}{4\pi \Delta p}$$
$$= \frac{6.626 \times 10^{-34} \text{ J s}}{4\pi \times 1 \times 10^{-7} \times 6.7 \text{ kg m s}^{-1}}$$
$$= 7.9 \times 10^{-29} \text{ m}$$

This is such a small number as to be of no consequence.

COMMENT

The uncertainty principle is negligible in the world of macroscopic objects but is very important for objects with small masses, such as the electron. Note that we used the equal sign rather than the "greater than" sign in Equation 11.12 to get the minimum value of the uncertainty.

Finally, we note that the Heisenberg uncertainty principle can be expressed also in terms of energy and time, as follows. Because

$$\text{momentum} = \text{mass} \times \text{velocity}$$

$$= \text{mass} \times \frac{\text{velocity}}{\text{time}} \times \text{time}$$

$$= \text{force} \times \text{time}$$

Thus,

$$\text{momentum} \times \text{distance} = \text{force} \times \text{distance} \times \text{time}$$

$$= \text{energy} \times \text{time}$$

or

$$\Delta x \Delta p = \Delta E \Delta t$$

where ΔE is the uncertainty in energy when the system is in a certain state, and Δt is the time interval during which the system is in the state. Equation 11.12 can now be written as

$$\Delta E \Delta t \geq \frac{h}{4\pi} \tag{11.13}$$

Thus, we cannot measure the (kinetic) energy of a particle with absolute precision (that is, to have $\Delta E = 0$) in a finite span of time. Equation 11.13 is particularly useful for estimating spectral line widths (see Section 14.1). In quantum mechanical language, momentum and position form a *conjugate pair*, as do energy and time. We shall return to this point in Chapter 14.

11.7 The Schrödinger Wave Equation

Bohr's theory of the hydrogen atom was one of the early triumphs of the quantum theory. It was soon found to be inadequate, however, for it could not account for the emission spectra of more complex atoms (like helium) or for the behavior of atoms in a magnetic field. Furthermore, the notion that the electron is circling the nucleus in a well-defined orbit is inconsistent with the uncertainty principle. A general equation was needed for submicroscopic systems, one comparable to Newton's equation for macroscopic bodies. In 1926, the Austrian physicist Erwin Schrödinger (1887–1961) furnished the necessary equation.

When expressed in one dimension (say x), Schrödinger's equation is given by

$$-\frac{h^2}{8\pi^2 m} \frac{d^2\psi}{dx^2} + V\psi = E\psi \tag{11.14}$$

where V is the potential energy, E is the total energy of the system, and h is the familiar Planck's constant. The particle properties are represented by mass m and the wave properties by the wave function ψ. Equation 11.14 does not contain time and is called the *time-independent Schrödinger equation*. The wave functions obtained from

Equation 11.14 are called *stationary-state wave functions* because they do not change with time.* In classical mechanics, the total energy (E) is given by the sum of kinetic energy (T) and potential energy (V):

$$T + V = E \tag{11.15}$$

The main difference between Equation 11.14 and 11.15 is that in the former, T is replaced with the kinetic-energy operator (see Appendix 1):

$$-\frac{h^2}{8\pi^2 m}\frac{d^2}{dx^2}$$

Keep in mind that Schrödinger's wave equation, like Newton's laws of motion, cannot be derived from first principles. Instead, it was deduced by analogy to classical mechanics and optics.

How should we interpret ψ? As a mathematical wave function, it has no physical meaning by itself. In fact, it may even be a complex function; that is, it may involve $i = \sqrt{-1}$. However, the German physicist Max Born (1882–1970) suggested in 1926 that for a one-dimensional system, for example, the probability of finding the particle between x and $x + dx$ is given by $\psi^2(x)dx$[†] The product $\psi^2(x)$ can be interpreted as probability density. Earlier, we mentioned that the intensity of light is proportional to the square of the amplitude of the wave. Analogously, if ψ represents the wave property of the particle, then the probability of locating the particle at some point in space is given by the value of ψ^2 at the same point.

For Equation 11.14 to apply to any system, ψ must be a "well-behaved" wave function, the conditions for which are

1. ψ must be single valued at any point.

2. ψ must be finite at any point.

3. ψ must be a smooth or continuous function of its coordinates, and its first derivatives with the coordinates also must be continuous.

Condition 1 means that there can be only one probability of finding the system (particle) at a certain point in space. Condition 2 is necessary because many mathematically acceptable solutions to the Schrödinger equation rise to infinity and are therefore physically unacceptable. Because the Schrödinger equation is a second-order differential equation, condition 3 requires that $d^2\psi/dx^2$ be well defined, meaning that ψ and $d\psi/dx$ must be continuous. Figure 11.16 shows some examples of unacceptable wave functions.

The Schrödinger wave equation marked the beginning of a new era in physics, which is referred to as the era of wave mechanics or quantum mechanics.

In general, ψ is a function of the x, y, and z coordinates.

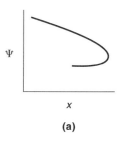

(a)

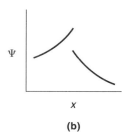

(b)

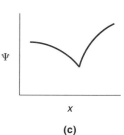

(c)

Figure 11.16
Unacceptable wave functions. (a) The wave function is not single valued. (b) The wave function is not continuous. (c) The slope of the wave function, $d\psi/dx$, is discontinuous.

---

*A more general Schrödinger equation contains a time dependence and can be applied to study spectroscopic transitions, for example. However, many problems of chemical interest can be described by the time-independent Schrödinger equation.

[†] Strictly speaking, this probability should be given by $\psi^*(x)\psi(x)dx$, where $\psi^*(x)$ is the *complex conjugate* of $\psi(x)$. The complex conjugate is found by changing i to $-i$ everywhere in $\psi(x)$. For example, if $\psi(x)$ is a complex function given by $a + ib$, then $\psi^*(x) = a - ib$ and $\psi^*(x)\psi(x) = (a + ib)(a - ib) = a^2 + b^2$. Thus, the product $\psi^*(x)\psi(x)$ is *always* positive and real. If $\psi(x)$ is a real function (that is, if it does not contain i), then $\psi^*(x)$ is the same as $\psi(x)$.

11.8 Particle in a One-Dimensional Box

The Schrödinger wave equation enables us to solve a particularly simple problem, the particle in a one-dimensional box. The situation is a model problem, one that may be applied to real situations of chemical and biological interest.

Suppose that we have a particle of mass m confined in a one-dimensional box of length L, and imagine that the particle is moving along a piece of straight wire. For simplicity, we assume that the particle has zero potential energy inside the box (or on the wire); that is, $V = 0$, so it has only kinetic energy. At each end of the box is a wall of infinite potential energy so that there is no probability of finding the particle either at the walls ($x = 0$ and $x = L$) or outside the box (Figure 11.17). Now Equation 11.14 can be written as

$$-\frac{h^2}{8\pi^2 m}\frac{d^2\psi}{dx^2} = E\psi \tag{11.16}$$

We are interested in knowing the values of E and ψ that the particle can possess. Equation 11.16 tells us that the wave function, ψ, is such that when it is differentiated twice with respect to x, the original function is obtained. Examples of such functions are trigonometric and exponential functions. As a trial solution, let ψ be

$$\psi = A\sin kx + B\cos kx \tag{11.17}$$

where A, B, and k are constants. This is the general solution to the second order differential Schrödinger equation. To proceed further and find a particular solution, we must determine the constants using the *boundary conditions*. Because the probability of finding the particle at either end of the box is zero, both $\psi(0)$ and $\psi(L)$ are zero. For $x = 0$, we have $\sin 0 = 0$ and $\cos 0 = 1$; therefore, B must be zero. Equation 11.17 is reduced to

$$\psi = A\sin kx \tag{11.18}$$

We now differentiate ψ with respect to x to get

$$\frac{d\psi}{dx} = kA\cos kx$$

$$\frac{d^2\psi}{dx^2} = -k^2 A\sin kx$$

$$= -k^2\psi \tag{11.19}$$

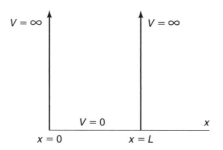

Figure 11.17
A one-dimensional box with infinite potential barriers.

From Equations 11.16 and 11.19, we obtain

$$k^2 = \frac{8\pi^2 mE}{h^2}$$

or

$$k = \left(\frac{8\pi^2 mE}{h^2}\right)^{1/2} \tag{11.20}$$

Substituting Equation 11.20 into Equation 11.18 gives

$$\psi = A \sin\left(\frac{8\pi^2 mE}{h^2}\right)^{1/2} x \tag{11.21}$$

Mathematically, an infinite number of solutions can satisfy Equation 11.21, because A can have any value. Physically, however, ψ must satisfy the following boundary conditions:

$$\text{at } x = 0, \quad \psi = 0$$

and

$$\text{at } x = L, \quad \psi = 0$$

The second condition, when applied to Equation 11.21, gives

$$0 = A \sin\left(\frac{8\pi^2 mE}{h^2}\right)^{1/2} L$$

Because $A = 0$ is a trivial solution, in general we have*

$$\left(\frac{8\pi^2 mE}{h^2}\right)^{1/2} L = n\pi \quad \text{where } n = 1, 2, 3, \ldots$$

Noting that

$$\sin \pi = \sin 2\pi = \sin 3\pi = \cdots = 0$$

we arrive at the result

$$E_n = \frac{n^2 h^2}{8mL^2} \tag{11.22}$$

where E_n is the energy for the nth level. Substituting Equation 11.22 ino Equation 11.21 gives

$$\psi_n = A \sin\frac{n\pi}{L} x \tag{11.23}$$

---

* Note that the condition $n = 0$ is eliminated because it leads to the result $(8\pi^2 mE/h^2)^{1/2} = 0$ and thus, from Equation 11.21, $\psi = 0$ for all values of x. This is a physically impossible result because it means that the probability of finding the particle anywhere in the box is zero.

The next step is to determine A. We start with the knowledge that because the particle must remain inside the box, the total probability of finding the particle between $x = 0$ and $x = L$ must be unity. Thus, we can carry out the *normalization* process by writing

$$\int_0^L \psi^2 \, dx = 1 \tag{11.24}$$

where $\psi^2 \, dx$ gives the probability of finding the particle between x and $x + dx$. Substituting the wave function, we have

$$A^2 \int_0^L \sin^2 \frac{n\pi}{L} x \, dx = 1$$

The definite integral above gives*

$$A^2 \frac{L}{2} = 1$$

or

$$A = \sqrt{\frac{2}{L}}$$

Finally, we have the *normalized* wave function

The quantity $\sqrt{\dfrac{2}{L}}$ is called the normalization constant.

$$\psi_n = \sqrt{\frac{2}{L}} \sin \frac{n\pi}{L} x \tag{11.25}$$

Plots of the allowed energy levels as well as ψ and ψ^2 are shown in Figure 11.18. Several important conclusions can be drawn from this model:

1. The (kinetic) energy of the particle is quantized according to Equation 11.22.

2. The lowest energy level is not zero but is equal to $h^2/8mL^2$. This *zero-point energy* can be accounted for by the Heisenberg uncertainty principle. If the particle could possess zero kinetic energy, its velocity would also be zero; consequently, there would be no uncertainty in determining its momentum. According to Equation 11.12, Δx would be infinite. If the box is of finite size, however, the uncertainty in determining the particle's position cannot exceed L; therefore, a zero energy would violate the Heisenberg uncertainty principle. Keep in mind that the zero-point energy means that the particle can never be at rest because its lowest energy is not zero.

3. Depending on the value of n, the wave behavior of the particle is described by Equation (11.25), but the probability is given by ψ_n^2, which is always positive. (In fact, the wave functions look just like the standing waves set

---

* This definite integral is evaluated by using the relation

$$\int \sin^2 ax \, dx = \frac{x}{2} - \frac{\sin 2ax}{4a}$$

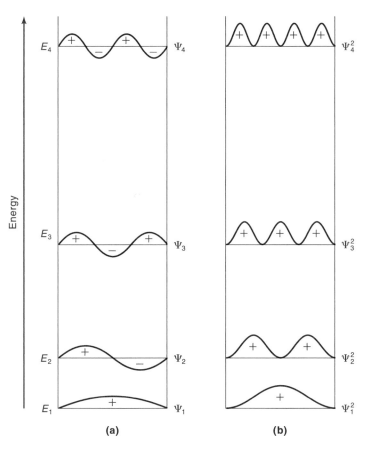

Figure 11.18
Plots of (a) ψ and (b) ψ^2 for the first four energy levels in a one-dimensional box with infinite potential barriers.

up in a vibrating string shown in Figure 11.14.) For $n = 1$, the maximum probability is at $x = L/2$; for $n = 2$, the maxima occur at $x = L/4$ and $x = 3L/4$, and the probability is zero at $x = L/2$. The point at which ψ (and hence ψ^2) is zero is called a node. Generally, the number of nodes increases with increasing energy. In classical mechanics, the probability of finding the particle is the same at all points along the box, regardless of its kinetic energy.

4. As Equation 11.22 shows, the energy of the system is inversely proportional to the mass of the particle. For macroscopic objects, m is very large, so the difference between successive energy levels would be exceedingly small. This means that the energy of the system is not quantized; instead, it can vary continuously. The inverse dependence of energy on L^2 means that if we confine a molecule in a container of macroscopic dimensions, its energy will also vary continuously rather than in a quantized fashion. We encountered this result earlier in the derivation of the translational kinetic energy of gases in Chapter 2. In summary, when we are dealing with systems of macroscopic magnitude, the quantum mechanical effects disappear and we have classical mechanical behavior.

Example 11.6

An electron is placed in a one-dimensional box that has a length of 0.10 nm (about the size of an atom). (a) Calculate the energy difference between the $n = 2$ and $n = 1$ states of the electron. (b) Repeat the calculation in (a) for a N_2 molecule in a container whose length is 10 cm. (c) Calculate the probability of finding the electron in (a) between $x = 0$ and $x = L/2$ in the $n = 1$ state.

ANSWER

(a) From Equation 11.22, we write the energy difference between the $n = 1$ and $n = 2$ states, ΔE, as

$$\Delta E = E_2 - E_1$$

$$= \frac{2^2 h^2}{8mL^2} - \frac{1^2 h^2}{8mL^2}$$

$$= \frac{(4-1)(6.626 \times 10^{-34} \text{ J s})^2}{8(9.109 \times 10^{-31} \text{ kg})[(0.10 \text{ nm})(1 \times 10^{-9} \text{ m}/1 \text{ nm})]^2}$$

$$= 1.8 \times 10^{-17} \text{ J}$$

This energy difference is of comparable magnitude to the difference between the $n = 1$ and $n = 2$ state for the hydrogen atom (see Equation 11.7).

(b) The mass of a N_2 molecule is 4.65×10^{-26} kg, so we write

$$\Delta E = E_2 - E_1$$

$$= \frac{(4-1)(6.626 \times 10^{-34} \text{ J s})^2}{8(4.65 \times 10^{-26} \text{ kg})[(10 \text{ cm})(1 \times 10^{-2} \text{ m}/1 \text{ cm})]^2}$$

$$= 3.5 \times 10^{-40} \text{ J}$$

This energy difference is some 23 orders of magnitude less than that in (a), meaning that the translational energy of the N_2 molecule varies essentially continuously. This result confirms our earlier statement that when molecules are confined to macroscopic systems, their translational motions are governed by classical mechanics.

(c) The probability (P) that the electron will be found between $x = 0$ and $x = L/2$ is

$$P = \int_0^{L/2} \psi^2 \, dx$$

Using the normalized wave function in Equation 11.25 and setting $n = 1$,

$$P = \frac{2}{L} \int_0^{L/2} \sin^2 \frac{\pi}{L} x \, dx$$

$$= \frac{2}{L} \left[\frac{x}{2} - \frac{\sin 2(\pi/L)x}{4(\pi/L)} \right]_0^{L/2}$$

$$= \frac{1}{2}$$

which is not an unexpected result, classically or quantum mechanically.

The problem of a particle in a one-dimensional box shows us that when a submicroscopic particle is in a *bound state* (that is, when its movement is restricted by potential barriers), its energy values should be quantized. This is precisely the case for electrons in atoms. Indeed, we can predict several atomic properties by using a particle in a three-dimensional box as a model. For example, the energies of an electron in a hydrogen atom must be quantized. Further, the electron should possess three quantum numbers (one for each dimension). We shall discuss this and related systems shortly.

Electronic Spectra of Polyenes

One application of the particle-in-a-one-dimensional-box model is the analysis of the electronic spectra of polyenes. Polyenes are important conjugated π systems (with alternating C–C and C=C bonds) that play a role in photosynthesis and vision (see Chapter 15). Consider the simplest polyene, butadiene,

$$H_2C=CH-CH=CH_2$$

which contains four π electrons. Although butadiene, like all other polyenes, is not linear in shape, we assume that the π electrons move along the molecule like particles moving in a one-dimensional box. The potential energy along the chain is constant but rises sharply at the ends. Thus, the energies of the π electrons are quantized. This assumption is called the *free electron model* and enables us to calculate the difference between energy levels and predict wavelengths associated with electronic transitions.

Figure 11.19 shows the π energy levels for butadiene. According to the Pauli exclusion principle (see Section 11.11), the electrons in each energy level have opposite spins. The electronic transition we are interested in is the one from the highest filled level to the lowest unfilled level (because it is the one that is usually measured experimentally)—that is, the $n = 2 \rightarrow 3$ transition. From Equation 11.22, we can derive a general expression for the wavelength for such a transition as follows. The number of filled energy levels is $N/2$, where N is the number of carbon atoms. This number $(N/2)$ is also equal to the quantum number of the highest occupied level. The transition, then, is from the $N/2$ level to the $(N/2) + 1$ level, and the energy difference is

$$\Delta E = \frac{[(N/2) + 1]^2 h^2}{8m_e L^2} - \frac{(N/2)^2 h^2}{8m_e L^2}$$

$$= \left[\left(\frac{N}{2} + 1 \right)^2 - \left(\frac{N}{2} \right)^2 \right] \frac{h^2}{8m_e L^2}$$

$$= (N + 1) \frac{h^2}{8m_e L^2} \tag{11.26}$$

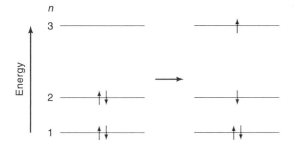

Figure 11.19
Pi energy levels in butadiene. The electronic transition is from the highest filled level to the lowest unfilled level.

Using the relations $c = \lambda v$ and $\Delta E = hv$, we arrive at the following expression for the wavelength:

$$\lambda = \frac{hc}{\Delta E} = \frac{8m_{\mathrm{e}}L^2 c}{h(N+1)} \tag{11.27}$$

For butadiene, we have $N = 4$. To calculate the value of L, the length of the molecule, we use bond lengths of 1.54 Å (154 pm) for C–C bonds and 1.35 Å (135 pm) for C=C bonds, plus the distance equal to a C atom radius at each end (0.77 Å or 77 pm). Thus, the length of the molecule is $(2 \times 135 \text{ pm}) + 154 \text{ pm} + (2 \times 77 \text{ pm}) = 578$ pm, or 5.78×10^{-10} m, so that

$$\lambda = \frac{8(9.1095 \times 10^{-31} \text{ kg})(5.78 \times 10^{-10} \text{ m})^2(3.00 \times 10^8 \text{ m s}^{-1})}{(6.626 \times 10^{-34} \text{ J s})(4+1)}$$

$$= 2.20 \times 10^{-7} \text{ m, or } 220 \text{ nm}$$

The experimentally measured wavelength is 217 nm. Considering the crudeness of the model, the agreement is remarkably good.

11.9 Quantum Mechanical Tunneling

What would happen if the potential walls surrounding the particle in the one-dimensional box were not infinitely high? The particle would escape when its kinetic energy became equal to or greater than the potential energy of the barrier. What is more surprising, however, is the fact that we might find the particle outside the box even if its kinetic energy is not sufficient to reach the top of the barrier! This phenomenon, known as *quantum mechanical tunneling*, has no analog in classical physics. It arises as a consequence of the wave nature of particles. Quantum mechanical tunneling has many profound consequences in chemistry and biology.*

The phenomenon of quantum mechanical tunneling was introduced by the Russian–American physicist George Gamow (1904–1968), among others, in 1928 to explain α decay, a process wherein a nucleus spontaneously decays by emitting an α particle, which is a helium nucleus (He^{2+}); for example,

$$^{238}_{92}\text{U} \rightarrow {}^{234}_{90}\text{Th} + \alpha \qquad t_{1/2} = 4.51 \times 10^9 \text{ yr}$$

The dilemma facing physicists was the following: For U-238 decay, the measured (kinetic) energy of the emitted α particle is about 4 MeV,[†] whereas the Coulombic barrier is on the order of 250 MeV. (Imagine the α particle being at the center of the nucleus. It is surrounded by other protons and therefore behaves like a particle trapped in a one-dimensional box. The potential barriers are the result of electrostatic repulsion due to other protons present. The barrier height can be calculated from the radius of the nucleus and the atomic number.) The question naturally arose as to how the α particle can overcome the barrier and leave the nucleus. Gamow suggested that the α particle, being a quantum mechanical object, has wavelike properties that allow

---

* See W. T. Scott, *J. Chem. Educ.* **48**, 524 (1971), for an interesting illustration of quantum mechanical tunneling.

† In nuclear physics and nuclear chemistry, the common unit for energy is the eV or MeV (1×10^6 eV), where 1 eV = 1.602×10^{-19} J.

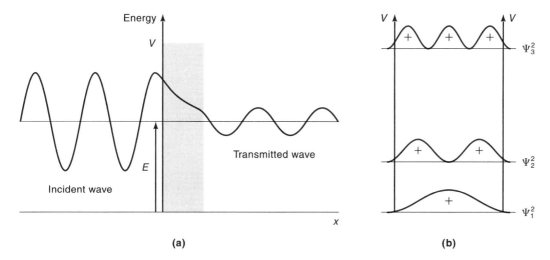

Figure 11.20
(a) Schematic diagram showing tunneling through a finite potential barrier. The particle is moving from left to right. Most of the incident wave of the particle is reflected back. A small part of the wave penetrates the barrier, emerging on the other side with diminished amplitude. (b) Plots of ψ^2 for a particle in a one-dimensional box with finite potential barriers. Note that the curves extend outside of the box.

it to penetrate a potential barrier, as shown in Figure 11.20. This explanation turned out to be correct. In general, for finite potential barriers, there is always some probability of finding the particle outside of the box.

The probability (P) of the particle tunneling through the barrier is proportional to the quantity*

$$P = \exp\left\{ -\frac{4\pi a}{h} [2m(V - E)]^{1/2} \right\} \quad V > E \qquad (11.28)$$

where exp means exponential, V is the potential barrier, E is the energy of the particle, and a is the thickness of the barrier. Unless $V = \infty$ or $a = \infty$, there is always a probability that the particle will escape, although P may be a very small number. This is certainly the case for the α decay of U-238, which has an exceedingly large half-life. Physically, the very small P value means that the α particles have to collide with the barrier many, many times before one of them can escape the nucleus. As Equation 11.28 shows, quantum mechanical tunneling is most likely to take place (for comparable V and a values) with light particles such as electrons, protons, and hydrogen atoms.

The energy profile for a chemical reaction is usually described in terms of reactant molecules acquiring sufficient energy to overcome the activation energy barrier to form products (see Figure 9.14). However, in some cases (for example, certain electron exchange reactions), reactions proceed even when an insufficient amount of energy is available. Such results have been attributed to quantum mechanical tunneling.

A practical application of the quantum mechanical tunneling effect is the scan-

---

* For the derivation, see Pilar, F. L. *Elementary Quantum Chemistry*, 2nd ed., McGraw-Hill, New York, 1990.

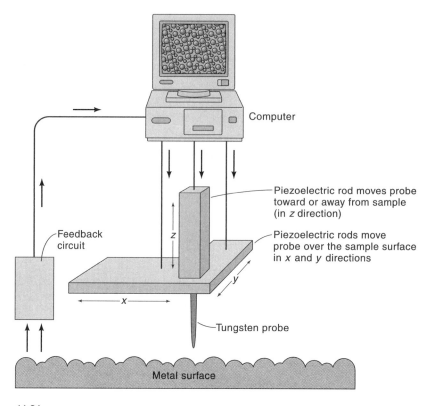

Figure 11.21
The scanning tunneling microscope. A tunneling current flows between the probe and the sample when there is a small voltage between them. A feedback circuit, which provides this voltage, senses the current and varies the voltage on a piezoelectric rod (the *z* drive) in order to keep the distance constant between the probe and the sample (a metal surface). A computer provides voltages to the *x* drive and *y* drive to move the probe over the surface of the metal.

ning tunneling microscope (STM), shown in Figure 11.21. The STM consists of a tungsten metal needle with a very fine point, the source of the tunneling electrons. A voltage is maintained between the needle and the surface of the sample to induce electrons to tunnel through space to the sample. As the needle moves over the sample, at a distance of a few atomic diameters from the surface, the tunneling current is measured. This current decreases with increasing distance from the sample. By using a feedback loop, the vertical position of the tip can be adjusted to a constant distance from the surface. The extent of these adjustments, which profile the sample, is recorded and displayed as a three-dimensional false-colored image. The STM is one of the most useful tools in chemical, biological, and materials science research. Large biological molecules such as DNA have been imaged by depositing them onto smooth surfaces.

11.10 The Schrödinger Wave Equation for the Hydrogen Atom

We are now ready to study the simplest atomic system, the hydrogen atom, which consists of one electron and one proton. This is a three-dimensional problem, so the wave function ψ for the electron will depend on the $x, y,$ and z coordinates. The

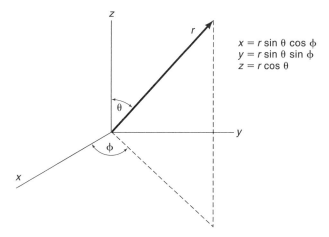

$$x = r \sin \theta \cos \phi$$
$$y = r \sin \theta \sin \phi$$
$$z = r \cos \theta$$

Figure 11.22
Relation between Cartesian coordinates and polar coordinates. For the hydrogen atom, the nucleus is at the origin ($r = 0$) and the electron is at the surface of a sphere of radius r.

Schrödinger wave equation is given by

$$\frac{\partial^2 \psi}{\partial x^2} + \frac{\partial^2 \psi}{\partial y^2} + \frac{\partial^2 \psi}{\partial z^2} + \frac{8\pi^2 m_e}{h^2}(E - V)\psi = 0 \qquad (11.29)$$

We assume ψ to be time independent.

where the potential energy (V) is the Coulombic interaction between the electron and the nucleus, given by $-e^2/4\pi\varepsilon_0 r$, where r is the distance between the electron and the nucleus, and ε_0 is the permittivity of free space. Because the attraction has spherical symmetry (depends only on r), Equation 11.29 is more conveniently solved in terms of *spherical polar coordinates*. Furthermore, this procedure will also make the results easier to interpret. The relation between Cartesian and spherical polar coordinates is shown in Figure 11.22. With this transformation, we can rewrite Equation 11.27 as

$$\frac{\partial^2 \psi}{\partial r^2} + \frac{2}{r}\frac{\partial \psi}{\partial r} + \frac{1}{r^2 \sin \theta}\frac{\partial[\sin\theta(\partial\psi/\partial\theta)]}{\partial\theta} + \frac{1}{r^2 \sin\theta}\frac{\partial^2\psi}{\partial\phi^2} + \frac{8\pi^2 m_e}{h^2}\left(E + \frac{e^2}{4\pi\varepsilon_0 r}\right)\psi = 0 \quad (11.30)$$

Fortunately, this fearsome-looking equation has already been solved, so we need be concerned only with the results. The main point is that the wave function can be expressed as a function of r, θ, and ϕ, as follows:

$$\Psi(r, \theta, \phi) = R(r)\Theta(\theta)\Phi(\phi) \qquad (11.31)$$

Thus, Ψ is given as a product of two independent quantities, the *radial part*, $R(r)$, and the *angular part*, $\Theta(\theta)\Phi(\phi)$, of the wave function, respectively.

The solution of Equation 11.30 gives rise to three quantum numbers, n, l, and m_l, which have the following properties.* The *principal quantum number*, n, determines the size of the wave function and the energy of the electron. The *azimuthal quantum number*, or the *angular momentum quantum number*, l, determines the shape of the wave function. Finally, the *magnetic quantum number*, m_l, determines the orientation of the wave function in space. We shall soon see how these quantum numbers are used to describe the electron in the hydrogen atom.

---

*See the standard physical chemistry texts listed on p. 6 for a detailed discussion of the solution of Equation 11.30.

For a given value of n (where $n = 1, 2, \ldots$), there are n values of l given by $l = 0, 1, 2, \ldots, (n-1)$; for a given l, there are $(2l + 1)$ values of m_l, given by $m_l = 0, \pm 1, \pm 2, \ldots, \pm l$. Thus, if $n = 3$, we have

$$l = 0, 1, \text{ and } 2$$

$$l = 0 \quad m_l = 0$$

$$l = 1 \quad m_l = 0, \pm 1$$

$$l = 2 \quad m_l = 0, \pm 1, \pm 2$$

The wave function for a single electron is called an *orbital*. Bohr's theory describes the orbits of an electron around a nucleus. In quantum mechanics, we speak of the position of an electron not in terms of its orbits, but in terms of its wave function or orbital ψ.

All the orbitals that have a given value of n form a single *shell* of the atom. The shells are referred to by letters:

$$
\begin{array}{ccccc}
n = & 1 & 2 & 3 & 4 \ldots \\
 & K & L & M & N \ldots
\end{array}
$$

The orbitals with the same value of n but different values of l form the *subshells* of a given shell. These subshells are generally designated by the letters $s, p, d, \ldots$ as follows:

l	0	1	2	3	4	5
Name of Subshell	s	p	d	f	g	h

Thus, if $n = 2$ and $l = 1$, we have a $2p$ subshell, and its three orbitals (corresponding to $m_l = +1, 0$, and -1, respectively) are called the $2p$ orbitals. The unusual sequence of letters (s, p, d, and f) has a historical origin. Physicists who studied atomic emission spectra tried to correlate the observed spectral lines with the particular energy states involved in the transitions. They noted that some of the lines were *s*harp; some were rather spread out, or *d*iffuse; and some were very strong and hence referred to as *p*rincipal lines. Subsequently, the initial letters of each characteristic were assigned to those energy states. However, starting with the letter f (for *f*undamental), the orbital designations follow alphabetical order.

The energy of the electron in a hydrogen atom is given by

$$E_n = -hcR_{\mathrm{H}} \left(\frac{1}{n^2} \right) \quad n = 1, 2, 3, \ldots$$

which is identical to Equation 11.5.

Atomic Orbitals

Table 11.2 lists the radial and angular wave functions for $n = 1, 2$, and 3. Let us first consider the radial part of ψ. Figure 11.23 shows the dependence of $R(r)$ and $R(r)^2$ on r for the hydrogen $1s$ orbital. The term $R(r)^2 dr$ gives the probability of finding the electron between r and $r + dr$ along a certain direction away from the nucleus. A more informative plot should give the total probability of finding the electron in a volume element between r and $r + dr$ in all directions. To do this, we

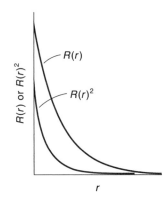

Figure 11.23
Plots of $R(r)$ and $R(r)^2$ versus r for the hydrogen $1s$ orbital.

Table 11.2
Hydrogen Atom Wave Functions for $n = 1, 2,$ and 3

n	l	m_l	$R(r)^a$	$\Theta(\theta)$	$\Phi(\phi)^b$
1	0	0	$\dfrac{2}{\sqrt{a_0^3}}e^{-\rho}$	$\dfrac{1}{\sqrt{2}}$	$\dfrac{1}{\sqrt{2\pi}}$
2	0	0	$\dfrac{1}{\sqrt{2a_0^3}}\left(1 - \dfrac{\rho}{2}\right)e^{-\rho/2}$	$\dfrac{1}{\sqrt{2}}$	$\dfrac{1}{\sqrt{2\pi}}$
2	1	0	$\dfrac{1}{\sqrt{24a_0^3}}\rho e^{-\rho/2}$	$\sqrt{\dfrac{3}{2}}\cos\theta$	$\dfrac{1}{\sqrt{2\pi}}$
2	1	± 1	$\dfrac{1}{\sqrt{24a_0^3}}\rho e^{-\rho/2}$	$\sqrt{\dfrac{3}{4}}\sin\theta$	$\dfrac{1}{\sqrt{2\pi}}e^{\pm i\phi}$
3	0	0	$\dfrac{2}{\sqrt{27a_0^3}}\left(1 - \dfrac{2}{3}\rho + \dfrac{2}{27}\rho^2\right)e^{-\rho/3}$	$\dfrac{1}{\sqrt{2}}$	$\dfrac{1}{\sqrt{2\pi}}$
3	1	0	$\dfrac{8}{27\sqrt{6a_0^3}}\rho\left(1 - \dfrac{\rho}{6}\right)e^{-\rho/3}$	$\sqrt{\dfrac{3}{2}}\cos\theta$	$\dfrac{1}{\sqrt{2\pi}}$
3	1	± 1	$\dfrac{8}{27\sqrt{6a_0^3}}\rho\left(1 - \dfrac{\rho}{6}\right)e^{-\rho/3}$	$\sqrt{\dfrac{3}{4}}\sin\theta$	$\dfrac{1}{\sqrt{2\pi}}e^{\pm i\phi}$
3	2	0	$\dfrac{4}{81\sqrt{30a_0^3}}\rho^2 e^{-\rho/3}$	$\sqrt{\dfrac{5}{8}}(3\cos^2\theta - 1)$	$\dfrac{1}{\sqrt{2\pi}}$
3	2	± 1	$\dfrac{4}{81\sqrt{30a_0^3}}\rho^2 e^{-\rho/3}$	$\dfrac{\sqrt{15}}{2}\sin\theta\cos\theta$	$\dfrac{1}{\sqrt{2\pi}}e^{\pm i\phi}$
3	2	± 2	$\dfrac{4}{81\sqrt{30a_0^3}}\rho^2 e^{-\rho/3}$	$\dfrac{\sqrt{15}}{4}\sin^2\theta$	$\dfrac{1}{\sqrt{2\pi}}e^{\pm 2i\phi}$

^{a} The variable $\rho = r/a_0$, where a_0 is the Bohr radius.
^{b} The symbol i denotes the complex number, $\sqrt{-1}$.

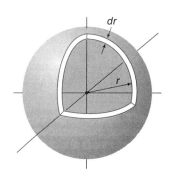

Figure 11.24
The radial distribution function gives the total probability of finding an electron in a spherical shell of thickness dr at a distance r from the nucleus. Note that the volume of the shell is proportional to r^2 and is zero when $r = 0$ (at the nucleus).

consider two concentric spheres of radii r and $r + dr$, shown in Figure 11.24. The volume between these two spheres is $4\pi r^2\,dr$,* and the probability of finding the electron within this spherical shell is $4\pi r^2 R(r)^2\,dr$. The function $r^2 R(r)^2$ is called the *radial distribution function*. Figure 11.25 shows the plot of the radial distribution function for $1s, 2s, 2p, 3s, 3p,$ and $3d$ orbitals. It is interesting that for the $1s$ orbital, the maximum value occurs at $0.529\,\text{Å}$ (52.9 pm), which is called the *Bohr radius*. From these plots, we see that an electron in any particular orbital does not have a well-defined position, and it is therefore more convenient to use the term *electron density* or *electron cloud* to describe the probability of locating the electron. Mathematically, the probability vanishes only when r approaches infinity. Physically, however, we need only consider each orbital over a relatively small distance (a few angstroms) because the function decreases rapidly with increasing r. The $2s$ orbital has two maxima. In this case, we can imagine two concentric spheres with a node somewhere in the spherical shell. The radial distribution function plots for the p and d orbitals are more complex in form, but they can be interpreted in a similar manner.

The angstrom (Å) unit is commonly used in spectroscopy and bond length.

* This function is obtained by taking the difference between the two volumes of radii r and $r + dr$: $(4\pi/3)(r + dr)^3 - (4\pi/3)r^3$ and neglecting the $(dr)^2$ and $(dr)^3$ terms.

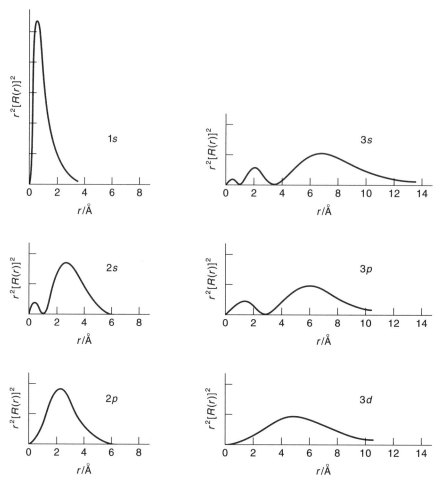

Figure 11.25
Radial distribution functions for the hydrogen $1s, 2s, 2p, 3s, 3p,$ and $3d$ orbitals. Note that the maximum for the $1s$ orbital occurs at $0.529\,\text{Å}$ (52.9 pm), which is the Bohr radius; that is, the radius of the $n = 1$ orbit.

The shapes of the orbitals are determined by the angular part of the wave functions, which are also listed in Table 11.2. The s orbitals contain only a constant; therefore, they are spherically symmetric. The p orbitals, on the other hand, depend on θ and ϕ and do not have spherical symmetry. Figure 11.26 shows the boundary surface diagrams for the three p orbitals along the x, y, and z coordinates. Each orbital consists of two adjacent regions in a dumbbell-like configuration. Furthermore, the wave function is positive in one region and negative in the other, with a nodal plane in between. The wave function is zero in this plane, which also contains the nucleus. These three orbitals are entirely equivalent except for their orientations. Thus, all three p orbitals have the same energy and are said to be *degenerate*. There is no physical significance in the signs by themselves. The only meaningful quantity is the probability of finding the electron given by the square of the wave function, which is always positive. The signs will be useful, however, when we consider the interaction between orbitals that occurs, for example, in chemical-bond formation. The five d orbitals are shown in Figure 11.27. These orbitals, too, are equivalent except for their orientations in space.

Although the d_{z^2} orbital looks different spatially, it is equivalent to the other four d orbitals.

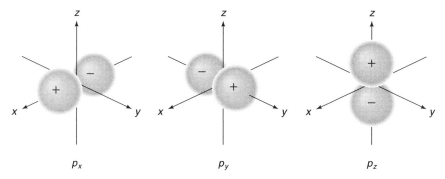

Figure 11.26
Plots of the angular parts of the real representation of the hydrogen wave functions for $l = 1$. These are the p orbitals.

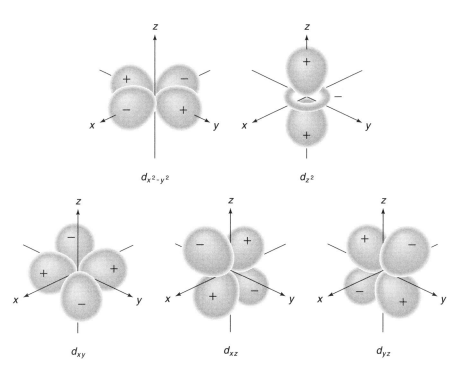

Figure 11.27
Plots of the angular parts of the real representation of the hydrogen wave functions for $l = 2$. These are the d orbitals.

Note that a complete wave function, according to Equation 11.31, is given by the product of the radial and angular parts; the size of the orbital is determined by the former and the shape by the latter. An orbital can be represented in several ways, as Figure 11.28 shows. The boundary-surface representation is the simplest to use, although it is also the least informative. The contour-surface and electron-density representations provide a more detailed description, but they take more time to draw.

The solution of the Schrödinger equation for the hydrogen atom provides us with three quantum numbers for an electron. The electron has yet a fourth quantum number associated with it, however. We know that each electron spins about its own

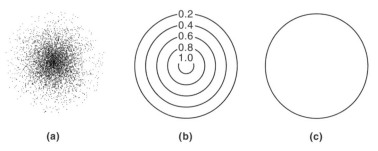

(a)   **(b)**   **(c)**

Figure 11.28
Representations of the hydrogen $1s$ orbital. (a) Charge cloud, (b) contour surfaces (the numbers represent the relative charge densities), and (c) boundary surface.

axis in either a clockwise or counterclockwise direction (Figure 11.29). (The spinning motion of a charged particle generates a magnetic field, so each electron behaves like a small magnet.) In quantum mechanics, we say that the electron has a spin, S, of $\frac{1}{2}$ and spin quantum numbers, $m_s = \pm\frac{1}{2}$. The value of m_s gives the orientation of the

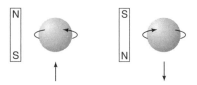

Figure 11.29
The two spinning motions of an electron. The ↑ and ↓ arrows are the symbols commonly employed to denote spin direction. The magnetic fields generated by the spinning motion are equivalent to those of two bar magnets.

magnetic dipole moment of the electron, which is a vector quantity showing the positions of the north and south poles of the magnet. Thus, m_s is analogous to m_l, which determines the orientation of the orbitals in space.

11.11 Many-Electron Atoms and the Periodic Table

The next simplest atom after hydrogen is helium. Helium contains two electrons and two protons (in the nucleus) so this is a three-body problem. The potential energy, V, is given by

$$V(r) = -\frac{2e^2}{4\pi\varepsilon_0 r_1} - \frac{2e^2}{4\pi\varepsilon_0 r_2} + \frac{e^2}{4\pi\varepsilon_0 r_{12}} \tag{11.32}$$

where r_1 and r_2 are the distances of the two electrons from the nucleus, and r_{12} is the distance between the electrons. The interelectronic repulsion term (the terms containing r_{12}) makes it impossible to obtain an exact solution of the Schrödinger equation for helium, as we can for a two-body system such as the hydrogen atom.

In quantum mechanics, the word "many" means two or more.

Solving the Schrödinger equation for helium and other many-electron atoms requires approximations. The usual approach to this problem is called the *self-consistent field (SCF) method*. Consider an atom of N electrons. A wave function is guessed for each electron except one. For example, we can guess the wave functions

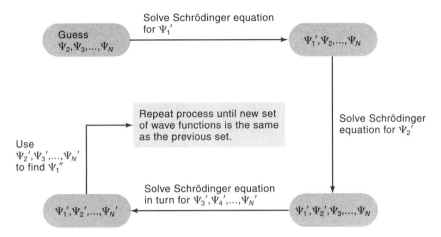

Figure 11.30
A schematic of the SCF method for obtaining the wave functions of a many-electron atom.

for electrons $2, 3, 4, \ldots, N$ to be $\psi_2, \psi_3, \psi_4, \ldots, \psi_N$. We can solve the Schrödinger equation for electron 1, which is moving in a potential field generated by the nucleus and electrons in orbitals $\psi_2, \psi_3, \psi_4, \ldots, \psi_N$. The repulsions between electron 1 and the rest of the electrons are calculated at each point in space from the sum of the average electron densities around that point. This procedure gives us the wave function for electron 1, which we call ψ_1'. Next, we do a similar calculation for electron 2, which is moving in a field of electrons described by the wave functions $\psi_1', \psi_3, \psi_4, \ldots, \psi_N$. This step yields a new function, ψ_2', for electron 2. We repeat this procedure for the rest of electrons until we obtain a set of new wave functions $\psi_1', \psi_2', \psi_3', \ldots, \psi_N'$ for all the electrons. This process is repeated as many times as necessary until we obtain a set of wave functions that is virtually identical to the previous set. At that point, a self-consistent field is achieved, and no more calculations are necessary. Figure 11.30 summarizes the procedure. The advent of high-speed computers has enabled us to calculate accurately the orbitals and energies of complex atoms. The results show that the orbitals of many-electron atoms are qualitatively similar to those of the hydrogen atom, and so we label them with the same quantum numbers used to describe the hydrogen atomic orbitals.

Electron Configurations

For the hydrogen atom and hydrogenlike ions, the energy of an electron depends only on the principal quantum number n (see Equation 11.5). Therefore, the orbitals can be arranged in order of increasing energy (decreasing stability) as follows:

$$1s < 2s = 2p < 3s = 3p = 3d < 4s = 4p = 4d < \cdots$$

For many-electron atoms, however, the electron's energy depends on both n and l, so that the order of increasing energy is given by

$$1s < 2s < 2p < 3s < 3p < 4s < 3d < 4p < 5s < 4d < \cdots$$

Figure 11.31 shows the order in which atomic subshells are filled in a many-electron atom. The difference between a hydrogen atom and a many-electron atom can be qualitatively explained as follows. Considering the $2s$ and $2p$ orbitals, we see from

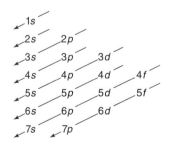

Figure 11.31
The order in which electrons fill energy levels in a many-electron atom.

Figure 11.25 that even though the most probable location of a $2p$ electron is closer to the nucleus than that of a $2s$ electron, the electron density close to the nucleus is actually greater for a $2s$ electron. Put another way, we say that an s electron is more penetrating than a p electron. Thus, the $2p$ electron is more shielded from the nucleus by the $2s$ electron than the other way around. Consequently, the energy of the $2p$ electron is higher than that of a $2s$ electron because the $2s$ electron shields it somewhat from the full attractive force of the nucleus. In general, for the same value of n, the penetrating power decreases as follows:

$$s > p > d > f > \cdots$$

The result of shielding means that each electron experiences a different effective nuclear charge and hence a different effective atomic number, Z_{eff}, given by

$$Z_{\text{eff}} = Z - \sigma \tag{11.33}$$

where Z is the atomic number of the atom, and σ is called the shielding constant. For carbon $(Z = 6)$, the effective atomic number for the $1s$ electrons is 5.7; for the $2s$ electrons it is 3.2; for the $2p$ electrons it is 3.1. A hydrogen atom or a hydrogenlike ion has only one electron; hence, no shielding occurs.

When the electron in the hydrogen atom is in the lowest possible energy level, the ground-state electron configuration is $1s^1$, meaning that there is one electron in the $1s$ orbital. The four quantum numbers $(n, l, m_l,$ and $m_s)$ of the electron can be either $\left(1, 0, 0, +\frac{1}{2}\right)$ or $\left(1, 0, 0, -\frac{1}{2}\right)$. In the absence of a magnetic field, the energy of the electron is the same whether $m_s = +\frac{1}{2}$ or $-\frac{1}{2}$. Helium has two electrons, so its ground-state electron configuration is $1s^2$. Helium atoms are diamagnetic, which means that the two electrons must have opposite spins, and the net magnetic field generated is zero. Thus, one electron must have $m_s = +\frac{1}{2}$ and the other $m_s = -\frac{1}{2}$; the other three quantum numbers are the same. This result is an illustration of the *Pauli exclusion principle* (after the Austrian physicist Wolfgang Pauli, 1900–1958), which states that no two electrons in an atom (or molecule) may have the same four quantum numbers.* According to the Pauli exclusion principle, the third electron in a lithium atom must enter the $2s$ orbital, so the electron configuration of lithium is $1s^2 2s^1$.

Note that the electrons in the outermost shell of an atom are called the *valence electrons* because they are largely responsible for the chemical bonds that the atom forms. Thus, hydrogen has one valence electron in the $1s$ orbital, and lithium has one valence electron in the $2s$ orbital.

As the procedure of filling orbitals continues across the second row of the periodic table, when we reach the carbon atom $(1s^2 2s^2 2p^2)$, we are faced with three different options for adding the two valence electrons to the three p orbitals:

Each square represents an orbital. None of the three arrangements shown violates the Pauli exclusion principle, so we must determine which one will give the carbon

---

*Another way of stating the Pauli exclusion principle is that the total wave function of the electrons (made up of a space part and a spin part) must change its sign when any two electrons are exchanged.

atom the greatest stability. The answer is provided by *Hund's rule* (after the German physicist Frederick Hund, 1896–1997), which states that when more than one electron enters a set of degenerate levels, the most stable arrangement is the one that has the greatest number of parallel spins. Hence, (c) is the most stable arrangement. The physical interpretation of Hund's rule is that the Pauli exclusion principle forbids two electrons with parallel spins in an atom to approach each other closely. Placing the two electrons in different orbitals decreases the electrostatic repulsion between them and results in a more stable arrangement. Indeed, experimentally we find that the ground-state carbon atom is *paramagnetic*, containing two unpaired electrons.

The stepwise procedure of writing the electron configuration of elements is based on the *Aufbau principle*, which states that as protons are added one by one to the nucleus to build up the elements, electrons are similarly added to the atomic orbitals. Table 11.3 lists the ground-state electron configurations of elements from H ($Z = 1$) through Ds ($Z = 110$). The electron configurations of all elements except hydrogen and helium are represented by a *noble gas core*, which shows in brackets the noble gas element that most nearly precedes the element, followed by the symbol for the highest filled subshells in the outermost shell. Notice that the electron configurations of the highest filled subshells in the outermost shells for the elements sodium ($Z = 11$) through argon ($Z = 18$) follow a pattern similar to those for lithium ($Z = 3$) through neon ($Z = 10$).

The $4s$ subshell is filled before the $3d$ subshell in a many-electron atom. Thus, the electron configuration of potassium ($Z = 19$) is $1s^2 2s^2 2p^6 3s^2 3p^6 4s^1$. Because $1s^2 2s^2 2p^6 3s^2 3p^6$ is the electron configuration of argon, we can simplify the electron configuration of potassium by writing [Ar]$4s^1$, where [Ar] denotes the "argon core." Similarly, we can write the electron configuration of calcium ($Z = 20$) as [Ar]$4s^2$. The placement of the outermost electron in the $4s$ orbital (rather than in the $3d$ orbital) of potassium is strongly supported by experimental evidence. The following comparison also suggests that this is the correct configuration. The chemistry of potassium is very similar to that of lithium and sodium, the first two alkali metals. The outermost electron of both lithium and sodium is in an s orbital (there is no ambiguity in assigning their electron configurations); therefore, we expect the last electron in potassium to occupy the $4s$ rather than the $3d$ orbital.

The elements from scandium ($Z = 21$) to copper ($Z = 29$) are transition metals. *Transition metals* either have incompletely filled d subshells or readily give rise to cations that have incompletely filled d subshells. Consider the first transition-metal series, from scandium through copper. In this series, additional electrons are placed in the $3d$ orbitals, according to Hund's rule. There are two irregularities, however. The electron configuration of chromium ($Z = 24$) is [Ar]$4s^1 3d^5$ and not [Ar]$4s^2 3d^4$, as we might expect. A similar break in the pattern is observed for copper, whose electron configuration is [Ar]$4s^1 3d^{10}$ rather than [Ar]$4s^2 3d^9$. The reason for these irregularities is that a slightly greater stability is associated with the half-filled ($3d^5$) and completely filled ($3d^{10}$) subshells. According to Hund's rule, the orbital diagram for Cr is

Having the d electrons in separate orbitals reduces the electrostatic repulsion. Thus, Cr has a total of six unpaired electrons. The orbital diagram for copper is

Aufbau is German for "building up."

Table 11.3
The Ground-State Electron Configurations of the Elements[a]

Atomic Number	Symbol	Electron Configuration	Atomic Number	Symbol	Electron Configuration	Atomic Number	Symbol	Electron Configuration
1	H	$1s^1$	38	Sr	$[Kr]5s^2$	75	Re	$[Xe]6s^24f^{14}5d^5$
2	He	$1s^2$	39	Y	$[Kr]5s^24d^1$	76	Os	$[Xe]6s^24f^{14}5d^6$
3	Li	$[He]2s^1$	40	Zr	$[Kr]5s^24d^2$	77	Ir	$[Xe]6s^24f^{14}5d^7$
4	Be	$[He]2s^2$	41	Nb	$[Kr]5s^14d^4$	78	Pt	$[Xe]6s^14f^{14}5d^9$
5	B	$[He]2s^22p^1$	42	Mo	$[Kr]5s^14d^5$	79	Au	$[Xe]6s^14f^{14}5d^{10}$
6	C	$[He]2s^22p^2$	43	Tc	$[Kr]5s^24d^5$	80	Hg	$[Xe]6s^24f^{14}5d^{10}$
7	N	$[He]2s^22p^3$	44	Ru	$[Kr]5s^14d^7$	81	Tl	$[Xe]6s^24f^{14}5d^{10}6p^1$
8	O	$[He]2s^22p^4$	45	Rh	$[Kr]5s^14d^8$	82	Pb	$[Xe]6s^24f^{14}5d^{10}6p^2$
9	F	$[He]2s^22p^5$	46	Pd	$[Kr]4d^{10}$	83	Bi	$[Xe]6s^24f^{14}5d^{10}6p^3$
10	Ne	$[He]2s^22p^6$	47	Ag	$[Kr]5s^14d^{10}$	84	Po	$[Xe]6s^24f^{14}5d^{10}6p^4$
11	Na	$[Ne]3s^1$	48	Cd	$[Kr]5s^24d^{10}$	85	At	$[Xe]6s^24f^{14}5d^{10}6p^5$
12	Mg	$[Ne]3s^2$	49	In	$[Kr]5s^24d^{10}5p^1$	86	Rn	$[Xe]6s^24f^{14}5d^{10}6p^6$
13	Al	$[Ne]3s^23p^1$	50	Sn	$[Kr]5s^24d^{10}5p^2$	87	Fr	$[Rn]7s^1$
14	Si	$[Ne]3s^23p^2$	51	Sb	$[Kr]5s^24d^{10}5p^3$	88	Ra	$[Rn]7s^2$
15	P	$[Ne]3s^23p^3$	52	Te	$[Kr]5s^24d^{10}5p^4$	89	Ac	$[Rn]7s^26d^1$
16	S	$[Ne]3s^23p^4$	53	I	$[Kr]5s^24d^{10}5p^5$	90	Th	$[Rn]7s^26d^2$
17	Cl	$[Ne]3s^23p^5$	54	Xe	$[Kr]5s^24d^{10}5p^6$	91	Pa	$[Rn]7s^25f^26d^1$
18	Ar	$[Ne]3s^23p^6$	55	Cs	$[Xe]6s^1$	92	U	$[Rn]7s^25f^36d^1$
19	K	$[Ar]4s^1$	56	Ba	$[Xe]6s^2$	93	Np	$[Rn]7s^25f^46d^1$
20	Ca	$[Ar]4s^2$	57	La	$[Xe]6s^25d^1$	94	Pu	$[Rn]7s^25f^6$
21	Sc	$[Ar]4s^23d^1$	58	Ce	$[Xe]6s^24f^15d^1$	95	Am	$[Rn]7s^25f^7$
22	Ti	$[Ar]4s^23d^2$	59	Pr	$[Xe]6s^24f^3$	96	Cm	$[Rn]7s^25f^76d^1$
23	V	$[Ar]4s^23d^3$	60	Nd	$[Xe]6s^24f^4$	97	Bk	$[Rn]7s^25f^9$
24	Cr	$[Ar]4s^13d^5$	61	Pm	$[Xe]6s^24f^5$	98	Cf	$[Rn]7s^25f^{10}$
25	Mn	$[Ar]4s^23d^5$	62	Sm	$[Xe]6s^24f^6$	99	Es	$[Rn]7s^25f^{11}$
26	Fe	$[Ar]4s^23d^6$	63	Eu	$[Xe]6s^24f^7$	100	Fm	$[Rn]7s^25f^{12}$
27	Co	$[Ar]4s^23d^7$	64	Gd	$[Xe]6s^24f^75d^1$	101	Md	$[Rn]7s^25f^{13}$
28	Ni	$[Ar]4s^23d^8$	65	Tb	$[Xe]6s^24f^9$	102	No	$[Rn]7s^25f^{14}$
29	Cu	$[Ar]4s^13d^{10}$	66	Dy	$[Xe]6s^24f^{10}$	103	Lr	$[Rn]7s^25f^{14}6d^1$
30	Zn	$[Ar]4s^23d^{10}$	67	Ho	$[Xe]6s^24f^{11}$	104	Rf	$[Rn]7s^25f^{14}6d^2$
31	Ga	$[Ar]4s^23d^{10}4p^1$	68	Er	$[Xe]6s^24f^{12}$	105	Db	$[Rn]7s^25f^{14}6d^3$
32	Ge	$[Ar]4s^23d^{10}4p^2$	69	Tm	$[Xe]6s^24f^{13}$	106	Sg	$[Rn]7s^25f^{14}6d^4$
33	As	$[Ar]4s^23d^{10}4p^3$	70	Yb	$[Xe]6s^24f^{14}$	107	Bh	$[Rn]7s^25f^{14}6d^5$
34	Se	$[Ar]4s^23d^{10}4p^4$	71	Lu	$[Xe]6s^24f^{14}5d^1$	108	Hs	$[Rn]7s^25f^{14}6d^6$
35	Br	$[Ar]4s^23d^{10}4p^5$	72	Hf	$[Xe]6s^24f^{14}5d^2$	109	Mt	$[Rn]7s^25f^{14}6d^7$
36	Kr	$[Ar]4s^23d^{10}4p^6$	73	Ta	$[Xe]6s^24f^{14}5d^3$	110	Ds	$[Rn]7s^25f^{14}6d^8$
37	Rb	$[Kr]5s^1$	74	W	$[Xe]6s^24f^{14}5d^4$			

[a] The symbol [He] is called the *helium core* and represents $1s^2$. [Ne] is called the *neon core* and represents $1s^22s^22p^6$. [Ar] is called the *argon core* and represents $[Ne]3s^23p^6$. [Kr] is called the *krypton core* and represents $[Ar]4s^23d^{10}4p^6$. [Xe] is called the *xenon core* and represents $[Kr]5s^24d^{10}5p^6$. [Rn] is called the *radon core* and represents $[Xe]6s^24f^{14}5d^{10}6p^6$. Elements 111–115 have been synthesized but have not yet been named.

The qualitative explanation for this configuration is as follows. Electrons in the same subshell have equal energy but different spatial distributions; consequently, their shielding of one another is relatively slight. Therefore, the effective nuclear charge increases as the actual nuclear charge increases, so that a completely filled subshell (d^{10}) has a high stability.

For elements Zn ($Z = 30$) through Kr ($Z = 36$), the $4s$ and $4p$ subshells fill in a straightforward manner. With rubidium ($Z = 37$), electrons begin to enter the $n = 5$ energy level. The electron configurations in the second transition-metal series [yttrium ($Z = 39$) to silver ($Z = 47$)] are also irregular, but we shall not discuss the details here.

The sixth period of the periodic table begins with cesium ($Z = 55$) and barium ($Z = 56$), whose electron configurations are [Xe]$6s^1$ and [Xe]$6s^2$, respectively. Next we come to lanthanum ($Z = 57$). The energies of the $5d$ and $4f$ orbitals are very close; in fact, for lanthanum, $4f$ is slightly higher in energy than $5d$. Thus, lanthanum's electron configuration is [Xe]$6s^2 5d^1$ and not [Xe]$6s^2 4f^1$. Following lanthanum are the 14 elements known as the *lanthanides*, or *rare earth series* [cerium ($Z = 58$) to lutetium ($Z = 71$)]. The rare earth metals have incompletely filled $4f$ subshells or readily give rise to cations that have incompletely filled $4f$ subshells. In this series, the added electrons are placed in $4f$ orbitals. After the $4f$ subshells are completely filled, the next electron enters the $5d$ subshell of lutetium. Note that the electron configuration of gadolinium ($Z = 64$) is [Xe]$6s^2 4f^7 5d^1$ rather than [Xe]$6s^2 4f^8$. Like chromium, gadolinium gains extra stability by having half-filled subshells ($4f^7$). The third transition-metal series, including lanthanum and hafnium ($Z = 72$) and extending through gold ($Z = 79$), is characterized by the filling of the $5d$ orbitals. The $6s$ and $6p$ subshells are filled next, bringing us to radon ($Z = 86$). The next row of elements is the *actinide series*, which starts at thorium ($Z = 90$) and have incompletely filled $5f$ subshells. Most of these elements are not found in nature but have been synthesized.

Finally, let us look at the procedure for writing electron configurations for ions. For cations, we first remove p valence electrons, then s valence electrons, and then as many d electrons as necessary to achieve the required charge. For example, the electron configuration of Mn is [Ar]$4s^2 3d^5$ so that the electron configuration of Mn^{2+} is [Ar]$3d^5$. The electron configurations of anions are derived by adding electrons to the atoms until the next noble gas core has been reached. Thus, for the oxide ion (O^{2-}), we add two electrons to [He]$2s^2 2p^4$, reaching [He]$2s^2 2p^6$, which is the same as the configuration for neon.

Variations in Periodic Properties

The periodic trends in electron configuration result in periodic trends in chemical and physical properties. Here, we consider a few of them: atomic radius, ionization energy, and electron affinity. The general periodic trends are that as we move across a period from left to right, the metallic properties decrease; down a particular group, the metallic character increases. These trends do not apply to transition elements that are all metallic and possess similar properties.

Atomic Radius. An atom does not have a definite size. Mathematically, the wave function of an atom extends to infinity. Therefore, we need to define it in a somewhat arbitrary manner. One way is to use the *covalent radii*, obtained from measurements of distances between the nuclei of atoms in molecules, as a measure of atomic size. Consider the second-period elements. From Li to Ne, atomic number increases, and electrons are added to the $2s$ and $2p$ orbitals. Because electrons in the same subshells do not shield each other well, the effective nuclear charge, Z_{eff}, increases, shrinking the electron density and hence the size of the atom. Within a periodic group, the atomic radius increases with increasing atomic number. In the alkali metals, for example, the outermost electron resides in the ns orbital. Because orbital size increases as the principal quantum number n increases, the size of the atoms increases from Li to Cs.

Ionization Energy. *Ionization energy* is the minimum energy required to remove an electron from a gaseous atom in its ground state,

$$\text{energy} + X(g) \rightarrow X^+(g) + e^-$$

where X represents an atom of any element. This measurement gives the first ionization energy. The process can be continued to give the second, third, ..., ionization energies as follows:

$$\text{energy} + X^+(g) \rightarrow X^{2+}(g) + e^-$$
$$\text{energy} + X^{2+}(g) \rightarrow X^{3+}(g) + e^-$$

Table 11.4 lists ionization energies for the first twenty elements, and Figure 11.32 plots the first ionization energy versus atomic number, which clearly shows periodic behavior. The effective nuclear charge increases across a period from left to right, and so the ionization energy also increases because the outermost electron is held more tightly. Down a group, the outermost electron is placed in successive outer shells, where it is effectively shielded by the inner electrons and can be removed more easily than the element above it.

Table 11.4
The Ionization Energies (kJ mol^{-1}) of the First 20 Elements

Z	Element	First	Second	Third	Fourth	Fifth	Sixth
1	H	1312					
2	He	2373	5251				
3	Li	520	7300	11815			
4	Be	899	1757	14850	21005		
5	B	801	2430	3660	25000	32820	
6	C	1086	2350	4620	6220	38000	47300
7	N	1400	2860	4580	7500	9400	53000
8	O	1314	3390	5300	7470	11000	13000
9	F	1680	3370	6050	8400	11000	15200
10	Ne	2080	3950	6120	9370	12200	15000
11	Na	495.9	4560	6900	9540	13400	16600
12	Mg	738.1	1450	7730	10500	13600	18000
13	Al	577.9	1820	2750	11600	14800	18400
14	Si	786.3	1580	3230	4360	16000	20000
15	P	1012	1904	2910	4960	6240	21000
16	S	999.5	2250	3360	4660	6990	8500
17	Cl	1251	2297	3820	5160	6540	9300
18	Ar	1521	2666	3900	5770	7240	8800
19	K	418.7	3052	4410	5900	8000	9600
20	Ca	589.5	1145	4900	6500	8100	11000

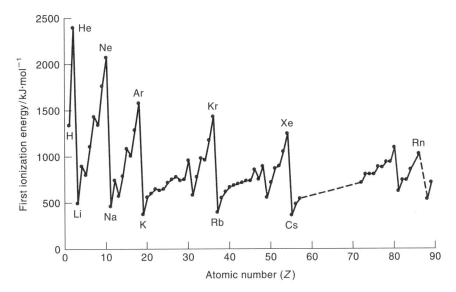

Figure 11.32
Plot of first ionization energy against atomic number.

Electron Affinity. *Electron affinity* is defined as the negative of the energy change that occurs when an electron is accepted by an atom in the gaseous state to form an anion,

$$X(g) + e^- \rightarrow X^-(g)$$

In contrast to ionization energies, electron affinities are difficult to measure. Table 11.5 lists the electron affinities of some representative elements. The more positive the electron affinity, the greater the tendency of the atom to accept an electron.

Electron affinity is positive if the reaction is exothermic and negative if the reaction is endothermic.

Table 11.5
Electron Affinities (kJ mol^{-1}) of Some Representative Elements and the Noble Gases[a]

1A	2A	3A	4A	5A	6A	7A	8A
H							He
73							<0
Li	Be	B	C	N	O	F	Ne
60	≤0	27	122	0	141	328	<0
Na	Mg	Al	Si	P	S	Cl	Ar
53	≤0	44	134	72	200	349	<0
K	Ca	Ga	Ge	As	Se	Br	Kr
48	2.4	29	118	77	195	325	<0
Rb	Sr	In	Sn	Sb	Te	I	Xe
47	4.7	29	121	101	190	295	<0
Cs	Ba	Tl	Pb	Bi	Po	At	Rn
45	14	30	110	110	?	?	<0

[a] The electron affinities of the noble gases, Be, and Mg have not been determined experimentally but are believed to be close to zero or negative.

Suggestions for Further Reading

Books

Atkins, P. W., *Quanta: A Handbook of Concepts*, Oxford University Press, New York, 1991.

Bell, R. P., *The Tunnel Effect in Chemistry*, Chapman and Hall, London, 1980.

Cropper, W. H., *The Quantum Physicists*, Oxford University Press, Inc., New York, 1970.

DeVault, D., *Quantum-Mechanical Tunneling in Biological Systems*, 2nd ed., Cambridge University Press, New York, 1984.

Herzberg, G., *Atomic Spectra and Atomic Structure*, Dover Publications, New York, 1944.

Hochstrasser, R. M., *Behavior of Electrons in Atoms*, W. A. Benjamin, Menlo Park, CA, 1964.

Karplus, M. and R. N. Porter, *Atoms and Molecules: An Introduction for Students of Physical Chemistry*, W. A. Benjamin, New York, 1970.

Articles

Quantum Theory

"The Limits of Measurement," R. Furth, *Sci. Am.* July 1950.

"The Quantum Theory," K. K. Darrow, *Sci. Am.* March 1952.

"What Is Matter?" E. Schrödinger, *Sci. Am.* September 1953.

"The Principle of Uncertainty," G. Gamow, *Sci. Am.* January 1958.

"The Bohr Atomic Model: Niels Bohr," A. B. Garrett, *J. Chem. Educ.* **39**, 534 (1962).

"Quantum Theory: Max Planck," A. B. Garrett, *J. Chem. Educ.* **40**, 262 (1963).

"Demonstration of the Uncertainty Principle," W. Laurita, *J. Chem. Educ.* **45**, 461 (1968).

"Particles, Waves, and the Interpretation of Quantum Mechanics," N. D. Christoudouleas, *J. Chem. Educ.* **52**, 573 (1975).

"The Mass of the Photon," A. S. Goldhaber and M. M. Nieto, *Sci. Am.* May 1976.

"The Spectrum of Atomic Hydrogen," T. W. Hänsch, A. L. Schawlow, and G. W. Series, *Sci. Am.* March 1979.

"Centrifugal Force and the Bohr Model of the Hydrogen Atom," B. L. Haendler, *J. Chem. Educ.* **58**, 719 (1981).

"Does Quantum Mechanics Apply to One or Many Particles?" F. Cactano, L. Lain, M. N. Sanchez Rayo, and A. Torre, *J. Chem. Educ.* **60**, 377 (1983).

"Illustrating the Heisenberg Uncertainty Principle," G. D. Peckham, *J. Chem. Educ.* **61**, 868 (1984).

"Dice Throwing as an Analogy for Teaching Quantum Mechanics," B. de Barros Neto, *J. Chem. Educ.* **61**, 1044 (1984).

"Perspectives on the Uncertainty Principle and Quantum Reality," L. S. Bartell, *J. Chem. Educ.* **62**, 192 (1985).

"On Introducing the Uncertainty Principle," G. M. Muha and D. W. Muha, *J. Chem. Educ.* **63**, 525 (1986).

"Exercises in Quantum Mechanics," F. Rioux, *J. Chem. Educ.* **64**, 789 (1987).

"Heisenberg, Uncertainty, and the Quantum Revolution," D. C. Cassidy, *Sci. Am.* May 1992.

"On a Relation Between the Heisenberg and de Broglie Principles," O. G. Ludwig, *J. Chem. Educ.* **70**, 28 (1993).

"The Duality in Matter and Light," B.-G. Englert, M. O. Scully, and H. Walther, *Sci. Am.* December 1994.

"Using Natural and Artificial Light Sources to Illustrate Quantum Mechanical Concepts," G. A. Rechtsteiner and J. A. Ganske, *Chem. Educator* **1998**, 3(4): S1430–4171 (98) 04230-7. Avail. URL: http://journals.springer-ny.com/chedr.

"One Hundred Years of Quantum Physics," D. Kleppner and R. Jackiw, *Science* **289**, 893 (2000).

Particle in a One-Dimensional Box

"A Particle in a Chemical Box," K. M. Jinks, *J. Chem. Educ.* **52**, 312 (1975).

"On the Momentum of a Particle in a Box," G. M. Muha, *J. Chem. Educ.* **63**, 761 (1986).

"How Do Electrons Get Across Nodes?" P. G. Nelson, *J. Chem. Educ.* **67**, 643 (1990).

"The Two-Dimensional Particle in a Box," G. L. Breneman, *J. Chem. Educ.* **67**, 866 (1990).

"More About the Particle-in-a Box System: The Confinement of Matter and the Wave-Particle Dualism," K. Volkamer and M. W. Lerom, *J. Chem. Educ.* **69**, 100 (1992).

Quantum Mechanical Tunneling

"Quantum Chemical Reactions in the Deep Cold," V. I. Goldanskii, *Sci. Am.* February 1980.

"The Scanning Tunneling Microscope," G. Binnig and H. Rohrer, *Sci. Am.* August 1985.

"Electron Transfer in Biology," R. J. P. Williams, *Molec. Phys.* **68**, 1 (1989).

"Electron Tunneling Pathways in Proteins," D. Beratan, J. N. Onuchic, J. R. Winkler, and H. B. Gray, *Science* **258**, 1740 (1992).

Atomic Structure

"The Exclusion Principle," G. Gamow, *Sci. Am.* July 1959.

"The Five Equivalent *d* Orbitals," R. E. Powell, *J. Chem. Educ.* **45**, 45 (1968).

"Five Equivalent *d* Orbitals," L. Pauling and V. McClure, *J. Chem. Educ.* **47**, 15 (1970).

"The Stability of the Hydrogen Atom," F. Rioux, *J. Chem. Educ.* **50**, 550 (1973).

"4*s* is Always Above 3*d*! or, How to tell the Orbitals from the Wavefunctions," F. L. Pilar, *J. Chem. Educ.* **55**, 2 (1978).

"Highly Excited Atoms," D. Kleppner, M. G. Littman, and M. L. Zimmerman, *Sci. Am.* May 1981.

"Teaching the Shapes of the Hydrogenlike and Hybrid Atomic Orbitals," R. D. Allendoerfer, *J. Chem. Educ.* **67**, 37 (1990).

"Relative Energies of 3*d* and 4*s* Orbitals," P. G. Nelson, *Educ. in Chem.* **29**, 84 (1992).

"Transition Metals and the Aufbau Principle," L. G. Vanquickenborne, K. Pierloot, and D. Devoghel, *J. Chem. Educ.* **71**, 469 (1994).

"Understanding Electron Spin," J. C. A. Boeyens, *J. Chem. Educ.* **72**, 412 (1995).

"Why the 4*s* is occupied Before the 3*d*," M. Melrose and E. R. Scerri, *J. Chem. Educ.* **74**, 498 (1996).

Periodic Trends

"Periodic Contractions Among the Elements: or, On Being the Right Size," J. Mason, *J. Chem. Educ.* **65**, 17 (1988).

"The Periodicity of Electron Affinity," R. T. Meyers, *J. Chem. Educ.* **67**, 307 (1990).

"Ionization Energies Revisited," N. C. Pyper and M. Berry, *Educ. Chem.* **27**, 135 (1990).

"Electron Affinities of the Alkaline Earth Metals and the Sign Convention for Electron Affinity," J. C. Wheeler, *J. Chem. Educ.* **74**, 123 (1997).

"The Evolution of the Periodic System," E. R. Scerri, *Sci. Am.* September 1998.

Problems

Quantum Theory

11.1 Calculate the energy associated with a quantum (photon) of light with a wavelength of 500 nm.

11.2 The threshold frequency for dislodging an electron from a zinc metal surface is 8.54×10^{14} Hz. Calculate the minimum amount of energy required to remove an electron from the metal.

11.3 Calculate the energies for the Bohr orbits with $n = 2$ and 3 for atomic hydrogen.

11.4 Calculate the frequency and wavelength associated with the transition from the $n = 5$ to the $n = 3$ level in atomic hydrogen.

11.5 What are the wavelengths associated with **(a)** an electron moving at 1.50×10^8 cm s^{-1}, and **(b)** a 60-g tennis ball moving at 1500 cm s^{-1}?

11.6 A photoelectric experiment was performed by separately shining a laser at 450 nm (blue light) and a laser at 560 nm (yellow light) on a clean metal surface and measuring the number and kinetic energy of the ejected electrons. Which light would generate more electrons? Which light would eject electrons with greater kinetic energy? Assume that the same number of photons is delivered to the metal surface by each laser and that the frequencies of the laser lights exceed the threshold frequency.

11.7 Explain how scientists are able to estimate the temperature on the surface of the sun. (*Hint:* Treat solar radiation like radiation from a blackbody.)

11.8 In a photoelectric experiment, a student uses a light source whose frequency is greater than that needed to eject electrons from a certain metal. After continuously shining the light on the same area of the metal for a long period of time, however, the student notices that the maximum kinetic energy of ejected electrons begins to decrease, even though the frequency of the light is held constant. How would you account for this behavior?

11.9 A proton is accelerated through a potential difference of 3.0×10^6 V, starting from rest. Calculate its final wavelength.

11.10 Suppose that the uncertainty in determining the position of an electron circling an atom in an orbit is 0.4 Å. What is the uncertainty in its velocity?

11.11 A person weighing 77 kg jogs at 1.5 m s^{-1}. **(a)** Calculate the momentum and wavelength of this person. **(b)** What is the uncertainty in determining his position at any given instant if we can measure his momentum to $\pm 0.05\%$? **(c)** Predict the changes that would take place in this problem if the Planck constant were 1 J s.

11.12 The diffraction phenomenon can be observed whenever the wavelength is comparable in magnitude to the size of the slit opening. To be "diffracted," how fast must a person weighing 84 kg move through a door 1 m wide?

11.13 Spectral lines of the Lyman and Balmer series do not overlap. Verify this statement by calculating the longest wavelength associated with the Lyman series and the shortest wavelength associated with the Balmer series (in nm).

11.14 The He^+ ion contains only one electron and is therefore a hydrogenlike ion. Calculate the wavelengths, in increasing order, of the first four transitions in the Balmer series of the He^+ ion. Compare these wavelengths with the same transitions in a H atom. Comment on the differences. (The Rydberg constant for He^+ is 8.72×10^{-18} J.)

11.15 An electron in an excited state in a hydrogen atom can return to the ground state in two different ways: first, via a direct transition in which a photon of wavelength λ_1 is emitted and second, via an intermediate excited state reached by the emission of a photon of wavelength λ_2. This intermediate excited state then decays to the ground state by emitting another photon of wavelength λ_3. Derive an equation that relates λ_1 to λ_2 and λ_3.

11.16 The retina of a human eye can detect light when radiant energy incident on it is at least 4.0×10^{-17} J. For light of 600-nm wavelength, how many photons does this correspond to?

11.17 A 368-g sample of water absorbs infrared radiation at 1.06×10^4 nm from a carbon dioxide laser. Suppose all the absorbed radiation is converted to heat. Calculate the number of photons at this wavelength required to raise the temperature of the water by 5.00°C.

11.18 Ozone (O_3) in the stratosphere absorbs the harmful radiation from the sun by undergoing decomposition: $O_3 \rightarrow O + O_2$. **(a)** Referring to Appendix 2, calculate the $\Delta_r H°$ value for this process. **(b)** Calculate the maximum wavelength of photons (in nm) that possess this energy to bring about the decomposition of ozone photochemically.

11.19 Scientists have found interstellar hydrogen atoms with quantum number n in the hundreds. Calculate the wavelength of light emitted when a hydrogen atom undergoes a transition from $n = 236$ to $n = 235$. In what region of the electromagnetic spectrum does this wavelength fall?

11.20 A student records an emission spectrum of hydrogen and notices that one spectral line in the Balmer series cannot be accounted for by the Bohr theory. Assuming that the gas sample is pure, suggest a species that might be responsible for this line.

11.21 In the mid-19th century, physicists studying the solar emission spectrum (a continuum) noticed a set of dark lines that did not match any of the emission lines (bright lines) on Earth. They concluded that the lines came from a yet unknown element. Later this element was identified as helium. **(a)** What is the origin of the dark lines? How were these lines correlated with the emission lines of helium? **(b)** Why was helium so difficult to detect in Earth's atmosphere? **(c)** Where is the most likely place to detect helium on Earth?

11.22 How many photons at 660 nm must be absorbed to melt 5.0×10^2 g of ice? On average, how many H_2O molecules does one photon convert from ice to water? (*Hint:* It takes 334 J to melt 1 g of ice at 0°C.)

Particle in a One-Dimensional Box

11.23 Show that Equation 11.22 is dimensionally correct.

11.24 According to Equation 11.22, the energy is inversely proportional to the square of the length of the box. How would you account for this dependence in terms of the Heisenberg uncertainty principle?

11.25 What is the probability of locating a particle in a one-dimensional box between $L/4$ and $3L/4$, where L is the length of the box? Assume the particle to be in the lowest level.

11.26 Derive Equation 11.22 using de Broglie's relation. (*Hint:* First you must express the wavelength of the particle in the nth level in terms of the length of the box.)

11.27 An important property of the wave functions of the particle in a one-dimensional box is that they are orthogonal; that is,

$$\int_0^L \psi_n \psi_m \, dx = 0 \quad m \neq n$$

Prove this statement using ψ_1 and ψ_2 and Equation 11.23.

11.28 Based on the particle-in-a-one-dimensional-box model, suggest where along the box the $n = 1 \rightarrow n = 2$ electronic transition would most likely take place. Explain your choice.

11.29 As stated in the chapter, the probability of locating a particle in a one-dimensional box is given by $\psi^2 \, dx$. Over a small distance, the probability can be calculated without integration. Consider an electron with $n = 1$ in a box of length 2.000 nm. Calculate the probability of locating the electron **(a)** between 0.500 nm and 0.502 nm and **(b)** between 0.999 nm and 1.001 nm. Comment on your results and on the validity of your approximation.

Electron Configuration and Atomic Properties

11.30 Obtain an expression for the most probable radius at which an electron will be found when it occupies the $1s$ orbital.

11.31 Use the $2s$ wave function given in Table 11.2 to calculate the value of r (other than $r = \infty$) at which this wave function becomes zero.

11.32 Write the ground-state electron configurations of the following ions, which play important roles in biochemical processes in our bodies: **(a)** Na^+, **(b)** Mg^{2+}, **(c)** Cl^-, **(d)** K^+, **(e)** Ca^{2+}, **(f)** Fe^{2+}, **(g)** Cu^{2+}, **(h)** Zn^{2+}.

11.33 Explain, in terms of their electron configurations, why Fe^{2+} is more easily oxidized to Fe^{3+} than Mn^{2+} is to Mn^{3+}.

11.34 Ionization energy is the energy required to remove a ground state ($n = 1$) electron from an atom. It is usually expressed in units of kJ mol^{-1}. **(a)** Calculate the ionization energy for the hydrogen atom. **(b)** Repeat the calculation, assuming in this case the electron is removed from the $n = 2$ state.

11.35 Plasma is a state of matter consisting of positive gaseous ions and electrons. In the plasma state, a mercury atom could be stripped of its 80 electrons and therefore would exist as Hg^{80+}. Calculate the energy required for the last ionization step—that is,

$$\text{Hg}^{79+}(g) \rightarrow \text{Hg}^{80+}(g) + e^-$$

11.36 A technique called photoelectron spectroscopy is used to measure the ionization energy of atoms. A sample is irradiated with UV light, which causes electrons to be ejected from the valence shell. The kinetic energies of the ejected electrons are measured. Knowing the energy of the UV photon and the kinetic energy of the ejected electron, we can write

$$h v = \text{IE} + \tfrac{1}{2} m_e v^2$$

where v is the frequency of the UV light, and m_e and v are the mass and velocity of the electron, respectively. In one experiment, the kinetic energy of the ejected electron from potassium is found to be 5.34×10^{-19} J using a UV source of wavelength 162 nm. Calculate the ionization energy of potassium. How can you be sure that this ionization energy corresponds to the electron in the valence shell (that is, the most loosely held electron)?

11.37 The energy needed for the following process is 1.96×10^4 kJ mol^{-1}:

$$\text{Li}(g) \rightarrow \text{Li}^{3+}(g) + 3e^-$$

If the first ionization of lithium is 520 kJ mol^{-1}, calculate the second ionization of lithium; that is, calculate the energy required for the process

$$\text{Li}^+(g) \rightarrow \text{Li}^{2+}(g) + e^-$$

11.38 Experimentally, the electron affinity of an element can be determined by using a laser light to ionize the anion of the element in the gas phase:

$$\text{X}^-(g) + h v \rightarrow \text{X}(g) + e^-$$

Referring to Table 11.5, calculate the photon wavelength (in nanometers) corresponding to the electron affinity for chlorine. In what region of the electromagnetic spectrum does this wavelength fall?

11.39 The standard enthalpy of atomization of an element is the energy required to convert 1 mole of an element in its most stable form at 25°C to 1 mole of monatomic gas. Given that the standard enthalpy of

atomization for sodium is 108.4 kJ mol^{-1}, calculate the energy in kilojoules required to convert 1 mole of sodium metal at 25°C to 1 mole of gaseous Na$^+$ ions.

11.40 Explain why the electron affinity of nitrogen is approximately zero, while the elements on either side, carbon and oxygen, have substantial positive electron affinities.

11.41 Calculate the maximum wavelength of light (in nanometers) required to ionize a single sodium atom.

11.42 The first four ionization energies of an element are approximately 738 kJ mol^{-1}, 1450 kJ mol^{-1}, 7.7×10^3 kJ mol^{-1}, and 1.1×10^4 kJ mol^{-1}. To which periodic group does this element belong? Why?

Additional Problems

11.43 When two atoms collide, some of their kinetic energy may be converted into electronic energy in one or both atoms. If the average kinetic energy $\left(\tfrac{3}{2} k_B T\right)$ is about equal to the energy for some allowed electronic transition, an appreciable number of atoms can absorb enough energy through an inelastic collision to be raised to an excited electronic state. **(a)** Calculate the average kinetic energy per atom in a gas sample at 298 K. **(b)** Calculate the energy difference between the $n = 1$ and $n = 2$ levels in hydrogen. **(c)** At what temperature is it possible to excite a hydrogen atom from the $n = 1$ level to the $n = 2$ level by collision?

11.44 Photodissociation of water,

$$\text{H}_2\text{O}(g) + h v \rightarrow \text{H}_2(g) + \tfrac{1}{2}\text{O}_2(g)$$

has been suggested as a source of hydrogen. The $\Delta_r H°$ value for the reaction, calculated from thermochemical data, is 285.8 kJ per mole of water decomposed. Calculate the maximum wavelength (in nm) that would provide the necessary energy. In principle, is it feasible to use sunlight as a source of energy for this process?

11.45 Based on the discussion of decay and quantum mechanical tunneling, suggest a relation between the energy of emitted α particles and the half-life for the radioactive decay.

11.46 Only a fraction of the electrical energy supplied to a tungsten light bulb is converted to visible light. The rest of the energy shows up as infrared radiation (that is, heat). A 75-W light bulb converts 15.0% of the energy supplied to it into visible light. Assuming a wavelength of 550 nm, how many photons are emitted by the light bulb per second? (1 W = 1 J s^{-1}.)

11.47 An electron in a hydrogen atom is excited from the ground state to the $n = 4$ state. State whether the following statements are true or false. **(a)** $n = 4$ is the first excited state. **(b)** It takes more energy to ionize (remove) the electron from the $n = 4$ state than from the ground state. **(c)** The electron is farther from the nucleus (on average) in the $n = 4$ state than in the ground state. **(d)** The wavelength of light emitted when the electron drops from $n = 4$ to $n = 1$ is longer than that from $n = 4$ to $n = 2$. (e) The wavelength the atom absorbs in going from $n = 1$ to $n = 4$ is the same as that emitted as it goes from $n = 4$ to $n = 1$.

11.48 The ionization energy of a certain element is 412 kJ mol^{-1}. When the atoms of this element are in the first excited state, however, the ionization energy is only 126 kJ mol^{-1}. Based on this information, calculate the wavelength of light emitted in a transition from the first excited state to the ground state.

11.49 The UV light responsible for sun tanning falls in the 320- to 400-nm region. Calculate the total energy (in joules) absorbed by a person exposed to this radiation for 2.0 hours, given that there are 2.0×10^{16} photons hitting Earth's surface per square centimeter per second over an 80-nm (320-nm to 400-nm) range and that the exposed body area is 0.45 m^2. Assume that only half of the radiation is absorbed and the other half is reflected by the body. (*Hint:* Use an average wavelength of 360 nm to calculate the energy of a photon.)

11.50 In 1996, physicists created an antiatom of hydrogen. In such an atom, which is the antimatter equivalent of an ordinary atom, the electrical charges of all the component particles are reversed. Thus, the nucleus of an antiatom is made of an antiproton, which has the same mass as a proton but bears a negative charge, and the electron is replaced by an antielectron (also called positron) with the same mass as an electron but bearing a positive charge. Would you expect the energy levels, emission spectra, and atomic orbitals of an antihydrogen atom to be different from those of a hydrogen atom? What would happen if an antiatom of hydrogen collided with a hydrogen atom?

11.51 A student carried out a photoelectric experiment by shining visible light on a clean piece of cesium metal. She determined the kinetic energy of ejected electrons by applying a retarding voltage such that the current due to the electrons read exactly zero. The condition was reached when $eV = (1/2)m_e v^2$, where e is electric charge and V is the retarding potential. Her results are as follows:

λ/nm	405	435.8	480	520	577.7	650
V/volt	1.475	1.268	1.027	0.886	0.667	0.381

Rearrange Equation 11.3 to read

$$v = \frac{\Phi}{h} + \frac{e}{h} V$$

Determine the values of h and Φ graphically.

11.52 Use Equation 2.28 to calculate the de Broglie wavelength of a N$_2$ molecule at 300 K.

11.53 Alveoli are tiny sacs of air in the lungs. Their average diameter is 5.0×10^{-5} m. Calculate the uncertainty in the velocity of an oxygen molecule (5.3×10^{-26} kg) trapped within a sac. (*Hint:* The maximum

uncertainty in the position of the molecule is given by the diameter of the sac.)

11.54 The sun is surrounded by a white circle of gaseous material called the corona, which becomes visible during a total eclipse of the sun. The temperature of the corona is in the millions of degrees Celsius, high enough to break up molecules and remove some or all of the electrons from atoms. One way astronomers have been able to estimate the temperature of the corona is by studying the emission lines of ions of certain elements. For example, the emission spectrum of Fe^{14+} ions has been recorded and analyzed. Knowing that it takes 3.5×10^4 kJ mol^{-1} to convert Fe^{13+} to Fe^{14+}, estimate the temperature of the sun's corona. (*Hint:* The average kinetic energy of 1 mole of a gas is $\frac{3}{2}RT$.)

11.55 Consider a particle in the one-dimensional box shown below.

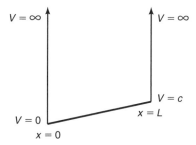

The potential energy of the particle is not zero everywhere in the box but varies according to its position, or x. Write the Schrödinger wave equation for this system.

11.56 The equation for calculating the energies of the electron in a hydrogen atom or a hydrogenlike ion is given by $E_n = -(2.18 \times 10^{-18} \text{ J})Z^2(1/n^2)$, where Z is the atomic number of the element. One way to modify this equation for many-electron atoms is to replace Z with $(Z - \sigma)$, where σ is a positive dimensionless quantity called the shielding constant. Consider the helium atom as an example. The physical significance of σ is that it represents the extent of shielding that the two $1s$ electrons exert on each other. Thus the quantity $(Z - \sigma)$ is appropriately called the "effective nuclear charge." Use the first ionization energy of helium in Table 11.4 to calculate the value of σ.

11.57 The radioactive Co-60 isotope is used in nuclear medicine to treat certain types of cancer. Calculate the wavelength and frequency of an emitted gamma particle having the energy of 1.29×10^{11} J mol^{-1}.

11.58 The figure shown below represents the emission spectrum of a hydrogenlike ion in the gas phase. All the lines result from the electronic transitions from the excited states to the $n = 2$ state. (a) What electronic transitions correspond to lines B and C? (b) If the wavelength of line C is 27.1 nm, calculate the wavelengths of lines A and B. (c) Calculate the energy needed to re-

move the electron from the ion in the $n = 4$ state.
(d) What is the physical significance of the continuum?

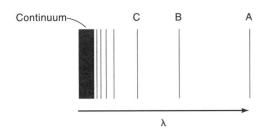

11.59 Use Equation 11.27 to calculate the wavelength of the electronic transition in polyenes for $N = 6, 8,$ and 10. Comment on the variation of λ with L, the length of the molecule.

11.60 Use Equation 11.22 to account for the difference in the standard molar entropies of He and Ne.

11.61 Use the Heisenberg uncertainty principle to show that an electron cannot be confined within a nucleus.

Repeat the calculation for a proton. Assume the radius of a nucleus to be 1×10^{-15} m.

11.62 The nitrogen atom has one electron in each of the three $2p$ orbitals. Referring to Table 11.2, show that the total electron density is spherically symmetric, that is, it is independent of θ and ϕ. (*Hint:* Take the squares of the angular wave functions.)

11.63 Alkali metals like Na and K dissolve in liquid ammonia to give a blue solution, which is due to the absorption of solvated electrons. As a crude model, we can treat the solvated electron as a particle in a three-dimensional cubical box of length L. Assuming that the absorbed light ($\lambda = 600$ nm) excites the electron from its ground state to the first excited state, calculate L. (*Hint:* Extend Equation 11.22 to include three quantum numbers $n_x, n_y,$ and n_z.)

11.64 Referring to Table 11.2, calculate the values of ρ for the $2s$ and $3s$ orbitals at which a node exists.

CHAPTER 12

The Chemical Bond

Using what we learned in Chapter 11 about quantum mechanics and the electronic structure of atoms, in this chapter we begin our study of molecules. How can we explain the fact that two hydrogen atoms will combine to form a stable H_2 molecule, but two helium atoms will not form a stable He_2 molecule? How do we account for the different bond lengths and bond strengths of various compounds? Why is the water molecule bent and the carbon dioxide molecule linear? The answers to all these questions, and many more, must come from quantum mechanics.

This chapter will survey the important theories of chemical bonding, some molecular properties, and the role of metal ions in biological systems.

12.1 Lewis Structures

Although the concept of molecules goes back to the 17th century, not until the early part of the 20th century did chemists begin to understand how and why molecules form. The first major breakthrough was made in 1916 by Lewis, who suggested that a chemical bond involves the sharing of electrons. He depicted the formation of a chemical bond in H_2 as

$$\text{H}\cdot + \cdot\text{H} \longrightarrow \text{H——H}$$

This type of electron pairing is an example of a *covalent bond*. In this so-called *Lewis structure*, the single line represents the shared pair of electrons. Other examples of Lewis structures are

$$:\!\ddot{\text{F}}\!-\!\ddot{\text{F}}\!: \qquad \text{H}\!-\!\ddot{\text{O}}\!-\!\text{H} \qquad \ddot{\text{O}}\!=\!\text{C}\!=\!\ddot{\text{O}} \qquad \begin{matrix} \text{H} & & \text{H} \\ & \diagdown\!\!\!\diagdown\text{C}\!=\!\text{C}\diagup & \\ \text{H} & & \text{H} \end{matrix}$$

$$:\!\text{N}\!\equiv\!\text{N}\!: \qquad \text{H}\!-\!\text{C}\!\equiv\!\text{C}\!-\!\text{H}$$

In a Lewis structure, the shared electron pairs of a covalent bond are represented by a line between two atoms, and lone pairs (nonbonding electrons) are shown as pairs of dots on individual atoms. Only valence electrons are shown in Lewis structures.

The guideline for drawing Lewis structures is the *octet rule*, which states that an atom other than hydrogen tends to form bonds until it is surrounded by eight valence electrons. In other words, a covalent bond forms when atoms individually do not have enough electrons for a complete octet. By sharing electrons in a covalent bond, the atoms can complete their octets.

Showing *formal charges* in Lewis structures is often useful. An atom's formal charge in a molecule is the difference between the number of valence electrons in an isolated atom and the number of electrons assigned to that atom in a Lewis structure. The equation for calculating formal charge is

$$\begin{array}{c}\text{formal}\\\text{charge}\end{array} = \begin{array}{c}\text{number of}\\\text{valence}\\\text{electrons in}\\\text{the free atom}\end{array} - \begin{array}{c}\text{number of}\\\text{nonbonding}\\\text{electrons}\end{array} - \frac{1}{2}\left(\begin{array}{c}\text{number of}\\\text{bonding electrons}\end{array}\right) \quad (12.1)$$

According to Equation 12.1, the formal charges on the ozone molecule, the carbon monoxide molecule, and the carbonate ion are

$$\overset{..}{O}=\overset{+}{\overset{..}{O}}-\overset{..}{\underset{..}{O}}:^{-} \qquad {}^{-}:C\equiv O:^{+} \qquad {}^{-}:\overset{..}{\underset{..}{O}}-\overset{\overset{\displaystyle:O:}{\|}}{C}-\overset{..}{\underset{..}{O}}:^{-}$$

Formal charges do not represent the actual charges in a molecule. Nevertheless, they provide information about charge distribution and can help us draw reasonable Lewis structures. For example, when two atoms carry formal charges in a molecule, the negative charge will most likely reside on the more electronegative atom. (The CO molecule shown above is an exception to this rule.)

Lewis's theory was a significant advance in our understanding of chemical bond formation. It soon turned out to be inadequate in several respects, however. Certain atoms in stable molecules are known to have fewer than eight valence electrons (the incomplete octet; for example, BeH_2 in the gas phase and BF_3), more than eight electrons (the expanded octet; for example, PCl_5 and SF_6), and an odd number of electrons (for example, NO and NO_2). Furthermore, the simple description of covalent bond formation does not provide information about bond lengths or bond strengths in molecules such as H_2 and F_2, both of which have a single covalent bond. A proper treatment of chemical bonding, then, must come from quantum mechanics. At present, two quantum mechanical theories are used to describe covalent bond formation and the electronic structure of molecules. *Valence bond (VB) theory* assumes that the electrons in a molecule occupy atomic orbitals of the individual atoms. It permits us to retain a picture of individual atoms taking part in the bond formation. The second theory, called *molecular orbital (MO) theory*, assumes the formation of molecular orbitals from atomic orbitals. Neither theory perfectly explains all aspects of bonding, but each has contributed much to our understanding of many observed molecular properties.

12.2 Valence Bond Theory

Consider the formation of H_2 from two H atoms. Figure 12.1 shows how the H $1s$ orbitals overlap as the two atoms approach each other; the electron on one atom is attracted by the nucleus of the other atom. This interaction continues until the atoms reach a minimum distance of separation, below which the repulsions between

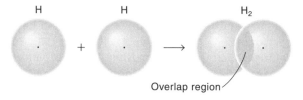

Figure 12.1
The formation of a covalent bond in H_2 resulting from the overlap of two hydrogen $1s$ orbitals.

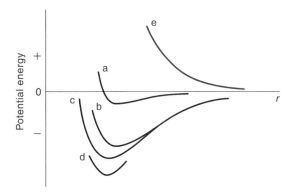

Figure 12.2
Potential energy curves for the H_2 molecule. Curves a–c represent successive improvements in the wave function, and curve d gives the experimental energy and bond distance. Curve e represents the nonbonding situation in which the two electrons have parallel spins. For simplicity, only truncated curves are shown.

the electrons and between the nuclei outweigh the electron–nucleus attraction. The buildup of electron density between the nuclei is the "glue" that holds the atoms together. This is the quantum mechanical description (in the VB framework) of Lewis's theory of covalent bond formation based on electron sharing.

As the two H atoms move toward each other, the potential energy changes. Initially, when the atoms are far apart, the potential energy is arbitrarily set to zero (Figure 12.2), and the wave function describing the two atoms is a product of the individual wave functions:

$$\psi = \psi_A(1)\psi_B(2) \tag{12.2}$$

where ψ_A and ψ_B are the hydrogen $1s$ wave functions for atoms A and B, and 1 and 2 denote the electrons. The potential-energy curve resulting from this wave function (curve a) has a shallow minimum (about 24 kJ mol^{-1}) at an internuclear distance of about 1 Å. These values are quite different from the experimentally measured values of 432 kJ mol^{-1} and 0.74 Å (curve d). Now, Equation 12.2 is inadequate to describe the bonding in H_2 because electrons are indistinguishable, so it is just as likely for electron 2 to be on atom A and electron 1 on atom B. An improved wave function is given by

The depth of the potential well is a measure of the bond dissociation energy.

$$\psi = \psi_A(1)\psi_B(2) + \psi_A(2)\psi_B(1) \tag{12.3}$$

which gives a potential-energy curve (curve b) with a minimum of about 300 kJ mol^{-1} and an equilibrium distance of about 0.9 Å. (Keep in mind that the two electrons in the H_2 molecule must have opposite spins in accord with the Pauli exclusion principle.) We can refine the wave function further by considering the possibility that both electrons might reside on the *same* atom; that is, we need to include both covalent and ionic structures, as follows (curve c):

$$\underset{\text{covalent}}{\text{H–H}} \qquad \underbrace{\text{H}^-\text{H}^+ \qquad \text{H}^+\text{H}^-}_{\text{ionic}}$$

The corresponding wave function is

$$\psi = \psi_A(1)\psi_B(2) + \psi_A(2)\psi_B(1) + \lambda[\psi_A(1)\psi_A(2) + \psi_B(1)\psi_B(2)] \tag{12.4}$$

where λ is a measure of the contribution of the ionic forms represented by $\psi_A(1)\psi_A(2)$ and $\psi_B(1)\psi_B(2)$ to the total wave function. Admittedly, the probability of finding both electrons on the same H atom is small because of electron–electron repulsion (that is, $\lambda \ll 1$); nevertheless, it does contribute to the overall properties of the molecule. Additional corrections include assigning an "effective" nuclear charge to the

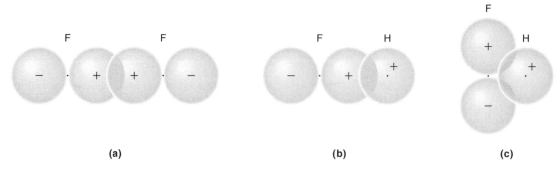

(a)   (b)   (c)

Figure 12.3
(a) The formation of the F_2 molecule results from the overlap of two $2p$ orbitals in a head-on position. (b) The covalent HF bond forms when a $2p$ orbital and a $1s$ orbital overlap. (c) The sideways overlap of a $2p$ orbital and a $1s$ orbital does not produce a stable bond because there is no net buildup of electron density between the nuclei.

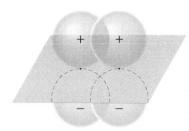

Figure 12.4
The sideways overlap of two p orbitals produces a π bond.

nuclei because each electron is now attracted to two nuclei rather than just one nucleus and taking into account that each hydrogen atom polarizes, or distorts, the electron cloud on the other hydrogen atom. With the aid of high-speed computers, we can now calculate the bond-dissociation energies and the equilibrium bond distances that agree with experimental values. Finally, we note that curve e represents a nonbonding state (the curve does not have a minimum). We can attribute this situation to two H atoms with parallel spins (called the triplet state). Such atoms cannot form a stable H_2 molecule because it would violate the Pauli exclusion principle.

We can apply the same approach to other diatomic molecules such as F_2 and HF. In F_2, the covalent bond is formed by the overlap of two $2p$ orbitals, whereas in HF, a hydrogen $1s$ orbital overlaps with a $2p$ orbital of fluorine to form a covalent bond (Figure 12.3). The process is more complex for molecules that contain multiple bonds. Consider the nitrogen molecule (N_2). The electron configuration of N is $1s^2 2s^2 2p^3$. Thus, there are three electrons in the $2p_x, 2p_y,$ and $2p_z$ orbitals on one N atom that can overlap with the same orbitals on the other N atom to form a triple bond. But these three bonds are not all equivalent. If we label the internuclear axis the z axis, then a bond is formed by the end-to-end overlap of the two $2p_z$ orbitals, as in F_2. Such a bond is called a *sigma (σ) bond* and is characterized by a high concentration of electron density between the nuclei of the bonding atoms. (The bonds in H_2, F_2, and HF are all σ bonds.) Because of their mutually perpendicular orientations, the $2p_x$ and $2p_y$ orbitals can overlap only in a sideways fashion (Figure 12.4), giving rise to two *pi (π) bonds*, in which the electron densities are concentrated above and below the plane of the nuclei of the bonding atoms.

12.3 Hybridization of Atomic Orbitals

The study of bonding in polyatomic molecules requires that we be able to account for their geometry as well. One widely used approach to this problem is *hybridization* of atomic orbitals, which is based on the VB theory.

Let us consider the carbon atom for which the electron configuration is $1s^2 2s^2 2p^2$. We might reasonably expect that carbon should be divalent (forming two covalent bonds), because it contains two unpaired electrons. Indeed, the molecule methylene (or carbene), CH_2, is known to exist, but it is a highly reactive species. Stable carbon compounds contain three types of chemical bonds best represented by the simplest molecules: methane, ethylene, and acetylene.

Methane (CH₄)

Methane is the simplest hydrocarbon. Physical and chemical studies show that all four C–H bonds are identical in length and strength, and the molecule has a tetrahedral geometry. The angle between each pair of C–H bonds is 109°28′. How can we explain carbon's tetravalency? A carbon atom in its ground state would not form four single bonds. Promoting a $2s$ electron into the empty $2p$ orbital would result in four unpaired electrons ($2s^1 2p^3$), which could form four C–H bonds. If this were the case, however, methane would contain three C–H bonds of one type and a fourth C–H bond of a different type. This configuration is contrary to experimental evidence. The fact that all four bonds are identical suggests that the bonding atomic orbitals of carbon are all equivalent, meaning that the s and p orbitals are mixed, so that hybridized, or *hybrid*, orbitals are formed. Because there are one s and three p orbitals, this process is called sp^3 hybridization.

Figure 12.5 shows the energy changes that occur in the process of hybridization. The state labeled $2s^1 2p^3$ is real and can be detected spectroscopically. The valence state (that is, the state in which the four equivalent hybrid orbitals are formed), is not real in the sense that it does not exist for an isolated carbon atom, but for us to imagine such a state just before the formation of the methane molecule is convenient. As Figure 12.5 shows, extra energy is needed to reach this state or to hybridize the atomic orbitals, but the investment is more than compensated for by the release of energy that results from bond formation.

Mathematically, the mixing of the atomic orbitals to form four hybrid orbitals t_1, t_2, t_3, and t_4 can be represented by*

$$t_1 = \tfrac{1}{2}(s + p_x + p_y + p_z)$$
$$t_2 = \tfrac{1}{2}(s + p_x - p_y - p_z)$$
$$t_3 = \tfrac{1}{2}(s - p_x + p_y - p_z) \qquad (12.5)$$
$$t_4 = \tfrac{1}{2}(s - p_x - p_y + p_z)$$

where s, p_x, p_y, and p_z represent the carbon $2s$ and $2p$ atomic orbitals and $\tfrac{1}{2}$ is the normalization constant. Note that the contributions of s and p character to each of

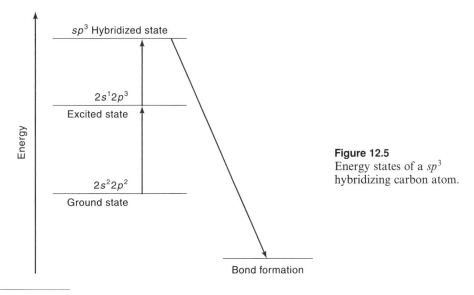

Figure 12.5
Energy states of a sp^3 hybridizing carbon atom.

---

*Each hybrid orbital is generated by taking the *linear combination* of the s and p orbitals. By linear combination, we mean that each orbital (s or p) is raised to the first power.

the hybrid orbitals are $\frac{1}{4}$ and $\frac{3}{4}$, respectively. Each sp^3 hybrid orbital has the shape shown in Figure 12.6; its direction is determined by the relative signs in Equation 12.5. A C–H σ bond can then be formed by the overlap between an sp^3 hybrid

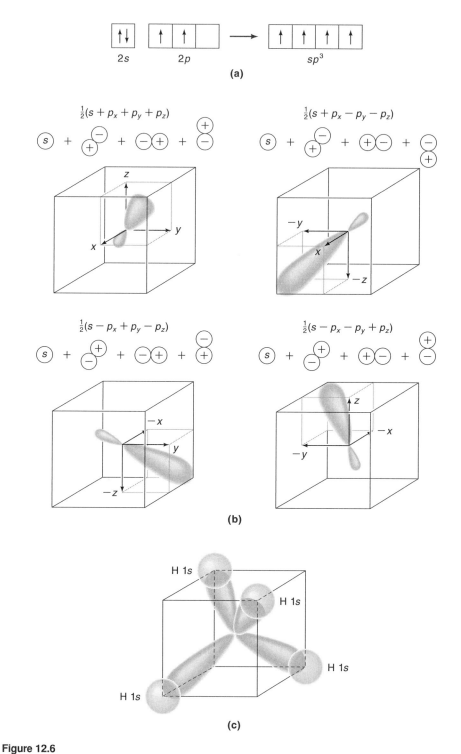

Figure 12.6
(a) Arrangement of carbon $2s$ and $2p$ electrons in sp^3 hybridization. (b) The four sp^3 hybrid orbitals in CH_4. (c) The formation of the C–H bonds between the sp^3 hybrid orbitals and the hydrogen $1s$ orbitals.

orbital and a hydrogen $1s$ orbital. Figure 12.7 shows the cross section of a computer-generated sp^3 hybrid orbital.

Ethylene (C₂H₄)

Ethylene is a planar molecule; the HCH angle is about 120°. In contrast to methane, each carbon atom is bonded to only three atoms. Both the geometry and bonding can be understood if we assume that each carbon atom is sp^2 hybridized. As Figure 12.8a shows, mixing the s electron with only two p electrons (say $2p_x$ and $2p_y$) produces three sp^2 hybrid orbitals (which all lie in the same xy plane), plus a pure p orbital (the $2p_z$ orbital). These three hybrid orbitals are then used to form two σ

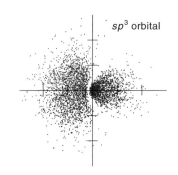

Figure 12.7
Cross section of an sp^3 orbital showing the electron probability distribution. (Generated from a program by Robert Allendoerfer. Adapted with permission from the *Journal of Chemical Education*, Project SERAPHIM, Division of Chemical Education, Inc.)

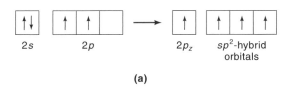

(a)

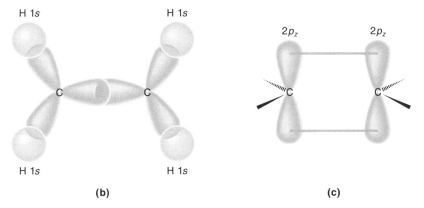

(b)   (c)

Figure 12.8
(a) Arrangement of carbon $2s$ and $2p$ electrons in sp^2 hybridization. (b) The three σ bonds between carbon and hydrogen and between the two carbon atoms in ethylene. (c) The π bond formed by the sideways overlap of the two $2p_z$ orbitals.

bonds with the two hydrogen atoms and a σ bond with the other carbon atom (Figure 12.8b). The $2p_z$ orbitals on the two carbon atoms can also overlap sideways to form a π bond (Figure 12.8c). These three hybrid orbitals are represented by

$$t_1 = \sqrt{\tfrac{1}{3}}(s + \sqrt{2}\,p_x)$$

$$t_2 = \sqrt{\tfrac{1}{3}}\left(s - \sqrt{\tfrac{1}{2}}\,p_x + \sqrt{\tfrac{3}{2}}\,p_y\right) \qquad (12.6)$$

$$t_3 = \sqrt{\tfrac{1}{3}}\left(s - \sqrt{\tfrac{1}{2}}\,p_x - \sqrt{\tfrac{3}{2}}\,p_y\right)$$

where $\sqrt{1/3}$ is the normalization constant.

Acetylene (C$_2$H$_2$)

Acetylene is a linear molecule. From Figure 12.9a we see that by mixing the $2s$ electron with only one p orbital (the $2p_z$ orbital), we obtain two sp hybrid orbitals plus two pure p orbitals (the $2p_x$ and $2p_y$ orbitals). Consequently, each carbon atom forms two σ bonds (one with hydrogen atom and one with the other carbon atom) and two π bonds (both with the other carbon atom), as shown in Figures 12.9b and 12.9c. The two hybrid orbitals are

$$t_1 = \sqrt{\tfrac{1}{2}}(s + p_z)$$
$$t_2 = \sqrt{\tfrac{1}{2}}(s - p_z)$$

(12.7)

Again, $\sqrt{1/2}$ is the normalization constant.

So far, we have discussed hybridization mathematically, but a physical interpretation is also possible. As the hydrogen atoms approach the carbon atom, electrostatic interactions between electrons and between electrons and nuclei cause the s and p orbitals of the carbon atom to become distorted. The s and p orbitals lose their distinct character, and each atomic orbital is more correctly described as partly resembling an s orbital and partly resembling a p orbital. For example, in methane the four sp^3 orbitals are identical in shape, although the three p orbitals are distorted to a different extent than the s orbital.

The concept of hybridization applies equally well to elements other than carbon. In ammonia, for example, each N–H bond points to the apex of a slightly irregular tetrahedron; the angle between any two N–H bonds is $107°20'$ (Figure 12.10). Because the electron configuration of nitrogen is $1s^2 2s^2 2p^3$, the bonding in NH$_3$ might be explained by assuming that the three p orbitals and the hydrogen $1s$ orbitals overlap. But if this were the case, the bond angles would be $90°$, because the p orbitals are all mutually perpendicular. It could also be argued that nitrogen is more electronegative than hydrogen, so that some charge separation may occur, resulting in a slightly negative nitrogen atom and slightly positive hydrogen atoms. Repulsion

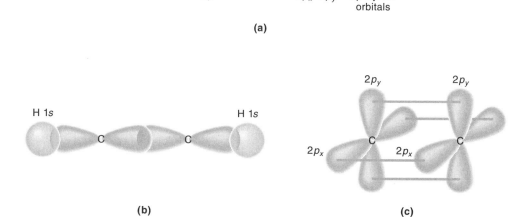

Figure 12.9
(a) Arrangement of carbon $2s$ and $2p$ electrons in sp hybridization. (b) The two σ bonds between carbon and hydrogen and between the two carbon atoms in acetylene. (c) The two π bonds between the two $2p_y$ orbitals and between the two $2p_z$ orbitals.

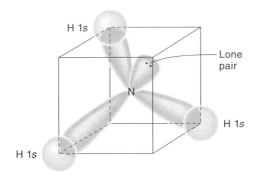

Figure 12.10
The N atom in NH_3 is sp^3 hybridized. The lone pair is placed in one of the hybrid orbitals.

between the hydrogen atoms would then increase the bond angles. Although this repulsion undoubtedly occurs, the effect is too small to account for the observed angle. The assumption that nitrogen is sp^3 hybridized in ammonia is more valid. One of the hybrid orbitals contains a lone pair. Repulsion between the lone pair electrons and those in the bonding orbitals changes the bond angle from $109°28'$ to $107°20'$. The lone pair is also responsible for the basicity of ammonia. When ammonia is dissolved in water, it readily accepts a proton to form the NH_4^+ ion, which possesses perfect tetrahedral symmetry.

Finally, we consider the water molecule. The HOH angle is $104°31'$, and there are two lone pairs on the O atom. The electron configuration of oxygen is $1s^2 2s^2 2p^4$. Often, the geometry of the water is explained by assuming that the O atom is sp^3 hybridized. Two of the sp^3 hybrid orbitals form σ bonds with the H atoms, and the two lone pairs reside in the other sp^3 hybrid orbitals. The strong repulsion between the lone pairs and the bond pairs reduces the HOH bond angle from $109°28'$ to $104°31'$. Spectroscopic evidence shows, however, that the O–H bond has no s character, suggesting that the O atom is unhybridized.* The reason is that the oxygen $2s$ and $2p$ electrons are so different in energy (the $2s$ level lies about 837 kJ mol^{-1} below the $2p$ level) that they do not readily interact to form sp^3 hybrid orbitals. If only the $2p$ orbitals are involved in bonding, we would expect the HOH angle to be $90°$. The ionic character of the O–H bonds causes them to repel each other, however, resulting in a larger angle of $104°31'$.

Although we have looked at hybridization only in terms of the s and p orbitals, the participation of d orbitals is also possible for the third-period elements and beyond. We shall return to this point in Section 12.8.

12.4 Electronegativity and Dipole Moment

In this section, we shall look at two properties that help us understand the ionic character of molecules and molecular geometry: electronegativity and dipole moment.

Electronegativity

In a homonuclear diatomic molecule such as H_2 or N_2, the electron density is evenly distributed between the two atoms. Such is not the case for heteronuclear diatomic molecules like HF or CO. The unequal distribution of electron density in these cases results directly from the difference in *electronegativity (X)*, which is the tendency of an atom to attract electrons to itself in a molecule. There are several ways to compare the electronegativity of elements. Here, we shall discuss a procedure intro-

---

* See M. Laing, *J. Chem. Educ.* **64**, 124 (1987).

1 1A	2 2A	3 3B	4 4B	5 5B	6 6B	7 7B	8	9 8B	10	11 1B	12 2B	13 3A	14 4A	15 5A	16 6A	17 7A	18 8A
H 2.1																	He
Li 1.0	Be 1.5											B 2.0	C 2.5	N 3.0	O 3.5	F 4.0	Ne
Na 0.9	Mg 1.2											Al 1.5	Si 1.8	P 2.1	S 2.5	Cl 3.0	Ar
K 0.8	Ca 1.0	Sc 1.3	Ti 1.5	V 1.6	Cr 1.6	Mn 1.5	Fe 1.8	Co 1.9	Ni 1.9	Cu 1.9	Zn 1.6	Ga 1.6	Ge 1.8	As 2.0	Se 2.4	Br 2.8	Kr 3.0
Rb 0.8	Sr 1.0	Y 1.2	Zr 1.4	Nb 1.6	Mo 1.8	Tc 1.9	Ru 2.2	Rh 2.2	Pd 2.2	Ag 1.9	Cd 1.7	In 1.7	Sn 1.8	Sb 1.9	Te 2.1	I 2.5	Xe 2.6
Cs 0.7	Ba 0.9	La–Lu 1.0–1.2	Hf 1.3	Ta 1.5	W 1.7	Re 1.9	Os 2.2	Ir 2.2	Pt 2.2	Au 2.4	Hg 1.9	Tl 1.8	Pb 1.9	Bi 1.9	Po 2.0	At 2.2	Rn
Fr 0.7	Ra 0.9	Ac 1.1	Rf	Db	Sg	Bh	Hs	Mt	Ds								

Figure 12.11
Electronegativity scale of the elements. The electronegativities of the actinides (Th–Lr) vary between 1.1–1.5.

duced by the American chemist Linus Pauling (1901–1994) in 1932. We define the electronegativity difference between atoms A and B to be

X is treated as a dimensionless quantity.

$$|X_A - X_B| = \sqrt{D_{AB} - (D_{A_2} D_{B_2})^{1/2}} \tag{12.8}$$

where D_{AB}, D_{A_2}, and D_{B_2} are the bond-dissociation energies of molecules AB, A_2, and B_2. Because only differences are obtained from Equation 12.8, one element must be assigned a specific electronegativity value, and then the values for the other elements can be calculated readily. By arbitrarily defining $X_F = 4.0$, Pauling developed the electronegativity scale shown in Figure 12.11.

Dipole Moment

A molecule possesses a permanent electric *dipole moment*, μ, if its center of positive charge does not coincide with its center of negative charge. The dipole moment is defined as

$$\mu = Q \times r \tag{12.9}$$

where Q is the charge (treated as a positive quantity), and r is the distance between the positive and negative centers. The SI units for dipole moment are coulomb meter (C m). If Q is one electron (1.602×10^{-19} C) and r is 1 Å (10^{-10} m), then μ is given by

$$\mu = 1.602 \times 10^{-19} \text{ C} \times 10^{-10} \text{ m}$$
$$= 1.602 \times 10^{-29} \text{ C m}$$
$$= 4.8 \text{ D}$$

where $1 \text{ D} = 3.3356 \times 10^{-30}$ C m. The unit D is called the *debye* in honor of Debye, who pioneered this field. Molecules that possess a permanent dipole moment are called *polar* molecules.

The dipole moment of a molecule can be readily measured. However, the value is given as a product of two quantities (that is, charge and distance), so that dipole-moment measurements cannot be used to obtain bond lengths and bond angles. In general, charge separation is difficult to estimate accurately. Yet the measurements are extremely useful in determining the *symmetry* of a molecule. An example is the CO_2 molecule, which in principle could be either linear or bent. The overall dipole moment of the molecule is determined by the vector sum of its bond moments, which measure the charge separation in each bond. (By convention, the symbol $\longmapsto$ indicates that the flow of electron density is away from the less electronegative atom toward the more electronegative atom.) As shown below, the bond moments for a linear CO_2 cancel by symmetry, whereas those for a bent CO_2 do not. Because CO_2 does not possess a permanent dipole moment, the molecule is linear:

$$O=C=O$$
$$\mu = 0$$

Resultant dipole
$$\mu \neq 0$$

Table 12.1
Dipole Moments of Some Polar Molecules

Molecule	Dipole Moment/D
HF	1.92
HCl	1.08
HBr	0.78
HI	0.38
H_2O	1.87
H_2S	1.10
NH_3	1.46
SO_2	1.60

The CO_2 example shows the usefulness of considering the dipole moment of a molecule in terms of its bond moments. The resultant dipole moment of a molecule can be estimated by vector addition of the individual bond moments. Table 12.1 lists the dipole moments of several molecules.

Dipole moments are useful for estimating the percent ionic character of a bond. If we know the bond distance in a diatomic molecule, we can calculate a dipole moment, μ_{ionic}, on the assumption that the atoms bear a full unit charge. The percent ionic character of the bond can then be calculated as

$$\% \text{ ionic character} = \frac{\mu_{exp}}{\mu_{ionic}} \times 100\% \tag{12.10}$$

where μ_{exp} is the experimental dipole moment. Figure 12.12 shows a nice correlation between ionic character and difference between the electronegativities of the two bonding atoms.

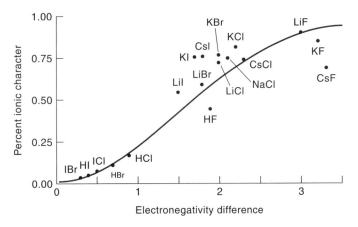

Figure 12.12
Relation between electronegativity difference and percent ionic character in several binary compounds. A purely ionic compound has 100% ionic character, whereas a purely covalent compound has 0% ionic character.

Example 12.1

The bond distance in HF is 0.92 Å (92 pm). Calculate the percent ionic character of the H–F bond.

ANSWER

From Table 12.1, we see that the dipole moment of HF is 1.91 D, or $1.91 \times 3.3356 \times 10^{-30}$ C m $= 6.37 \times 10^{-30}$ C m. Thus, we write

$$\text{\% ionic character} = \frac{6.37 \times 10^{-30} \text{ C m}}{(1.602 \times 10^{-19} \text{ C})(92 \times 10^{-12} \text{ m})} \times 100\%$$

$$= 43.2\%$$

As expected, the H–F bond is quite polar because of the large difference between the electronegativities of H and F.

COMMENT

No compound has 100% ionic character.

12.5 Molecular Orbital Theory

The second theory of chemical bonding, molecular orbital (MO) theory, assumes that two atomic orbitals merge to become molecular orbitals. To illustrate, let us again consider the formation of a H_2 molecule from two H atoms. By analogy with the interference phenomenon discussed in Section 11.1, the two $1s$ orbitals can interact either constructively or destructively, depending on whether their wave functions add or subtract in the region of "overlap." These interactions then lead to the formation of a σ *(sigma) bonding molecular orbital* and a σ* *(sigma star) antibonding molecular orbital* (Figure 12.13). In the bonding σ orbital, there is a buildup of electron density between the two nuclei; in the antibonding σ* orbital, there is a decrease in electron density. Furthermore, σ and σ* orbitals correspond to the potential-energy curves with and without a minimum, respectively, as shown in Figure 12.14.

The σ and σ* molecular orbitals arise from the linear combination of the atomic orbitals. In this *LCAO–MO* (*linear combination of atomic orbitals–molecular orbitals*) model, the wave functions for the molecular orbitals are given by[†]

$$\psi(\sigma) = N(\psi_A + \psi_B) \tag{12.11}$$

$$\psi(\sigma^*) = N(\psi_A - \psi_B) \tag{12.12}$$

where the plus sign denotes the bonding σ molecular orbital and the minus sign denotes the antibonding σ* molecular orbital. Here N is the normalization constant.

The concept of molecular orbitals is a natural extension of atomic orbitals. Each electron in the atomic case is localized in the $1s$ orbital. We can also think of the paired electrons in H_2 as being in the bonding molecular orbital. Squaring Equation 12.11, we get

$$\psi(\sigma)^2 = N^2(\psi_A^2 + \psi_B^2 + 2\psi_A\psi_B) \tag{12.13}$$

[†] It is important not to confuse Equations 12.11 or 12.12 with Equation 12.5. In the latter case, the hybrid orbitals are still atomic orbitals, because they all arise from the *same* carbon atom.

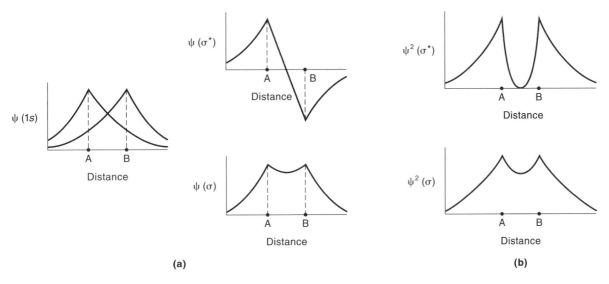

Figure 12.13
(a) Molecular wave functions resulting from the constructive and destructive interactions between the hydrogen $1s$ wave functions. These molecular wave functions correspond to the bonding and antibonding σ molecular orbitals. The black dots denote the positions of the nuclei. (b) The square of the molecular wave functions give the electron probability distribution in the bonding and antibonding σ molecular orbitals. In the bonding molecular orbital, there is a buildup of electron density between the nuclei; in the antibonding molecular orbital, there is a decrease in electron density between the nuclei.

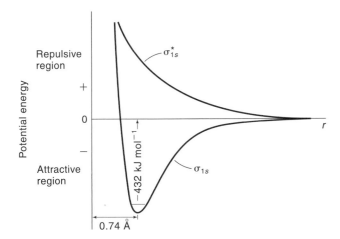

Figure 12.14
Potential energy curves for the H_2 bonding and antibonding molecular orbitals. Note that the antibonding curve has no minimum.

Equation 12.13 says that the probability of finding an electron near nucleus A or B is still given by ψ_A^2 or ψ_B^2 (although modified by the N^2 term), and there is also a buildup of electron density between A and B. This buildup is the result of the overlap of the two atomic orbitals, and its magnitude is given by the product $2\psi_A\psi_B$. The main difference between $\psi(\sigma)$ and $\psi(\sigma^*)$ is that for the antibonding molecular orbital, the overlap is given by $-2\psi_A\psi_B$, which actually corresponds to a decrease in electron density. This description explains the "antibonding" label.

The electron configuration of molecules, like that of atoms, must satisfy both the Pauli exclusion principle and Hund's rule. The electron configuration of H_2 is simply $(\sigma_{1s})^2$, where σ_{1s} denotes the sigma molecular orbital formed by the $1s$ orbitals, and the superscript 2 represents two electrons. The electrons in a bonding σ molecular

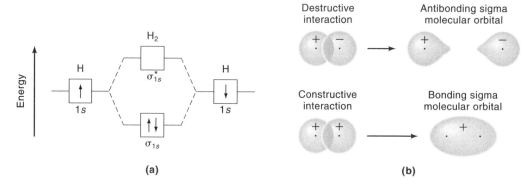

Figure 12.15
(a) Molecular orbital energy level diagram of H_2. (b) Constructive and destructive interactions of the $1s$ orbitals.

orbital give rise to a σ bond. The relative energies of the σ_{1s} and σ_{1s}^* molecular orbitals are shown in Figure 12.15.

The stability of a diatomic molecule can be estimated by calculating the number of bonds joining the atoms, or the *bond order*, given by

$$\text{bond order} = \frac{1}{2}\left(\begin{array}{c}\text{number of electrons}\\ \text{in the bonding MO}\end{array} - \begin{array}{c}\text{number of electrons}\\ \text{in the antibonding MO}\end{array}\right) \quad (12.14)$$

The H_2 molecule has two electrons in the bonding molecular orbital and none in the antibonding molecular orbital, so the bond order is $\frac{1}{2}(2) = 1$. A bond order of 1 means that there is one covalent bond and the H_2 molecule is stable. Note that the bond order can be a fraction, but a bond order of 0 (or a negative value) means the bond has no stability and the molecule cannot exist. Bond order can be used only qualitatively for the purpose of comparison.

12.6 Diatomic Molecules

We shall now apply the MO theory to several diatomic molecules, paying particular attention to their ground-state electron configurations, stability (as measured by their bond orders), and magnetic properties. We shall study both homonuclear and heteronuclear diatomic molecules.

Homonuclear Diatomic Molecules of the Second-Period Elements

Li$_2$. The electron configuration of Li is $1s^2 2s^1$, so in Li_2 the four $1s$ electrons are paired in the σ_{1s} and σ_{1s}^* molecular orbitals. In addition, we also have σ_{2s} and σ_{2s}^* orbitals. Because there are only two $2s$ electrons, the electron configuration for Li_2 is given by

$$(\sigma_{1s})^2(\sigma_{1s}^*)^2(\sigma_{2s})^2$$

The molecule has a bond order of 1 and is diamagnetic.

By convention, we take the z axis to be the internuclear axis in a diatomic molecule.

B$_2$. The electron configuration of B is $1s^2 2s^2 2p^1$. The presence of a p orbital suggests that two types of interaction are possible. In Figure 12.16a, the linear overlap of the p orbitals gives rise to a σ molecular orbital; in Figure 12.16b, a π molecular orbital is

Destructive interaction

Antibonding sigma molecular orbital

Constructive interaction

Bonding sigma molecular orbital

(a)

Destructive interaction

Antibonding pi molecular orbital

Constructive interaction

Bonding pi molecular orbital

(b)

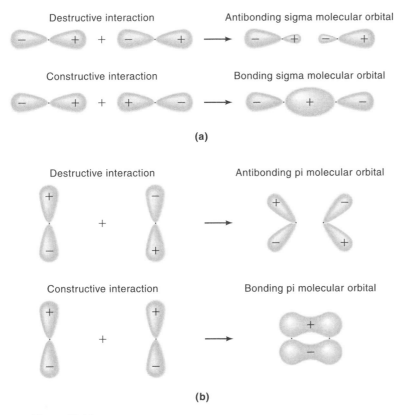

Figure 12.16
Formation of (a) σ and (b) π molecular orbitals from two p orbitals.

formed. As mentioned earlier, because each atom has three p orbitals, we can see by simple symmetry that one σ and two π molecular orbitals will result from the interaction. In isolated cases, a σ molecular orbital is more stable than a π molecular orbital because of the greater extent to which the $2p$ orbitals overlap. The situation is more complicated for a molecule in which molecular orbitals are formed by $2s$ and $2p$ orbitals. There are two types of molecular orbital energy-level diagrams; Figure 12.17a applies to Li_2 through N_2, and Figure 12.17b applies to O_2 and F_2. Both diagrams indicate that the σ_{2s} and σ_{2s}^* are the lowest-energy molecular orbitals because the $2s$ atomic orbitals lie below the $2p$ atomic orbitals. They differ, however, in the order of the σ_{2p} and π_{2p} orbitals. This difference is the result of the interaction of the $2s$ orbital on one of the atoms with the $2p$ orbitals on the other. This so-called s–p mixing affects the energies of the σ_{2s} and σ_{2p} molecular orbitals in such a way that these MOs move further apart in energy, with the σ_{2s} falling and σ_{2p} rising in energy. For B_2, C_2, and N_2, the interaction is strong enough so that the σ_{2p} MO is above the π_{2p} MOs in energy. For O_2 and F_2, the σ_{2p} MO is below the π_{2p} MOs. Thus, the electron configuration of B_2 is

$$(\sigma_{1s})^2(\sigma_{1s}^*)^2(\sigma_{2s})^2(\sigma_{2s}^*)^2(\pi_x)^1(\pi_y)^1$$

where π_x and π_y represent the π molecular orbitals formed from $2p_x$ and $2p_y$. Hund's rule dictates that the two electrons enter the degenerate orbitals with parallel spins. Thus, the B_2 molecule is paramagnetic and has a bond order of 1.

Electron configurations are confirmed by spectroscopic and magnetic measurements.

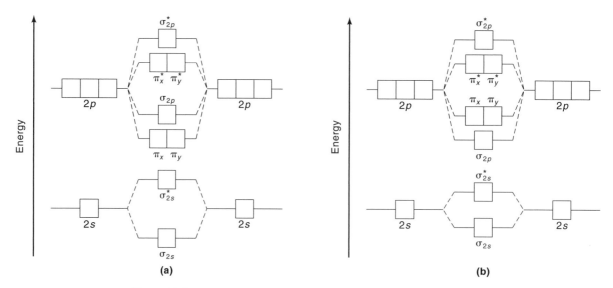

Figure 12.17
Molecular orbital energy level diagrams for homonuclear diatomic molecules of the second-period elements. (a) Diagram for Li_2, B_2, C_2, and N_2. (b) Diagram for O_2 and F_2. For simplicity, we omit the σ_{1s} and σ_{1s}^* orbitals.

N$_2$. The electron configuration of N is $1s^2 2s^2 2p^3$. Referring to Figure 12.17a, we write the electron configuration of N_2 as

$$(\sigma_{1s})^2 (\sigma_{1s}^*)^2 (\sigma_{2s})^2 (\sigma_{2s}^*)^2 (\pi_x)^2 (\pi_y)^2 (\sigma_{2p})^2$$

where σ_{2p} is the molecular orbital formed by the overlap of the $2p_z$ orbitals. This electron configuration is consistent with what we know—namely, that N_2 is a diamagnetic molecule, having a bond order of 3 (a triple bond).

O$_2$. It has long been known that the O_2 molecule is paramagnetic (Figure 12.18), but its Lewis structure does not account for this property:

$$\ddot{O}{=}\ddot{O}$$

Figure 12.18
Liquid oxygen is held between the poles of a magnet. Paramagnetic substances are attracted into a magnetic field. (Reprinted with permission from Photo Researchers, Inc.)

The electron configuration of O is $1s^2 2s^2 2p^4$. Based on Figure 12.17b, the electron configuration for O_2 is

$$(\sigma_{1s})^2 (\sigma_{1s}^*)^2 (\sigma_{2s})^2 (\sigma_{2s}^*)^2 (\sigma_{2p})^2 (\pi_x)^2 (\pi_y)^2 (\pi_x^*)^1 (\pi_y^*)^1$$

The two unpaired electrons in the π_x^* and π_y^* orbitals account for the paramagnetic properties. Furthermore, the O_2 molecule has a bond order of 2. The successful explanation of the bonding and magnetic properties of O_2 was one of the early triumphs of the MO theory.

Table 12.2 lists the electron configurations and other properties of all the homonuclear diatomic molecules formed by the second-period elements.

Table 12.2
Electron Configuration and Bond Properties of Homonuclear Diatomic Molecules of the Second-Period Elements

Molecule	Electron Configuration[a]	Bond Order	Bond Enthalpy kJ · mol^{-1}	Bond Length/Å
H_2	$(\sigma_{1s})^2$	1	436.4	0.74
Li_2	$KK(\sigma_{2s})^2$	1	104.6	2.67
B_2	$KK(\sigma_{2s})^2(\sigma_{2s}^*)^2(\pi_x)^1(\pi_y)^1$	1	288.7	1.59
C_2	$KK(\sigma_{2s})^2(\sigma_{2s}^*)^2(\pi_x)^2(\pi_y)^2$	2	627.6	1.31
N_2	$KK(\sigma_{2s})^2(\sigma_{2s}^*)^2(\pi_x)^2(\pi_y)^2(\sigma_{2p})^2$	3	941.4	1.10
O_2	$KK(\sigma_{2s})^2(\sigma_{2s}^*)^2(\sigma_{2p})^2(\pi_x)^2(\pi_y)^2(\pi_x^*)^1(\pi_y^*)^1$	2	498.8	1.21
F_2	$KK(\sigma_{2s})^2(\sigma_{2s}^*)^2(\sigma_{2p})^2(\pi_x)^2(\pi_y)^2(\pi_x^*)^2(\pi_y^*)^2$	1	150.6	1.42

[a]The notation KK represents the electron configuration $(\sigma_{1s})^2(\sigma_{1s}^*)^2$, which is a He_2 molecule.

Heteronuclear Diatomic Molecules of the First- and Second-Period Elements

For heteronuclear diatomic molecules containing elements of the first two periods, the approach is basically the same as that for homonuclear diatomic molecules, except that the molecular orbitals are no longer symmetrically positioned relative to the atomic orbitals because the atoms are different. In particular, the bonding molecular orbitals will lie closer to the atomic orbitals of the more electronegative element. The opposite is true for the antibonding molecular orbitals: they lie closer to the atomic orbitals of the less electronegative element.

Hydrogen Fluoride. The Lewis structure of hydrogen fluoride contains a single bond and three lone pairs on the F atom:

$$H-\ddot{\underset{\cdot\cdot}{F}}:$$

Figure 12.19 shows the relative energies of atomic and molecular orbitals in HF.

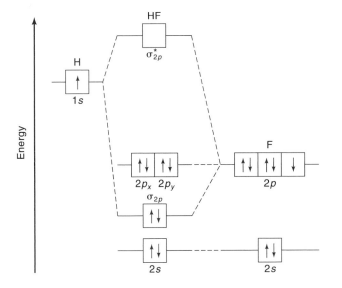

Figure 12.19
Molecular orbital energy-level diagram for the HF molecule. The placements of hydrogen 1s and carbon 2p orbitals are based on their first ionization energy values (see Table 11.4). For simplicity, the fluorine 1s orbital is omitted.

Several features are worth noting. First, the $1s$ orbital of H overlaps with the $2p_z$ orbital of F to form a σ molecular orbital that contains a large component of fluorine's $2p_z$ orbital. This is consistent with the fact that the H–F bond is polar, and the greater electron density is in the vicinity of fluorine. The σ^* molecular orbital lies closer to the hydrogen $1s$ orbital and resembles mostly that orbital. The $1s$ orbital has no net interaction with the $2p_x$ and $2p_y$ orbitals of fluorine. Consequently, these two p orbitals are nonbonding; that is, two of the three lone pairs reside in them. The energy of the fluorine $2s$ orbital is too low to interact with the hydrogen $1s$ orbital. For this reason, it is also nonbonding and holds the remaining lone pair on the F atom.

Mathematically, the wave functions for the bonding and antibonding σ molecular orbitals are

For simplicity, we omit the normalization constants.

$$\psi(\sigma) = c_1\psi_{1s} + c_2\psi_{2p_z} \tag{12.15}$$

$$\psi(\sigma^*) = c_3\psi_{1s} - c_4\psi_{2p_z} \tag{12.16}$$

where the coefficients c_1 and c_2 give the relative contributions of the hydrogen $1s$ and the fluorine $2p_z$ orbital in the σ molecular orbital. The large difference between the electronegativities of H and F means that $c_2 \gg c_1$, so the electron density is much greater near F in the HF molecule. The opposite is true for the σ^* orbital, which is empty but could be occupied in certain excited electronic states. This orbital has largely hydrogen $1s$ character—that is, $c_3 \gg c_4$.

Carbon Monoxide. Carbon monoxide is *isoelectronic* (i.e., has the same number of electrons) with N_2. Indeed, the energies of the $2s$ orbitals and the $2p$ orbitals of both C and O are close enough that the way they interact is similar to the way in which the $2s$ and $2p$ orbitals interact in N_2. Figure 12.20 shows the molecular orbital energy-level diagram for CO. Again, the placement of the atomic orbitals of the C and O

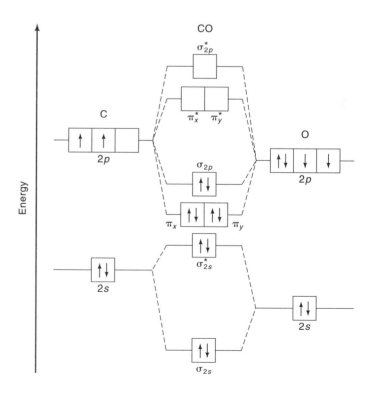

Figure 12.20
Molecular orbital energy-level diagram for the CO molecule. The placements of carbon and oxygen $2p$ orbitals are based on their first ionization energy values (see Table 11.4). For simplicity, the $1s$ orbitals are omitted.

atoms is determined by the values of the first ionization energy. As for HF, the orbital coefficients for the more electronegative atom (O) will be larger in the more stable molecular orbitals (the bonding orbitals). The less electronegative atom (C) will have larger coefficients for its orbitals in the less stable molecular orbitals (the antibonding orbitals). The electron configuration for CO is analogous to that for N_2 (see p. 462).

Example 12.2

Nitric oxide (NO) takes part in photochemical smog formation. Recently, it has also been found to act as a neurotransmitter and to play a role in controlling blood pressure and in gene regulation. (a) Write its electron configuration, (b) calculate its bond order, (c) predict its magnetic properties, and (d) draw two Lewis structures for the molecule, including formal charges.

ANSWER

(a) The electronegativity of N is close to that of O (see Figure 12.11); therefore, they interact to form a set of molecular orbitals similar to those in N_2 and CO. Nitric oxide has one more electron than N_2, so its electron configuration is

$$(\sigma_{1s})^2(\sigma_{1s}^*)^2(\sigma_{2s})^2(\sigma_{2s}^*)^2(\pi_x)^2(\pi_y)^2(\sigma_{2p})^2(\pi_x^*)^1$$

(b) The bond order is given by

$$\text{bond order} = \tfrac{1}{2}(6-1) = 2.5$$

Note that we do not need to include the inner electrons (in the σ orbitals) because their net contribution to the bond order is zero.
(c) The molecule is paramagnetic because it has one unpaired electron.
(d) The Lewis structures are

$$\cdot \ddot{N}=\ddot{O} \qquad {}^-\ddot{N}=\ddot{O}^+$$

Note that we cannot draw a Lewis structure showing a triple bond between N and O. Both of these structures show a double bond between N and O.

12.7 Resonance and Electron Delocalization

The advantage of the concept of orbital hybridization is that, in addition to explaining the geometry of molecules, it enables us to continue thinking of a chemical bond as the pairing of electrons. The properties of a molecule, however, cannot always be completely represented by a single structure. A case in point is the carbonate ion:

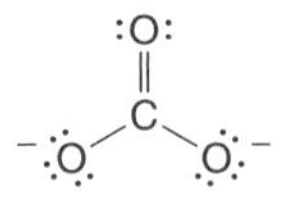

The ion has a planar structure and the OCO bond angle is 120°, which can be readily explained if we assume that the carbon atom is sp^2 hybridized. But because experimental studies indicate that all three carbon-to-oxygen bonds have equal length and equal strength, the structure shown above is inadequate to describe the ion. The position of the C=O double bond is chosen arbitrarily, so we must consider the following three structures:

The two-headed arrow indicates that we are looking at *resonance structures* of the carbonate ion. The term *resonance* means the use of two or more Lewis structures for a molecule (or ion) that cannot be described fully with a single Lewis structure. The point is that none of the individual resonance structures accurately represents the carbonate ion; the ion is best represented by the superposition of all three resonance structures. Thus, the character of each carbon-to-oxygen bond is somewhere between a single and a double bond, in accord with experimental observations. No evidence whatsoever indicates that these three structures actually oscillate back and forth. In fact, each of the three resonance structures is of a nonexistent ion. The model above merely enables us to solve the dilemma of trying to explain properties such as bond length and bond strength.*

In VB theory, we can write the wave function for the carbonate ion as

$$\psi = c_A \psi_A + c_B \psi_B + c_C \psi_C$$

where ψ_A, ψ_B, and ψ_C are the wave functions for the three individual resonance structures and c_A, c_B, and c_C are the coefficients that determine the weight or importance of the resonance structures. Here we have three equivalent structures, so c_A, c_B, and c_C are all equal.

The concept of resonance is applied most often to aromatic hydrocarbons. In 1865, the German chemist August Kekulé (1829–1896) first proposed the ring structure for benzene. Since then, considerable progress has been made in the study of these molecules. The measured carbon-to-carbon distance in benzene is 1.40 Å, which is somewhere between the single C–C bond (1.54 Å) and the double C=C bond (1.33 Å). It is more realistic to describe the resonance between two Kekulé structures as follows:

MO theory offers an alternative approach to explaining the properties discussed above. Instead of the resonance concept, MO theory employs the delocalization of electrons in the molecular orbitals. In the ethylene molecule, for instance, two overlapping $2p_z$ orbitals give rise to two molecular orbitals—one bonding and one antibonding (Figure 12.21). The two electrons are placed in the π bonding molecular orbital, which extends over the two carbon atoms. The more stable σ molecular orbitals lie lower on the energy scale. Similarly, for the benzene molecule, the six overlapping $2p_z$ orbitals give rise to six molecular orbitals, of which three are bonding and three are antibonding (Figure 12.22). A pair of electrons is assigned to each of the bonding molecular orbitals. These electrons are free to move within the boundaries denoted

---

*An interesting analogy has been given for resonance. A medieval European traveler returns home from a journey to Africa and describes a rhinoceros as a cross between a griffin and a unicorn. Thus, a real animal is described in terms of two familiar (in concept) but imaginary animals. Similarly, the real chemical species, the carbonate ion, is described in terms of the three familiar-looking but nonexistent resonance structures.

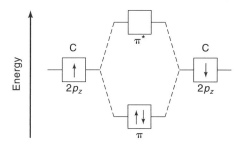

Figure 12.21
Molecular orbital energy-level diagram for the π bonding and π^* antibonding molecular orbitals in ethylene.

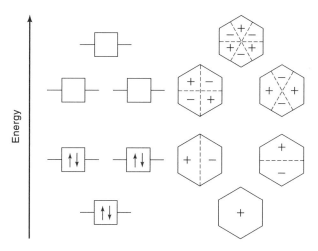

Figure 12.22
Energy-level diagram for the three bonding and three antibonding molecular orbitals in benzene, together with the π electron wave functions (top view). The dashed lines represent the nodal planes.

by the dashed lines in Figure 12.22. For example, the lowest-energy bonding molecular orbital is formed by taking the linear combination of the six $2p_z$ orbitals $(\psi_1 + \psi_2 + \psi_3 + \psi_4 + \psi_5 + \psi_6)$.[†] The number of nodes increases as the energy of the molecular orbital increases, a situation analogous to the particle-in-a-box case discussed in Chapter 11. Within the framework of MO theory, the benzene molecule is often represented as

in which the circle indicates that the π bonds between carbon atoms are not confined to individual pairs of atoms; rather, the π electron densities are evenly distributed throughout the benzene molecule. This description also accounts for the equal carbon-to-carbon bond length and equal bond strength in benzene.

Both MO theory and VB theory are useful in studying chemical bonding, but each has strong and weak points. The former tends to emphasize electron delocalization; the latter emphasizes the electron pair bond. Likewise, the concept of resonance is helpful, although one needs considerable chemical intuition when applying it to large molecules. Hybridization appears to fit VB theory, dealing as it does with the electron pair concept rather than electron delocalization. In fact, hybridization can be accounted for by MO theory equally well, for hybrid orbitals can also be used to construct molecular orbitals.

The Peptide Bond

Peptide bonds (that is, the bonds between the C and N atoms in the amide group, –CO–NH–) play an important role in determining the structure of proteins. X-ray diffraction and other experimental techniques show that all four atoms (C, O, N, and H) lie in a plane, indicating a restricted rotation about the C–N bond. The carbon-to-nitrogen bond length is about 1.33 Å, which is somewhere between the length of

---

[†] Higher-energy π molecular orbitals are formed by taking different linear combinations of the $2p_z$ orbitals, in which some of the coefficients are negative.

the C–N bond (1.49 Å) and that of the C=N bond (1.27 Å). The planarity of the amide group can be accounted for in terms of the following resonance structures:

$$
\begin{array}{ccc}
\overset{\textstyle :\!O\!:}{\underset{\textstyle \underset{H}{|}}{\overset{\textstyle \|}{-C-\ddot{N}-}}} & \longleftrightarrow & \overset{\textstyle :\!\ddot{O}\!:^{-}}{\underset{\textstyle \underset{H}{|}}{\overset{\textstyle |}{-C=\overset{+}{N}-}}}
\end{array}
$$

The partial double bond character restricts the rotation about the carbon-to-nitrogen bond and locks the four atoms in the same plane.

The sigma bonds are formed between the $2p_x$ orbital on O and an sp^2 orbital on C and between the sp^2 orbitals on C and N.

In MO theory, we assume that the C and N atoms are sp^2 hybridized. There is one electron in each of the unhybridized $2p_z$ orbitals on C and O and two electrons in the unhybridized $2p_z$ orbital on N. The three unhybridized $2p_z$ orbitals overlap to form three molecular orbitals whose wave functions are (Figure 12.23)

$$\psi_1 = c_{11}\psi(O) + c_{12}\psi(C) + c_{13}\psi(N)$$
$$\psi_2 = c_{21}\psi(O) + c_{22}\psi(C) + c_{23}\psi(N)$$
$$\psi_3 = c_{31}\psi(O) + c_{32}\psi(C) + c_{33}\psi(N)$$

where ψ_1, ψ_2, and ψ_3 are the wave functions for the lowest-to-highest-energy molecular orbitals and $\psi(O), \psi(C)$, and $\psi(N)$ are the $2p_z$ wave functions of the atoms. By analogy to the first three states for a particle in a one-dimensional box (see Figure 11.18), we can make the following predictions about the coefficients:

ψ_1: c_{11}, c_{12}, and c_{13} are all positive.

ψ_2: c_{21} is positive, c_{22} is 0, and c_{23} is negative.

ψ_3: c_{31} and c_{33} are positive, and c_{32} is negative.

The delocalized molecular orbital over the atoms imparts double-bond character to the carbon-to-nitrogen bond and so accounts for the geometry of the amide group.

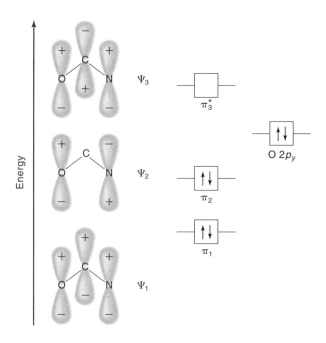

Figure 12.23
The overlapping $2p_z$ orbitals on the N, C, and O atoms generate three π molecular orbitals of which one is bonding (ψ_1), one is nonbonding (ψ_2), and one is anti-bonding (ψ_3). The ψ_1 and ψ_2 orbitals each contain two electrons, and the ψ_3 orbital is empty. The bonding ψ_1 molecular orbital imparts some double bond character to the carbon-to-nitrogen linkage and is responsible for the planarity of the amide group. The lone pairs on O are placed in the $2p_y$ and $2s$ orbitals.

12.8 Coordination Compounds

A *coordination compound* typically consists of a *complex ion*, a metal ion with one or more attached ligands, and counterions, which are anions or cations as needed to produce a compound with no net charge. Examples are $[Co(NH_3)_5Cl]Cl_2$ and $K_3[Fe(CN)_6]$. Some coordination compounds do not contain complex ions; an example is $Fe(CO)_5$. The reaction between a ligand and a metal ion is often classified as a Lewis acid–base interaction, in which the base (the ligand) donates an electron pair to the acid (the metal ion) to form a coordinate covalent bond.

The American chemist Ralph Gottfrid Pearson (1919–) proposed that the central metal ion—or acid—in a complex be designated either hard or soft, according to its ability to attach various ligands. Hard acids are small, compact, and less easily polarized metal ions (that is, their electron density is not easily distorted). Soft acids, on the other hand, are large, fairly polarizable metal ions. Similarly, we can classify ligands as hard or soft according to how tightly the lone pair is held by the donor atom—the more electronegative the atom, the harder the base. Table 12.3 lists several so-called hard and soft acids and bases. The significance of this designation is

Table 12.3
List of Hard and Soft Acids and Bases in Ligand–Metal Reactions

	Acids	Bases[a]
Hard	H^+, Li^+, Na^+, K^+ Mg^{2+}, Ca^{2+}, Mn^{2+}, Al^{3+} Ga^{3+}, Co^{3+}, Cr^{3+}, Fe^{3+}	H_2O, NH_3, RNH_2 OH^-, F^-, Cl^-, NO_3^- CO_3^{2-}, CH_3COO^-, PO_4^{3-}
Intermediate	Fe^{2+}, Co^{2+}, Ni^{2+}, Cu^{2+} Zn^{2+}, Sn^{2+}	N_2, Br^-, NO_2^-, SO_3^{2-} N_3^-, imidazole
Soft	Cu^+, Au^+, Pt^{2+}, Ag^+ Cd^{2+}, Hg^{2+}, Pb^{2+}	H^-, CN^-, SCN^-, I^- RSH, RS^-, R_3P, CO

[a] R is an alkyl or aryl group.

that generally hard acids preferentially bind to hard bases and soft acids to soft bases. This correlation helps us understand and predict many chemical and biological processes. Figure 12.24 shows some of the common geometries of coordination compounds. These geometries enable us to define the *coordination number* (CN), which

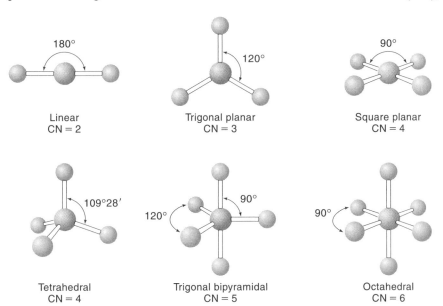

Linear
CN = 2

Trigonal planar
CN = 3

Square planar
CN = 4

Tetrahedral
CN = 4

Trigonal bipyramidal
CN = 5

Octahedral
CN = 6

Figure 12.24
Common geometries of complex ions. The red sphere denotes the metal ion and the gray spheres the ligands. CN stands for coordination number.

Table 12.4
Electron Configurations and Other Properties of the First-Row Transition Metals

	Sc	Ti	V	Cr	Mn	Fe	Co	Ni	Cu
	Electron Configurations								
M	$4s^2 3d^1$	$4s^2 3d^2$	$4s^2 3d^3$	$4s^1 3d^5$	$4s^2 3d^5$	$4s^2 3d^6$	$4s^2 3d^7$	$4s^2 3d^8$	$4s^1 3d^{10}$
M^{2+}	—	$3d^2$	$3d^3$	$3d^4$	$3d^5$	$3d^6$	$3d^7$	$3d^8$	$3d^9$
M^{3+}	[Ne]	$3d^1$	$3d^2$	$3d^3$	$3d^4$	$3d^5$	$3d^6$	—	—
	Ionization Energy/kJ $\cdot$ mol^{-1}								
First	631	658	650	652	717	759	760	736	745
Second	1235	1309	1414	1591	1509	1561	1645	1751	1958
Third	2388	2650	2828	2986	3251	2956	3231	3393	3578
	Radius/Å								
M	1.44	1.36	1.22	1.17	1.17	1.16	1.16	1.15	1.17
M^{2+}	—	0.90	0.88	0.85	0.80	0.77	0.75	0.72	0.72
M^{3+}	0.81	0.77	0.74	0.68	0.66	0.63	0.64	—	—

is the number of donor atoms (in ligands) surrounding the central metal atom in a complex.

Many of the metal ions or atoms in coordination compounds are transition metals. Table 12.4 gives the ground-state electron configurations for the first-row transition metals. Two configurations, d^5 and d^{10}, are preferred because their half-filled and completely filled arrangements provide extra stability (see Section 11.11).

Any theory that explains bonding in coordination compounds must account for their color, magnetic properties, geometry, and other characteristics. Currently, three theories are applied to the study of these compounds, as described below.

Crystal Field Theory

In an isolated transition-metal atom or ion, the five d orbitals all have the same energy regardless of their orientations. This is not the case when the atom or ion is surrounded by ligands. *Crystal field theory* focuses on the electrostatic interaction between the metal and the ligands. There are two types of electrostatic interaction: (1) the attraction between the metal cation and either negatively charged ligands such as Cl^- or the negative ends of polar ligands, such as H_2O, and (2) the repulsion between the valence electrons of the metal and the lone pair on the ligands. The electric field resulting from these interactions is called crystal field, because the theory was originally applied to the study of ions in crystals.

Octahedral Complexes. Consider a complex ion with six ligands octahedrally surrounding a metal ion (Figure 12.25). The magnitude of the electrostatic interaction between the metal ion and the lone pairs on the ligands depends on the orientation of the d orbital involved. Take the $d_{x^2-y^2}$ orbital as an example. We see that the lobes of this orbital point toward the corners of the octahedron along the x and y axes, where the lone pair electrons are positioned. Thus, an electron residing in this orbital would experience a greater repulsion from the ligands than an electron would in, say, the d_{xy} orbital. For this reason, the energy of the $d_{x^2-y^2}$ orbital is higher than that of the d_{xy}, d_{yz}, and d_{xz} orbitals. The d_{z^2} orbital's energy is also greater, because its lobes are pointed at the ligands along the z axis.

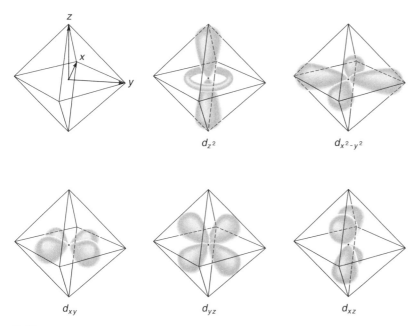

Figure 12.25
The five *d* orbitals in an octahedral environment. The metal ion is at the center of the octahedron, and the six lone pairs on the donor atoms of the ligands are at the corners.

Figure 12.26 shows the effect of an octahedral crystal field on the energies of the *d* orbitals. The energy difference between the two sets of *d* orbitals is called the *crystal field splitting*, denoted by Δ. The *crystal field stabilization energy* (CFSE) is the net decrease in energy of the *d* orbital (hence extra stabilization) in a complex ion relative to the *d* orbital energy in a spherically symmetric crystal field. As can be seen from Figure 12.26, the CFSE depends on the magnitude of Δ (which depends on the nature of the ligands) and the number of electrons present. It can be calculated as follows:

$$\text{CFSE} = n(e_g)(0.6\Delta) - n(t_{2g})(0.4\Delta) \tag{12.17}$$

where $n(e_g)$ and $n(t_{2g})$ are the number of electrons in the e_g and t_{2g} orbitals, respec-

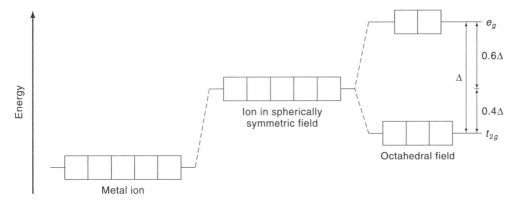

Figure 12.26
Effect of a crystal field on the *d* orbitals in an octahedral environment. The crystal field stabilization energy is given by Equation 12.17. The symbols e_g and t_{2g} are group theory (a mathematical theory of symmetry) symbols for the two sets of orbitals.

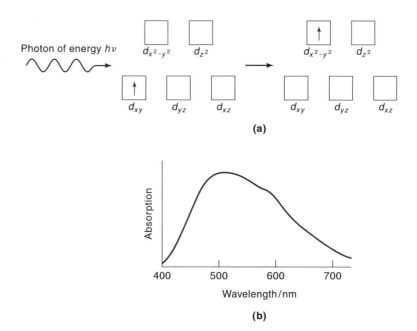

Figure 12.27
(a) An electronic transition in $Ti(H_2O)_6^{3+}$. The condition for absorption is $\Delta = h\nu$. (b) The absorption spectrum of $Ti(H_2O)_6^{3+}$, with a maximum at 498 nm.

tively. This equation tells us that, because the ground-state electron configuration always has more electrons in the t_{2g} orbitals than the e_g orbitals, the CFSE is either negative (indicating that an ion is more stable in an octahedral field than in a spherically symmetric field) or zero. In the latter case, no stabilization occurs when the e_g and t_{2g} orbitals are either completely empty (d^0) or completely filled (d^{10}).

Ligands can be classified as *strong-* or *weak-field ligands*, depending on the magnitude of Δ that results from the interaction. The larger Δ is, the stronger the ligand is. A quantitative measure of Δ can be obtained from the absorption spectrum of the complex ion in solution. For example, when $TiCl_3$ dissolves in water, the hydrated Ti^{3+} ion ($3d^1$) exists as $Ti(H_2O)_6^{3+}$. The lone d electron in the ion must reside in one of the t_{2g} orbitals. When irradiated with light of a specific wavelength, this electron can be excited to one of the e_g orbitals. Figure 12.27 shows the absorption spectrum of $Ti(H_2O)_6^{3+}$, which has a maximum at 498 nm. From the condition for a spectroscopic transition ($\Delta = h\nu$) and the relation $\nu = c/\lambda$, we write

$$\Delta = \frac{hc}{\lambda} = \frac{(6.626 \times 10^{-34} \text{ J s})(3.00 \times 10^8 \text{ m})}{498 \text{ nm} \times 10^{-9} \text{ m/nm}}$$

$$= 3.99 \times 10^{-19} \text{ J}$$

This is the energy required to excite one ion. For 1 mole of ions, we find that $\Delta = 240$ kJ mol^{-1}, which is the crystal splitting due to the water ligands. Thus, one way to compare the strength of the ligands is to use the same metal ion and vary the ligands. From the absorption spectra of these complex ions, we can measure Δ and obtain the *spectrochemical series*, a list of ligands arranged in increasing order of their abilities to split the d orbital energy levels:

$$I^- < Br^- < SCN^- < Cl^- < F^- < OH^- < H_2O < NH_3 < en < NO_2^- < CN^- < CO$$

where the abbreviation *en* denotes the ligand ethylenediamine. Both CO and CN$^-$

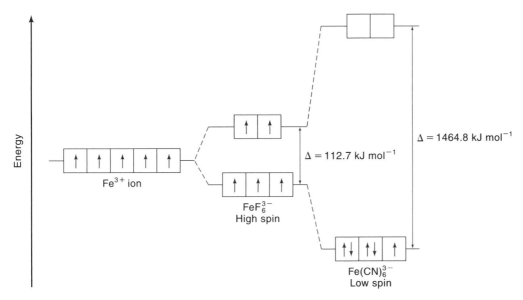

Figure 12.28
Energy-level diagrams for FeF_6^{3-} and $Fe(CN)_6^{3-}$.

are strong-field ligands because they produce the largest values of Δ; the halides and hydroxide ions are weak-field ligands.

Knowledge of the strength of ligands helps us understand the optical and magnetic properties of complex ions. For complex ions of metals with more than three and fewer than eight d electrons, there are two ways of filling the t_{2g} and e_g orbitals. Consider, for example, the octahedral complexes FeF_6^{3-} and $Fe(CN)_6^{3-}$ (Figure 12.28). The electron configuration of Fe^{3+} is $[Ar]3d^5$, and there are two possible ways to distribute the five d electrons. According to Hund's rule, maximum stability is reached when the electrons are placed in five separate orbitals with parallel spins. But this arrangement can be achieved only at a cost: two of the five electrons must be promoted to the higher-energy e_g orbitals. Note that placing all five electrons in the t_{2g} orbitals suffers from a greater electrostatic repulsion due to the proximity of the electrons in the *same d* orbitals. The actual electron configuration, then, is a compromise between these opposing effects. Because F^- is a weak-field ligand with a small value of Δ, Hund's rule prevails, and FeF_6^{3-} is a *high-spin complex*. The opposite holds true for the strong-field ligand CN^-. In $Fe(CN)_6^{3-}$ the electrons are preferentially paired in the t_{2g} orbitals, and a *low-spin complex* results. In general, we can reliably predict the magnetic properties of a complex ion from the electron configuration and the nature of the ligands.

For the same metal ion, the high-spin complex contains a greater number of unpaired spins.

The color of a complex ion also depends on the type of ligands because of the equation $\Delta = h\nu$ governing the electronic transition; the larger the Δ value, the greater the frequency and the shorter the wavelength of light. Table 12.5 shows the approximate relationship between the wavelength of light absorbed and the wavelength transmitted. For example, the $Ti(H_2O)_6^{3+}$ ion absorbs mostly in the yellow-green region (see Figure 12.27), so it appears violet.

Tetrahedral and Square-Planar Complexes. The splitting of the d orbital energy levels in two other types of complexes—tetrahedral and square-planar—can also be accounted for by crystal field theory. In fact, the splitting pattern for a tetrahedral ion is just the opposite of that for octahedral complexes (Figure 12.29). In this case, the d_{xy}, d_{yz}, and d_{xz} orbitals are more closely directed at the ligands and therefore have more energy than the $d_{x^2-y^2}$ and d_{z^2} orbitals. Because tetrahedrally arranged

Table 12.5
Relationship of Wavelength to Color

Wavelength Absorbed/nm	Color Observed
400 (violet)	Greenish yellow
450 (blue)	Yellow
490 (blue-green)	Red
570 (yellow-green)	Violet
580 (yellow)	Dark blue
600 (orange)	Blue
650 (red)	Green

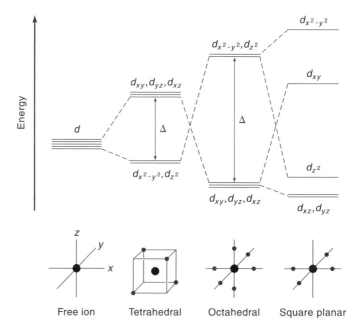

Figure 12.29
Splitting of d orbitals in tetrahedral, octahedral, and square-planar environments. Crystal field splitting can be defined only for tetrahedral and octahedral complexes.

ligands do not differentiate the d orbitals as much as the ligands do in the octahedral case, the crystal field splitting is smaller. In fact, for the same metal ion and ligands, the tetrahedral splitting is $\frac{4}{9}$ that of the octahedral splitting—that is, $\Delta_{\text{tet}} = (4/9)\Delta_{\text{oct}}$. As a result, most tetrahedral complexes are high-spin complexes.

As Figure 12.29 shows, the splitting pattern for square-planar complexes is the most complicated. The $d_{x^2-y^2}$ orbital possesses the highest energy (as in the octahedral case), and the d_{xy} orbital is the next highest. Next comes d_{z^2}, which has a significant band of electron density in the xy plane. The d_{xz} and the d_{yz} form the lowest-lying orbitals. This pattern of splitting does not permit us to define crystal field splitting in the same way as in the octahedral and tetrahedral cases.

Example 12.3

Calculate the CFSE (in terms of Δ) and predict the magnetic properties of the $Mn(CN)_6^{4-}$ ion.

ANSWER

The metal ion is Mn^{2+}, which has the electron configuration $[Ar]3d^5$. Because CN^- is a strong-field ligand, all five d electrons are in the t_{2g} orbitals. According to Equation 12.17, we have

$$CFSE = (0)(0.6\Delta) - (5)(0.4\Delta)$$
$$= -2.0\Delta$$

Four of the five electrons are paired, giving us a paramagnetic but low-spin complex.

COMMENT

Note that the actual CFSE is smaller (more positive) than -2.0Δ because we ignored the repulsion due to the pairing of electrons in the t_{2g} orbitals.

Molecular Orbital Theory

Although crystal field theory is conceptually easy to understand and provides an explanation for the spectral and magnetic properties of some complexes, it has a serious defect. By considering only the electrostatic interaction, it completely neglects the covalent character of the metal–ligand bonds. A more satisfactory treatment that takes this property into consideration is MO theory.[†]

Let us start with the idea that atomic orbitals that are close in energy will interact more strongly than those widely separated in energy. Consider, for example, an octahedral complex ion with the general formula ML_6^{n+} (or ML_6^{n-}, depending on the charge on the metal and whether L carries a negative charge). As mentioned (see Figure 12.25), the $d_{x^2-y^2}$ and d_{z^2} orbitals point at the ligands; therefore, they will form σ molecular orbitals with the ligand lone pair orbitals. On the other hand, the d_{xy}, d_{yz}, and d_{xz} orbitals point between the ligands and thus will not be involved in the σ bonding with the ligands. The $4s$ orbital has spherical symmetry and will overlap with all of the ligands' lone pair orbitals. The three $4p$ orbitals will overlap with the ligand lone pair orbitals on the x, y, and z axes. Thus, the $d_{x^2-y^2}$ and $d_{z^2}, 4s, 4p_x, 4p_y$, and $4p_z$ orbitals will be involved in the σ molecular orbital formations and the d_{xy}, d_{yz}, and d_{xz} orbitals will be the nonbonding molecular orbitals.

Figure 12.30 shows the molecular orbital energy-level diagram for the FeF_6^{3-} ion. Note that the antibonding e_g^* orbitals are primarily composed of $d_{x^2-y^2}$ and d_{z^2} atomic orbitals, with relatively little contribution from the ligand orbitals. This lack of mixing is due to the large energy difference between the ligand orbitals and the metal $3d$ orbitals. Thus, MO theory predicts the same type of d orbital splitting as does the crystal field theory. In addition, MO theory provides a sounder basis for bonding and offers a clearer picture of crystal field splitting. For example, a ligand with a very electronegative donor atom, say F, will have lone pair orbitals with very low energy (because electrons are strongly held by the atom). Consequently, these orbitals do not mix well with the metal orbitals, and crystal field splitting is minimal.

We assume M to be a first-row transition metal.

[†] An extended treatment of the MO theory for metal complexes, which includes delocalized molecular orbitals, is called ligand field theory.

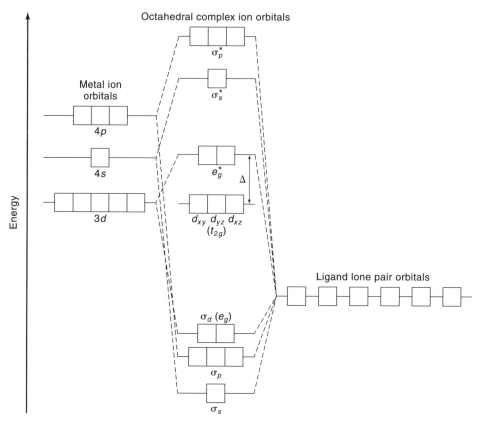

Figure 12.30
Molecular orbital energy-level diagram for an octahedral complex ion ML_6^{n+} (or ML_6^{n-}).
The e_g molecular orbitals are essentially pure $d_{x^2-y^2}$ and d_{z^2} orbitals of the metal ion.

For this reason, FeF_6^{3-} is a high-spin complex.

Another advantage of MO theory is that it can also explain π bonding interactions between the ligand and metal ion (see Section 12.9). It is, however, generally much more complicated to apply than crystal field theory.

Valence Bond Theory

A third approach assumes that each ligand donates a pair of electrons to the metal ion to form a coordinate covalent bond. In an octahedral complex, the metal ion has six vacant orbitals available for bond formations. We can visualize the configuration by assuming that the metal ion is hybridized. Our earlier discussion of hybridization did not include the d orbitals, although their participation is expected for the third-period elements and beyond. The criterion is that the d orbitals must be close to the s and p orbitals on the energy scale. Thus, the trigonal bipyramid shape of PCl_5 can be accounted for if we assume that the P atom is dsp^3 hybridized. The electron configuration of phosphorus is $[Ne]3s^2 3p^3$. Promoting an s electron into a low-lying vacant $3d$ orbital followed by mixing generates five dsp^3 hybrid orbitals. Table 12.6 summarizes different geometries in terms of hybridization.

In octahedral complexes such as $Fe(CN)_6^{3-}$, the metal ion is d^2sp^3 hybridized. This process involves the $3d_{x^2-y^2}, 3d_{z^2}, 4s$, and the three $4p$ orbitals. Upon accepting 12 electrons from the six cyanide groups, the metal ion has a total of 17 electrons

Table 12.6
Hybridization and Geometry of Molecules

Hybridization of Central Atom	Shape	Bond Angle	Example
sp	Linear	180°	HCN, C_2H_2
sp^2	Planar	120°	BF_3, C_2H_4
sp^3	Tetrahedral	109°28′	CH_4, NH_4^+
dsp^2	Square planar	90°	$Ni(CN)_4^{2-}$, $PtCl_4^{2-}$
dsp^3	Trigonal bipyramid	90° (axial-equatorial)[a] 120° (equatorial-equatorial)	PCl_5
d^2sp^3	Octahedral	90°	$Ti(H_2O)_6^{3+}$, SF_6

[a] The atoms that are above and below the triangular plane are said to occupy axial positions, and those that are in the triangular plane are said to occupy equatorial positions.

(Fe^{3+} is $3d^5$) and the electron configuration $3d^{10}4s^24p^5$. After hybridization, we have 12 electrons in the six hybrid orbitals and five more in the unhybridized d_{xy}, d_{yz}, and d_{xz} orbitals. The major disadvantage of VB theory is that it does not satisfactorily account for crystal field splitting and therefore cannot readily explain color and magnetism of complex ions. Nevertheless, the use of hybridization to account for geometry is helpful, so this aspect of the theory has been retained.

12.9 Coordination Compounds in Biological Systems

Metals play a central role in many biological processes. A comprehensive survey of different metals and their functions is not possible to give here. Instead, we shall discuss only a few metal ions and examine their structural aspects in proteins and other biological molecules. Because of iron's importance to all living organisms, we shall focus mainly on the bioinorganic chemistry of the metal.

Iron

It is interesting, although not surprising from an evolutionary point of view, that in proteins such as myoglobin, hemoglobin, and the cytochromes, and in enzymes such as catalase and peroxidase, iron is situated at the center of a planar porphyrin system, as shown in Figure 12.31. Like aromatic hydrocarbons, the porphyrin molecule is extensively delocalized. In both myoglobin and hemoglobin, iron is in the +2 oxidation state. It forms four σ bonds with the nitrogen atoms in the porphyrin ring. This leaves us two more ligands to account for in an octahedral complex. The fifth ligand in these cases is provided by the histidine group, which is part of the protein chain. In the absence of oxygen, the sixth ligand is a water molecule, which binds the Fe^{2+} ion on the other side of the ring to complete the octahedral complex (Figure 12.32). In oxyhemoglobin, where the water molecule is replaced by molecular oxygen, three different orientations have been proposed for the O_2 molecule (Figure 12.33). It turns out that the O_2 molecule assumes a bent configuration, as shown in Figure 12.33c. Although the end-on configuration shown in Figure 12.33b would seem more reasonable, because it allows for a greater degree of orbital overlap, steric hindrance due to a neighboring histidine reside—called the *distal* histidine—that is not bonded to the heme group, forces the O_2 molecule to tilt at an angle (the FeOO angle is about 120°). The nature of binding in oxymyoglobin and oxyhemoglobin is

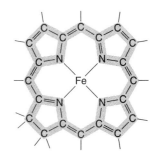

Figure 12.31
Iron–porphyrin system. The electron-delocalized portion of the molecule is shown in color. The chlorophyll molecule also contains the porphyrin molecule, but the metal ion there is Mg^{2+}.

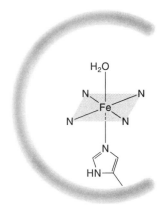

Figure 12.32
The heme group in hemoglobin and myoglobin. The fifth ligand is histidine, which is called *proximal histidine* to distinguish it from another nearby histidine, called *distal histidine*. In the absence of oxygen, the sixth ligand is water.

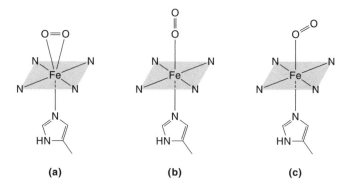

Figure 12.33
Three possible ways for O_2 to bind to the heme group in hemoglobin. The structure shown in (a) would have a coordination number of 7, which is considered unlikely for iron complexes. Although the end-on arrangement in (b) seems most reasonable, experimental evidence points to the structure in (c) as the correct one.

essentially the same, yet their affinity for oxygen is drastically different. The cooperative effect, discussed in Chapter 11, is responsible for this difference.*

In Chapter 10 (p. 392), we saw that deoxyhemoglobin is paramagnetic and oxyhemoglobin is diamagnetic. The transition from high-spin Fe^{2+} to low-spin Fe^{2+} causes the ionic radius to shrink, so that the iron atom can slide into the porphyrin ring to initiate the cooperative effect. Figure 12.34 shows the differences in spin states of deoxyhemoglobin and oxyhemoglobin. Upon dissociation, an O_2 molecule is released and the low-spin Fe^{2+} ion reverts to the high-spin state with H_2O as the sixth ligand.

Besides water and oxygen, several other ligands, such as CO and N_3^-, bind to the Fe^{2+} ion. In fact, the binding of CO is some 50 times stronger than O_2 in myoglobin

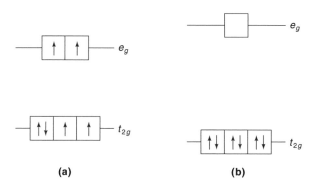

Figure 12.34
Diagram showing spin states of deoxyhemoglobin (high spin) and oxyhemoglobin (low spin). Because low-spin Fe^{2+} is smaller, it is pulled into the porphyrin ring, which is partly responsible for the cooperative effect discussed in Chapter 10 (see Figure 10.23).

---

*In the presence of oxygen, the Fe^{2+} ion in an isolated heme group in solution is readily oxidized to Fe^{3+}, which lacks the ability to bind O_2. The mechanism involves an intermediate in which an O_2 molecule is sandwiched between two heme groups. In both myoglobin and hemoglobin, the steric hindrance of the protein's three-dimensional structure prevents the close approach of any two heme groups, so the Fe^{2+} ion remains unchanged.

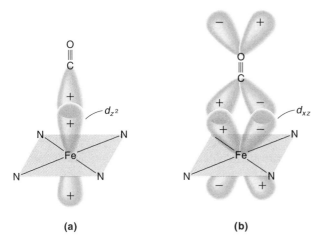

Figure 12.35
The formation of (a) a σ bond and (b) a π bond between the carbon monoxide ligand and Fe^{2+} ion in the heme group. Note that the extent of overlap in both (a) and (b) will decrease if the CO molecule assumes a bent geometry like that shown in Figure 12.33c for the O_2 molecule.

and 200 times stronger in hemoglobin. The enhanced affinity of CO for Fe^{2+} is due to the d_π–p_π interaction. The carbon atom in CO is sp hybridized. The sp hybrid orbital containing the lone pair electrons overlaps with the d_{z^2} orbital of iron to form a σ bond. In addition, the d_{xz} orbital donates electron density to an empty π^* orbital on CO (Figure 12.35). As a result of this interaction, the linkage between Fe^{2+} and CO has appreciable double-bond character and is therefore harder to break than the typical single bond. The strength of the Fe–C bond depends on the extent of the orbital overlap. Fortunately, the CO molecule assumes a bent configuration in carboxyhemoglobin, as does the O_2 molecule in oxyhemoglobin. If the CO molecule were to bind the Fe^{2+} ion in an end-on fashion, the CO affinity for Fe^{2+} in myoglobin and hemoglobin would be a thousand times or more greater than that of O_2, and a short drive on a freeway during rush hour, say, could prove fatal to most people! Although the cyanide ion is isoelectronic with carbon monoxide, cyanide does not complex strongly with the Fe^{2+} ion in hemoglobin due to repulsion between the d electrons of Fe and the negative charge on the cyanide group. The toxicity of the cyanide ion lies in its ability to attack cytochrome oxidase, a respiratory enzyme.[†] Table 12.7 shows the spin states of some important heme proteins—that is, proteins that contain the heme group.

Table 12.7
Spin States for Some Heme Proteins

Heme Protein	Oxidation State for Fe	Sixth Ligand	Number of Unpaired Electrons	Spin State
Hemoglobin	+2	H_2O	4	High
Oxyhemoglobin	+2	O_2	0	Low
Myoglobin	+2	H_2O	4	High
Cyanohemoglobin	+2	CN^-	0	Low
Carboxyhemoglobin	+2	CO	0	Low
Methemoglobin[a]	+3	H_2O	5	High
Cyanomethemoglobin	+3	CN^-	1	Low

[a] Methemoglobin is hemoglobin in which the iron is in the ferric (Fe^{3+}) state.

[†] See R. Chang and L. E. Vickery, *J. Chem. Educ.* **51**, 800 (1974); D. A. Labianca, *ibid*, **56**, 789 (1979).

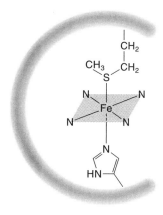

Figure 12.36
The heme group in cytochrome c. In this case, the sixth ligand, a methionine residue, cannot be replaced by other ligands.

Another heme protein is cytochrome c, which differs from hemoglobin in that it carries electrons instead of oxygen. This compound is present on the electron-transport chains in both photosynthetic and respiratory systems. The redox reaction is

$$\text{cyc}Fe^{3+} + e^- \rightleftharpoons \text{cyc}Fe^{2+}$$

Again, iron is at the center of the porphyrin ring and histidine is the fifth ligand. In this case, however, the sixth ligand is a methionine segment (Figure 12.36). The iron–sulfur bond is strong enough to prevent the replacement of the methionine ligand by oxygen.

Rubredoxin and ferredoxin are examples of another type of iron protein that does not contain the heme group. These proteins are found in all green plants, algae, and photosynthetic bacteria. They also play an important role in nitrogen fixation. Like the cytochromes, the ferredoxins function as electron carriers. X-ray studies have revealed that the iron atom is tetrahedrally bonded to sulfur atoms in these compounds.

Copper

The copper ion is also present in a variety of proteins and enzymes, including cytochrome oxidase, an enzyme in respiratory chains; hemocyanin, an oxygen-carrying protein; plastocyanin, an electron-carrying protein in photosynthesis; and tyrosinase, an enzyme that catalyzes the conversion of the amino acid tyrosine to melanin pigment in plants and animals. The electron configuration of Cu^+ is $[Ar]3d^{10}$, so it is not a transition metal ion. Most Cu(I) complexes are colorless and tetrahedral. In the +2 oxidation state, copper usually forms square-planar and octahedral complexes that are green or blue. For example, deoxyhemocyanin is a colorless Cu(I) species that turns blue when it combines with oxygen, indicating the presence of Cu(II). There are two copper atoms per molecule of hemocyanin, and the protein-to-oxygen binding ratio is 1 : 1.

Manganese, Cobalt, and Nickel

Manganese. Manganese has many oxidation states, but only Mn(II) and Mn(III) are important in biological systems. In the +2 oxidation state, manganese usually forms high-spin octahedral complexes, whereas Mn^{3+} is unstable in aqueous solution unless it is complexed. Photosynthesis cannot occur in many higher plants, such as spinach, without the presence of manganese. The reason is that oxygen evolution in these plants (that is, the oxidation of water to molecular oxygen) is catalyzed by a Mn(III) protein complex.

A coenzyme is a substance that activates enzymes. Most coenzymes are vitamins.

Cobalt. Like copper, cobalt has two common oxidation states in solution, Co(II) and Co(III). The biological activity of cobalt is largely confined to its role in the vitamin B_{12} series of coenzymes. In vitamin B_{12}, the cobaltic ion is situated at the center of a conjugated corrin ring that is similar to the porphyrin structure (Figure 12.37). The corrin system provides four nitrogen atoms, whose lone pairs form sigma bonds with cobalt. The fifth and sixth ligands are benzimadazole and a cyanide ion. The cyanide complex itself does not function as a coenzyme, but the cyanide ion can be replaced by a number of other ligands, many of which are biologically active.

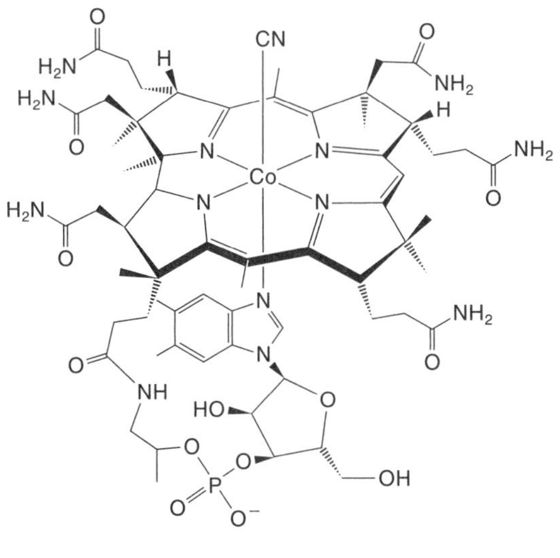

Figure 12.37
Structure of vitamin B_{12}.

Nickel. As mentioned in Chapter 10, urease, the enzyme that catalyzes the conversion of urea to carbamate and ammonium ion, contains nickel ions:

$$(NH_2)_2CO + H_2O \rightarrow NH_2COO^- + NH_4^+$$

In fact, there are two Ni^{2+} ions within 4 Å at the active site. There are at least three other nickel-containing enzymes known. Hydrogenases catalyze the oxidation of hydrogen to protons ($2H_2 + O_2 \rightarrow 2H_2O$). This redox process involves the I, II, and III oxidation states of nickel. Carbon monoxide dehydrogenase catalyzes the oxidation of CO to CO_2 and methyl–coenzyme M reductase catalyzes the conversion of CO_2 to CH_4.

The ability of certain transition metal ions to form complexes has found useful application in protein purification. In the metal–chelate affinity chromatography (MCAC) technique, metal ions such as Ni^{2+}, Cu^{2+}, and Zn^{2+} are bound to solid matrices that bind to the exposed imidazole (histidine residue) and thiol (cysteine reside) groups within the protein molecule. Of particular value is the binding between the Ni^{2+} ion and the histidine residue. In a typical procedure, a solution containing an impure protein solution is poured down a column containing the nickel-based solid matrix. Under neutral or slightly alkaline conditions, the N atom of the imidazole ring becomes deprotonated (see Figure 8.6), which can form a coordinate bond with Ni^{2+}. In this way the target protein becomes attached to the column and separated from the solution. Eventually the protein is freed by eluting the column with a low-pH buffer or a solution of imidazole, which competes for the Ni^{2+} binding site. Even in cases where the protein does not possess exposed histidine residues, it is possible to tag a small peptide of histidine (six is the chosen number) to the C or N terminal of the protein. This added portion can then be cleaved from the parent protein after purification.

> Urease was the first protein to be isolated and crystallized in 1926. It took another 50 years for chemists to realize that it contains Ni^{2+} ions.

> From Table 12.3, we see that Ni^{2+} should bind strongly to the imidazole group.

Zinc

The electron configuration of Zn^{2+} is $[Ar]3d^{10}$; therefore, it is not a transition metal. This is consistent with the fact that zinc compounds are largely colorless and diamagnetic. Zinc performs both a structural and a catalytic role in proteins. For example, liver alcohol dehydrogenase catalyzes the reaction

$$CH_3CH_2OH + NAD^+ \rightarrow CH_3CHO + NADH + H^+$$

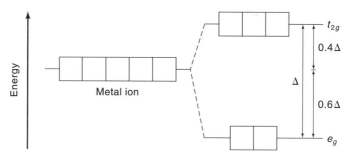

Figure 12.38
Ligand-field stabilization energy for a tetrahedral complex ion.

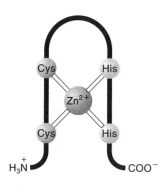

Figure 12.39
Schematic diagram showing the zinc ion tetrahedrally bonded to two cysteine and two histidine residues to form the zinc-finger domain. The zinc finger makes contact with and identifies DNA sequences.

Each subunit of the enzyme binds one NAD^+ and two functionally very different Zn^{2+} ions. The geometry of zinc complexes is tetrahedral. One of the Zn^{2+} ions is bound to four cysteine ligands and is inaccessible to solvent; its function is undoubtedly structural. The other Zn^{2+} ion is bound to two cysteines, one histidine, and a water molecule; together they make up the catalytic domain.

As mentioned, the CFSE for an octahedral complex with d^{10} configuration is zero. This conclusion also applies to tetrahedral complexes (Figure 12.38). The fact that the CFSE is zero for *any* geometry for ions with d^{10} configuration helps explain the specificity and stability of Zn^{2+} in proteins. Recent studies have found "structural" Zn^{2+} ions complexed with proteins that bind to DNA. In one case, Zn^{2+} tetrahedrally binds to two cysteine and two histidine residues to form a loop, or "finger," that fits into the major groove of DNA (Figure 12.39). For this reason, this portion of the structure is called a *zinc finger*. The thermodynamic stability of the zinc finger can be readily appreciated if we examine the energetics of replacing Zn^{2+} with another ion. Suppose we have Co^{2+} ($3d^7$) in solution as $Co(H_2O)_6^{2+}$. Because H_2O is a weak-field ligand, the crystal field splitting Δ is 111 kJ mol^{-1} and the CFSE value, given by Equation 12.17, is $-(4/5)\Delta$, or -89 kJ mol^{-1}. To replace Zn^{2+}, Co^{2+} must assume a tetrahedral geometry with a CFSE value of $-(6/5)\Delta$.* For a zinc-finger site containing two cysteines and two histidine ligands, Δ is 59 kJ mol^{-1}, so the CFSE value is $-(6/5)(59$ kJ mol$^{-1})$ or -71 kJ mol^{-1}. Thus, a Co^{2+} ion will lose $89 - 71 = 18$ kJ mol^{-1} in CFSE during the transition from an octahedral site in solution to the tetrahedral site in a zinc-finger domain. This comparison at least partly accounts for the stability of many zinc complexes and for the difficulty of replacing zinc with other metals.

Toxic Heavy Metals

Cadmium, mercury, and lead have no known biological functions but are extremely poisonous to living systems. The toxicities of these heavy metals are well known and understood because they are used in industrial processes, as well as in batteries, and they are present in old buildings in paint and water pipes. In solution, these metals exist mostly as divalent cations (Cd^{2+}: $[Kr]4d^{10}$; Hg^{2+}: $[Xe]4f^{14}5d^{10}$; and Pb^{2+}: $[Xe]6s^24f^{14}5d^{10}$). Being soft acids, they have a great affinity for soft bases, particularly the sulfhydryl group on the cysteine residue and, to a lesser extent, the imidazole group on histidine. Consequently, they can inactivate enzymes and disrupt

---

* The CFSE for a tetrahedral complex is given by:

$$\text{CFSE} = n(t_{2g})(0.4\Delta) - n(e_g)(0.6\Delta)$$

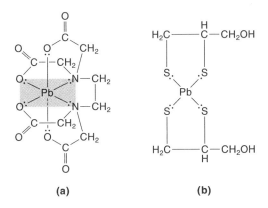

Figure 12.40
(a) EDTA complex of lead. The complex ion bears a net charge of -2, since each O donor atom has one negative charge and the lead is present as Pb^{2+}. Note the octahedral geometry around Pb^{2+}. (b) BAL complex of lead. This complex ion also bears a net charge of -2, because each S atom has one negative charge. The Pb^{2+} ion is tetrahedrally bonded to the four S atoms.

protein functions in cellular energy output and oxygen transport. Because their d orbitals are completely filled, CFSE plays no role in the geometry of complexes containing these ions. Cd^{2+} preferentially forms tetrahedral complexes, whereas Hg^{2+} forms octahedral as well as linear and tetrahedral complexes. More dangerous than inorganic mercury compounds are organomercury compounds such as $(CH_3)_2Hg$ and CH_3HgCl because biomembranes are more permeable to them.* They concentrate in blood and have a more immediate and permanent effect on the brain and central nervous system, no doubt by binding to the $-SH$ groups in proteins. Lead poisoning affects the kidney and the liver, and inhibits the synthesis of heme, resulting in anemia.

Metal poisoning is commonly treated by administering a chelating agent. The two most effective chelating compounds are ethylenediaminetetraacetate (EDTA) and 2,3-dimercaptopropanol, which is more familiarly called BAL (British anti-Lewisite), shown in Figure 12.40. Interestingly, virtually all mammal organs contain metallothionein proteins, which act as the first line of defense against heavy-metal poisoning. Although quite small (molar mass about 6,500 daltons), nearly 30% to 35% of the amino acids in these proteins are cysteine residues, whose main function is to protect cells against the toxic effects of the heavy metals. Tests have shown that nonlethal doses of cadmium, mercury, and lead in animals can induce the synthesis of metallothionein.

---

* X-ray absorption spectroscopy shows that the chemical form of mercury present in fish is a compound in which Hg is coordinated with a methyl group and a cysteinyl residue (see Table 8.6). This structure is in accord with the fact that Hg^{2+} is a soft acid and the RS^- group of cysteine is a soft base (see Table 12.3). The cysteine is likely to be part of a larger peptide or protein. This compound is considered to be much less toxic than either $(CH_3)_2Hg$ or CH_3HgCl. [See "The Chemical Form of Mercury in Fish," H. H. Harris, I. J. Pickering, and G. N. George, *Science* **301**, 1203 (2003).]

Suggestions for Further Reading

BOOKS

Companion, A. L., *Chemical Bonding*, 2nd ed., McGraw-Hill, New York, 1979.

Dekock, R. L. and H. B. Gray, *Chemical Structure and Bonding*, 2nd ed., University Science Books, Sausalito, CA, 1989.

Fraústo da Silva, J. J. R. and R. J. P. Williams, *The Biological Chemistry of the Elements*, Clarendon Press, Oxford, England, 1991.

Karplus, M. and R. N. Porter, *Atoms and Molecules: An Introduction for Students of Physical Chemistry*, W. A. Benjamin, New York, 1970.

Lippard, S. J. and J. M. Berg, *Principles of Bioinorganic Chemistry*, University Science Books, Sausalito, CA, 1994.

McWeeny, R., *Coulson's Valence*, 3rd ed., Oxford University Press, New York, 1979.

Richards, W. G. and P. R. Scott, *Energy Levels in Atoms and Molecules*, Oxford University Press, New York, 1994.

Wulfsberg, G., *Inorganic Chemistry*, University Science Books, Sausalito, CA, 2000.

ARTICLES

Chemical Bonding

"Molecular Orbital Theory for Transition Metal Complexes," H. B. Gray, *J. Chem. Educ.* **41**, 2 (1964).

"Ligand Field Theory," F. A. Cotton, *J. Chem. Educ.* **41**, 466 (1964).

"Kekulé and Benzene," C. A. Russell, *Chem. Bri,* **1**, 141 (1965).

"Hard and Soft Acids and Bases," R. G. Pearson, *J. Chem. Educ.* **45**, 581, 681 (1968).

"The Shape of Organic Molecules," J. B. Lambert, *Sci. Am.* January 1970.

"Chemistry by Computer," A. C. Wahl, *Sci. Am.* April 1970.

"A Simple, Quantitative Molecular Orbital Theory," W. F. Cooper, G. A. Clark, and C. R. Hare. *J. Chem. Educ.* **48**, 247 (1971).

"Size and Shape of a Molecule," M. J. Demchik and V. C. Demchik, *J. Chem. Educ.* **48**, 770 (1971).

"Molecular Orbitals and Air Pollution," B. M. Fung, *J. Chem. Educ.* **49**, 26 (1972). See also p. 654, same volume.

"Molecular Oxygen Adducts of Transition Metal Complexes: Structure and Mechanism," L. Klevan, I. Peone, Jr., and S. K. Madan, *J. Chem. Educ.* **50**, 670 (1973).

"Predicting Chemistry from Topology," D. H. Rouvray, *Sci. Am.* September 1986.

"No Rabbit Ears on Water. The Structure of the Water Molecule: What Should We Tell the Students?" M. Laing, *J. Chem. Educ.* **64**, 124 (1987).

"Localized and Spectroscopic Orbitals: Squirrel Ears on Water," R. B. Martin, *J. Chem. Educ.* **65**, 668 (1988).

"The Relative Energies of Molecular Orbitals for Second-Row Homonuclear Diatomic Molecules: The Effect of *s–p* Mixing," A. Haim, *J. Chem. Educ.* **68**, 737 (1991).

"Why Aromatic Compounds Are Stable," J.-I. Aihara, *Sci. Am.* March 1992.

"How Should Chemists Think?" R. Hoffmann, *Sci. Am.* February 1993.

Bioinorganic Chemistry

"The Role of Chelation in Iron Metabolism," P. Saltman, *J. Chem. Educ.* **42**, 682 (1965).

"Chelation in Medicine," J. Schubert, *Sci. Am.* May 1966.

"The Biochemistry of Copper," E. Frieden, *Sci. Am.* May 1968.

"Lead Poisoning," J. I. Chisolm, Jr., *Sci. Am.* February 1971.

"Mercury Poisoning," L. E. Strong, *J. Chem. Educ.* **49**, 28 (1972).

"Environmental Bioinorganic Chemistry," E.-I. Ochiai, *J. Chem. Educ.* **51**, 25 (1974).

"Iron and Susceptibility to Infectious Disease," E. D. Weinberg, *Science* **184**, 952 (1974).

"The Role of Metal Ions in Proteins and Other Biological Molecules," E. W. Ainscough and A. M. Brodie, *J. Chem. Educ.* **53**, 156 (1976).

"Biochemical Effects of Excited State Molecular Oxygen," J. Bland, *J. Chem. Educ.* **53**, 274 (1976).

"Therapeutic Chelating Agents," M. M. Jones and T. H. Pratt, *J. Chem. Educ.* **53**, 342 (1976).

"Metals, Models, Mechanisms, Microbes, and Medicine," H. A. O. Hill, *Chem. Bri.* **12**, 119 (1976).

"Hemocyanin: The Copper Blood," N. M. Senozan, *J. Chem. Educ.* **53**, 684 (1977).

"Principles in Bioinorganic Chemistry," E.-I. Ochiai, *J. Chem. Educ.* **55**, 631 (1978).

"Hemoglobin Structure and Respiratory Transport," M. F. Perutz, *Sci. Am.* December 1978.

"Inorganic Elements in Biology and Medicine," R. J. P. Williams, *Chem. Bri.* **15**, 506 (1979).

"Chemical Toxicology: Part II. Metal Toxicity," D. E. Carter and Q. Fernando, *J. Chem. Educ.* **56**, 491 (1979).

"Cytochrome *c* and the Evolution of Energy Metabolism," R. E. Dickerson, *Sci. Am.* March 1980.

"Modeling Coordination Sites in Metallobiomolecules," J. A. Ibers and R. H. Holm, *Science* **209**, 223 (1980).

"Hemoglobin: Its Occurrence, Structure, and Adaptation," N. M. Senozan and R. L. Hunt, *J. Chem. Educ.* **59**, 173 (1982).

"Bacterial Resistance to Mercury," N. L. Brown, *Trends Biochem. Sci.* **10**, 400 (1985).

"Methemoglobinemia: An Illness Caused by the Ferric

State," M. M. Senozan, *J. Chem. Educ.* **62**, 181 (1985).

"New Perspectives on the Essential Trace Elements," E. Frieden, *J. Chem. Educ.* **62**, 917 (1985).

"Zinc Enzyme," I. Bertini, C. Luchinat, and R. Monnanni, *J. Chem. Educ.* **62**, 924 (1985).

"Uniqueness of Zinc as a Bioelement," E.-I. Ochiai, *J. Chem. Educ.* **65**, 943 (1988).

"Copper Precipitation in the Human Body," R. P. Csintalau and N. M. Senozan, *J. Chem. Educ.* **68**, 365 (1991).

"Biological Roles of Nitric Oxide," S. H. Snyder and D. S. Bredt, *Sci. Am.* May 1992.

"Nitric Oxide in Cells," J. R. Lancaster, Jr., *Am. Sci.* **80**, 248 (1992).

"Zinc Fingers," D. Rhodes and A. Klug, *Sci. Am.* February 1993.

"Toxicity of Heavy Metals and Biological Defense," E.-I. Ochiai, *J. Chem. Educ.* **72**, 479 (1995).

"CO, N$_2$, NO, and O$_2$—Their Bioinorganic Chemistry," E.-I. Ochiai, *J. Chem. Educ.* **73**, 130 (1996).

"Carbon Monoxide Poisoning," N. M. Senozan and J. A. Devore, *J. Chem. Educ.* **73**, 767 (1996).

"Iron as Nutrient and Poison," N. M. Senozan and M. P. Christiano, *J. Chem. Educ.* **74**, 1060 (1997).

"Analysis of Iron in Ferritin, the Iron-Storage Protein," M. J. Donlin, R. F. Frey, C. Putnam, J. K. Proctor, and J. K. Baskin, *J. Chem. Educ.* **75**, 437 (1998).

"Bacterial Solutions to the Iron-Supply Problems," V. Braun and H. Killman, *Trends Biochem. Sci.* **24**, 104 (1999).

"Nitric Oxide Moves Myoglobin Centre Stage," M. Brunori, *Trends Biochem. Sci.* **26**, 209 (2001).

Problems

Lewis Theory and Related Topics

12.1 Which of the following molecules has the shortest nitrogen-to-nitrogen bond? Explain.

$$N_2H_4 \quad N_2O \quad N_2 \quad N_2O_4$$

12.2 The chlorine nitrate molecule ($ClONO_2$) is believed to be involved in the destruction of ozone in the Antarctic stratosphere. Draw a plausible Lewis structure for this molecule.

12.3 Draw resonance structures for N_2O. The atomic arrangement is NNO. Show formal charges. How would you distinguish this structure from the NON structure?

12.4 Carbon monoxide has a rather small dipole moment ($\mu = 0.12$ D) even though the electronegativity difference between C and O is rather large ($X_C = 2.5$ and $X_O = 3.5$). How would you explain this fact in terms of resonance structures?

12.5 The resonance concept is sometimes described by analogy to a mule, which is a cross between a horse and a donkey. Compare this analogy with the description of a rhinoceros as a cross between a griffin and a unicorn. Which description is more appropriate? Why?

12.6 Comment on the appropriateness of using the following resonance structure for O_2 intended to explain its paramagnetism.

$$\cdot\ddot{O}-\ddot{O}\cdot$$

12.7 Consider the following Lewis structure for boron trifluoride (BF_3):

Does it satisfy the octet rule? If not, draw additional resonance structures that do satisfy the octet rule. Suggest an experimental measurement that would enable you to show the relative importance of the resonance structures.

Valence Bond Theory and Hybridization

12.8 Disulfide bonds play an important role in the three-dimensional structure of protein molecules. Discuss the nature of the –S–S– bond.

12.9 Consider the HCl molecule. Let ψ_{1s} be the hydrogen $1s$ wave function and ψ_{3p} be the chlorine $3p$ wave function. Write the VB wave function of HCl assuming that **(a)** the bond is purely covalent, **(b)** the bond is purely ionic, and the electron is transferred from H to Cl, and **(c)** the bond is polar.

12.10 The unstable molecule carbene or methylene (CH_2) has been isolated and studied spectroscopically. Suggest two types of bonding that might be present in this molecule. How would you determine which type of bond is present in CH_2?

12.11 Describe the bonding in CO_2 and C_3H_4 (allene) in terms of hybridization. Draw diagrams to show the formation of σ bonds and π bonds in allene.

Molecular Orbital Theory, Resonance, and Electron Delocalization

12.12 Describe the bonding scheme in the following species in terms of molecular orbital theory: H_2^+, H_2, He_2^+, and He_2. List the species in order of decreasing stability.

12.13 Which molecule would have the longer bond length, F_2 or F_2^+? Explain in terms of molecular orbital theory.

12.14 Which of the following species has the longest bond: CN^+, CN, or CN^-?

12.15 Borazine ($B_3N_3H_6$) is isoelectronic with benzene. Describe qualitatively the bonding in this molecule in terms of **(a)** resonance and **(b)** molecular orbital theory.

12.16 Which of the following two molecules has a greater degree of π-electron delocalization: naphthalene or biphenyl?

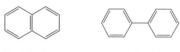

Naphthalene   Biphenyl

12.17 Use MO theory to describe the bonding in NO^+, NO, and NO^-. Compare their bond energies and bond lengths.

12.18 Compare the MO theory description for the H_2 molecule, where the wave function is given by

$$\psi = [\psi_A(1) + \psi_B(1)][\psi_A(2) + \psi_B(2)]$$

with the VB theory treatment given by Equation 12.4. Under what condition do they become identical?

12.19 Acetylene (C_2H_2) has a tendency to lose two protons (H^+) and form the carbide ion (C_2^{2-}), which is present in several ionic compounds, such as CaC_2 and MgC_2. Describe the bonding scheme in the C_2^{2-} ion in terms of molecular orbital theory. Compare the bond order in C_2^{2-} with that in C_2.

12.20 Describe the bonding in the nitrate ion, NO_3^-, in terms of delocalized molecular orbitals.

12.21 A single bond is usually a σ bond, and a double bond is usually made up of a σ bond and a π bond. Can you identify the exceptions in the homonuclear diatomic molecules of the second period?

Crystal Field Theory

12.22 Draw energy-level diagrams to show the low- and high-spin octahedral complexes of the transition-metal ions that have the electron configurations d^4, d^5, d^6, and d^7.

12.23 The $Ni(CN)_4^{2-}$ ion, which has square-planar geometry, is diamagnetic, whereas the $NiCl_4^{2-}$ ion, which has tetrahedral geometry, is paramagnetic. Show the crystal field splitting diagrams for those two complexes.

12.24 Predict the number of unpaired electrons in the following complex ions: **(a)** $Cr(CN)_6^{4-}$ and **(b)** $Cr(H_2O)_6^{2+}$.

12.25 Transition metal complexes containing CN^- ligands are often yellow in color, whereas those containing H_2O ligands tend to be green or blue. Explain.

12.26 The absorption maximum for the complex ion $Co(NH_3)_6^{3+}$ occurs at 470 nm. **(a)** Predict the color of the complex, and **(b)** calculate the crystal field splitting in kJ mol^{-1}.

12.27 The label of a certain brand of mayonnaise lists EDTA as a food preservative. How does EDTA prevent the spoilage of mayonnaise?

12.28 Hydrated Mn^{2+} ions are practically colorless even though they possess five $3d$ electrons. Explain. (*Hint:* Electronic transitions in which there is a change in the number of unpaired electrons do not occur readily.)

12.29 Oxyhemoglobin is bright red, whereas deoxy-hemoglobin is purple. Explain the difference in color in terms of the electron configurations of iron in these two complexes.

Additional Problems

12.30 Although both carbon and silicon are in Group 4A, very few Si=Si bonds are known. Account for the instability of silicon-to-silicon double bonds in general. (*Hint:* Compare the covalent radii of C and Si.)

12.31 Compare the bonding in FeF_6^{3-} and $Fe(CN)_6^{3-}$ in terms of hybridization.

12.32 Chemical analysis shows that hemoglobin is 0.34% Fe by mass. What is the minimum possible molar mass of hemoglobin? The actual molar mass of hemoglobin is about 65,000 g. How do you account for the discrepancy between your minimum value and the actual value?

12.33 Use the molecular orbital energy-level diagram for O_2 to show that the following Lewis structure corresponds to an excited state:

$$\ddot{O}=\ddot{O}$$

12.34 Co binds better to the heme group than Fe, and Co^{2+} has less of a tendency to be oxidized to Co^{3+} than Fe^{2+} does to Fe^{3+}. Why is Fe the metal in hemoglobin and myoglobin rather than Co?

12.35 Suffocation victims usually look purple, but a person poisoned by carbon monoxide often has rosy cheeks. Explain.

12.36 The dipole moment of *cis*-dichloroethylene is 1.81 D at 25°C. On heating, its dipole moment begins to decrease. Give a reasonable explanation for this observation.

12.37 Although the hydroxyl radical (OH) is present only in a trace amount in the atmosphere, it plays an important role in atmospheric chemistry because it is a strong oxidant and can react with many pollutants. Assume that the radical is analogous to the HF molecule and that the molecular orbitals result from the overlap of an oxygen $2p_x$ orbital and a hydrogen $1s$ orbital. **(a)** Draw pictures of the σ and σ^* molecular orbitals in OH. **(b)** Which of the two molecular orbitals has more hydrogen $1s$ character? **(c)** Draw a molecular orbital energy-level diagram, and write the electron configuration for the radical. Note that the electrons in the non-bonding orbitals on oxygen have π character and should be assigned as such. **(d)** Estimate the bond order of OH. Compare this value with that for OH^+.

12.38 The H_3^+ ion is the simplest polyatomic molecule. It has equilateral geometry. **(a)** Draw three resonance structures to represent this species. **(b)** Use MO theory to describe the bonding molecular orbital for this ion. Write the wave function for the lowest-energy molecular orbital. Is it a σ or π delocalized molecular orbital? **(c)** Given that $\Delta_r H = -849$ kJ mol^{-1} for the reaction $2H + H^+ \rightarrow H_3^+$ and that $\Delta_r H = 436.4$ kJ mol^{-1} for $H_2 \rightarrow 2H$, calculate the value of $\Delta_r H$ for the reaction $H^+ + H_2 \rightarrow H_3^+$. Comment on the magnitude of $\Delta_r H$.

12.39 A novel electron-transport protein is found to contain only zinc as the metal. Comment on this finding.

12.40 Does the molecule $HBrC=C=CHBr$ have a dipole moment?

12.41 What is the state of hybridization of the central O atom in O_3? Describe the bonding in O_3 in terms of delocalized molecular orbitals.

12.42 Oxalic acid, $H_2C_2O_4$, is sometimes used to remove rust stains from sinks and bathtubs. Explain the chemistry underlying this cleaning action.

12.43 Use the particle-in-a-box model to explain the difference between the potential energy curves that result from Equations 12.2 and 12.3.

12.44 Draw qualitative diagrams for the crystal-field splitting in **(a)** a linear complex ion ML_2, **(b)** a trigonal-planar complex ion ML_3, and **(c)** a trigonal-bipyramidal complex ion ML_5.

12.45 You are given two solutions containing $FeCl_2$ and $FeCl_3$ at the same concentration. One solution is light yellow and the other one is brown. Identify these solutions based only on color.

12.46 The geometries discussed in this chapter all lend themselves to fairly straightforward elucidation of bond angles. The exception is the tetrahedron, because its bond angles are hard to visualize. Consider the CCl_4 molecule, which has tetrahedral geometry and is nonpolar. By equating the bond moment of a particular C–Cl bond to the resultant bond moments of the other three C–Cl bonds in opposite directions, show that the bond angles are all equal to $109.5°$.

12.47 Plastocyanin, a copper-containing protein found in photosynthetic systems, is involved in electron transport with the copper ion switching between the +1 and +2 oxidation states. The copper ion is coordinated with two histidine residues, a cysteine residue, and a methionine residue in a tetrahedral manner. How does the crystal-field splitting (Δ) change between these two oxidation states?

12.48 Cu^{2+} ions coordinated with S atoms tend to form tetrahedral complexes whereas those coordinated with N atoms tend to form octahedral complexes. Explain.

12.49 $[Pt(NH_3)_2Cl_2]$ is found to exist in two geometric isomers designated I and II, which react with oxalic acid as follows:

$$I + H_2C_2O_4 \rightarrow [Pt(NH_3)_2C_2O_4]$$
$$II + H_2C_2O_4 \rightarrow [Pt(NH_3)_2(HC_2O_4)_2]$$

Comment on the structures of I and II.

12.50 Which has a lower first ionization energy, O or O_2? Explain.

12.51 The porphine group shown here gives rise to the metal-porphyrin complex ion in many biological systems.

(a) Treating the porphyrin group as a two-dimensional square box of length L, write the expression for the energies of an electron in the box. (*Hint:* Extend Equation 11.22 to include two quantum numbers n_x and n_y.) **(b)** Consider a M^{2+}–porphyrin complex that contains 26 π electrons, where M^{2+} is Fe^{2+} as in the heme group or Mg^{2+} as in chlorophyll (see Figures 12.31 and 15.3), sketch an energy level diagram for this system. Note that a number of the levels are degenerate in energy. **(c)** Given that the box length is 10 Å, calculate the wavelengths for the first two lowest energy electronic transitions in nm.

12.52 How many geometric isomers can the square-planar platinum complex Pt(abcd) have? Each letter represents a monodentate ligand.

CHAPTER 13

Intermolecular Forces

In Chapter 12 we discussed the covalent bond and related the force holding atoms together to the overlapping of atomic orbitals. Interactions between molecules are best explained by several types of intermolecular forces, forces responsible for phenomena such as the liquefaction of gases and the stability of proteins. A special type of interaction, hydrogen bonding, plays an important role in determining the structure and properties of DNA and water.

13.1 Intermolecular Interactions

When two molecules approach each other, various interactions between the electrons and nuclei of one molecule and the electrons and nuclei of the other molecule generate potential energy. At a very large distance of separation, where there is no intermolecular interaction, we can arbitrarily set the potential energy of the system at zero. As the molecules approach each other, electrostatic attractions outweigh electrostatic repulsions, so the molecules are pulled toward each other and the potential energy of interaction is negative. This trend continues until the potential energy reaches a minimum. Beyond this point (that is, as the distance of separation decreases further), repulsive forces predominate and the potential energy rises (becoming more positive).

A simple demonstration of intermolecular forces is to ask why the far end of a walking cane rises when you lift the handle.

For a discussion of molecular interaction, it is useful to distinguish between force and potential energy. In mechanics, work done is force times distance. The work done (dw) in moving two interacting molecules apart by an infinitesimal distance (dr) is given by

$$dw = -F \, dr \qquad (13.1)$$

The sign convention for work is that if the molecules attract each other (i.e., F is negative) and dr is positive (i.e., molecules are moved farther apart), then dw is positive. An expression for the potential energy (V) of the molecules separated by distance r can be obtained as follows. Imagine that one molecule is fixed in position, and we want to know how much work is done in bringing another molecule from infinite separation to a distance r from this molecule. This work is the potential energy acquired by the system, given by

$$V = \int_{\infty}^{r} dw$$

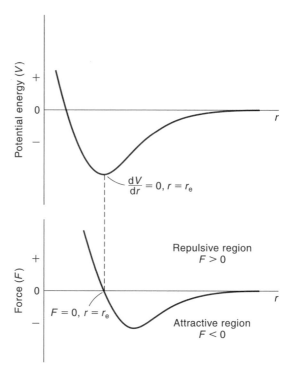

Figure 13.1
Relation between potential energy and force. Because $F = -dV/dr$, at the minimum of the $V(r)$ versus r curve, $F = 0$. Note that at $r < r_e$, where the potential energy is still negative, force becomes repulsive.

From Equation 13.1,

$$V = -\int_{\infty}^{r} F \, dr \qquad (13.2)$$

Equation 13.2 relates the potential energy of interaction to the force between the two molecules. Differentiating Equation 13.2 with respect to r, we get

$$F = -\frac{dV}{dr} \qquad (13.3)$$

Thus, force is the negative of the slope of the curve that describes the dependence of V on r. Figure 13.1 shows the relationship between potential energy and force.

13.2 The Ionic Bond

Before we study the interaction between molecules, we should first examine the bond between a pair of ions for comparison purposes. At elevated temperatures, ionic compounds such as NaCl vaporize to form ion pairs. The dipole moments of this and other similar alkali halide ion pairs are large, about 10 times that of hydrogen halides, indicating that the character of the bond is largely ionic. The potential energy due to the attraction between a pair of Na^+ and Cl^- ions is given by Coulomb's law (see p. 156):

$$V = -\frac{q_{Na^+}q_{Cl^-}}{4\pi\varepsilon_0 r} \tag{13.4}$$

We assume air is the medium so the dielectric constant is 1 in all cases.

where r is the distance of separation and only the magnitude (not the signs) of the charges are shown. However, we must also include a term representing the repulsion between electrons and between nuclei on these ions. This term usually takes the form of be^{-ar} or b/r^n, where a and b are constants that are specific for a given ion pair and n is an integer between 8 and 12. Using the latter term, we write the complete equation for potential energy as

$$V = -\frac{q_{Na^+}q_{Cl^-}}{4\pi\varepsilon_0 r} + \frac{b}{r^n} \tag{13.5}$$

We can solve for b by realizing that, at the minimum of the potential energy curve (see Figure 13.1), $dV/dr = 0$ so that

$$\frac{dV}{dr} = 0 = \frac{q_{Na^+}q_{Cl^-}}{4\pi\varepsilon_0 r_e^2} - \frac{nb}{r_e^{n+1}}$$

or

$$b = \frac{q_{Na^+}q_{Cl^-}}{4\pi\varepsilon_0 n} r_e^{n-1} \tag{13.6}$$

where r_e is the equilibrium bond length of the ion pair. Substituting Equation 13.6 into 13.5, we write

$$V_0 = -\frac{q_{Na^+}q_{Cl^-}}{4\pi\varepsilon_0 r_e} + \frac{q_{Na^+}q_{Cl^-}}{4\pi\varepsilon_0 n r_e}$$

$$= -\frac{q_{Na^+}q_{Cl^-}}{4\pi\varepsilon_0 r_e}\left(1 - \frac{1}{n}\right) \tag{13.7}$$

Note that V_0 denotes the potential energy associated with the most stable separation (r_e). Using the NaCl(g) bond length of 2.36 Å (236 pm), 1.602×10^{-19} C for unit charge, and $n = 10$ for the repulsion, we write, for 1 mole of Na^+ and Cl^- ion pairs,

$$V_0 = -\frac{(1.602 \times 10^{-19}\ \text{C})^2(6.022 \times 10^{23}\ \text{mol}^{-1})(1 - 0.1)}{4\pi(8.854 \times 10^{-12}\ \text{C}^2\ \text{N}^{-1}\ \text{m}^{-2})(236 \times 10^{-12}\ \text{m})}$$

$$= -5.297 \times 10^5\ \text{N m mol}^{-1}$$

$$= -529.7\ \text{kJ mol}^{-1}$$

This is the energy given off when 1 mole of NaCl ion pairs forms from Na^+ and Cl^- ions in the gaseous state:

$$Na^+(g) + Cl^-(g) \rightarrow NaCl(g)$$

However, the ground state of the dissociated system consists of atoms rather than ions—that is,

$$NaCl(g) \rightarrow Na(g) + Cl(g)$$

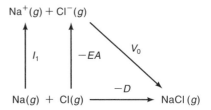

Figure 13.2
Born–Haber cycle for the formation of a gaseous NaCl ion pair.

To calculate the bond-dissociation enthalpy for this process, we apply the *Born–Haber cycle* shown in Figure 13.2. Based on Hess's law, this procedure breaks the formation of NaCl(g) into separate steps involving the ionization energy of Na and electron affinity of Cl. Let D be the bond-dissociation enthalpy of NaCl(g) into atoms so that

$$-D = I_1 - EA + V_0 \tag{13.8}$$

where I_1 is the first ionization energy of Na and EA is the electron affinity of Cl. Using data from Tables 11.4 and 11.5, we write

$$-D = 495.9 \text{ kJ mol}^{-1} - 349 \text{ kJ mol}^{-1} - 529.7 \text{ kJ mol}^{-1}$$

$$= -383 \text{ kJ mol}^{-1}$$

Therefore, the bond-dissociation enthalpy for NaCl into Na and Cl atoms is 383 kJ mol^{-1}. This value differs somewhat from the experimentally measured value (414 kJ mol^{-1}) because the repulsion term is inexact, and the actual bond has some covalent character.

13.3 Types of Intermolecular Forces

We are now ready to survey the different types of intermolecular forces. For the sake of completeness, we shall discuss interactions between molecules as well as between ions and molecules.

Dipole–Dipole Interaction

An intermolecular interaction of the *dipole–dipole* type occurs between polar molecules, which possess permanent dipole moments. Consider the electrostatic interaction between the two dipoles μ_A and μ_B separated by distance r. In extreme cases, these two dipoles can be aligned as shown in Figure 13.3. For the top example, the potential energy of interaction is given by

$$V = -\frac{2\mu_A \mu_B}{4\pi\varepsilon_0 r^3} \tag{13.9}$$

and for the bottom pair we have

$$V = -\frac{\mu_A \mu_B}{4\pi\varepsilon_0 r^3} \tag{13.10}$$

where the negative sign indicates that the interaction is attractive; that is, energy is released when these two molecules interact. Reversing the charge signs of one of the

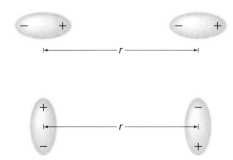

Figure 13.3
Schematic drawing showing the extreme orientations
of two permanent dipoles for attractive interaction.

dipoles makes V a positive quantity. Then the interaction between the two molecules
is repulsive.

Example 13.1

Two HCl molecules ($\mu = 1.08$ D) are separated by 4.0 Å (400 pm) in air. Calculate the
dipole–dipole interaction energy in kJ mol^{-1} if they are oriented end-to-end—that is,
H–Cl H–Cl.

ANSWER

We need Equation 13.9. The data are

$$\mu = 1.08 \text{ D} = 3.60 \times 10^{-30} \text{ C m} \qquad \text{(see Section 12.4)}$$
$$r = 4.0 \text{ Å} = 4.0 \times 10^{-10} \text{ m}$$

The potential energy due to this interaction is

$$V = -\frac{2(3.60 \times 10^{-30} \text{ C m})(3.60 \times 10^{-30} \text{ C m})}{4\pi(8.854 \times 10^{-12} \text{ C}^2 \text{ N}^{-1} \text{ m}^{-2})(4.0 \times 10^{-10} \text{ m})^3}$$
$$= -3.6 \times 10^{-21} \text{ N m}$$
$$= -3.6 \times 10^{-21} \text{ J}$$

To express the potential energy on a per mole basis, we write

$$V = (-3.6 \times 10^{-21} \text{ J})(6.022 \times 10^{23} \text{ mol}^{-1})$$
$$= -2.2 \text{ kJ mol}^{-1}$$

COMMENT

We have assumed that the dielectric constant of air is 1. In general, the dielectric
constant (ε) of the medium through which the dipoles interact appears in the
denominator of Equation 13.9 (see Section 5.7).

In a macroscopic system where all possible orientations of the dipoles are present, we might expect that the mean value of V would be zero, because there would
be as many repulsions as attractions. But even under conditions of free rotation in

a liquid or gaseous state, orientations giving rise to a lower potential energy are favored over those resulting in a higher potential energy, in accordance with the Boltzmann distribution law (see Chapter 2). A rather elaborate derivation shows that the average or net energy of interaction of permanent dipoles is given by

$$V = -\frac{2}{3}\frac{\mu_A^2\mu_B^2}{(4\pi\varepsilon_0)^2 r^6}\frac{1}{k_B T}$$ (13.11)

where k_B is Boltzmann's constant and T is absolute temperature. Note that V is inversely proportional to the sixth power of r, so that the energy of interaction falls off rapidly with distance. Also, V is inversely proportional to T, because at higher temperatures the average kinetic energy of the molecules is greater, a condition unfavorable to aligning dipoles for attractive interaction. In other words, the dipole–dipole interaction will gradually average out to zero with increasing temperature.

Ion–Dipole Interaction

The interaction between an ion and polar molecules was first discussed in Chapter 5 in relation to ionic hydration. The potential energy of interaction between an ion of charge q at a distance r from a dipole μ is given by

$$V = -\frac{q\mu}{4\pi\varepsilon_0 r^2}$$ (13.12)

Equation 13.12 holds only when the ion and the dipole lie along the same axis. This attractive interaction is mainly responsible for the dissolution of ionic compounds in polar solvents.

Example 13.2

A sodium ion (Na^+) is situated in air at a distance of $4.0\,\text{Å}$ (400 pm) from a HCl molecule with a dipole moment of 1.08 D. Use Equation 13.12 to calculate the potential energy of interaction in kJ mol^{-1}.

ANSWER

The data are

$$\mu = 1.08\,\text{D} = 3.60 \times 10^{-30}\,\text{C m}$$

$$r = 4.0\,\text{Å} = 4.0 \times 10^{-10}\,\text{m}$$

From Equation 13.12,

$$V = -\frac{(1.602 \times 10^{-19}\,\text{C})(3.60 \times 10^{-30}\,\text{C m})}{4\pi(8.854 \times 10^{-12}\,\text{C}^2\,\text{N}^{-1}\,\text{m}^{-2})(4.0 \times 10^{-10}\,\text{m})^2}$$

$$= -3.2 \times 10^{-20}\,\text{J}$$

$$= -19\,\text{kJ mol}^{-1}$$

Ion–Induced Dipole and Dipole–Induced Dipole Interactions

In a neutral nonpolar species such as the helium atom, the electron charge density is spherically symmetrical about the nucleus. If an electrically charged object, such as a positive ion, is brought near the helium atom, electrostatic interaction will cause a redistribution of the charge density (Figure 13.4). The atom will then acquire a dipole moment induced by the charged particle. The magnitude of the *induced dipole moment*, μ_{ind}, is directly proportional to the strength of the electric field, E:

$$\mu_{ind} \propto E$$
$$= \alpha' E \tag{13.13}$$

where α', the proportionality constant, is called the *polarizability*. The potential energy of interaction is given by the work done in bringing the helium atom from infinite distance ($E = 0$) to distance r ($E = E$); that is,

$$V = -\int_0^E \mu_{ind}\, dE$$
$$= -\int_0^E \alpha' E\, dE$$
$$= -\frac{1}{2}\alpha' E^2 \tag{13.14}$$

The electric field exerted by the ion of charge q on the atom is (see Appendix 5.1)

$$E = \frac{q}{4\pi\varepsilon_0 r^2}$$

Substituting the expression for E in Equation 13.14, we get

$$V = -\frac{1}{2}\frac{\alpha' q^2}{\left(4\pi\varepsilon_0\right)^2 r^4} \tag{13.15}$$

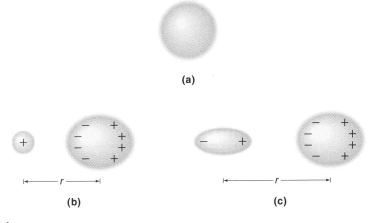

(a)

(b)

(c)

Figure 13.4
(a) An isolated helium atom has spherically symmetrical electron density. (b) Induced dipole moment in helium due to a cation. (c) Induced dipole moment in helium due to a permanent dipole. The plus and minus signs in helium represent shifts in electron density.

Qualitatively, polarizability measures how easily the electron density in an atom or a molecule can be distorted by an external electric field. Unsaturated bonds such as those in C=C, C=N, the nitro group ($-NO_2$), the phenyl group ($-C_6H_5$), the base pairs in DNA, and negative ions are highly polarizable groups. Generally, the larger the number of electrons and the more diffuse the electron charge cloud in the molecule, the greater its polarizability. As defined in Equation 13.13, however, α' has the rather awkward units of C m^2 V^{-1}. For this reason, using the polarizability α in units m^3 is more convenient, where

$$\alpha = \frac{\alpha'}{4\pi\varepsilon_0}$$

Equation 13.15 can now be expressed as

$$V = -\frac{1}{2}\frac{\alpha q^2}{4\pi\varepsilon_0 r^4} \qquad (13.16)$$

A permanent dipole can also induce a dipole moment in a nonpolar molecule (see Figure 13.4). The potential energy of interaction for the dipole–induced dipole interaction is given by

$$V = -\frac{\alpha'\mu^2}{(4\pi\varepsilon_0)^2 r^6} = -\frac{\alpha\mu^2}{4\pi\varepsilon_0 r^6} \qquad (13.17)$$

where α is the polarizability of the nonpolar molecule, and μ is the dipole moment of the polar molecule. Note that both Equations 13.16 and 13.17 are independent of temperature. This is so because the dipole moment can be induced instantaneously so that the value of V is unaffected by the thermal motion of the molecules. Table 13.1 lists the polarizability values of some atoms and simple molecules.

In general, both ion–induced dipole and dipole–induced dipole interactions are fairly weak compared with ion–dipole interactions. This is the reason that ionic compounds like NaCl and polar molecules like alcohols are not soluble in nonpolar solvents such as benzene or carbon tetrachloride.

Table 13.1
Polarizabilities of Some Atoms and Molecules

Atom	$\alpha/10^{-30}$ m^3	Molecule	$\alpha/10^{-30}$ m^3
He	0.20	H_2	0.80
Ne	0.40	N_2	1.74
Ar	1.66	CO_2	2.91
Kr	2.54	NH_3	2.26
Xe	4.15	CH_4	2.61
I	4.96	C_6H_6	10.4
Cs	42.0	CCl_4	11.7

Example 13.3

A sodium ion (Na^+) is situated in air at a distance of 4.0 Å (400 pm) from a nitrogen molecule. Use Equation 13.16 to calculate the potential energy of ion–induced dipole interaction in kJ mol^{-1}.

ANSWER

The data are (see Table 13.1)

$$\alpha(N_2) = 1.74 \times 10^{-30} \text{ m}^3$$

$$r = 4.0 \text{ Å} = 4.0 \times 10^{-10} \text{ m}$$

From Equation 13.16,

$$V = -\frac{1}{2} \frac{(1.74 \times 10^{-30} \text{ m}^3)(1.602 \times 10^{-19} \text{ C})^2}{4\pi(8.854 \times 10^{-12} \text{ C}^2 \text{ N}^{-1} \text{ m}^{-2})(4.0 \times 10^{-10} \text{ m})^4}$$

$$= -7.8 \times 10^{-21} \text{ J}$$

$$= -4.7 \text{ kJ mol}^{-1}$$

Dispersion, or London, Interactions

The cases considered thus far consist of at least one charged ion or one permanent dipole among the interacting species, and they can be satisfactorily treated by classical physics. We must now ask the following question: Because nonpolar gases such as helium and nitrogen can be condensed, what kind of attractive interaction exists between atoms and between nonpolar molecules?

When we speak of the spherical symmetry of the charge density in helium, we mean that averaged over a certain period of time (for example, a time long enough for us to carry out a physical measurement on the system), the electron density at a fixed distance away from the nucleus is the same in every direction. If we could take snapshots of the instantaneous configuration of each individual helium atom, we would most likely find varying degrees of deviation from spherical symmetry, owing to interactions among the atoms. Nevertheless, the temporary dipole created at every instant can induce a dipole in its neighboring atom(s), so an attractive interaction will result. We expect this interaction to be a weak attraction; indeed, the low boiling point of helium (4 K) suggests that very weak forces hold the atoms together in the liquid state. For molecules with large polarizabilities, however, this interaction can be comparable to or even greater than dipole–dipole and dipole–induced dipole interactions. For example, carbon tetrachloride (CCl_4), a nonpolar molecule, has a large polarizability (see Table 13.1) and a considerably higher boiling point ($76.5°C$) than methyl fluoride (CH_3F), a polar molecule ($-141.8°C$).

A quantum mechanical treatment for interactions between nonpolar molecules was given in 1930 by the German physicist Fritz London (1900–1954), who showed that the potential energy arising from the interaction of two identical atoms or nonpolar molecules is given by

$$V = -\frac{3}{4} \frac{\alpha'^2 I}{(4\pi\varepsilon_0)^2 r^6}$$

$$V = -\frac{3}{4} \frac{\alpha^2 I}{r^6} \tag{13.18}$$

where I is the first ionization energy of the atom or molecule. For unlike atoms or molecules A and B, Equation 13.18 becomes

$$V = -\frac{3}{2}\frac{I_A I_B}{I_A + I_B}\frac{\alpha'_A \alpha'_B}{(4\pi\varepsilon_0)^2 r^6}$$

$$V = -\frac{3}{2}\frac{I_A I_B}{I_A + I_B}\frac{\alpha_A \alpha_B}{r^6} \qquad (13.19)$$

The forces that arise from this kind of interaction are called *dispersion*, or *London*, *forces*.

Example 13.4

Calculate the potential energy of interaction between two argon atoms separated by 4.0 Å in air.

ANSWER

The data are

$$\alpha = 1.66 \times 10^{-30} \text{ m}^3 \quad \text{(Table 13.1)}$$
$$I = 1521 \text{ kJ mol}^{-1} \quad \text{(Table 11.4)}$$

From Equation 13.18,

$$V = -\frac{3}{4}\frac{(1.66 \times 10^{-30} \text{ m}^3)^2(1521 \text{ kJ mol}^{-1})}{(4.0 \times 10^{-10} \text{ m})^6}$$

$$= -0.77 \text{ kJ mol}^{-1}$$

Dipole–dipole, dipole–induced dipole, and dispersion forces are collectively called *van der Waals forces*. These forces are responsible for the deviation of gas behavior from ideality discussed in Chapter 2.

Repulsive and Total Interactions

In addition to the attractive forces discussed so far, atoms and molecules must repel one another; otherwise, they would eventually fuse. Fusion is prevented by strong repulsive forces between electron clouds and between nuclei.* The potential energy of repulsion is extremely short-range. It is proportional to $1/r^n$, where n is between 8 and 12. The British physicist Sir John Edward Lennard-Jones (1894–1954) proposed the following expression to represent the attractive and repulsive interactions in nonionic systems:

$$V = -\frac{A}{r^6} + \frac{B}{r^{12}} \qquad (13.20)$$

where A and B are constants for two interacting atoms or molecules. The first term in

*The repulsion between atoms or molecules is a direct consequence of the Pauli exclusion principle, which prevents electrons from sharing the same region in space.

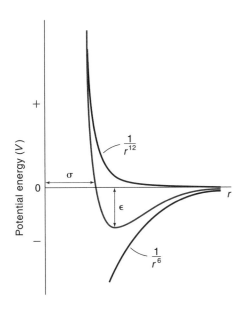

Figure 13.5
The potential energy curve between two molecules or two nonbonded atoms is the sum of the $1/r^6$ (attraction) and $1/r^{12}$ (repulsion) terms. The depth of the well is given by ε and σ gives the distance between the centers of the molecules at $V = 0$.

Equation 13.20 represents attraction. (As we have seen, the dipole–dipole, dipole–induced dipole, and dispersion interactions all have $1/r^6$ dependence.) The second term, which is very short-range (depends on $1/r^{12}$), describes repulsion between molecules. A more common form of Equation 13.20, called the *Lennard-Jones 6–12 potential*, is given by

$$V = 4\varepsilon\left[\left(\frac{\sigma}{r}\right)^{12} - \left(\frac{\sigma}{r}\right)^{6}\right] \tag{13.21}$$

Figure 13.5 shows the Lennard-Jones potential between two molecules. For a given pair of molecules, the quantity ε measures the depth of the potential well, and σ is the separation at which $V = 0$. Table 13.2 lists ε and σ values for a few atoms and molecules. Although the Lennard-Jones potential has been used extensively in calculations, the $1/r^{12}$ term is a poor representation of the repulsive potential. For accurate work, a more satisfactory term is $e^{-ar/\sigma}$, where a is a constant.

Several points are worth noting about the total interaction potential. First, the second virial coefficient B (see Section 2.4) can be related to the intermolecular potential. According to Equation 2.14, this coefficient can be measured experimentally from the slope of a plot of Z against P (if we ignore the third and higher virial coefficients). Therefore, knowledge of B enables us to determine the intermolecular potential. Second, the value of σ gives a measure of how closely two nonbonded atoms

Table 13.2
Lennard-Jones Parameters for Atoms and Molecules

Particle	$\varepsilon/\text{kJ} \cdot \text{mol}^{-1}$	$\sigma/\text{Å}$
Ar	0.997	3.40
Xe	1.77	4.10
H_2	0.307	2.93
N_2	0.765	3.92
O_2	0.943	3.65
CO_2	1.65	4.33
CH_4	1.23	3.82
C_6H_6	2.02	8.60

Table 13.3
van der Waals Radii of Atoms
and the Methyl Group

Atom	Radius/Å
H	1.2
C	1.5
N	1.5
O	1.4
P	1.9
S	1.85
F	1.35
Cl	1.80
Br	1.95
I	2.2
$-CH_3$	2.0

can approach each other. Called the *van der Waals radius*, this measure is one-half the internuclear distance between two nonbonded atoms in a crystal. For example, in solid Cl_2, the average distance between two adjacent, nonbonded Cl atoms is 3.60 Å, so the van der Waals radius of Cl is taken to be 3.60 Å/2, or 1.80 Å. This value is considerably larger than the covalent radius of Cl, which is 1.01 Å. Table 13.3 lists the van der Waals radii of several atoms and the methyl group. Note that the size of atoms in space-filling models is based on their van der Waals radii. Third, we can make an interesting comparison of the relative magnitudes of intermolecular potential with intramolecular potential. Consider the interaction between a pair of H_2 molecules and the potential energy of a H_2 molecule. In the former case, the depth of the potential well is about 0.3 kJ mol^{-1}, and the "bond length" is 3.4 Å (340 pm). The corresponding quantities for H_2 are 432 kJ mol^{-1} and 0.74 Å (74 pm), respectively (see Figure 12.14). Thus, the stability of a normal chemical bond is 2 to 3 orders of magnitude greater than that of species held together by intermolecular forces, and bonded atoms are much closer in a molecule.

The Role of Dispersion Forces in Sickle-Cell Anemia

Dispersion forces play an important role in protein structures. In lipoprotein membranes, the interaction between hydrocarbon tails of lipid molecules is mainly due to dispersion and other van der Waals forces. Dispersion forces are partly responsible for holding the heme group in the "pocket" formed by the side chains in hemoglobin and myoglobin.

As mentioned in Chapter 1, the replacement of glutamic acid by valine in the sixth position in each of the β chains in hemoglobin (called hemoglobin S) causes the disease known as sickle-cell anemia. The nonpolar alkyl groups in the valine residues on the surface can fit into a nonpolar pocket in the deoxygenated form of another hemoglobin molecule through dispersion forces. Because there is a conformational change in oxyhemoglobin, the pocket is sufficiently altered and no such interaction occurs. The hemoglobin concentration in the erythrocytes is very high, about 350 mg mL^{-1}, and the average distance between individual hemoglobin molecules is only 10 Å. The proximity of the molecules facilitates the hydrophobic interaction among the deoxgenated hemoglobin molecules. Eventually, the aggregated hemoglobin S molecules, in the form of fibers, precipitate out of solution. The precipitate causes the normally disc-shaped red blood cells to assume a warped crescent or sickle shape (Figure 13.6). These deformed cells clog the narrow capillaries, restricting

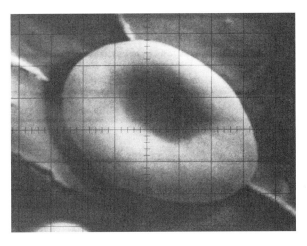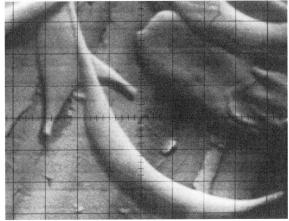

Figure 13.6
Electron micrographs showing a normal red blood cell (left) and a sickled red blood cell (right). (Courtesy of Phillips Electronic Instruments, Inc.)

blood flow to vital organs. The usual symptoms of sickle-cell anemia are swelling, severe pain, and other complications. Sickle-cell anemia was described by Pauling as a molecular disease because the destructive action is understood at the molecular level and the disease is, in effect, due to a molecular defect.*

Despite the intensive research effort currently underway, no cure for this disease in known. Treatment has been based largely on antisickling agents such as urea and the cyanate ion:

$$H_2N-\underset{\underset{O}{\|}}{C}-NH_2 \qquad O=C=N^-$$

Urea   Cyanate ion

The intravenous administration of urea in 10% sugar solution has met with limited success. The cyanate ion is a more effective antisickling agent, but it has rather serious toxic side effects. Another promising therapeutic agent is hydroxyurea:

$$HO-NH-\underset{\underset{O}{\|}}{C}-NH_2$$

Hydroxyurea

These molecules break up the dispersion forces between different hemoglobin S molecules to reverse the sickling of the red blood cells.

---

* As an aside, it is interesting to note how the sickle-cell anemia disease gives increased resistance to malaria. The malarial parasite spends part of its life cycle within the erythrocyte of its host. The presence of parasites lowers the pH of the intracellular fluid slightly, making the cell more prone to sickling. (The Bohr effect discussed on p. 386 shows that the formation of deoxyhemoglobin is favored at low pHs.) When sickling occurs, the cell becomes more permeable to K^+ ions, which leak out into the surroundings. This depletion in K^+ ion concentration eventually kills the malarial parasite. This intriguing mechanism explains the survival of heterozygotes (that is, people who have received a defective gene for hemoglobin S from one parent but a normal gene from the other parent) exposed to malaria, especially in Africa.

13.4 Hydrogen Bonding

Table 13.4 summarizes different types of intermolecular interactions, including hydrogen bonding, which we shall discuss here. The hydrogen bond is a special type of

Table 13.4
Interactions Between Molecules

Type of Interaction	Distance Dependence	Example	Order of Magnitude (kJ mol^{-1})[a]
Covalent bond[b]	No simple expression	H–H	200–800
Ion–ion	$\dfrac{q_A q_B}{4\pi\varepsilon_0 r}$	Na$^+$Cl$^-$	40–400
Ion–dipole	$\dfrac{q\mu}{4\pi\varepsilon_0 r^2}$	Na$^+$(H$_2$O)$_n$	5–60
Dipole–dipole	$\dfrac{2}{3}\dfrac{\mu_A^2 \mu_B^2}{(4\pi\varepsilon_0)^2 r^6}\dfrac{1}{k_B T}$	SO$_2$ SO$_2$	0.5–15
Ion–induced dipole	$\dfrac{1}{2}\dfrac{\alpha q^2}{4\pi\varepsilon_0 r^4}$	Na$^+$ C$_6$H$_6$	0.4–4
Dipole–induced dipole	$\dfrac{\alpha\mu^2}{4\pi\varepsilon_0 r^6}$	HCl C$_6$H$_6$	0.4–4
Dispersion	$\dfrac{3}{4}\dfrac{\alpha^2 I}{r^6}$	CH$_4$ CH$_4$	4–40
Hydrogen bond	No simple expression	H$_2$O$\cdots$H$_2$O	4–40

[a] The actual value depends on distance of separation, charge, dipole moment, polarizability, and the dielectric constant of the medium.
[b] This is listed for comparison purposes only.

interaction between molecules; it forms whenever a polar bond containing the hydrogen atom (for example, O–H or N–H) interacts with an electronegative atom such as oxygen, nitrogen, or fluorine. This interaction is represented as A–H$\cdots$B, where A and B are the electronegative atoms and the dotted line denotes the hydrogen bond.* Figure 13.7 shows several examples of hydrogen bonding. Although hy-

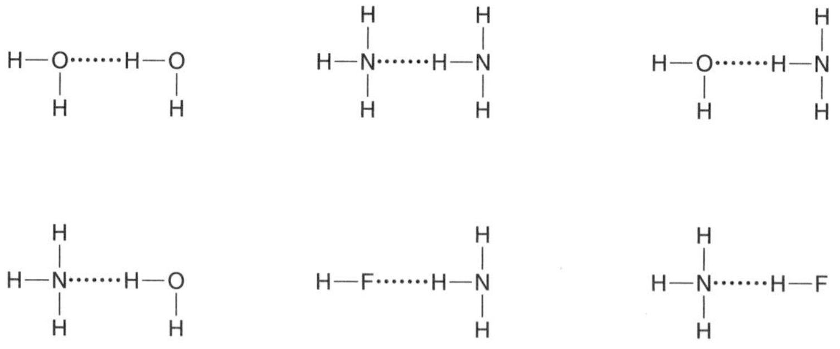

Figure 13.7
Some examples of hydrogen bonding. The dotted red lines represent hydrogen bonds.

* Detailed X-ray study of ice shows that the O$\cdots$H hydrogen bond, and presumably other types of strong hydrogen bonds as well, has considerable covalent character. This conclusion is supported by quantum mechanical calculations. See E. D. Isaacs, A. Shukla, P. M. Platzmann, et al., *Phys. Rev. Lett.* **82**, 600 (1999). Also see *Science* **283**, 614 (1999).

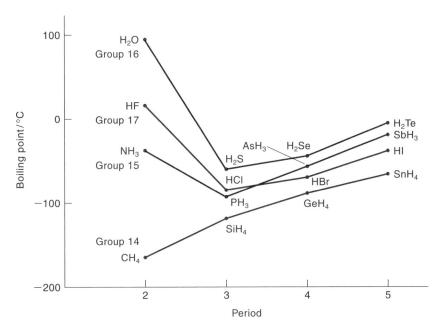

Figure 13.8
Boiling points of the hydrogen compounds of Groups 14, 15, 16, and 17 elements. Although normally we expect the boiling point to increase as we move down a group, we see that NH_3, H_2O, and HF behave differently, as a result of intermolecular hydrogen bonding.

drogen bonds are relatively weak (about 40 kJ mol^{-1} or less), they play a central role in determining the properties of many compounds.

Early evidence of hydrogen bonding came from the study of the boiling points of compounds. Normally, the boiling points of a series of similar compounds containing elements in the same periodic group increase with increasing molar mass (and hence increasing polarizability). But, as Figure 13.8 shows, the binary hydrogen compounds of the elements in Groups 15, 16, and 17 do not follow this trend. In each of these series, the lightest compounds (NH_3, H_2O, HF) have the *highest* boiling points, because there is extensive hydrogen bonding between molecules in these compounds.

This type of bonding is unique to hydrogen primarily because the hydrogen atom has only one electron. When that electron is used to form a covalent bond with an electronegative atom, the hydrogen nucleus becomes partially unshielded. Consequently, its proton can interact directly with another electronegative atom on a different molecule. Depending on the strength of the interaction, such bonding can exist in the gas phase as well as in the solid and liquid phases. In solids and liquids, HF forms a polymeric chain as follows:

For maximum stability, the donor pair (AH) and the acceptor (B) are usually co-linear (that is, $\angle AHB = 180°$), but deviations of up to 30° are known.

Hydrogen bonding is largely responsible for the stability of protein conformations. Intramolecular hydrogen bonds between the $>C=O$ and $>N-H$ groups of a polypeptide chain result in the α helix. On the other hand, intermolecular hydrogen bonds between two polypeptide chains account for β pleated-sheet structures. We shall postpone the discussion of these structures until Chapter 16. Here, let us consider the importance of hydrogen bonding in DNA.

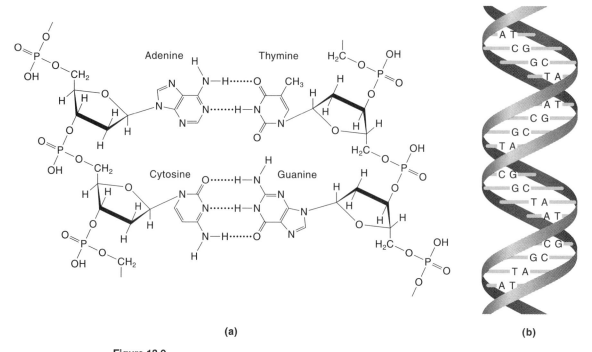

Figure 13.9
(a) Base-pair formation between adenine (A) and thymine (T) and between cytosine (C) and guanine. (b) The most common structure of DNA, which is a right-handed double helix. The two strands are held together by hydrogen bonds and other intermolecular forces.

DNA molecules are polymers that have molar masses in the millions to tens of billions of grams. They consist of three parts: phosphate groups, sugar groups (deoxyribose), and purine and pyrimidine bases (adenine, cytosine, guanine, and thymine). Figure 13.9 shows the Watson–Crick model of the DNA double helix, after the American biologist James Dewey Watson (1928–) and the British biologist Francis Harry Compton Crick (1916–2004). The molecule's backbone contains alternating sugar and phosphate residues. Each sugar residue is attached to a purine or pyrimidine base. Hydrogen bonds form between bases on two strands of DNA, generating the double-helical structure. The bases are roughly perpendicular to the axis of the helix; each one can form a strong hydrogen bond with only one of the four bases available. This specificity of base pairing gives the DNA structure the stability required for its function as the storage site for the genetic code.

Energetically, the most favorable pairings in DNA molecules are adenine (A) to thymine (T) and cytosine (C) to guanine (G), and as Figure 13.9b shows, the two strands are complementary. Although the amount of energy required to break a hydrogen bond is rather small (about 5 kJ mol^{-1}), the double-helical structure of DNA is stable under normal physiological conditions. The stability of the molecule rests on the cooperative nature of hydrogen bond formation. Consider the pairing of two nucleotides, C and G, in solution at room temperature. The ratio of free bases to hydrogen-bonded base pair can be calculated from the Boltzmann distribution law:

The cytosine–guanine base pair has three hydrogen bonds.

$$\frac{(C, G)_{\text{free bases}}}{(C-G)_{\text{base pair}}} = e^{-\Delta E/RT} = \exp\left(\frac{-3 \times 5000 \text{ J mol}^{-1}}{8.314 \text{ J K}^{-1} \text{ mol}^{-1} \times 300 \text{ K}}\right)$$

$$= 0.00244$$

Thus, there are 409 pairs of hydrogen-bonded bases to one pair of free bases. For dinucleotides in which the strands are made of two cytosines and guanines, respectively, the complexed form is favored over free bases by a factor of 409×409, or 1.67×10^5. Clearly, then, in a polynucleotide containing thousands of bases, equilibrium overwhelmingly favors the hydrogen-bonded structure.

So far, our discussion of hydrogen bond formation has focused on the very electronegative atoms N, O, and F. Ample evidence suggests that hydrogen bonds also exist in compounds not containing these atoms. Shown below are examples of "weak" hydrogen bonds, so-called because of their diminished magnitude compared with hydrogen bonds in H_2O, NH_3, or HF.

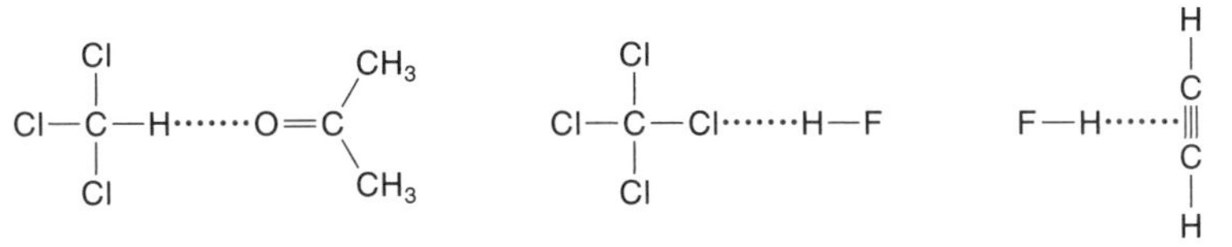

Interestingly, the electron-rich triple bond in acetylene can form a hydrogen bond with hydrogen fluoride. Similarly, the delocalized π electrons in benzene can form weak hydrogen bonds.

13.5 The Structure and Properties of Water

Water is so common a substance that we often overlook its unique properties. For example, given its molar mass, water should be a gas at room temperature, but due to hydrogen bonding, it has a boiling point of 373.15 K at 1 atm. In this section, we shall study the structure of ice and liquid water and consider some biologically significant aspects of water.

The Structure of Ice

To understand the behavior of water, we must first investigate the structure of ice. There are nine known crystalline forms of ice; most of them stable only at high pressures. Ice I, the familiar form, has been studied thoroughly. It has a density of 0.924 g mL^{-1} at 273 K and 1 atm pressure.

There is a significant difference between H_2O and other polar molecules, such as NH_3 and HF. The number of hydrogen atoms in a water molecule that can form the positive ends of hydrogen bonds is equal to the number of lone pairs on the oxygen atom that can form the negative ends:

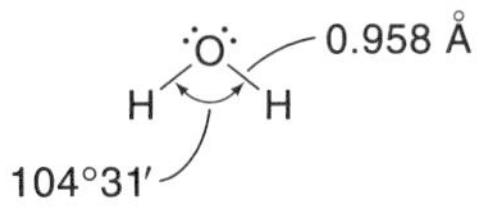

The result is an extensive three-dimensional network in which each oxygen atom is bonded tetrahedrally to four hydrogen atoms by means of two covalent bonds and two hydrogen bonds. This equality in number of protons and lone pairs is not characteristic of NH_3 and HF. Consequently, NH_3 and HF can form only rings or chains and not an extensive three-dimensional structure.

Figure 13.10 shows the structure of ice I. The distance between adjacent oxygen atoms is 2.76 Å. The O–H distance is between 0.96 Å and 1.02 Å, and the O$\cdots$H distance is between 1.74 Å and 1.80 Å. Because of its open lattice, ice has a lower density than water, a fact that has profound ecological significance. Were it not for this unique type of hydrogen bonding, ice, like most other solid substances, would be

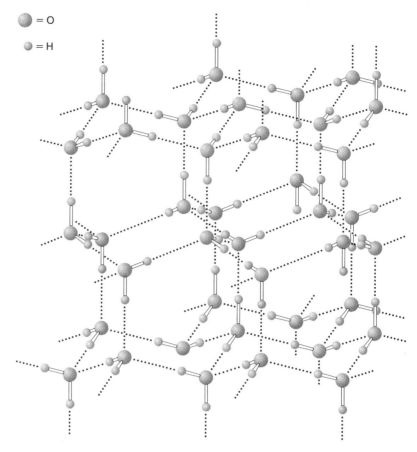

Figure 13.10
Structure of ice. The red dotted lines represent hydrogen bonds.

denser than the corresponding liquid. On freezing, it would sink to the bottom of a lake or pond, causing all the water to freeze gradually and killing most live organisms within it. Fortunately, water reaches its maximum density at 277.15 K (4 K above freezing). Cooling below 277.15 K decreases the density of water and allows it to rise to the surface, where freezing occurs. An ice layer formed on the surface does not sink; just as important, it acts as a thermal insulator to protect the biological environment beneath it.

The Structure of Water

Although using the word *structure* may seem strange when discussing liquids, most liquids possess short-range order. A convenient way to study the structure of liquids is to use the *radial distribution function*, $g(r)$. This function is defined so that $4\pi r^2 g(r)dr$ gives the probability that a molecule will be found in a spherical shell of width dr at distance r from the center of another molecule.* For a crystalline solid, a plot of $g(r)$ versus r gives a series of sharp lines because crystals have long-range order. In contrast, as Figure 13.11 shows, the radial distribution curve for liquid water at 4°C produces a major peak at 2.90 Å, with weaker peaks at 3.50 Å, 4.50 Å,

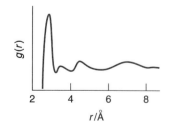

Figure 13.11
Experimental radial distribution curve for water at 4°C. The peaks become broader at higher temperatures.

---

* This radial distribution function is similar to the one applied to the hydrogen atom in Chapter 11 (see p. 429). The distribution function can be constructed from the intensity of the X-ray diffraction patterns of the liquid.

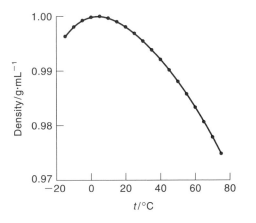

Figure 13.12
Plot of density versus temperature for liquid water. The maximum density of water is reached at 4°C. The density of ice at 0°C is 0.9167 g cm^{-3}.

and 7.00 Å. Beyond 7.00 Å, the function is essentially zero, meaning that the local order does not extend beyond this distance. X-ray diffraction studies of ice I show that the O–O distance is 2.76 Å. The strong peak at 2.90 Å suggests a very similar tetrahedral arrangement in the liquid. The peak at 3.50 Å does not correspond to any bond distance in ice I, which does, however, have interstitial sites at a distance of 3.50 Å from each O atom. Therefore, when ice melts, some of the water molecules break loose and become trapped in these interstitial sites, which are responsible for the peak at 3.50 Å. The peaks at 4.50 Å and 7.00 Å are also consistent with the tetrahedral arrangement.

The above discussion suggests that the extensive three-dimensional hydrogen-bonded structure that characterizes ice I is largely intact in water, although the bonds may become bent and distorted. On melting, monomeric water molecules occupy holes in the remaining "icelike" lattice, causing the density of water to be greater than that of ice. As temperature increases, more hydrogen bonds are broken, but at the same time the kinetic energy of molecules increases. Consequently, more water molecules are trapped, but the elevated kinetic energy decreases the density of water because the molecules occupy a greater volume. Initially, the trapping of monomeric water molecules outweighs the expansion in volume due to the increase in kinetic energy, so the density rises from 0°C to 4°C. Beyond this temperature, the expansion predominates, so the density decreases with increasing temperature (Figure 13.12).

Some Physiochemical Properties of Water

Table 13.5 lists some important physiochemical properties of water. The abnormally high values of several of its properties make water a unique solvent, particularly suited to the support of living systems. The reasons are briefly discussed below.

1. Water has one of the highest dielectric constants of all liquids (see Table 5.3). This property makes water an excellent solvent for ionic compounds. In addition, its ability to form hydrogen bonds enables it to dissolve carbohydrates, carboxylic acids, and amines.

2. Due to its extensive hydrogen bonds, water has a very high heat capacity. Because $\Delta H = C_P \Delta T$, $\Delta T = \Delta H / C_P$, which means that a large amount of heat is needed to raise the temperature of an aqueous solution by one kelvin. This property is important in regulating the temperature of a cell from the heat generated by metabolic processes. From an environmental point of view, the ability of water to absorb a lot of heat with little

Table 13.5
Some Physiochemical Properties of Water

Melting point	0°C (273.15 K)
Boiling point	100°C (373.15 K)
Density of water	0.99987 g mL^{-1} at 0°C
Density of ice	0.9167 g mL^{-1} at 0°C
Molar heat capacity	75.3 J K^{-1} mol^{-1}
Molar heat of fusion	6.01 kJ mol^{-1}
Molar heat of vaporization	40.79 kJ mol^{-1} at 100°C
Dielectric constant	78.54 at 25°C
Dipole moment	1.82 D
Viscosity	0.001 N s m^{-2}
Surface tension	0.07275 N m^{-1} at 20°C
Diffusion coefficient	2.4×10^{-9} m^2 s^{-1} at 25°C

temperature rise greatly influences Earth's climate. Our lakes and oceans absorb solar radiation or give up large amounts of heat with only a small change in temperature. For this reason, the climate close to the ocean is more moderate than that inland.

3. Water has a high molar heat of vaporization (41 kJ mol^{-1}). Thus, sweating is an effective way to regulate body temperature. On average, a 60-kg person generates 1×10^7 J of heat daily from metabolism. If sweating were the only mechanism for cooling, then one would need to vaporize $(1 \times 10^7$ J/41 × 1000 J mol$^{-1}) = 244$ mol, or nearly 4.4 liters of water, to maintain a constant temperature. Normally a person does not sweat this much (unless, for example, she is training for a marathon race in Houston, Texas, in mid-July). Part of the excess heat is radiated to the cooler surroundings. Although water's molar heat of fusion is not unusually high (6 kJ mol^{-1}), it is still sizable and helps protect the body against freezing.

4. The ecological significance of water's higher density than ice has been discussed.

5. Again, strong intermolecular interaction due to hydrogen bonding gives water a high surface tension. This property forces biological organisms to produce detergentlike compounds (called surfactants) to lower surface tension that would otherwise inhibit certain functions. For example, lung surfactants are needed to decrease the work required to open alveolar spaces and allow efficient respiration to take place.

6. The viscosity of water, unlike many of its properties, is comparable to that of many other liquids. Because the presence of macromolecules (proteins and nucleic acids, for example) appreciably increases the viscosity of a solution, the fact that water is not a viscous fluid facilitates blood flow and the diffusion of molecules and ions in the medium.

13.6 Hydrophobic Interaction

Experience tells us that oil and water do not mix. At first glance, the reason seems to be that the dipole–induced dipole and dispersion forces between water and nonpolar oil molecules are weak. From this observation, we might conclude that the enthalpy

Table 13.6
Thermodynamic Quantities for the Transfer of Nonpolar Solutes from Organic Solvents to Water at 25°C

Process	$\Delta H / \text{kJ} \cdot \text{mol}^{-1}$	$\Delta S / \text{J} \cdot \text{K}^{-1} \cdot \text{mol}^{-1}$	$\Delta G / \text{kJ} \cdot \text{mol}^{-1}$
$CH_4(CCl_4) \rightarrow CH_4(H_2O)$	-10.5	-75.8	12.1
$CH_4(C_6H_6) \rightarrow CH_4(H_2O)$	-11.7	-75.8	10.9
$C_2H_6(C_6H_6) \rightarrow C_2H_6(H_2O)$	-9.2	-83.6	15.7
$C_6H_{14}(C_6H_{14}) \rightarrow C_6H_{14}(H_2O)$	0.0	-95.3	28.4
$C_6H_6(C_6H_6) \rightarrow C_6H_6(H_2O)$	0.0	-57.7	17.2

of mixing (ΔH) is positive, which causes ΔG to be positive $(\Delta G = \Delta H - T\Delta S)$. Thus, the solubility of oil in water is very low. But this explanation is incorrect. The unfavorable interaction is primarily due to the *hydrophobic interaction* (also called the *hydrophobic effect* or *hydrophobic bond*), a term coined by the American chemist Walter Kauzmann (1916–). Hydrophobic interaction describes the influences that cause nonpolar substances to cluster together to minimize their contacts with water. This interaction forms the basis for many important chemical and biological phenomena, including the cleaning action of soaps and detergents, the formation of biological membranes, and the stabilization of protein structure.

Table 13.6 shows the thermodynamic quantities for the transfer of small nonpolar molecules from nonpolar solvents to water. The most striking feature of this table is that ΔS is negative for all of the compounds. When nonpolar molecules enter the aqueous medium, some hydrogen bonds must be broken to make room or create a cavity for the solutes. This part of the interaction is endothermic because the broken hydrogen bonds are much stronger than the dipole–induced dipole and dispersion interactions. Each solute molecule is now trapped in an icelike cage structure, referred to as the clathrate cage model, which consists of a specific number of water molecules held together by hydrogen bonds (Figure 13.13). The formation of the clathrate has two important consequences. First, the newly formed hydrogen bonds (an exothermic process) can partly or totally compensate for the hydrogen bonds that were broken initially to make the cavity. This explains why ΔH could be negative, zero, or positive for the overall process. Furthermore, because the cage structure is highly ordered (a decrease in the number of microstates), there is an appreciable decrease in entropy, which far outweighs the increase in entropy due to the mixing of solute and water molecules, so that ΔS is negative. Thus, the immiscibility of nonpolar molecules and water, or the hydrophobic interaction, is entropy driven rather than enthalpy driven.*

Hydrophobic interaction has a profound effect on the structure of proteins. When the polypeptide chain of a protein folds into a three-dimensional structure in solution, the nonpolar amino acids (for example, alanine, phenylalanine, proline, tryptophan, and valine) are in the interior of the macromolecule and have little or no contact with water, while the polar amino acid residues (such as arginine, aspartic acid, glutamic acid, and lysine) are on the exterior. An insight into this entropy-driven process can be gained by considering just two nonpolar molecules in aqueous

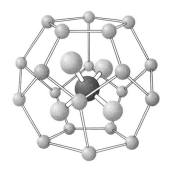

Figure 13.13
A probable structure for methane in water, called methane hydrate. The methane molecule is trapped in a cage of water molecules (gray spheres) held together by hydrogen bonds.

---

* Comparing the solubility of nonpolar molecules with ionic compounds in water is instructive. In the latter, there is a large decrease in enthalpy $(\Delta H < 0)$ due to the strong ion–dipole interaction, which outweighs the decrease in entropy $(\Delta S < 0)$ when water becomes more organized around the charged ions, so that $\Delta G < 0$. Note that the structure of the hydration sphere surrounding ions is different from the clathrate cage structure.

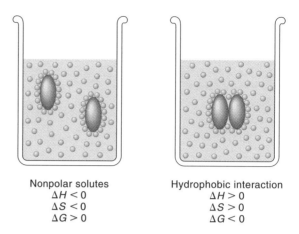

Nonpolar solutes
$\Delta H < 0$
$\Delta S < 0$
$\Delta G > 0$

Hydrophobic interaction
$\Delta H > 0$
$\Delta S > 0$
$\Delta G < 0$

Figure 13.14
Left: The dissolution of nonpolar molecules in water is unfavorable because of the large decrease in entropy resulting from clathrate formation, even though the process is exothermic ($\Delta H < 0$). Consequently, $\Delta G > 0$. Right: As a result of hydrophobic interaction, the nonpolar molecules come together, releasing some of the ordered water molecules in the clathrate structure and thus increasing entropy. This is a thermodynamically favorable process ($\Delta G < 0$), even though it is endothermic ($\Delta H > 0$) because more hydrogen bonds are broken than are made.

solution (Figure 13.14). The hydrophobic interaction causes the nonpolar molecules to come together into a single cavity to reduce the unfavorable interactions with water by decreasing the surface area. This response destroys part of the cage structure, resulting in an increase in ΔS and hence a decrease in ΔG. Moreover, enthalpy increases ($\Delta H > 0$) because some of the hydrogen bonds in the original cage structures are broken. Similarly, the folding of a protein is an example of this phenomenon because it minimizes the exposure of nonpolar surfaces to water. More will be said of the stability of proteins in Chapter 16.

Suggestions for Further Reading

BOOKS

Eisenberg, D. and W. Kauzmann, *The Structure and Properties of Water*, Oxford University Press, New York, 1969.

Franks, F., *Water: A Matrix of Life*, Cambridge University Press, Cambridge, 2000.

Jeffrey, G. A., *An Introduction to Hydrogen Bonding*, Oxford University Press, New York, 1997.

Jeffrey, G. A. and W. Saenger, *Hydrogen Bonding in Biological Structures*, Springer-Verlag, New York, 1994.

Kavanau, J. L., *Water and Water-Solute Interactions*, Holden-Day, San Francisco, 1964.

Pimentel, G. C. and A. L. McClellan, *The Hydrogen Bond*, W. H. Freeman, San Francisco, 1960.

Rigby, M., E. B. Smith, W. A. Wakeham, and G. C. Maitland, *The Forces Between Molecules*, Clarendon Press, Oxford, 1986.

Vinogrador, S. N. and R. H. Linnell, *Hydrogen Bonding*, Van Nostrand Reinhold, New York, 1971.

ARTICLES
General

"The Force between Molecules," B. V. Derjaguin, *Sci. Am.* July 1960.

"The Human Thermostat," T. H. Benzinger, *Sci. Am.* January 1961.

"A Molecular Theory of General Anesthesia," L. Pauling, *Science* **134**, 15 (1961).

"Inclusion Compounds," J. F. Brown, Jr., *Sci. Am.* July 1962.

"Clathrates: Compounds in Cages," M. M. Hagan, *J. Chem. Educ.* **40**, 643 (1963).

"Chemical Forces," H. H. Jaffé, *J. Chem. Educ.* **40**, 649 (1963).

"Early Views on Forces between Atoms," L. Holliday, *Sci. Am.* May 1970.

"The Role of van der Waals Forces in Surface and Colloid Chemistry," P. C. Hiemenz, *J. Chem. Educ.* **49**, 164 (1972).

"Why Does a Stream of Water Deflect in an Electric Field?" G. K. Vemulapalli and S. G. Kukolich, *J. Chem. Educ.* **73**, 887 (1996).

Sickle-Cell Anemia
"Cyanate and Sickle-Cell Disease," A. Cerami and C. M. Peterson, *Sci. Am.* April 1975.
"Non-Covalent Interactions," E. Frieden, *J. Chem. Educ.* **52**, 754 (1975).
"Sickle-Cell Anemia: Molecular and Vellular Bases of Therapeutic Approaches," J. Dean and A. N. Schechter, *New Engl. J. Med.* **299**, 752, 804, 863 (1978).
"Rational Approaches to Chemotherapy: Antisickling Agents," I. M. Klotz, D. N. Havey, and L. C. King, *Science* **213**, 724 (1981).

The Hydrogen Bond/Water Structure
"On Hydrogen Bonds," J. Donohue, *J. Chem. Educ.* **40**, 598 (1963).
"Ice," L. K. Runnels, *Sci. Am.* December 1966.
"The Significance of Hydrogen Bonds in Biological Structure," A. L. McClellan, *J. Chem. Educ.* **44**, 547 (1967).

"The Structure of Ordinary Water," H. S. Frank, *Science* **169**, 635 (1970).
"Hydrogen Bonding and Proton Transfer," M. D. Joesten, *J. Chem. Educ.* **59**, 362 (1982).
"Water Clusters," K. Liu, J. D. Cruzan, and R. J. Saykally, *Science* **271**, 929 (1996).
"Simulating Water and the Molecules of Life," M. Gerstein and M. Levitt, *Sci. Am.* November 1998.
"Hydrogen Bonds Involving Transition Metal Centers Acting As Proton Acceptors," A. Martin, *J. Chem. Educ.* **76**, 578 (1999).
"More Hydrogen Bonds for the (Structural) Biologists," M. S. Weiss, M. Brandl, J. Sühnel, D. Pal, and R. Hilgenfeld, *Trends Biochem. Sci.* **26**, 521 (2001).

Hydrophobic Interactions
"Hydrophobic Interactions," G. Némethy, *Angew. Chem. Intl. Ed.* **6**, 195 (1967).
"The Hydrophobic Effect," E. M. Huque, *J. Chem. Educ.* **66**, 581 (1989).
"How Protein Chemists Learned About the Hydrophobic Effect," C. Tanford, *Protein Science* **6**, 1358 (1997).
"The Real Reasons Why Oil and Water Don't Mix," T. P. Silverstein, *J. Chem. Educ.* **75**, 116 (1998).

Problems

Intermolecular Forces

13.1 List all the intermolecular interactions that take place in each of the following kinds of molecules: Xe, SO_2, C_6H_5F, and LiF.

13.2 Arrange the following species in order of decreasing melting points: Ne, KF, C_2H_6, MgO, H_2S.

13.3 The compounds Br_2 and ICl have the same number of electrons, yet Br_2 melts at $-7.2°C$, whereas ICl melts at $27.2°C$. Explain.

13.4 If you lived in Alaska, which of the following natural gases would you keep in an outdoor storage tank in winter: methane (CH_4), propane (C_3H_8), or butane (C_4H_{10})? Explain.

13.5 List the types of intermolecular forces that exist between molecules (or basic units) in each of the following species: **(a)** benzene (C_6H_6), **(b)** CH_3Cl, **(c)** PF_3, **(d)** NaCl, **(e)** CS_2.

13.6 The boiling points of the three different structural isomers of pentane (C_5H_{12}) are $9.5°C$, $27.9°C$, and $36.1°C$. Draw their structures, and arrange them in order of decreasing boiling points. Justify your arrangement.

13.7 Two water molecules are separated by 2.76 Å in air. Use Equation 13.9 to calculate the dipole–dipole interaction. The dipole moment of water is 1.82 D.

13.8 Coulombic forces are usually referred to as long-range forces (they depend on $1/r^2$), whereas van der Waals forces are called short-range forces (they depend on $1/r^7$). **(a)** Assuming that the forces (F) depend only on distances, plot F as a function of r at $r = 1$ Å, 2 Å, 3 Å, 4 Å, and 5 Å. **(b)** Based on your results, explain the fact that although a $0.2\,M$ nonelectrolyte solution usually behaves ideally, nonideal behavior is quite noticeable in a $0.02\,M$ electrolyte solution.

13.9 Calculate the induced dipole moment of I_2 due to a Na^+ ion that is 5.0 Å away from the center of the I_2 molecule. The polarizability of I_2 is 12.5×10^{-30} m^3.

13.10 Differentiate Equation 13.21 with respect to r to obtain an expression for σ and ε. Express the equilibrium distance, r_e, in terms of σ, and show that $V = -\varepsilon$.

13.11 Calculate the bond enthalpy of LiF using the Born–Haber cycle. The bond length of LiF is 1.51 Å. See Tables 11.4 and 11.5 for other information. Use $n = 10$ in Equation 13.7.

13.12 **(a)** From the data in Table 13.2, determine the van der Waals radius for argon. **(b)** Use this radius to determine the fraction of the volume occupied by 1 mole of argon at $25°C$ and 1 atm.

Hydrogen Bonding

13.13 Diethyl ether ($C_2H_5OC_2H_5$) has a boiling point of 34.5°C, whereas 1-butanol (C_4H_9OH) boils at 117°C. These two compounds have the same type and number of atoms. Explain the difference in their boiling points.

13.14 If water were a linear molecule, **(a)** would it still be polar and **(b)** would the water molecules still be able to form hydrogen bonds with one another?

13.15 Which of the following compounds is a stronger base: $(CH_3)_4NOH$ or $(CH_3)_3NHOH$? Explain.

13.16 Explain why ammonia is soluble in water but nitrogen trichloride is not.

13.17 Acetic acid is miscible with water, but it also dissolves in nonpolar solvents such as benzene or carbon tetrachloride. Explain.

13.18 Which of the following molecules has a higher melting point? Explain your answer.

13.19 What type of chemical analysis is needed to test the A–T and C–G pairing in DNA?

13.20 Assume the energy of hydrogen bonds per base pair to be 10 kJ mol^{-1}. Given two complementary strands of DNA containing 100 base pairs each, calculate the ratio of two separate strands to hydrogen-bonded double helix in solution at 300 K.

Additional Problems

13.21 The term "like dissolves like" has often been used to describe solubility. Explain what it means.

13.22 List all the intra- and intermolecular forces that could exist between hemoglobin molecules in water.

13.23 A small drop of oil in water usually assumes a spherical shape. Explain.

13.24 Which of the following properties indicates very strong intermolecular forces in a liquid? **(a)** A very low surface tension, **(b)** a very low critical temperature, **(c)** a very low boiling point, or **(d)** a very low vapor pressure.

13.25 Figure 13.9 shows that the average distance between base pairs measured parallel to the axis of a DNA molecule is 3.4 Å. The average molar mass of a pair of nucleotides is 650 g mol^{-1}. Estimate the length in cm of a DNA molecule of molar mass 5.0×10^9 g mol^{-1}. Roughly how many base pairs are contained in this molecule?

13.26 Using values listed in Table 13.1 and a handbook of chemistry, plot the polarizabilities of the noble gases versus their boiling points. On the same graph, also plot their molar masses versus boiling points. Comment on the trends.

13.27 Given the following general properties of water and ammonia, comment on the problems that a biological system (as we know it) would have in developing in an ammonia medium.

	H_2O	NH_3
Boiling point	373.15 K	239.65 K
Melting point	273.15 K	195.3 K
Molar heat capacity	75.3 J K^{-1} mol^{-1}	8.53 J K^{-1} mol^{-1}
Molar heat of vaporization	40.79 kJ mol^{-1}	23.3 kJ mol^{-1}
Molar heat of fusion	6.0 kJ mol^{-1}	5.9 kJ mol^{-1}
Dielectric constant	78.54	16.9
Viscosity	0.001 N s m^{-2}	0.0254 N s m^{-2} (at 240 K)
Surface tension	0.07275 N m^{-1} (293 K)	0.0412 N m^{-1} (at 244 K)
Dipole moment	1.82 D	1.46 D
Phase at 300 K	Liquid	Gas

13.28 The HF_2^- ion exists as

$$\left[:\ddot{F}\!-\!H\cdots\cdots:\ddot{F}: \right]^-$$

The fact that both HF bonds are the same length suggests that proton tunneling occurs. **(a)** Draw resonance structures for the ion. **(b)** Give a molecular orbital description (with an energy-level diagram) of hydrogen bonding in the ion.

13.29 **(a)** Draw a potential-energy curve for two atoms based on a hard-sphere model. **(b)** A potential intermediate between the hard-sphere and the Lennard-Jones potentials is the square-well potential, defined by $V = \infty$ for $r < \sigma$, $V = -\varepsilon$ for $\sigma \le r \le a$, and $V = 0$ for $r > a$. Sketch this potential.

13.30 The potential energy of the helium dimer (He_2) is given by

$$V = \frac{B}{r^{13}} - \frac{C}{r^6}$$

where $B = 9.29 \times 10^4$ kJ Å^{13} (mol dimer)$^{-1}$ and $C = 97.7$ kJ Å^6 (mol dimer)$^{-1}$. **(a)** Calculate the equilibrium distance between the He atoms. **(b)** Calculate the binding energy of the dimer. **(c)** Would you expect the dimer to be stable at room temperature (300 K)?

13.31 The internuclear distance between two closest Ar atoms in solid argon is about 3.8 Å. The polarizability of argon is 1.66×10^{-30} m^3, and the first ionization energy is 1521 kJ mol^{-1}. Estimate the boiling point of argon. [*Hint:* Calculate the potential energy due to the dispersion interaction for solid argon, and equate this quantity to the average kinetic energy of 1 mole of argon gas, which is $(3/2)RT$.]

CHAPTER 14

Spectroscopy

Spectroscopy is the study of the interaction between electromagnetic radiation and matter; it is a phenomenon of quantum mechanics. Detailed information about structure and bonding, as well as about various intra- and intermolecular processes, can be obtained from the analysis of atomic and molecular spectra. In this chapter, we introduce the vocabulary of spectroscopy and discuss several spectroscopic techniques and optical activity.

14.1 Vocabulary

Here we shall become acquainted with some common terms in spectroscopy.

Absorption and Emission

There are two categories of spectroscopy: absorption and emission. These processes were briefly mentioned in Chapter 11. The fundamental equation for both absorption and emission is

$$\Delta E = E_2 - E_1 = h\nu \qquad (14.1)$$

where E_1 and E_2 are the energies of the two quantized energy levels involved in a transition (see Figure 11.11). Microwave spectroscopy, infrared spectroscopy, electronic spectroscopy, nuclear magnetic resonance, and electron spin resonance are usually studied in the absorption mode. Fluorescence and phosphorescence are emission processes. Laser is a special type of emission called stimulated emission.

Units

The position of a spectral line corresponds to the difference in energy between two levels involved in a transition. This position can be measured in several different units.

1. *Wavelength.* Wavelength (λ) can be measured in meters (m), micrometers (μm), or nanometers (nm), where

$$1 \ \mu m = 1 \times 10^{-6} \ m$$

$$1 \ nm = 1 \times 10^{-9} \ m$$

2. *Frequency.* Frequency (ν) is given by s^{-1} or Hz.

3. *Wavenumber.* Wavenumber ($\tilde{v}$) is the number of waves per centimeter,

$$\tilde{v} = \frac{1}{\lambda} = \frac{v}{c} \tag{14.2}$$

where c is the speed of light. Note that energy is directly proportional to wavenumber. (Energy is also directly proportional to frequency, but unlike wavenumbers, frequencies are too high and therefore impractical to use.)

Depending on the particular type of spectroscopy, any one of these units can be used to label the lines in a spectrum. No confusion should arise as long as we remember the fundamental equation, $c = \lambda v$.

Regions of the Spectrum

Figure 14.1 summarizes the features of the major branches of spectroscopy that analyze molecular motion. The spectroscopic techniques that are most commonly used to study chemical and biological systems are infrared, visible and UV, nuclear magnetic resonance, and fluorescence. To gain a better overview, however, we shall also discuss microwave spectroscopy, electron spin resonance, and phosphorescence. The advent of laser technology has revolutionized spectroscopy, so we shall devote a section to this topic.

Line Width

Every spectral line has a finite, nonzero width, which is usually defined as the full width at half-height of the peak. If the two states involved in a spectroscopic transition have precisely well-defined energies, then their energy difference must also be an exactly measurable quantity. In this case, we would observe a line of no width. In reality, however, many phenomena cause every spectral line to have a definite width

	γ-ray	X-ray	Ultraviolet	Visible	Infrared	Microwave	Radio frequency
Wavelength/ nm	0.0003 0.03		10 30	400 800	1000 3×10^5 3×10^7		3×10^{11} 3×10^{13}
Frequency/ Hz	1×10^{21} 1×10^{19}		3×10^{16} 1×10^{16} 8×10^{14}	4×10^{14} 3×10^{14}	1×10^{12} 1×10^{10}	1×10^6	1×10^4
Wavenumber/ cm^{-1}	3×10^{10} 3×10^8		1×10^6 3×10^5 3×10^4 1.3×10^4 1×10^4		33 3		3×10^{-5} 3×10^{-7}
Energy/ (kJ mol^{-1})	4×10^8 4×10^6 1.2×10^4 4×10^3		330	170 125	0.4 4×10^{-3}	4×10^{-7}	4×10^{-9}
Phenomenon observed		Nuclear transitions	Inner electronic transitions $\sigma \to \sigma^*$	Outer electronic transitions $\pi \to \pi^*, n \to \pi^*$	Molecular vibration	Molecular rotation, electron spin resonance	Nuclear magnetic resonance
Type of spectroscopy	Mössbauer	UV	UV, Visible		IR	Microwave, ESR	NMR

Figure 14.1
Types of spectroscopy. Mössbauer and Raman spectroscopy are not discussed in this text.

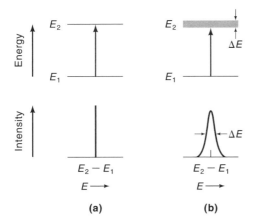

Figure 14.2
(a) A hypothetical absorption line having no width. (b) An actual absorption line having a width, ΔE, at half-height. The lifetime of the ground state is very long so that its energy is well defined.

(Figure 14.2). We shall discuss the three most basic mechanisms below.

The Natural Line Width. The so-called natural line width of a spectral line is a consequence of the Heisenberg uncertainty principle, which can be expressed as follows (see Equation 11.13):

$$\Delta E \Delta t \geq \frac{h}{4\pi}$$

This equation relates the uncertainty in the energy and the lifetime of the system in a particular state. It says that the longer it takes to measure the energy of a system in that state (the larger the Δt is), the more accurately this energy can be determined (the smaller ΔE is). In an absorption process, the final state (the excited state) has a finite lifetime, t; therefore, the uncertainty in determining the lifetime, Δt, cannot exceed t. Consequently, the uncertainty in the energy of that state is given by

$$\Delta E \geq \frac{h}{4\pi\Delta t}$$

Because $E = h\nu$, we have $\Delta E = h\Delta\nu$, or

$$\Delta\nu = \frac{1}{4\pi\Delta t}$$

Note that we use the equals sign here to get the minimum value of $\Delta\nu$. The uncertainty in energy (expressed in Hz) shows up as the width of the spectral line. This width is called the natural line width because it is inherent in the system and cannot be influenced (decreased) by external parameters such as temperature or concentration. Natural line widths depend strongly on the type of transition. For example, for transitions between rotational energy levels, a typical lifetime is about 10^3 s, which translates into a natural line width of about 8×10^{-5} Hz. On the other hand, an electronic excited state has a lifetime on the order of 10^{-8} s, resulting in an uncertainty in frequency of about 8×10^6 Hz!

The Doppler Effect. Experimentally, the widths of spectral lines are invariably much greater than those predicted solely for the lifetimes of excited states. Therefore, other mechanisms must also be responsible for broadening the lines. The Doppler broadening of spectral lines is an interesting result of the *Doppler effect* (after the Austrian physicist Christian Doppler, 1803–1853). When radiation is emitted, its frequency depends on the velocity of the atom or molecule relative to the detector.* For the same reason, the whistle of a train traveling toward you seems to have a frequency higher than it really is, and when the train is moving away from you the whistle sounds lower in frequency than it is. The equation describing the Doppler effect is

$$v = v_0 \left(1 \pm \frac{v}{c} \right) \tag{14.3}$$

where v_0 is the frequency of the emitting molecule, v is the frequency registered by the detector, v is the average speed of the molecules in the sample, and c is the speed of light. The $\pm$ sign indicates that some molecules are moving toward the detector ($+$) and others are moving away from it ($-$).

It is possible to estimate how much broader a line is as a result of the Doppler effect. For N_2 molecules at 300 K, we can use Equation 2.23 to calculate the root-mean-square speed, 517 m s^{-1}. Substituting this value into Equation 14.3, we get

$$v = v_0 \left(1 \pm \frac{517 \text{ m s}^{-1}}{3.00 \times 10^8 \text{ m s}^{-1}} \right)$$

$$= v_0 (1 \pm 1.72 \times 10^{-6})$$

Using a typical frequency of electronic transition of 1×10^{15} Hz for $N_2(v_0)$, we find a total frequency shift (plus and minus) of about 2×10^9 Hz, which is about 400 times that of the natural line width. Line width broadening due to the Doppler effect increases with temperature, because there is a larger spread in molecular speeds. To minimize this effect, spectra should be obtained from cold samples.

The Pressure Effect. Another influence on the width of a spectral line is called pressure, or collisional, broadening. Molecular collisions can deactivate excited states, thereby shortening their lifetimes. If τ is the mean time between collisions, and each collision results in a transition between two states, then according to the Heisenberg uncertainty principle there is a broadening Δv, given by $1/4\pi\tau$. Referring to Section 2.7, we see that $\tau = 1/Z_1$, where Z_1 is the collision frequency. Because Z_1 is directly proportional to pressure, it follows that an increase in pressure will lead to a broader spectral line. Figure 14.3 shows the electronic absorption spectra of benzene in the vapor and liquid states. Because the collision frequency is greater in liquid, the spectral lines are much broader. To minimize collisional broadening, spectra should be recorded in the vapor state (if possible) at low pressures.

Finally, we note that rate processes such as dissociation, rotation, and electron- and proton-transfer reactions can also cause line broadening. We shall see some examples of these effects later.

---

*The same conclusion is obtained for an absorption process; that is, the Doppler effect influences (increases) the width of both emission and absorption spectral lines.

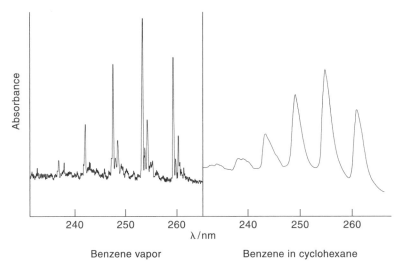

Figure 14.3
Electronic absorption spectrum of benzene. Left: vapor; right: in cyclohexane.
(By permission of Varian Associates, Palo Alto, CA.)

Resolution

Related to line width is the separation of one spectral line from another, called *resolution* (Figure 14.4). In all of spectroscopic techniques, the *resolving power* (R) of an instrument is a measure of the ability of a spectrometer to distinguish closely spaced lines from one another. An instrument of high resolving power will show two closely spaced lines separately, whereas a low-resolution instrument will merge them. If $\Delta\lambda$ is the wavelength separation of the closest lines that can be seen to be two lines, the resolving power is given by

$$R = \frac{\lambda}{\Delta\lambda} \tag{14.4}$$

where λ is the average wavelength of the lines. An analogous equation, expressed in terms of frequencies, is $R = \nu/\Delta\nu$.

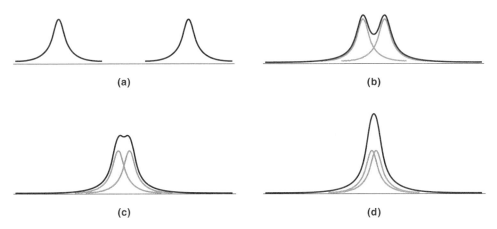

Figure 14.4
(a) Two well-resolved lines. (b)–(d) Two overlapping lines. From (b) through (d), the observed line shape is the sum of the two overlapping lines.

Intensity

Several factors affect the intensity of an absorption line, which is related to the number of molecules participating in the spectroscopic transition. Here, we shall discuss the treatment presented by Einstein in 1917. Consider a two-state system separated by $\Delta E = E_n - E_m$. When molecules are exposed to radiation with frequency, v, such that $\Delta E = hv$, they undergo a transition from the lower state, m, to the higher state, n. The rate of transition, N_{mn}, to the higher state is proportional to the number of molecules, N_m, in the lower state and also to the radiation density $\rho(v)$ at this frequency. Thus,

$$N_{mn} \propto N_m \rho(v)$$
$$= B_{mn} N_m \rho(v) \tag{14.5}$$

where B_{mn} is the *Einstein coefficient of stimulated absorption*. Einstein realized that radiation can also induce molecules in the excited state to undergo a transition to the lower state. The rate of this stimulated emission, N_{nm}, is

$$N_{nm} = B_{nm} N_n \rho(v) \tag{14.6}$$

where B_{nm} is the *Einstein coefficient of stimulated emission*, and N_n is the number of molecules in the excited state. Note that only radiation of the *same* frequency as the transition can stimulate the shift from the excited state to the lower state. The two coefficients, B_{mn} and B_{nm}, are equal. In addition, molecules in an excited state can lose energy by spontaneous emission at a rate that is independent of the radiation frequency. This rate is given by $A_{nm} N_n$, where A_{nm} is the *Einstein coefficient of spontaneous emission*. These three situations are summarized in Figure 14.5. At equilibrium, the number of molecules going from the m to the n state is equal to the number going from the n to the m state, so

$$N_m B_{mn} \rho(v) = N_n B_{nm} \rho(v) + N_n A_{nm}$$

or

$$\rho(v) = \frac{A_{nm}}{B_{nm}} \frac{N_n}{N_m - N_n} = \frac{A_{nm}}{B_{nm}} \frac{1}{\dfrac{N_m}{N_n} - 1} \tag{14.7}$$

because $B_{mn} = B_{nm}$. The ratio N_n / N_m is given by the Boltzmann distribution law (see

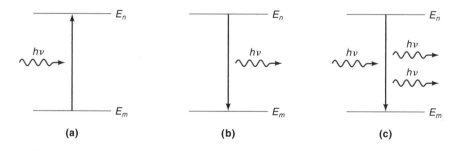

Figure 14.5
(a) Stimulated absorption. (b) Spontaneous emission. (c) Stimulated emission. The energy of the incoming or emitted photon is hv.

Equation 3.22),

$$\frac{N_n}{N_m} = e^{-h\nu/k_B T}$$

so that

$$\frac{N_m}{N_n} = e^{h\nu/k_B T} \tag{14.8}$$

Substituting Equation 14.8 into 14.7, we get

$$\rho(\nu) = \frac{A_{nm}}{B_{nm}} \frac{1}{e^{h\nu/k_B T} - 1} \tag{14.9}$$

The radiation density was shown by Planck to be

$$\rho(\nu) = \frac{8\pi h\nu^3}{c^3} \frac{1}{e^{h\nu/k_B T} - 1} \tag{14.10}$$

Substituting Equation 14.10 into Equation 14.9 gives

$$A_{nm} = B_{nm} \frac{8\pi h\nu^3}{c^3} \tag{14.11}$$

Note the dependence of A_{nm} on frequency. In electronic spectroscopy, ν is a large number, so the probability of spontaneous emission is usually much higher than it is for stimulated emission. (This explains the short lifetime of the excited electronic states mentioned earlier.) When the frequencies are much smaller, as in microwave or magnetic resonance spectroscopies, the stimulated emission predominates. We shall return to the phenomenon of stimulated emission in the section on lasers.

Keep in mind that any type of spectrum (absorption or emission) is actually a superposition of numerous transitions from individual molecules. Most spectrometers are not designed to detect the energy absorbed or emitted by a single molecule. Further, the interaction between a photon of electromagnetic radiation and a molecule can give rise to only one transition and hence one line. Any spectrum containing more than one line, as most do, is actually the statistical sum of all the transitions.

Selection Rules

Transitions do not take place between any two levels in an atom or molecule just because the frequency of radiation is appropriate to the resonance condition ($\Delta E = h\nu$). Generally, a transition has to obey certain *selection rules*, which are theoretical conditions obtained by time-dependent quantum mechanical calculations. Transitions, then, are classified as *allowed* (having a high probability) or *forbidden* (having a low probability), depending on how they occur according to selection rules.

Theoretically, we predict two types of transitions as being forbidden: spin-forbidden transitions and symmetry-forbidden transitions.

Spin-Forbidden Transitions. Spin-forbidden transitions involve a change in *spin multiplicity*. The spin multiplicity is given by $(2S + 1)$, where S is the spin quantum number of the system (Table 14.1). The numerical value gives us the number of different ways in which the unpaired spins can line up in an external magnetic field. The

Table 14.1
Spin Multiplicity of Atoms and Molecules

Number of Unpaired Electrons	Electron Spin S	$2S+1$	Multiplicity
0	0	1	Singlet
1	$\frac{1}{2}$	2	Doublet
2	1	3	Triplet
3	$\frac{3}{2}$	4	Quartet
.	.	.	.
.	.	.	.
.	.	.	.

selection rule is that the spin multiplicity must not change in a transition; that is, we must have $\Delta S = 0$. Normally, for example, a transition from a singlet to a triplet, or vice versa, is strongly forbidden.

Symmetry-Forbidden Transitions. A quantitative measure for the intensity of a transition is provided by the *transition dipole moment*, μ_{ij}, given by

$$\mu_{ij} = \int \psi_i \mu \psi_j \, d\tau \tag{14.12}$$

where ψ_i and ψ_j are the wave functions for the ith and jth state, and μ is the dipole moment vector connecting these two states. The integration is taken over all coordinates and $d\tau$ represents the volume element ($d\tau = dxdydz$). For an allowed transition, $\psi_i \mu \psi_j$ must be an even function.* Because μ depends only on the first power of the coordinates and is therefore odd, ψ_i and ψ_j must have different symmetry with respect to each other (even–odd or odd–even) so that the product will be even.

To gain insight into the transition dipole moment, consider the electronic transition in the hydrogen atom. The physical significance of the dipole-moment vector is that it denotes the electron charge migration during the transition. Figure 14.6 shows

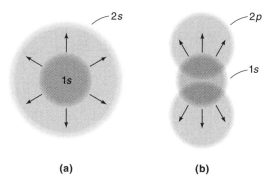

(a)   (b)

Figure 14.6
(a) There is no dipole moment associated with a $1s \rightarrow 2s$ transition because the electric charge migrates spherically. Consequently, this transition is symmetry forbidden. (b) During a $1s \rightarrow 2p$ transition, there is a dipole associated with the charge migration. This is an allowed transition.

---

* An even function, $f(x)$, has the property that it is unchanged when we reverse the sign of x; that is, $f(x) = f(-x)$. The opposite holds true for an odd function where $f(x) = -f(-x)$. Thus, x^2 is an even function, and x^3 is an odd function.

the charge migrations accompanying the transitions from hydrogen $1s$ to $2s$ and to $2p$ states. As we can see, the $1s \rightarrow 2s$ transition is forbidden, because the charge redistribution remains spherically symmetrical, whereas the $1s \rightarrow 2p$ transition is allowed because the charge redistribution is dipolar. We can generalize these observations with the selection rule $\Delta l = \pm 1$, where l is the angular momentum quantum number.

Mechanisms too complex to list here cause various degrees of breakdown in the selection rules. Consequently, transitions predicted as forbidden may appear as weak lines.

Signal-to-Noise Ratio

A recorded spectrum, because of the manner in which the signals are detected, always contains random fluctuations of electronic signals called *noise*. The sensitivity of detection of any signal due to the sample being studied depends on how easily we can distinguish it from the noise. An effective way to increase the signal-to-noise ratio is by signal averaging—that is, by repeatedly recording the same spectrum and adding the signals. Theoretically, if we signal-average a spectrum N times, the intensity will increase by a factor N, and the noise will increase by a factor $\sqrt{N}$. Thus, the signal-to-noise ratio will increase by $N/\sqrt{N}$, or $\sqrt{N}$, times, so that scanning the same spectrum 10 times will enhance the signal-to-noise ratio by a factor of $\sqrt{10}$, or 3.2. The Fourier-transform technique (see p. 547) enables us to scan a spectrum rapidly, so accumulating hundreds or even thousands of scans in a reasonable amount of time is now practical. This procedure has greatly enhanced our ability to study weak transitions or samples at low concentrations.

The Beer–Lambert Law

A useful equation for studying the quantitative aspects of absorption is the Beer–Lambert law. Consider the passage of a monochromatic beam of radiation (that is, radiation of one wavelength) through a homogeneous medium, say a solution. Let I_0 and I be the intensity* of the incident and transmitted light and I_x the intensity of the light at distance x (Figure 14.7). The incremental decrease in intensity, $-dI_x$, is proportional to $I_x \, dx$; that is,

$$-dI_x \propto I_x \, dx$$
$$= k I_x \, dx \qquad (14.13)$$

where I_x is the intensity at x, and k is a constant whose value depends on the nature of the absorbing medium. Rearranging Equation 14.13, we obtain

$$\frac{dI_x}{I_x} = -k \, dx$$

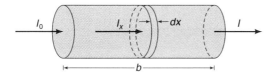

Figure 14.7
Absorption of light by a uniform medium of pathlength b.

---

* The intensity of light is determined by the number of photons $cm^{-2} \, s^{-1}$.

Upon integration,

$$\ln I_x = -kx + C$$

where C is a constant of integration. At $x = 0$, $I_x = I_0$, so $C = \ln I_0$. If we consider the entire length of the absorbing medium, I_x can be replaced by I (which is the intensity of the emerging light) and x by b. Hence,

$$\ln I = -kb + \ln I_0$$

$$-\ln \frac{I}{I_0} = kb$$

or

$$-\log \frac{I}{I_0} = k'b \qquad (14.14)$$

where $k = 2.303k'$. The ratio I/I_0 is called the *transmittance* (T); it measures the amount of light transmitted after passing through the medium. Equation 14.14 can be expressed in a more convenient form as follows:

$$-\log T = A = \varepsilon bc$$

or

$$A = \varepsilon bc \qquad (14.15)$$

where A is the *absorbance*, ε is the *molar absorptivity* (also called *molar extinction coefficient*), b is the *pathlength* (cm), and c is the concentration in mol L^{-1}.

Equation 14.15 is known as the Beer–Lambert law, after the German astronomer Wilhelm Beer (1797–1850) and the German mathematician Johann Heinrich Lambert (1728–1777). We see that the absorbance is equal to the negative log base 10 of transmittance. Thus, the smaller the transmittance (and hence the larger the $-\log T$), the greater the absorption of light. The absorbance is a dimensionless quantity, so ε has the units L mol^{-1} cm^{-1}. Theoretically, A can have any positive value; in practice, for UV and visible spectrometers, A normally varies between zero and one.

The Beer–Lambert law forms the quantitative basis for all types of absorption spectroscopy. It holds as long as there is no interaction between solute molecules, such as dimerization or ion pair formation. An interesting application of Equation 14.15 is the pulse oximeter, a device used to monitor the extent of oxygenation of hemoglobin molecules of patients without drawing their blood. Light is directed through a fingertip, and the absorbances of the arterial blood at different wavelengths are measured at different times. Using the known molar absorptivities of oxyhemoglobin and deoxyhemoglobin, the instrument determines the fraction of oxyhemoglobin in the blood.

14.2 Microwave Spectroscopy

Microwave spectroscopy is concerned with the rotational motion of molecules. Consider a heteronuclear diatomic molecule such as HCl or CO, which may be treated as a rigid rotor; that is, it behaves like a dumbbell. The *moment of inertia* (I) of the

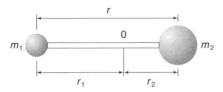

Figure 14.8
Diatomic molecule as a rigid rotor. The zero indicates the center of gravity.

molecule about the center of gravity is given by

$$I = m_1 r_1^2 + m_2 r_2^2 \qquad (14.16)$$

where m_1 and m_2 are the masses of the atoms, and r_1 and r_2 are the distances from the nuclei to the center of gravity (Figure 14.8). Also, the center of gravity requires that

$$m_1 r_1 = m_2 r_2 \qquad (14.17)$$

Therefore, we write

$$r_1 = \frac{m_2}{m_1} r_2 = \frac{m_2}{m_1}(r - r_1) \qquad (14.18)$$

From Equation 14.18 it follows that

$$r_1 = \frac{m_2}{m_1 + m_2} r \qquad (14.19)$$

and

$$r_2 = \frac{m_1}{m_1 + m_2} r \qquad (14.20)$$

Substituting Equations 14.19 and 14.20 into 14.16, we get

$$I = m_1 \left(\frac{m_2}{m_1 + m_2}\right)^2 r^2 + m_2 \left(\frac{m_1}{m_1 + m_2}\right)^2 r^2$$

$$= \frac{m_1 m_2}{m_1 + m_2} r^2 = \mu r^2 \qquad (14.21)$$

where μ, the *reduced mass*, is defined by

$$\frac{1}{\mu} = \frac{1}{m_1} + \frac{1}{m_2} \qquad (14.22)$$

so that

$$\mu = \frac{m_1 m_2}{m_1 + m_2}$$

With the introduction of the reduced mass, we can treat the rotation of the molecule as a single particle of mass μ moving in a circle of radius r.

From the solution of the Schrödinger equation, we obtain the following quantized energies for rotation:

$$E_{\text{rot}} = \frac{J(J+1)h^2}{8\pi^2 I} = BJ(J+1)h \qquad (14.23)$$

where B is the *rotational constant*, given by $h/8\pi^2 I$ and J is the rotational quantum number $(J = 0, 1, 2, \ldots)$. We see that the lowest rotational energy level has zero energy.*

Transition from a lower energy level to a higher one can be induced by irradiating a sample of molecules with the appropriate microwave frequency. Not all the transitions are allowed because of the selection rule $\Delta J = \pm 1$. From Equation 14.23, we see that the energy change, ΔE, for a transition from the $J = 0$ to the $J = 1$ level is given by

$$\Delta E_{\text{rot}} = E_1 - E_0$$

$$= 2Bh$$

For the $J = 1$ to $J = 2$ transition, ΔE_{rot} is given by

$$\Delta E_{\text{rot}} = Bh[2(2+1) - 1(1+1)] = 4Bh$$

and so on. We can generalize the results as follows. Let J' and J'' be the rotational quantum numbers for the upper and lower levels, respectively. The energy difference is given by

$$\Delta E_{\text{rot}} = BJ'(J'+1)h - BJ''(J''+1)h$$

$$= Bh[J'(J'+1) - J''(J''+1)]$$

Because $J' - J'' = 1$, the above equation becomes

$$\Delta E_{\text{rot}} = 2BhJ' \quad J' = 1, 2, 3, \ldots \qquad (14.24)$$

Thus, absorptions $J = 0 \to 1, 1 \to 2, 2 \to 3, \ldots$ will have energy differences of $2Bh, 4Bh, 6Bh, \ldots$, and a set of equally spaced lines with separations of $2Bh$ will be obtained (Figure 14.9).

Under high resolution, the spacing between adjacent lines in a rotational spectrum is found to decrease with increasing values of J. At higher energy levels, a molecule rotates faster, so the internuclear bond is stretched somewhat by the centrifugal force. An increase in bond length, r, increases the moment of inertia, I, causing E_{rot} to decrease (see Equation 14.23). For better accuracy, this effect can be corrected by adding a term to Equation 14.23 as follows:

$$\Delta E_{\text{rot}} = BJ(J+1)h - D[J(J+1)]^2 h \qquad (14.25)$$

where D, the *centrifugal constant*, is about 1,000 times smaller than B. Thus, the second term can usually be neglected unless J is a large number.

Microwave spectroscopy is an important tool for determining molecular geometry. From the separation between adjacent lines, we obtain the rotational constant B and hence the moment of inertia and the internuclear distance, r.

---

*A zero rotational energy (and momentum) does not violate the Heisenberg uncertainty principle because the rigid rotor can have infinitely many angular orientations. In other words, there is complete uncertainty in the angular orientation.

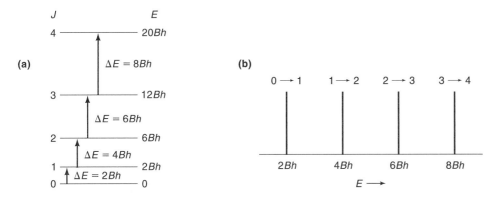

Figure 14.9
(a) Allowed resonance conditions for microwave transitions for a diatomic molecule. (b) Equally spaced rotational lines. In practice, these lines are of unequal intensities.

Example 14.1

The microwave spectrum of carbon monoxide consists of a series of lines separated by 1.15×10^{11} Hz. Calculate the bond length of CO.

ANSWER

Figure 14.9 shows that the spacing between successive lines is $2Bh$, which is equal to ΔE_{rot}. Therefore, the difference in frequencies, Δv, is given by

$$\Delta v = \frac{\Delta E_{rot}}{h} = 2B = 2 \times \frac{h}{8\pi^2 I}$$

Solving for I, we get

$$I = \frac{h}{4\pi^2 \Delta v}$$

$$= \frac{6.626 \times 10^{-34} \text{ J s}}{4\pi^2 (1.15 \times 10^{11} \text{ s}^{-1})}$$

$$= 1.46 \times 10^{-46} \text{ kg m}^2$$

From Equation 14.21,

$$I = \frac{m_1 m_2}{m_1 + m_2} r^2$$

$$1.46 \times 10^{-46} \text{ kg m}^2$$

$$= \frac{(12.01 \text{ amu})(16.00 \text{ amu})(1.661 \times 10^{-27} \text{ kg amu}^{-1})}{(12.01 \text{ amu} + 16.00 \text{ amu})} r^2$$

$$r^2 = 1.28 \times 10^{-20} \text{ m}^2$$

$$r = 1.13 \times 10^{-10} \text{ m}$$

$$= 1.13 \text{ Å}$$

COMMENT

Many bond lengths have been determined to four or five decimal-place precision by microwave spectroscopy.

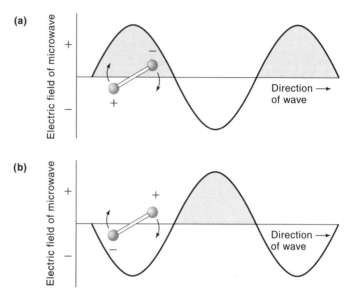

Figure 14.10

Interaction between the electric field component of microwave radiation and an electric dipole. (a) The negative end of the dipole follows the propagation of the wave (the positive region) and rotates in a clockwise direction. (b) If, after the molecule has rotated to the new position, the radiation has also moved along to its next cycle, the positive end of the dipole will move into the negative region of the wave while the negative end will be pushed up. Thus, the molecule will rotate faster. No such interaction can occur with nonpolar molecules.

The analysis of the microwave spectra of polyatomic molecules can be quite complex and will not be described here. Instead, we shall briefly discuss a linear triatomic molecule, carbonyl sulfide, OCS. Like CO, OCS gives a series of equally spaced lines in its microwave spectrum. Because there are two bond lengths, however, the moment of inertia is expressed in terms of two unknowns, r_{CO} and r_{CS}. We can overcome this difficulty by making the good assumption that isotopic substitution does not alter bond lengths. Then we can record the microwave spectra of $^{16}O^{12}C^{32}S$ and $^{16}O^{12}C^{34}S$ to get two moments of inertia corresponding to these species and solve a system of two equations and two unknowns to determine the two bond lengths.

Nonpolar molecules (for example, homonuclear diatomic molecules such as N_2 and O_2) do not absorb radiation in the microwave region and are said to be micro-wave inactive. To see why a polar molecule such as CO behaves differently, let us consider the interaction between a dipole and an oscillating electric field from the electromagnetic wave, shown in Figure 14.10. In Figure 14.10a, the negative end of the dipole follows the propagation of the wave (the positive region) and rotates in a clockwise direction. If, after the molecule has rotated by 180°, the wave has moved along to its next half cycle (Figure 14.10b), the positive end of the dipole will move into the negative region of the wave, and the negative end will be pushed up. Unless the frequency of the radiation is equal to that of the molecular rotation, the dipole cannot absorb energy from the radiation to rotate faster. This classical wave-nature description supplements the quantum mechanical picture for a transition from a lower rotational energy level to a higher one. It also explains why nonpolar molecules are microwave inactive, because they cannot interact with the electric field component of the radiation.

Finally, we note that microwave spectroscopy generally applies only to molecules in the gas phase. In solution, the frequency of molecular collision is much greater than the frequency of molecular rotation. Consequently, molecules cannot execute full rotational motion, and hence no microwave spectra can be obtained.

14.3 Infrared Spectroscopy

Infrared (IR) spectroscopy concerns the vibrational motion of molecules. To study molecular vibration, we will start with a system that behaves according to classical (that is, Newtonian) mechanics. Figure 14.11 shows such a system. An object of mass m is attached to a spring. According to Hooke's law (after the English natural philosopher, Robert Hooke, 1635–1703), the force, f, acting on the object is proportional to the displacement, x, from the equilibrium position:

$$f \propto -x$$
$$= -kx \qquad (14.26)$$

where k is the *force constant*, a characteristic of the stiffness of the spring measured in N m^{-1}. The negative sign means that the force acts in the direction opposite to x; if x is positive (the spring is stretched), f is negative, meaning that there is a restoring force pulling the object upward. The reverse holds true for a compressed spring. If we pull the object downward and let go, it will undergo a periodic vibrational motion known as *simple harmonic motion*, for which the displacement, x, at any time, t, is given by a sinusoidal function,

$$x = A \sin \alpha t \qquad (14.27)$$

where A is the amplitude of vibration and α is a constant. The frequency of vibration, v, is

$$v = \frac{1}{2\pi} \sqrt{\frac{k}{m}} \quad (\alpha = 2\pi v) \qquad (14.28)$$

Because every term on the right side of Equation 14.28 is a constant, this equation says that there is one characteristic, or natural, frequency of vibration. During the complete period of a vibration, the kinetic energy of the particle is being converted to the potential energy of the spring, and vice versa. The potential energy of the system is given by

$$V = \tfrac{1}{2} k x^2 \qquad (14.29)$$

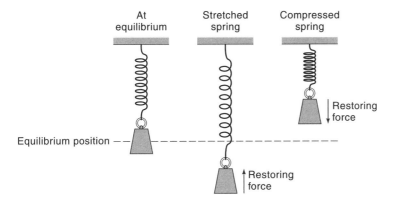

Figure 14.11
A weight attached to a spring exhibits simple harmonic motion when it is pulled downward and then released.

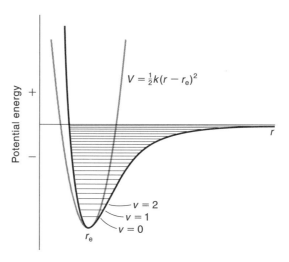

Figure 14.12

Potential energy curve for a diatomic molecule. The symmetric curve is given by Equation 14.31; the other curve represents the actual behavior of the molecule. The vertical distance from $v = 0$ to the r axis is the bond dissociation energy of the molecule. The rotational energy levels associated with each vibrational level are not shown.

This analysis of simple harmonic motion can be applied to the vibration of a diatomic molecule. As in rotational motion, we can treat the system as being comprised of one particle by using the reduced mass, μ, so that

$$v = \frac{1}{2\pi}\sqrt{\frac{k}{\mu}} \tag{14.30}$$

where v is now the fundamental frequency of vibration of the molecule. If a molecule behaves as a harmonic oscillator, then its potential energy is given by

$$V = \tfrac{1}{2}k(r - r_e)^2 \tag{14.31}$$

where r is the distance between the two atoms, and r_e is the equilibrium bond distance. Figure 14.12 compares the potential energy curve based on Equation 14.31 with that of a real molecule, such as nitric oxide, NO. At small displacements from r_e, the two curves match quite well, and the oscillation is harmonic. As r increases due to more energetic vibrations, however, appreciable deviation occurs. The character of the vibrations now is said to be *anharmonic*. We shall return to this point later.

The solution of the Schrödinger equation for the harmonic oscillator gives the vibrational energies as

$$E_{\text{vib}} = \left(v + \tfrac{1}{2}\right)hv \tag{14.32}$$

where v is the *vibrational quantum number* $(v = 0, 1, 2, \ldots)$. Thus, the vibrational energies are quantized. Furthermore, the lowest vibrational energy $(v = 0)$ is not zero but is equal to $\tfrac{1}{2}hv$. This means that a molecule will execute vibrational motion even at the absolute zero of temperature. This *zero-point* energy is in accord with the Heisenberg uncertainty principle. If a molecule did not vibrate, then its energy and the momentum associated with the motion would be zero. In this case, the uncertainty in the momentum would also be zero, which means the uncertainty in the position (in locating the atoms) would be infinite. But the atoms are separated by a finite distance, so the uncertainty should be comparable to the bond distance and not infinite.

As the following example shows, the fundamental frequency determined from an IR spectrum provides information about the strength of the bond.

Example 14.2

The fundamental frequency of vibration for the $H^{35}Cl$ molecule is 2886 cm^{-1}. Calculate the force constant of the molecule. (Atomic mass of $^{35}Cl = 34.97$ amu.)

ANSWER

First, we need to convert cm^{-1} to Hz:

$$v = c\tilde{v}$$
$$= (3.00 \times 10^{10} \text{ cm s}^{-1})(2886 \text{ cm}^{-1})$$
$$= 8.66 \times 10^{13} \text{ Hz}$$

Next, we calculate the reduced mass of the molecule, given by

$$\mu = \frac{m_H m_{Cl}}{m_H + m_{Cl}}$$
$$= \frac{(1.008 \text{ amu})(34.97 \text{ amu})(1.661 \times 10^{-27} \text{ kg amu}^{-1})}{(1.008 \text{ amu} + 34.97 \text{ amu})}$$
$$= 1.627 \times 10^{-27} \text{ kg}$$

Rearranging Equation 14.30,

$$k = 4\pi^2 v^2 \mu$$
$$= 4\pi^2 (8.66 \times 10^{13} \text{ s}^{-1})^2 (1.627 \times 10^{-27} \text{ kg})$$
$$= 4.82 \times 10^2 \text{ kg s}^{-2}$$
$$= 4.82 \times 10^2 \text{ kg m s}^{-2} \text{ m}^{-1}$$
$$= 4.82 \times 10^2 \text{ N m}^{-1}$$

COMMENT

The force constant measures how much force is needed to stretch a bond by unit length (per meter or per angstrom). As expected, triple bonds have larger force constants than double bonds, which have larger values than single bonds. For example, k is approximately 450 N m^{-1} for the C–C bond, 930 N m^{-1} for the C=C bond, and 1600 N m^{-1} for the C≡C bond.

Molecules do not behave exactly like harmonic oscillators. For example, as r increases, the chemical bond weakens, and eventually dissociation takes place. A more realistic description of molecular vibration is presented by the asymmetric curve in Figure 14.12. Each horizontal line represents a vibrational level. The spacing between successive levels decreases with increasing v, due to the anharmonic character of the vibration. As a correction, we rewrite Equation 14.32 as

$$E_{vib} = \left(v + \tfrac{1}{2}\right)hv - x\left(v + \tfrac{1}{2}\right)^2 hv \tag{14.33}$$

where x is the *anharmonicity constant*. As for the centrifugal distortion of rotation, x can be ignored except for large values of v.

The selection rule for transitions between vibrational energy levels is $\Delta v = \pm 1$. Because the spacing between energy levels is large, most of the molecules reside in the ground level at room temperature. Therefore, the absorption of IR radiation almost always involves the $v = 0 \rightarrow 1$ transition (called *the fundamental band*). If the molecule behaves as a harmonic oscillator, then the $v = 1 \rightarrow 2$ transition, called a *hot band* (because its intensity increases with increasing temperature), will occur at the same frequency as the fundamental band. If there is appreciable anharmonicity, however, then the hot band can be distinguished from the fundamental band by its slightly lower frequency in the spectrum. Another consequence of anharmonicity is the breakdown of the selection rule, so that it is possible to have $v = 0 \rightarrow 2, 0 \rightarrow 3, \ldots$ transitions, which are called *overtones*. Note that the first overtone ($v = 0 \rightarrow 2$) band does not appear at exactly twice the frequency of the fundamental band. As stated in Example 14.2, the fundamental band of $H^{35}Cl$ occurs at 2886 cm^{-1}. The first overtone is observed at 5668 cm^{-1}, which is somewhat less than 2×2886 cm^{-1}, or 5772 cm^{-1}. The difference is due to anharmonicity.

> The intensity of the overtone bands is much weaker than that of the fundamental band.

For a particular vibration to absorb IR radiation or to be IR active, we must have

$$\frac{d\mu}{dr} > 0$$

that is, the electric dipole moment must change with bond length during a vibration. Again, using the more intuitive wave-nature picture of radiation, we see that for a vibrational mode to be excited, the frequency of the oscillating electric field of the IR radiation must match that of the bond vibration. Keep in mind that when energy is absorbed by the molecule, the frequency of the bond vibration does not change. It is the amplitude that increases.* The requirement that the dipole moment must change with bond distance during vibration rules out IR activity in all homonuclear diatomic molecules.

According to the *equipartition of energy theorem*, the energy of a molecule is equally divided among all types of motion or *degrees of freedom*. For monatomic gases, each atom needs three coordinates (x, y, and z) to completely define its position. It follows that an atom has three translational degrees of freedom. Because a molecule also executes internal rotational and vibrational motions, there must be additional degrees of freedom to describe them. For a molecule containing N atoms, we would need a total of $3N$ coordinates to describe its motions. Three coordinates are required to describe the translational motion (for example, the motion of the center of mass of the molecule), leaving $(3N - 3)$ coordinates for rotation and vibration. Three angles are needed to define the rotation of the molecule about the three mutually perpendicular axes through its center mass, leaving $(3N - 6)$ degrees of freedom for vibration. If the molecule is linear, two angles will suffice. Because rotation of the molecule about the internuclear axis does not change the positions of the nuclei, this motion does not constitute a rotation. Thus, a linear molecule such as HCl or CO_2 has $(3N - 5)$ degrees of freedom for vibration. A nonlinear molecule like H_2O has $(3 \times 3 - 6)$ or three, degrees of freedom for vibration. The apparently complex vibrational motion of H_2O can then be analyzed in terms of three fundamental frequencies, shown in Figure 14.13. We expect a total of three lines from this

---

*An analogy of this effect is the situation of pushing someone on a swing. You must push "in phase" with the oscillation of the swing, and your energy goes into increasing the amplitude, not the frequency, of the swinging motion.

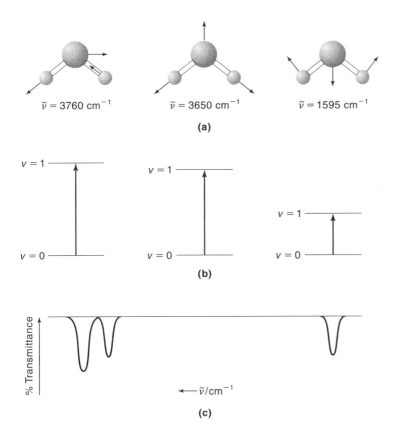

Figure 14.13
(a) The three fundamental modes of vibration of H_2O, all of which are IR active. From the symmetry of the molecule we determine the details of the vibrations (that is, the movement of individual atoms) by using a mathematical procedure called group theory. Note that in each case, the center of gravity of the molecule remains unchanged during the course of a vibration. (b) Energy levels. (c) IR transmission spectrum.

scheme if we consider only the $v = 0 \rightarrow 1$ transition for each vibrational mode.

As another example, let us consider the linear molecule, CO_2, which has four vibrational degrees of freedom $(3N - 5 = 4)$ (Figure 14.14). Although CO_2 does not possess a permanent electric dipole moment, three of the four vibrations are IR active

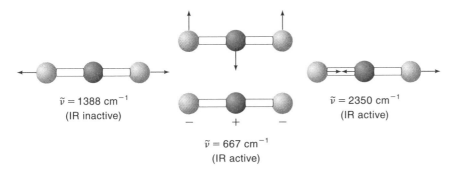

Figure 14.14
The four fundamental modes of vibration of CO_2. The middle two vibrations have the same frequency and are said to be degenerate. The "+" and "−" signs indicate the movement of the atoms into and out of the plane of the paper. The frequency of the IR inactive stretching mode is determined by another spectroscopic technique (Raman) not discussed in text.

because there is a change in dipole moment with respect to bond distance. Two of the four vibrations are called degenerate because they have the same frequency.

Simultaneous Vibrational and Rotational Transitions

Associated with any given vibrational state v is a set of rotational levels. Thus, a molecule is constantly executing both rotational and vibrational motions. The rotational energy levels shown in Figure 14.9 are assumed to be associated with the $v = 0$ level. Generally, then, a $v = 0 \rightarrow 1$ transition is accompanied by a simultaneous transition between two rotational levels associated with the lower and upper vibrational states. The selection rule $\Delta J = \pm 1$ still holds in this case. In solution, however, collisions between molecules effectively prevent molecular rotation. The vibrations of molecules, on the other hand, are little affected by their neighbors because vibrational frequencies are greater than the frequency of collision. The situation is different for molecules in the gas phase, where simultaneous rotational and vibrational energy changes can occur. For example, it is possible to obtain a high-resolution spectrum for diatomic molecules in the gas phase because the rotational fine structure can be observed (Figure 14.15).

Starting with Equations 14.23 and 14.32, we can write the energy difference for a *simultaneous* transition between vibrational and rotational levels as

$$\Delta E = \left(v' + \tfrac{1}{2}\right)h\nu + BJ'(J' + 1)h - \left(v'' + \tfrac{1}{2}\right)h\nu - BJ''(J'' + 1)h \qquad (14.34)$$

where v' and v'' represent the higher and lower vibrational states, and J' and J'' the rotational levels in the v' and v'' states, respectively.* For the $v = 0 \rightarrow 1$ transition (that is, $v' - v'' = 1$), we obtain

$$\Delta E = h\nu + Bh[J'(J' + 1) - J''(J'' + 1)] \qquad (14.35)$$

or, expressed in terms of wavenumbers,

$$\Delta \tilde{\nu} = \tilde{\nu} + \frac{B}{c}[J'(J' + 1) - J''(J'' + 1)] \qquad (14.36)$$

In many cases, an IR spectrum can be divided into two branches, called P and R,

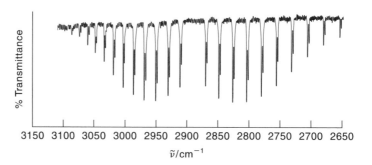

Figure 14.15
IR spectrum of HCl gas. The more intense line of each doublet is produced by $H^{35}Cl$ (75%); the weaker line is due to $H^{37}Cl$ (25%). [From J. L. Hollenberg, *J. Chem. Educ.* **47**, 2 (1970).]

---

*This equation holds if there is no interaction between rotation and vibration, or if the molecule behaves as a perfect rigid rotor. In practice, the rotational constant B is affected by the vibrational state the molecule is in, so a correction is needed for accurate work.

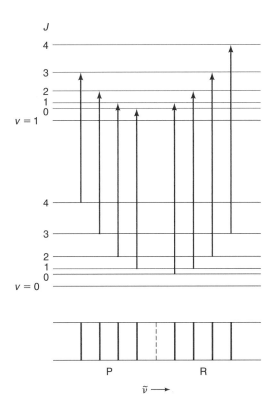

Figure 14.16
Simultaneous rotational energy level transitions accompanying the $v = 0 \rightarrow 1$ transition for a diatomic molecule.

according to the following conditions (Figure 14.16):

P branch: $J' = J'' - 1$ $\Delta \tilde{v} = \tilde{v} - \dfrac{2BJ''}{c}$ $J'' = 1, 2, 3, \ldots$ (14.37)

R branch: $J' = J'' + 1$ $\Delta \tilde{v} = \tilde{v} + \dfrac{2B(J'' + 1)}{c}$ $J'' = 0, 1, 2, \ldots$ (14.38)

Infrared spectroscopy is a highly useful technique for chemical analysis. The complexity of molecular vibration virtually assures that any two different molecules cannot produce identical IR spectra. Matching the IR spectrum of an unknown with that of a standard compound, a procedure known as *fingerprinting*, is an unequivocal method of identification. So far, over 200,000 reference spectra have been recorded and stored for fingerprinting. The details of an IR spectrum reveals much useful information about the structure and bonding of the molecule. Table 14.2 shows some characteristic frequencies and Figure 14.17 shows the IR spectrum of a relatively simple molecule, 2-propenenitrile, $CH_2 = CHCN$, and the assignment of its major peaks.

Table 14.2
Some Bond Stretching Frequencies

Bond	$\tilde{v}/\text{cm}^{-1}$	Bond	$\tilde{v}/\text{cm}^{-1}$
O–H	2500–3600	C=C	1600–1700
N–H	3300–3500	C=O	1700–1800
C–H	2800–3300	C≡C	2100–2200
C–O	1000–1200	C≡N	2200–2300

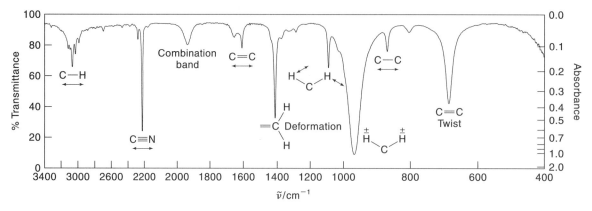

Figure 14.17
IR spectrum of 2-propenenitrile.

14.4 Electronic Spectroscopy

The absorption of radiation in the visible and UV regions involves electronic transitions and gives rise to electronic spectra. There is a major difference in the appearance of the electronic spectra of diatomic molecules and polyatomic molecules. We shall discuss diatomic molecules first because their spectra are better resolved.

Figure 14.18 shows potential-energy curves for the ground state and an excited state of a diatomic molecule and their respective vibrational energy levels, v'' and v'. According to the Boltzmann distribution, at room temperature practically all the transitions originate in the ground vibrational state. Two features of electronic transitions are worth noting. First, the selection rules $\Delta v = \pm 1$, which hold for vibrational transitions within a given electronic state, are not valid. Thus, Δv can assume any value (see Figure 14.18). Second, we can use the *Franck–Condon principle* (after the German physicist James Franck, 1882–1964, and the American physicist Edward Uhler Condon, 1902–1974) to predict the relative intensities of the bands. This principle states that because it takes much more time for a molecule to execute a vibration (about 10^{-12} s) than it does for an electronic transition (about 10^{-15} s), the nuclei do not appreciably alter their positions during an electronic transition. Therefore, the most probable (most intense) transitions are those for which the internuclear distance remains unchanged, as indicated in Figure 14.18.

In the gas phase, the electronic spectra of diatomic molecules, at high resolution, show both vibrational bands and rotational fine structure. Although these spectra are quite complex, consisting of hundreds or even thousands of lines, these lines have been assigned to specific vibrational and rotational transitions for many molecules. Such an analysis enables us to determine bond lengths of homonuclear diatomic molecules such as N_2 and I_2. An interesting note is that the bond length of I_2 is 2.67 Å in the ground electronic state, but it changes to 3.02 Å in the first excited electronic state because the transition involves the promotion of an electron to an antibonding molecular orbital.

The situation is quite different for polyatomic molecules. Large moments of inertia can make it difficult to resolve the rotational fine structure of these molecules. In solution, they usually produce electronic spectra with broad, unresolved bands. We shall briefly discuss the electronic spectra of three types of polyatomic molecules: organic molecules, transition-metal complexes, and molecules that undergo charge-transfer interactions.

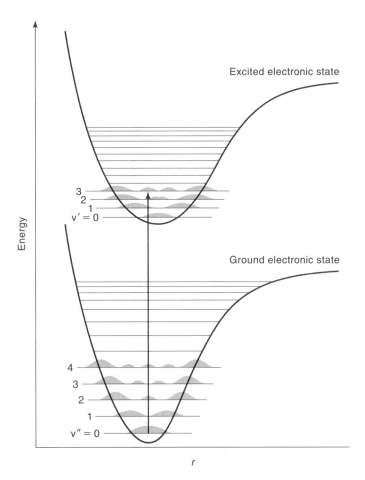

Figure 14.18
Diagram showing the most probable electronic transition of a diatomic molecule. A transition is favored (that is, it will give rise to a strong line) when the internuclear distance is such that the transition connects probable states of the molecule. In other words, the transition starts in the ground state at some point along r where the vibrational probability density function (ψ^2) is large and ends in the excited state at some point along r where the value of ψ^2 is also appreciable.

Organic Molecules

In saturated organic molecules, such as the alkanes, electronic transitions are of the $\sigma \rightarrow \sigma^*$ type. Aromatic molecules and compounds that contain $>$C=C$<$ and $>$C=O groups also have $\pi \rightarrow \pi^*$, $\sigma \rightarrow \pi^*$, and $n \rightarrow \pi^*$ transitions, where n denotes the nonbonding orbital. The $\sigma \rightarrow \pi^*$ and $n \rightarrow \pi^*$ transitions are weaker transitions because they are symmetry forbidden. Analogous to the group vibrational frequencies in IR, electronic spectra can often be characterized by special groups of atoms called *chromophores*. Table 14.3 lists absorption wavelengths of some common chromophores. The actual maximum absorption wavelength for these chromophores depends not only on the compound involved but also on the environment, because it is affected by changes in variables such as solvent and temperature.

The electronic spectra of most amino acids arise from the $\sigma \rightarrow \sigma^*$ transitions that occur in the far UV range, below 230 nm. The exceptions are phenylalanine, tryptophan, and tyrosine, all of which contain the phenyl ($-C_6H_5$) chromophore and absorb strongly above 250 nm (Figure 14.19). Absorbance at 280 nm, due mainly to the

Table 14.3
Some Common Chromophores and Their Approximate Maximum Absorption Wavelengths

Chromophore	$\lambda_{max}/$nm
\C=C/	190
\C=C—C=C/	210
(benzene ring)	190
	260
\C=O	190
	280
—C≡N	160
—COOH	200
—N=N—	350
—NO$_2$	270

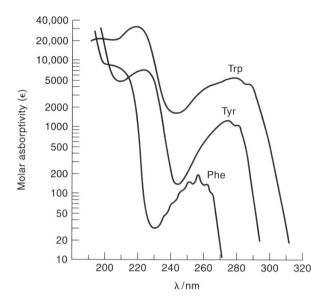

Figure 14.19
UV spectra of phenylalanine (Phe), trytophan (Trp), and tyrosine (Tyr). [From D. C. Neckers, *J. Chem. Educ.* **50**, 164 (1973).]

$\pi \rightarrow \pi^*$ transition in the tryptophan and tyrosyl residues, is useful in measuring the concentration of protein solutions.

Figure 14.20 shows the electronic spectra of purines (adenine and guanine) and pyrimidines (cytosine, thymine, and uracil), which are the components of DNA and RNA. Both DNA and RNA exhibit an interesting phenomenon called *hypochrom-*

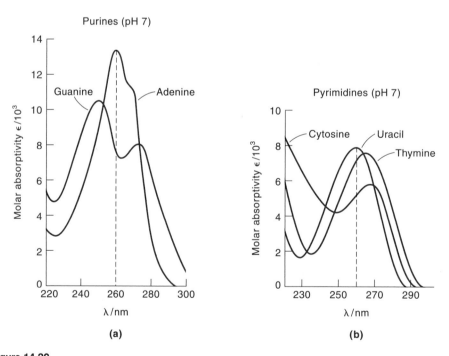

Figure 14.20
UV spectra of purines and pyrimidines. The concentration of nucleic acid solutions is determined by the absorbance at 260 nm. (From A. L. Lehninger, *Biochemistry*, 2nd ed., Worth Publishers, New York, 1975.)

ism. In general, the molar absorptivity of intact DNA is some 20% to 40% lower than we would expect it to be given the number of nucleotides present. For example, molar absorptivity of calf thymus DNA at 260 nm increases from about 6500 to 9500 L mol^{-1} cm^{-1} when the polymer undergoes thermal denaturation. Although the theory of hypochromism is beyond the scope of this text, the phenomenon is attributed to Coulombic interactions between electric dipoles induced by light absorption in the base pairs. This interaction depends on the orientation of the dipoles relative to one another. A random orientation would have little or no interaction and thus no effect on the absorption spectrum. In the native state, the dipoles are stacked parallel on top of one another, leading to a *decrease* in the absorbance. This property has been successfully employed to monitor the helix–coil transition in DNA. Figure 14.21 shows a *melting curve* of a DNA solution. The term *melting* here refers to the unwinding of the double-helical structure. The melting temperature T_m corresponds to the inflection point (the point at which the second derivative of the curve is zero); its value depends on the base-pair composition of the DNA. Because a G–C pair forms three hydrogen bonds compared to two hydrogen bonds for an A–T pair, it follows that the higher the ratio (G–C/A–T) in a polynucleotide, the larger the T_m. Note that the helix–coil transition occurs over a very narrow temperature range, which is characteristic of a cooperative transition. Thus, the double helical structure of the DNA does not unwind bit by bit. Instead, the whole structure breaks apart at the appropriate temperature.

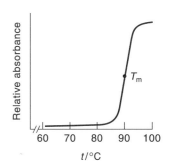

Figure 14.21
The relative absorbance of DNA at 260 nm as a function of temperature. The melting point (T_m) is about 90°C.

Transition-Metal Complexes

The color and absorption spectra of transition-metal complexes were discussed in Section 12.8. The electronic transitions occur between *d* orbital energy levels as a result of crystal-field splitting. For this reason, these transitions are commonly referred to as *d–d* transitions.

Molecules That Undergo Charge-Transfer Interactions

A special type of electronic spectrum arises from *charge-transfer* interaction between a pair of molecules. When tetracyanoethylene [(CN)$_2$C=C(CN)$_2$], an electron acceptor, dissolves in carbon tetrachloride, the resulting solution is colorless. The reason is that the $\pi \rightarrow \pi^*$ transition occurs in the UV region. Upon the addition to the solution of a small amount of an electron donor (an aromatic hydrocarbon such as benzene or toluene), the solution immediately turns yellow (Figure 14.22). Many similar reactions, including that between iodine and benzene, have been observed. In 1952, the American chemist Robert Mulliken (1896–1986) proposed the following scheme to explain the spectra of charge-transfer complexes:

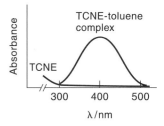

Figure 14.22
Visible absorption spectrum of the tetracyanoethylene–toluene charge-transfer complex in carbon tetrachloride. (TCNE = tetracyanoethylene.)

$$D + A \rightleftharpoons \underset{\text{ground state}}{[(D, A)]} \xrightarrow{h\nu} \underset{\text{excited state}}{[(D^+, A^-)]^*}$$

where D and A are the donor and acceptor molecules and (D, A) and (D^+, A^-) represent the covalent and ionic resonance structures of the charge-transfer complex, respectively. In the ground state, van der Waals forces hold the molecules together, and there is little, if any, actual transfer of charge from D to A. When the complex is excited at a suitable wavelength, however, a large charge transfer takes place, and the ionic structure makes a major contribution to the excited state. If the exciting wavelength falls in the visible region, the solution will appear colored. There is an interesting difference between this electronic transition and the normal absorption. In this case, an electron is excited from a lower level (bonding molecular orbital) in the donor molecule to a higher level (antibonding molecular orbital) in the acceptor molecule. The tendency for charge-transfer formation generally depends on the ionization energy of the donor and the electron affinity of the acceptor. Many transition-metal complexes also produce charge-transfer spectra. In such cases, the absorption process is accompanied by the transfer of an electron from the ligands to the metal, or from the metal to the ligands. These charge-transfer transitions often give rise to intense bands, but they can be distinguished from the d–d transitions because they fall in the far UV region, whereas most of the d–d transitions take place in the visible region (hence the color of these complex ions).

Application of the Beer–Lambert Law

The UV–visible technique is not as reliable for compound identification as is the IR technique, because an electronic spectrum generally does not possess the fine details of an IR spectrum. Electronic spectroscopy is a useful tool in quantitative analysis, however. The concentration of a solution can be determined readily using the Beer–Lambert law by measuring the absorbance (if the molar absorptivity is known). A commonly encountered case is the analysis of a solution containing two species X and Y whose absorption bands overlap. The absorbance (A) at a wavelength λ is additive so that, from the Beer–Lambert law, we have

$$A = A^X + A^Y = \varepsilon_\lambda^X b[X] + \varepsilon_\lambda^Y b[Y]$$
$$= b(\varepsilon_\lambda^X[X] + \varepsilon_\lambda^Y[Y]) \tag{14.39}$$

where ε is the molar absorptivity and b the path lenth. If the molar absorptivities of X and Y are known at two different wavelengths λ_1 and λ_2, then the absorbances are given by

$$A_1 = b(\varepsilon_1^X[X] + \varepsilon_1^Y[Y]) \tag{14.40}$$
$$A_2 = b(\varepsilon_2^X[X] + \varepsilon_2^Y[Y]) \tag{14.41}$$

From Equations 14.40 and 14.41, we can solve for [X] and [Y] to get

$$[X] = \frac{1}{b}\frac{\varepsilon_2^Y A_1 - \varepsilon_1^Y A_2}{\varepsilon_1^X \varepsilon_2^Y - \varepsilon_2^X \varepsilon_1^Y} \tag{14.42}$$

and

$$[Y] = \frac{1}{b}\frac{\varepsilon_1^X A_2 - \varepsilon_2^X A_1}{\varepsilon_1^X \varepsilon_2^Y - \varepsilon_2^X \varepsilon_1^Y} \tag{14.43}$$

Now consider a situation in which at some wavelength of the overlapped region the molar absorptivities of the two species are equal. Then the sum of the molar concentrations of these two compounds in solution is held constant and there will be *no* change in absorbance at this wavelength as the ratio of these two compounds is varied. This invariant point is called the *isosbestic point*. The existence of one or more isosbestic points in a system is a good indication of chemical equilibrium between two compounds. Figure 14.23 shows the absorption spectra of tyrosine at various pH values. There are two isosbestic points, at 267 nm and 277 nm, due to the following equilibrium process:

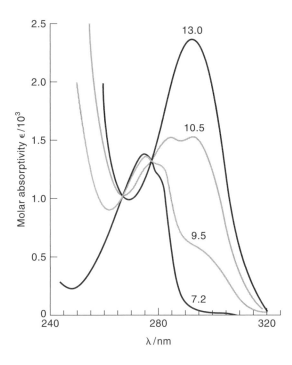

Figure 14.23
Absorption spectrum of tyrosine at pH values indicated. Note the isosbestic points at 267 nm and 277.5 nm. [From D. Schugar, *Biochem. J.* **52**, 142 (1952).]

14.5 Nuclear Magnetic Resonance Spectroscopy

Some nuclei have a spinning motion and therefore a magnetic moment associated with them. The nuclear spin, I, may have one of the following values:

$$I = 0, \tfrac{1}{2}, 1, \tfrac{3}{2}, 2, \dots$$

A value of zero means the nucleus has no spin. Table 14.4 gives rules for determining nuclear spin on the basis of atomic number and the number of neutrons present. A nucleus with spin I can adopt $(2I + 1)$ spin orientations, which are degenerate in the absence of a magnetic field. Consider the proton (^{1}H), for which $I = \tfrac{1}{2}$. The two values of its nuclear spin quantum number, m_I, are $+\tfrac{1}{2}$ and $-\tfrac{1}{2}$. When an external magnetic field is applied, the degeneracy is removed. The energy of a given spin state, E_I, is directly proportional to the value of m_I and the magnetic field strength, B_0:

$$E_I = -m_I B_0 \frac{\gamma h}{2\pi} \tag{14.44}$$

Table 14.4
Rules for Predicting Nuclear Spin

Number of Protons (Z)	Number of Neutrons[a]	Nuclear Spin, (I)
Even	Even	0
Even	Odd	$\tfrac{1}{2}$ or $\tfrac{3}{2}$ or $\tfrac{5}{2}\cdots$
Odd	Even	$\tfrac{1}{2}$ or $\tfrac{3}{2}$ or $\tfrac{5}{2}\cdots$
Odd	Odd	1 or 2 or 3 $\cdots$

[a] In the only case in which the nucleus has no neutrons (that is, the ^{1}H isotope), "0" is treated as an even number, and $I = \tfrac{1}{2}$.

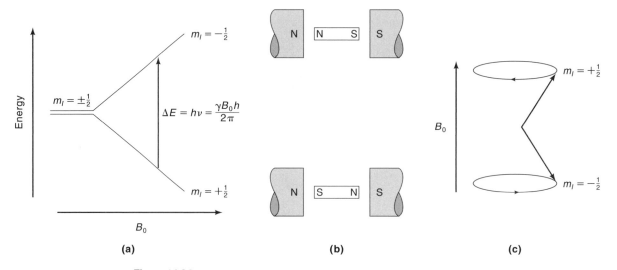

Figure 14.24
(a) Difference in nuclear spin energy levels in an external magnetic field B_0. (b) Classical description of the alignment of the nuclear spins parallel and antiparallel to the external magnetic field. (c) The precession of the nuclear spins at their Larmor frequency.

where γ is the *gyromagnetic* ratio (also called the *magnetogyric ratio*), which is a constant characteristic of the nucleus being studied. The minus sign follows the convention that a positive m_I value corresponds to a lower (negative) energy than that of its negative counterpart. Figure 14.24a shows the splitting of the nuclear spin energy levels relative to magnetic field strength for $I = \frac{1}{2}$. The energy difference, ΔE, is given by

$$\Delta E = E_{-1/2} - E_{+1/2}$$

$$= -B_0 \left[\left(-\frac{1}{2} \right) - \left(+\frac{1}{2} \right) \right] \frac{\gamma h}{2\pi}$$

$$\Delta E = \frac{\gamma B_0 h}{2\pi} \tag{14.45}$$

A nuclear magnetic resonance (NMR)—that is, a transition from the $m_I = +\frac{1}{2}$ level to the $m_I = -\frac{1}{2}$ level—can be observed by varying either the frequency, ν, of the applied radiation (given by $\Delta E/h$ or $\gamma B_0/2\pi$) or the intensity of the magnetic field, B_0, until the resonance condition ($\Delta E = h\nu$) is satisfied.* The selection rule for nuclear spin energy-level transitions is $\Delta m_I = \pm 1$.

The two magnetic moments associated with $m_I = \pm\frac{1}{2}$ are not aligned statically with or against the external magnetic field; rather, they wobble around (like a spinning top), or precess about, the axis of the applied field (Figure 14.24c). The frequency of this precession, called the *Larmor frequency* (ω), is given by

$$\omega = \gamma B_0 \tag{14.46}$$

The Larmor frequency is given in radians per second (rad s^{-1}), but it can be con-

---

*Note that unlike microwave, IR, and electronic spectroscopies, NMR spectroscopy focuses on the interaction between the magnetic-field component of electromagnetic radiation and the magnetic moment of the nuclei.

Table 14.5
Gyromagnetic Ratios, NMR Frequencies (in a 4.7-T Field), and Natural Abundances of Isotopes

Isotope	I	$\gamma/10^7$ T$^{-1} \cdot$ s^{-1}	ν/MHz	Natural Abundance (%)
^{1}H	$\frac{1}{2}$	26.75	200	99.985
^{2}H	1	4.11	30.7	0.015
^{13}C	$\frac{1}{2}$	6.73	50.3	1.108
^{14}N	1	1.93	14.5	99.63
^{15}N	$\frac{1}{2}$	2.71	20.3	0.37
^{17}O	$\frac{5}{2}$	3.63	27.2	0.037
^{19}F	$\frac{1}{2}$	25.17	188.3	100
^{31}P	$\frac{1}{2}$	10.83	81.1	100
^{33}S	$\frac{3}{2}$	2.05	15.3	0.76

verted into linear frequency, ν, as follows (see Appendix 1)

$$\nu_{\text{precession}} = \frac{\omega}{2\pi} = \frac{\gamma B_0}{2\pi} \tag{14.47}$$

This precession frequency is independent of m_I, so all spin orientations of a given nucleus precess at this frequency in a magnetic field. Note that Equation 14.47 looks the same as the frequency for observing nuclear magnetic resonance mentioned above. The reason is that the frequency of the applied radiation must be equal to the Larmor frequency for resonance to occur.

The strength of a magnetic field is measured in *tesla* (T), after the Serbian engineer and inventor Nikola Tesla (1856–1944), where

$$1 \text{ T} = 10^4 \text{ gauss}$$

The gyromagnetic ratio has the units T^{-1} s^{-1}. Table 14.5 lists the gyromagnetic ratios, NMR frequencies (in a 4.7-T field), and natural abundances of several isotopes. These frequencies are in the radiofrequency region. For a given nucleus, the larger the γ, the easier it is to detect the corresponding NMR signal. Thus, the most readily studied nuclei are ^{1}H, ^{19}F, and ^{31}P. With modern instrumentation, however, the NMR of the ^{13}C nucleus, which has a small γ value and very low natural abundance but is of great importance in organic chemistry and biochemistry, can be studied with ease.

Example 14.3

Calculate the magnetic field, B_0, that corresponds to a precession frequency of 400 MHz for ^{1}H.

ANSWER

From Equation 14.47 and Table 14.5,

$$B_0 = \frac{2\pi\nu}{\gamma}$$

$$= \frac{2\pi(400 \times 10^6 \text{ s}^{-1})}{26.75 \times 10^7 \text{ T}^{-1} \text{ s}^{-1}}$$

$$= 9.40 \text{ T}$$

The Boltzmann Distribution

Because NMR is a branch of absorption spectroscopy, its sensitivity is governed by the Boltzmann distribution. Consider a sample of ^1H nuclei in a magnetic field measuring 9.40 T at 300 K. From Equation 14.45, we have

$$\Delta E = \frac{(26.75 \times 10^7 \text{ T}^{-1}\text{ s}^{-1})(9.40 \text{ T})(6.626 \times 10^{-34}\text{ J s})}{2\pi}$$

$$= 2.65 \times 10^{-25}\text{ J}$$

and $k_B T = 4.14 \times 10^{-21}$ J. Thus, the ratio of the number of nuclear spins in the upper energy level to that in the lower energy level is

$$\frac{N_{-1/2}}{N_{+1/2}} = e^{-\Delta E/k_B T}$$

$$= \exp\left(\frac{-2.65 \times 10^{-25}\text{ J}}{4.14 \times 10^{-21}\text{ J}}\right)$$

$$= 0.99994$$

This number is very close to one, meaning that the two levels are almost equally populated.* This distribution is the result of strong thermal motion in the sample, which overwhelms the tendency to orient the spins in a magnetic field. Nevertheless, even the slight excess of spins in the lower level is sufficient to give rise to detectable NMR signals.

Chemical Shifts

The discussion so far may seem to suggest that all protons resonate at the same frequency, but this is not the case. In reality, at a particular magnetic field strength, the resonance frequency for any given ^1H nucleus depends on its position in the molecule under study. This effect, called the *chemical shift*, is what makes NMR spectroscopy so useful.

Figure 14.25a shows the proton NMR spectrum of ethanol (CH_3CH_2OH). The three peaks of relative areas $1:2:3$ correspond to the hydroxyl, methylene, and methyl protons, respectively. The fact that three separate peaks are observed means that the local magnetic field present at each type of nucleus is different from the external magnetic field B_0. According to electromagnetic theory, an electric current moving through a wire generates a magnetic field. In an analogous way, the external magnetic field (B_0) causes the electrons surrounding the nucleus to circulate through the orbitals in such a way as to generate an induced magnetic field, which is *opposed*

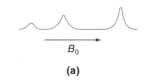

(a)

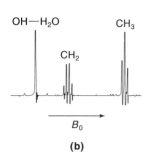

(b)

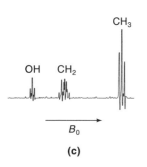

(c)

Figure 14.25
(a) Low-resolution and (b) high-resolution proton NMR spectrum of ethanol. (c) NMR spectrum of anhydrous pure ethanol. [Parts (b) and (c) from G. Glaros and N. H. Cromwell, *J. Chem. Educ.* **48**, 202 (1971).]

---

*This ratio is much smaller (favoring the absorption process) for IR and electronic spectroscopy because of the appreciably larger separation in energy levels.

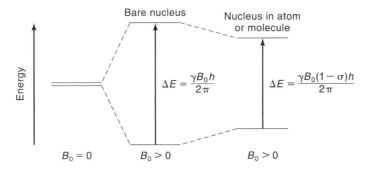

Figure 14.26
Effect of electron shielding on the nuclear magnetic resonance condition.

to the external field. The net magnetic field experienced by the nucleus (B) is given by

$$B = B_0(1 - \sigma) \tag{14.48}$$

where σ, a dimensionless constant, is called the *screening constant* or the *shielding constant*, whose value depends on the electronic structure around the nucleus. As a result of this shielding, the resonance frequency for a given nucleus becomes

σ is a small number for protons, about 10^{-5}.

$$\nu = \frac{\gamma B_0(1 - \sigma)}{2\pi} \tag{14.49}$$

Figure 14.26 shows the modified resonance condition. Because the protons in ethanol are in three different chemical environments and hence experience three different values of B, we see three signals. Keep in mind that the greater the shielding of the nucleus (the larger the σ), the lower will be its resonance frequency and the farther to the right it will appear in an NMR spectrum. For ethanol, then, the methyl protons are the most shielded and the hydroxyl proton the least shielded.

In general, we are interested not in the absolute shifts of the NMR peaks from that expected for a bare proton, but in the relative positions of the peaks. Thus, it is common practice to define the chemical shift by the difference in resonance frequencies between a nucleus of interest (ν) and a reference nucleus (ν_{ref}) in terms of the chemical shift parameter (δ), where

$$\delta = \frac{\nu - \nu_{ref}}{\nu_{spec}} \times 10^6 \tag{14.50}$$

where ν_{spec} is the spectrometer frequency. Because the difference between ν and ν_{ref} is typically on the order of hundreds of hertz while ν_{spec} is typically hundreds of megahertz, the ratio is multiplied by 10^6 to make δ a convenient number with which to work. For this reason, chemical shifts are expressed in units of ppm (parts per million). Note that the frequency difference ($\nu - \nu_{ref}$) is divided by ν_{spec}. This means that δ is *independent* of the magnetic field used to measure it. The reference compound chosen for most organic systems is tetramethylsilane (TMS), $(CH_3)_4Si$, because it has the following advantages: (1) it contains 12 protons of the same type so only a small amount is needed as an internal reference, (2) it is chemically inert, and (3) its protons have a smaller resonance frequency than that observed for most other protons, so their chemical shifts can be assigned positive values. Table 14.6 shows the chemical shifts of some protons relative to TMS in ppm.

By convention, the chemical shift of TMS is 0 ppm.

Table 14.6
Chemical Shift Ranges for Protons*

Type of Proton	δ/ppm	Type of Proton	δ/ppm
C–H	0–2	ArOH	4.0–9.5
X–C–H	1.6–6.0	C=C–H	4.5–8.0
H–C≡C	1.8–3.0	ArH	6.0–9.0
ROH, RNH_2	0.4–5.0	RCHO	9.5–10.5
$RCONH_2$	4.0–9.5	RCOOH	9.7–12

* Based on $\delta = 0$ ppm for TMS.

Conventionally, NMR spectra are plotted with ν (and δ) increasing from right to left. Sometimes chemists refer to chemical shifts as "upfield" or "downfield," meaning "more shielded" or "less shielded," respectively. Chemical shifts can be readily converted back to the frequencies separating the sample and reference peaks using Equation 14.50. For example, the chemical shift of benzene is about 7.3 ppm so that, if the spectrometer operates at 200 MHz frequency,

$$\nu_{benzene} - \nu_{TMS} = \delta \times \nu$$
$$= (7.3 \times 10^{-6})(200 \times 10^6 \text{ Hz})$$
$$= 1.46 \times 10^3 \text{ Hz}$$

You can see that if the signals were recorded with a 400-MHz spectrometer, the difference in frequencies would be 2.92×10^3 Hz. So the separation between peaks in a given NMR spectrum is directly proportional to the spectrometer frequency (or the magnetic field), but δ is independent of the frequency. For this reason, high-field NMR (approaching 1000 MHz as of 2004) is becoming increasingly popular in the study of protein solutions, for which separation of the many overlapping peaks observed is critical to meaningful analysis.

Another advantage of high-field NMR is the increase in sensitivity due to the Boltzmann distribution.

Spin–Spin Coupling

At high resolution, the spectrum of ethanol is as shown in Figure 14.25b. The –CH_2 and –CH_3 peaks actually consist of four and three lines, respectively, with relative intensities of $1:3:3:1$ and $1:2:1$. The spacing between each group of lines is *independent* of the spectrometer frequency. Therefore, it cannot be a chemical-shift effect as discussed above. How can we explain this observation? Each nucleus with $I \neq 0$ has a nuclear magnetic moment, and the magnetic field generated by this nucleus can affect the magnetic field experienced by a neighboring nucleus, thereby slightly changing the frequency at which the neighboring nucleus will undergo NMR absorption. In the liquid or gas phase, where rapid molecular rotation occurs, the direct nuclear spin–spin interaction, called dipole interaction, averages to zero.

There is, however, an additional, indirect interaction between the nuclear spins that is transmitted through the bonding electrons. This interaction is unaffected by molecular rotation and causes splitting of the NMR peaks. In ethanol, with two possible orientations for each nuclear spin in the methylene group, the methyl peak is split into two lines by the magnetic field generated by the first methylene proton. Each of these two lines is then further split by the second methylene proton, and a total of four lines result. We see only three lines for –CH_3 because two of the lines fall on top of each other, giving rise to the observed intensity pattern of $1:2:1$. Similarly, four lines are obtained for the –CH_2 group due to splitting by the methyl protons (Figure 14.27). The separation between the lines in each group gives the *spin–spin*

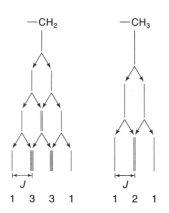

Figure 14.27
Spin–spin splitting between –CH_2 and –CH_3 groups in ethanol that gives rise to the triplet and quartet in the NMR spectrum. The coupling constant (J) is the same in both cases.

Table 14.7
The Coefficients of the Binomial Distribution*

n			Intensity Ratio				Multiplicity
0				1				singlet
1			1		1			doublet
2		1		2		1		triplet
3		1	3		3		1	quartet
4	1		4	6		4	1	quintet
5	1	5	10		10	5	1	sextet

*The coefficients are generated by the equation $(1 + x)^n$. The intensity ratio starts with the number 1, and is followed by the coefficients of $x, x^2, \ldots$.

coupling constant (J), whose magnitude is determined by the extent of this magnetic interaction. The following points are worth noting about spin–spin coupling:

1. Nuclei must be magnetically nonequivalent to produce spin–spin coupling. For example, the protons in the methyl group in ethanol are magnetically equivalent, so they do not interact with each other. They cause the splitting of the methylene peak only because the methylene protons are magnetically nonequivalent to the methyl protons.

2. Spin–spin coupling is observed only for two nuclei separated by no more than three bonds.

3. For ^1H (or any nucleus with $I = \frac{1}{2}$), the splitting of a line by a group of n equivalent protons is governed by the $(n + 1)$ rule, and the intensities are given by the *binomial distribution* (Table 14.7). The binomial distribution satisfactorily explains the NMR splitting patterns in ethanol and other hydrogen-containing compounds.

NMR and Rate Processes

To finish our discussion of the ethanol spectrum, we must account for the absence of spin–spin interaction between the methylene and hydroxyl groups. Actually, in pure ethanol, the hydroxyl peak is indeed split into a $1:2:1$ triplet by the methylene group, and each of the four lines in the methylene group is further split into a doublet of equal intensity by the hydroxyl proton (Figure 14.25c). There is no observable splitting between the –OH and –CH_3 groups because these protons are separated by more than three bonds. In the presence of a small amount of water, a rapid proton-exchange reaction between the –OH group and H_2O and between C_2H_5OH and protonated C_2H_5OH effectively eliminates the spin–spin interaction between the –OH and –CH_2 groups:

$$C_2H_5OH' + H_3O^+ \rightleftharpoons C_2H_5OHH'^+ + H_2O$$

$$C_2H_5OHH'^+ + C_2H_5OH \rightleftharpoons C_2H_5OH + C_2H_5OHH'^+$$

We use H' to show the proton involved in the exchange reaction.

In fact, NMR spectroscopy is convenient for studying the rates of proton-exchange reactions and of many other chemical processes such as rotation about a chemical bond and ring inversion. Consider, for example, the conformational

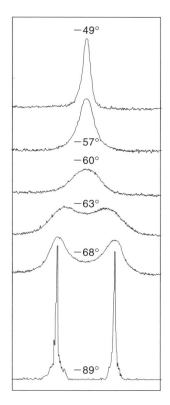

Figure 14.28
Proton NMR spectra of deuterated cyclohexane ($C_6D_{11}H$) at various temperatures (degrees Celsius). (From F. A. Bovey, *Nuclear Magnetic Resonance Spectroscopy*, Academic Press, Inc., New York, 1969.)

change, or "ring inversion," that occurs in cyclohexane:

The NMR spectrum of cyclohexane is rather complex because of spin–spin interactions. By using a deuterated compound—$C_6D_{11}H$—and applying a procedure known as *spin decoupling*, however, we eliminate these interactions, which leaves only two lines for observation, one representing the proton in the axial position and the other corresponding to the equatorial position (Figure 14.28). At $-89°C$, the ring-inversion rate is very slow so that two lines are observed corresponding to a state in which only half of the protons in the sample are in the axial position and half are in the equatorial position. Warming the sample causes the peaks to broaden. At $-60°C$, the peaks merge into a single line, which sharpens as temperature is increased further. This chemical exchange (between the axial and equatorial positions) can be understood in terms of the Heisenberg uncertainty principle:

$$\Delta E = \frac{h}{4\pi\tau}$$

or

$$\Delta v = \frac{1}{4\pi\tau}$$

where τ is the mean lifetime of the proton in a particular magnetic environment, and Δv is the width of an NMR line. The exchange process causes the lifetime to shorten, leading to a large Δv and hence a broadening of the line that exceeds its natural line width (lifetime broadening). The ring-inversion rate increases rapidly with temperature. When the exchange rate, $1/\tau$, is large compared with the frequency difference between the two lines, the spectrum collapses into a single line. At a still higher rate of inversion, the two protons change sites so rapidly that the system behaves as though there were only one type of proton present. The spectrum observed is called the exchange-narrowing region (the peak recorded at $-49°C$). From an analysis of the change in line width with temperature, the energy of activation for the cyclohexane ring inversion is found to be 42 kJ mol^{-1}.

NMR of Nuclei Other Than ^1H

Proton magnetic resonance is the most frequently encountered form of NMR, but other nuclei are also important to investigations of chemical and biological systems. As a NMR nucleus, ^{13}C is second in popularity only to ^1H, thanks to the development of Fourier-transform spectroscopy. The spin–spin coupling discussed for protons also occurs between the ^{13}C nucleus and any protons attached to it. Thus, the ^{13}C NMR spectrum of a methylene group is observed as a triplet due to the interaction of ^{13}C with the two protons. The low natural abundance of the ^{13}C isotope does have an advantage: we do not observe ^{13}C–^{13}C splitting because the probability of two ^{13}C atoms being bonded to each other is very small. In practice, ^{13}C NMR spectra are *proton-decoupled*, a procedure that eliminates all ^{13}C–^1H splitting so that the spectra are more easily analyzed.* Moreover, the chemical shifts of ^{13}C NMR span a range about 250 ppm in magnitude, which is an order of magnitude larger than the range for protons.

In addition to ^{13}C, the isotopes ^{15}N, ^{19}F, and ^{31}P are important in NMR spectroscopy because these elements are found in many chemical and biological compounds. As an interesting example, note that in the 1990s chemists detected spin–spin coupling between two nitrogen-15 nuclei taking part in hydrogen bonding—that is, N–H$\cdots$N. This finding was significant for it provides strong evidence that this hydrogen bond, and indeed all hydrogen bonds in general, possess some covalent character. We saw earlier that spin–spin coupling is transmitted through bonding electrons; therefore, such an interaction could not arise if hydrogen bonds were purely electrostatic attractions. Thus, there must be some overlap of the wave functions between the donor group (N–H) and the acceptor (N).

The NMR finding is consistent with the X-ray experiment discussed on p. 502.

Figure 14.29 shows the ^{1}H, ^{13}C, and ^{31}P NMR spectra of a small but important biomolecule, adenosine-5′-triphosphate (ATP).

Fourier-Transform NMR

In recent years, the technique of Fourier transforms has had a profound impact on several branches of spectroscopy. Here we shall discuss Fourier-transform NMR (FT-NMR).

As mentioned in Section 14.5, nuclear magnetic resonance can be observed either by keeping the external magnetic field constant and varying the radio frequency (rf) of the applied radiation or by keeping the radio frequency constant and sweeping the magnetic field until the resonance condition is satisfied. Technically it is simpler to maintain a constant radio frequency and vary the magnetic field. In either case, however, the spectrometer operates in the cw (continuous wave) mode because radiation is supplied continuously to the sample. As a result, it takes minutes to record an NMR spectrum with such an instrument.

To understand how FT-NMR works, consider a sample consisting of many identical spins of $I = \frac{1}{2}$. When placed in a strong external magnetic field B_0, the nuclei align themselves parallel or antiparallel to the applied field, with a very slight excess in parallel alignment, which corresponds to the lower energy level. The net magnetization, M, then precesses around B_0 (the z axis) at the Larmor frequency (Figure 14.30). In a pulsed NMR experiment, a single, short, intense burst of magnetic field of strength B_1 along the x axis is applied to the sample. As a result, the net magnetization is rotated through an angle α given by

$$\alpha = \gamma B_1 t_{\mathrm{p}} \tag{14.51}$$

where γ is the gyromagnetic ratio and t_{p} is the duration of the applied pulse, which is on the order of microseconds. The appropriate choice of t_{p} causes the magnetization to rotate from the z axis to the y axis. (This is called a 90° pulse because $\alpha = 90°$.) The NMR signal is measured by a detecting coil along the y axis. Immediately after the pulse (when B_1 is turned off), the magnetization vector begins to rotate in the xy plane at the Larmor frequency. Relaxation mechanisms subsequently cause the magnetization to decrease along the y axis, and eventually, at thermal equilibrium, it

---

*Decoupling is achieved by irradiating the sample at the ^{1}H resonance frequency while the ^{13}C spectrum is being recorded. The ^{13}C NMR spectrum of a C–H group shows a doublet (assuming no other magnetic nuclei are present). When the power at the decoupling frequency is large enough, the rate of change of the ^{1}H spin orientations becomes much greater than the coupling constant, and the doublet collapses into a single peak.

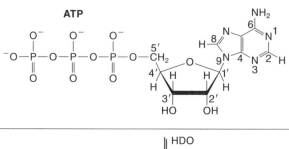

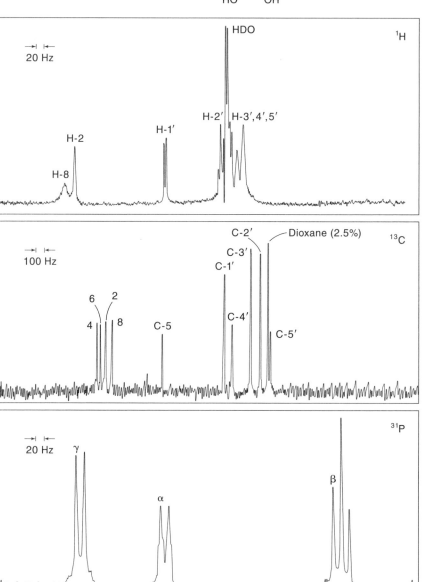

Figure 14.29
^{1}H, ^{13}C, and ^{31}P NMR spectra of adenosine-5′-triphosphate (ATP). Note that the ^{13}C spectrum is proton-decoupled so that only the chemical shifts of different types of carbon atoms are shown. (By permission of Varian Associates, Palo Alto, CA.)

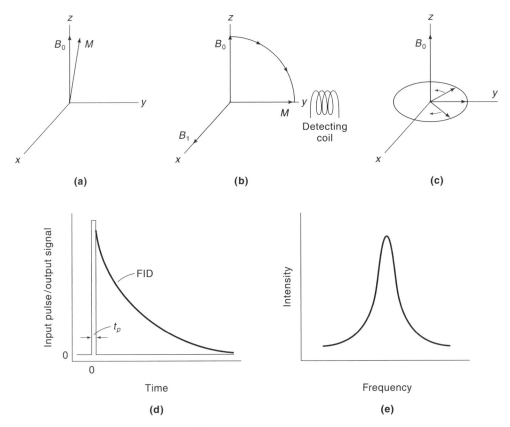

Figure 14.30
(a) Precession of the net magnetization of a collection of spins about an external magnetic field B_0 along the z axis. (b) A 90° pulse (by a rf field B_1 along the x axis) flips the magnetization vector along the y axis, where the detecting coil is located. (c) Right after the pulse, the spins begin to precess in the xy plane. As the precession continues, the magnetization vector gives a signal with alternating maxima and minima. (d) The time sequence between the 90° pulse and the subsequent free induction decay. (e) Fourier transform of the FID $[f(t)]$ produces an intensity versus frequency spectrum.

is restored along the z axis. The decay in NMR signal as a function of time is called the *free induction decay* (FID). Finally, by applying a Fourier transformation (see below), the FID is converted to an absorption peak.

Figure 14.30d applies to the situation in which the nuclei are identical and the rf radiation (B_1) is chosen to match the Larmor frequency. More often than not, we study nuclei that differ in Larmor frequency as a result of chemical shifts and spin–spin coupling. In those cases, different groups of nuclei precess at different frequencies, and interference effects can occur so that the FID will have a much more complex appearance.

An essential feature of pulsed NMR is that we can excite simultaneously nuclei with different chemical shifts even if we apply an rf field at just one frequency. Consider a pulse of 10 μs (1 μs $= 10^{-6}$ s) duration. From the Heisenberg uncertainty principle,

$$\Delta E \Delta t = \frac{h}{4\pi}$$

Because $\Delta E = h\Delta v$, we have

$$\Delta v = \frac{1}{4\pi\Delta t}$$

$$= \frac{1}{4\pi(10 \times 10^{-6}\ \text{s})}$$

$$\approx 8 \times 10^3\ \text{s}^{-1}$$

$$= 8\ \text{kHz}$$

This frequency range is broad enough to cover most proton chemical shifts. Because of the difficulty in interpreting $f(t)$, the function that describes the variation in intensity of the signal with time associated with the FID, we must convert it to the more recognizable form $I(v)$, which is a function of variation in intensity with frequency. These two spectral functions are related by the Fourier transformation as follows:

$$f(t) = \int_{-\infty}^{+\infty} I(v)\cos(vt)dv \tag{14.52}$$

$$I(v) = \int_{-\infty}^{+\infty} f(t)\cos(vt)dt \tag{14.53}$$

Figure 14.31a shows the FID for acetaldehyde (CH_3CHO). The FID curve is so complex because it arises from the precession of a magnetization vector that is composed of six components (from the six peaks), each of which precesses with a characteristic frequency. By performing the Fourier transformation using Equation 14.53,

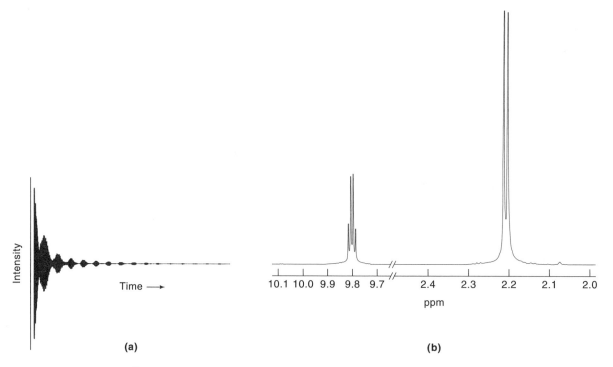

(a)   (b)

Figure 14.31
The FID (a) and its Fourier-transform spectrum (b) of acetaldehyde (CH_3CHO).

we obtain the conventional NMR spectrum of acetaldehyde in which the intensities of the peaks are plotted against frequency (Figure 14.31b).

Because of the rapidity with which data can be collected and processed, FT-NMR enables one to record literally hundreds or thousands of similar spectra over a relatively short time period for signal averaging. In addition, the advances in instrumentation in recent years have made NMR one of the most powerful and versatile spectroscopic techniques.

Magnetic Resonance Imaging (MRI)

MRI is a noninvasive technique for obtaining cross-sectional pictures through the human body without exposing the patient to ionizing radiation, as in X-ray computerized tomography, or CT scanning. Figure 14.32 illustrates the basic idea behind MRI. Two water samples are separated in space (as they might be for water in two different regions of the body). The normal NMR spectrum of water shows a single peak, because the two protons are magnetically equivalent. Suppose that, in addition to the normal magnetic field B_0, a gradient magnetic field, G_x, is also applied along the x direction. Because the gradient field varies with distance, the field at the left cylinder will be slightly different from that at the second cylinder. After a pulsed NMR experiment, the Fourier transform of the FID will now show two peaks

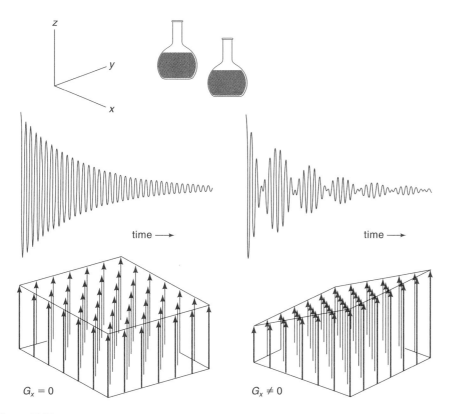

Figure 14.32
Top: Two small water-filled cylinders. Left: In the absence of a gradient, the B_0 field (along the z axis) is the same at all points in the cylinders as depicted by the set of vectors. Application of a short rf pulse yields an FID, which gives rise to a single line. Right: Here, a gradient field G_x is applied along the x axis. The water molecules in the two cylinders are now exposed to different magnetic fields. Consequently, the FID will give rise to two signals.

instead of one. To make a three-dimensional image, however, we need to apply gradient fields along the x, y, and z axis. Furthermore, the signals are shown as intensity versus distance rather than intensity versus frequency.

14.6 Electron Spin Resonance Spectroscopy

Electron spin resonance (ESR), also called electron paramagnetic resonance (EPR), is very similar to NMR in theory. The electron has a spin of $S = \frac{1}{2}$. The spinning motion of the electron generates a magnetic field, and the orientation of the electron magnetic moment in an external magnetic field (B_0) is characterized by the electron spin quantum numbers, $m_s = \pm\frac{1}{2}$. The resonance condition is given by

$$\Delta E = h\nu = g\beta B_0 \tag{14.54}$$

where g is a dimensionless constant called the Landé g factor, which, for a free electron, is equal to 2.0023*; β is the Bohr magneton, given by $eh/2\pi m_e c$, where e and m_e are the electronic charge and mass, respectively; and c is the speed of light (Figure 14.33a).

Because the magnetic moment for an electron is about 600 times greater than that for a proton, ESR measurements are usually carried out in a magnetic field of about 0.34 T at a frequency of 9.5×10^9 Hz, or 9.5 GHz, which falls in the microwave region. Most spectrometers are designed to present the ESR lines as the first derivative of the absorption lines (Figure 14.34).

Although isolated electrons, or electrons trapped in a matrix, undergo only one transition and therefore only one line is observed, the ESR spectrum of the hydrogen

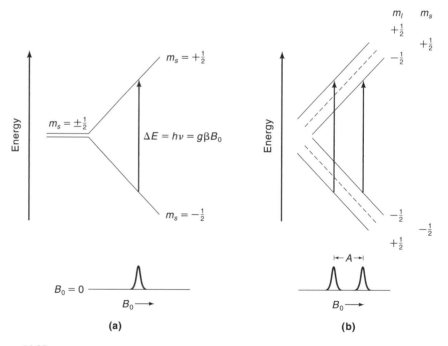

Figure 14.33
(a) Resonance condition for an electron. (b) Resonance conditions for the electron in a hydrogen atom. A is the hyperfine splitting constant.

---

*The g factor is the ratio of the electron's magnetic moment to its spin angular momentum.

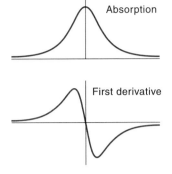

Figure 14.34
Relation between an absorption line
and its first derivative.

atom consists of two lines of equal intensity, as shown in Figure 14.33b. This *hyper-fine splitting* results from the magnetic interaction between the unpaired electron and the nucleus, which is analogous to the spin–spin interaction discussed earlier for NMR. Only two transitions are allowed, however, because of the selection rules $\Delta m_s = \pm 1$ and $\Delta m_I = 0$. One interpretation of the selection rules is that the motion of a nucleus is much slower than that of an electron, so that during the time it takes for an electron to change its orientation, the nuclear spin has no time to reorient. The separation between these two lines is the *hyperfine splitting constant* (A). In general, the number of hyperfine lines can be predicted by the quantity $(2nI + 1)$, where n is the number of equivalent nuclei, and I is the nuclear spin. As in NMR, the intensity of the lines arising from proton hyperfine splitting is given by the binomial distribution.

In most molecules, electrons regularly are paired with electrons that have opposite spins, as required by the Pauli exclusion principle; hence, ESR experiments cannot be performed on them. A few molecules, including NO, NO_2, ClO_2, and O_2, do contain one or more unpaired electrons in their electronic ground states. The ESR spectra for these molecules have been studied. It is also possible to reduce diamagnetic molecules by chemical or electrochemical means, converting them to anion radicals. For example, when benzene and naphthalene are dissolved in an inert organic solvent, such as tetrahydrofuran, and are treated with potassium metal in the absence of oxygen and water, the benzene and naphthalene anion radicals are generated (Figure 14.35a,b):

$$C_6H_6 + K \rightarrow C_6H_6^- K^+$$
$$C_{10}H_8 + K \rightarrow C_{10}H_8^- K^+$$

An important class of stable, neutral radicals are the nitroxides. In these molecules, the unpaired electron is localized on the nitrogen and oxygen atoms. An example is the di-*tert*-butyl nitroxide radical:

$$(CH_3)_3C \diagdown \underset{\underset{O\cdot}{|}}{N} \diagup C(CH_3)_3$$

Because ^{16}O has no magnetic moment $(I = 0)$, the hyperfine splittings are due solely to the ^{14}N nucleus; a total of three lines of equal intensity are observed (Figure 14.35c).* Because of their stability and the simplicity of their ESR spectra, nitroxides

The natural abundance of ^{14}N is 99.63%.

---

* The nuclear spin of ^{14}N is 1, so according to the $(2nI + 1)$ rule, we have $2 \times 1 \times 1 + 1 = 3$.

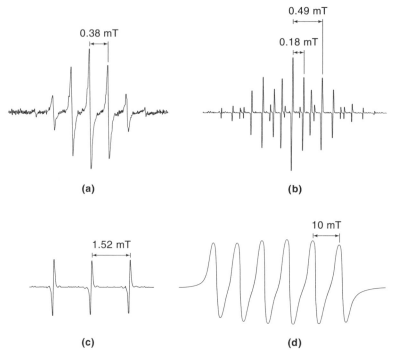

Figure 14.35
ESR spectra: (a) benzene anion radical; (b) naphthalene anion radical; (c) di-*tert*-butyl nitroxide radical; and (d) Mn^{2+} in water. Note that there are two types of protons in naphthalene and therefore two different coupling constants.

have been used extensively as spin labels to probe the structure and dynamics of proteins.

Many transition-metal ions contain unpaired d electrons and are particularly suited for ESR studies.* Of particular interest are the Cu^{2+}, Co^{2+}, Fe^{3+}, Ni^{3+}, and Mn^{2+} ions because they occur in biological systems. The ESR spectrum of Mn^{2+} $\left(I = \frac{5}{2}\right)$ gives rise to six equally spaced lines (Figure 14.35d). The Cu^{2+} ion is specially amenable to ESR investigation because it has only one unpaired electron. Due to hyperfine interaction (both ^{63}Cu and ^{65}Cu have nuclear spin $I = \frac{3}{2}$), its spectrum gives rise to a four-line pattern.

14.7 Fluorescence and Phosphorescence

If the excitation of a molecule by light—the basis for the spectroscopic techniques considered so far—does not lead to a chemical reaction, such as dissociation or rearrangement, or to energy transfer by collision, then the molecule will eventually return to the ground state through the release of a photon of energy $h\nu$. The liberated photon shows up as luminescent emission. In luminescence, there are two paths for energy depletion in the excited molecule, fluorescence and phosphorescence.

Fluorescence

Fluorescence is the emission of radiation that causes the transition of an electron from an excited state to a ground state without any change in spin multiplicity. Be-

---

* According to a theorem by the Dutch physicist Hendrik Kramers (1894–1952), the most suitable ions for ESR study are those containing an *odd* number of electrons.

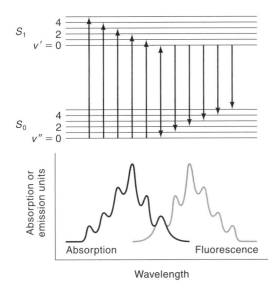

Figure 14.36
Relation between absorption and fluorescence.
(Adapted from D. A. McQuarrie and J. D. Simon,
Physical Chemistry, University Science Books,
Sausalito, CA, 1997.)

cause electrons in molecules are paired according to the Pauli exclusion principle and most molecules have no net electron spin, the initial absorption is from the ground-state singlet, S_0, to the first excited singlet, S_1 (or some higher singlet level). At first, fluorescence might appear to be exactly the reverse of the absorption process. This is true on the atomic level, but a comparison of the absorption and emission spectra of molecules shows them not to be superimposed. Instead, they usually form the mirror image of each other, and the emission spectrum is displaced toward the longer wavelengths (Figure 14.36). Because the time required for vibrational energy transfer (about 10^{-13} s) is much shorter than the decay or mean lifetime of the fluorescent state (about 10^{-9} s), most of the excess vibrational energy dissipates to the surroundings as heat, and the electronically excited molecules will decay from their ground vibrational levels.

We define the quantum yield of fluorescence emission, Φ_F, as the ratio of photons emitted through fluorescence to the total number of photons originally absorbed. The maximum value of Φ_F is 1, although it can be appreciably smaller than 1 if other processes are present to deactivate the excited molecules, as is usually the case. The intensity of radiation emitted after the exciting light is turned off is given by

$$I = I_0 e^{-t/\tau} \tag{14.55}$$

where I_0 is the intensity at $t = 0$, I is the intensity at time t, and τ is the mean lifetime of the fluorescent state. The mean lifetime is equal to the time it takes for the original intensity to decrease to $1/e$, or 0.368, of its original value. Thus, when $t = \tau$, $I = I_0/e$. Equation 14.55 shows that the decay of fluorescence obeys first-order kinetics (see Equation 9.7), and the rate constant, k, for the decay is given by $1/\tau$.

Liquid Scintillation Counting. The fluorescence technique, besides yielding information about the electronic structure of molecules in excited states, is also useful in chemical and biochemical analysis. For example, it is employed in *liquid scintillation counting*, a common method of assaying radioactive tracer-labeled compounds containing ^3H, ^{14}C, ^{32}P, and ^{35}S. Scintillators are compounds that can be excited either in the solid state or in solution. The intensity of the subsequent fluorescence of these compounds is related to the amount of the exciting source present. The general procedure in liquid scintillation counting is to first dissolve the scintillator (called a fluor) in a solvent (toluene or dioxane, depending on the nature of the sample to be studied), giving what is referred to as a "cocktail." Then, the radioactive sample is added to the cocktail, and the following sequence of events takes place: (1) The solvent molecules are excited by bombardment with β particles emitted from the radioactive nuclei; (2) the excited solvent molecules transfer the energy to the scintillator; (3) the fluorescence of the scintillator molecules is measured; and (4) the amount of radioactive nuclei present in the original sample is determined from previously calibrated fluorescence versus concentration measurements.

The fluor, characterized by a large Φ_F value, is excited by energy from the solvent (say, toluene), received via a nonradiative mechanism such as collision.* This singlet–singlet energy transfer is represented by

$$D(S_1) + A(S_0) \rightarrow D(S_0) + A(S_1)$$

Here the donor molecule (D) is the excited toluene, and the acceptor molecule (A) is the fluor. If the wavelengths of the photons emitted by the fluor are not in the region of highest sensitivity of the detector, a second fluor is added. The secondary fluor absorbs the photons emitted by the primary fluor and re-emits them as fluorescence at a longer wavelength, which is better suited for the detector. The most commonly used primary fluor is 2,5-diphenyloxazole (PPO), and the most commonly used secondary fluor is 1,4-bis-2-(5-phenyloxazole) benzene (POPOP):

PPO   POPOP

Phosphorescence

Phosphorescence offers a different path for the return of an excited molecule to the ground state with the emission of light. Phosphorescence can be distinguished readily from fluorescence by two characteristics. First, phosphorescence has a much longer decay period than fluorescence, about 10^{-3} s to several seconds. Second, a molecule in the phosphorescent state is paramagnetic, containing two unpaired electrons; that is, it is in a triplet state. The relation between the excited singlet and triplet electronic states is conveniently illustrated by the *Jablonski diagram* (after the Polish physicist Alexander Jablonski, 1898–1980), shown in Figure 14.37. Initially, an electron is promoted from S_0 to S_1. The promotion is followed by a process called *radiationless transition*, in which the electron flips its spin and drops from S_1 to T_1, the lowest triplet level, without the emission of light. In the end, a radiative transition from T_1 to S_0 occurs. This step is called *phosphorescence*. Because the transition involves a change in spin multiplicity (from triplet to singlet), it is spin forbidden and therefore has a low probability, accounting for the long lifetimes observed. Because the excited state (T_1) is easily deactivated by collision due to its long lifetime, phosphorescence, unlike fluorescence, cannot be studied in a liquid phase. Phosphorescence is best studied when the sample is in a frozen clear glass at or below the temperature of liquid nitrogen (77 K).

---

*Another mechanism, which is responsible for long-range intermolecular transfers of energy, depends on the overlap between the emission band of the donor molecule and the absorption band of the acceptor molecule. The excited donor interacts with the ground-state acceptor through a dipole–dipole mechanism.

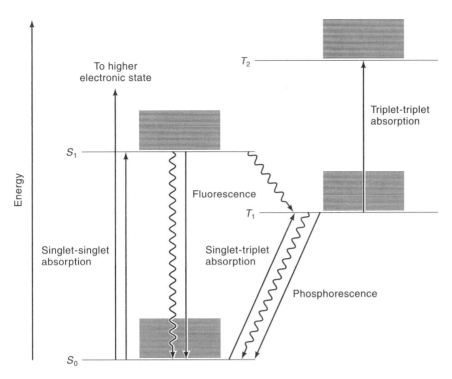

Figure 14.37
Jablonski diagram showing absorption, fluorescence, and phosphorescence. The wavy lines indicate radiationless transitions; the closely spaced lines represent the vibrational levels.

14.8 Lasers

Laser is an acronym for ***light amplification by stimulated emission of radiation***. It is a special type of emission involving either atoms or molecules. Consider first a two-level system. Suppose that N molecules are irradiated by light at intensity $\rho(v)$. In Section 14.1, we saw that the rate of stimulated absorption is given by $B_{mn}\rho(v)N(1 - x)$, and the rate of stimulated emission is $B_{nm}\rho(v)Nx$, where x is the fraction of the molecules in the excited state. In addition, an excited molecule undergoes spontaneous emission, the rate of which is $A_{nm}Nx$. At equilibrium, the rates of absorption and emission are equal, so that

$$B_{mn}\rho(v)N(1 - x) = B_{nm}\rho(v)Nx + A_{nm}Nx \qquad (14.56)$$

or

$$x = \frac{B_{nm}\rho(v)}{2B_{nm}\rho(v) + A_{nm}} \qquad (14.57)$$

(Recall that $B_{mn} = B_{nm}$.) It follows, therefore, that the maximum value of x is 0.5, reached only as $\rho(v)$ approaches infinity.

The above discussion means that the condition $x > 0.5$ can never be achieved in a two-level system. If we can somehow populate the upper level without using the normal radiation process, we might be able to bring about a *population inversion*, causing x to exceed 0.5. In such a case, intense emission could be induced by irradiating the system with photons of the appropriate frequency. In fact, this objective can

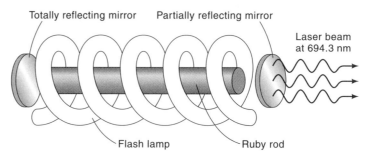

Figure 14.38
Emission of light from a ruby laser.

be reached by using a three- or four-level system, on which most laser actions are based.

The first successful laser was the creation of the American physicist and engineer Theodore Harold Maiman (1927–). In 1960, Maiman constructed a ruby rod by doping synthetic sapphire (Al_2O_3) with about 0.5% Cr_2O_3 so that Cr^{3+} replaced some of the Al^{3+} ions in the crystal lattice. A schematic diagram of a ruby laser is shown in Figure 14.38. The energy levels for Cr^{3+} ions are shown in Figure 14.39. At first, the laser system is subjected to a short, intense irradiation, known as *optical pumping* (by discharging a flashlamp, for example), which causes transitions from the E_0 to E_2 and E_3 levels of Cr^{3+}. The excited states then decay to the E_1 state by radiationless transitions. The lifetime of the E_1 state is rather long—about 0.003 s at room temperature—because the $E_1 \rightarrow E_0$ transition is spin forbidden. If the pumping is effective, the population in the E_1 state will exceed that in the E_0 state, and a laser transition can be effected by stimulating the transition with photons of wavelength 694.3 nm.

A variety of different lasers are now available. Operating in the solid, liquid, and gas states, they emit radiation ranging from infrared to ultraviolet and X ray. We will not survey the types of lasers here, except to point out that the mechanism for achieving population inversions can be very different for different lasers. For example, in the helium–neon laser, which is an example of an atomic gas laser, the He atoms are first excited by electrons to the higher electronic states, which are then deactivated by collisions with the Ne atoms. The populations in the upper excited

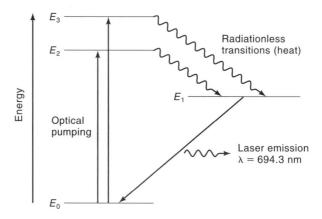

Figure 14.39
Energy-level diagram for a Cr^{3+} ion used in a ruby laser.

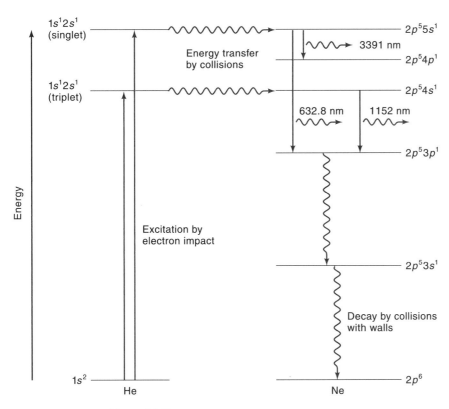

Figure 14.40
Energy-level diagram for a helium–neon laser.

states in Ne build up to exceed those of the lower excited states, so laser transitions can occur (Figure 14.40).

Table 14.8 summarizes the properties of common lasers. Lasers are operated in one of two different modes: continuous wave (cw) or pulsed. As the names suggest, in the cw mode, the laser light is emitted continuously, whereas in the pulsed mode the

Table 14.8
Some Common Laser Systems

Laser	Emitted Wavelengths/nm	Mode[a]
Ruby	694.3	pulsed
He–Ne(g)	632.8	cw
	1152	
	3391	
Ar$^+$(g)	457	cw
	488	
	514.5	
N$_2(g)$	337.1	pulsed
Nd/YAG[b]	1064.1	cw/pulsed
CO$_2(g)$	10,600	cw/pulsed

[a] cw, continuous wave.
[b] This laser system is made of neodymium ions (Nd^{3+}) trapped in *y*ttrium *a*luminum *g*arnet crystal (Y$_3$Al$_5$O$_{12}$). A procedure called *frequency doubling* makes it possible to convert the 1064-nm light to 532 nm and 266 nm, which are more applicable for research in photochemistry and photobiology.

light comes out in pulses, some of which may be as short as 1×10^{-14} s, or 10 fs (1 femtosecond $= 10^{-15}$ s). The actual mode depends on the system, the method of pumping, and the design of the apparatus. For example, if the rate of pumping is less than the decay rate from the upper laser level, then a population inversion cannot be sustained, and pulsed operation must be used, and the pulse duration is governed by the decay kinetics. A laser can operate continuously if the heat it generates is easily dissipated and population inversion can be sustained.

Properties and Applications of Laser Light

A laser beam is characterized by three properties: high intensity, high coherence, and high monochromaticity. We shall briefly discuss these properties and applications based on them.

Intensity. Laser light has the highest intensity of any light on Earth. As an example, consider a Nd^{3+}: YAG laser (see Table 14.8) that produces 7.0×10^{15} photons at 1064.1 nm during a pulse lasting 150 ps (1 ps $= 10^{-12}$ s). Because $E = h\nu = hc/\lambda$, the total energy output is given by

$$E = \left(\frac{hc}{\lambda} \text{ photon}^{-1}\right)(7.0 \times 10^{15} \text{ photons})$$

$$= \frac{(6.626 \times 10^{-34} \text{ J s})(3.00 \times 10^8 \text{ m s}^{-1})(7.0 \times 10^{15})}{1064.1 \times 10^{-9} \text{ m}}$$

$$= 1.3 \times 10^{-3} \text{ J}$$

Now, 1.3×10^{-3} J may not seem like much, but it is generated in an extremely short period of time. We can calculate the peak power associated with such a laser beam as follows:

$$\text{power} = \frac{\text{energy}}{\text{time}} = \frac{1.3 \times 10^{-3} \text{ J}}{150 \times 10^{-12} \text{ s}} = 8.7 \times 10^6 \text{ J s}^{-1}$$

$$= 8.7 \times 10^6 \text{ W}$$

An ordinary 60-watt light bulb has a continuous output of 60 W or 60 J s^{-1}.

(The unit of power is the watt, denoted by W, where 1 W $= 1$ J s^{-1}.) This is the power output during the pulse of laser action. When such a laser beam is focused on a small target of area 0.01 cm^2, say, the *power flux density* is given by

$$\text{power flux density} = \frac{\text{power}}{\text{area}} = \frac{8.7 \times 10^6 \text{ W}}{0.01 \text{ cm}^2} = 8.7 \times 10^8 \text{ W cm}^{-2}$$

Shorter pulses and smaller areas yield higher power flux densities.

Intense laser beams have been used to cut and weld metals and even to produce nuclear fusion. Medically, lasers are used in surgery. For example, a pulsed argon laser is employed to "spot-weld" a detached retina back onto its support (the choroid). This procedure has some advantages over the traditional treatment in that it is noninvasive and does not require the administration of anesthetics.

The high intensity of a laser beam also allows "multiphoton absorption," a process in which a molecule absorbs two or more photons during a spectroscopic transition. In a conventional spectroscopic measurement, an atom or molecule absorbs a photon of the same energy as that separating the ground state and the excited state. This is the normal one-photon process. When a system is irradiated with a high-power laser beam of frequency ν_l, however, it may undergo a two-photon process in

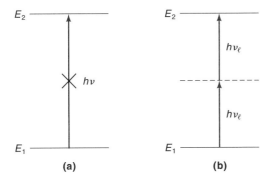

Figure 14.41
(a) A forbidden transition in a one-photon absorption process. (b) The same transition becomes allowed in a two-photon absorption process. Note that $\nu = 2\nu_l$.

which the system attains the excited state by absorbing two photons in a concerted fashion (Figure 14.41). A two-photon process requires a very intense laser beam because the two photons must pass essentially simultaneously through the region of space occupied by one molecule. An interesting consequence of multiphoton spectroscopy is that the selection rules are different, so that transitions that are strictly forbidden for one-photon absorption occur in a two-photon process.* In addition, transitions that take place in the UV region can be probed using visible laser light because of the additivity of frequencies.

Coherence. By *coherent* we mean that the photons in a laser are emitted *in phase* with one another. The high degree of coherence arises from the fact that the stimulated emission synchronizes the radiation of the individual molecules, so that the photon emitted from one molecule stimulates another molecule to emit a photon of the same wavelength that is exactly in phase with the first photon, and so on. One application of laser coherence is *holography*, a technique for producing three-dimensional images. This process produces a *hologram*, which contains information not only on the intensity (as in a conventional two-dimensional photograph), but also on the phase of light reflected from the subject. Subsequent illumination of the hologram reconstructs a three-dimensional image.

Monochromaticity. Laser light is highly monochromatic (having the same wavelength) because all the photons are emitted as a result of a transition between the same two atomic or molecular energy levels. Therefore, they possess the same frequency and wavelength. In the ruby laser, for example, the emitted light is centered at 694.3 nm over a width less than 0.01 nm. Although narrow line widths also can be obtained from ordinary light (from an incandescent light bulb, say) and a *monochromator* (a prism or diffraction grating which separates light of different wavelengths), the intensity of a laser beam at a particular wavelength can be six orders of magnitude greater than that from a conventional source.

The highly monochromatic nature of laser light enables us to induce and identify specific transitions between electronic, vibrational, and even rotational energy levels for many molecules and therefore to obtain well-resolved absorption spectra. The

---

* As mentioned in Section 14.1, for an allowed transition, ψ_i and ψ_j must have different symmetries with respect to each other. For the two-photon process, we can imagine the presence of an intermediate state between the initial and final states; that is, the transition takes place in two steps. Thus, the initial and final states must have the *same* symmetry (even–even or odd–odd).

systems discussed so far, however, are all fixed-frequency lasers, which provide emission of light at one or a few discrete wavelengths. Thus, they are not suitable for the usual absorption methods, which require scanning over a continuous range of wavelengths. Organic dye lasers are appropriate for this application because they are *tunable*; that is, they provide wavelengths over a continuous range. One of the most widely used organic dyes is Rhodamine 6G ($C_{28}H_{31}N_2O_3BF_4$), a molecule that has many modes of vibration. The electronic spectrum of Rhodamine 6G in solution shows a broad peak due to strong molecular interaction in the liquid state. (Collisions with solvent molecules broaden the vibrational structure of the transitions into unresolved bands.) Consequently, the fluorescence of the dye, which occurs at a longer wavelength, also appears as a broad peak. By pumping the solution with a laser, we can bring about a population inversion in Rhodamine 6G and subsequent laser action. With a proper diffraction grating, it is possible to tune a particular wavelength for use. For example, a dye laser using Rhodamine 6G in methanol is continuously tunable over the range of 570 nm to 660 nm. Using different organic dyes, spectroscopists have expanded the tunable range of dye lasers to between 200 nm and 1000 nm. This technique has greatly broadened the scope of high-resolution spectroscopy.

Finally, we note that the high intensity, high monochromaticity, and tunability of laser light enable us to excite atoms and molecules and monitor the subsequent fluorescence of these species. The advantage of the so-called *laser-induced fluorescence* (LIF) over fluorescence produced by a conventional light source is that it is highly sensitive. In elemental analysis, for example, a sample solution is atomized (decomposed into atomic species) in a furnace or a flame and then irradiated with a laser beam to induce atomic fluorescence. By this procedure, spectroscopists can detect a concentration (of the original solution) in the 10^{-11} g mL^{-1} range. The high monochromaticity and narrow linewidth of a laser source enable one to excite molecules to specific rotational and vibrational levels in excited electronic states and to observe the subsequent fluorescence. If the fluorescence intensity is recorded as a function of the frequency or wavelength of the exciting laser, a line spectrum will be obtained. At every frequency for which the resonance condition is satisfied for an allowed transition from the lower to upper state, molecules will be excited and will subsequently fluoresce. This method yields valuable information about the electronic structure of excited states and is particularly useful for studying small transient species, such as radicals generated in a flame.

14.9 Optical Rotatory Dispersion and Circular Dichroism

In this section we shall discuss some aspects of optical activity, which has great significance for biological molecules such as proteins and nucleic acids.

Molecular Symmetry and Optical Activity

Molecular symmetry plays a central role in the study of optical activity. A molecule is *optically active* (or *chiral*) if it can rotate the plane of polarized light (see below). The general criterion for optical activity is that the molecule and its mirror image must not be superimposable. However, this rule is difficult to apply on complex molecules without actually constructing three-dimensional models. In terms of molecular symmetry, any molecule with a plane of symmetry or center of symmetry is not optically active, and therefore is said to be *achiral*. [A plane of symmetry bisects a molecule so that one-half of the molecule is the mirror image of the other half. A molecule possesses a center of symmetry if the coordinates (x, y, z) of every atom in the molecule are changed into $(-x, -y, -z)$ and the molecule is indistinguishable from the original configuration.] A chiral molecule in organic and biological systems contains at least one asymmetric carbon atom; that is, it contains an atom that is bonded to four different atoms or different groups of atoms. Keep in mind that chirality may also be the result of a molecule lacking a plane of symmetry, even if no asymmetric carbon atoms are present in the molecule.

An asymmetric carbon atom is also referred to as a *stereogenic center*.

Polarized Light and Optical Rotation

As we saw in Chapter 11, an electromagnetic wave has an electric-field component (E) and a magnetic-field component (B) perpendicular to it. Both components oscillate in space with the same wavelength and frequency. Light is a *transverse* wave; that is, the planes of E and B are perpendicular to the direction of propagation. For ordinary, unpolarized light, the direction of E and B change rapidly and randomly (Figure 14.42a). When a beam of unpolarized light passes through a polarizer, it

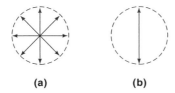

(a)   (b)

Figure 14.42
Viewed toward the light source. (a) Unpolarized light. The electric-field vector of the light oscillates in all directions. (b) Plane-polarized light. The electric field vector is confined to one plane. In both cases, the double-headed arrow shows the oscillation between the maximum and minimum of the wave.

emerges polarized, meaning that its electric field components are confined to a single plane, as shown in Figure 14.42b. A polarizer can be a Nicol prism, which is a crystal of calcite ($CaCO_3$), or a Polaroid sheet made of polyvinyl alcohol stained with iodine. These polarizers produce what is called *linearly* or *plane-polarized* light.

A plane-polarized light beam can be resolved into two component vectors, E_L and E_R, which correspond to the left and right *circularly polarized* waves, respectively. Circularly polarized light can be produced by passing a beam of plane-polarized light through an optical device known as a *quarter-wave plate*. Figure 14.43 shows the variation of E as the resultant of the two rotating vectors (the addition of E_L and E_R gives E at every point). In an ordinary medium, E_L and E_R rotate at the

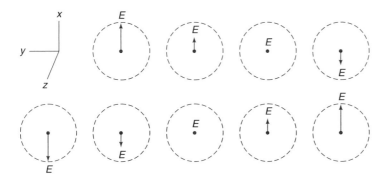

Figure 14.43
Electric-field vector E is the resultant of two rotating vectors E_L and E_R. [Adapted from C. Djerassi, *Optical Rotatory Dispersion*, © 1960 by McGraw-Hill Book Company. Used by permission of McGraw-Hill Company.]

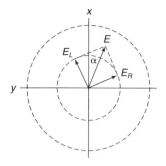

Figure 14.44
Rotation of E (measured by α) when E_L and E_R make unequal angles with the x axis. [From C. Djerassi, *Optical Rotatory Dispersion*, © 1960 by McGraw-Hill Book Company. Used by permission of McGraw-Hill Company.]

same speed, so E is confined to the xz plane. In an optically active medium, E_L and E_R rotate at different speeds, and although the light is polarized, the plane containing E makes an angle α with the x axis (Figure 14.44). The plane of polarization rotates because in an optically active medium the *refractive index* (n)—the ratio of the speed of light in vacuum to the speed of light in the medium—is different for left and right circularly polarized light and this difference results in different speeds of rotation of E_L and E_R. Such a medium is said to be *circularly birefringent*.

A *polarimeter* is used to study optical rotation (Figure 14.45). In a polarimeter, the polarizer and analyzer (another polarizer) are first aligned so that no light passes through the analyzer. If the substance in the polarimeter tube is optically active, some light is transmitted because the medium causes the plane of polarization to rotate. The angle of rotation, α, can be measured by turning the analyzer until again no light is transmitted. A useful quantity for characterizing a chiral compound is the *specific rotation*, defined as

$$[\alpha]_\lambda^T = \frac{\alpha}{lc} \tag{14.58}$$

where l is the pathlength in dm and c is the concentration of the optically active substance in g cm^{-3}. Because specific rotation is a function of the wavelength of light and temperature, it is customary to denote both of these factors as shown in Equation 14.58. Note that T is in degrees Celsius. The units of specific rotation are deg dm^{-1} cm^3 g^{-1}. (In SI units, specific rotation is expressed in deg m^2 kg^{-1}.) For a

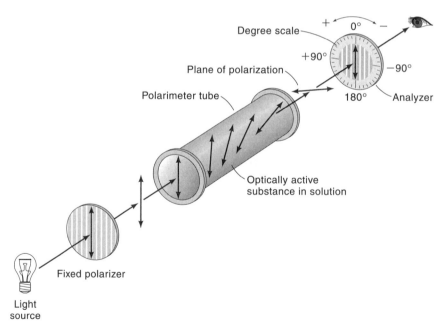

Figure 14.45
Operation of a polarimeter. Initially the sample tube is filled with an achiral solvent. The analyzer is rotated so that its plane of polarization is perpendicular to that of the polarizer. Under this condition, no light reaches the observer. Next, a solution containing a chiral compound is placed in the tube as shown. The plane of the polarized light is rotated as it travels through the length of the tube so that some light now reaches the observer. Rotating the analyzer (either to the left or to the right) until no light reaches the observer again allows the angle of optical rotation to be measured.

pure liquid, Equation 14.58 becomes

$$[\alpha]_\lambda^T = \frac{\alpha}{ld} \tag{14.59}$$

where d is the density of the liquid. Another related quantity is the *molar rotation*, $[\Phi]_\lambda^T$, given by

$$[\Phi]_\lambda^T = \frac{[\alpha]_\lambda^T \mathcal{M}}{100} \tag{14.60}$$

where $\mathcal{M}$ is the molar mass of the optically active compound. The units of molar rotation are deg dm^{-1} cm^3 mol^{-1}. (In SI units, the molar rotation has the units deg m^2 mol^{-1}.) Both specific rotation and molar rotation are *independent* of the concentration of solution and path length of the cell.

If a medium causes the plane of polarization to rotate to the right, it is called *dextrorotatory*, denoted by $(+)$; if the rotation is to the left, it is *levorotatory*, denoted by $(-)$.

Example 14.4

The specific rotation of L-lysine ($C_6H_{14}N_2O_2$) is $+13.5°$ dm^{-1} cm^3 g^{-1} at 589.3 nm and 25°C. (a) Calculate the optical rotation of a lysine solution (concentration: 0.148 g cm^{-3}) in a 10-cm cell. (b) What is the molar rotation, given that the molar mass of lysine is 146.2 g mol^{-1}.

ANSWER

(a) From Equation 14.58 we have

$$c = 0.148 \text{ g cm}^{-3}$$
$$l = 10 \text{ cm} = 1.0 \text{ dm}$$

Hence

$$+13.5° \text{ dm}^{-1} \text{ cm}^3 \text{ g}^{-1} = \frac{\alpha}{1 \text{ dm} \times 0.148 \text{ g cm}^{-3}}$$
$$\alpha = +2.0°$$

(b) From Equation 14.60

$$[\Phi]_\lambda^T = \frac{13.5° \text{ dm}^{-1} \text{ cm}^3 \text{ g}^{-1} \times 146.2 \text{ g mol}^{-1}}{100}$$
$$= 19.7° \text{ dm}^{-1} \text{ cm}^3 \text{ mol}^{-1}$$

Optical Rotatory Dispersion (ORD) and Circular Dichroism (CD)

The refractive index of a medium is not constant but depends on the wavelength of light employed. It follows that optical rotation, which depends on the refractive index, must also vary with wavelength. Figure 14.46 shows two types of optical rotatory dispersion (ORD) curves, which demonstrate how optical rotation varies with wavelength. The instrument that measures ORD curves is called a *spec-*

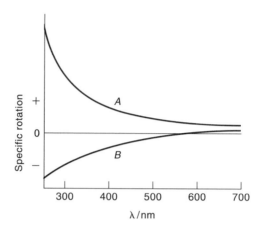

Figure 14.46
Optical rotatory dispersion curves: plain positive (*A*) and plain negative (*B*).

tropolarimeter, which differs from a polarimeter in that the latter employs only one wavelength. These ORD curves are called *plain* dispersion curves because they lack maxima and minima. For compounds that do not absorb in the visible region, rotation is quite small—a major experimental difficulty at a time when optical rotation was measured only at a single wavelength of sodium D-line.[†]

The optical activity of most organic and biological compounds generally increases toward the shorter-wavelength region because these compounds possess optically active absorption bands in the UV ($\pi \rightarrow \pi^*$, $n \rightarrow \pi^*$, and $\sigma \rightarrow \sigma^*$ transitions). The chromophores for these bands are either intrinsically asymmetric (that is, they do not contain an asymmetric carbon atom) or become asymmetric as a result of the interaction with an asymmetric environment. As mentioned above, optical rotation is a function of wavelength. Within the spectral region of the optically active absorption band, an anomaly is observed. This phenomenon is known as the *Cotton effect* (after the French physicist Aimé Cotton, 1869–1951). A positive Cotton effect is characterized by an initial rise of the ORD curve with decreasing wavelength reaching a maximum (called a *peak*) at a longer wavelength than the maximum of the absorption band. Beyond this point, the curve changes its slope, reaching a minimum (called a *trough*) at a shorter wavelength than the maximum of the absorption band (Figure 14.47). Just the opposite holds for a *negative* Cotton effect. The vertical distance between the peak and the trough is a measure of the amplitude of the Cotton effect while the horizontal distance between the same two points gives the breadth of the curve.

Optical activity is also manifest in small differences in the molar absorptivity, ε_L and ε_R, of circularly polarized components. Then the medium is said to exhibit *cir-*

[†] Actually, the familiar yellow color of the sodium D-line is a closely spaced doublet at 589.0 nm and 589.6 nm. The symbol D indicates that the wavelength is about 589.3 nm.

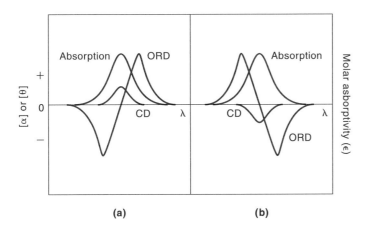

Figure 14.47
The Cotton effect of an enantiomeric pair (a) and (b). The Cotton effect is positive in (a) and negative in (b), as discussed in the text. Note also the different signs of the CD peaks. In general, the CD peaks are several orders of magnitude smaller than absorption peaks. For an idealized optically active transition, the full width at half maximum of the CD band should be the same as that for an absorption band.

cular dichroism (CD). Because ε_L is not equal to ε_R, ε will no longer oscillate along a single line but will trace out the ellipse shown in Figure 14.48. The difference in ε values is measured according to the equation

$$[\theta] = 3300(\varepsilon_L - \varepsilon_R) \tag{14.61}$$

where $[\theta]$, the molar ellipticity, is in deg cm^2 dmol^{-1} of the optically active compound. Note that in contrast to an absorption band, a CD band is signed because the difference $(\varepsilon_L - \varepsilon_R)$ can be either positive or negative (see Figure 14.47).

Both ORD and CD are important tools for studying the conformation of biomolecules in solution. Much of the early work was carried out on model compounds, synthetic polypeptides such as a poly-γ-benzyl-L-glutamate and poly-L-proline. There are two different contributions to optical activity in these molecules, as well as in proteins: the presence of L-amino acids and the folding of the polypeptide chains, for example, into α helices. The helix can be either left-handed or right-handed, as shown in Figure 14.49. Protein helices are exclusively right-handed and rotate plane-

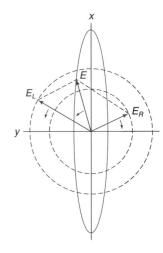

Figure 14.48
Variation of E for circular dichroism. The E vector traces out an ellipse. [From C. Djerassi, *Optical Rotatory Dispersion,* © 1960 by McGraw-Hill Book Company. Used by permission of McGraw-Hill Company.]

Left-handed   Right-handed

Figure 14.49
Left-handed and right-handed helices. As a reference, it is useful to remember that standard metal screws, which insert when turned clockwise, are right-handed. Note that a helix preserves the same handedness when it is turned upside down. The α helix of polypeptides has a right-handed helical structure.

polarized light in the opposite direction from the L-amino acids. Studying the optical rotation of many protein molecules enables biochemists to estimate the percent of helical structure. From the change in rotation with temperature, pH, and so on, we can also monitor the helix–coil transitions.

ORD has largely been supplanted by CD for the conformational studies of proteins and nucleic acids since CD instrumentation became available commercially. The reason is that CD has much greater resolving power. As Figure 14.47 shows, the CD band associated with an electronic transition drops off sharply as one moves away from λ_{max}, while for ORD, the contribution of a transition decreases much more slowly. This slow change means that closely spaced transitions are extremely difficult to resolve in ORD. The only advantage ORD has over CD is that one can study compounds for which CD measurements are impossible because the regions in which they absorb are outside the range of CD instrumentation or are obscured by solvent absorption.

Suggestions for Further Reading

BOOKS

Andrews, D. L., *Lasers in Chemistry*, 3rd ed., Springer-Verlag, New York, 1997.

Barrow, G. M., *The Structure of Molecules*, W. A. Benjamin, New York, 1963.

Barrow, G. M., *Molecular Spectroscopy*, McGraw-Hill, New York, 1962.

Crabbé, P., *ORD and CD in Chemistry and Biochemistry*, Academic Press, New York, 1972.

Djerassi, C., *Optical Rotatory Dispersion*, McGraw-Hill, New York, 1960.

Dunford, H. B., *Elements of Diatomic Molecular Spectra*, Addison-Wesley, Reading, MA, 1968.

Fasman, G. D., Ed. *Circular Dichroism and the Conformational Analysis of Biomolecules*, Plenum Press, New York, 1996.

Gibilisco, S., *Understanding Lasers*, Tab Books, Blue Ridge Summit, PA, 1989.

Hore, P. J., *Nuclear Magnetic Resonance*, Oxford University Press, New York, 1995.

Jirgensons, B., *Optical Rotatory Dispersion of Proteins and Other Macromolecules*, Springer-Verlag, New York, 1969.

Lakowicz, J. R., *Principles of Fluorescence Spectroscopy*, Kluwer Academic Publishers, Norwell, MA, 1999.

Macomber, R. S., *A Complete Introduction to Modern NMR Spectroscopy*, John Wiley & Sons, New York, 1998.

Roger, A. and B. Norden, *Circular Dichroism and Linear Dichroism*, Oxford University Press, New York, 1997.

Straughan, B. P. and S. Walker, Eds. *Spectroscopy*, Vols. 1, 2, and 3, John Wiley & Sons, New York, 1976.

Stuart, B., *Biological Applications of Infrared Spectroscopy*, John Wiley & Sons, New York, 1997.

Swartz, H. M., J. R. Bolton, and D. C. Borg, *Biological Applications of Electron Spin Resonance*, Wiley-Interscience, New York, 1972.

Wertz, J. E. and J. R. Bolton, *Electron Spin Resonance: Elementary Theory and Practical Applications*, Chapman & Hall, New York, 1986.

Yoder, C. H. and C. D. Schaeffer, Jr. *Introduction to Multinuclear NMR*, Benjamin/Cummings, Reading, MA, 1987.

Zare, R. N., *Laser: Experiments for Beginners*, University Science Books, Sausalito, CA, 1995.

ARTICLES

General

"Light," G. Feinberg, *Sci. Am.* September 1968.

"How Light Interacts With Matter," V. F. Weisskopf, *Sci. Am.* September 1968.

"Chemistry and Light Generation," N. Slagg, *J. Chem. Educ.* **45**, 103 (1968).

"The Chemical Origin of Color," M. V. Orna, *J. Chem. Educ.* **55**, 478 (1978).

"The Causes of Color," K. Nassau, *Sci. Am.* October 1980.

"Demonstration of Maxwell Distribution Law of Velocity by Spectral Line Shape Analysis," C. L. Berg and R. Chang, *Am. J. Phys.* **52** (1), 80 (1984).

"A Time Scale for Fast Events," D. Onwood, *J. Chem. Educ.* **63**, 680 (1986).

"Band Breadth of Electronic Transitions and the Particle-in-a-Box Model," L. F. Olsson, *J. Chem. Educ.* **63**, 756 (1986).

"Radiationless Relaxation and Red Wine," H. D. Burrows and A. C. Cardoso, *J. Chem. Educ.* **64**, 995 (1987).

"The Fourier Transform," R. N. Bracewell, *Sci. Am.* June 1989.

"The Early History of Spectroscopy," N. C. Thomas, *J. Chem. Educ.* **68**, 631 (1991).

"The Beer-Lambert Law Revisited," P. Lykos, *J. Chem. Educ.* **69**, 730 (1992).

"Using Fourier Transform to Understand Spectral Line Shape," E. Grunwald, J. Herzog, and C. Steel, *J. Chem. Educ.* **72**, 210 (1995).

"Why Do Spectral Lines Have Linewidth?" V. B. E. Thomsen, *J. Chem. Educ.* **72**, 616 (1995).

"A Unified Approach to Absorption Spectroscopy at the Undergraduate Level," R. S. Macomber, *J. Chem. Educ.* **74**, 65 (1997).

Nuclear Magnetic Resonance

"NMR Imaging in Medicine," I. L. Pykett, *Sci. Am.* May 1982.

"NMR Spectroscopy in Living Cells," R. G. Shulman, *Sci. Am.* January 1983.

"The NMR Time Scale," R. G. Bryant, *J. Chem. Educ.* **60**, 933 (1983).

"Atomic Memory," R. G. Brewer and E. L. Hahn, *Sci. Am.* December 1984.

"Sensitivity Enhancement by Signal Averaging in Pulsed/Fourier Transform NMR Spectroscopy," D. L. Rabenstein, *J. Chem. Educ.* **61**, 909 (1984).

"A Primer on Fourier Transform NMR," R. S. Macomber, *J. Chem. Educ.* **62**, 213 (1985).

"A Step-by-Step Picture of Pulsed (Time-Domain) NMR," L. J. Schwartz, *J. Chem. Educ.* **65**, 959 (1988).

"Spin-Lattice Relaxation Times in ^{1}H NMR Spectrscopy," D. J. Wink, *J. Chem. Educ.* **66**, 810 (1989).

"Nuclear Magnetic Resonance in Biochemistry," S. Cheatham, *J. Chem. Educ.* **66**, 111 (1989).

"A Comparison of FTNMR and FTIR Techniques," M. K. Ahn, *J. Chem. Educ.* **66**, 802 (1989).

"Protein Structure Determination in Solution by NMR Spectroscopy," K. Wüthrich, *Science* **243**, 45 (1989).

"A Demonstration of Imaging on an NMR Spectrometer," L. A. Hull, *J. Chem. Educ.* **67**, 782 (1990).

"NMR of Whole Body Fluids," B. Faust, *Educ. Chem.* **32**, 22 (1995).

"Structural Determination by Nuclear Magnetic Resonance Spectroscopy," A. M. Chippendall in *Encyclopedia of Applied Physics*, G. L. Trigg, Ed., VCH Publishers, New York, 1997, Vol. 20, p. 119.

Other Branches of Spectroscopy

"Vibration-Rotation Spectrum of HCl," F. E. Stafford, C. W. Holt, and G. L. Paulson, *J. Chem. Educ.* **40**, 245 (1963).

"The Ultraviolet Spectra of Aromatic Molecules," P. E. Stevenson, *J. Chem. Educ.* **41**, 234 (1964).

"The Fates of Electronic Excitation Energy," H. H. Jaffé and A. L. Miller, *J. Chem. Educ.* **43**, 469 (1966).

"Applications of Absorption Spectroscopy in Biochemistry," G. R. Penzer, *J. Chem. Educ.* **45**, 692 (1968).

"The Triplet State," N. J. Turro, *J. Chem. Educ.* **46**, 2 (1969).

"Liquid Scintillation Counting," W. Yang and E. K. C. Lee, *J. Chem. Educ.* **46**, 277 (1969).

"Infrared Spectrometry," D. R. Johnson and C. T. Moynihan, *J. Chem. Educ.* **46**, 431 (1969).

"Progress in Our Understanding of the Optical Properties of Nucleic Acids," A. M. Lesk, *J. Chem. Educ.* **46**, 821 (1969).

"Free Radicals in Biological Systems," W. A. Pryor, *Sci. Am.* August 1970.

"Energy States of Molecules," J. L. Hollenberg, *J. Chem. Educ.* **47**, 2 (1970).

"Vibrational Frequencies of Sulfur Dioxide," A. G. Briggs, *J. Chem. Educ.* **47**, 391 (1970).

"ESR Study of Electron Transfer Reactions," R. Chang, *J. Chem. Educ.* **47**, 563 (1970).

"The Spectroscopy of Supercooled Gases," D. H. Levy, *Sci. Am.* February 1984.

"Introduction to the Interpretation of Electron Spin Resonance Spectra of Organic Radicals," N. J. Bunce, *J. Chem. Educ.* **64**, 907 (1987).

"Introduction to a Quantum Mechanical Harmonic Oscillator Using a Modified Particle-in-a-Box Problem," H. F. Blanck, *J. Chem. Educ.* **69**, 98 (1992).

"Molecular Spectroscopy," D. A. Ramsay in *Encyclopedia of Applied Physics*, G. L. Trigg, Ed., VCH Publishers, New York, 1994, Vol. 10, p. 491.

"The Fundamental Rotational-Vibrational Band of CO and NO," H. H. R. Schor and E. L. Teixeira, *J. Chem. Educ.* **71**, 771 (1994); see also **75**, 258 (1998).

Lasers

"Photography by Laser," E. N. Leith and J. Upatnieks, *Sci. Am.* June 1965.

"Laser Chemistry," D. L. Rousseau, *J. Chem. Educ.* **43**, 566 (1966).

"Advances in Holography," K. S. Pennington, *Sci. Am.* February 1968.

"Laser Light," A. L. Schawlow, *Sci. Am.* September 1968.

"Applications of Laser Light," D. R. Harriott, *Sci. Am.* September 1968.

"Organic Lasers," P. Sorokin, *Sci. Am.* February 1969.

"Laser Spectroscopy," M. S. Feld and V. S. Letokhov, *Sci. Am.* December 1973.

"Applications of Lasers to Chemical Research," S. R. Leone, *J. Chem. Educ.* **53**, 13 (1976).

"Laser Separation of Isotopes," R. N. Zare, *Sci. Am.* February 1977.

"Laser Chemistry," A. M. Ronn, *Sci. Am.* May 1979.

"Laser—An Introduction," W. F. Coleman, *J. Chem. Educ.* **59**, 441 (1982).

"Laser-Induced Fluorescence in Spectroscopy, Dynamics, and Diagnostics," D. R. Crosley, *J. Chem. Educ.* **59**, 446 (1982).

"Lasers: A Valuable Tool for Chemists," E. W. Findsen and M. R. Ondrias, *J. Chem. Educ.* **63**, 479 (1986).

"Cooling and Trapping of Atoms," W. D. Phillips and H. J. Metcalf, *Sci. Am.* March 1987.

"Free Electron Lasers," A. M. Sessler and D. Vaughan, *Am. Sci.* **75**, 34 (1987).

"Detecting Individual Atoms and Molecules With Laser," V. S. Letokhov, *Sci. Am.* September 1988.

"Soft X-ray Lasers," D. L. Matthews and M. D. Rosen, *Sci. Am.* December 1988.

"The Birth of Molecules," A. H. Zewail, *Sci. Am.* December 1990.

"Laser Surgery," M. W. Berns, *Sci. Am.* June 1991.

"Laser Trapping of Neutral Molecules," S. Chu, *Sci. Am.* February 1992.

"Medical Use of Lasers," I. Itzkan and J. A. Izatt in *Encyclopedia of Applied Physics*, G. L. Trigg, Ed., VCH Publishers, New York, 1994, Vol. 10, p. 33.

"Laser Control of Chemical Reactions," P. Brumer and M. Shapiro, *Sci. Am.* March 1995.

"Using Lasers to Demonstrate the Concept of Polarizability," G. R. van Hecke, K. K. Karukstis, and J. M. Underhill, *Chemical Educator* **1997**, 2(5): S1430-4171(97)05147-X. Avail. URL: http://journals.springer-ny.com/chedr.

"Innovative Laser Technique in Chemical Kinetics," L. J. Kovalenko and S. R. Leone, *J. Chem. Educ.* **65**, 681 (1998).

"Freezing Atoms in Motion: Principles of Femtochemistry and Demonstration by Laser Spectroscopy," J. S. Baskin and A. H. Zewail, *J. Chem. Educ.* **78**, 737 (2001).

Optical Rotatory Dispersion and Circular Dichroism
"A Model for Optical Rotation," L. L. Jones and H. Eyring, *J. Chem. Educ.* **38**, 601 (1961).

"Optical Rotation and the Conformation of Polypeptides and Proteins," P. Urnes and P. Doty, *Adv. Protein Chem.* **16**, 421 (1961).

"Absorption, Dispersion, Circular Dichroism, and Rotatory Dispersion," J. G. Foss, *J. Chem. Educ.* **40**, 593 (1963).

"A Brief History of Polarimetry," R. E. Lyle and G. G. Lyle, *J. Chem. Educ.* **41**, 308 (1964).

"An Experiment in Optical Rotatory Dispersion," J. P. Schelz and W. C. Purdy, *J. Chem. Educ.* **41**, 645 (1964).

"Optical Activity and Molecular Dissymmetry," S. F. Mason, *Chem. Brit.* **1**, 245 (1965).

"Symmetry," C. A. Coulson, *Chem. Brit.* **4**, 113 (1968).

"Criteria for Optical Activity in Organic Molecules," D. F. Mowery, Jr., *J. Chem. Educ.* **46**, 269 (1969).

"The Natural Origin of Optically Active Compounds," W. E. Elias, *J. Chem. Educ.* **49**, 448 (1972).

"Spontaneous Generation of Optical Activity," R. E. Pincock and K. R. Wilson, *J. Chem. Educ.* **50**, 455 (1973).

"Optical Rotation and the DNA Helix to Coil Transition," G. L. Baker and M. E. Alden, *J. Chem. Educ.* **51**, 591 (1974).

"Optical Rotatory Dispersion and Circular Dichroism," K. P. Wong, *J. Chem. Educ.* **51**, A573 (1974); **52**, A9 (1975).

"Optical Activity," C. D. Mickey, *J. Chem. Educ.* **57**, 442 (1980).

"Rotation of Plane-Polarized Light," R. R. Hill and B. G. Whatley, *J. Chem. Educ.* **57**, 467 (1980).

"Optical Activity: Biot's Bequest," J. Applequist, *Am. Sci.* **75**, 58 (1987).

"Molecular Structure and Chirality," D. J. Brand and J. Fisher, *J. Chem. Educ.* **64**, 1035 (1987); **67**, 358 (1990).

"Polarized Light and Rates of Chemical Reactions," J. J. Weir, *J. Chem. Educ.* **66**, 1035 (1989).

"Symmetry Elements and Molecular Achirality," G. Q. Chen, *J. Chem. Educ.* **69**, 159 (1992).

"Chiroptical Spectroscopy," J. E. Gurst, *J. Chem. Educ.* **72**, 827 (1995).

"Polarized Starlight and the Handedness of Life," S. Clark, *Am. Sci.* **87**, 336 (1999).

"Demonstration of Optical Rotatory Dispersion of Sucrose," S. M. Mahurin, R. N. Compton, and R. N. Zare, *J. Chem. Educ.* **76**, 1234 (1999).

Problems

General

14.1 Convert 15,000 cm^{-1} to wavelength (nm) and frequency.

14.2 Convert 450 nm to wavenumber and frequency.

14.3 Convert the following percent transmittance to absorbance: **(a)** 100%, **(b)** 50%, and **(c)** 0%.

14.4 Convert the following absorbance to percent transmittance: **(a)** 0.0, **(b)** 0.12, and **(c)** 4.6.

14.5 The absorption of radiation energy by a molecule results in the formation of an excited molecule. Given enough time, it would seem that all of the molecules in a sample would have been excited and no more absorption would occur. Yet in practice we find that the absorbance of a sample at any wavelength remains unchanged with time. Why?

14.6 The mean lifetime of an electronically excited molecule is 1.0×10^{-8} s. If the emission of the radiation occurs at 610 nm, what are the uncertainties in frequency ($\Delta \nu$) and wavelength ($\Delta \lambda$)?

14.7 The familiar yellow D-lines of sodium is actually a doublet at 589.0 nm and 589.6 nm. Calculate the difference in energy (in J) between these two lines.

14.8 The resolution of visible and UV spectra can usually be improved by recording the spectra at low temperatures. Why does this procedure work?

14.9 Assuming that the width of a spectral line is the result solely of lifetime broadening, estimate the lifetime of a state that gives rise to a line of width **(a)** 1.0 cm^{-1}, **(b)** 0.50 Hz.

14.10 What is the molar absorptivity of a solute that absorbs 86% of a certain wavelength of light when the beam passes through a 1.0-cm cell containing a 0.16 M solution?

14.11 The molar absorptivity of a benzene solution of an organic compound is 1.3×10^2 L mol^{-1} cm^{-1} at 422 nm. Calculate the percentage reduction in light intensity when light of that wavelength passes through a 1.0-cm cell containing a solution of concentration 0.0033 M.

14.12 A single NMR scan of a dilute sample exhibits a signal-to-noise (S/N) ratio of 1.8. If each scan takes 8.0 minutes, calculate the minimum time required to generate a spectrum with a S/N ratio of 20.

Microwave, IR, and Electronic Spectroscopies

14.13 Which of the following molecules are microwave active? C_2H_2, CH_3Cl, C_6H_6, CO_2, H_2O, HCN.

14.14 What is the degeneracy of the rotational energy level with $J = 7$ for a diatomic rigid rotor? [The degeneracy is given by $(2J + 1)$.]

14.15 The $J = 3 \rightarrow 4$ transition for a diatomic molecule occurs at 0.50 cm^{-1}. What is the wavenumber for the $J = 6 \rightarrow 7$ transition for this molecule? Assume the molecule is a rigid rotor.

14.16 The equilibrium bond length in nitric oxide ($^{14}N^{16}O$) is 1.15 Å. Calculate **(a)** the moment of inertia of NO, and **(b)** the energy for the $J = 0 \rightarrow 1$ transition. How many times does the molecule rotate per second at the $J = 1$ level?

14.17 Which of the following molecules are IR active? **(a)** N_2, **(b)** HBr, **(c)** CH_4, **(d)** Xe, **(e)** H_2O_2, **(f)** NO.

14.18 Give the number of normal vibrational modes of **(a)** O_3, **(b)** C_2H_2, **(c)** CBr_4, **(d)** C_6H_6.

14.19 Draw a vibrational mode of the BF_3 molecule that is IR inactive.

14.20 A 500-g object suspended from the end of a rubber band has a vibrational frequency of 4.2 Hz. Calculate the force constant of the rubber band.

14.21 The fundamental frequency of vibration for carbon monoxide is 2143.3 cm^{-1}. Calculate the force constant of the carbon–oxygen bond.

14.22 If molecules did not possess zero-point energy, would they be able to undergo the $v = 0 \rightarrow 1$ transition?

14.23 Under what conditions can one observe a hot band in the IR spectrum?

14.24 Show all the fundamental vibration modes of **(a)** carbon disulfide (CS_2), and **(b)** carbonyl sulfide (OCS), and indicate which ones are IR active.

14.25 Calculate the number of vibrational degrees of freedom of the hemoglobin molecule, which contains 9272 atoms.

14.26 Which of the following molecules has the highest fundamental frequency of vibration? H_2, D_2, HD.

14.27 The fundamental frequency of vibration for $D^{35}Cl$ is given by $\tilde{\nu} = 2081.0$ cm^{-1}. Calculate the force constant, k, and compare this value with the force constant obtained for $H^{35}Cl$ in Example 14.2. Comment on your result.

14.28 Anthracene is colorless, but tetracene is light orange. Explain.

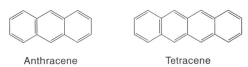

Anthracene   Tetracene

14.29 Use the particle-in-a-one-dimensional-box model to calculate the longest-wavelength peak in the electronic absorption spectrum of hexatriene. (*Hint:* See Equation 11.27.)

14.30 Many aromatic hydrocarbons are colorless, but their anion and cation radicals are often strongly colored. Give a qualitative explanation for this phenomenon. (*Hint:* Consider only the π molecular orbitals.)

14.31 Referring to Figure 14.21, explain why the value of T_m increases as the mole percent (C + G) increases.

NMR and ESR Spectroscopies

14.32 The NMR signal of a compound is found to be 240 Hz downfield from the TMS peak using a spectrometer operating at 60 MHz. Calculate its chemical shift in ppm relative to TMS.

14.33 Both NMR and ESR spectroscopy differ from other branches of spectroscopy discussed in this chapter in one important respect. Explain.

14.34 What is the field strength (in tesla) needed to generate a ^1H frequency of 600 MHz?

14.35 Suppose the NMR spectrum of acetaldehyde (see Figure 14.31) is recorded at 200 MHz and 400 MHz. State whether each of the following quantities remains unchanged or is different from 200 MHz to 400 MHz: **(a)** sensitivity of detection, **(b)** $|\delta_{CH_3} - \delta_H|$, **(c)** $|\nu_{CH_3} - \nu_H|$, **(d)** J.

14.36 For an applied field of 9.4 T (used in a 400-MHz spectrometer), calculate the difference in frequencies for two protons whose δ values differ by 2.5.

14.37 For each of the following molecules, state how many proton NMR peaks occur and whether each peak is a singlet, doublet, triplet, etc. **(a)** CH_3OCH_3, **(b)** $C_2H_5OC_2H_5$, **(c)** C_2H_6, **(d)** CH_3F, **(e)** $CH_3COOC_2H_5$.

14.38 Sketch the NMR spectrum of isobutyl alcohol [$(CH_3)_2CHCH_2OH$], given the following chemical shift data: $-CH_3$: 0.89 ppm, $-C-H$: 1.67 ppm, $-CH_2$: 3.27 ppm, and $-O-H$: 4.50 ppm.

14.39 The toluene proton NMR spectrum, consisting of two peaks due to the methyl and aromatic protons, has been recorded at 60 MHz and 1.41 T. **(a)** What would be the magnetic field at 300 MHz? **(b)** At 60 MHz, the resonance frequencies are: methyl, 140 Hz; aromatic, 430 Hz. What would the frequencies be if recorded by a 300-MHz spectrometer? **(c)** Calculate the chemical shifts (δ) of the two signals, using both the 60 MHz and 300 MHz data.

14.40 The methyl radical has a planar geometry. How many lines would you observe in the ESR spectrum of $\cdot CH_3$? Of $\cdot CD_3$?

14.41 Account for the number of lines observed in the ESR spectra of benzene and naphthalene anion radicals shown in Figure 14.35. How would you use isotopic substitution to assign the two hyperfine splitting constants in naphthalene?

14.42 One way to study membrane structure (see Section 5.7) is to use a spin label, which is a nitroxide radical that has the following structure:

$$R-O-\overset{\overset{\displaystyle O}{\|}}{\underset{\underset{\displaystyle O_-}{|}}{P}}-O-CH_2-CH_2-\overset{+}{N}\!$$

where R represents the hydrophobic tail part of the phosphatidic acid derivative. The ESR spectrum of this spin label, like that of di-*tert*-butyl nitroxide, shows three lines of equal intensities. The ESR signals disappear rapidly when the nitroxide comes in contact with a reducing agent such as ascorbate. In one experiment, these spin-label molecules were incorporated in the membrane lipid bilayer structure at about 5% concentration. The amplitude of the nitroxide ESR signals decreased to 35% of the initial value within a few minutes of the addition of ascorbate. The amplitude of the residual spectrum decayed exponentially with a half-life of about 7 h. Explain these observations.

Fluorescence, Phosphorescence, and Lasers

14.43 List some important differences between fluorescence and phosphorescence.

14.44 The lowest triplet state in naphthalene ($C_{10}H_8$) is about 11,000 cm^{-1} below the lowest excited singlet electronic level at 77 K. Calculate the ratio of the populations in these two states. [*Hint:* The Boltzmann equation is given by $N_2/N_1 = (g_2/g_1)\exp(-\Delta E/k_B T)$, where g_1 and g_2 are the degeneracies for levels 1 and 2.]

14.45 The luminescent first-order decay of a certain organic molecule yields the following data:

t/s	0	1	2	3	4	5	10
I	100	43.5	18.9	8.2	3.6	1.6	0.02

where I is the relative intensity. Calculate the mean lifetime, τ, for the process. Is it fluorescence or phosphorescence?

14.46 Give a qualitative explanation as to why POP absorbs light at a shorter wavelength than that for POPOP (see p. 556).

14.47 The fluorescence of a protein is due to tryptophan, tyrosine, and phenylalanine (assuming that the protein does not contain a prosthetic group that is fluorescent). Iodide ions are known to quench the fluorescence of tryptophan. If a protein is known to contain only one tryptophan residue and iodide fails to quench its fluorescence, what can you conclude about the location of the tryptophan residue?

14.48 Name three characteristic properties of laser.

14.49 Explain why we cannot produce laser light with a two-level system.

14.50 For a three-level laser system, the wavelength for an absorption from level A to level C is found to be 466 nm, and the wavelength for a transition between levels B and C is 752 nm. What is the wavelength for a transition between levels A and B?

14.51 What is the difference between a one-photon and a multiphoton process? Why does the use of lasers make it favorable to observe, say, a two-photon process?

14.52 How many unpaired electrons are in a molecule in a quartet state?

Optical Rotatory Dispersion and Circular Dichroism

14.53 Is 1,3-dichloroallene ($C_3H_2Cl_2$) chiral?

14.54 The optical rotation of a sucrose solution (concentration: 9.6 g in 100-mL soln) is $+0.34°$ when measured in a 10-cm cell with the sodium D-line at room temperature. Calculate the specific rotation and molar rotation for sucrose.

14.55 The optical rotation of an L-leucine solution (concentration: 6.0 g in 100 mL solution) is $+1.81°$ in a 20-cm polarimeter cell at 589.3 nm and 25°C. **(a)** Calculate the specific rotation and **(b)** the molar rotation, given that the molar mass of L-leucine is 131.2 g mol^{-1}.

14.56 Two optical isomers, A and B, having specific rotations of $+27.6°$ and $-19.5°$, respectively, are in equilibrium in solution. If the specific rotation of the mixture is 16.2°, calculate the equilibrium constant for the process A $\rightleftharpoons$ B.

14.57 Two substances, A and B, have identical absorption spectra and identical CD curves except that one CD curve is positive and the other is negative. What is the structural relationship between A and B?

14.58 In an optical-rotation measurement, the angle measured for a solution of concentration c in a cell of pathlength l is $-12.7°$. How can you be certain that this is the correct rotation and not $(-12.7° + 360°)$ or 347.3°, considering that clockwise rotation by 347.3° is equivalent to counterclockwise rotation by 12.7°?

14.59 In what wavelength range would you expect to find the CD spectrum of tryptophan? (*Hint:* See Figure 14.19.)

14.60 The CD of a protein solution changes appreciably upon the addition of a certain achiral compound. What might have happened?

14.61 The optical rotation of a sample of α-D-glucose is +112.2° and that of β-D-glucose is +18.7°. A mixture of these two sugars has an optical rotation of 56.8°. Calculate the composition of the mixture.

14.62 Winemakers often use a pocket polarimeter to check the maturity of grapes in their vineyards. Explain how it works.

14.63 Sucrose ($C_{12}H_{22}O_{11}$) is known as cane sugar. In the confectionery industry, sucrose is hydrolyzed to glucose and fructose by dilute acids or the enzyme invertase as follows:

$$C_{12}H_{22}O_{11} \rightarrow C_6H_{12}O_6 + C_6H_{12}O_6$$
$$+66.48° \qquad +112.2° \qquad -132°$$

The specific rotations are all measured at 25°C with the sodium D-line. Both glucose and fructose have the same molecular formula; fructose has the negative specific rotation. One reason for the breakdown of sucrose is that fructose is the sweetest sugar known. **(a)** Why is the sugar manufactured from this process called "invert sugar"? **(b)** What additional advantage does a mixture of fructose and glucose have over pure sucrose in making candy?

14.64 The optical rotation of the *d*-form of α-piene ($C_{10}H_{16}$), measured at 20°C with the sodium D-line in a cell of pathlength 1.0 cm, is +4.4°. Given that the density of the liquid is 0.859 g mL^{-1}, calculate the specific rotation of α-piene. What does the positive sign of the rotation mean?

14.65 Using the results of Problem 14.64, calculate the molar rotation of α-piene.

14.66 The rotation of a solution of L-ribulose measured at 25°C with the sodium D-line is −3.8°. The pathlength of the cell is 10 cm and $[\alpha]_D^{25}$ is −16.6° dm^{-1} cm^3 g^{-1}. What is the concentration of the solution in g mL^{-1}?

Additional Problems

14.67 The typical energy differences for transitions in the microwave, IR, and electronic spectroscopies are 5×10^{-22} J, 0.5×10^{-19} J, and 1×10^{-18} J, respectively. Calculate the ratio of the number of molecules in two adjacent energy levels (for example, the ground level and the first excited level) at 300 K in each case. (*Hint:* See Problem 14.14.)

14.68 The molar absorptivity of a solute at 664 nm is 895 L mol^{-1} cm^{-1}. When light at that wavelength is passed through a 2.0-cm cell containing a solution of the solute, 74.6% of the light is absorbed. Calculate the concentration of the solution.

14.69 The frequency of molecular collision in the liquid phase is about 1×10^{13} s^{-1}. Ignoring all other mechanisms contributing to line width, calculate the width (in Hz) of vibrational transitions if **(a)** every collision is

effective in deactivating the molecule vibrationally, and **(b)** that one collision in 40 is effective.

14.70 Consider the 2-propenenitrile molecule whose IR spectrum is shown in Figure 14.17. Which of the following types of energy has the largest number of energy levels appreciably occupied at 300 K? Electronic, C–H stretching vibration, C=C stretching vibration, HCH bending motion, or rotational.

14.71 Analysis of lines broadened by the Doppler effect shows that the width at half-height, $\Delta\lambda$, is given by

$$\Delta\lambda = 2\left(\frac{\lambda}{c}\right)\left(\frac{2k_BT}{m}\right)^{1/2}$$

where c is the speed of light, T is the temperature (in kelvin), and m is the mass of the species involved in the transition. The corona of the sun emits a spectral line at about 677 nm due to the presence of an ionized ^{57}Fe atom (molar mass: 0.0569 kg mol^{-1}). If the line has a width of 0.053 nm, what is the temperature of the corona?

14.72 Derive an expression for the value of J corresponding to the most populous rotational energy level of a rigid diatomic rotor at temperature T. Evaluate the expression for HCl ($B = 10.59$ cm^{-1}) at 25°C. (*Hint:* See Problem 14.14.)

14.73 Analyze the ^{31}P NMR spectrum of ATP shown in Figure 14.29.

14.74 An aqueous solution contains two species, A and B. The absorbance at 300 nm is 0.372, and at 250 nm is 0.478. The molar absorptivities of A and B are:

$$A: \varepsilon_{300} = 3.22 \times 10^4 \text{ L mol}^{-1} \text{ cm}^{-1}$$
$$\varepsilon_{250} = 4.05 \times 10^4 \text{ L mol}^{-1} \text{ cm}^{-1}$$
$$B: \varepsilon_{300} = 2.86 \times 10^4 \text{ L mol}^{-1} \text{ cm}^{-1}$$
$$\varepsilon_{250} = 3.76 \times 10^4 \text{ L mol}^{-1} \text{ cm}^{-1}$$

If the pathlength of the cell is 1.00 cm, calculate the concentrations of A and B in mol L^{-1}.

14.75 A molecule XY_2 is known to be linear, but it is not clear whether it is Y–X–Y or X–Y–Y. How would you use IR spectroscopy to determine its structure?

14.76 The NMR spectrum of *N,N′*-dimethylformamide shows two methyl peaks at 25°C. When heated to 130°C, there is only one peak due to the methyl protons. Explain.

14.77 This problem deals with the amplitude of molecular vibration of a diatomic molecule in its ground vibrational state. **(a)** When the molecule is stretched by an extent x from the equilibrium position, the increase in the potential energy is given by the integral

$$\int_0^x kx\,dx$$

where k is the force constant. Evaluate this integral.
(b) To calculate the amplitude of vibration, we equate the potential energy with the vibrational energy in the ground state. Use x_{max} to represent the maximum displacement. **(c)** Given that the force constant for $H^{35}Cl$ is 4.84×10^2 N m^{-1}, calculate the amplitude of vibration in the $v = 0$ state. **(d)** What is the percent of the amplitude compared to the bond length (1.27 Å)? **(e)** Repeat the calculations in **(c)** and **(d)** for carbon monoxide, given that the force constant is 1.85×10^3 N m^{-1} and the bond length is 1.13 Å. (^{35}Cl: 34.97 amu.)

14.78 The IR spectrum of the carbon monoxide–hemoglobin complex shows a peak at about 1950 cm^{-1}, which is due to the carbonyl stretching frequency.
(a) Compare this value with the fundamental frequency of free CO, which is 2143.3 cm^{-1}. Comment on the difference. **(b)** Convert this frequency to kJ mol^{-1}.
(c) What conclusion can you draw from the fact that there is only one band present?

14.79 True or false? **(a)** To be IR active, a polyatomic molecule must possess a permanent dipole moment.
(b) The moment of inertia of a diatomic molecule measured from its microwave spectrum provides information about the force constant of the bond. **(c)** The fluorescence spectrum of a molecule occurs at a shorter wavelength than the absorption spectrum of the molecule. **(d)** A 600-MHz NMR spectrometer is more sensitive than a 400-MHz spectrometer. **(e)** Phosphorescence is a spin-forbidden process. **(f)** To observe hyperfine splittings in an ESR spectrum, the nucleus involved must have $I \neq 0$. **(g)** Whenever a molecule goes from one energy level to another, it must emit or absorb a photon whose energy is equal to the energy difference between the two levels.

14.80 The Beer–Lambert law (see Equation 14.15) predicts a linear dependence of absorbance on concentration. The law often breaks down at very high concentrations. Why?

14.81 The wavelengths of absorption of chromophores in electronic spectra are often influenced by the solvent. For example, polar solvents stabilize the ground state of $n \rightarrow \pi^*$ transitions more than the excited state. On the other hand, for $\pi \rightarrow \pi^*$ transitions, the excited state is more stabilized. Sketch diagrams to show the changes in energy levels involved in the electronic transition when a chromophore changes from a nonpolar to a polar solvent environment and predict the shift in wavelength in each case.

14.82 (a) Calculate the energy difference between the two spin states of ^1H and of ^{13}C in a magnetic field of 4.70 T. **(b)** What is the precession frequency of a ^1H nucleus at this magnetic field? Of a ^{13}C nucleus? **(c)** At what magnetic field do protons precess at a frequency of 500 MHz?

14.83 Oxygen is an effective quencher of fluorescence because it is a triplet in its electronic ground state. The unpaired spins of O_2 can induce the excited state of the fluorescent molecule to undergo intersystem crossing from the singlet state to the triplet state, that is, $S_1 \rightarrow T_1$ (see Figure 14.37). **(a)** How would you verify the mechanism experimentally? **(b)** Assume that the quenching rate constant (that is, the rate constant for the collision between O_2 and fluorescent molecules) is 1.0×10^{10} M^{-1} s^{-1}. How many collisions s^{-1} on average does each fluorescent molecule in solution experience? The concentrations are $[O_2] = 3.4 \times 10^{-4}$ M and $[F] = 0.5$ M, where F is the fluorescent molecule. **(c)** The fluorescence lifetime of pyrene, a molecule that is often used to probe biological systems, is 500 ns while that of tryptophan is about 5 ns. Explain why under normal atmospheric conditions O_2 can interfere only with the fluorescence of pyrene but not with that of tryptophan. **(d)** A quantitative relationship of fluorescence quenching is the Stern–Volmer equation

$$\frac{I_0}{I} = 1 + k_Q \tau_0 [Q]$$

where I_0 and I are the fluorescence intensities in the absence and presence of the quencher, k_Q is the quenching rate constant, τ_0 is the mean lifetime of the fluorescent state in the absence of the quencher, and $[Q]$ is the concentration of the quencher. Use this equation to support your conclusion in **(c)**.

14.84 The absorbance at 260 nm of a DNA solution is 0.120 at 25°C and below (all double helices) and 0.142 at 90°C and above (totally denatured). Calculate the fraction of the double helix remaining at 70°C if the absorbance is 0.131.

14.85 Ants can sometimes survive in a microwave oven during heating by moving around. Explain.

CHAPTER 15

Photochemistry and Photobiology

A photochemist is interested in the fate of an electronically excited molecule. Depending on the system and conditions under which photoexcitation is carried out, such a molecule can undergo one of several processes. It can lose energy in collisions with other molecules, liberating heat. It can return to the ground state by emitting a photon; that is, it can fluoresce or phosphoresce. Alternatively, it can undergo a chemical reaction, such as isomerization, dissociation, or ionization. Chapter 14 dealt with the phenomena of fluorescence and phosphorescence. In this chapter we shall introduce the vocabulary of photochemical reactions and discuss two important photobiological processes: photosynthesis and vision. We shall also examine the biological effects of radiation.

15.1 Introduction

We begin our study of photochemical and photobiological events with an introduction to some of the terms used in this chapter.

Thermal Versus Photochemical Reactions

Chemical reactions can be categorized as thermal or photochemical. *Thermal reactions*, discussed in Chapter 9, involve atoms and molecules in their electronic *ground* state. By definition, a *photochemical reaction* takes place in the presence of light, which usually means radiation from the visible and UV region or high-energy radiation such as X rays and γ rays.

If we take 4×10^{-19} J as the typical energy of an electronically excited state, then, using the Boltzmann distribution law (see Equation 3.22), we can show that at room temperature (25°C) $N_2/N_1 \approx 6 \times 10^{-43}$, and so only a negligible fraction of molecules are electronically excited. To achieve a mere 1% concentration of excited molecules would require a temperature of about 6000°C! At that temperature, practically all of the molecules would undergo rapid thermal decomposition in their ground electronic state, and it would be impossible to produce appreciable concentrations of electronically excited molecules.

On the other hand, if molecules absorb radiation at 500 nm, which roughly corresponds to the wavelength required for the electronic transition, then electronic excitation must occur. The concentration of the excited molecules depends on several factors, including the intensity of irradiation and the rate at which the excited molecules return to the ground state. Further, if the electronic excitation energy can somehow be harnessed for bond breaking, then chemical change may occur. Thus, the energy of excitation for a photochemical reaction is analogous to the activation energy for a thermal reaction.

Primary Versus Secondary Processes

Photochemical reactions are subclassified as *primary* or *secondary processes*. Primary processes include vibrational relaxation, or loss of vibrational energy, by collision with other molecules; fluorescence; phosphorescence; isomerization; and dissociation. Dissociation of excited molecules may provide reactive intermediates that can undergo secondary processes of a thermal nature.

Let us illustrate the primary and secondary processes with the decomposition of hydrogen iodide in the gas phase. The overall reaction is

$$2HI \rightarrow H_2 + I_2$$

When light of the appropriate wavelength is applied, the reactions are

$$HI \xrightarrow{h\nu} H + I \quad \text{photochemical reaction (primary process)}$$
$$H + HI \rightarrow H_2 + I \quad \text{thermal reactions (secondary process)}$$
$$I + I \rightarrow I_2$$
$$H + H \rightarrow H_2$$
$$\text{overall:} \quad 2HI \rightarrow H_2 + I_2$$

where $h\nu$ represents the energy of the photon absorbed.

Quantum Yields

A useful ratio in the study of photochemical reactions is the *quantum yield* (Φ), which is the ratio of the number of molecules of product formed (or reactant molecules consumed) to the number of light quanta absorbed

$$\Phi = \frac{\text{number of molecules produced}}{\text{number of photons absorbed}} \tag{15.1}$$

Equation 15.1 can be expressed in molar quantities as

$$\Phi = \frac{\text{number of moles of product formed}}{\text{number of einsteins absorbed}} \tag{15.2}$$

where an *einstein* is equal to 1 mole of photons.

The quantum yield of photochemical reactions varies greatly from one system to another, and the value of Φ often reveals the mechanism involved in the process. For the hydrogen iodide reaction discussed above, the quantum yield is 2 because the absorption of one photon leads to the removal of two reactant molecules (HI). When irradiated with UV light at about 280 nm, acetone forms a methyl and an acetyl radical with high yield:

$$(CH_3)_2CO \xrightarrow{h\nu} CH_3\cdot + CH_3CO\cdot$$

In the liquid phase, however, these radicals are likely to recombine because of the solvent cage effect (see p. 346). Therefore, the overall quantum yield for this reaction is below 0.1.

A mixture of gaseous hydrogen and chlorine is stable at room temperature. When exposed to visible light (about 400 nm), the gases react explosively to form hydrogen chloride. The mechanism is

$$Cl_2 \overset{hv}{\to} Cl + Cl$$

$$Cl + H_2 \to HCl + H \qquad\qquad\qquad (a)$$

$$H + Cl_2 \to HCl + Cl \qquad\qquad\qquad (b)$$

This is a *chain reaction* in which the propagation steps are (a) and (b). The quantum yield of this reaction is about 10^5! In general, a quantum yield greater than 2 is evidence of a chain mechanism.

Alternatively, a photochemical reaction can be analyzed in terms of rate constants. Consider the following situation:

$$A \overset{hv}{\to} A^*$$

$$A^* \overset{k_1}{\to} A$$

$$A^* \overset{k_2}{\to} product$$

where A is the reactant, and A^* is an electronically excited molecule. Assuming steady-state conditions, we write

$$\text{rate of formation of } A^* = \text{rate of removal of } A^*$$

$$= k_1[A^*] + k_2[A^*]$$

The quantum yield of product formation is given by

$$\Phi_P = \frac{\text{rate of product formation}}{\text{total rate of removal of } A^*}$$

$$= \frac{k_2[A^*]}{k_1[A^*] + k_2[A^*]} = \frac{k_2}{k_1 + k_2} \qquad\qquad (15.3)$$

Note that Φ and rate are not fundamentally related. Two reactions may have very similar Φ values but differ greatly in their rate constants. Consider the following photochemical decompositions:

$$C_6H_5COCH_2CH_2CH_3 \overset{k}{\to} C_6H_5COCH_3 + CH_2{=}CH_2$$

$$\Phi = 0.40 \quad k = 3 \times 10^6 \text{ s}^{-1}$$

$$CH_3COCH_2CH_2CH_2CH_3 \overset{k}{\to} CH_3COCH_3 + CH_2{=}CHCH_3$$

$$\Phi = 0.38 \quad k = 1 \times 10^9 \text{ s}^{-1}$$

For an insight into the difference between Φ and rate, we need to look at the factors affecting the rate of a photochemical reaction, which can be expressed as

$$\text{rate} = IFf\Phi_P \qquad\qquad (15.4)$$

where I is the rate of absorption of light, F is the fraction of the total incident light that is absorbed, f is the fraction of absorbed light that produces the reactive state, and Φ_P is the quantum yield of product formation. Now we see why two reactions can have similar Φ values but very different rates if the reactants have different values of I, F, and f.

Measurement of Light Intensity

Regardless of the mechanism involved, the rate of a photochemical reaction should be proportional to the rate of absorption of light. Thus, kinetic studies of photochemical reactions require accurate measurements of the intensity of light employed. Light intensity is measured with a chemical *actinometer*—a chemical system whose photochemical behavior is quantitatively understood. One of the most useful solution-phase actinometers is the potassium ferrioxalate system. When a sulfuric acid solution of $K_3Fe(C_2O_4)_3$ is irradiated with light in the range of 250 to 470 nm, the reduction of iron from Fe(III) to Fe(II) and oxidation of the oxalate ion to carbon dioxide occur simultaneously. The simplified equation for this process is

$$2Fe(C_2O_4)_3^{3-} \rightarrow 2Fe^{2+} + 5C_2O_4^{2-} + 2CO_2$$

This reaction has been carefully studied, and its quantum yields are known at various wavelengths. The amount of Fe^{2+} ions formed can be readily determined from the formation of the red 1,10-phenanthroline–Fe^{2+} complex ion whose molar absorptivity is known. In this way, the amount of photons absorbed in a given time period can be determined.

Example 15.1

A 35-mL solution of $K_3Fe(C_2O_4)_3$ is irradiated with monochromatic light at 468 nm for 30 min. The solution is then titrated with 1,10-phenanthroline to form the red complex of 1,10-phenanthroline–Fe^{2+}. The absorbance of this complex ion measured in a 1-cm cell at 510 nm is 0.65 ($\varepsilon_{510} = 1.11 \times 10^4$ L mol^{-1} cm^{-1}). Assume that the quantum yield for the decomposition at this wavelength is 0.93, and calculate the number of einsteins absorbed per second and the total energy absorbed.

ANSWER

From Equation 14.15

$$c = \frac{A}{\varepsilon b} = \frac{0.65}{(1.11 \times 10^4 \text{ L mol}^{-1} \text{ cm}^{-1})(1 \text{ cm})}$$

$$= 5.86 \times 10^{-5} \; M$$

The number of einsteins absorbed is given by (see Equation 15.2)

$$\frac{\text{number of moles of } Fe^{2+} \text{ produced}}{\text{quantum yield}} = \frac{(5.86 \times 10^{-5} \text{ mol/L})(1 \text{ L}/1000 \text{ mL})(35 \text{ mL})}{0.93}$$

$$= 2.2 \times 10^{-6} \text{ mol}$$

$$= 2.2 \times 10^{-6} \text{ einstein}$$

The rate of absorption is given by

$$\frac{2.2 \times 10^{-6} \text{ einstein}}{30 \times 60 \text{ s}} = 1.2 \times 10^{-9} \text{ einstein s}^{-1}$$

Finally,

$$\text{total energy absorbed} = \text{number of photons} \times h\nu$$
$$= (2.2 \times 10^{-6} \text{ mol})(6.022 \times 10^{23} \text{ mol}^{-1})$$
$$\times (6.626 \times 10^{-34} \text{ J s})\left(\frac{3.00 \times 10^8 \text{ m s}^{-1}}{468 \times 10^{-9} \text{ m}}\right)$$
$$= 0.56 \text{ J}$$

COMMENT

Light intensity is measured in photons cm^{-2} s^{-1} (or J cm^{-2} s^{-1}). In photochemistry, we are more interested in the amount of light energy that is deposited in the sample, which is called absorbed intensity. Absorbed intensity is energy input into the reacting system per unit volume per unit time and has the units J cm^{-3} s^{-1}. In our example, the absorbed intensity is

$$\frac{0.56 \text{ J}}{(35 \text{ cm}^3)(30 \times 60 \text{ s})} = 8.9 \times 10^{-6} \text{ J cm}^{-3} \text{ s}^{-1}.$$

Action Spectrum

Often, very useful information regarding the species responsible for photochemical and photobiological processes can be obtained if we measure the response or the effectiveness of the system as a function of the wavelength of the light employed. This procedure gives the *action spectrum*. In general, if a simple system contains only one type of molecule, the action spectrum should and does resemble the absorption spectrum closely. In a complex biological system, there are usually several different compounds that strongly absorb the incident radiation over the range of wavelength of interest. The molecules responsible for the photochemical reaction may be present in very low concentrations, so that their absorption spectra cannot always be detected. Their presence can be revealed, however, by recording the action spectrum instead of the usual absorption spectrum (Figure 15.1).

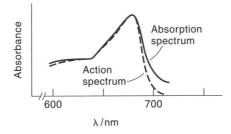

Figure 15.1
Comparison of the absorption spectrum for the unicellular alga chlorella. The photosynthetic efficiency (measured by oxygen evolution) of light of different wavelengths (action spectrum) closely parallels the absorption spectrum of chlorophyll molecules. The discrepancy at about 700 nm is known as the "red drop" (to be discussed later). This comparison strongly suggests that chlorophyll plays a key role in photosynthesis. (From *Light and Living Matter*, Vol. 2 by R. K. Clayton. Copyright 1971 by McGraw-Hill Book Company. Used with permission of McGraw-Hill Company.)

15.2 Photosynthesis

Photosynthesis is the most important of all photobiological reactions. It is the process by which plants and other organisms capture solar energy and convert it to carbohydrates and other complex molecules. Although green plant photosynthesis involves many very complex steps, the overall change can be represented as

$$6CO_2 + 6H_2O \xrightarrow{h\nu} C_6H_{12}O_6 + 6O_2 \quad \Delta_r G^\circ = 2879 \text{ kJ mol}^{-1}$$

This thermodynamically highly unfavorable reaction is driven by light. (Note that this reaction is just the opposite of oxidative carbohydrate metabolism.) In a simplified form, we write

$$CO_2 + H_2O \xrightarrow{h\nu} (CH_2O) + O_2 \qquad \Delta_r G^\circ = 480 \text{ kJ mol}^{-1}$$

The formation of oxygen as a by-product is not universal among photosynthetic organisms. For example, in photosynthetic bacteria, the reaction is

$$2H_2S + CO_2 \xrightarrow{h\nu} (CH_2O) + H_2O + 2S$$

Thus, a general equation for photosynthesis is

$$2H_2A + CO_2 \rightarrow (CH_2O) + H_2O + 2A$$

where H_2A acts as the hydrogen donor, and CO_2 acts as the hydrogen acceptor.

The study of photosynthesis encompasses a wide range of science, including chemical physics and molecular biology. Research completed over the past 50 years has given us a fairly complete description of the overall process. Here, we shall discuss only the initial process of light absorption and some of the reactions that immediately follow it. Photosynthesis occurs in two main stages. The first stage, called the *light reaction*, involves the absorption of photons, which takes place in picoseconds (1 ps = 10^{-12} s). This step is followed by a series of chemical transformations that are sometimes called *dark reactions* because they occur in the absence of light. Light reactions use light energy to generate NADPH and ATP, which the dark reactions require to drive the synthesis of carbohydrate from CO_2 and H_2O.

Actually, the dark reactions are reactions that are just not directly affected by light.

The Chloroplast

The site of photosynthesis in eukaryotes (algae and higher plants) is the *chloroplast*. A typical plant cell contains between 50 and 200 chloroplasts, which are about 5000 Å long (Figure 15.2). In addition to the two outer membranes, chloroplasts have interior membranes that form a highly laminated structure consisting of individual sacs, or *thylakoids*, piled into cylinders called *grana*. The light reactions occur in the grana, and the dark reactions take place in the *stroma*, a concentrated solution of enzymes, RNA, DNA, and ribosomes that direct protein synthesis.

In 1937, the American biochemist Robert Hill (1899–1991) discovered that isolated chloroplasts evolve oxygen when illuminated with light in the presence of a suitable electron acceptor, such as the ferricyanide ion $[Fe(CN)_6]^{3-}$. There is a concomitant production of O_2. This experiment was significant because it showed that the oxygen comes from water rather than CO_2, because no CO_2 was present. Subsequently, Hill's conclusion was confirmed by an isotope-label experiment:

$$CO_2 + H_2{}^{18}O \xrightarrow{h\nu} (CH_2O) + {}^{18}O_2$$

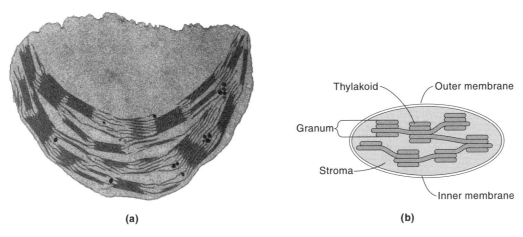

Figure 15.2
(a) Electron micrograph of a chloroplast from a citrus leaf. (b) Schematic drawing of a chloroplast. The closely packed membranes, which are packed into stacks, are called grana. They conduct the initial steps in photosynthesis—the light reaction that traps the photons. The black particles in part (a) are called plastoglobuli and consist of lipids. They appear black as a result of the staining technique used in electron microscopy. (Electron micrograph courtesy of Kenneth R. Miller.)

All of the heavy oxygen isotopes (^{18}O) from the labeled water turned up in O_2.

Chlorophyll and Other Pigment Molecules

Chlorophyll molecules are the light-absorbing chromophores in photosynthetic organisms. There are different types of chlorophyll molecules, and an organism may contain two or more of them. In green plants, we find chlorophyll a and chlorophyll b, shown in Figure 15.3. The highly conjugated porphyrin system in chlorophyll results in their absorption of light in the visible region. As Figure 15.4 shows, chlorophyll a absorbs strongly in the blue and red regions but transmits light in the green, yellow, and orange regions. Consequently, it has a characteristic green color. The peak molar absorptivities of chlorophyll a and b are among the highest for organic compounds, on the order of 10^5 cm^{-1} L mol^{-1}.

Like the chlorophylls, other pigment molecules also serve as *antenna* molecules, helping to gather, or harvest, solar energy and relay it to the site where the light reactions occur. Two of these molecules are phycoerythrin and β-carotene, also shown in Figure 15.3. As we can see from Figure 15.4, together these pigments (plus others not shown) absorb most of the visible light in the solar emission spectrum.

The Reaction Center

The initial steps of photosynthesis involve the absorption of light and the transfer of this energy to the *reaction center*, where the light reactions take place. Knowledge of this process was provided by the American biologist Robert Emerson (1903–1959), who studied oxygen production by the green algae *Chlorella*. Emerson exposed the chlorella cells to light flashes lasting a few microseconds (1 μs = 10^{-6} s). With weak flashes, he found that about one O_2 molecule was generated per eight photons absorbed. Emerson expected to find that, as the intensity of light increased, the yield of O_2 would increase with each flash until each chlorophyll molecule absorbed a pho-

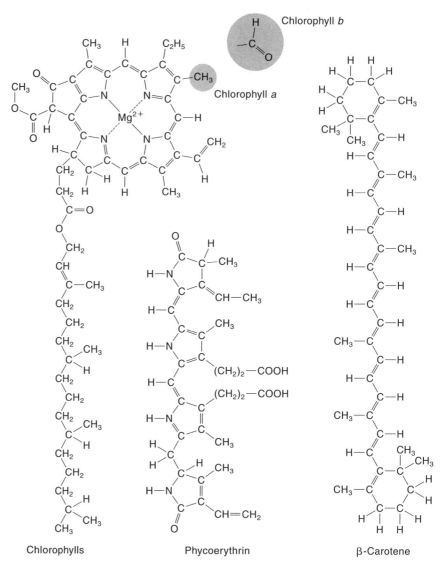

Figure 15.3
Structures of some photosynthetic pigments. Chlorophylls and β-carotene are present in higher plants; phycoerythrin is found only in certain algaes.

Figure 15.4
Absorption spectra of several photosynthetic pigments. Together, they effectively cover the solar emission spectrum in the visible region.

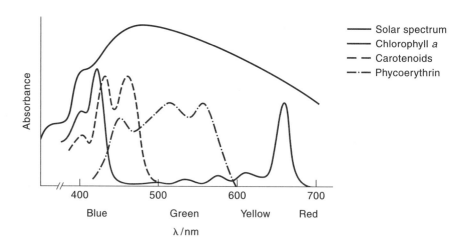

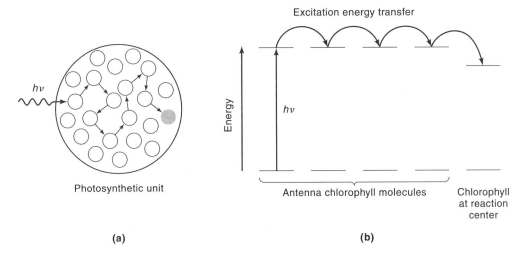

Figure 15.5
(a) The photosynthetic unit. Initially, an antenna chlorophyll molecule is photoexcited. By a nonradiative transfer mechanism (the path is totally random), the energy is finally captured by the chlorophyll dimer (gray sphere) at the reaction center. (b) Energy-level diagram illustrating excitation transfer. The excitation energy is trapped by the reaction center because the excited state of its chlorophyll has a lower energy than that of the antenna molecules. More than 95% of the light absorbed is transferred to the reaction center.

ton, which would then be used in dark reactions. Instead, the measurements showed that each flash of saturating intensity produced only one O_2 per 2500 chlorophyll molecules present. Because at least eight photons must be absorbed to liberate one O_2 molecule, these results suggest that the reaction center must contain $2500/8 \approx 300$ chlorophyll molecules.

It is unlikely that all 300 chlorophyll molecules would participate in the photochemical process. Indeed, subsequent experiments showed that most of the chlorophylls act as antenna molecules that do not take part in light reactions. Figure 15.5a shows a schematic diagram of a *photosynthetic unit*, which consists of the antenna molecules and the reaction center. The mechanism of energy transfer is a nonradiative process requiring that the emission spectrum of the donor molecule overlaps with the absorption spectrum of the acceptor molecule. The efficiency of this energy transfer depends on $1/r^6$, where r is the distance between the donor and the acceptor, and on the relative orientations of these molecules.* Eventually, the excitation energy is trapped at the reaction center by chlorophylls that, although chemically identical to antenna chlorophylls, have slightly lower excited-state energies because of their different environments (Figure 15.5b).

Photosystems I and II

In 1956, Emerson and his colleagues showed that photosynthesis is more complex than Figure 15.5 suggests. They measured the quantum efficiency of photosynthesis in an isolated chloroplast as a function of wavelength and discovered a curious effect. For a single kind of photoreceptor, the quantum efficiency is expected to be independent of wavelength over its entire absorption spectrum. Starting from the short wavelength of 400 nm, the efficiency was found to remain fairly constant and close to the maximum value until 680 nm is reached. Beyond this point, the efficiency rapidly drops to zero (called the *red drop*) even though the chlorophyll *a* molecules

Quantum efficiency is the ratio of oxygen molecules produced to number of photons absorbed.

---

*For an analogy of this energy-transfer mechanism, see R. Chang, *The Physics Teacher*, p. 593, December 1983.

still absorb in the range from 680 nm to 700 nm (see Figure 15.4). Efficiency can be restored if the chloroplast is simultaneously irradiated with light of a shorter wavelength, say 600 nm. Furthermore, the efficiency under simultaneous irradiation at 600 nm and 700 nm is greater than the sum of the efficiencies under separate irradiations at these two wavelengths. This phenomenon, now called the *Emerson enhancement effect*, suggests that there are two separate photochemical systems called *photosystem I* (PSI) and *photosystem II* (PSII). Both of these systems are driven by light of wavelength shorter than 680 nm, but only one is driven by light of a longer wavelength. Each photosystem has its own photosynthetic unit, containing roughly 300 light-harvesting chlorophylls and other pigments. The ratio of chlorophyll a to chlorophyll b is much greater in PSI than in PSII.

Photosynthesis in plants depends on the interplay of PSI and PSII. PSI, which can be excited by light of $\lambda \leq 700$ nm, produces a strong reductant capable of reducing $NADP^+$ to NADPH, and, concomitantly, a weak oxidant. PSII, which requires light of $\lambda \leq 680$ nm, generates a strong oxidant capable of oxidizing water to oxygen, and, concomitantly, a weak reductant. Electron flow from PSII to PSI, as well as electron flow within each photosystem, generates a transmembrane proton gradient that drives the formation of ATP. This process is called *photophosphorylation.*[†]

How are these two photosystems related to each other, and what are their roles in the overall process of photosynthesis? Let us start first with PSII. Our knowledge of PSII has been greatly enhanced by the structural study of a similar photosystem in purple bacteria. X-ray crystallographic analysis of its reaction center reveals detailed information about the type of compounds present and their spatial orientations along the electron-transfer chain. The reaction center in PSII contains a chlorophyll dimer whose maximum absorption is at 680 nm, so it is called pigment 680, or P680. Upon receiving a photon, P680 becomes electronically excited, which is denoted by P680*. Being a reductant, P680* transfers an electron to a neighboring bound pheophytin (Ph, a molecule like chlorophyll a except that it does not contain a Mg^{2+} ion)

$$P680^* + Ph \rightarrow P680^+ + Ph^-$$

The Ph^- ion then transfers the extra electron to plastoquinone (Q), eventually reducing it to the hydroquinone form (QH_2). $P680^+$ is a strong oxidizing agent that can oxidize water to oxygen:

$$2H_2O \rightarrow O_2 + 4H^+ + 4e^-$$

This reaction is catalyzed by a manganese-containing enzyme.

The transfer of electrons from water to quinone is an uphill process, as shown by the following standard reduction potentials:

$$O_2 + 4H^+ + 4e^- \rightarrow 2H_2O \qquad E^{\circ\prime} = 0.82 \text{ V}$$
$$Q + 2H^+ + 2e^- \rightarrow QH_2 \qquad E^{\circ\prime} = 0.1 \text{ V}$$

[†] As mentioned in Chapter 7, the principles by which ATP synthesis takes place in photophosphorylation are nearly identical with those for oxidative phosphorylation. In 1966 the American biochemist André Jagendorf (1926–) provided strong evidence for the Mitchell chemiosmotic hypothesis. In his experiment, thylakoids were suspended in an acidic medium (pH 4 buffer) and were allowed to equilibrate. They were then rapidly transferred to a pH 8 buffer solution. As a result, a pH difference of four units was established across the thylakoid membrane, with the inside acidic relative to the outside. By adding ADP and inorganic phosphate (P_i) to the medium, Jagendorf found that large amounts of ATP were formed in the *absence* of light or electron transport. This simple yet elegant experiment demonstrates that it is the proton gradient that provides the driving force for ATP synthesis.

We see that QH_2 is a stronger reducing agent than water, so it is more easily oxidized. The driving force for the uphill electron-transfer reaction comes from the absorbed photons. The energy of a photon of wavelength 680 nm is $E = h\nu$, or 3×10^{-19} J. Because 1 eV $= 1.6 \times 10^{-19}$ J, the energy of the photon in electron volts is 3×10^{-19} J/$(1.6 \times 10^{-19}$ J eV$^{-1})$, or 1.9 eV, which is more than enough to change the potential of an electron by 0.72 V (from 0.82 V to 0.1 V) under standard-state conditions.

Continuing our journey along the electron-transfer chain, we find that the flow of electrons from QH_2 to PSI is an energetically downhill process. Some of the electron carriers along the path are cytochrome b_6 and cytochrome f (called the cytochrome bf complex), and the copper protein plastocyanin.

Now we come to PSI. Like PSII, PSI contains in its reaction center a chlorophyll dimer that absorbs maximally at 700 nm, so it is called pigment 700 or P700. After the absorption of a photon, P700* transfers an electron to an acceptor chlorophyll molecule, and thus becomes P700$^+$. The acceptor chlorophyll molecule passes the electron along the chain containing a quinone, three iron–sulfur clusters, and ferredoxin. Then, two reduced ferredoxins pass the electron to NADP$^+$ to form NADPH. (The structure of NADP$^+$ is quite similar to that of NAD$^+$, shown in Figure 6.20.) Finally, P700$^+$ receives an electron from reduced plastocyanin and reverts to P700, ready to absorb another photon. The overall process is summarized by a diagram called the Z scheme, because the redox diagram from P680 to P700* resembles the letter Z (Figure 15.6). The net reaction is

$$2H_2O + 2NADP^+ \rightarrow O_2 + 2NADPH + 2H^+$$

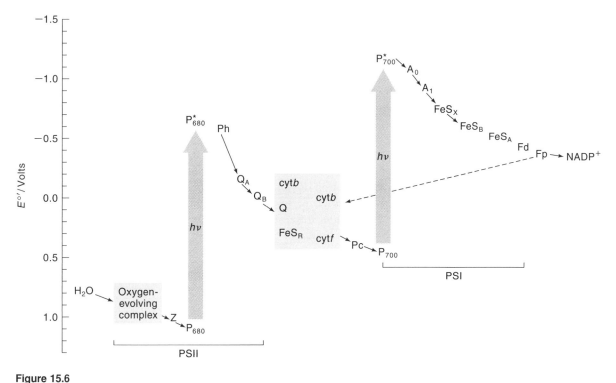

Figure 15.6
The Z scheme tracing the flow of electrons from H_2O to NADP$^+$. The abbreviations are: Z, donor to photosystem II; Ph, pheophytin; Q_A, Q_B, Q, different types of quinones; cyt b and cyt f, cytochromes b and f; FeS$_R$, Rieske iron–sulfur protein; Pc, plastocyanin; A_0 and A_1, early acceptors of PSI; FeS$_X$, FeS$_B$, FeS$_A$, iron–sulfur proteins; Fd, ferredoxin; and Fp, flavoprotein. The dashed line indicates cyclic electron flow around PSI. [After R. E. Blankenship and R. C. Prince, *Trends Biochem. Sci.* **10**, 383 (1985).]

A few points are worth noting regarding the Z scheme. The electrons from fer-redoxin may be returned to the cytochrome *bf* complex in a *cyclic* process rather than to $NADP^+$. The electrons then flow back to $P700^+$ through plastocyanin, causing the protons to be pumped across the membrane. The resulting proton gradient drives the synthesis of ATP according to the chemiosmotic theory discussed in Section 7.6. PSII does not participate in a cyclic process.

In the noncyclic process, proton gradients are established at two sites—where water is oxidized to molecular oxygen in PSII and in the cytochrome *bf* complex. Altogether, about 12 protons are pumped across the membrane per O_2 molecule produced by noncyclic electron transfer. Estimates show that one ATP is synthesized for every four protons transported, resulting in the production of $12/4 = 3$ molecules of ATP for every O_2 generated.

The Z scheme also enables us to explain the red drop and the Emerson enhancement effect. As we have seen, PSI is driven most efficiently by far-red light (680 nm or longer). In the absence of short-wavelength radiation to drive PSII, the flow of electrons along the Z scheme soon ceases because water molecules are not oxidized to moleculer oxygen. Irradiation at a wavelength shorter than 680 nm alone does increase the efficiency, but maximum efficiency is obtained only when both wavelengths are employed.

Finally, we look briefly at photosynthesis in bacterial systems. As mentioned earlier, the reaction center in purple bacteria is analogous to that in PSII of green plants. Many photosynthetic bacteria do not produce oxygen. Instead of the directional (noncyclic) electron flow from water to $NADP^+$, these bacteria utilize a single cyclic photosystem. The reaction center contains a bacteriochlorophyll dimer that is similar to the chlorophyll dimer except for small structural differences that shift the absorption maximum to 870 nm. Therefore, it is called P870. The basic mechanism of light-induced electron transfer remains the same, but the donor and acceptor chains are connected so that no net oxidation or reduction occurs. Some of the energy of the photon is conserved in ATP formation, which in turn drives NAD^+ reduction, using electrons from H_2S or organic acids. The NADH thus formed, along with additional light-generated ATP, is used in a variety of cellular reactions.

Dark Reactions

The initial electron-transfer steps in photosynthesis are extremely rapid, on the order of picoseconds. The subsequent steps, the dark reactions, are much slower. They involve the incorporation of CO_2 from the atmosphere using ATP and NADPH synthesized from light energy, to form part of a glucose molecule. This process is called *carbon dioxide fixation*. Interested readers should consult a standard biochemistry text for a detailed discussion of the dark reactions.

15.3 Vision

Like photosynthesis, the first step in vision is absorption of light energy. The chromophore that absorbs visible light is vitamin A aldehyde, or retinal. The retina of the eye has some 100 million rod-shaped cells and 5 million cone-shaped cells. Between the cells and the nerve fibers leading to the brain are synapses, or junctions (Figure 15.7). Retinal is associated with a protein called *opsin*. Four different types of opsin exist, one in the rod-shaped cells and three in the cone-shaped cells. The chromophore–opsin complexes in the rods and cones are called *rhodopsin* and *iodopsin*, respectively. As a result of research done by the American biologist George Wald (1906–1997) and others, we now understand the basic mechanism of vision fairly well. The changes that occur after excitation by light are basically the same in

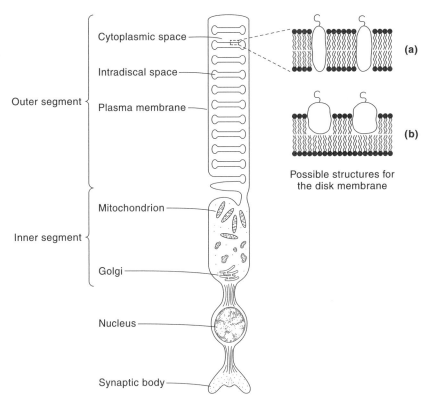

Figure 15.7
Schematic representation of a vertebrate rod cell. The volume of the intradiscal space is greatly exaggerated; in the normal bovine outer segment there are approximately 1500 discs within an outer segment 500,000 Å in length. In the magnified views (a) and (b) showing possible structures of the disc membrane, the wiggly lines represent the hydrocarbon chains of the phospholipids, and the circles the polar head groups. The "S" on the rhodopsin molecule signifies the carbohydrate portion. Light from outside falls on the top portion of the rod cells. [Reprinted with permission from W. L. Hubbell, *Acc. Chem. Res.* **8**, 85 (1975). Copyright by the American Chemical Society.]

rhodopsin and iodopsin, namely, a *cis* to *trans* isomerization of the chromophore.

Figure 15.8 shows the geometric isomers 11-*cis*-retinal and all-*trans*-retinal. Actually, a total of six geometric isomers are possible, but only these two isomers are important in the vision process. Light serves only to initiate the isomerization of 11-*cis*-retinal to all-*trans*-retinal. Here is the fundamental difference between the action of light in vision and its role in photosynthesis. In photosynthesis, light energy is used for the chemical work of boosting electrons against an electrochemical gradient and

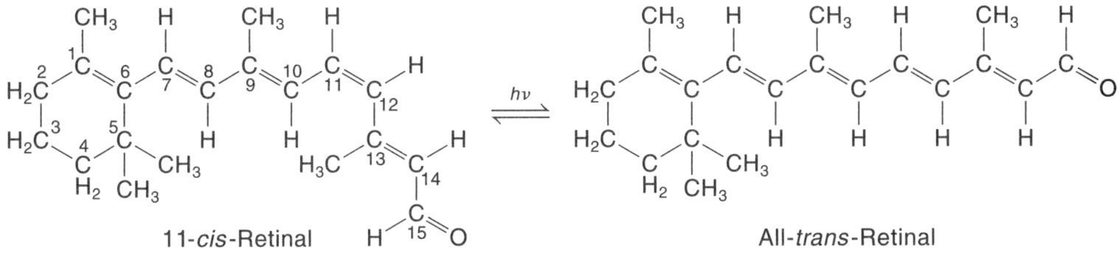

Figure 15.8
Structures of 11-*cis*-retinal and all-*trans*-retinal.

synthesizing ATP and NADPH molecules. In vision, no evidence suggests that chemical syntheses are brought about by light energy. Nerve fibers upon which light acts are ready to discharge because they were previously charged by chemical reactions totally unrelated to excitation of the chromophore. Light is needed only to trigger their discharge.

Structure of Rhodopsin

A Schiff base contains the functional group C=N (imine) formed between a carbonyl group and a primary amine.

Opsin is a protein with a molar mass of about 38,000 daltons. In rhodopsin, the aldehyde group of 11-*cis*-retinal forms a Schiff base with the amino group of lysine on opsin:

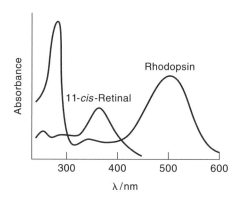

As we can see from Figure 15.9, the formation of a protonated Schiff base (in going from 11-*cis*-retinal to rhodopsin) shifts the absorption from λ_{max} at about 380 nm for 11-*cis*-retinal to about 500 nm. The molar absorptivity of rhodopsin is about 40,000 L mol^{-1} cm^{-1}, which is due to the extensive π electron conjugation in the retinal molecule.

Figure 15.9
Absorption spectra of 11-*cis*-retinal and rhodopsin. The absorption at 280 nm is due to the protein opsin. The molar absorptivity of rhodopsin at 500 nm is about 40,000 L mol^{-1} cm^{-1}.

Mechanism of Vision

Early *in vitro* studies showed that when the retina is exposed to light, rhodopsin decomposes into opsin and the all-*trans* isomer of retinal. Because rhodopsin can be regenerated by binding opsin to 11-*cis*-retinal in the dark, the conclusion was that light causes the isomerization of retinal about its C-11-to-C-12 double bond (Figure 15.10). Experiments using short laser pulses have shown that the initial excitation of rhodopsin produces a species called *bathorhodopsin*, which is a strained all-*trans* form of retinal. This event generates a series of intermediates leading to the release of all-*trans*-retinal (Figure 15.11). The different absorption characteristics of these transient species are due to their different conformations and charge distributions. Finally, in an enzyme-catalyzed process, the all-*trans*-retinal is isomerized in the dark to 11-*cis*-retinal, which then forms the Schiff base to regenerate rhodopsin.

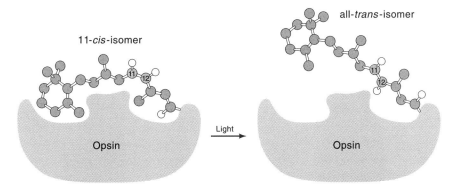

11-*cis*-isomer

all-*trans*-isomer

Light

Opsin

Opsin

Figure 15.10
Schematic drawing showing the isomerization process in rhodopsin.

The sensation of seeing occurs at the formation of metarhodopsin II, which triggers an enzymatic cascade, leading to the closing of cation-specific channels and the generation of a nerve signal. This signal is transmitted to the brain, where it is processed and transformed into a visual image.

Rotation About the C=C Bond

To gain a better understanding of the isomerization of 11-*cis*-retinal to all-*trans*-retinal, let us examine the energetics of the process. To begin with, we note that rotation about a C–C single bond, such as that in ethane, is quite free. There is only a small barrier to rotation, about 10 kJ mol^{-1}, due to the steric hindrance between the hydrogen atoms on adjacent carbon atoms. For molecules containing C=C bonds, however, rotation is limited by the presence of a π bond between the two carbon atoms, in addition to the σ bond. In this case, the restricted rotation gives rise to the phenomenon of geometric isomerism. Figure 15.12 shows the bond-breaking and bond-making steps in *cis*-to-*trans* isomerization. The activation energy for such a reaction is typically on the order of 120 kJ mol^{-1}. Heating, irradiation, or chemical catalysis are the usual means to bring about a geometric isomerization.

Photoisomerization can cause a change from *cis* to *trans* or *trans* to *cis*. Prolonged irradiation of either isomer therefore produces a steady state of [*cis*] to [*trans*] ratio. The actual value of this ratio depends on the molar absorptivity of each isomer

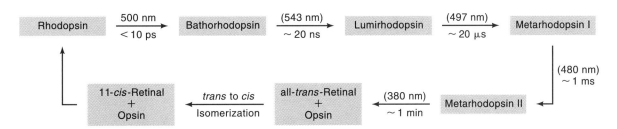

Figure 15.11
The vision cycle. Initially the rhodopsin molecule absorbs a photon at 500 nm. (This is the only step what involves the absorption of light.) Within 10 ps, rhodopsin is transformed into bathorhodopsin. Both retinal and the protein continue to change their conformations, as shown by the series of intermediates, which are characterized by their absorption maxima (shown in parentheses) and mean lifetimes. Eventually the all-*trans*-retinal dissociates from opsin and is isomerized in the dark to 11-*cis*-retinal to form a Schiff base linkage with another opsin molecule.

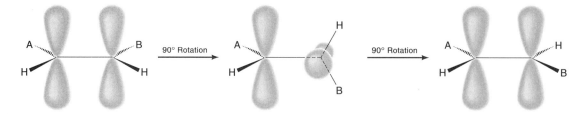

Figure 15.12
The breaking and remaking of a π bond in a *cis*-to-*trans* isomerization process. Note that in the intermediate state, the two *p* orbitals are perpendicular to each other. The breaking of the π bond makes rotation about the C–C bond possible.

and the wavelength of radiation. Photoexcitation of the C=C bond results in the $\pi \rightarrow \pi^*$ transition. The molecule may be in the first excited singlet state or the lowest triplet state. Figure 15.13 shows a simplified potential-energy diagram for the ground and excited states of rhodopsin, as a function of the rotation about the bonds between carbon 11 and carbon 12. A rotational angle of 0° corresponds to the 11-*cis*-retinal in rhodopsin prior to the absorption of light, and 180° corresponds to the all-*trans* form (bathorhodopsin). Note the relative potential energies of these two species. In solution in the free state, 11-*cis*-retinal is less stable than the all-*trans* isomer because of the steric interaction between the methyl group on carbon 13 and the hydrogen atom on carbon 10 in the *cis* isomer. In rhodopsin, the interaction between 11-*cis*-retinal and opsin is more favorable. The potential energy of the excited state is minimal when the rotational angle is 90° (also see Figure 15.12). Finally, the excited state relaxes back to the ground state, which can be either bathorhodopsin or the original 11-*cis*-retinal. Measurements show that two-thirds of the time, the singlet formed after the initial excitation is bathorhodopsin.

Interestingly, in the absence of light, the *cis*-to-*trans* isomerization of retinal takes place only once in a thousand years at the physiological temperature of 37°C. Thus, there is practically no background "noise" level in our perception of light. This is important because the sensitivity of the human eye is impressively high—only 5 to 6 photons are required to elicit the sensation of seeing.

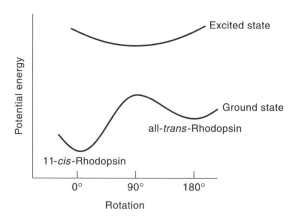

Figure 15.13
Potential energy diagram of the ground electronic states of 11-*cis*-retinal and all-*trans*-retinal and their common excited state. A rotation of 0° corresponds to 11-*cis*-retinal whereas a rotation angle of 180° means the molecule has isomerized to the all-*trans* form. The minimum energy of the excited state occurs at a rotational angle of about 90°.

Rhodopsin functions at low light intensity, for example, at nighttime. It cannot distinguish colors, however, because it has only one pigment. Three types of iodopsin, which contain pigments that absorb light at λ_{max} at 426 nm (blue), 530 nm (green), and 560 nm (yellow), are responsible for color vision in the cone cells. The pigment with its maximum at 560 nm extends its sensitivity into the longer wavelength region to allow visual sensation of red as well. Cone cells are much less sensitive to light than rod cells, so in dim light all objects appear in shades of gray. In addition to humans, primates, bony fishes, and birds possess cone cells and can perceive at least some colors. The retinas of cats and cattle, on the other hand, contain predominantly rod cells. These animals are therefore color-blind.

We conclude, therefore, that a matador does not really need to use a red cape to provoke a bull.

15.4 Biological Effects of Radiation

Both photosynthesis and vision are natural photobiological phenomena. Radiation has also been used successfully to treat disease. In this section, we shall discuss both the harmful and beneficial effects of radiation.

Sunlight and Skin Cancer

In the United States, about 1 million new cases of skin cancer occur annually, rivaling the incidence of all other types of cancer combined. Of these, approximately 40,000 are malignant melanoma, which has an 18% fatality rate. In the vast majority of cases, skin cancer is attributable to solar radiation.

The harmful radiation from the sun is mainly in the UV range, which is divided into three regions called UV-C (200–280 nm), UV-B (280–320 nm), and UV-A (320–400 nm). The most harmful type is UV-C. Fortunately, most UV-C radiation is absorbed by the ozone layer in the stratosphere. UV-B reaches Earth's surface in small amounts and is responsible for the redness and blistered, peeling skin associated with sunburn. (The redness is due to increased flow of blood vessels beneath the skin, which widen in response to the radiation.) UV-B rays are blamed for skin cancer. The least energetic radiation, UV-A, causes what we call a "suntan."

When UV-A or UV-B strikes the pigment-producing melanocyte cells beneath the skin, they produce a UV-absorbing dark pigment called *melanin*. This substance screens out part of the radiation and helps to minimize damage to the underlying layers of skin. In addition, the melanocytes start dividing more rapidly than usual to replace damaged cells in the outer layer. Normally it takes a few weeks for the new cells to reach the surface, where they are shed as part of the skin's renewal cycle. Prolonged exposure to the sun speeds up this process, so that a large number of melanin-containing cells arrive at the surface in a few days, giving the skin a sun-tanned appearance.

To understand sunlight-induced cancer, we must look at the effect of UV radiation on DNA. DNA molecules absorb radiation strongly between 200 nm and 300 nm, with a maximum at about 260 nm (see Figure 14.20). Unlike the pyrimidines, the purines (adenine and guanine) are much less sensitive to UV light. Experiments suggest that the dimerization of thymine is the most important photochemical reaction to occur in DNA molecules. Normally, a thymine solution is relatively insensitive to UV light, but when a frozen solution of thymine is irradiated with UV, thymine dimer is formed in high yield. The fact that thymine dimers are formed only in the frozen state shows that the reaction requires not only that two thymine molecules be close to each other, but also that they be held in a certain orientation. Two adjacent thymine base pairs are both close and fixed in position on the *same* strand of a DNA molecule. We would then expect thymine dimers to form when DNA molecules are

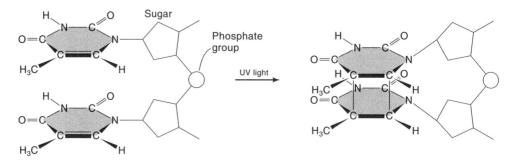

Figure 15.14
Dimerization of adjacent thymine bases on the same strand of a DNA molecule.

exposed to UV radiation, and this is indeed the case (Figure 15.14). This is probably the first step in the mutation of specific genes within skin cells. For example, a cell may reproduce excessively if the mutation turns a normal gene into a growth promoter (an oncogene). Alternatively, mutation may inactivate a gene that normally limits cell growth (a tumor suppresser gene).

The thymine dimer can be restored to its monomeric form through photoreactivation, a process in which light-absorbing enzymes called photoreactivating enzymes, or DNA photolyases, repair DNA by utilizing the energy of visible light to break the cyclobutane ring of the dimer. Photolyases are monomeric proteins with two flavin cofactors that act as chromophores. A photolyase enzyme binds the DNA substrate in a light-independent reaction. Then one chromophore of the bound enzyme absorbs a visible photon and, by dipole–dipole interaction, transfers energy to the second flavin, which, in turn, transfers an electron to the thymine dimer in DNA. Subsequently, the dimer breaks up. Back-electron transfer restores the functional form of the chromophore flavin, and the enzyme is ready for a new cycle of catalysis. There is no net redox change in the dimer splitting. Interestingly, photolyases are the only light-driven enzymes that are not involved in photosynthesis.

Photomedicine

Photomedicine is the application of the principles of photochemistry and photobiology to the diagnosis and therapy of disease. Interest in this subject dates back to the 19th century, when it was found that facial lesions resulting from tuberculosis could be cured by irradiation with UV light. It was reinforced by the discovery that UV light kills microorganisms and that sunlight is effective in the treatment and prevention of vitamin D deficiency (rickets). We shall briefly discuss two examples below.

Photodynamic Therapy. Photodynamic therapy utilizes light to generate a reactive species that can destroy cancerous cells. A patient is intravenously injected with a solution containing a compound called a *photosensitizer* (S). After a day or so, the solution has been distributed throughout the body. Specially designed fiber-optic probes are then inserted in the region containing affected cells, and the photosensitizer is irradiated with a dye laser. The following reactions take place:

$$S_0 \xrightarrow{hv} S_1 \qquad \text{singlet–singlet excitation}$$

$$S_1 \rightarrow S_0 + hv \qquad \text{fluorescence}$$

$$S_1 \rightarrow T_1 \qquad \text{intersystem crossing}$$

$$T_1 + {}^3O_2 \rightarrow S_0 + {}^1O_2 \quad \text{energy transfer to produce singlet oxygen}$$

Intersystem crossing is the radiationless transition of a molecule from one electronic state into another with a different spin multiplicity.

where S_0 and S_1 are the ground and first excited singlet state, and T_1 is the lowest triplet state of the photosensitizer (see Figure 14.37), and ${}^3O_2$ and ${}^1O_2$ are the triplet state and singlet state of molecular oxygen.* Singlet oxygen is a highly reactive species, and it has the ability to destroy the neighboring tumor cells.

To be successful as an agent for photodynamic therapy, a photosensitizer must satisfy three requirements. First, it must be nontoxic and preferably water soluble. Second, it should absorb strongly in the red region of the visible spectrum or in the near-IR region. The reason is that after injection, the solution containing the photosensitizer is distributed throughout the body, including the skin. If the compound absorbs appreciably in the shorter wavelengths of the visible or UV light, the patient becomes sensitized to photodamage by sunlight, a side effect that is clearly undesirable. Third, to minimize damage to healthy tissues, the photosensitizer should be selectively retained by the cancerous cells. In this respect, the location of the compound can be monitored by studying its fluorescence.

The prospect of photodynamic therapy is promising. Besides treating cancer, photosensitizers also appear to be highly effective in killing bacteria. At present, much effort is being expended to the synthesis of photosensitizers (mostly compounds with complex structures containing the porphyrin ring system) with suitable photochemical and chemical properties for clinical applications.

Light-Activated Drugs. The ancient Egyptians recognized that a common plant called *Ammi majus* possesses medicinal properties that are elicited by light. *Ammi majus* is a weed that grows on the banks of the Nile. The physicians of the time found that people became unusually prone to sunburn after ingesting the plant. Consequently, the plant was used to treat certain skin disorders. Chemical analysis has shown that the active ingredients in the plant belong to a class of compounds called psoralens, an example of which is 8-methoxpsoralen, or 8-MOP:

8-Methoxypsoralen (8-MOP)

Clinical studies have shown that 8-MOP is an effective anticancer drug that can be activated by light.

Cutaneous T-cell lymphoma (CTCL) is a malignancy of white blood cells; it has a poor prognosis. Treatment with 8-MOP and light, however, has yielded very promising results in CTCL. In a typical procedure, about 500 mL of blood (roughly the same volume as that given in a single blood donation) is drawn from a patient. Centrifugation separates the blood into three components: erythrocytes (red blood cells), leukocytes (white blood cells), and plasma (liquid portion of the blood). The leukocytes and plasma are combined with a saline solution to which 8-MOP is added. This solution is then irradiated with high-intensity UV-A light. After irradiation, the erythrocytes are recombined with the remainder of the original, drawn blood, which is then transfused into the patient. Without light, 8-MOP is inert and is totally harmless to the body.

---

* According to Hund's rule, the lowest, or the ground, electronic state of O_2 is a triplet, with two unpaired electrons.

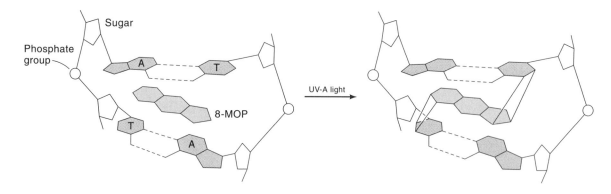

Figure 15.15
Schematic diagram showing chemical bond formation between 8-MOP and the thymine molecules on different strands of a DNA molecule. This linkage keeps the strands from unwinding for replication.

Figure 15.15 shows that the size and shape of 8-MOP allow it to slide between base pairs of DNA molecules in the cell nucleus of a leukocyte. Upon irradiation, it forms chemical bonds with the bases on *both* strands. The strong chemical bonds now prevent the DNA from replicating, resulting in cell death. The treatment is non-specific and damages both malignant cells and healthy ones. Interestingly, when the damaged malignant cells are returned to the patient's bloodstream, they somehow induce the immune system to destroy the malignant cells that have *not* been treated with 8-MOP and radiation. Although more work needs to be done to find drugs that have greater affinity for DNA and ways to activate them inside the body, there is little doubt that light-activated drugs will play an important role as therapeutic agents against cancer and other diseases.

Suggestions for Further Reading

BOOKS

Blankenship, R. E., *Molecular Mechanisms of Photosynthesis*, Blackwell Science, Oxford, 2002.

Clayton, R. K., *Light and Living Matter*, Vols. 1 and 2, McGraw-Hill, New York, 1971.

Clayton, R. K., *Photosynthesis: Physical Mechanisms and Chemical Patterns*, Cambridge University Press, New York, 1980.

Cramer, W. A. and D. B. Knaff, *Energy Transduction in Biological Membranes*, Springer-Verlag, New York, 1991.

Hall, D. O. and K. K. Rao, *Photosynthesis*, 5th ed., Cambridge University Press, New York, 1994.

Harm, W., *Biological Effects of Ultraviolet Radiation*, Cambridge University Press, New York, 1980.

Nicholls, D. G. and S. J. Ferguson, *Bioenergetics*, 2nd ed., Academic Press, New York, 1992. (Chapter 6 discusses photosynthesis.)

Suppan, P., *Chemistry and Light*, Royal Society of Chemistry, London, 1994.

Turro, N. J. *Modern Molecular Photochemistry*, University Science Books, Sausalito, CA, 1991.

ARTICLES
General

"The Fates of Electronic Excitation Energy," H. H. Jaffé and A. L. Miller, *J. Chem. Educ.* **43**, 469 (1966).

"Photochemical Reactivity," N. J. Turro, *J. Chem. Educ.* **44**, 536 (1967).

"The Chemical Effects of Light," G. Oster, *Sci. Am.* September 1968.

"Photochemistry of Organic Compounds," J. S. Swenton, *J. Chem. Educ.* **46**, 7, 217 (1969).

"Photochemical Reactions of Natural Macromolecules," D. C. Neckers, *J. Chem. Educ.* **50**, 164 (1973).

"Photochemical Reactions of Tris(oxalato)Iron(III)," A. D. Baker, A. Casadavell, H. D. Gafney, and M. Gellender, *J. Chem. Educ.* **57**, 317 (1980).

"Photochemistry and Beer," A. Vogler and H. Kunkely, *J. Chem. Educ.* **59**, 25 (1982).

"Photochemistry in Organized Media," J. H. Fendler, *J. Chem. Educ.* **60**, 872 (1983).

"Atmospheric Physics," F. W. Taylor in *Encyclopedia of Applied Physics*, G. L. Trigg, Ed., VCH Publishers, New York, 1994, Vol. 1, p. 489.

"Laser Photochemistry," P. Engelking in *Encyclopedia of Applied Physics*, G. L. Trigg, Ed., VCH Publishers, New York, 1994, Vol. 8, p. 283.

"Reactions Induced by Light," K. L. Stevenson and O. Horváth in *Encyclopedia of Applied Physics*, G. L. Trigg, Ed., VCH Publishers, New York, 1996, Vol. 16, p. 117.

"Does a Photochemical Reaction Have a Reaction Order?" S. R. Logan, *J. Chem. Educ.* **74**, 1303 (1997).

Photosynthesis

"The Absorption of Light in Photosynthesis," Govindjee and R. Govindjee, *Sci. Am.* December 1974.

"The Photosynthetic Membrane," K. R. Miller, *Sci. Am.* April 1979.

"Electrode Potential Diagrams and Their Use in the Hill-Bendall or Z-Scheme for Photosynthesis," P. Borrell and D. T. Dixon, *J. Chem. Educ.* **61**, 83 (1984).

"Energy Conversions in Photosynthesis," C. L. Bering, *J. Chem. Educ.* **62**, 659 (1985).

"Excited State Redox Potentials and the Z Scheme for Photosynthesis," R. E. Blankenship and R. C. Prince, *Trends Biochem. Sci.* **10**, 382 (1985).

"Photosynthesis and Carbon Dioxide Fixation," M. B. Bishop and C. B. Bishop, *J. Chem. Educ.* **64**, 302 (1987).

"Molecular Mechanism of Photosynthesis," D. C. Yonvan and B. L. Marrs, *Sci. Am.* June 1987.

"How Plants Make Oxygen," Govindjee and W. J. Coleman, *Sci. Am.* February 1990.

"Photosynthesis," R. E. Blankenship, Chapter in *Encyclopedia of Inorganic Chemistry*, John Wiley & Sons, New York, 1994, p. 3828.

"Revealing the Blueprint of Photosynthesis," J. Barber and B. Andersson, *Nature* **370**, 31 (1994).

"Photosynthesis: Why Does It Occur?" J. J. MacDonald, *J. Chem. Educ.* **72**, 1113 (1995).

"Photosynthesis," J. Whitmarsch in *Encyclopedia of Applied Physics*, G. L. Trigg, Ed., VCH Publishers, New York, 1995, Vol. 13, p. 513.

"Artificial Photosynthesis," N. Lewis, *Am. Sci.* **83**, 534 (1995).

"The Origin and Evolution of Oxygenic Photosynthesis," R. E. Blankenship and H. Hartman, *Trends Biochem. Sci.* **23**, 94 (1998).

"How Plants Produce Dioxygen," V. A. Szalai and G. W. Brudvig, *Am. Sci.* **86**, 542 (1998).

"Mimicking Natural Photosynthesis," M. Freemantle, *Chem. Eng. News* **76**(43), 37 (1998).

"Chance, Luck and Photosynthesis Research: An Inside Story," A. T. Jagendorf, *Photosynthesis Research* **57**, 215 (1998).

"Photosynthesis: The Light Reactions," R. E. Blankenship, Chapter in *Plant Physiology*, L. Taiz and E. Zeiger, Eds., Benjamin Cummings, Redwood City, CA, 1999.

"The Heart of Photosynthesis in Glorious 3D," A. W. Rutherford and P. Fuller, *Trends Biochem. Sci.* **26**, 341 (2001).

Vision

"Molecular Isomers in Vision," R. Hubbard and A. Kropf, *Sci. Am.* June 1967.

"How Light Interacts With Living Matter," S. B. Hendricks, *Sci. Am.* September 1968.

"The Process of Vision," U. Weisser, *Sci. Am.* September 1968.

"The Functional Architecture of the Retina," R. H. Masland, *Sci. Am.* December 1986.

"How Photoreceptor Cells Respond to Light," J. L. Schnapf and D. A. Baylor, *Sci. Am.* April 1987.

"The Molecules of Visual Excitation," L. Stryer, *Sci. Am.* July 1987.

"Rhodospin: Structure, Function, and Genetics," J. Nathans, *Biochemistry* **31**, 4923 (1992).

"How the Retina Works," H. Kolb, *Am. Sci.* **91**, 28 (2003).

Biological Effects of Radiation

"Ultraviolet Radiation and Nucleic Acid," R. A. Deering, *Sci. Am.* December 1962.

"The Repair of DNA," P. C. Hanawalt and R. H. Haynes, *Sci. Am.* February 1967.

"The Chemical Effects of Light," G. Oster, *Sci. Am.* September 1968.

"The Effects of Light on the Human Body," R. J. Wurtman, *Sci. Am.* July 1975.

"Biochemical Effects of Excited State Molecular Oxygen," J. Bland, *J. Chem. Educ.* **53**, 274 (1976).

"Inducible Repair of DNA," P. Howard-Flanders, *Sci. Am.* November 1981.

"Radiation Sensitization in Cancer Therapy," C. L. Greenstock, *J. Chem. Educ.* **58**, 156 (1981).

"The Biological Effects of Low-Level Ionizing Radiation," A. C. Upton, *Sci. Am.* February 1982.

"Phototherapy and the Treatment of Hyperbilirubinemia: A Demonstration of Intra- Versus Intermolecular Hydrogen Bonding," A. C. Wilbraham, *J. Chem. Educ.* **61**, 540 (1984).

"Light Activated Drugs," R. L. Edelson, *Sci. Am.* August 1988.

"Effect of UV Irradiation on DNA as Studied by Its Thermal Denaturation," C. M. Lovett, Jr., T. N. Fitsgibbon, and R. Chang, *J. Chem. Educ.* **66**, 526 (1989).

"DNA, Sunlight, and Skin Cancer," J. S. Taylor, *J. Chem. Educ.* **67**, 835 (1990).

"A Simple UV Experiment of Environmental Significance," D. W. Daniel, *J. Chem. Educ.* **71**, 83 (1994).

"Structure and Function of DNA Photolyase," A. Sancar, *Biochemistry*, **33**(1), 2 (1994).

"Sunlight and Skin Cancer," D. J. Leffell and D. E. Brash, *Sci. Am.* July 1996.

"The Photochemistry of Sunscreens," D. R. Kimbrough, *J. Chem. Educ.* **74**, 51 (1997).

"The Spectrophotometric Analysis and Modeling of Sunscreens," C. Walters, A. Keeney, C. T. Wigal, C. R. Johnston, and R. D. Cornelius, *J. Chem. Educ.* **74**, 99 (1997).

"Let There Be Light and Let It Heal," A. M. Rouhi, *Chem. & Eng. News* **76**(44), 22 (1998).

"Photodynamic Therapy: The Sensitization of Cancer Cells to Light," J. Miller, *J. Chem. Educ.* **76**, 592 (1999).

"Photochemotherapy: Light-Dependent Therapies in Medicine," E. P. Zovinka and D. R. Sunseri, *J. Chem. Educ.* **79**, 1331 (2002).

"New Light on Medicine," N. Lane, *Sci. Am.* January 2003.

Problems

General

15.1 In a photochemical reaction, 428.3 kJ mol^{-1} of energy input is required to break a chemical bond. What wavelength must be employed in the irradiation?

15.2 Convert 450 nm to kJ einstein^{-1}.

15.3 Design an experiment that would allow you to measure the rate of absorption of light by a solution.

15.4 An organic molecule absorbs light at 549.6 nm. If 0.031 mole of the molecule is excited by 1.43 einsteins of light, what is the quantum efficiency for this process? Also, calculate the total energy taken up in the process.

15.5 In the photochemical decomposition of a certain compound, light intensity of 5.4×10^{-6} einsteins s^{-1} was employed. Assuming the most favorable conditions, estimate the time needed to decompose 1 mole of the compound.

15.6 The first-order rate constants for the fluorescence and phosphorescence of naphthalene ($C_{10}H_8$) are 4.5×10^7 s^{-1} and 0.50 s^{-1}, respectively. Calculate how long it takes for 1.0% of fluorescence and phosphorescence to occur following termination of excitation.

15.7 Photochemical reactions often produce different products than thermal reactions with the same reactants. Why?

15.8 Why does one have to irradiate a sample for hours or even days to achieve acceptable yields in some photochemical reactions even though the lifetimes of excited electronic states are on the order of micro- or nanoseconds? Assume that the rate of light absorption is 2.0×10^{19} photons s^{-1}.

15.9 Suppose that an excited singlet, S_1, can be deactivated by three different mechanisms whose rate constants are $k_1, k_2,$ and k_3. The rate of decay is given by $-d[S_1]/dt = (k_1 + k_2 + k_3)[S_1]$. **(a)** If τ is the mean lifetime—that is, the time required for $[S_1]$ to decrease to $1/e$ or 0.368 of the original value—show that $(k_1 + k_2 + k_3)\tau = 1$. **(b)** The overall rate constant, k, is given by

$$\frac{1}{\tau} = k = k_1 + k_2 + k_3 = \frac{1}{\tau_1} + \frac{1}{\tau_2} + \frac{1}{\tau_3}$$

Show that the quantum yield Φ_i is given by

$$\Phi_i = \frac{k_i}{\sum_i k_i} = \frac{\tau}{\tau_i}$$

where i denotes the ith decay mechanism. **(c)** If $\tau_1 = 10^{-7}$ s, $\tau_2 = 5 \times 10^{-8}$ s, and $\tau_3 = 10^{-8}$ s, calculate the lifetime of the singlet state and the quantum yield for the path that has τ_2.

15.10 Consider the photochemical isomerization $A \rightleftharpoons B$. At 650 nm, the quantum yields for the forward and reverse reactions are 0.73 and 0.44, respectively. If the molar absorptivities of A and B are 1.3×10^3 L mol^{-1} cm^{-1} and 0.47×10^3 L mol^{-1} cm^{-1}, respectively, what is the ratio [B]/[A] in the photostationary state?

Photobiology

15.11 What is the biological significance of the fact that λ_{max} occurs at around 500 nm in the solar emission spectrum?

15.12 In the sea, light intensity decreases with depth. For example, at a depth of 20 m below the surface, light intensity is one-half of that at the sea level. In practice, total darkness sets in when 99% of the light is absorbed by water. Explain why green algae are found near the surface, but red algae are located as deep as 100 m.

15.13 Transition metals such as Fe, Cu, Co, and Mn are necessary for respiration and photosynthesis but nontransition metals such as Zn, Ca, and Na are not. Explain.

15.14 In photosynthesis, the term *quantum requirement* refers to the number of photons required to reduce one CO_2 molecule to (CH_2O):

$$H_2O + CO_2 \rightarrow (CH_2O) + O_2$$

The efficiency of this process depends on the wavelength of light employed. Assuming a quantum requirement of 8, calculate the efficiency under standard-state conditions for the synthesis of 1 mole of glucose if the wavelength of light employed is **(a)** 400 nm and **(b)** 700 nm.

15.15 At low light intensities, the rate of photosynthesis increases linearly with intensity. At high intensities, however, the rate is constant (saturation rate). Suggest an interpretation at the molecular level. The saturation rate varies with temperature. Explain.

15.16 Calculate the number of moles of ATP that can be synthesized at 80% efficiency by a photosynthetic organism upon the absorption of 2.1 einsteins of photons at 650 nm. (*Hint:* $\Delta_r G^{\circ\prime}$ for the synthesis of ATP from ADP and P_i is 31.4 kJ mol^{-1}.)

15.17 What is the advantage of the broad absorption spectrum of rhodopsin (see Figure 15.9)?

15.18 In solution, 11-*cis*-retinal absorbs maximally at 380 nm. In rhodopsin, the maximum absorption occurs at 500 nm. Explain.

15.19 A light source of power 2×10^{-16} W is sufficient to be detected by the human eye. Assuming the wavelength of light is at 550 nm, calculate the number of photons that must be absorbed by rhodopsin per second. (*Hint:* Vision persists for only 1/30 of a second.)

15.20 DCMU (dichlorophenyldimethylurea) is a herbicide (weed killer). Its function is to interfere with photophosphorylation and oxygen evolution. However, in the presence of an artificial electron acceptor, it does not inhibit oxygen evolution. Suggest a site for the inhibitory action of DCMU. Would DCMU affect a plant's ability to perform cyclic photophosphorylation? (*Hint:* See Figure 15.6.)

15.21 Estimate the minimum pH gradient required to synthesize ATP from ADP and P_i ($\Delta_r G^{\circ\prime} = 31$ kJ mol^{-1}) in photophosphorylation. Assume that the ratio [ATP] to [ADP][P_i] is 1×10^3 and $T = 298$ K. (*Hint:* As stated on p. 586, four protons are transported for every ATP synthesized.)

CHAPTER 16

Macromolecules

A macromolecule is a chemical species distinguished by a high molar mass (10^4 to 10^{10} g mol^{-1}). The chemistry of macromolecules, which are called *polymers*, differs greatly from the chemistry of small, ordinary molecules. Studying the properties of these giant molecules requires special techniques.

Macromolecules are divided into two classes: natural and synthetic. Examples of *natural macromolecules* are proteins, nucleic acids, polysaccharides (cellulose), and polyisoprene (rubber). Most *synthetic macromolecules* are organic polymers, such as polyhexamethylene adipamide (nylon), polyethylene terephthalate (Dacron, Mylar), and polymethylmethacrylate (Lucite, Plexiglas).

In this chapter, we examine some of the methods used to characterize macromolecules and discuss their structure and conformation. We shall also study protein stability and protein folding.

16.1 Methods for Determining Size, Shape, and Molar Mass of Macromolecules

Molar mass has a special meaning when applied to macromolecules. In a sucrose solution, all solute molecules have the same molar mass; different methods of determining molar mass yield the same value. The same statement applies to molecules of hemoglobin and other proteins in solution, assuming no dissociation of the solute into subunits. Such is not the case, however, for polystyrene, DNA, fibrous proteins such as collagen, rubber, and other substances composed of polymers. In these systems, the molecules are not all identical, and distribution of molar masses is uneven. A polymeric system whose molecules all have the same molar mass is said to be *monodisperse*; if the molecules do not have identical molar masses, the polymer is said to be *polydisperse*.

Molar Mass of Macromolecules

The molar mass of a polymer can be defined in various ways. The two most common definitions are the number-average molar mass and the weight-average molar mass.

Number-Average Molar Mass ($\bar{\mathcal{M}}_n$). Consider a sample of N polymer molecules containing n_1 molecules of molar mass $\mathcal{M}_1$, n_2 molecules of molar mass $\mathcal{M}_2$, and so on.

The *number-average molar mass* is defined as

$$\bar{\mathscr{M}}_n = \frac{n_1 \mathscr{M}_1 + n_2 \mathscr{M}_2 + \cdots}{n_1 + n_2 + \cdots} = \frac{\sum_i n_i \mathscr{M}_i}{\sum_i n_i}$$

$$\bar{\mathscr{M}}_n = \frac{\sum_i n_i \mathscr{M}_i}{N} \qquad (16.1)$$

where $\sum_i n_i = N$. Thus, $\bar{\mathscr{M}}_n$ is simply the arithmetic mean of all the molar masses.

Weight-Average Molar Mass ($\bar{\mathscr{M}}_w$). This is defined as

$$\bar{\mathscr{M}}_w = \frac{n_1 \mathscr{M}_1^2 + n_2 \mathscr{M}_2^2 + \cdots}{n_1 \mathscr{M}_1 + n_2 \mathscr{M}_2 + \cdots} = \frac{\sum_i n_i \mathscr{M}_i^2}{\sum_i n_i \mathscr{M}_i} \qquad (16.2)$$

For a polydisperse system, we have $\bar{\mathscr{M}}_w > \bar{\mathscr{M}}_n$; for a monodisperse system, $\bar{\mathscr{M}}_n = \bar{\mathscr{M}}_w$. It follows, therefore, that the determination of molar mass by two different methods can, in principle, test the homogeneity of the system under investigation.

Sedimentation in the Ultracentrifuge

Everyday experience tells us that particles suspended in a solution are pulled downward by Earth's gravitational force. This movement is partially offset by the buoyancy of the particle. Because Earth's gravitational field is weak, a solution containing macromolecules is usually homogeneous, due to the random thermal motion of the molecules.

Moving-Boundary Sedimentation. Consider a solution under the influence of a strong gravitational field, such as a solution being spun in a centrifuge tube. The centrifugal force acting on a solute particle of mass m is $m\omega^2 r$, where ω is the angular velocity of the rotor in radians per second (the relation between angle and radian is given in Appendix 1), r is the distance from the center of rotation to the particle, and $\omega^2 r$ is the centrifugal acceleration of the rotor (Figure 16.1).

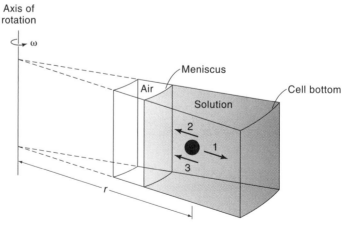

Figure 16.1
Forces acting on a molecule in a sector-shaped cell being spun in an ultracentrifuge at angular velocity ω. The labels are: 1, centrifugal force; 2, buoyancy force; and 3, frictional force. The distance from the molecule to the center of rotation is r. Initially, the solution is homogeneous.

In addition to centrifugal force, we must also consider the particle's buoyancy due to the displacement of the solvent molecules by the particle. This buoyancy reduces the force on the particle by $\omega^2 r$ times the mass of the displaced solvent. Thus, the net force acting on the particle is given by

$$\text{net force} = \text{centrifugal force} - \text{buoyancy force}$$

$$= \omega^2 rm - \omega^2 rm_s$$

$$= \omega^2 rm - \omega^2 rv\rho \tag{16.3}$$

The buoyancy force is equal to the centrifugal acceleration times the mass of displaced solvent, m_s, and v and ρ are the volume of the particle and the density of the solution, respectively.

According to Newton's second law of motion, a net force acting on a particle causes it to accelerate. At the same time, the medium exerts on the particle a frictional force, which is proportional to the sedimentation velocity, dr/dt. The frictional force is equal to the product of the frictional coefficient, f (in units N m^{-1} s), and the sedimentation velocity. It acts in the opposite direction to the net force. At steady state, then, the frictional force is equal to the net force and the molecule moves with velocity dr/dt toward the bottom of the cell:

$$f\frac{dr}{dt} = \omega^2 rm - \omega^2 rv\rho \tag{16.4}$$

Because measuring the volume of the particle is difficult, for convenience we use a term called *partial specific volume*, $\bar{v}$, defined as the increase in volume when 1 g of dry solute is dissolved in a large volume of solvent. The quantity $m\bar{v}$ is the incremental volume increase when one molecule of mass m is added to the solvent; that is, it is equal to the volume v of the particle. For most proteins, $\bar{v}$ has a value of about 0.74 mL g^{-1}. Equation 16.4 can now be written as

$$f\frac{dr}{dt} = \omega^2 rm - \omega^2 rm\bar{v}\rho$$

$$= \omega^2 rm(1 - \bar{v}\rho) \tag{16.5}$$

Rearranging Equation 16.5 gives

$$s = \frac{dr/dt}{\omega^2 r} = \frac{m(1 - \bar{v}\rho)}{f}$$

$$s = \frac{\mathscr{M}}{N_A}\frac{1 - \bar{v}\rho}{f} \tag{16.6}$$

where $\mathscr{M}$ is the molar mass of the solute and N_A is Avogadro's number.

The quantity on the left side in Equation 16.6 is called the *sedimentation coefficient* (s), which has the unit *svedberg*, named after the Swedish chemist Theodor Svedberg (1884–1971), who pioneered ultracentrifuge studies. One svedberg is equal to 10^{-13} s. The British physicist Sir George Gabriel Stokes (1819–1903) showed that the fricitional coefficient for a spherical solute particle of radius r_s is given by

$$f = 6\pi\eta r_s \tag{16.7}$$

A more refined treatment deals with the nonspherical and solvated solute molecules.

where η is the viscosity of the medium. From Equation 16.6, we have

$$\mathcal{M} = \frac{sN_A f}{1 - \bar{v}\rho} = \frac{sN_A(6\pi\eta r_s)}{1 - \bar{v}\rho} \tag{16.8}$$

The ease of movement of the solute molecule depends on its thermal energy, represented by $k_B T$ (where k_B is the Boltzmann constant and T is the absolute temperature), and its frictional coefficient, f. The greater the $k_B T$, the more energetic its motion, and the larger the f, the greater the frictional force the medium exerts on the solute. According to Einstein, the ratio of these two opposing factors gives the diffusion coefficient, D, of the solute:

$$D = \frac{k_B T}{f} \tag{16.9}$$

Therefore, the value of D (in units $m^2\ s^{-1}$ or $cm^2\ s^{-1}$) tells us the ease with which the solute molecules diffuse in solution.* Rearranging Equation 16.9, we get

$$f = \frac{k_B T}{D} \tag{16.10}$$

and substitution of Equation 16.10 into Equation 16.8 gives

$$\mathcal{M} = \frac{sN_A k_B T}{D(1 - \bar{v}\rho)} = \frac{sRT}{D(1 - \bar{v}\rho)} \tag{16.11}$$

where $N_A k_B = R$, the gas constant.

Because D and $\bar{v}$ can be determined by separate experiments, the only other quantity we need to measure in order to determine molar mass is the sedimentation coefficient. By definition,

$$s = \frac{dr/dt}{\omega^2 r}$$

or

$$s\,dt = \frac{1}{\omega^2} \frac{dr}{r}$$

Integration over the distance traveled by the particle from $r = r_0\ (t = 0)$ to $r = r\ (t = t)$ gives

$$\int_0^t s\,dt = \frac{1}{\omega^2} \int_{r_0}^r \frac{dr}{r}$$

$$s = \frac{1}{t\omega^2} \ln\frac{r}{r_0}$$

or

$$\ln(r/r_0) = s\omega^2 t \tag{16.12}$$

---

* See the physical chemistry texts listed in Chapter 1 (p. 6) for a discussion of diffusion.

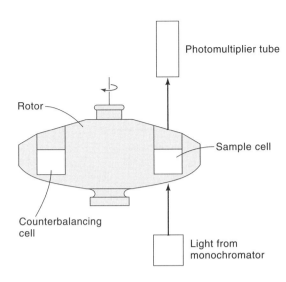

Figure 16.2
Schematic representation of an analytical
ultracentrifuge. The rotor is placed in a
vacuum chamber to reduce resistance.

A plot of $\ln r$ versus t gives a straight line with the slope $s\omega^2$, which enables us to calculate the value of s.

Figure 16.2 shows a schematic diagram of an analytical ultracentrifuge designed to measure the sedimentation coefficient. The rotor spins up to 70,000 rpm (revolutions per minute) in an evacuated chamber at constant temperature. Initially, a homogeneous solution containing the protein is placed in a sector-shaped cell. During centrifugation, the protein molecules move toward the bottom of the cell, and a solvent–solution boundary is established (Figure 16.3). By monitoring the move-

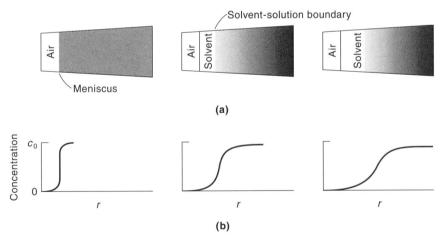

Figure 16.3
Sedimentation velocity measurement. (a) The distribution of protein in the cell as a function
of centrifugation time. Immediately after centrifugation has started, a solvent–solution
boundary appears and moves steadily from left to right. (b) Concentration as a function of
the distance of the solvent–solution boundary from the center of rotation. The concentration
of the original protein solution is c_0.

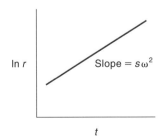

Figure 16.4
Plot of $\ln r$ versus t gives a
straight line with a slope equal
to $s\omega^2$ (see Equation 16.12).

ment of the boundary, we obtain a series of plots of absorbance (or concentration) versus r. From the positions of the boundary at different time intervals, we can plot $\ln r$ versus t to obtain s (from knowledge of ω), as described above (Figure 16.4).

Note that the sedimentation coefficient for a given molecule is *independent* of the angular velocity of the rotor. As $\omega^2 r$ increases, so does dr/dt. Therefore, the ratio

Table 16.1
Some Physical Properties of Proteins in Water at 293 K[a]

Protein	$s_{20,w}$ $\overline{10^{-13} \text{ s}}$	D $\overline{10^{-7} \text{ cm}^2 \cdot \text{s}^{-1}}$	$\bar{v}$ $\overline{\text{mL} \cdot \text{g}^{-1}}$	$\mathcal{M}$ $\overline{\text{g} \cdot \text{mol}^{-1}}$
Cytochrome c_1 (bovine heart)	1.71	11.40	0.728	13,370
Lysozyme (chicken egg white)	1.91	11.20	0.703	13,930
Ribonuclease (bovine pancrease)	2.00	13.10	0.707	12,640
Myoglobin	2.04	11.3	0.74	16,890
Human serum albumin	4.60	6.10	0.733	68,460
Alcohol dehydrogenase (horse liver)	4.88	6.50	0.751	73,050

[a] From *Handbook of Biochemistry*, H. A. Sober, Ed., Copyright by The Chemical Rubber Co., 1968. Used by permission of The Chemical Rubber Co.

remains constant. From Equation 16.11, we see that knowledge of s enables us to calculate the molar mass, $\mathcal{M}$. Measuring molar mass this way provided the first conclusive proofs that biploymers are giant molecules. This approach has been used less frequently in recent years because determining diffusion coefficients is inconvenient. The advent of fast, accurate, laser scattering methods for measuring D will likely revive this technique for molar-mass determination.

Table 16.1 lists several properties of proteins for the ultracentrifuge experiment. Each protein has a characteristic sedimentation coefficient. However, because s depends on $\bar{v}, \rho$, and f, all of which in turn depend on temperature and the medium in which the protein is dissolved, values obtained under different experimental conditions must be corrected to standard conditions for comparison purposes. The standard conditions are chosen to correspond to a hypothetical sedimentation in pure water at 20°C. This standardized sedimentation coefficient is symbolized $s_{20,w}$.

Example 16.1

At 20°C, catalase (horse liver) has a diffusion coefficient of 4.1×10^{-11} m^2 s^{-1} and a sedimentation coefficient of 11.3×10^{-13} s. The density of water at 20°C is 0.998 g mL^{-1}. Calculate the molar mass of catalase, assuming a partial specific volume of 0.715 mL g^{-1}.

ANSWER

From Equation 16.11, we write

$$\mathcal{M} = \frac{sRT}{D(1 - \bar{v}\rho)}$$

$$= \frac{(11.3 \times 10^{-13} \text{ s})(8.314 \text{ J K}^{-1} \text{ mol}^{-1})(293 \text{ K})}{(4.1 \times 10^{-11} \text{ m}^2 \text{ s}^{-1})[1 - (0.715 \text{ mL g}^{-1})(0.998 \text{ g mL}^{-1})]}$$

$$= 236 \text{ kg mol}^{-1}$$

$$= 2.36 \times 10^5 \text{ g mol}^{-1}$$

Conversion factor: 1 J = 1 kg m^2 s^{-2}.

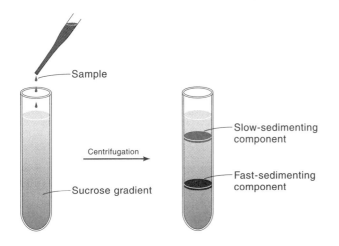

Figure 16.5
A sample is layered onto a sucrose density gradient. Centrifugation results in the appearance of different bands that correspond to the different masses of the proteins present. These proteins can be collected separately by puncturing the bottom of the centrifuge tube and allowing the solution to run into different collecting tubes.

Zonal Sedimentation. The moving-boundary technique has the disadvantage that it cannot satisfactorily separate complex mixtures. Faster molecules (those with greater *s* values) always sediment through a solution of slower ones, so that the former is contaminated by the latter. An ideal technique would allow each component to sediment only through the solvent. *Zonal* (also called *band*) ultracentrifugation comes close to achieving this goal. The first step in zonal centrifugation is to form a density gradient in a celluloid centrifuge tube by mixing sucrose solutions of different densities. A small volume of the solution containing a mixture of proteins is layered on top of the gradient (Figure 16.5). As the rotor is spun, proteins move through the density gradient and separate according to their sedimentation coefficients. Centrifugation is stopped before the fastest protein reaches the bottom of the tube. The separate bands of proteins can be collected by piercing the bottom of the tube and allowing the solution to drip into different tubes. The different components then can be assayed for catalytic, binding, and other properties.

Sedimentation Equilibrium. We now consider a technique called *sedimentation equilibrium*, which makes possible an accurate and direct determination of the molar mass of macromolecules.

A concentration gradient is created in a sedimentation velocity experiment. When the rotor speed is great enough, all solute molecules will eventually collect in the bottom of the cell. Now let us suppose that the rotor speed is lowered to about 10,000 rpm, instead of the 60,000 rpm or so required for a sedimentation velocity experiment. A perfect balance between sedimentation and diffusion can be achieved in this way. In diffusion, solute molecules move from a higher concentration to a lower one, while sedimentation reverses this process. When an equilibrium is established, no net flow occurs.

The German physiologist Adolf Eugen Fick (1829–1901) showed that in a normal diffusion process, the *flux* (J)—that is, the net amount of solute that diffuses through a unit area per unit time—is proportional to the concentration gradient:

$$J \propto -\frac{dc}{dr}$$

$$= -D\frac{dc}{dr} \qquad (16.13)$$

where (dc/dr) is the concentration gradient (change of concentration c along direction r) and D is the diffusion coefficient. Because concentration gradient is negative in the direction of diffusion, a negative sign is needed to make the flux J a positive quantity. In the sedimentation equilibrium experiment, however, the concentration gradient actually *increases* with increasing r value. Therefore, from Equation 16.13 we write

$$J = D\frac{dc}{dr}$$

$$= \frac{k_B T}{f}\frac{dc}{dr} = \frac{RT}{fN_A}\frac{dc}{dr} \tag{16.14}$$

According to Equation 16.5, the sedimentation rate for solute molecules in a solution of concentration c is

$$\frac{dr}{dt} = \frac{\omega^2 rm}{f}(1 - \bar{v}\rho)$$

or

$$c\frac{dr}{dt} = \frac{c\omega^2 rm}{f}(1 - \bar{v}\rho) \tag{16.15}$$

At equilibrium, the diffusion rate is equal to the sedimentation rate, so that

$$c\frac{dr}{dt} = \frac{RT}{fN_A}\frac{dc}{dr}$$

or

$$c\omega^2 rm(1 - \bar{v}\rho) = \frac{RT}{N_A}\frac{dc}{dr} \tag{16.16}$$

Rearranging, we obtain

$$\frac{dc}{c} = \frac{\mathcal{M}\omega^2 r(1 - \bar{v}\rho)}{RT}dr \tag{16.17}$$

where $\mathcal{M}$ is the molar mass, given by mN_A. Integration between $r_1(c_1)$ and $r_2(c_2)$ yields

$$\int_{c_1}^{c_2}\frac{dc}{c} = \frac{\mathcal{M}\omega^2(1 - \bar{v}\rho)}{RT}\int_{r_1}^{r_2} r\,dr$$

$$\ln\frac{c_2}{c_1} = \frac{\mathcal{M}\omega^2(1 - \bar{v}\rho)}{2RT}(r_2^2 - r_1^2) \tag{16.18}$$

Again, optical techniques enable us to measure the concentrations of the solute concentrations c_1 and c_2 at r_1 and r_2, and so if $\bar{v}, \rho$, and ω are known, we can calculate the value of $\mathcal{M}$. In contrast to the sedimentation velocity method, this technique does not require any knowledge of the shape of the molecule or its diffusion coefficient. It is, therefore, one of the most accurate methods of molar-mass determination.

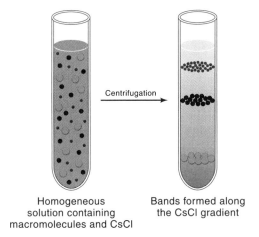

Figure 16.6
A homogeneous solution of macromolecules in 6 M CsCl is spun in an ultracentrifuge. At equilibrium, the molecules separate into bands along the CsCl density gradient according to their molar masses.

Homogeneous solution containing macromolecules and CsCl

Bands formed along the CsCl gradient

Density-Gradient Sedimentation. One improvement has greatly widened the scope of the ultracentrifuge technique: the cesium chloride sedimentation equilibrium method. For this procedure, a macromolecular solution mixed with a concentrated (about 6 M) CsCl solution is spun in a celluloid centrifuge tube until equilibrium is reached (Figure 16.6). At equilibrium, a CsCl density gradient has formed along the tube. At some value of r, the density of the solution, ρ, is equal to the reciprocal of the partial specific volume of the macromolecule; that is, $\rho = 1/\bar{v}$. This means that $(1 - \bar{v}\rho) = 0$, and from Equation 16.5, we see that the sedimentation rate, dr/dt, is zero. Consequently, a band will form for each type of macromolecule at some point along the density gradient. Generally, the bands are narrow and well resolved. Above the band (nearer the center of rotation), we have $\rho < 1/\bar{v}$, and $(1 - \bar{v}\rho)$ is positive. Sedimentation will drive the macromolecules down toward the band. Below the band (toward the bottom of the tube) $\rho > 1/\bar{v}$, and $(1 - \bar{v}\rho)$ is negative. At this point, buoyancy will drive the macromolecules up toward the band. Thus, the presence of the CsCl density gradient acts as a stabilizing force to prevent intermixing due to changes in temperature and mechanical disturbances. This technique has enabled scientists to separate DNA molecules that differ only in their ^{14}N and ^{15}N isotope contents (see Section 16.3). The ordinary sedimentation equilibrium method does not produce such well-resolved bands because of diffusion.

Another advantage of the CsCl sedimentation equilibrium method is that once equilibrium has been reached, the tube can be removed from the centrifuge, and different bands can be separated from one another by puncturing the bottom of the celluloid tube and collecting each portion into a different tube. In this respect, the method is like the sucrose gradient technique. The difference between these two procedures, however, is that in the CsCl method, the bands form at the *same* time the density gradient is established, so no additional movement of macromolecules is necessary.

Viscosity

The *viscosity* of a fluid is an index of its resistance to flow. A relatively simple apparatus for measuring viscosity is the Ostwald viscometer [devised by the German chemist Wolfgang Ostwald (1883–1943)] shown in Figure 16.7. It consists of a bulb (A) with markings x and y, attached to a capillary tube B and a reservoir bulb C. A definite volume of the liquid under study is introduced into C, sucked into A, and the time it takes for the liquid to flow between x and y is recorded. In practice, the viscosity of a solution is determined by comparison with a standard, which is usually the

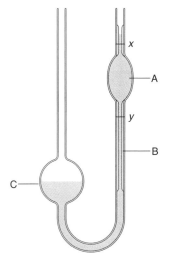

Figure 16.7
An Ostwald viscometer. The time a liquid takes to flow between markings x and y is measured and compared with that of a reference liquid. A: bulb; B: capillary tube; and C: reservoir bulb.

solvent. In such a case, we measure the *relative* viscosity, η_{rel}, given by

$$\eta_{rel} = \frac{\eta}{\eta_0} = \frac{\rho}{\rho_0}\frac{t}{t_0} \qquad (16.19)$$

where η and η_0 are the viscosities of the solution and the standard, respectively; t and t_0 are the respective flow times; and ρ and ρ_0 are the respective densities. Because the presence of solute molecules normally disturbs the streamline flow of liquids, thus causing an increase in viscosity, η_{rel} is usually greater than unity. Additional definitions of viscosity follow:

$$\text{Specific viscosity:}\quad \eta_{sp} = \eta_{rel} - 1 \qquad (16.20)$$

$$\text{Reduced viscosity:}\quad \eta_{red} = \frac{\eta_{sp}}{c} \qquad (16.21)$$

$$\boxed{\text{Intrinsic viscosity:}\quad [\eta] = \lim_{c \to 0}\frac{\eta_{sp}}{c}} \qquad (16.22)$$

where c is the concentration in $g\ mL^{-1}$ or $g\ (100\ mL)^{-1}$.

Relative viscosity is a measure of the change in the solution's viscosity compared with the viscosity of the pure solvent. Specific viscosity measures the increase in viscosity over one. To take the concentration into account—that is, to find out how large the specific viscosity per unit concentration of the solute is—we divide η_{sp} by c to give η_{red}. Further, η_{sp} itself depends on concentration, so we still need another quantity, $[\eta]$, the intrinsic viscosity, which is obtained by the extrapolation shown in Figure 16.8.

A useful equation relating intrinsic viscosity to the molar mass is given by

$$\boxed{[\eta] = K\mathscr{M}^\alpha} \qquad (16.23)$$

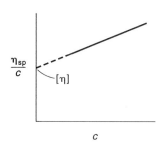

Figure 16.8
Determination of intrinsic viscosity.

where $\mathscr{M}$ is the molar mass of the polymer, and K and α are empirical constants. The value of α depends on the shape, or geometry, of the macromolecule: $\alpha = 0$ for a sphere, $\alpha = 0.5$ for a random coil, and $\alpha \approx 1.8$ for a long rigid rod. For a macromolecule of known K and α values, Equation 16.23 provides a quick estimate of its molar mass from the intrinsic viscosity measurement. On the other hand, if the molar mass of the macromolecule is known, then Equation 16.23 enables us to deduce the shape of the molecule.

Intrinsic viscosity has often been used to follow the conformational changes of proteins. For example, when a globular protein (to be discussed later) unfolds to form a random coil, its intrinsic viscosity increases, but when a rod-shaped protein such as collagen or myosin unfolds, the intrinsic viscosity decreases because of a decrease in the asymmetry of the molecule. Table 16.2 lists the intrinsic viscosity of various macromolecules.

Electrophoresis

Electrophoresis is the migration of ions under the influence of an applied electric field. As for sedimentation, solute molecules in electrophoresis move under the influence of an external field. Electrophoresis, however, depends primarily on the charge and not on the molar mass of the solute. It is a useful technique for separating proteins in a mixture and for determining molar mass.

An electric field, E, is applied across a solution. The force acting on the charged solute molecules is given by zeE, where z is the number of charges on the molecule,

Table 16.2
Intrinsic Viscosity of Some Macromolecules

Molecule	Shape	$\dfrac{[\eta]}{\text{mL} \cdot \text{g}^{-1}}$	$\dfrac{\mathscr{M}}{\text{g} \cdot \text{mol}^{-1}}$
Myoglobin	Globular	3.1	17,800
Myoglobin[a]	Random coil	21	17,800
Hemoglobin	Globular	3.6	64,450
Hemoglobin[a]	Random coil	19	64,450
Ribonuclease	Globular	3.4	13,683
Bovine serum albumin	Globular	3.7	67,500
Bovine serum albumin[a]	Random coil	51	67,500
Myosin	Rod	217	440,000
Tobacco mosaic virus	Rod	37	39,000,000
Collagen	Rod	1150	350,000
Polystyrene[b]	Random coil	130	500,000

[a] Denatured.
[b] In toluene.

and e is the electronic charge. As for sedimentation, each ion accelerates for a very brief time immediately after the field is turned on. Then, a steady state is reached as the electrostatic force is balanced by the frictional force exerted by the solvent medium. At this point, the ions are moving at a constant velocity, v, and

$$zeE = fv$$
$$= 6\pi\eta r_{s}v \tag{16.24}$$

Hence,

$$v = \frac{zeE}{f} \tag{16.25}$$

Defining the *electrophoretic mobility*, u, as the velocity per unit electric field, we write

$$u = \frac{v}{E} = \frac{ze}{f}$$

$$= \frac{ze}{6\pi\eta r_{s}} \tag{16.26}$$

Electrophoretic mobility has the units $\text{m}^2 \text{ s}^{-1} \text{ V}^{-1}$.

We see from Equation 16.26 that a particular ion's ease of movement depends on its charge and is inversely proportional to its size and the viscosity of the medium. This equation is oversimplified, for it assumes the ion is spherical. The equation also neglects the influence of ionic atmosphere (see Chapter 5) on the movement of the ions. Nevertheless, it enables us to estimate electrophoretic mobility and therefore suggests a convenient means for separating a mixture of macromolecules into pure components.

One of the simplest measurements of electrophoretic mobility utilizes the *moving-boundary method*. Figure 16.9 shows the setup for this procedure. The solution under investigation is poured into the bottom of the U tube, and buffer solutions are then carefully added to obtain sharp boundaries. Electrodes are inserted in the side arms

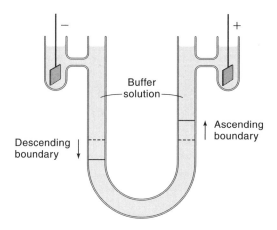

Figure 16.9
A schematic of a moving-boundary electrophoresis apparatus. The dashed horizontal lines indicate the positions before the electric field was turned on.

Table 16.3
Isoelectric Points of Some Proteins[a]

Protein	pI
Bovine serum albumin	4.9
β-Lactoglobulin	5.2
Carboxypeptidase	6.0
Hemoglobin	6.7
Hemoglobin S[b]	6.9
Ribonuclease	9.5
Cytochrome c	10.7
Lysozyme	10.7

[a] The precise value depends on the temperature and the ionic strength of the solution.
[b] This is sickle-cell hemoglobin.

to prevent products formed from electrolysis from falling into the boundary regions. The entire apparatus is then immersed in a thermostat. If the solution contains protein molecules bearing excess charges on their surfaces, the boundaries of the layered buffer solutions will move toward the electrode of the opposite sign. It follows that the direction of movement of protein molecules depends on the pH of the medium. At a pH above its isoelectric point (see Section 8.7), a protein is negatively charged and the boundaries will move toward the anode; at a pH below its isoelectric point, the net charge on the protein is positive and the boundaries will move toward the cathode. At the isoelectric point, the net charge on the protein is zero and the boundaries remain stationary (Figure 16.10). The separation of proteins is possible if they have different isoelectric points. Table 16.3 lists the isoelectric points of some common proteins. The moving-boundary method was pioneered by the Swedish biochemist Arne Tiselius (1902–1971) in the 1930s. Its major shortcoming is that convective mixing of migrating proteins can occur with this technique. This method has now been supplanted by *zone electrophoresis*, in which a sample is constrained to move in a solid support such as filter paper, cellulose, or a gel.

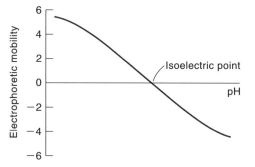

Figure 16.10
Determination of the isoelectric point of a protein by plotting its electrophoretic mobility as a function of pH. At a pH below its isoelectric point, the protein is positively charged and moves toward the cathode. At a pH above its isoelectric point, the protein is negatively charged and moves toward the anode. At the isoelectric point, its mobility is zero.

Gel Electrophoresis. *Gel electrophoresis* is one of the most useful methods for separating macromolecules. The gels in common use are polyacrylamide and agarose. Polyacrylamide gels have the following advantages: (1) they are chemically inert, (2) they suppress convective currents caused by temperature changes, and (3) their pore size can be controlled when they are synthesized from the monomer acrylamide and *N,N'*-methylenebisacrylamide (a cross-linking agent). Thus, in addition to providing a medium for electrophoresis, the gels also act as molecular sieves that retard large molecules relative to smaller ones. Consequently, a better separation of proteins results.

A variation of the gel electrophoresis technique, called *sodium dodecyl sulfate–polyacrylamide gel electrophoresis* (SDS–PAGE), enables us to determine the molar mass of proteins under denaturing conditions. This method requires treatment of the protein with the denaturing agent sodium dodecyl sulfate (SDS) and *β*-mercaptoethanol or dithiothreitol, both of which act to reduce the number of disulfide bonds in proteins:

$$CH_3-(CH_2)_{10}-CH_2OSO_3^- Na^+ \quad HOCH_2CH_2SH \quad HSCH_2(CHOH)_2CH_2SH$$
sodium dodecyl sulfate (SDS)   *β*-mercaptoethanol   dithiothreitol

The denaturing agent SDS binds strongly to most proteins (about 1.4 g of SDS to 1 g of protein), and *β*-mercaptoethanol ruptures disulfide linkages in proteins. The surface charge on the SDS–protein complex is almost entirely due to the exposed sulfate ions. The complex is not a completely random coil; rather, it assumes the shape of a long rod of constant width. Its length is a function of the protein's molar mass. The important point is that the surface charge per unit length of the complex tends to be constant, regardless of the charge on the polypeptide chain. Thus, its electrophoretic mobility depends only on its size (that is, length) rather than the net charge. By carrying out an electrophoresis experiment using polyacrylamide gel as the supporting medium, scientists can obtain well-resolved bands of various SDS–protein complexes (Figure 16.11).

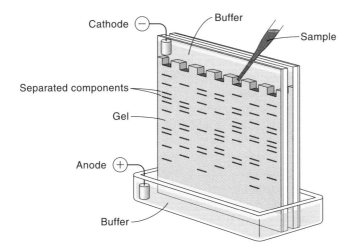

Figure 16.11
Schematic representation of the apparatus for SDS–PAGE. Several samples are electrophoresed on one flat polyacrylamide gel. The negatively charged SDS–protein complexes migrate downward toward the anode. In time, bands develop in parallel lanes corresponding to the proteins present in the original samples.

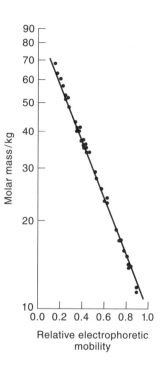

Figure 16.12
A logarithmic plot of the molar masses of many proteins versus their relative electrophoretic mobilities on an SDS–polyacrylamide gel. [After K. Weber and M. Osborn, *J. Biol. Chem.* **244**, 4406 (1969).]

A plot of the logarithm of the molar mass of the proteins versus the electrophoretic mobility of their SDS complexes produces a straight line that has a negative slope (Figure 16.12). The molar mass of an unknown protein can then be readily determined from the electrophoretic mobility of its SDS complex and the standard calibration curve. An additional advantage of this technique is that SDS dissociates oligomeric proteins into their individual polypeptide chains. Thus, from the molar-mass determination, we can often deduce the number of subunits present. For example, the molar mass of glyceraldehyde-3-phosphate dehydrogenase is found to be 140,000 g by the ultracentrifugation technique. The SDS–PAGE technique, however, measures the molar mass of the same compound to be only 36,500 g. We may conclude that this enzyme has four subunits.

Isoelectric Focusing. In the *isoelectric focusing technique*, a pH gradient is first established between the electrodes so that different proteins form stationary bands along the gradient at points where the pH is equal to their isoelectric points. The pH gradient is established in a special way. First, low-molar-mass polyampholytes that cover a wide range of isoelectric points are dissolved in water. Before the application of an electric field, the pH of the solution is the same throughout (the pH is the average value of all the polyampholytes in solution). After the field has been turned on, the polyampholytes start to migrate toward the electrodes. As a consequence of their own buffering capacities, a pH gradient is gradually established between the electrodes. Eventually, each type of polyampholyte will come to rest in the self-established gradient at the point corresponding to its own isoelectric point. If a mixture of proteins is introduced to the medium, each type of protein will migrate to the position corresponding to its isoelectric point so that several bands will form. The bands can then be separated from one another for characterization. Note that, in principle, the isoelectric focusing technique is analogous to the CsCl density equilibrium method discussed earlier.

Isoelectric focusing can be combined with SDS–PAGE to obtain very high-resolution separations. First, a single sample is subjected to isoelectric focusing, producing a series of bands displayed vertically according to different p*I* values

Polyampholytes are a mixture of low-molar-mass (300–6000) substances containing aliphatic amino and carboxylic acid groups.

Table 16.4
Summary of Different Methods for Molar-Mass Determination

Method	Approximate Range	Type of Molar Mass Determined
Freezing-point depression	$\lesssim 500$	$\bar{\mathscr{M}}_n$
Osmotic pressure	$\lesssim 100,000$	$\bar{\mathscr{M}}_n$
Viscosity	Unrestricted	$\bar{\mathscr{M}}_v$
Ultracentrifugation	$\gtrsim 5,000$	$\bar{\mathscr{M}}_w$
X-ray diffraction[a]	Unrestricted	$\bar{\mathscr{M}}_n$
Gel electrophoresis	$\gtrsim 5,000$	$\bar{\mathscr{M}}_w$
Electron microscopy[a]	$\gtrsim 100,000$	$\bar{\mathscr{M}}_n$
Light scattering[a]	$\gtrsim 5,000$	$\bar{\mathscr{M}}_w$

[a] Not discussed in this text.

of the proteins. Next, this single-lane gel is placed horizontally on top of an SDS–polyacrylamide slab. The proteins are electrophoresed vertically downward to yield a two-dimensional pattern of spots. Because two different proteins are highly unlikely to have the same pI value and the same molar mass, each spot represents a *single* protein. This two-dimensional electrophoresis technique has enabled biochemists to separate more than a thousand proteins from the bacterium *Escherichia coli* (*E. coli*) in a single experiment!

Table 16.4 summarizes various techniques for molar-mass determination.

16.2 Structure of Synthetic Polymers

Although determining the molar mass and overall shape of a macromolecule is relatively straightforward, discovering the three-dimensional arrangement of individual atoms in such a system is more complex. X-ray diffraction provides the complete structural details of a macromolecule, but only for the solid state. The solution conformation of the macromolecule may be appreciably different from the crystalline structure. To probe different structural aspects of the macromolecule, chemists rely on several recently developed techniques. By assembling pieces of information like parts of a jigsaw puzzle, we can create an overall picture. In this section, we shall briefly consider the stereochemistry of macromolecules.

The NMR technique now enables chemists to study proteins with molar masses as high as 100,000 daltons in solution.

Configuration and Conformation

We need to distinguish between two terms that are used in describing molecular structure: *configuration* and *conformation*. The *configurations* of a molecule are stereo arrangements that are related to one another by symmetry but cannot be converted into one another without rupturing bonds. For example, *cis*- and *trans*-dichloroethylene are the two configurations of the $C_2H_2Cl_2$ molecule:

cis-Dichloroethylene   *trans*-Dichloroethylene

Other examples are optical isomers and derivatives of alicyclic compounds such as 1,2-*cis*- and 1,2-*trans*-dibromocyclopropane. On the other hand, *conformations* are arrangements of atoms in three-dimensional space. In ethane, the molecule can exist in one of an infinite number of conformations, the most stable of which is called the staggered conformation and the least stable the eclipsed conformation:

Staggered   Eclipsed

The Random-Walk Model

Imagine a long polymer chain made up of identical units, say $-CH_2-$, dissolved in some solvent. For simplicity, let us ignore all the solute–solvent and solute–solute interactions. The question we ask is: What shape does the polymer assume? In one extreme, the chain would be completely stretched out; in the other extreme, the chain would be wound up on itself like a ball of string. Generally, the actual structure is something in between.

If the repeating units of a polymer have no preferred orientation, we can apply the *random-walk model* to determine its structure. Place a drunken person in the center of a large room, and instruct him to take a succession of steps of equal length. No restriction is placed on his direction, so each step is completely uncorrelated with the previous one. It turns out that the lines joining the successive steps quite accurately represent the arrangement of a polymer chain. The analogy is not exact, however, because polymers have three-dimensional structure, whereas the random walk occurs on a two-dimensional floor.

Figure 16.13 shows a two-dimensional representation of a freely jointed polymer chain containing 50 identical units. The quantity of interest is the distance between the two ends of the chain (r), which gives us some idea about the size of the molecule. Statistical calculations averaged over many different random walks show that the average of the square of the distance r is given by

$$\overline{r^2} = nl^2 \tag{16.27}$$

where n is the number of bonds, and l is the length of each bond. The root-mean-

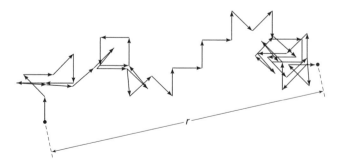

Figure 16.13
A two-dimensional, 50-step, random-walk model. (Reprinted from P. J. Flory, *Principles of Polymer Chemistry.* Copyright 1953 by Cornell University. Used by permission of Cornell University Press.)

square distance is

$$r_{rms} = \sqrt{\overline{r^2}} = l\sqrt{n} \tag{16.28}$$

For example, a chain containing 1000 C–C bonds, each having a length of 1.54 Å, would give

$$r_{rms} = 1.54\,\text{Å} \times \sqrt{1000} = 48.7\,\text{Å}$$

which is considerably shorter than a completely stretched out chain of length $1000 \times 1.54\,\text{Å} = 1540\,\text{Å}$.

We can also estimate the volume occupied by our polymer chain. If we take the diameter as 48.7 Å, the volume is given by $\frac{4}{3}\pi(48.7\,\text{Å}/2)^3 = 6.0 \times 10^4\,\text{Å}^3$. The actual volume occupied by the chain is only a small fraction of this volume. What, then, is the significance of the volume of this sphere? In solution, the polymer chain does not remain stationary but is constantly changing its shape and size because of thermal motion. The calculated volume represents the space within which the polymer is contained, on the average.

Another important value is the probability, $P(r)$, of finding the two ends of a polymer chain separated by distance r. Determining this probability is a well-known mathematical problem; the exact solution takes the form

$$P(r) = Ar^2 \exp\left(-\frac{3r^2}{2nl^2}\right) \tag{16.29}$$

where A is a constant. A plot of $P(r)$ versus r is shown in Figure 16.14. As the curve shows, the probability of two ends meeting at $r = 0$ and $r \rightarrow \infty$ is zero. The value of r for the maximum value of $P(r)$ is called the most probable distance, r_{mp}.

The results obtained from Equation 16.29 give a fairly good picture of what a polymer chain in solution might look like. The random-walk model, besides being a two-dimensional representation of a three-dimensional object,* is oversimplified in other ways. First, the bonds in the chain are not free to take any orientation with respect to one another. For example, in polyethylene, all the bond angles must be close to 109°:

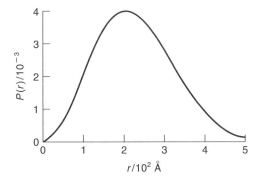

Figure 16.14
Plot of $P(r)$ versus r according to Equation 16.29, assuming $n = 1 \times 10^4$ and $l = 2.5\,\text{Å}$. (Reprinted from P. J. Flory, *Principles of Polymer Chemistry.* Copyright 1953 by Cornell University. Used by permission of Cornell University Press.)

---

* Equation 16.29 does apply to a three-dimensional model.

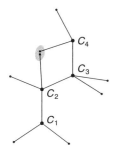

Figure 16.15
Steric interaction between hydrogen atoms (shaded region) on neighboring carbon atoms in polyethylene.

Second, we must also take into account the steric interaction between hydrogen atoms on adjacent carbon atoms (Figure 16.15). Both of these effects tend to reduce the compactness of the chain and increase the value of r_{rms}. Fortunately, the corrections can be made easily as follows:

$$\overline{r^2} = Cnl^2 \qquad (16.30)$$

where C is a constant for a particular type of polymer. Its value lies between 2 and 10 in most cases. Third, while the drunk on a random-walk model recrosses his path as many times as he desires, no two atoms in a polymer chain can occupy the same space at the same time. This *volume exclusion effect* must also be taken into account. For our purposes, Equation 16.30 is adequate for most calculations.

The actual conformation of the polymer must depend on the nature of the solvent. In a "good" solvent—that is, one that has a zero or negative heat of mixing with the polymer (an exothermic process)—the chain will be in the extended form. In a "poor" solvent—that is, one that has a positive heat of mixing with the polymer (an endothermic process)—the chain tends to roll up like a ball of string.

16.3 Structure of Proteins and DNA

The simplest polymeric systems are those that have identical units, such as polyethylene. Many naturally occurring polymers possess structures far more complex than that of polyethylene. Theoretical treatment of these systems is quite formidable. Two important groups of complex polymers are proteins and DNA.

Proteins

Proteins can be thought of as amino acid polymers. The first step in the synthesis of a protein molecule is a condensation reaction between the amino group on one amino acid and the carboxyl group on another amino acid. The molecule formed from the two amino acids is called a *dipeptide*, and the bond joining them is a *peptide bond*:

$$H_3\overset{+}{N}-\underset{R}{\overset{H}{\underset{|}{\overset{|}{C}}}}-\overset{O}{\overset{\|}{C}}-O^- \ + \ H_3\overset{+}{N}-\underset{R}{\overset{H}{\underset{|}{\overset{|}{C}}}}-\overset{O}{\overset{\|}{C}}-O^- \ \longrightarrow \ H_3\overset{+}{N}-\underset{R}{\overset{H}{\underset{|}{\overset{|}{C}}}}-\overset{O}{\overset{\|}{C}}-\underset{H}{\overset{H}{\underset{|}{\overset{|}{N}}}}-\underset{R}{\overset{H}{\underset{|}{\overset{|}{C}}}}-\overset{O}{\overset{\|}{C}}-O^- \ + \ H_2O$$

where R represents a H atom or some other group; –CO–NH– is called the *amide group*. The condensation reaction can continue to form a *tripeptide*, and so on until the complete *polypeptide* chain is formed, as shown in Figure 16.16. Twenty amino acids are found in proteins (see Table 8.6). All except glycine are optically active, having the L configuration. Even for a small protein, such as insulin, which contains only 50 amino acid residues, the number of chemically different structures that could be formed is 20^{50} or 10^{65}. This would be an enormously large group, considering that Avogadro's number is only 6×10^{23}. The actual number of proteins found to date, however, is far smaller than this figure.

In the 1930s, Linus Pauling and Robert Corey (American chemist, 1897–1971), systematically investigated protein structure, relying mainly on the X-ray diffraction technique. They studied amino acids, dipeptides, and tripeptides as model compounds, and they arrived at the following conclusions:

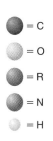

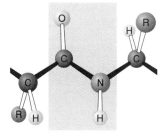

Figure 16.16
Condensation of amino acids to form dipeptides and polypeptides.

1. The amide group is essentially planar due to structural resonance:

   where C_α denotes the carbon atom adjacent to the carbon atom in the amide group. The C–N bond has about 30% to 40% double-bond character; consequently, rotation about this bond is restricted by an activation energy of about 40 to 80 kJ mol^{-1}. Rotation about the C_α–C and C_α–N bonds are not restricted. Figure 16.17 shows the planar amide group.

Figure 16.17
The planar amide group.

2. The *trans* configuration is more stable than the *cis* configuration because of the steric interaction between the groups of the C_α atoms in the *cis* isomers.

   cis   *trans*

3. The backbone of the polypeptide, the –C–C–N–C–C–N– linkage, is more important than the side chains in determining the structure. Further, all residues are considered equivalent.

4. Hydrogen bonds play an essential role in stabilizing the polypeptide chain conformation. In all cases studied, the hydrogen atom was found to lie within 30° of the N $\cdots$ O line. In the most stable forms, all four atoms are colinear:

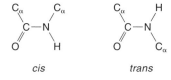

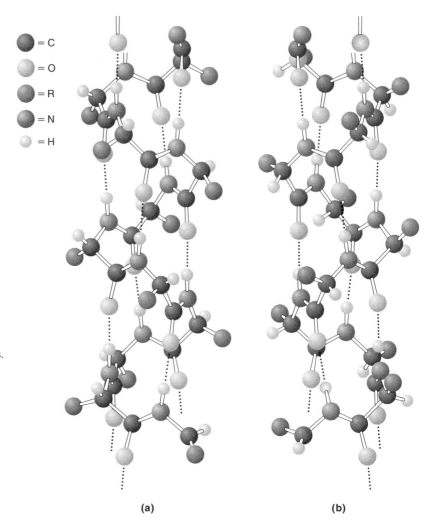

= C
= O
= R
= N
= H

Figure 16.18
(a) Left-handed and (b) right-handed α helices.
The spheres labeled R represent the amino
acid side chains.

(a)   (b)

From these findings, Pauling and Corey predicted that a stable arrangement for the polypeptide chain is the α-helix structure, because it allows for a maximum number of hydrogen bonds and introduces the least distortion of bond length and bond angle (Figure 16.18).

Because the amide group is essentially planar, the polypeptide chain has only two degrees of rotational freedom, which are the rotation about the C_α–N bond, characterized by angle ϕ, and the rotation about the C_α–C bond, characterized by angle ψ (Figure 16.19). In fact, the three-dimensional structure of the polypeptide chain can be completely defined in terms of these two angles. For a right-handed α helix, $\phi = -57°$ and $\psi = -47°$. Not all possible combinations of ϕ and ψ yield stable conformations. For example, severe steric hindrance between amide hydrogen atoms would result if $\phi = 0°$ and $\psi = 0°$.

Another important structure predicted from X-ray diffraction studies is the β-pleated structure, or simply the β structure. This structure results from *intermolecular* hydrogen bonding between extended polypeptide chains and intramolecular hydrogen bonding within a protein molecule. In contrast to the α helix, the hydrogen bonds in the β structure are oriented roughly perpendicularly to the long axis of the polypeptide chain. Two different arrangements, called *parallel* and *antiparallel*, are known (Figure 16.20).

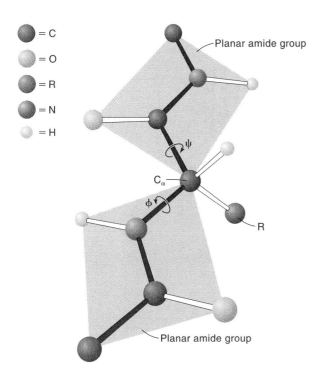

= C
= O
= R
= N
= H

Planar amide group

ψ

C_α

ϕ

R

Planar amide group

Figure 16.19
The angles ϕ and ψ are both defined as 180° when the polypeptide chain is in its planar, fully extended conformation (as shown here) and increase for a clockwise rotation when viewed from the C_α atom.

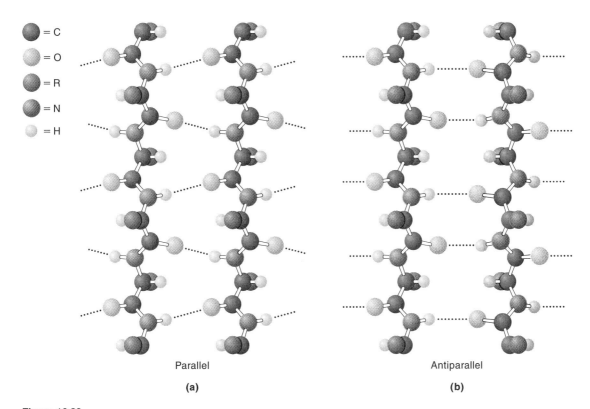

= C
= O
= R
= N
= H

Parallel

(a)

Antiparallel

(b)

Figure 16.20
(a) Parallel β structure: the polypeptide chains all run in the same direction. (b) Antiparallel β structure: adjacent polypeptide chains run in opposite directions.

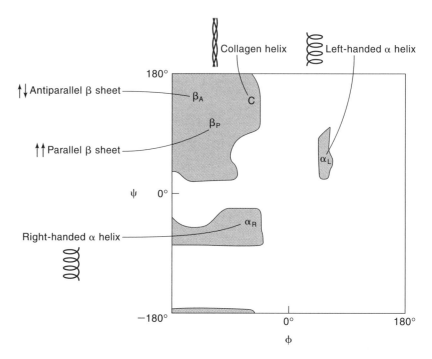

Figure 16.21
Ramachandran plot showing allowed values of ϕ and ψ for poly-L-alanine (shaded regions).
All proteins are right-handed α helices. The left-handed α helices are energetically less
favored.

One way to study the conformational stability of the polypeptide chain is to use a
Ramachandran plot, after the Indian chemist G. N. Ramachandran (1922–), shown
in Figure 16.21. This plot lays out the values of ϕ and ψ between $-180°$ and $+180°$ so
that all possible conformations of the polypeptide chain can be described in terms of
the combination of the angles. The enclosed areas represent possible combinations of
ϕ and ψ that yield the least steric hindrance. We see that there are three separate
allowed regions, namely, the left- and right-handed α helix and the β structure.

Proteins are arranged in both the α helix and the β structure to varying degrees,
confirming Pauling and Corey's theory. Generally, proteins are divided into two
categories, *globular* and *fibrous*. Globular proteins are characterized by their compactness. In these molecules, the polypeptide chain is folded up to fill most of the
space within its domain, leaving relatively little empty volume. Both myoglobin and
hemoglobin contain more than 75% α-helical structure. This is not true of all globular
proteins, however. Lysozyme has only about 40% α-helical content, papain, 20%, and
chymotrypsin, practically none. Clearly, factors other than Pauling and Corey's criteria contribute to protein conformation. We shall return to this point in the next
section.

Fibrous proteins are present in wool, hair, and silk. The fibrous proteins in wool
and hair are called *keratins*. The α helix in keratins accounts for the flexible and
elastic properties of wool. Because interactions do not occur among different chains,
wool fibers are not very strong. Silk possesses the β structure. Since the polypeptide
chains are already in extended forms, silk lacks elasticity and extendibility, but it is
quite strong because of its intermolecular hydrogen bonds. Collagen is another example of fibrous protein. It accounts for about one-third of all proteins in the human
body. Collagen is the most important component of connective tissues, such as car-
tilage, ligament, and tendon. The fundamental structural unit of collagen is tropo-
collagen, an elongated triple helix consisting of three polypeptide chains intertwined

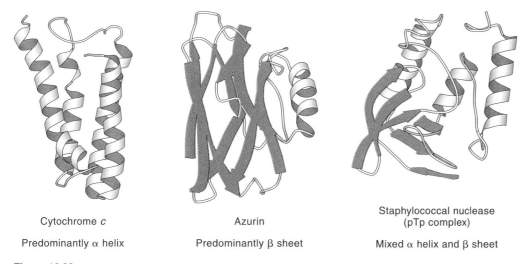

Cytochrome *c*	Azurin	Staphylococcal nuclease (pTp complex)
Predominantly α helix	Predominantly β sheet	Mixed α helix and β sheet

Figure 16.22
Ribbon diagrams for three proteins having different amounts of α helix and β structure. The α helices are shown as coils and the arrows of the β structure strands indicate parallel or antiparallel arrangements. (Adapted from S. J. Lippard and J. M. Berg, *Principles of Bioinorganic Chemistry*, University Science Books, Sausalito, California, 1994. Used with permission.)

to form a superhelix with a diameter of about 15 Å and a length of 3,000 Å. The most notable properties of collagen are its rigidity and resistance to deformation, both of which are vital to transmitting the mechanical force generated by muscles.

The structure of proteins is customarily divided into four levels of organization. The *primary structure* refers to the unique amino acid sequence of the polypeptide chain. The *secondary structure* refers to those parts of the polypeptide chain that are stabilized by hydrogen bonds—for example, the α helix. The term *tertiary structure* is given to the three-dimensional structure stabilized by the disulfide bonds between cysteine residues, as well as to such noncovalent forces as van der Waals forces, hydrogen bonding, and electrostatic forces. The hydrogen bonds in the tertiary structure join residues that are far apart in the polypeptide chain but are brought into proximity as a result of the folding of the chain. Figure 16.22 shows the tertiary structures of three proteins that contain different amount of α helices and β-pleated sheets. A *quaternary structure* results from interaction between polypeptide chains; for example, the four polypeptide chains or four subunits in hemoglobin. These chains are held together by noncovalent forces such as dispersion forces and electrostatic forces (Figure 16.23).

DNA

DNA molecules do not exhibit the structural complexity of proteins because they contain only four base pairs compared to the 20 amino acids in proteins. Consequently, DNA has a limited secondary structure and no comparable tertiary or quaternary structures. The structure of DNA proposed by Watson and Crick was shown in Figure 13.9. In this structure, now known as B-DNA, two polynucleotide chains run in opposite directions coiled around a common axis to form a right-handed double helix. In recent years, detailed X-ray analysis of DNA crystals has revealed two other structural variations called A-DNA and Z-DNA (Figure 16.24).

DNA molecules are huge. Depending on the source, they contain from hundreds to millions of base pairs and range in length from 10^4 Å to 10^{10} Å, or 1 m! In com-

DNA also exists in circular and supercoiled forms.

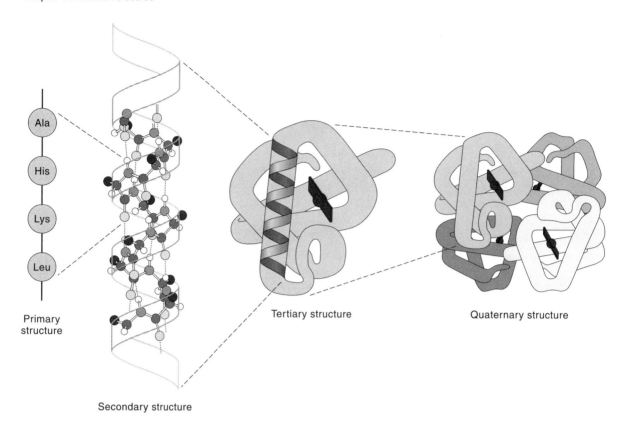

Primary
structure

Secondary structure

Tertiary structure

Quaternary structure

Figure 16.23
The primary, secondary, tertiary, and quaternary structure of the hemoglobin molecule.

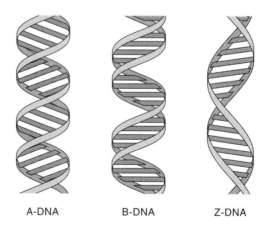

A-DNA   B-DNA   Z-DNA

Figure 16.24
The three forms of DNA. Both A-DNA and B-DNA are right-handed double helices, whereas Z-DNA is a left-handed double helix. The B-DNA is the preferred conformation *in vivo*, although double-stranded RNA does adopt the A conformation. It is not known if Z-DNA occurs in nature. (From S. J. Lippard and J. M. Berg, *Principles of Bioinorganic Chemistry*, University Science Books, Sausalito, CA, 1994. Used with permission.)

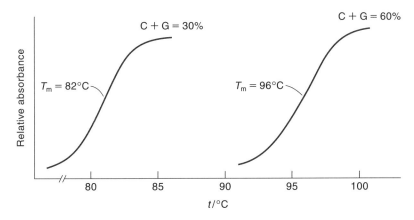

Figure 16.25
The melting curves of two DNAs containing different percentages of C + G.

parison, the roughly spherical hemoglobin molecule has a diameter of about 65 Å, and collagen, one of the longest proteins, is about 3000 Å in length. The highly elongated shape of double-helical DNA, together with its stiffness, make it susceptible to mechanical damage outside the cell's protective environment. Heat disrupts the hydrogen bonds between the base pairs and causes the two strands of the molecule to come apart. This process is called *melting*. Figure 16.25 shows the melting curves of two DNA molecules that contain different percentages of C + G. The temperature at which half the helical structure is destroyed is called the melting temperature (T_m). A DNA molecule denatured by heating can be renatured if it is carefully cooled—a process called *annealing*. If cooling takes place too rapidly, the complementary strands will not have sufficient time to find each other before the partially base-paired structures freeze.

DNA Replication. The Watson–Crick base-pairing model immediately suggests that the sequence of bases carries the genetic information. During replication, each strand of a double helix can act as a *template* for the synthesis of a new complementary strand. This is the *semiconservative model*, so called because the daughter double strands arising from replication each contain one old (conserved) strand and one new strand. A critical test of this hypothesis was carried out by the American biochemists Matthew Meselson (1930–) and Franklin Stahl (1929–) in 1958.* By growing *E. coli* bacteria in a ^{15}N-rich medium for many generations, they produced DNA with a greater than normal density because of the incorporation of the heavier N isotope. The labeled bacteria were then abruptly transferred to a ^{14}N-containing medium, and the cells were allowed to go through one or more rounds of replication.

The results of Meselson and Stahl's experiment are summarized in Figure 16.26. DNA from the cells before and after replication were isolated and analyzed by the CsCl density-gradient technique described earlier. Before replication, a single band of DNA containing ^{15}N was observed. After one generation, each of two daughters consisted of one ^{14}N strand and a ^{15}N strand, so again a single band, displaced upward from the pure ^{15}N band, was observed. Two bands appeared after two generations, corresponding to DNA containing only ^{14}N and DNA made up of a hybrid of ^{14}N and ^{15}N. After three generations, six of the eight daughters contained only ^{14}N, and the other two contained a hybrid of ^{14}N and ^{15}N. These findings are entirely consistent with the Watson–Crick model.

Meselson and Stahl's work has been described as the most beautiful experiment in biology.

* M. Meselson and F. Stahl, *Proc. Natl. Acad. Sci. U.S.A.* **44**, 671 (1958).

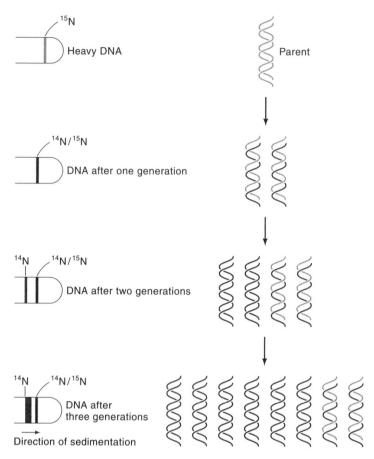

Figure 16.26
The Meselson–Stahl experiment, showing the semiconservative nature of DNA replication.

16.4 Protein Stability

Pauling and Corey's work was a great triumph in protein chemistry, for their theory successfully predicted the structure of a number of proteins. But many other proteins, such as cytochrome c, were found to possess very little or no α helix or β structure. Chemists now know that Pauling and Corey overemphasized the importance of hydrogen bonding and the backbone polypeptide chain and that a complete description of protein structure must include other types of interaction, such as electrostatic forces and van der Waals forces (discussed in Chapter 13). The nature of side-chain residues also appears to play an important role in the overall stability of proteins. Figure 16.27 shows many of the noncovalent interactions present in a protein molecule. In addition, the cysteine residues can form disulfide bonds in an oxidative environment:

$$2-CH_2SH + \tfrac{1}{2}O_2 \rightarrow -CH_2-S-S-CH_2- + H_2O$$

The disulfide bridges help to stabilize the three-dimensional structure of proteins.

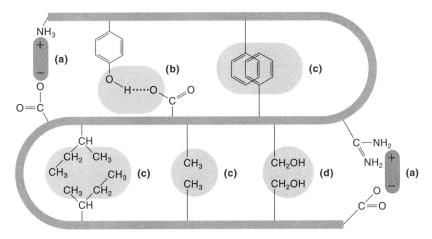

Figure 16.27
Various noncovalent interactions present in a protein molecule. The interactions are (a) electrostatic forces, (b) hydrogen bonding, (c) dispersion forces, and (d) dipole–dipole forces.

Hydrophobic Interaction

Another factor that influences protein conformation is hydrophobic interaction, which differs from the other types of intermolecular forces because it does not arise from any interaction between a pair of atoms or two groups of a molecule; instead, it involves numerous water molecules (see p. 508).

When we study protein structure, one of the first things we notice is that the nonpolar groups in valine, leucine, isoleucine, tryptophan, and phenylalanine are usually located in the interior of the protein, where they have little or no contact with water. This is the arrangement found in native proteins—that is, proteins in their normally functioning state. What would happen if the polypeptide chain unfolded, so that the nonpolar groups came in contact with water molecules? As an estimate, let us consider the following thermodynamic changes for the transfer of 1 mole of a hydrocarbon from a nonpolar solvent, such as benzene, to water at 298 K. For methane, we have $\Delta S = -75.8$ J K^{-1}, $\Delta H = -11.7$ kJ, and $\Delta G = 10.9$ kJ (see Table 13.6). Because ΔG is positive, the process is nonspontaneous. The process is exothermic and therefore enthalpy-driven. What makes the process nonspontaneous, then, is the large decrease in entropy. In Section 13.6, we discussed the formation of clathrate compounds when nonpolar hydrocarbons are dissolved in water. The ordering of the water molecules around the hydrocarbons is what results in the large negative ΔS values. Therefore, hydrophobic interaction is largely an entropy-driven process. The reverse process—that is, the interaction between two nonpolar groups, R_1 and R_2, in aqueous solution,

$$\text{R}_1(\text{H}_2\text{O})_n \;+\; \text{R}_2(\text{H}_2\text{O})_n \;\xrightarrow{\text{Hydrophobic interaction}}\; \text{R}_1\text{R}_2 \;+\; 2n\text{H}_2\text{O}$$

liberates the ordered water molecules that originally surrounded R_1 and R_2; consequently, the entropy of the system increases. Because the binding of water by individual R groups gives off some heat, the process described above, which releases water molecules, is usually endothermic; that is, ΔH is positive. At room temperature, the $T\Delta S$ term predominates so that $\Delta G \, (= \Delta H - T\Delta S)$ is negative, and the reaction is spontaneous from left to right. At lower temperatures, the $T\Delta S$ term decreases, and the sign of ΔG may be determined by ΔH. In this case, ΔG could become positive at a certain temperature, breaking up the hydrophobic interaction between R_1 and R_2 and causing the protein to unfold. Certain cold-sensitive enzymes—enzymes that lose their activity at low temperatures, such as pyruvate carboxylase—seem to fit this description.

Denaturation

A useful key to understanding the stability of proteins in solution is the manner in which these molecules denature. Protein denaturation describes any process that results in a change in the three-dimensional structure from the native conformation to some other conformation. By native conformation, we mean the conformation of the protein in its normal, physiological state. Strictly speaking, the native state of a protein is found only *in vivo*, for example, in a cell. Because we cannot isolate a protein without changing its environment somewhat, any protein *in vitro* has probably undergone some degree of denaturation.

In principle, a protein has many possible conformations. In one extreme, the native state, the protein possesses the maximum number of noncovalent interactions and disulfide linkages. In the other extreme, the completely denatured state, most of the noncovalent interactions and the disulfide bonds are broken, and the molecule has the shape of a random coil. Depending on the protein and environmental conditions, the protein may also have other, intermediate conformations.

Protein denaturation occurs in various ways. Changes in temperature, pH, and ionic strength, as well as the addition of organic solvents and reagents such as 8 M urea, 6 M guanidine hydrochloride, and detergents to a protein solution, will cause the protein molecule to denature. The action of urea probably disrupts both hydrogen bonds and hydrophobic interactions, whereas organic solvents, such as ethanol and ethylene glycol, form hydrogen bonds with polar residues on the surface of the protein molecule. The compound β-mercaptoethanol specifically cleaves disulfide bonds as follows:

$$R_1-S-S-R_2 \; + \; 2HOCH_2CH_2SH \; \longrightarrow \; R_1-SH \; + \; R_2-SH \; + \; \begin{array}{c} S-CH_2CH_2OH \\ | \\ S-CH_2CH_2OH \end{array}$$

Studies of protein denaturation have yielded significant information on protein stability. The work of the American biochemist Christian Anfinsen (1916–) on ribonuclease is particularly interesting. Ribonuclease isolated from bovine pancreas is an enzyme that has a molar mass of 13,700 g. It contains 124 amino acid residues and four disulfide linkages. Its three-dimensional structure has been determined by X-ray diffraction (Figure 16.28). The specific action of ribonuclease is to catalyze the hydrolysis of the phosphodiester bonds in ribonucleic acids.

Ribonuclease in solution is denatured by adding β-mercaptoethanol and 8 M urea. Under these conditions, all the disulfide bonds are reduced to the sulfhydryl group; that is,

$$-S-S- \; \xrightarrow{\text{2H}} \; 2-SH$$

and most or all of the secondary and tertiary structure is destroyed (Figure 16.29). In this form, the enzyme becomes completely inactive. When the denatured enzyme is oxidized with oxygen in the presence of the denaturants, a mixture of products is formed. If we assume that the probability of forming a disulfide bond between any two cysteine residues is the same, then, statistically, the total number of structurally different isomers formed from eight cysteine residues is given by $7 \times 5 \times 3 = 105$.

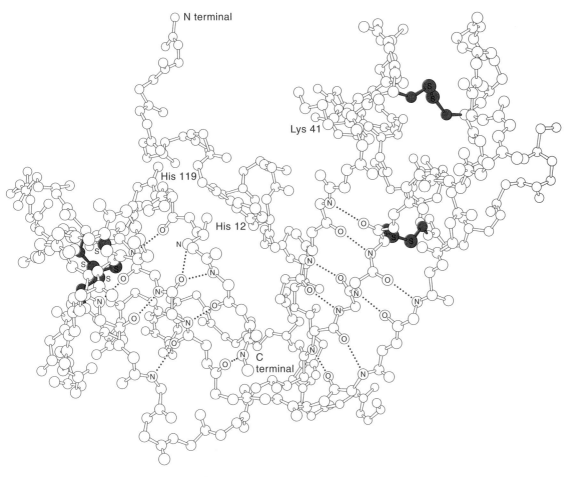

Figure 16.28
Structure of ribonuclease. The four disulfide bonds are shown in red. The hydrogen atoms taking part in hydrogen bonds are not shown. (Illustration copyright by Irving Geis.)

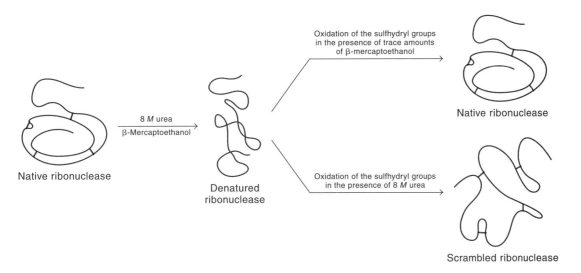

Figure 16.29
Denaturation and renaturation of ribonuclease. Depending on the conditions for renaturation, we obtain either native ribonuclease or scrambled ribonuclease. The disulfide bonds are shown as red lines.

Note that the first cysteine residue has seven choices in forming an S–S bond, the next cysteine residue has only five choices, and so on. This relationship can be generalized to $(N-1)(N-3)(N-5)\cdots 1$, where N is the total (even) number of cysteine residues present (see Problem 16.30). The observed activity of the mixture—the "scrambled protein"—is less than 1% of that of the native enzyme $(1/105 < 0.01)$. This finding is consistent with the fact that only one out of every 105 possible structures corresponds to the original state. If, on the other hand, the oxidation is carried out in the absence of urea, but a trace amount of β-mercaptoethanol is retained to promote disulfide interchanges, a homogeneous product is obtained that is *indistinguishable* from the native ribonuclease in activity and every other respect.

An important conclusion can be drawn from Anfinsen's work: the *thermodynamic hypothesis*, which states that the native state of a protein has a minimum Gibbs energy. The conformation of the native protein is determined by its overall intramolecular interactions, which in turn depend on the primary structure—that is, its amino acid sequence. Thus, although 105 different conformations are all kinetically accessible, native ribonuclease folds in such a way to give only one specific form that has the greatest thermodynamic stability. Similar observations have been obtained on other systems, such as lysozyme and proinsulin.

Another interesting result has emerged from the ribonuclease study. Contrary to earlier beliefs, disulfide bonds do not play an essential role in the folding process. Instead, their function is to stabilize the three-dimensional structure once it has formed. In the experiment described above, folding of the protein is entirely governed by noncovalent forces; correct cysteine pairs are brought into proximity for disulfide bond formation *after* a three-dimensional network has been established.

The denaturation process is generally cooperative; that is, the transition from the native state, say helix, to random-coil conformation occurs over a narrow range of temperature, pH, or denaturant concentration. Experimentally, this transition can be studied most conveniently by viscosity, optical rotation, or optical absorption measurements. Figure 16.30 shows a plot of the fraction of unfolded ribonuclease with temperature as monitored by several different physical techniques. The fact that the changes closely parallel each other, and that the transition is the same whether we proceed from low to high temperature or in the reverse direction, suggests that only two conformational states are present in appreciable amounts. This situation is an example of the *two-state model*. The equilibrium constant for the transition in proteins,

$$N \underset{k_r}{\overset{k_f}{\rightleftharpoons}} RC$$

is given by

$$K = \frac{k_f}{k_r} = \frac{[RC]}{[N]} = e^{-\Delta G^\circ/RT} = e^{-(\Delta H^\circ - T\Delta S^\circ)/RT} \tag{16.31}$$

where N and RC represent the native and random coil conformations, respectively. The forward and reverse rate constants, k_f and k_r, can be measured by suitable means, including the relaxation techniques discussed in Chapter 9. From the equilibrium constant, K, measured either directly or from the ratio k_f/k_r, we can calculate the Gibbs energy change for the transition.

The changes in thermodynamic properties that result from protein denaturation are interesting to examine. When ribonuclease is denatured at 30°C in a solution whose pH is 2.5, we have $\Delta G^\circ = 3.8$ kJ mol^{-1}, $\Delta H^\circ = 238$ kJ mol^{-1}, and $\Delta S^\circ = 774$ J K^{-1} mol^{-1}. Similar values have been obtained for the denaturation of other proteins, such as chymotrypsinogen and myoglobin. The large positive value of ΔS° may seem strange at first because it contradicts our earlier discussion of the importance of

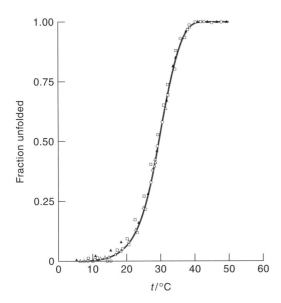

Figure 16.30
Thermal denaturation of bovine ribonuclease as measured by the increase in viscosity (□), the decrease in optical rotation at 365 nm (○), and in UV absorbance at 287 nm (Δ). The ▲ show measurements of a second melting after cooling from 41°C for 16 h. (From *Proteins: Structures and Molecular Properties*, 2/e, by Thomas E. Creighton. © 1984 and 1993 by W. H. Freeman and Company. Used with permission.)

the hydrophobic interaction in holding proteins in the native state. (When protein is denatured, the nonpolar residues in the interior are exposed to water, which should lead to a decrease in entropy, as indicated by the data in Table 13.6.) We must realize that there are actually two factors that contribute to $\Delta S°$. The disruption of hydrophobic interaction does lead to a negative $\Delta S°$ value. Whereas each protein has only one conformation in its native state, however, a denatured protein may possess any one of the many possible conformations. As we saw in Chapter 4, any process in an isolated system that goes from a less probable state to a more probable state will result in an increase in the number of microstates and hence in entropy. Thus, the sign of $\Delta S°$ depends on the magnitude of these two opposing effects. For ribonuclease, the entropy increase for the "order-to-disorder" transition is greater than the entropy decrease when the nonpolar residues are exposed to water; therefore, $\Delta S°$ is positive.

Another interesting result of denaturation is the large value of $\Delta \bar{C}_P$, which is given by

$$\Delta \bar{C}_P = [(\bar{C}_P)_{\text{denatured}} - (\bar{C}_P)_{\text{native}}]$$

For ribonuclease, $\Delta \bar{C}_P = 8.4$ kJ K^{-1} mol^{-1}. In a solution containing denatured ribonuclease, additional heat is required to melt the clathrate structure (made of water molecules) around the nonpolar residues (see p. 509). This contribution to heat capacity in denatured (but not native) solution is responsible for the large value of $\Delta \bar{C}_P$.

In many cases, protein denaturation cannot be described by a simple two-state model because the transition may involve several stages, and several intermediates could be present. A many-state model is more difficult to handle theoretically and experimentally.

Protein Folding

One of the most important problems in biology centers on protein folding, the mechanism by which the linear information contained in the amino acid sequence of an unfolded polypeptide gives rise to the unique three-dimensional structure of the native protein. Although Anfinsen's work demonstrates that a completely unfolded protein will spontaneously refold to its native conformation, it does not explain how

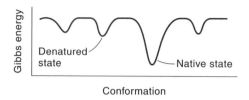

Figure 16.31
Schematic diagram comparing the relative stabilities of a native protein and the same protein in its denatured states.

this happens. A rough calculation should convince us that proteins cannot fold to their native states by a random search of all possible conformations. Suppose we have a protein with N residues. In the native state, each residue is spatially fixed relative to the others in a unique three-dimensional structure. If we assume that the spatial orientation of each residue can be described by three geometric parameters (say three rotational degrees of freedom about the covalent bonds), then we need to specify $3N$ internal geometric parameters. We further assume that each geometric parameter can have only two possible values, such as two angles (a gross underestimate). Therefore, the total possible conformations for the protein is given by 2^{3N}, or 2^{300} for a protein containing 100 residues. If the protein were to randomly sample all of the conformations at the rate that single bonds can reorient, which is about 10^{13} s^{-1}, then the time it would take for the protein to explore all the conformations is $2^{300}/10^{13} \text{ s}^{-1} = 2 \times 10^{77}$ s, or 6×10^{69} years. Considering that the age of the universe is only about 10 billion years (10^{10} years), this is an almost unimaginably long time!

For our study of protein folding, it is useful to compare the relative stability of the native conformation with that of denatured states (Figure 16.31). Thermodynamic measurements indicate that the native protein is only 20 to 40 kJ mol^{-1} more stable than the denatured one. This surprisingly small value has the same order of magnitude as a typical hydrogen bond. The fact that native proteins are only marginally stable (compared to denatured ones) enables them to change their conformations rapidly, as in allosteric interaction (see Section 10.6). Proteins that can change their conformations easily can also diffuse across cell membranes more readily.

Many small proteins (fewer than 100 residues) do fold according to the two-state model; that is, they do not go through an intermediate state.

Most proteins fold to their native conformations in a matter of seconds or less.* Therefore, the folding process must go through a kinetic pathway of unstable intermediates to avoid sampling a huge number of irrelevant conformations. In many cases, the initial stage of folding is the collapse of the flexible disordered polypeptide chain into a partly organized structure, called the *molten globule (MG)* (Figure 16.32). The molten globule has most of the secondary structure of the native protein and may even possess α helices and β sheets. It is less compact than the native structure, however, and the interior side chains may be mobile. Some proteins fold through one major pathway (one molten globule) whereas others may fold through multiple pathways (two or more molten globules).

It is instructive to probe the mechanism of protein folding in some detail. The initial step is believed to be bringing the hydrophobic residues out of contact with water and into contact with one another. This step, which is accompanied by a large decrease in Gibbs energy, vastly reduces the number of possible conformations that need to be sampled. This partial folding also results in the formation of α helices and

---

*In Anfinsen's experiment, the refolding of ribonuclease to the native conformation took about 10 hours. In the presence of the enzyme *protein disulfide isomerase*, which catalyzes the exchange of disulfide linkages, the time for renaturation was reduced to about two minutes. Keep in mind that these are *in vitro* experiments.

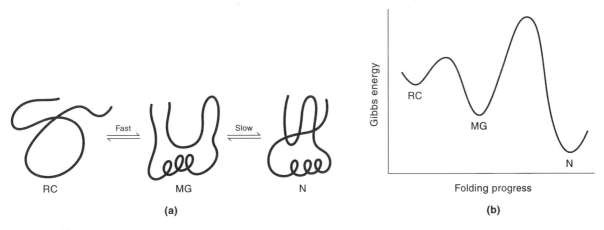

Figure 16.32
(a) The folding of a protein involving a molten globule as an intermediate. (b) Gibbs energy diagram for the same folding process. RC = random coil; MG = molten globule; and N = native protein.

β sheets. Thus, hydrogen bonding in these secondary structures is a consequence and not a driving force in forming the molten globule. During folding, the enzyme *protein disulfide isomerase* catalyzes disulfide exchange to remove intermediates with incorrectly formed disulfide linkages. Earlier we saw that the *trans* configuration is more stable for the peptide groups (see p. 617). In fact, the relative stability is about 1000 to 1, favoring the *trans* form. On the other hand, *cis*-proline is only four times less stable than *trans*-proline. Thus, although most proline residues in proteins have the *trans* configuration, the *cis* form is also known to exist. In the unfolded state, there is an equilibrium between these two isomers. When the polypeptide chain folds, a fraction of the molecule may therefore have one or more proline in the wrong configuration. Unaided *cis–trans* isomerization occurs too slowly for corrective measure; the rate of this process is enhanced by the enzyme *peptidyl propyl isomerase*.

Small polypeptides generally fold spontaneously in the test tube, but within the cell, efficient folding of many newly synthesized (nascent) proteins is thought to depend on a class of special molecules called *molecular chaperones*. Molecular chaperones are proteins that assist in the noncovalent assembly of protein-containing structures *in vivo*. Unfolded proteins formed in the complex and highly concentrated intracellular environment contain many hydrophobic regions exposed to water. Consequently, they have a great tendency to aggregate, either among themselves or with other components, leading to precipitation or incorrectly folded structures, or both. Like their human counterparts, molecular chaperones inhibit inappropriate interactions among the nascent proteins and facilitate more favorable association.

Only a few molecular chaperones tend to assist in the folding and assembly of numerous different proteins; therefore, their binding to proteins seems to be fairly nonspecific. No evidence suggests that molecular chaperones catalyze the correct folding of the bound proteins. Rather, their main function seems to be sequestering the unfolded polypeptide to keep it from aggregating. Release of the protein from a molecular chaperone is energy dependent and requires ATP hydrolysis.

Experimental Study of Protein Folding. A number of physical techniques and chemical methods enable us to study protein structures in considerable detail. Several branches of spectroscopy, discussed in Chapter 14, as well as ORD and CD, are particularly useful in monitoring conformational changes in proteins. For example,

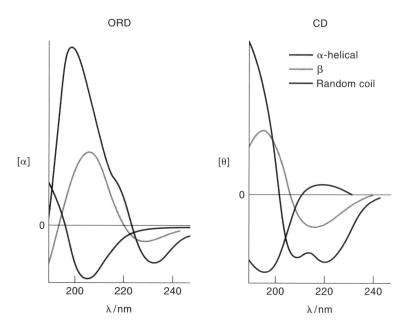

Figure 16.33
ORD and CD spectra for poly-L-lysine in the α-helical, β, and random coil conformations. [From N. Greenfield and G. Fasman, *Biochemistry* **8**, 4108 (1969).]

the ORD and CD spectra of a polypeptide are distinctly different for different conformations (Figure 16.33). We shall briefly discuss three cases to illustrate the approaches employed by chemists to study protein folding.

CASE 1. CYTOCHROME *c*. Cytochrome *c* is an electron-transfer protein. Initially, the protein exists in the unfolded state in a 4.3 *M* guanidine hydrochloride (the denaturant) solution. In a stopped-flow CD experiment, the protein solution is rapidly mixed with a buffer solution to lower the concentration of the denaturant, and the rate of folding is monitored at 222 nm (for α helix) and 289 nm (for aromatic side chains). Measurements at 222 nm show that 44% of the total change associated with refolding occurs within the dead time (the time required for mixing) of the stopped-flow experiment, indicating that a significant amount of helical secondary structure is formed in less than 4 ms. The remaining changes in the ellipticity (see p. 567) at this wavelength occur in about 40 ms and 0.7 s, respectively. The aromatic CD band monitored at 289 nm, which is indicative of the formation of a tightly packed core, only begins to appear in a 400-ms step and is completed in a final 10-s phase.*

CASE 2. BOVINE PANCREATIC TRYPSIN INHIBITOR. This protein inactivates trypsin (a digestive enzyme) in the pancreas to prevent it from digesting the organ before it is secreted. It is a 58-residue protein with three disulfide bonds. In the presence of a reducing agent, all the disulfide linkages are converted to the sulfhydryl (–SH) groups, and the protein exists in the unfolded form. Folding is initiated by removal of the denaturant, which enables the –S–S– bonds to reform. As the protein folds, intermediates that have disulfide bonds that do not exist in the final molecule appear and disappear. The folding pathway can be tracked by trapping the intermediates at different time intervals with iodoacetate, which can block the sulfhydryl

---

*G. Elöve, A. F. Chaffotte, H. Roder, and M. E. Goldberg, *Biochemistry* **31**, 6879 (1992).

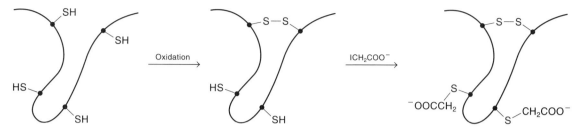

Figure 16.34
The trapping of sulfhydryl groups by iodoacetate during protein folding.

groups (Figure 16.34):

$$-SH + ICH_2COO^- \rightarrow -SCH_2COO^- + HI$$

The intermediates are separated and analyzed. Knowing the positions of the –SH groups, chemists are able to get a clear picture of the folding sequence.*

CASE 3. RIBONUCLEASE. The denatured protein is placed in heavy water (D_2O), so that all the labile H atoms are replaced by D atoms. Then the denaturant is removed. As folding proceeds, what would have been hydrogen bonds ($>N-H \cdots O-$) become "deuterium" bonds ($>N-D \cdots O-$). At some point, ordinary water (H_2O) is added to the sample, and any deuterium atoms that are not protected—that is, those that are exposed to the solvent—are replaced by H atoms. Using NMR spectroscopy, chemists can detect the individual NH resonances and hence identify the regions of the compact molecule that contain protected deuterium. In this way, they can determine which parts folded before the others. By varying the time at which they add H_2O, they also learn the order in which several different intermediates form. Results with ribonuclease show that the β-sheet part of the enzyme, found in the middle of the molecule, forms early.

---

*T. E. Creighton, *Biochem. J.* **270**, 12 (1990).

Suggestions for Further Reading

BOOKS

Blake, R., *The Informational Biopolymers of Genes and Gene Expression*, University Science Books, Sausalito, CA, 2005.

Bloomfield, V. A. and R. E. Harrington, Eds. *Biophysical Chemistry*, W. H. Freeman, San Francisco, 1975.

Branden, C. and J. Tooze, *Introduction to Protein Structure*, Garland Publishing, New York, 1999.

Calladine, C. and H. Drew, *Understand DNA*, 2nd ed., Academic Press, New York, 1997.

Cantor, C. R. and P. R. Schimmel, *Biophysical Chemistry*, W. H. Freeman, New York, 1980.

Creighton, T. E., *Proteins: Structures and Molecular Properties*, 2nd ed., W. H. Freeman, New York, 1993.

Fersht, A., *Structure and Mechanism in Protein Science*, W. H. Freeman, New York, 1999.

Freifelder, D., *Physical Biochemistry: Applications to Biochemistry and Molecular Biology*, W. H. Freeman, New York, 1982.

Kyte, J., *Structure in Protein Chemistry*, Garland Publishing, New York, 1994.

Perutz, M. F., *Protein Structure: New Approaches to Disease and Therapy*, W. H. Freeman, New York, 1992.

Robyt, J. F. and B. J. White, *Biochemical Techniques: Theory and Practice*, Waveland Press, Prospect Heights, IL, 1987.

Saenger, W., *Principles of Nucleic Acid Structure*, Springer-Verlag, New York, 1984.

Schultz, G. E. and R. H. Schirmer, *Principles of Protein Structure*, Springer-Verlag, New York, 1979.

ARTICLES

General

"How Giant Molecules Are Measured," P. J. W. Debye, *Sci. Am.* September 1957.

"The Viscosity of Macromolecules in Relation to Molecular Conformation," J. T. Yang, *Adv. Protein Chem.* **16**, 323 (1961).

"Average Quantities in Colloid Science," J. T. Bailey, W. H. Beattie, and C. Booth, *J. Chem. Educ.* **39**, 196 (1962).

"Errors in Representing Structures of Proteins and Nucleic Acids," R. A. Day and E. J. Ritter, *J. Chem. Educ.* **44**, 761 (1967).

"Hydrophobic Interaction," G. Némethy, *Angew. Chem. Int. Ed.* **6**(3), 195 (1967).

"Conformation of Peptides," S. Lande, *J. Chem. Educ.* **45**, 587 (1968).

"Molecular Weight Distributions of Polymers," A. Rudin, *J. Chem. Educ.* **46**, 595 (1969).

"Conformation of Macromolecules," D. H. Napper, *J. Chem. Educ.* **46**, 305 (1969).

"Polymer Models," C. E. Carraher, Jr., *J. Chem. Educ.* **47**, 581 (1970).

"Some Stereochemical Principles from Polymers," C. C. Price, *J. Chem. Educ.* **50**, 744 (1973).

"SDS-Polyacrylamide Gel Electrophoresis," J. Svasti and B. Panijpan, *J. Chem. Educ.* **54**, 560 (1977).

"Macromolecular Crystals," A. McPherson, *Sci. Am.* March 1989.

"A Polymer Viscosity Experiment With No Right Answers," L. C. Rosenthal, *J. Chem. Educ.* **67**, 78 (1990).

"Visualizing Biological Molecules," A. J. Olson and D. S. Goodsell, *Sci. Am.* November 1992.

"Viscosity and Shapes of Macromolecules," J. L. Richards, *J. Chem. Educ.* **70**, 685 (1993).

"Molecular Properties of Polymers," C. M. Guttman and B. Fanconi in *Encyclopedia of Applied Physics*, G. L. Trigg, Ed., VCH Publishers, New York, 1996, Vol. 14, p. 549.

DNA

"DNA Helix to Coil Transition: A Simplified Model," G. L. Baker, *Am. J. Phys.* **44**, 599 (1976).

"The Buoyant Density of DNA and the G + C Content," B. Panijpan, *J. Chem Educ.* **54**, 172 (1977).

"DNA Topoisomerases," J. C. Wang, *Sci. Am.* July 1982.

"The DNA Helix and How It is Read," R. E. Dickerson, *Sci. Am.* December 1983.

"DNA," G. Felsenfeld, *Sci. Am.* October 1985.

"Supercoiled DNA: Biological Significance," R. R. Sinden, *J. Chem. Educ.* **64**, 294 (1987).

"The Unusual Origin of the Polymerase Chain Reaction," K. B. Mullis, *Sci. Am.* April 1990.

"Conformational Analysis," S. S. Zimmerman in *Encyclopedia of Applied Physics*, G. L. Trigg, Ed., VCH Publishers, New York, 1992, Vol. 4, p. 229.

"Nucleic Acids," M. D. Frank-Kamenetskii in *Encyclopedia of Applied Physics*, G. L. Trigg, Ed., VCH Publishers, New York, 1995, Vol. 12, p. 19.

"Computing with DNA," L. M. Adleman, *Sci. Am.* August 1998.

Proteins

"Collagen," J. Gross, *Sci. Am.* May 1961.

"Three-Dimensional Structure of a Protein," J. C. Kendrew, *Sci. Am.* December 1961. (Myoglobin)

"The Hemoglobin Molecule," M. F. Perutz, *Sci. Am.* November 1964.

"The Three-Dimensional Structure of an Enzyme Molecule," D. C. Phillips, *Sci. Am.* November 1966. (Lysozyme)

"Keratins," R. D. B. Fraser, *Sci. Am.* August 1969.

"The Structure and History of an Ancient Protein," R. E. Dickerson, *Sci. Am.* April 1972. (Cytochrome *c*)

"Principles That Govern the Folding of Protein Chains," C. B. Anfinsen, *Science* **181**, 223 (1973).

"Cyanate and Sickie-Cell Disease," A. Cerami and C. M. Peterson, *Sci. Am.* April 1975.

"Amino Acid Sequence Diversity in Proteins," D. Blackman, *J. Chem. Educ.* **54**, 170 (1977).

"Electrostatic Effects in Proteins," M. F. Perutz, *Science* **201**, 1187 (1978).

"Cytochrome *c* and the Evolution of Energy Metabolism," R. E. Dickerson, *Sci. Am.* March 1980.

"Proteins," R. F. Doolittle, *Sci. Am.* April 1985.

"The Dynamics of Proteins," M. Karplus and J. A. McCammon, *Sci. Am.* April 1986.

"How Does Protein Folding Get Started?" R. L. Baldwin, *Trends Biochem. Sci.* **14**, 291 (1989).

"Effect of 'Crowding' in Protein Solutions," G. B. Ralston, *J. Chem. Educ.* **67**, 857 (1990).

"The Protein Folding Problem," F. M. Richards, *Sci. Am.* January 1991.

"Protein Dynamics," G. U. Nienhaus and R. D. Young in *Encyclopedia of Applied Physics*, G. L. Trigg, Ed., VCH Publishers, New York, 1996, Vol. 15, p. 163.

"How Protein Chemists Learned About the Hydrophobic Effect," C. Tanford, *Protein Science*, **6**, 1358 (1997).

"Simulating Water and the Molecules of Life," M. Gerstein and M. Levitt, *Sci. Am.* November 1998.

"Consideration of Lewis Acidity in the Context of Heme Biochemistry: A Molecular Visualization Exercise," T. E. Elgren, *Chem. Educator* **1998** 3(3): S1430-4171(98)03206-7. Avail. URL: http://journals.springer-ny.com/chedr.

"The Evolution of Hemoglobin," R. Hardison, *Am. Sci.* **87**, 126 (1999).

"Is Protein Folding Hierarchic? I. Local Structure and Peptide Folding," R. L. Baldwin and G. D. Rose, *Trends Biochem. Sci.* **24**, 26 (1999).

"Is Protein Folding Hierarchic? II. Folding Intermediates and Transition States," R. L. Baldwin and G. D. Rose, *Trends Biochem. Sci.* **24**, 77 (1999).

"Protein Unfolding of Metmyoglobin Monitored by Spectroscopic Techniques," C. M. Jones, *Chem. Educator* **1999** 4(3): S1430-4171(99)032998-5. Avail. URL: http://journals.springer-ny.com/chedr.

"An Unfolding Story," A. P. Capaldi and S. E. Radford, *Trends Biochem. Sci.* **26**, 753 (2001).

"Membrane Proteins Shaping Up," C. N. Chiu, G. von Heijne, and J. W. L. de Gier, *Trends Biochem. Sci.* **27**, 231 (2002).

"The Chaperon Folding Machine," H. R. Saibil and N. A. Ranson, *Trends Biochem. Sci.* **27**, 627 (2002).

"Protein Structures: From Famine to Feast," H. M. Berman, D. S. Goodsell, and P. E. Bourne, *Am. Sci.* **90**, 350 (2002).

"Protein Folding and Misfolding," J. King, C. Hasse-Pettingell, and D. Gossard, *Am. Sci.* **90**, 445 (2002).

"Is There a Unifying Mechanism for Protein Folding?," V. Daggett and A. R. Fersht, *Trends Biochem. Sci.* **28**, 18 (2003).

"Disulfide Bonds as Switches for Protein Function," P. J. Hogg, *Trends Biochem. Sci.* **28**, 210 (2003).

Problems

Molar-Mass Determination

16.1 A polydisperse solution has the following distribution:

Number of Molecules	Molar Mass $g \cdot mol^{-1}$
10	25,000
7	17,000
24	31,000
16	49,000

Calculate the values of both $\bar{\mathcal{M}}_n$ and $\bar{\mathcal{M}}_w$ and the polydispersity of solution. (Polydispersity is defined as $\bar{\mathcal{M}}_w / \bar{\mathcal{M}}_n$.)

16.2 Ceruloplasmin is a protein present in the blood plasma. It contains 0.33% copper by weight. **(a)** Calculate its minimum molar mass. **(b)** The actual molar mass of ceruloplasmin is 150,000 g mol^{-1}. How many copper atoms does each protein molecule contain?

16.3 Depending on experimental conditions, the measurement of the molar mass of hemoglobin in an aqueous solution may show that the solution is monodisperse or polydisperse. Explain.

16.4 An ultracentrifuge is spinning at 60,000 rpm. **(a)** Calculate the value of ω in radians s^{-1}. **(b)** Calculate the centrifugal acceleration, a, given by $\omega^2 r$, at a point 7.4 cm from the center of rotation. **(c)** How many "g's" is this acceleration equivalent to?

16.5 How does the sedimentation coefficient, s, depend on the mass of a protein (assumed to be spherical)? Compare the rates of sedimentation of two proteins with molar masses of 70,000 g and 35,000 g, respectively.

16.6 A protein with $\bar{v} = 0.74$ mL g^{-1} is sedimented in water at 20°C. If $s_{20,w} = 3.0 \times 10^{-13}$ s and $D = 1.5 \times$ 10^{-6} cm^2 s^{-1}, what is the molar mass of the protein? The density of the solution is 0.998 g mL^{-1}.

16.7 In a sedimentation equilibrium experiment carried out at 293 K, the following data were obtained for a certain protein molecule: $\omega = 19,000$ rpm, $s = 2.15 \times 10^{-13}$ s, $\bar{v} = 0.71$ mL g^{-1}, and $\rho = 1.1$ g mL^{-1}. The relative concentrations at distances r_1 and r_2 from the center of rotation are $c_1 = 4.72$ ($r_1 = 5.95$ cm) and $c_2 = 12.98$ ($r_2 = 6.23$ cm). What is the molar mass of the protein?

Viscosity and Electrophoresis

16.8 What are the units for the various viscosities defined in Equations 16.19 to 16.22?

16.9 Will dissolving 1×10^{-3} g of glucose result in a greater relative viscosity in water or in a 10% glycerol solution?

16.10 The intrinsic viscosity of ribonuclease is 3.4 at 20°C and 6 at 50°C. What can you say about the change in structure?

16.11 Show that the units for zeE in Equation 16.25 are those for force.

16.12 At pH 6.5, the electrophoretic mobility of carboxyhemoglobin is 2.23×10^{-5} cm^2 s^{-1} V^{-1}, and that of sickle-cell carboxyhemoglobin is 2.63×10^{-5} cm^2 s^{-1} V^{-1}. Calculate how long it will take to separate these two proteins by 1.0 cm if the potential gradient is 5.0 V cm^{-1}.

16.13 In an electrophoretic study of an aqueous protein solution, two species were found with molar masses of 60,000 and 30,000, respectively. The solution contains 1.85% protein by weight. If the fraction of the larger protein is found to be 70%, calculate the values of $\bar{\mathcal{M}}_w$ and $\bar{\mathcal{M}}_n$.

16.14 The relative electrophoretic mobilities of several protein–SDS complexes in a polyacrylamide gel are as follows:

Protein	Molar Mass $g \cdot mol^{-1}$	Relative Mobility
Myoglobin	17,200	0.95
Trypsin	23,300	0.82
Aldolase	40,000	0.59
Fumarase	49,000	0.50
Carbonic anhydrase	29,000	0.73

Plot log (molar mass) versus relative mobility. The relative mobility of creatine kinase is 0.60. What is its molar mass? Compare your result with the molar mass of 80,000 obtained by ultracentrifugation. What conclusions can you draw?

16.15 The extent to which a protein molecule behaves like a spherical molecule can be tested by the frictional ratio, f/f_0, where f_0 is the frictional coefficient in Stokes law (Equation 16.7), and f is the frictional coefficient obtained from the diffusion coefficient (Equation 16.9). For spherical molecules, $f/f_0 = 1$; deviations from unity can be used as a measure of the nonspherical shape of the molecule. Consider myoglobin and human fibrinogen (molar mass 339,700, $s = 7.63 \times 10^{-13}$ s, $D = 1.98 \times 10^{-7}$ cm^2 s^{-1}, and $\bar{v} = 0.725$ mL g^{-1}). What conclusions can you draw about the shape of the molecules? (The radius of an assumed spherical molecule, r, can be obtained from the equation $\mathcal{M} = 4\pi N_A r^3 / 3\bar{v}$, where $\mathcal{M}$ is the molar mass.) Assume $T = 298$ K and the viscosity of water is 0.00101 N s m^{-2}.

Proteins and DNA

16.16 Referring to Figure 16.28, state whether the β structure in ribonuclease is parallel or antiparallel.

16.17 How does hydrophobic interaction differ from both covalent and noncovalent bonds? What role does it play in protein structure and stability?

16.18 The average pitch (the distance between successive turns) of an α helix is 5.4 Å. Assuming this pitch to be the same for human hair and that hair grows at the rate of 0.6 inch month^{-1}, how many turns of the α helix are generated each second? (Assume 1 month = 30 days.)

16.19 Hair contains keratins made of α helixes coiled to form a superhelix. The disulfide bonds linking the α helixes together are largely responsible for the shape of the hair. Based on this information, explain how "permanent waves" are formed.

16.20 The α helical structure of poly-L-lysine is formed at pH 10, whereas that of the random coil is formed at pH 7. Account for the pH-dependent structural change.

16.21 Proteases such as trypsin and carboxypeptidase catalyze the hydrolysis of peptide bonds of proteins (as in digestion). Explain how such an enzyme might bind to the substrate protein so that its main chain becomes fully extended in the vicinity of the targeted peptide bond. What kind of structure would it resemble?

16.22 Proteins denatured by compounds such as urea or SDS can be renatured when the denaturants are removed by dialysis. On the other hand, thermal denaturation of proteins is often irreversible. Explain. (*Hint:* Consider the *trans* configuration of the peptide bond.)

16.23 Proteins generally have widely different structures, whereas nucleic acids have quite similar structures. Explain.

16.24 A compact disc (CD) stores about 4.0×10^9 bits of information. This information is stored as a binary code; that is, every bit is either 0 or 1. **(a)** How many bits would it take to specify each nucleotide pair in a DNA sequence? **(b)** How many CDs would it take to store the information in the human genome, which consists of 3×10^9 nucleotide pairs?

16.25 Referring to Figure 16.26, if after one generation, the DNA sample was thermally denatured and then the sample was annealed, how many bands would appear in the CsCl gradient?

16.26 The following T_m data were obtained for double-stranded DNA in certain buffer solutions:

Sample	Percent (C + G)	$T_m/°C$
1	40.0	86.6
2	49.0	90.0
3	62.0	95.0
4	71.0	98.4

(a) Derive an equation relating the percent (C + G) to T_m. **(b)** Calculate the percent (C + G) content for a sample whose $T_m = 88.3°C$.

16.27 The enthalpy change in the denaturation of a certain protein is 125 kJ mol^{-1}. If the entropy change is 397 J K^{-1} mol^{-1}, calculate the minimum temperature at which the protein would denature spontaneously.

16.28 Consider the formation of a dimeric protein

$$2P \rightarrow P_2$$

At 25°C, we have $\Delta_r H° = 17$ kJ mol^{-1} and $\Delta_r S° = 65$ J K^{-1} mol^{-1}. Is the dimerization favored at this temperature? Comment on the effect of lowering the temperature. What general conclusion can you draw about the so-called cold labile enzymes?

16.29 The cause and properties of sickle-cell hemoglobin (HbS) were discussed in Chapter 13. Under certain conditions, the aggregates of the HbS molecules formed at body temperature break apart as the temperature is lowered. Explain.

16.30 In this chapter, an expression was derived showing that the number of ways N sulfhydryl groups can form a linkage is $(N-1)(N-3)(N-5)\cdots 1$ if N is even. Derive an expression if N is odd.

16.31 What assumptions must be made in the study of protein folding using the hydrogen–deuterium exchange technique?

16.32 A denatured protein contains 10 cysteine residues. Upon oxidation, what fraction of the molecules will randomly form the correct linkages if the native protein has **(a)** five disulfide bonds, and **(b)** three disulfide bonds?

16.33 As mentioned in the chapter, the *trans* configuration is more stable (about 1000 times) than the *cis* configuration for the amide group. The only exception occurs for the proline residue. In this case the *cis* configuration is only about four times less stable than the *trans* configuration. For this reason, *cis*-prolines are found in some polypeptide chains. Sketch the *cis* and *trans* amide groups in which one of the C_α atoms is part

of the proline residue and show that the difference in steric interactions between these two configurations is not as pronounced as for other residues.

16.34 At what pH would you expect the conformation of poly-L-glutamate to be random coil? α helix?

16.35 An effective remedy to deodorize a dog that has been sprayed by a skunk is to rub the affected areas with a solution of hydrogen peroxide and sodium bicarbonate. What is the chemical basis for this action? (*Hint:* An odiferous component of a skunk's secretion is 2-butene-1-thiol, $CH_3CH=CHCH_2SH$.)

16.36 In density-gradient sedimentation experiments, EDTA is usually added to the DNA solution. Without EDTA, the results may vary from one experiment to another. Explain.

Review of Mathematics

This appendix will briefly review some of the basic equations and formulas that are useful in physical chemistry.

Exponents and Powers

Many numbers are more conveniently expressed as powers of 10. For example,

$$1 = 10^0$$

$$0.1 = 10^{-1}$$

$$0.00023 = 2.3 \times 10^{-4}$$

$$100 = 10^2$$

$$100,000 = 10^5$$

$$3.1623 = 10^{0.5}$$

In general, we write a^n, where a is called the *base* and n the *exponent*. This expression is read as "a to the power of n." The following relations are useful:

Operation	Example
$a^m \times a^n = a^{m+n}$	$10^{0.2} \times 10^3 = 10^{3.2}$
$(a^m)^n = a^{m \times n}$	$(10^4)^2 = 10^8$
$\dfrac{a^m}{a^n} = a^{m-n}$	$\dfrac{10^3}{10^7} = 10^{-4}$

Note that a^0 (a to the power of zero) is equal to unity for all values of a except for $a = 0$; that is, $0^n = 0$ (for all values of n). Further, we have $1^n = 1$ for all values of n.

Logarithm. The concept of logarithm is a natural extension of exponents. The logarithm to the base a of a number x is equal to the exponent y to which the base number a must be raised so that $x = a^y$. Thus, if

$$x = a^y$$

then

$$y = \log_a x$$

For example, because $3^4 = 81$, we have

$$4 = \log_3 81$$

Similarly, for logarithm to the base 10, we write

Logarithm	Exponent
$\log_{10} 1 = 0$	$10^0 = 1$
$\log_{10} 2 = 0.3010$	$10^{0.301} = 2$
$\log_{10} 10 = 1$	$10^1 = 10$
$\log_{10} 100 = 2$	$10^2 = 100$
$\log_{10} 0.1 = -1$	$10^{-1} = 0.1$

The logarithm to the base 10 is called the *common logarithm*. By convention, we use the notation $\log a$ instead of $\log_{10} a$ to denote the common logarithm of a.

Because the logarithms of numbers are exponents, they have the same properties as exponents. For simplicity, we express the following relations in terms of common logarithms:

Logarithm	Exponent
$\log AB = \log A + \log B$	$10^A \times 10^B = 10^{A+B}$
$\log \dfrac{A}{B} = \log A - \log B$	$\dfrac{10^A}{10^B} = 10^{A-B}$
$\log A^n = n \log A$	

Logarithms taken to the base e are known as *natural logarithms*. The quantity e is a number given by

$$e = 1 + \frac{1}{1!} + \frac{1}{2!} + \frac{1}{3!} + \cdots$$

$$= 2.71828182845\ldots$$

$$\approx 2.7183$$

where the symbol ! is called "factorial." For example, 3! is given by $3 \times 2 \times 1$ and by definition, $0! = 1$.

In physical chemistry, the exponential function $y = e^x$ is of great importance. Taking the natural logarithm on both sides, we get

$$\ln y = x \ln e = x$$

where "ln" represents $\log_e$. The relation between natural logarithm and common logarithm is as follows. We start with the equation

$$y = e^x$$

Taking the common logarithm on both sides, we obtain

$$\log y = x \log e$$

$$= \ln y \log e$$

because $x = \ln y$. Now $\log e = \log 2.7183 = 0.4343$; thus,

$$\log y = 0.4343 \ln y$$

or

$$2.303 \log y = \ln y$$

Simple Equations

Linear Equation.   A linear equation is represented by

$$y = mx + b$$

A plot of y versus x gives a straight line with slope m and an intercept (on the y axis; that is, at $x = 0$) b.

Quadratic Equation.   A quadratic equation takes the form

$$y = ax^2 + bx + c$$

where a, b, and c are constants and $a \neq 0$. A plot of y versus x gives a parabola.
   Let us consider a particular quadratic equation,

$$y = 3x^2 - 5x + 2$$

A plot of y versus x is shown in Figure 1. The curve intercepts the x axis $(y = 0)$ twice at $x = 1$ and $x = 0.67$. Alternatively, we can solve the equation as follows. By setting the equation to be zero (that is, $y = 0$), we get

$$3x^2 - 5x + 2 = 0$$

$$x = \frac{-b \pm \sqrt{b^2 - 4ac}}{2a}$$

$$= \frac{5 \pm \sqrt{25 - 4 \times 3 \times 2}}{2 \times 3}$$

$$= 1.00 \text{ or } 0.67$$

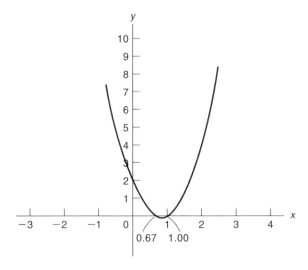

Figure 1

Mean Values

If we repeat a measurement of an experiment, we often obtain a value that is different from the previous reading, and it is appropriate to represent the result as a mean of these two numbers. The most common mean value is the *arithmetic mean*. For two readings a and b, the arithmetic mean is given by $(a+b)/2$. There are occasions when the readings do not vary randomly. In such cases, we may try the *geometric mean*. The geometric mean of two numbers a and b is given by $\sqrt{ab}$.

Series and Expansions

Arithmetic Series

$$1, 2, 3, 4, \ldots$$

or

$$a, 2a, 3a, 4a, \ldots$$

Geometric Series

$$1, 2, 4, 8, \ldots$$

or

$$a, 2a, 4a, 8a, \ldots$$

Binomial Expansion

$$(1+x)^n = 1 + nx + \frac{n(n-1)}{2!}x^2 + \frac{n(n-1)(n-2)}{3!}x^3 + \cdots$$

Exponential Expansion

$$e^{\pm x} = 1 \pm \frac{x}{1!} \pm \frac{x^2}{2!} \pm \frac{x^3}{3!} \pm \cdots$$

$$e^{\pm ax} = 1 \pm \frac{ax}{1!} \pm \frac{(ax)^2}{2!} \pm \frac{(ax)^3}{3!} \pm \cdots$$

Trigonometric Expansions

$$\sin x = x - \frac{x^3}{3!} + \frac{x^5}{5!} - \frac{x^7}{7!} + \cdots$$

$$\cos x = 1 - \frac{x^2}{2!} + \frac{x^4}{4!} - \frac{x^6}{6!} + \cdots$$

Logarithmic Expansion

$$\ln(1+x) = x - \frac{x^2}{2} + \frac{x^3}{3} - \frac{x^4}{4} + \cdots$$

Angles and Radians

The common unit of angular measure is the *degree*, which is defined as $\frac{1}{360}$ of a complete circle. Often in physical chemistry, we find it more convenient to use another unit, called the *radian* (rad). The relation between angle and radian can be understood as follows. Consider a certain portion of the circumference of a circle of radius r. The length of the arc (s) is proportional to the angle θ and the radius r, so that

$$s = r\theta$$

where θ is measured in radians. Thus, 1 radian is defined to be the angle subtended when the arc length, s, is exactly equal to the radius.

If we consider the entire circle as the arc, then

$$s = 2\pi r = r\theta$$

or

$$2\pi = \theta$$

This means that $\theta = 2\pi$ radians corresponds to $\theta = 360°$. Thus,

$$1 \text{ rad} = \frac{360°}{2\pi} \approx \frac{360°}{2 \times 3.1416} = 57.3°$$

On the other hand,

$$1° = \frac{2\pi}{360°} \approx \frac{2 \times 3.1416}{360°} = 0.0175 \text{ rad}$$

Keep in mind that although the radian is a unit of angular measure, it does not have physical dimensions. For example, the circumference of a circle of radius 5 cm is given by $2\pi(\text{rad}) \times 5$ cm $= 31.42$ cm.

Areas and Volumes

Triangle.   Consider a triangle with sides a, b, and c and height h (with side a as base). The semiperimeter s is given by

$$s = \frac{a+b+c}{2}$$

The area (A) of the triangle is

$$A = \tfrac{1}{2}ah = \sqrt{s(s-a)(s-b)(s-c)}$$
$$= \tfrac{1}{2}ab \sin C$$

where angle C is opposite side c. If a, b, and c are the sides of a right-angled triangle, c being the hypotenuse, then

$$c^2 = a^2 + b^2$$

which is the Pythagorean theorem.

Rectangle.   The area of a rectangle of sides a and b is ab.

Parallelogram.   The area of a parallelogram of sides a and b is ah, where h is the perpendicular distance between the two sides whose lengths are a.

Circle.   The circumference of a circle is $2\pi r$, and the area of the circle is πr^2, where r is the radius.

Sphere.   The area of the curved surface of a sphere of radius r is $4\pi r^2$, and the volume of the sphere is $\frac{4}{3}\pi r^3$.

Cylinder.   The area of the curved surface of a cylinder of radius r and length h is $2\pi rh$, and the volume of the cylinder is $\pi r^2 h$.

Cone.   The area of the curved surface of a cone is πrl, where r is the radius of the base and l is the slant height. The volume of the cone is $\frac{1}{3}r^2 h$, where h is the vertical height (from the apex to the base).

Operators

In Section 11.7, we mentioned the use of operators. An operator is a mathematical symbol that tells us specifically what to do to a number or a function. Some examples of operators are as follows:

Operator	Function or Number	Final Form
log	24.1	$\log 24.1 = 1.382$
$\sqrt{}$	974.2	$\sqrt{974.2} = 31.21$
sin	61.9°	$\sin 61.9° = 0.882$
cos	x	$\cos x$
$\dfrac{d}{dx}$	e^{kx}	$\dfrac{de^{kx}}{dx} = ke^{kx}$

Differential and Integral Calculus

Functions of Single Variables.   The following are derivatives of some common functions.

$y = f(x)$	dy/dx
x^n	nx^{n-1}
e^x	e^x
e^{kx}	ke^{kx}
$\sin x$	$\cos x$
$\sin(ax+b)$	$a\cos(ax+b)$
$\cos x$	$-\sin x$
$\cos(ax+b)$	$-a\sin(ax+b)$
$\ln x$	$1/x$
$\ln(ax+b)$	$\dfrac{a}{ax+b}$

Some Useful Integrals

$$\int x^n \, dx = \frac{1}{n+1}x^{n+1} + C \qquad \int \cos x \, dx = \sin x + C$$

$$\int \frac{dx}{x} = \ln x + C \qquad \int \ln x \, dx = x \ln x - x + C$$

$$\int \frac{dx}{ax + b} = \frac{1}{a} \ln(ax + b) + C \qquad \int e^x \, dx = e^x + C$$

$$\int \sin x \, dx = -\cos x + C \qquad \int e^{kx} \, dx = \frac{e^{kx}}{k} + C$$

Because all these integrals are indefinite integrals, a constant term C must be added to the results.

Total and Partial Differentiation

If y is a function of x, then the derivative of $y(x)$ at some value of x is defined as

$$\frac{dy}{dx} = \lim_{\Delta x \to 0} \frac{y(x + \Delta x) - y(x)}{\Delta x}$$

The quantity dy/dx, called the total derivative, gives the rate of change of y with respect to x—that is, how fast y changes as x is changing at a point on the y versus x plot.

In thermodynamics, we often deal with functions of two or more variables. For example, the pressure P of an ideal gas depends on temperature, volume, and number of moles ($P = nRT/V$). We can express P in terms of the three independent variables as

$$P = f(n, T, V)$$

Consider a function z of two independent variables, x and y:

$$z = f(x, y)$$

This function can be differentiated with respect to x while keeping y constant:

$$\left(\frac{\partial z}{\partial x} \right)_y = \left[\frac{\partial f(x, y)}{\partial x} \right]_y$$

Thus, if

$$z = 3x^2 - 4xy + 6y^2 \tag{1}$$

then

$$\left(\frac{\partial z}{\partial x} \right)_y = 6x - 4y$$

Note that we use the symbol "∂" to denote partial differentiation, and the subscript y reminds us that y is kept constant.

Similarly, the partial derivative of z with respect to y at constant x is given by

$$\left(\frac{\partial z}{\partial y} \right)_x = -4x + 12y$$

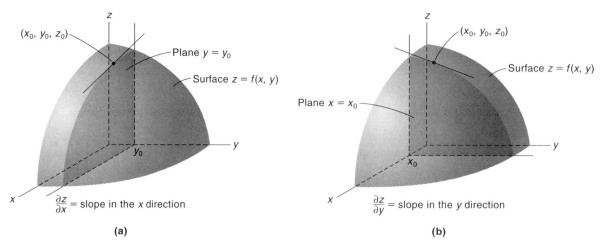

Figure 2
Geometric interpretation of partial derivatives. (a) $(\partial z/\partial x)_y$. For each fixed number, y_0, the points (x, y_0, z) form a vertical plane whose equation is $y = y_0$. If $z = f(x, y)$ and if y is kept fixed at $y = y_0$, then the corresponding points $[x, y_0, f(x, y_0)]$ form a curve in three-dimensional space that is the intersection of the surface $z = f(x, y)$ with the plane $y = y_0$. At each point on this curve, the partial derivative $(\partial z/\partial x)$ is simply the slope of the line in the plane $y = y_0$ that is tangent to the curve at that point in question. That is, $(\partial z/\partial x)$ is the slope of the tangent in the x direction. (b) $(\partial z/\partial y)_x$. The partial derivative can be similarly interpreted.

Figure 2 shows the geometric interpretation of these quantities.

For a fixed amount of an ideal gas (n is constant), we can write the partial derivative of P $(= nRT/V)$ with respect to T and V as follows:

$$\left(\frac{\partial P}{\partial T}\right)_V = \frac{nR}{V} \quad \text{and} \quad \left(\frac{\partial P}{\partial V}\right)_T = -\frac{nRT}{V^2}$$

In general, the order of differentiation is not important. For Equation 1, the second-order derivative can be shown to be the same whether we differentiate with respect to x first or with y first; that is,

$$\left(\frac{\partial^2 z}{\partial x \partial y}\right) = \left(\frac{\partial^2 z}{\partial y \partial x}\right) = -4$$

We do not indicate which variable is held constant because they vary with each differentiation.

Finally, if we have a function y such that

$$y = f(x_1, x_2, x_3, \ldots)$$

then we must write the partial derivative of y with respect to x_1 as

$$\left(\frac{\partial y}{\partial x_1}\right)_{x_2, x_3, \ldots}$$

Sources

"Some Comments on Partial Derivatives in Thermodynamics," E. W. Anacker, S. E. Anacker, and W. J. Swartz, *J. Chem. Educ.* **64**, 670 (1987).
"Thermodynamic Partial Derivatives and Experimentally Measurable Quantities," G. A. Estévez, K. Yang, and B. B. Dasgupta, *J. Chem. Educ.* **66**, 890 (1989).

Exercise 1

Derive the partial derivatives $(\partial P/\partial V)_T$ and $(\partial P/\partial T)_V$ for one mole of a van der Waals gas. (*Hint:* First rearrange Equation 2.13 to show P as a function of V and T.)

Exact and Inexact Differentials

Consider a function z of two variables x and y:

$$z = f(x, y)$$

If there is an infinitesimal change in x at constant y, the corresponding change in z is dz, given by $dz = (\partial z/\partial x)_y dx$. Similarly, for an infinitesimal change dy while keeping x constant, we have $dz = (\partial z/\partial y)_x dy$. Now if both x and y undergo infinitesimal changes, the change in z is the sum of the changes due to dx and dy:

$$dz = \left(\frac{\partial z}{\partial x}\right)_y dx + \left(\frac{\partial z}{\partial y}\right)_x dy \tag{2}$$

Here, dz is called the *total differential* because it is expressed in terms of both dx and dy.

Consider the following example of a total differential. The pressure of one mole of a gas is a function of volume and temperature:

$$P = f(V, T)$$

We can write the total differential, dP, as

$$dP = \left(\frac{\partial P}{\partial V}\right)_T dV + \left(\frac{\partial P}{\partial T}\right)_V dT \tag{3}$$

From Exercise 1 shown above, we find that, for a van der Waals gas,

$$\left(\frac{\partial P}{\partial V}\right)_T = -\frac{RT}{(V-b)^2} + \frac{2a}{V^3} \quad \text{and} \quad \left(\frac{\partial P}{\partial T}\right)_V = \frac{R}{V-b}$$

Substituting these expressions in Equation 3 we obtain the total differential, dP, for a van der Waals gas:

$$dP = \left[-\frac{RT}{(V-b)^2} + \frac{2a}{V^3}\right]dV + \frac{R}{V-b}dT \tag{4}$$

Total differentials are either *exact* or *inexact*, with important differences. A differential of the type

$$dz = M(x, y)dx + N(x, y)dy$$

is said to be an exact differential if the following condition is satisfied:

$$\left(\frac{\partial M}{\partial y}\right)_x = \left(\frac{\partial N}{\partial x}\right)_y$$

This test is known as Euler's theorem (after the Swiss mathematician Leonhard Euler, 1707–1783). Suppose that we have a function given by

$$dz = (y^2 + 3x)dx + e^x\,dy$$

where $M(x, y) = y^2 + 3x$ and $N(x, y) = e^x$. Applying Euler's theorem, we write

$$\left(\frac{\partial M}{\partial y}\right)_x = \left[\frac{\partial(y^2 + 3x)}{\partial y}\right]_x = 2y$$

and

$$\left(\frac{\partial N}{\partial x}\right)_y = \left(\frac{\partial e^x}{\partial x}\right)_y = e^x$$

Therefore, dz is not an exact differential.

On the other hand, dP in Equation 4 is an exact differential because

$$\left[\frac{\partial\left(-\dfrac{RT}{(V-b)^2} + \dfrac{2a}{V^3}\right)}{\partial T}\right]_V = -\frac{R}{(V-b)^2}$$

and

$$\left[\frac{\partial\left(\dfrac{R}{V-b}\right)}{\partial V}\right]_T = -\frac{R}{(V-b)^2}$$

The significance of exact differentials is that if df is an exact differential (where f is a function of x and y), then the value of the following integral depends only on the limits of integration; that is,

$$\int_1^2 df = f_2 - f_1$$

However, if $đf$ is an inexact differential, then

$$\int_1^2 đf \neq f_2 - f_1$$

Note that we have used the symbol $đ$ to denote an inexact differential. Unless the functional relationship between the variables x and y is known, the integral $\int_1^2 đf$ cannot be carried out. We saw earlier that dU and dH are exact differentials whereas $đw$ and $đq$ are inexact differentials. This means that the amount of work done or heat exchanged in a process depends on the path and not just on the initial and final states of the system. Thus, if any thermodynamic function X is a state function, then dX is an exact differential.

Exercise 2

For a given amount of an ideal gas, we can write $V = f(P, T)$. Prove that dV is an exact differential.

Exercise 3

Using the result from Exercise 1, show that $đw$, expressed as $đw = -P\,dV$, is an inexact differential.

Thermodynamic Data

Thermodynamic Data for Selected Elements and Inorganic Compounds at 1 bar and 298 K^a

Substance	$\Delta_f \bar{H}^\circ / \text{kJ} \cdot \text{mol}^{-1}$	$\Delta_f \bar{G}^\circ / \text{kJ} \cdot \text{mol}^{-1}$	$\bar{S}^\circ / \text{J} \cdot \text{K}^{-1} \cdot \text{mol}^{-1}$	$\bar{C}_P^\circ / \text{J} \cdot \text{K}^{-1} \cdot \text{mol}^{-1}$
$Ag(s)$	0	0	42.7	25.49
$Ag^+(aq)$	105.9	77.11	72.68	37.66
$AgBr(s)$	−99.5	−95.94	107.11	52.38
$AgCl(s)$	−127.0	−109.72	96.2	50.79
$AgI(s)$	−62.4	−66.3	114.2	54.43
$AgNO_3(s)$	−129.4	−33.41	140.9	93.05
$Al(s)$	0	0	28.32	24.34
$Al^{3+}(aq)$	−531.0	−485	−321.7	
$AlCl_3(s)$	−704.2	−628.8	110.67	91.84
$Al_2O_3(s)$	−1669.8	−1576.4	50.99	78.99
$Ar(g)$	0	0	154.8	20.79
$Ba(s)$	0	0	62.8	26.36
$Ba^{2+}(aq)$	−537.6	−560.8	10	
$BaO(s)$	−553.5	−525.1	70.3	47.45
$BaCl_2(s)$	−858.6	−810.9	123.7	75.31
$BaSO_4(s)$	−1464.4	−1353.1	132.2	101.75
$Be(s)$	0	0	9.54	17.82
$BeO(s)$	−610.9	−581.6	14.1	25.4
$Br_2(l)$	0	0	152.23	75.69
$Br^-(aq)$	−121.6	−103.96	82.4	
$HBr(g)$	−36.4	−53.45	198.7	29.12
$C(graphite)$	0	0	5.7	8.52
$C(diamond)$	1.90	2.87	2.4	6.11
$CO(g)$	−110.5	−137.3	197.9	29.14
$CO_2(g)$	−393.5	−394.4	213.6	37.1
$CO_2(aq)$	−413.8	−386.0	117.6	
$CO_3^{2-}(aq)$	−677.1	−527.8	−56.9	
$HCO_3^-(aq)$	−692.0	−586.8	91.2	
$H_2CO_3(aq)$	−699.65	−623.1	187.4	
$HCN(g)$	135.1	124.7	201.8	35.9
$CN^-(aq)$	151.6	172.4	94.1	
$Ca(s)$	0	0	41.42	25.31
$Ca^{2+}(aq)$	−542.8	−553.6	53.1	
$CaO(s)$	−635.6	−604.2	39.8	42.8
$Ca(OH)_2(s)$	−986.6	−896.8	83.4	84.52
$CaCl_2(s)$	−795.8	−748.1	104.6	72.63
$CaCO_3(calcite)$	−1206.9	−1128.8	92.9	81.9
$Cl_2(g)$	0	0	223.0	33.93
$HCl(g)$	−92.3	−95.3	186.5	29.12

Thermodynamic Data for Selected Elements and Inorganic Compounds at 1 bar and 298 K[a]
(*continued*)

Substance	$\Delta_f \bar{H}°/\text{kJ} \cdot \text{mol}^{-1}$	$\Delta_f \bar{G}°/\text{kJ} \cdot \text{mol}^{-1}$	$\bar{S}°/\text{J} \cdot \text{K}^{-1} \cdot \text{mol}^{-1}$	$\bar{C}_P°/\text{J} \cdot \text{K}^{-1} \cdot \text{mol}^{-1}$
$Cl^-(aq)$	−167.2	−131.2	56.5	
$Cr(s)$	0	0	23.77	23.35
$Cr_2O_3(s)$	−1128.4	−1046.8	81.2	118.74
$CrO_4^{2-}(aq)$	−881.2	−727.8	50.2	
$Cr_2O_7^{2-}(aq)$	−1490.3	−1301.1	261.9	
$Cu(s)$	0	0	33.15	24.47
$Cu^+(aq)$	71.67	49.98	40.6	
$Cu^{2+}(aq)$	64.77	65.49	−99.6	
$Cu_2O(s)$	−168.6	−146.0	93.14	
$CuO(s)$	−157.3	−129.7	42.63	44.35
$CuS(s)$	−48.53	−48.95	66.53	47.82
$CuSO_4(s)$	−771.36	−661.8	109	100.8
$F_2(g)$	0	0	202.8	31.3
$F^-(aq)$	−329.11	−276.5	−13.8	
$HF(g)$	−271.1	−273.2	173.5	29.08
$Fe(s)$	0	0	27.2	25.23
$Fe^{2+}(aq)$	−89.1	−86.3	−137.7	
$Fe^{3+}(aq)$	−48.5	−4.7	−315.9	
$Fe_2O_3(s)$	−824.2	−742.2	90.0	104.6
$H(g)$	218.2	203.3	114.7	
$H_2(g)$	0	0	130.6	28.8
$H^+(aq)$	0	0	0	
$OH^-(aq)$	−229.6	−157.3	−10.75	
$H_2O(g)$	−241.8	−228.6	188.7	33.6
$H_2O(l)$	−285.8	−237.2	69.9	75.3
$H_2O_2(l)$	−187.8	−120.4	109.6	89.1
$He(g)$	0	0	126.1	20.79
$Hg(l)$	0	0	77.4	27.98
$Hg^{2+}(aq)$	171.1	164.4	−32.2	
$HgO(\text{red})$	−90.8	−58.5	70.29	44.06
$I_2(s)$	0	0	116.13	54.44
$I^-(aq)$	55.19	51.57	111.3	
$HI(g)$	26.48	1.7	206.3	29.16
$K(s)$	0	0	64.18	29.58
$K^+(aq)$	−252.38	−283.27	102.5	
$KOH(s)$	−424.8	−379.1	78.9	
$KCl(s)$	−436.8	−409.1	82.59	51.3
$KClO_3(s)$	−391.2	−289.9	142.97	100.3
$KNO_3(s)$	−492.7	−393.1	132.9	96.3
$Kr(g)$	0	0	164.08	20.79
$Li(s)$	0	0	28.03	23.64
$Li^+(aq)$	−278.5	−293.8	14.23	
$LiOH(s)$	−487.2	−443.9	50.21	
$Mg(s)$	0	0	32.68	23.89
$Mg^{2+}(aq)$	−466.9	−454.8	−138.1	
$MgO(s)$	−601.8	−569.6	26.78	37.41
$MgCO_3(s)$	−1095.8	−1012.1	65.7	75.5
$MgCl_2(s)$	−641.3	−591.8	89.62	71.3
$N_2(g)$	0	0	191.6	29.12
$NH_3(g)$	−46.3	−16.6	192.5	35.66

Thermodynamic Data for Selected Elements and Inorganic Compounds at 1 bar and 298 K^a
(*continued*)

Substance	$\Delta_f \bar{H}°/\text{kJ} \cdot \text{mol}^{-1}$	$\Delta_f \bar{G}°/\text{kJ} \cdot \text{mol}^{-1}$	$\bar{S}°/\text{J} \cdot \text{K}^{-1} \cdot \text{mol}^{-1}$	$\bar{C}_P°/\text{J} \cdot \text{K}^{-1} \cdot \text{mol}^{-1}$
$NH_4^+(aq)$	-132.5	-79.3	113.4	
$NH_4Cl(s)$	-314.4	-202.87	94.6	
$N_2H_4(l)$	50.63	149.4	121.2	139.3
$NO(g)$	90.4	86.7	210.6	29.86
$NO_2(g)$	33.9	51.84	240.5	37.9
$N_2O_4(g)$	9.7	98.29	304.3	79.1
$N_2O(g)$	81.56	103.6	220.0	38.7
$HNO_3(l)$	-174.1	-80.7	155.6	109.87
$HNO_3(aq)$	-207.6	-111.3	146.4	
$Na(s)$	0	0	51.21	28.41
$Na^+(aq)$	-240.12	-261.9	59.0	46.4
$NaBr(s)$	-361.06	-348.98	86.82	52.3
$NaCl(s)$	-411.15	-384.14	72.13	50.5
$NaI(s)$	-287.78	-286.06	98.53	54.39
$Na_2CO_3(s)$	-1130.9	-1047.7	135.98	110.5
$NaHCO_3(s)$	-947.7	-851.9	102.1	87.6
$NaOH(s)$	-425.61	-379.49	64.46	59.54
$Ne(g)$	0	0	146.3	20.79
$O(g)$	249.4	231.73	161.0	
$O_2(g)$	0	0	205.0	29.4
$O_3(g)$	142.7	163.4	237.7	38.2
$P(\text{white})$	0	0	41.09	23.22
$PO_4^{3-}(aq)$	-1277.4	-1018.7	-221.8	
$P_4O_{10}(s)$	-2984.0	-2697.0	228.86	211.7
$PCl_3(g)$	-287.0	-267.8	311.78	71.84
$PCl_5(g)$	-374.9	-305.0	364.6	112.8
$S(\text{rhombic})$	0	0	31.88	22.59
$S(\text{monoclinic})$	0.30	0.10	32.55	23.64
$SO_2(g)$	-296.1	-300.1	248.5	39.79
$SO_3(g)$	-395.2	-370.4	256.2	50.63
$SO_4^{2-}(aq)$	-909.3	-744.5	20.1	
$H_2S(g)$	-20.63	-33.56	205.8	33.97
$H_2SO_4(l)$	-814.0	-690.0	156.9	
$H_2SO_4(aq)$	-909.27	-744.53	20.1	
$SF_6(g)$	-1209	-1105.3	291.8	97.3
$Si(s)$	0	0	18.83	19.87
$SiO_2(s)$	-910.9	-856.6	41.84	44.43
Xe	0	0	169.6	20.79
$Zn(s)$	0	0	41.63	25.06
$Zn^{2+}(aq)$	-153.9	-147.1	-112.1	
$ZnO(s)$	-348.3	-318.3	43.64	40.25
$ZnS(s)$	-202.9	-198.3	57.74	45.19
$ZnSO_4(s)$	-978.6	-871.6	124.9	117.2

^{a}Data are mostly from the NBS *Tables of Chemical Thermodynamic Properties* (1982). The values for ions in aqueous solution (1 M), such as $Li^+(aq)$, are based on the convention that all the properties listed for $H^+(aq)$ are equal to zero.

Thermodynamic Data for Selected Organic Compounds at 1 bar and 298 K[a]

Compound	State	$\Delta_f \bar{H}°/kJ \cdot mol^{-1}$	$\Delta_f \bar{G}°/kJ \cdot mol^{-1}$	$\bar{S}°/J \cdot K^{-1} \cdot mol^{-1}$
Acetic acid (CH_3COOH)	l	−484.2	−389.9	159.8
	aq	−485.8	−396.5	178.7
Acetate (CH_3COO^-)	aq	−486.01	−369.3	86.6
Acetaldehyde (CH_3CHO)	l	−192.3	−128.1	160.2
Acetone (CH_3COCH_3)	l	−248.1	−155.4	200.4
Acetylene (C_2H_2)	g	226.6	209.2	200.8
Benzene (C_6H_6)	l	49.04	124.5	172.8
Benzoic acid (C_6H_5COOH)	s	−385.1	−245.3	167.6
Ethanol (C_2H_5OH)	l	−277.0	−174.2	161.0
Ethane (C_2H_6)	g	−84.7	−32.9	229.5
Ethylene (C_2H_4)	g	52.3	68.12	219.5
Formic acid ($HCOOH$)	l	−424.7	−361.4	129.0
Formate ($HCOO^-$)	aq	−410.0	−334.7	91.6
α-D-Glucose ($C_6H_{12}O_6$)	s	−1274.5	−910.6	210.3
D-Glucose ($C_6H_{12}O_6$)	s	−1268.1	−908.9	228.0
Glycine ($C_2H_5O_2N$)	s	−537.2	−377.7	103.5
	$eq\ buf$	−523.0	−379.9	158.6
Glycylglycine ($C_4H_8O_3N_2$)	s	−746.0	−491.5	190.0
	$eq\ buf$	−734.3	−493.1	
Lactic acid ($C_3H_6O_3$)	s	−694.0	−523.3	143.5
Lactate ion ($C_3H_5O_3^-$)	aq	−686.6	−516.7	146.4
Methane (CH_4)	g	−74.85	−50.79	186.2
Methanol (CH_3OH)	l	−238.7	−166.3	126.8
2-Propanol (C_3H_7OH)	l	−317.9	−180.3	180.6
Pyruvic acid ($C_3H_4O_3$)	l	−585.8	−463.4	179.5
Pyruvate ($C_3H_3O_3^-$)	aq	−596.2	−472.4	171.5
Sucrose ($C_{12}H_{22}O_{11}$)	s	−2221.7	−1544.3	360.2
Urea [$(NH_2)_2CO$]	s	−333.5	−197.3	104.6
	aq	−317.7	−202.7	173.9

[a] Data mostly from the NBS *Tables of Chemical Thermodynamic Properties* (1982). The abbreviation *eq buf* is for an equilibrium mixture of species in an aqueous solution buffered to pH 7. The concentration of all solutions is 1 *M*.

Glossary*

A

absolute zero Theoretically, the lowest attainable temperature. (2.3)

absolute zero temperature scale A temperature scale that uses the absolute zero of temperature as the lowest temperature. (2.3)

actinides Elements that have incompletely filled $5f$ subshells or readily give rise to cations that have incompletely filled $5f$ subshells. (11.11)

action potential A transient change in electric potential at the surface of a nerve or muscle cell occurring at the moment of excitation. (7.7)

action spectrum An absorption spectrum that displays the photochemical response or effectiveness of a system as a function of the wavelength of light employed. (15.1)

activated complex An energetically excited state that is intermediate between reactants and products in a chemical reaction. Also called the transition state. (9.7)

activation energy The minimum energy required to initiate a chemical reaction. (9.5)

active site The site on an enzyme molecule where the substrate binds and the catalytic reaction is facilitated. (10.1)

active transport The energy-requiring movement of molecules across a membrane against a concentration gradient. (5.10)

activity The activity is an effective thermodynamic concentration that takes into account deviation from ideal behavior. (5.5)

activity coefficient A characteristic of a quantity expressing the deviation of a solution from ideal behavior. It relates activity to concentration. (5.5)

adiabatic process A process in which there is no heat exchange between the system and its surroundings. (3.4)

allosteric interaction The long-range interaction between spatially distant ligand-binding sites mediated by the structure of a protein molecule. (1.1, 10.6)

alpha (a) helix Secondary structure of proteins in which the polypeptide chain assumes a helical form. (16.3)

amphoterism The ability to behave both as an acid and a base. (8.2)

antenna molecule A molecule that absorbs light energy and transfers it to a reaction center during photosynthesis. (15.2)

antibonding molecular orbital A molecular orbital with an energy higher than that of its constituent atomic orbitals. (12.5)

Arrhenius equation An equation that relates the rate constant to the pre-exponential factor (A) and the activation energy (E_a), $k = A \exp(-E_a/RT)$. (9.5)

Aufbau principle The principle stating that as protons are added one at a time to the nucleus to build up the elements, electrons similarly are added to the atomic orbitals. (11.11)

Avogadro's law The law stating that at constant pressure and temperature, the volume of a gas is directly proportional to the number of moles of the gas present. (2.3)

Avogadro's number 6.022×10^{23}; the number of particles in a mole. (1.3)

B

Beer–Lambert law An equation that relates the absorbance (A) at a particular wavelength to the concentration of the solution (c) and pathlength of the cell (b); that is, $A = \varepsilon bc$, where ε is the molar absorptivity of the light-absorbing species at that wavelength. (14.1)

beta (β) sheet Secondary structure of proteins in which portions of a polypeptide chain in extended conformation are held together by hydrogen bonds. β sheets also exist between molecules such as β-keratin. (16.3)

bimolecular reaction An elementary step that involves two molecules. (9.3)

bioenergetics The study of energy transformations in living organisms. (6.6)

Bohr effect The effect of pH on the oxygen equilibrium of hemoglobin. A decrease in pH decreases hemoglobin's affinity for oxygen. (10.6)

Bohr radius The radius of the smallest orbit of the hydrogen atom. It is equal to $0.529\,\text{Å}$. (11.4)

Boltzmann distribution law The law that expresses the population (the number of molecules) in a state of energy E_i in a system at thermal equilibrium at temperature T. It is often used to calculate the ratio of populations (N_2/N_1) in two states of energy, E_1 and E_2:

$$\frac{N_2}{N_1} = \exp\left(-\frac{E_2 - E_1}{k_B T}\right) \quad (3.3)$$

bond dissociation enthalpy The enthalpy change that accompanies the breaking of a chemical bond in a diatomic molecule. (3.7)

bond enthalpy The enthalpy change that accompanies the breaking of a chemical bond in a polyatomic molecule. (3.7)

bond moment The degree of polarity of a chemical bond. For a diatomic molecule, the bond moment is equal to the dipole moment and is given by the product of charge (magnitude only) and the distance between the charges. (12.4)

bond order The difference between the number of electrons in bonding molecular orbitals and antibonding molecular orbitals, divided by two. (12.5)

bonding molecular orbital A molecular orbital with an energy lower than that of its constituent atomic orbitals. (12.5)

boundary-surface diagram A diagram of the region of an atomic orbital containing a substantial amount of electron density (about 90%). (11.10)

---

*The number in parentheses indicates the number of the chapter section or chapter appendix in which the term first appears.

Boyle's law   The law stating that the volume of a fixed amount of gas maintained at constant temperature is inversely proportional to the gas pressure. (2.3)

buffer capacity   An index of a buffer solution's resistance to change in pH as a result of the addition of an acid or a base. (8.5)

buffer solution   A solution of (a) a weak acid or base and (b) its salt; both components must be present. The solution has the ability to resist the changes in pH upon the addition of small amounts of either acid or base. (8.5)

C

capacitance   The ratio of the charge on either of a pair of conductors of a capacitor to the potential difference between the conductors. (Chapter Appendix 5.1)

carbon dioxide fixation   In photosynthesis, the incorporation of carbon dioxide and its ultimate conversion to carbohydrates. (15.2)

Carnot cycle   A hypothetical cycle that consists of four reversible processes in succession: an isothermal expansion with the absorption of heat, an adiabatic expansion, an isothermal compression with the release of heat, and an adiabatic compression. In such a cycle, some of the heat absorbed is converted to work. The Carnot cycle is used to show that entropy is a state function and to derive an expression for the thermodynamic efficiency. (4.2)

catalyst   A substance that increases the rate of a reaction without itself being consumed. (10.1)

catalytic rate constant (k_{cat})   *See* turnover number.

chain reaction   A reaction in which an intermediate generated in one step attacks another species to produce another intermediate, and so on. (9.4)

Charles' law   The law stating that the volume of a fixed amount of gas maintained at constant pressure is directly proportional to the absolute temperature of the gas. (2.3)

chemical potential (μ)   Partial molar Gibbs energy. The chemical potential of the ith component in a mixture is defined by

$$\mu_i = \left(\frac{\partial G}{\partial n_i}\right)_{T,P,n_j}$$

Chemical potential is used to predict the direction of a spontaneous process in a mixture just as Gibbs energy is for a pure component. (5.2)

chemical shift   The difference between the NMR resonance frequency of the nucleus in question and that of a reference standard, divided by the resonance frequency of the standard. (14.5)

chemiosmotic theory   The theory stating that ATP synthesis is coupled to electron transport by an electrochemical proton gradient across a membrane. (7.6)

chiral   The property of a substance that means it is not superimposable on its mirror image. The term *chiral* is used synonymously with *optically active* in describing a compound. (14.9)

chloroplasts   The organelles in plants and algae cells that carry out photosynthesis. (15.2)

chromophore   A part of a molecule that absorbs light of a specific wavelength. (14.4)

circular birefringence   A substance is circularly birefringent if it has different refractive indices for left- and right-circularly polarized light. (14.9)

circular dichroism   Describes the phenomenon in which a substance absorbs left- and right-circularly polarized light with different molar absorptivities. (14.9)

circularly polarized light   Light in which the electric field vector rotates in a plane perpendiclar to the propagation direction with constant angular velocity. (14.9)

citric acid cycle   The biochemical pathway that degrades the acetyl group of acetyl CoA to CO_2 and H_2O as three molecules of NAD^+ and one molecule of FAD are reduced. (7.6)

Clapeyron equation   A relation between the changes in pressure and temperature for two phases at equilibrium:

$$\frac{dP}{dT} = \frac{\Delta \bar{H}}{T \Delta \bar{V}}$$

It expresses the slope of a phase boundary in a phase diagram. (4.9)

Clausius–Clapeyron equation   An approximate form of the Clapeyron equation in which one phase is condensed and the other is a gas that is treated as ideal. (4.9)

closed system   A system that allows the exchange of energy (usually in the form of heat) but not mass with its surroundings. (2.1)

coenzyme   A small organic molecule required in the catalytic mechanisms of certain enzymes. (12.9)

colligative properties   Properties of solutions that depend on the number of solute particles in solution and not on the nature of the solute particles. (5.6)

collision frequency (Z_1)   Number of collisions made by a molecule per unit time. (2.8)

complex ion   A metal ion with one or more ligands. (12.8)

compressibility factor (Z)   A quantity given by ($P\bar{V}/RT$). Deviation of Z from 1 indicates nonideal behavior of the gas. (2.4)

concerted model   A model employed to explain the cooperative nature of hemoglobin binding to oxygen. It assumes the protein exists in two states at equilibrium, called the T state and R state. In the absence of oxygen, the T state, which has a low affinity for oxygen, is favored. When oxygen molecules are bound to hemoglobin, the equilibrium shifts to the R state, which has a higher affinity for oxygen. This model can also be applied to allosteric enzymes. (10.6)

configuration   Stereo arrangements of atoms in a molecule that are related to one another by symmetry but cannot be converted into one another without rupturing bonds. Examples are geometric isomers. (16.2)

conformation   Arrangements of atoms in three-dimensional space. In general, a molecule possesses an infinite number of conformations such as those generated by the rotation of the C–C bond in ethane (C_2H_6). (16.2)

consecutive reactions   Consecutive reactions are reactions of the type A $\rightarrow$ B $\rightarrow$ C . . . . (9.4)

cooperativity   A mechanism in which binding one ligand to a protein promotes the binding of other ligands. (10.6)

coordination compound   A compound that typically consists of a complex ion and counter ion(s). Some coordination compounds such as $Fe(CO)_5$ do not contain complex ions. (12.8)

Cotton effect   The characteristic wavelength dependence of the optical rotatory dispersion curve in the vicinity of an absorption band. (14.9)

coupled reaction   A process in which an endergonic reaction is made to proceed by coupling it to an exergonic reaction. Biological coupled reactions are usually mediated with the aid of enzymes. (6.6)

covalent bond   A chemical bond formed by the sharing of one or more pairs of electrons. (12.1)

critical temperature   The temperature above which a gas will not liquefy. (2.5)

crystal-field stabilization energy   The net decrease in energy of the d orbital (hence extra stabilization) in a complex ion relative to the d orbital energy in a spherically symmetric crystal field. (12.8)

crystal-field theory   The theory that assumes that the ligands of a coordination compound are the sources of negative charges that, through electrostatic interaction, remove the degeneracy of the d orbitals. The theory is called "crystal field" because the metal ion is subject to an electric field resulting from the presence of the ligands that is analogous to the electric field within an ionic crystalline lattice. (12.8)

D

Dalton's law of partial pressures   The law stating that the total pressure of a mixture of gases is just the sum of the pressures that each gas would exert if it were alone. (2.3)

dark reaction   Steps in photosynthesis that do not depend directly on light energy; for example, the synthesis of carbohydrate from CO_2 and H_2O. (15.2)

Debye–Hückel limiting law   A mathematical expression for calculating the mean activity coefficient of an electrolyte solution in regions of low ionic strength. (5.8)

degeneracy   Two or more states are said to be degenerate if they have the same energy. (11.10)

degrees of freedom   The number of ways a molecule can execute its kinetic motion (translation, rotation, and vibration). (14.8) In phase rule, it is the number of intensive variables (pressure, temperature, and composition) that can be changed independently without disturbing the number of phases in equilibrium. (4.9)

delocalized molecular orbital   A molecular orbital that extends over more than two atoms. (12.5)

dialysis   The process by which low-molar-mass solutes are added to or removed from a solution by means of diffusion across a semipermeable membrane. (6.5)

diamagnetic   A diamagnetic substance contains only paired electrons and is slightly repelled by a magnet. (11.11)

dielectric constant (ε)   The dielectric constant of a medium is the ratio of the capacitance (C) of a capacitor when the region between the plates filled with the material to the capacitance (C_0) of the same capacitor when the region between the plates is a vacuum; that is, $\varepsilon = C/C_0$. (5.7)

diffusion   The gradual mixing of molecules of one gas with the molecules of another by virtue of their kinetic properties. (2.9)

diffusion-controlled reaction   A reaction whose rate-determining step is the rate of diffusion of reactant molecules in an encounter to form a product. (9.10)

dipolar ion   A species carrying equal numbers of positive and negative charges. Also called a zwitterion. (8.7)

dipole–dipole interaction   The electrostatic interaction between the electric dipoles of two polar molecules. (13.3)

dipole–induced dipole interaction   The electrostatic interaction between the electric dipole of a polar molecule and the induced electric dipole (by the polar molecule) of a nonpolar molecule. (13.3)

dipole moment   The vector sum of the bond moments in a molecule. It is a measure of the polarity of the molecule. (12.4)

dispersion interaction   Attractive interactions between molecules due to the fluctuating electron distributions in them. Also called London interaction. (13.3)

distribution diagram   A plot showing the fractions of acids and their conjugate bases present in solution as a function of pH. (8.4)

disulfide bond   A covalent bond formed between the sulfhydryl groups in two cysteine residues. (16.4)

Donnan effect   The unequal equilibrium distribution of small diffusible ions on the two sides of a membrane that is freely permeable to these ions but impermeable to macromolecular ions, in the presence of a macromolecular electrolyte on one side of the membrane. (5.9)

Doppler effect   The change in the observed frequency of an acoustic or electromagnetic wave due to relative motion of source and observer. (14.1)

E

effector   An activator of an allosteric enzyme. (10.6)

effusion   The process by which a gas under pressure escapes from one compartment of a container to another by passing through a small opening. (2.9)

einstein   A unit for one mole of photons. (15.1)

elastic collision   In an elastic collision, there is no transfer of energy from translational motion into internal modes of motion such as rotation and vibration. (2.6)

electromagnetic radiation   Radiation that is emitted or absorbed in the form of electromagnetic waves. (11.1)

electromagnetic wave   A wave that has an electric field and a mutually perpendicular magnetic field. (11.1)

electron affinity   The negative of the energy change when an electron is accepted by an atom in the gaseous state to form an anion. (11.11)

electron configuration   Describes the distribution of electrons among the various orbitals in an atom or molecule. (11.11)

electronegativity   The ability of an atom to attract electrons toward itself in a chemical bond. (12.4)

electrophoresis   A process in which a macromolecule with a net electric charge migrates in a solution under the influence of an electric field. It is used to purify or separate proteins and nucleic acids. (16.1)

electrophoretic mobility   Ionic mobility per unit electric field. (16.1)

elementary step   A reaction that represents the progress at the molecular level. (9.3)

enantiomers   Stereoisomers that are nonsuperimposable

mirror images of each other. Also called optical isomers. (14.9)

endergonic process   A process that is accompanied by a positive change in Gibbs energy ($\Delta G > 0$) and therefore is thermodynamically not favored. (4.6)

endothermic reaction   A reaction that absorbs heat from the surroundings. (3.2)

energy   The capacity to do work or to produce change. (1.2)

energy level   An energy level is an allowed energy of a quantized system. The level is said to be degenerate if several states possess the same energy. (11.4)

enthalpy (H)   A thermodynamic quantity used to describe heat changes taking place at constant pressure. It is defined by the equation $H = U + PV$, where U is the internal energy, and P and V are the pressure and volume of the system, respectively. (3.2)

entropy (S)   A thermodynamic quantity that expresses the degree of disorder or randomness in a system at the molecular level. The thermodynamic definition is the change in entropy defined as $dS = dq_{rev}/T$, where dq_{rev} is the infinitesimal quantity of energy supplied reversibly as heat to the system at temperature T. The statistical definition of entropy is given by the Boltzmann equation, $S = k_B \ln W$, where k_B is the Boltzmann constant, and W is the total number of ways in which molecules of the system can be arranged to achieve the same overall energy. (4.2)

enzyme   A biological catalyst that is either a protein or an RNA molecule. (10.1)

equation of state   For gases, an equation that provides the mathematical relationships among the properties that define the state of the system, such as n, P, T, and V. (2.3)

equilibrium vapor pressure   The vapor pressure of a liquid in equilibrium with its vapor at a particular temperature. Frequently, it is referred to simply as vapor pressure. (2.5)

exergonic process   A process that is accompanied by a negative change in Gibbs energy ($\Delta G < 0$) and therefore is thermodynamically favored. (4.6)

exothermic reaction   A reaction that gives off heat to the surroundings. (3.2)

extensive property   A property that depends on how much matter is being considered. (2.1)

F

Faraday constant (F)   The charge carried by one mole of electrons; it has the value 96,485 C mol^{-1}. (7.1)

fibrous protein   A protein composed of polypeptides arranged in long sheets or fibers. (16.3)

first law of thermodynamics   The law that states that energy can be converted from one form to another but cannot be created or destroyed. In chemistry, the first law is usually expressed by the equation $\Delta U = q + w$, where U is the internal energy of the system, q is the heat exchange between the system and its surroundings, and w is the work done on the system by the surroundings or by the system on the surroundings. (3.2)

first-order reaction   A reaction whose rate depends on the reactant concentration raised to the first power. (9.2)

fluorescence   The emission of electromagnetic radiation by a substance while the substance is illuminated.

Fluorescence is characterized by a short lifetime (about 10^{-9} s) and by the fact that the emissive state and the ground state have the same spin multiplicity (usually singlet state). (14.7)

force   According to Newton's second law of motion, force is mass times acceleration. (1.2)

force constant (k)   The force constant of a harmonic oscillator is the constant of proportionality between the restoring force and the displacement of a body (x) that obeys Hooke's law; that is, $f = -kx$. It is a measure of the strength of chemical bonds. (14.3)

formal charge   The difference between the valence electrons in an isolated atom and the number of electrons assigned to that atom in a Lewis structure. (12.1)

Franck–Condon principle   The principle stating that in any molecular system, the transition from one electronic state to another is so rapid that the nuclei of the atoms in the molecule can be considered stationary during the transition. (14.4)

fugacity (f)   The fugacity of a gas is its effective thermodynamic pressure. (6.1)

fugacity coefficient (γ)   A quantity that relates the fugacity of a gas (f) to its pressure (P): $f = \gamma P$. (6.1)

G

gas constant (R)   The universal constant that appears in the ideal gas equation. It has the value 0.08206 L atm K^{-1} mol^{-1}, or 8.314 J K^{-1} mol^{-1}. (2.3)

gel electrophoresis   A type of electrophoresis in which the supporting medium is a thin slab of gel. (16.1)

Gibbs energy (G)   A thermodynamic quantity defined by the equation $G = H - TS$, where H, T, and S are enthalpy, temperature, and entropy, respectively. The change in Gibbs energy of a system in a process at constant temperature and pressure is given by $\Delta G = \Delta H - T\Delta S \leq 0$, where the equal sign denotes equilibrium and the "less than" sign denotes a spontaneous process. (4.6)

Gibbs–Helmholtz equation   An equation that expresses the temperature dependence of the Gibbs energy of a system in terms of its enthalpy. (4.8)

globular protein   A protein that assumes a compact rounded shape. (16.3)

glycolysis   The enzymatic pathway that converts a glucose molecule to two molecules of pyruvate. This anaerobic process generates energy in the form of two ATP molecules and two NADH molecules. (6.6)

Graham's law of diffusion   The law stating that the rate of diffusion of gas molecules is inversely proportional to the square root of the molar mass of the gas at constant temperature and pressure. (2.9)

Graham's law of effusion   The law stating that the rate of effusion of gas molecules from a particular orifice is inversely proportional to the square root of the molar mass of the gas at constant temperature and pressure. (2.9)

H

half-life   The time required for the concentration of a reactant to decrease to half of its initial concentration. (9.2)

harmonic oscillator   A body that obeys Hooke's law. In the study of their vibrational motions, molecules are treated (to a good approximation) as harmonic oscillators. (14.3)

heat A process in which energy is transferred from one system to another as a result of a temperature difference between them. (3.1)

heat capacity The amount of energy required to raise the temperature of a given quantity of the substance by one degree Celsius. (3.3)

heat capacity ratio (γ) The ratio given by $\bar{C}_P / \bar{C}_V$. (3.4)

Heisenberg uncertainty principle The principle stating that it is impossible to know simultaneously both the momentum and the position of a particle with certainty. (11.6)

Helmholtz energy (A) A thermodynamic quantity defined by the equation $A = U - TS$. The change in Helmholtz energy of a system in a process at constant temperature and volume is given by $\Delta A = \Delta U - T\Delta S \leq 0$, where the equal sign denotes equilibrium and the "less than" sign denotes a spontaneous process. (4.6)

Henry's law The law stating that the solubility of a gas in a liquid is proportional to the pressure of the gas over the solution. (5.4)

Hess's law The law stating that when reactants are converted to products, the change in enthalpy is the same whether the reaction takes place in one step or a series of steps. (3.6)

Hill coefficient (n) A measure of cooperativity. (10.6)

Hooke's law The law stating that the restoring force (f) on a body is proportional to its displacement (x) from the equilibrium position: $f = -kx$, where k is the constant of proportionality, called the force constant. (14.3)

Hund's rule The rule that says the most stable arrangement of electrons in subshells is the one with the greatest number of parallel spins. (11.11)

hybrid orbitals Atomic orbitals obtained when two or more nonequivalent atomic orbitals combine. (12.3)

hybridization The process of mixing the atomic orbitals in an atom that have similar energies to generate a set of new atomic orbitals with different spatial distributions. (12.3)

hydration number The number of water molecules associated with a solute molecule or an ion in aqueous solution. (5.7)

hydrogen bond A special type of dipole–dipole and covalent interaction between the hydrogen atom bonded to an atom of an electronegative element and another atom of an electronegative element. (13.4)

hydrophobic interaction Influences that cause nonpolar substances to cluster together so as to minimize their contact with water. (13.6)

hypertonic solution A concentrated solution with a high osmotic pressure. (5.6)

hypochromism A reduction in the absorbance of UV light at 260 nm that accompanies the transition from denatured DNA strands to a double-strand helix. This phenomenon is used to monitor the process of denaturation or renaturation of DNA molecules. (14.4)

hypotonic solution A dilute solution with a low osmotic pressure. (5.6)

I

Ideal solution A solution in which both the solvent and the solute obey Raoult's law. (5.4)

ideal dilute solution A solution in which the solvent obeys Raoult's law and the solute obeys Henry's law. (5.5)

ideal gas equation An equation expressing the relationships among pressure, volume, temperature, and amount of an ideal gas ($PV = nRT$, where R is the gas constant). (2.3)

inhibitor A substance that is capable of stopping or retarding an enzyme-catalyzed reaction. (10.5)

intensive properties A property that does not depend on how much matter is being considered. (2.1)

intermediate A species that appears in the mechanism of the reaction (that is, the elementary steps) but not in the overall balanced equation. (9.3)

internal energy The internal energy of a system is the total energy of all its components. It consists of translational, rotational, vibrational, electronic, and nuclear energies, as well as energy resulting from intermolecular interactions. (3.2)

international system of units (SI) A particular choice of metric units that was adopted by the General Conference of Weights and Measures. (1.2)

intersystem crossing The radiationless transition of a molecule from one electronic state into another with a different spin multiplicity. (14.7)

ion–dipole interaction The electrostatic interaction between an ion and the electric dipole of a molecule. (13.3)

ion–induced dipole The electrostatic interaction between an ion and the induced electric dipole (by the ion) of a nonpolar molecule. (13.3)

ionic atmosphere A sphere of opposite charge surrounding each ion in an aqueous solution. (5.8)

ionic strength (I) A characteristic of an electrolyte solution, defined by

$$I = \frac{1}{2}\sum_i m_i z_i^2$$

where m_i is the molality of the ith ion, and z_i is its charge. (5.8)

ionization energy The minimum energy required to remove an electron from an isolated atom (or ion) in its ground electronic state. (11.11)

ionophore A compound that has a high affinity for certain ions and the ability to transport them across membranes. (5.10)

irreversible inhibition An inhibition in which the inhibitor forms a covalent bond with the enzyme at the active site. The inhibitor cannot be removed by dialysis. (10.5)

isoelectric focusing A technique in which a mixture of protein molecules is resolved into its components by subjecting the mixture to an electric field in a supporting gel medium that has a previously established pH gradient. (16.1)

isoelectric point (pI) The pH at which the net charge on a macromolecule such as a protein is zero. Consequently, the protein does not move in an electric field. (8.7)

isoelectric A term describing ions, atoms and ions, or molecules and ions that possess the same number of electrons. (12.6)

isolated system A system that does not allow the transfer of either mass or energy to or from its surroundings. (2.1)

isosbestic point   The point of equal absorbance that occurs when two species are in equilibrium and both contribute to the absorbance at a certain wavelength. The absorbance does not depend on their relative concentrations. (14.4)

isotherm   A plot of pressure versus volume of a gas at the same temperature. (2.2)

isothermal process   A process that occurs at constant temperature. (3.2)

isotonic solution   Solutions with the same concentration and hence the same osmotic pressure. (5.6)

J

Jablonsky diagram   A schematic display of the relative energies of the electronic states of molecules and the vibrational levels associated with each state. The diagram also shows both the radiative and nonradiative transitions between electronic states. (14.7)

K

kinetic isotope effect   The reduction in the rate of a reaction by the replacement of an atom by a heavier isotope. The effect is due to the lowering of the zero-point energy. (9.8)

kinetic salt effect   The effect of ionic strength on the rate of a reaction in solution. (9.9)

Kirchhoff's law   The law stating that the difference between the enthalpies of a reaction at two different temperatures, T_1 and T_2 ($T_2 > T_1$), is just the difference in the enthalpies of heating the products and reactants from T_1 to T_2. (3.6)

L

lanthanides   Elements that have incompletely filled $4f$ subshells or that readily give rise to cations that have incompletely filled $4f$ subshells. Also called rare earth metals. (11.11)

laser   An acronym for Light Amplification by the Stimulated Emission of Radiation. The operation of a laser requires a population inversion—that is, a greater number of atoms (or molecules) in an upper energy level than in some lower level. (14.8)

lattice energy   The enthalpy change when one mole of a solid is converted to a vapor. (5.7)

Le Chatelier's principle   The principle stating that if an external stress is applied to a system at equilibrium, the system adjusts in such a way that the stress is partially offset as it tries to reestablish equilibrium. (6.4)

Lewis structure   A representation of covalent bonding in which shared electrons are shown either as lines or as pairs of dots between two atoms, and lone pairs are shown as pairs of dots on individual atoms. (12.1)

ligand-field theory   An extended treatment of the molecular orbital theory for metal complexes that includes delocalized molecular orbitals. (12.8)

light reaction   The step in photosynthesis that depends directly on light energy. It results in the synthesis of ATP by photophosphorylation and the reduction of $NADP^+$ to NADPH via the oxidation of water. (15.2)

line spectra   Spectra produced when radiation is absorbed or emitted by substances only at certain wavelengths. (11.4)

linear birefringence   A substance is linearly birefringent if it has different refractive indices for left- and right-linearly polarized light. (14.9)

linearly polarized light   *See* plane-polarized light.

liquid scintillation counting   A method of assaying radioactive-tracer-labeled compounds by fluorescence. (14.7)

London interaction   *See* dispersion interaction.

lone pairs   Valence electrons that are not involved in covalent bond formation. (12.1)

M

macrostate   The state of a system as described by the macroscopic properties. (4.5)

maximum rate (V_{max})   The rate of an enzyme-catalyzed reaction when all the enzymes are bound to substrate molecules. (10.2)

Maxwell distribution of speed   A theoretical relationship that predicts the relative number of molecules at various speeds for a sample of gas at thermal equilibrium at a particular temperature. (2.7)

Maxwell distribution of velocity   A theoretical relationship that predicts the relative number of molecules at various velocities for a sample of gas at thermal equilibrium at a particular temperature. (2.7)

mean activity coefficient   A quantity that describes the deviation from ideality in the behavior of ions in solution. (5.8)

mean free path (λ)   Average distance traveled by a molecule between successive collisions. (2.8)

melting temperature   The temperature at which denaturation occurs for half of the helical strands of a DNA molecule. (16.3)

membrane potential   A voltage difference that exists across a membrane due to differences in the concentrations of ions on either side of the membrane. (7.7)

Michaelis–Menton kinetics   A mathematical treatment that assumes the initial step in enzyme cayalysis is a pre-equilibrium between the substrate and the enzyme, followed by the conversion of the enzyme–substrate complex to product. (10.2)

microstate   The state of a system as specified by the actual properties of each individual component (atoms or molecules). (4.5)

molar absorptivity   The constant of proportionality in the Beer–Lambert law. It measures the ability of a compound to absorb the electromagnetic radiation at a particular wavelength. (14.1)

molar mass   The mass (in grams or kilograms) of one mole of atoms, molecules, or other particles. (1.3)

mole   The amount of substance that contains as many elementary entities (atoms, molecules, or other particles) as there are atoms in exactly 12 grams (or 0.012 kilograms) of the carbon-12 isotope. (1.3)

mole fraction   The ratio of the number of moles of one component of a mixture to the total number of moles of all components in the mixture. (2.3)

molecular chaperone   A molecule that assists in protein folding. (16.4)

molecular orbital   A wave function describing an electron in a molecule. (12.5)

molecular orbital theory   A theory of electronic structures of molecules. The electrons are considered to occupy the molecular orbitals that extend over the entire molecule. The molecular orbitals are generated by the linear combination of atomic orbitals of the atoms in the molecule. (12.5)

molecularity   The number of molecules reacting in an elementary step. (9.3)

moment of inertia (I)   The moment of inertia of a body composed of point masses, m_i, at a perpendicular distance, r_i, from a specified line (usually a line through its center of mass) is given by

$$I = \sum_i m_i r_i^2 \quad (14.2)$$

monodispersity   A polymer system that is homogeneous in molar mass. (16.1)

most probable speed   The speed possessed by the largest number of molecules of a collection of molecules. (2.7)

N

negative cooperativity   A mechanism in which the binding of one ligand to a protein molecule decreases the likelihood of subsequent ligand binding. (10.6)

Nernst equation   An equation that expresses the cell potential (E) in terms of the standard cell potential ($E°$) and the reaction quotient for the cell reaction:

$$E = E° - \frac{RT}{vF} \ln Q$$

where v is the stoichiometric coefficient, and Q is the reaction quotient. (7.3)

neurotransmitter   A molecule released at a nerve terminal that binds to and influences the function of other nerve or muscle cells. (7.7)

noble gas core   The electron configuration of the noble gas element that most nearly precedes the element being considered. (11.11)

node   A point at which the wave function is zero. (11.5)

nonvolatile   Does not have a measurable vapor pressure. (5.4)

nucleic acid   A macromolecule composed of nucleotides. (16.3)

O

octet rule   The rule stating that an atom other than hydrogen tends to form bonds until it is surrounded by eight valence electrons. (12.1)

open system   A system that can exchange mass and energy (usually in the form of heat) with its surroundings. (2.1)

optical isomers   *See* enantiomers.

optical rotatory dispersion   Optical rotation as a function of wavelength. (14.9)

orbital   A one-electron wave function in an atom or molecule. (11.10)

oscillating reaction   A reaction that shows a periodic variation in space or time of the concentration of one or more intermediates. (9.11)

osmosis   The net movement of solvent molecules through a semipermeable membrane from a pure solvent or from a dilute solution to a more concentrated solution. (5.6)

osmotic pressure   The pressure required to stop osmosis. (5.6)

oxidative phosphorylation   The synthesis of ATP (the phosphorylation of ADP to ATP) that occurs in

conjunction with the electron transport across the inner mitochondrial membrane. (6.6)

P

paramagnetic   A paramagnetic substance contains one or more unpaired electrons and is attracted by a magnet. (11.11)

partial molar volume   The partial molar volume of the ith component of a mixture is defined by

$$\bar{V}_i = \left(\frac{\partial V}{\partial n_i}\right)_{T,P,n_j}$$

It measures the change in the volume of a solution upon the addition of n_i moles of the component at constant temperature, pressure, and moles of other components. Such a function is necessary because the volume of a solution is not, in general, the sum of the volumes of the individual components due to unequal intermolecular interactions upon mixing. (5.2)

partial specific volume ($\bar{v}$)   The increase in volume when 1 g of solute is dissolved in a large volume of solvent. (16.1)

protein denaturation   The disruption of protein structure caused by exposure to heat or chemicals, leading to loss of native three-dimensional structure and biological activity. (16.4)

Q

quantum-mechanical tunneling   The penetration of the wave function of a particle through a potential barrier into a classically forbidden region when the kinetic energy of the particle is less than the height of the potential barrier. (11.9)

quantum mechanics   The modern theory of matter, electromagnetic radiation, and the interaction between matter and radiation. It applies mainly to atomic and molecular systems. (11.7)

quantum number   An integer or half-integer used to characterize an electron in a certain state in an atom or molecule. (11.10)

quantum yield (Φ)   Ratio of the number of molecules of product formed to the number of light quanta absorbed. (15.1)

quaternary structure   A structure that results from the association of two or more folded polypeptides to form a functional protein. (16.3)

R

radial distribution function   The radial distribution, $r^2 R(r)^2$, that gives the probability of finding a particle in the range r and $r + dr$ regardless of direction, where r is the radius. (11.10)

Ramanchandran plot   A map that shows all possible backbone configurations for an amino acid in a polypeptide. (16.3)

random-walk model   A mathematical model for studying the three-dimensional structure of a polymer. In particular, it provides information about the distance between the beginning and the end of a polymeric chain. (16.2)

Raoult's law   The law stating that the partial pressure of a component of a solution is given by the product of the

mole fraction of the component and the vapor pressure of the pure component. (5.4)

rare earth metals   *See* lanthanides.

rate constant   Constant of proportionality between the reaction rate and the concentrations of reactants. (9.2)

rate-determining step   The slowest elementary step in the sequence of steps leading to formation of products. (9.3)

rate law   An expression relating the rate of a reaction to the rate constant and the concentrations of the reactants. (9.2)

reaction center   The location in a photosynthesizing cell that mediates the conversion of light energy into chemical energy. (15.2)

reaction mechanism   The sequence of elementary steps that leads to product formation. (9.3)

reaction order   The sum of the powers to which all reactant concentrations appearing in the rate law are raised. (9.2)

reaction quotient   A number equal to the ratio of product concentrations to reactant concentrations, each raised to the power of its stoichiometric coefficient at some point other than equilibrium. (6.1)

reduced mass (μ)   The reduced mass of two particles of masses m_1 and m_2 is defined by

$$\frac{1}{\mu} = \frac{1}{m_1} + \frac{1}{m_2} \quad (14.2)$$

residue   An individual amino acid unit in a polypeptide chain. (16.3)

resonance   The use of two or more Lewis structures to represent a particular molecule. (12.7)

resonance structure   One of two or more Lewis structures for a single molecule that cannot be described fully with only one Lewis structure. (12.7)

reverse osmosis   A process in which a solvent is forced by a pressure greater than the osmotic pressure to flow through a semipermeable membrane from a concentrated solution to a more dilute one. (5.6)

reversible inhibition   An inhibition in which the inhibitor can be removed by dialysis. (10.5)

reversible process   In a reversible process, a system is always infinitesimally close to equilibrium. Such a process is of theoretical interest but can never be realized in practice. (3.1)

root-mean-square velocity   The square-root of the sum of all the squares of the speeds divided by the total number of molecules present. (2.7)

S

salting-in effect   The increase in solubility of an electrolyte at high ionic strengths. (5.8)

salting-out effect   The decrease in solubility of an electrolyte at high ionic strengths. (5.8)

Schrödinger wave equation   The fundamental equation in quantum mechanics. It calculates the wave functions and energies of atoms and molecules. (11.7)

second law of thermodynamics   The law that says the entropy of an isolated system increases in an irreversible process and remains unchanged in a reversible process; it can never decrease. The mathematical statement of the second law, for a finite change, is

$$\Delta S_{univ} = \Delta S_{sys} + \Delta S_{surr} \geq 0$$

where ΔS_{univ}, ΔS_{sys}, and ΔS_{surr} are the changes in the entropy of the universe, system, and surroundings, respectively. (4.3)

second-order reaction   A reaction whose rate depends on the reactant concentration raised to the second power or on the concentrations of two different reactants, each raised to the first power. (9.2)

secondary structure   The folding of a polypeptide chain into local patterns such as the α helix or β sheet. Secondary structure is maintained by hydrogen bonds between the amide hydrogen and the carbonyl oxygen of the peptide bond. (16.3)

sedimentation coefficient (s)   A quantity that determines the velocity at which a particular particle will sediment during centrifugation. (16.1)

sedimentation equilibrium   The equilibrium situation in which the forward movement of sedimentation of a particle during centrifugation is balanced by the reverse movement due to its tendency to diffuse to lower concentration. (16.1)

selection rule   The rule that predicts which changes of state may occur in a specific type of spectroscopic transition. (14.1)

self-consistent field method   An iterative procedure for obtaining wave functions of a many-electron atom. (11.11)

semiconservative model   Replication of DNA by separation of the two complementary strands of the molecule, each being conserved and acting as a template for synthesis of a new complementary strand. (16.3)

semipermeable membrane   A membrane that allows solvent and certain solute molecules to pass through but blocks the movement of other solute molecules. (5.6)

sequential model   A model employed to explain the cooperative nature of hemoglobin binding to oxygen. When an oxygen molecule binds to a vacant site on one of its four subunits, the interaction causes the site to change its conformation, which in turn affects the binding constants of the three sites that are still vacant. This model can also be applied to allosteric enzymes. (10.6)

SI system   *See* international system of units. (1.2)

sigma (σ) bond   A covalent bond formed by orbitals overlapping end-to-end; its electron density is concentrated between the nuclei of the bonding atoms. (12.2)

singlet state   A state of an atom or molecule with zero electronic spin ($S = 0$). (14.1)

specific heat   The amount of energy required to raise the temperature of one gram of the substance by one degree Celsius. (3.3)

specific rotation   The calculated rotation of plane-polarized light passing through a solution containing a chiral solute as related to the concentration of the solution, pathlength, and optical rotation at a given wavelength and temperature. (14.9)

spectrochemical series   A list of ligands arranged in order of their abilities to split the d-orbital energy levels. (12.8)

spin multiplicity   Spin multiplicity is given by ($2S + 1$), where S is the spin quantum number. (14.1)

spin–spin coupling   A coupling between nuclear spins that gives rise to the fine structure of an NMR spectrum. (14.5)

spontaneous process A process that occurs on its own accord under a given set of conditions. (4.1)

standard enthalpy of reaction The enthalpy change at a certain temperature when reactants in their standard states are converted to products in their standard states. (3.6)

standard hydrogen electrode An electrode involving the reversible half-reaction

$$H^+(1\ M) + e^- \rightleftharpoons \tfrac{1}{2}H_2(g)$$

It is assigned a zero electrode potential when the gas is at 1 bar pressure and the concentration of the H^+ ions is at 1 M at 298 K. (7.2)

standard molar enthalpy of formation The enthalpy change when 1 mole of a compound is synthesized from its elements in their standard states of 1 bar at some temperature. (3.6)

standard molar Gibbs energy of formation Gibbs energy change when 1 mole of a compound is synthesized from its elements in their standard states of 1 bar at some temperature. (4.7)

standard reduction potential The electrode potential of a substance for the reduction half-reaction

$$Ox + ve^- \rightarrow Rd$$

where Ox and Rd are the oxidized and reduced forms of the substance, and v is the stoichiometric coefficient. The Ox and Rd forms are in their standard states, and the measured potential is based on the standard hydrogen electrode reference scale. (7.2)

standard state A reference state with respect to which thermodynamic quantities are defined. The standard state of a solid or liquid is the most stable form of the solid or liquid at 1 bar and the specified temperature. The standard state of an ideal gas is the pure gas at 1 bar and the specified temperature. (3.6)

state The condition of a system that is specified as completely as possible by observations of a specific nature. An example is the thermodynamic state, which is described by thermodynamic properties such as temperature, pressure, and composition. (2.1)

state function A property that is determined by the state of a system. The change in any state function in a process is path independent. (3.1)

state of the system The values of all pertinent macroscopic variables (for example, composition, volume, pressure, and temperature) of a system. (2.1)

steady-state approximation Assumes that the concentration of a reaction intermediate remains constant during the main part of the reaction. (9.4)

supercritical fluid The state of a substance above its critical temperature. (2.5)

surroundings The rest of the universe outside the system. (2.1)

svedberg (s) A unit used for sedimentation coefficient; it is equal to 10^{-13} s. (16.1)

system Any specific part of the universe that is of interest to us. (2.1)

T

termolecular reaction An elementary step that involves three molecules. (9.3)

tertiary structure The specific three-dimensional shape into which the entire polypeptide chain is folded in a protein molecule. (16.3)

thermal motion The random, chaotic molecular motion. The more energetic the thermal motion, the higher the temperature. (2.6)

thermal reaction Reaction that occurs with the reactant molecules in their electronic ground states. The rate of the reaction is governed by the thermal motion of the reactant molecules. (15.1)

thermochemistry The study of heat changes in chemical reactions. (3.6)

thermodynamic efficiency The ratio of the work done by a heat engine to the heat absorbed by the engine. It is expressed as $1 - (T_1/T_2)$, where T_2 is the temperature at which the heat engine absorbs heat, and T_1 is the temperature of the cold reservoir that receives the heat that has not been transformed into work. (4.2)

thermodynamic equilibrium constant The equilibrium constant expression in which concentration terms are expressed either as activities (for solutes in solution) or fugacities (for gases). (6.1)

thermodynamics The scientific study of the interconversion of heat and other forms of energy. (3.1)

third law of thermodynamics The law stating that every substance has a finite positive entropy, but at the absolute zero of temperature the entropy may become zero, and it does in the case of a pure, perfect crystalline substance. (4.5)

threshold frequency The minimum frequency of light required to eject an electron from a metal's surface. (11.3)

transition dipole moment A measure of the dipolar character of the shift in electronic charge that occurs during a spectroscopic transition. The transition is allowed if the transition dipole moment is nonzero. (14.1)

transition metals Elements that have incompletely filled d subshells or readily give rise to cations that have incompletely filled d subshells. (11.11)

transition state *See* activated complex.

triplet state A state of an atom or molecule in which the total spin quantum number is $S = 1$. Thus, the spin multiplicity of this state is $(2S + 1) = 3$. (14.1)

turnover number The number of substrate molecules processed by an enzyme molecule per second when the enzyme is saturated with the substrate. Also referred to as k_{cat}, the catalytic rate constant. (10.2)

U

unimolecular reaction An elementary step that involves one molecule. (9.3)

V

valence bond theory A theory of electronic structures of molecules. It describes each bond as being formed by spin-pairing of electrons in atomic orbitals. (12.2)

valence electron The outer electrons of an atom, which are those involved in chemical bonding. (11.11)

van der Waals equation An equation of state that describes P, V, T, and n of a real gas. (2.4)

van der Waals forces The weak attractive forces: dipole–dipole, dipole–induced dipole, and dispersion forces. (13.3)

van't Hoff equation   An equation that shows the temperature dependence of the equilibrium constant in terms of the enthalpy of reaction:

$$\left(\frac{\partial \ln K}{\partial T}\right)_P = \frac{\Delta_r H^\circ}{RT^2} \quad (6.4)$$

van't Hoff factor   The ratio of the actual number of ionic particles in solution after dissociation to the number of formula units initially dissolved in solution. (5.9)

virial equation of state   An equation of state for real gases. The equation is expressed as an expansion in powers of the molar volume, or pressure. (2.4)

volatile   A term meaning that a substance has a measurable vapor pressure. (5.4)

W

Wien effect   An increase in the conductance of an electrolyte at very high potential gradients. (5.8)

work   In mechanics, work is force times distance. In thermodynamics, the most common forms of work are gas expansion (or compression) and electrical work carried out in an electrochemical cell. (3.1)

Z

Z scheme   A mechanism whereby electrons flow between PSII and PSI during photosynthesis. (15.2)

zero-order reaction   A reaction whose rate is independent of the concentrations of the reactants. (14.2)

zero-point energy   The minimum energy that a system may possess. (11.8)

zeroth law of thermodynamics   The law stating that if system A is in thermal equilibrium with system B, and system B is in thermal equilibrium with system C, then system C is also in thermal equilibrium with system A. (2.2)

zwitterion   *See* dipolar ion.

Chapter 2

2.6 32.0 g mol^{-1}
2.8 2.98 g L^{-1}
2.10 (a) 1.1×10^{-7} mol L^{-1}.
   (b) 18 ppm
2.12 (a) 0.85 L
2.14 (a) 4.9 L. (b) 6.0 atm.
   (c) 0.99 atm
2.16 N_2O
2.18 3.2×10^7 molecules; 2.5×10^{22}
   molecules
2.20 O_2: 28%; N_2: 72%
2.24 N_2: 88.9%; H_2: 11.1%
2.26 13 days
2.28 349 mmHg
2.30 0.45 g
2.32 4.8%
2.38 $P_T = 1.02$ atm, $P_{Ar} = 0.30$ atm,
   $P_{He} = 0.720$ atm, $x_{Ar} = 0.29$,
   $x_{He} = 0.71$
2.42 $a = 18.7$ atm L^2 mol^{-2}
   $b = 0.120$ L mol^{-1}
2.52 0.29 atm
2.54 6.07×10^{-21} J; 3.65×10^3
   J mol^{-1}
2.56 460 K
2.58 42.6 K
2.62 N_2: 1.33×10^4 m; He: 9.31×10^4 m
2.64 $c_{rms} = 2.8$ m s^{-1}, $\bar{c} = 2.7$ m s^{-1}
2.68 $c_{mp} = 406$ m s^{-1}; 0.0427
2.72 3.53×10^{-8} m; 7.70×10^{31}
   collisions L^{-1} s^{-1}
2.74 12.0 K
2.76 At 1.0 atm: $Z_1 = 3.4 \times 10^9$
   collisions s^{-1}, $Z_{11} = 4.0 \times 10^{34}$
   collisions m^{-3} s^{-1}. At 0.10 atm:
   $Z_1 = 3.4 \times 10^8$ collisions s^{-1},
   $Z_{11} = 4.0 \times 10^{32}$ collisions
   m^{-3} s^{-1}
2.78 16.0 g mol^{-1}; CH_4
2.80 CO: 54%; CO_2: 46%
2.82 H_2: 0.5857; D_2: 0.4143
2.84 5.27×10^{18} kg
2.86 1.07×10^3 mmHg
2.92 (b) 509 K
2.94 (a) 61.3 m s^{-1}. (b) 4.57×10^{-4} s. (c) 328 m s^{-1}
2.96 16.3
2.100 NO
2.102 30%

Chapter 3

3.4 (a) -112 J. (b) -230 J
3.6 -2.27×10^3 J
3.10 $\Delta U = 0$, $q = -20$ J
3.14 $\Delta U = 0$, $\Delta H = 0$ for both (a)
   and (b)
3.20 0.71 atm
3.22 50.8°C
3.26 $\bar{C}_V = 22$ J K^{-1} mol^{-1};
   $\bar{C}_P = 30$ J K^{-1} mol^{-1}
3.34 (a) 207 K. (b) 226 K
3.36 24.8 kJ g^{-1}; 603 kJ mol^{-1}
3.38 25.0°C
3.40 (a) -2905.6 kJ mol^{-1}.
   (b) 1452.8 kJ mol^{-1}.
   (c) -1276.8 kJ mol^{-1}
3.42 (a) -167.2 kJ mol^{-1}.
   (b) -229.6 kJ mol^{-1}
3.44 -337 kJ mol^{-1}
3.46 -23.2 kJ mol^{-1}
3.48 500 J mol^{-1}
3.50 -197 kJ mol^{-1}
3.52 1.9 kJ mol^{-1}
3.54 -238.7 kJ mol^{-1}
3.56 0
3.58 (b) 75.3 kJ mol^{-1}
3.60 -2758 kJ mol^{-1}; -3119.4
   kJ mol^{-1}
3.66 2.8×10^3 g
3.68 1.19×10^4 K
3.72 (a) -65.2 kJ mol^{-1}. (b) -9.4
   kJ mol^{-1}
3.74 47.8 K; 4.1×10^3 g
3.76 7.60%
3.78 9.90×10^8 J; 305°C
3.86 56°C
3.88 7.6×10^4 K

Chapter 4

4.8 $\Delta U = 43.34$ kJ, $\Delta H = 46.44$ kJ,
   $\Delta S = 126.2$ J K^{-1}
4.10 4.5 J K^{-1}. Same.
4.12 (a) 75.0°C. (b) $\Delta S_A = 22.7$
   J K^{-1}, $\Delta S_B = -20.79$ J K^{-1},
   $\Delta S_T = 1.9$ J K^{-1}
4.14 0.36 J K^{-1}
4.16 (a) $\Delta S_{sys} = 5.8$ J K^{-1},
   $\Delta S_{surr} = -5.8$ J K^{-1}, $\Delta S_{univ} = 0$.
   (b) $\Delta S_{sys} = 5.8$ J K^{-1},
   $\Delta S_{surr} = -4.15$ J K^{-1},
   $\Delta S_{univ} = 1.7$ J K^{-1}

4.18 0
4.20 (a) -543.8 J K^{-1} mol^{-1}.
   (b) -117.0 J K^{-1} mol^{-1}.
   (c) 284.4 J K^{-1} mol^{-1}. (d) 19.4
   J K^{-1} mol^{-1}
4.22 (a) $\Delta S_{sys} = 5.8$ J K^{-1},
   $\Delta S_{surr} = -5.8$ J K^{-1}, $\Delta S_{univ} = 0$.
   (b) $\Delta S_{sys} = 5.8$ J K^{-1}, $\Delta S_{surr} = -3.4$ J K^{-1}, $\Delta S_{univ} = 2.4$ J K^{-1}
4.24 $\Delta_r S° = 24.6$ J K^{-1} mol^{-1},
   $\Delta S_{surr} = -607$ J K^{-1} mol^{-1},
   $\Delta S_{univ} = -582$ J K^{-1} mol^{-1}
4.28 (a) $w = -1.5 \times 10^3$ J, $q = 1.5 \times 10^3$ J, $\Delta U = 0$, $\Delta H = 0$,
   $\Delta S = 5.3$ J K^{-1}, $\Delta G = -1.5 \times 10^3$ J. (b) $\Delta U, \Delta S$, and ΔG the
   same as (a). $w = -6.3 \times 10^2$ J,
   $q = 6.3 \times 10^2$ J
4.30 -222.7 kJ mol^{-1}
4.32 -75.9 kJ mol^{-1}
4.34 (a) $\Delta_r H° = 1.90$ kJ mol^{-1},
   $\Delta_r S° = -3.3$ J K^{-1} mol^{-1}.
   (b) 1.4×10^4 bar
4.42 -2.20 K
4.50 88.9 torr
4.54 0.20 J K^{-1}
4.56 $\Delta U = -1.25 \times 10^3$ J,
   $\Delta H = -2.08 \times 10^3$ J,
   $\Delta S = -15.1$ J K^{-1}
4.58 352°C
4.62 $S_{sys} = 162$ J K^{-1}, $S_{surr} = -158$ J K^{-1}, $\Delta S_{univ} = 4$ J K^{-1}
4.66 $\Delta H° = -11.5$ kJ mol^{-1},
   $\Delta G° = 12.0$ kJ mol^{-1}
4.76 -6.24×10^3 J
4.80 1.20 kJ K^{-1} mol^{-1}; 9.84 J K^{-1}
   mol amino acid^{-1}

Chapter 5

5.2 2.28 M
5.4 5.0×10^2 m; 18.3 M
5.8 10 m
5.10 (a) 11.53 J K^{-1}. (b) 50.45 J K^{-1}
5.22 2.7×10^{-3}
5.28 3.91×10^3 mL. Yes
5.30 10.2 atm
5.32 O_2: 4.6×10^{-6}; N_2: 8.9×10^{-6}
5.34 $\Delta P = 6.147 \times 10^{-5}$ mmHg;
   $\Delta T_f = 2.67 \times 10^{-4}$ K; $\Delta T_b = 7.3 \times 10^{-5}$ K; $\pi = 2.67$ torr
5.36 39.7 g mol^{-1}

5.38 No

5.40 A: 1.2×10^2 g mol^{-1};
B: 2.5×10^2 g mol^{-1}. The
compound dimerizes in benzene.

5.42 In solution: $x_A = 0.65$,
$x_B = 0.35$. In vapor: $x_A = 0.81$,
$x_B = 0.19$

5.44 A: 0.19; B: 0.81

5.46 0.5114 K m^{-1}

5.50 22.3 J mol^{-1}

5.52 0.63

5.54 (a) 0.10 m; 0.69. (b) 0.030 m;
0.67. (c) 1.0 m; 9.1×10^{-3}

5.56 $m_{\pm} = 0.32\ m$, $a_{\pm} = 0.041$,
$a = 7.0 \times 10^{-5}$

5.58 24.8 Å

5.60 (a) $\gamma_+ = 0.70$, $\gamma_- = 0.91$.
(b) $\gamma_{\pm} = 0.83$

5.64 7.2×10^5 g mol^{-1}

5.66 -14 kJ

5.68 22.2 kJ mol^{-1}

5.72 168 m

5.74 (a) 2.5×10^{-3} g L^{-1}. (b) 3.9×10^{-4} g L^{-1}

5.76 (a) 5.1×10^{-5} M. (b)
$[\text{Oxa}^{2-}] = 3.0 \times 10^{-7}$ M,
$[\text{Ca}^{2+}] = 0.010$ M

5.78 $[\text{Co}(\text{NH}_3)_5\text{Cl}]\text{Cl}_2$

5.80 (a) 0.150 atm. (b) 0.072 atm

5.82 -35.8 kJ

Chapter 6

6.2 (a) 0.49. (b) 0.23. (c) 0.036.
(d) >0.036 mol

6.4 (b)(i) 1.4×10^5. (ii) CH$_4$: 2 atm;
H$_2$O: 2 atm; CO: 13 atm; H$_2$:
38 atm

6.6 (a) $K_c = 1.07 \times 10^{-7}$,
$K_P = 2.67 \times 10^{-6}$. (b) 22 mg m^{-3}

6.8 1.74×10^5 J mol^{-1}

6.10 2.59×10^4 J mol^{-1}

6.12 $\Delta_r G^\circ$/kJ mol^{-1}: 23, 11, 0, -11,
-23

6.14 $K_P = 0.116$, $\Delta_r G^\circ = 5.33 \times 10^3$
J mol^{-1}

6.16 575.3 K

6.20 NO$_2$: 0.96 bar; N$_2$O$_4$: 0.21 bar

6.28 1.1×10^{-5}

6.30 -29.2 kJ mol^{-1}

6.32 29.5 kJ mol^{-1}; 6.72×10^{-6}

6.34 $\Delta_r H^\circ = -3.9 \times 10^4$ J mol^{-1},
$\Delta_r S^\circ = -1.3 \times 10^2$ J K^{-1} mol^{-1}

6.36 (a) 1.59×10^{-4}. (b) 4.00×10^{-5}

6.38 9.1 kJ mol^{-1}

6.40 -7.3×10^3 J mol^{-1}

6.42 Isobutane: 70%; butane: 30%

6.46 4; 1.7×10^{-4}

Chapter 7

7.2 1.125 V; 1.115 V

7.6 0.531 V

7.8 (a) $E^\circ = 0.913$ V, $\Delta_r G^\circ = -1.76 \times 10^5$ J mol^{-1}; 7.34×10^{30}. (b) 0.824 V, -1.59×10^5
J mol^{-1}, 7.12×10^{27}. (c) 0.736 V,
-3.55×10^5 J mol^{-1}; 1.60×10^{62}

7.10 0.50 bar

7.12 2.55; -2.32×10^3 J mol^{-1}

7.14 0.010 V

7.16 -0.142 V

7.18 1.45×10^4

7.20 117 mol

7.22 2.97×10^4 J mol^{-1}; 6.17×10^{-6}

7.24 (a) -0.197 V, (b) -0.241 V

7.26 1.4 mol ATP

7.28 -18 mV

7.30 4.81×10^{-13}

7.34 1.124×10^{-4} V K^{-1}; 1.098 V

7.36 1.01×10^{-14}

7.38 (a) X^{2+}/X is negative and
Y^{2+}/Y is positive. (b) 0.59 V

7.42 3.3×10^2 bar

7.44 -86.3 kJ mol^{-1}

7.46 1.09 V

Chapter 8

8.6 8.4×10^{-4}

8.8 (a) 0.32. (b) 0.34

8.12 $[\text{HSO}_4^-] = 0.16$ M, $[\text{H}^+] = 4.5 \times 10^{-2}$ M, $[\text{SO}_4^{2-}] = 4.5 \times 10^{-2}$ M

8.14 (c)

8.16 2.7×10^{-3} g

8.18 2.8×10^{-2}

8.24 5.13

8.28 5.70

8.30 1.3 M

8.32 (a) 12.6. (b) 8.8×10^{-6} M

8.36 9.25; 9.18

8.38 6.8

8.40 (a) H$_2$PO$_4^-$/HPO$_4^{2-}$. (b) 6.76

8.44 10.5

8.48 13 mL

8.54 Lysine: 9.74; valine: 5.97

8.58 52 kJ mol^{-1}

8.62 3.783

8.64 38 kJ mol^{-1}

8.68 21.4 kJ mol^{-1}

8.70 1×10^{-4}

8.74 (b) 5.2×10^3. (c) 7.93

8.76 9.25; for pure NH$_3$, pH changes
from 11.28 to 10.26

8.78 (a) 0.054 M H$^+$ (pH unit)$^{-1}$
(b) 0.014 M H$^+$ (pH unit)$^{-1}$

8.80 7.83

Chapter 9

9.2 6.2×10^{-6} M s^{-1}

9.4 0.99

9.6 (a) 1.21×10^{-4} yr^{-1}. (b) 2.1×10^4 yr

9.8 Second order. $k = 0.42$
M^{-1} min^{-1}

9.10 1.19×10^{-4} s^{-1}

9.12 3.6 s

9.18 47.95 g

9.22 (a) Rate $= k[\text{NO}]^2[\text{H}_2]$. (b) 0.38
M^{-2} s^{-1}

9.26 10^{11} s^{-1}; 4.5×10^{10} s^{-1};
2.0×10^2 s^{-1}

9.30 $A = 3.38 \times 10^{16}$ s^{-1}, $E_a = 100$
kJ mol^{-1}

9.32 371°C

9.34 298 K

9.36 $E_a = 13.4$ kJ mol^{-1},
$\Delta H^{\circ\ddagger} = 10.8$ kJ mol^{-1},
$\Delta S^{\circ\ddagger} = -29.3$ J K^{-1} mol^{-1},
$\Delta G^{\circ\ddagger} = 19.8$ kJ mol^{-1}

9.46 (a) 22.5 cm^2. (b) 44.9 cm^2

9.48 First order

9.50 86 kJ mol^{-1}

9.52 2.7×10^{14}

9.54 (b) 3.8×10^{-3} M^{-1} s^{-1}

9.58 (a) 1.11×10^{-2} min^{-1}

9.62 4×10^{-4} M

9.64 6.4×10^9 M^{-1} s^{-1}

Chapter 10

10.6 $V_{\max} = 1.3 \times 10^{-5}$ M min^{-1},
$K_M = 1.2 \times 10^{-3}$ M, $k_2 = 3.3$
min^{-1}. Both plots give the same
values.

10.8 (a) 898 K. (b) 76.4 kJ mol^{-1}

10.10 8.5×10^3 g mol^{-1} (minimum
molar mass)

10.12 (a) 16 μM min^{-1}. (b) 1.7
μM min^{-1}. (c) 1.8 μM min^{-1}

10.14 $[\text{I}] = 6.5 \times 10^{-5}$ M,
$[\text{S}] = 1.4 \times 10^{-2}$ M

Chapter 11

11.2 5.66×10^{-19} J

11.4 2.34×10^{14} s^{-1}

11.10 1×10^6 m s^{-1}

11.12 7.9×10^{-36} m s^{-1}

11.14 He$^+$(H): n_3: 164 nm (656 nm);
n_4: 122 nm (486 nm); n_5: 109 nm
(434 nm); n_6: 103 nm (410 nm)

11.16 1.2×10^2

11.18 (a) 106.7 kJ mol^{-1}. (b) 1.12×10^3 nm

11.22 1.67×10^{25}; 30

11.34 (a) 1.313×10^3 kJ mol^{-1}.
(b) 3.282×10^2 kJ mol^{-1}

11.36 4.173×10^2 kJ mol^{-1}

11.38 343 nm

11.44 419 nm

11.46 3.1×10^{19}

11.48 419 nm

11.52 2.75×10^{-11} m

11.54 2.8×10^6 K

11.56 0.649

11.58 **(a)** B: $n_i = 4 \rightarrow n_f = 2$; C: $n_i = 5 \rightarrow n_f = 2$. **(b)** A: 41.0 nm; B: 30.4 nm. **(c)** 2.18×10^{-18} J. **(d)** The continuum denotes ionization.

11.64 **(a)** 2 **(b)** 1.9 and 7.1

Chapter 12

12.26 **(b)** 255 kJ mol^{-1}

12.32 1.6×10^4 g mol^{-1}; 4 Fe/hemoglobin

12.38 **(c)** -413 kJ mol^{-1}

Chapter 13

13.12 **(a)** 1.70 Å. **(b)** 5.1×10^{-4}

13.20 0

13.30 **(a)** 2.98 Å. **(b)** 7.60×10^{-2} kJ

Chapter 14

14.2 2.2×10^4 cm^{-1}; 6.7×10^{14} s^{-1}

14.4 **(a)** 100%. **(b)** 76%. **(c)** 0.0025%

14.6 8.0×10^6 s^{-1}; 9.9×10^{-6} nm

14.10 5.3 L mol^{-1} cm^{-1}

14.12 16 h

14.14 15

14.16 1.02×10^{11} s^{-1}

14.18 **(a)** 3. **(b)** 7. **(c)** 9. **(d)** 30

14.20 3.5×10^2 N m^{-1}

14.26 H_2

14.32 4.0 ppm

14.34 14.1 T

14.36 1.0×10^3 Hz

14.44 1.6×10^{-90}

14.50 1225 nm

14.54 $[\alpha]_D^{25} = 3.5$ deg dm^{-1} cm^3 g^{-1}; $[\Phi]_D^{25} = 12$ deg dm^{-1} cm^3 mol^{-1}

14.56 0.319

14.64 51 deg dm^{-1} cm^3 g^{-1}

14.66 0.23 g mL^{-1}

14.68 3.3×10^{-4} M

14.72 3

14.74 A: 6.04×10^{-6} M; B: 6.21×10^{-6} M

14.78 **(b)** 23.3 kJ mol^{-1}

14.82 **(a)** ^{1}H: 1.33×10^{-25} J; ^{13}C: 3.34×10^{-26} J. **(b)** ^{1}H: 200 MHz; ^{13}C: 50.3 MHz. **(c)** 11.7 T

14.84 0.50

Chapter 15

15.2 266 kJ einstein^{-1}

15.4 0.022; 3.11×10^5 J

15.6 2.01×10^{-2} s

15.10 4.6

15.14 **(a)** 20.0%. **(b)** 35.1%

15.16 9.9 mol

Chapter 16

16.2 **(a)** 1.9×10^4 g mol^{-1}. **(b)** 8

16.4 **(a)** 6283 rad s^{-1}. **(b)** 2.9×10^6 m s^{-2}. **(c)** 3.0×10^5 g

16.6 19 kg mol^{-1}

16.12 14 h

16.14 3.90×10^4 g mol^{-1}; dissociates into 2 subunits

16.18 11 turns s^{-1}

16.24 **(a)** 2 bits. **(b)** 2 CDs

16.26 **(b)** 44.5%

16.32 **(a)** 1/945. **(b)** 1/3150

16.34 Random coil above pH 7 and α helical at pH below 4.25

Index

A

Absolute entropy, 96
Absolute zero, 9
Absolute zero temperature scale, 9
Absorbance, 522
Absorption spectroscopy, 513
Acetaldehyde, 344, 550
Acetylcholine, 260, 384
Acetylcholinesterase, 260
Acetyl coenzyme A, 249
Acetylene, 454
Acid-base indicator, 287
Acid-base titration, 286
    Proteins, 292
Acids, 267
    amino, 288, 289t
    Brønsted, 267
    diprotic, 276
    dissociation constants, 270, 273t
    hard, 469
    Lewis, 268
    polyprotic, 276
    soft, 469
Actinides, 437
Actinometer, 578
Action potential, 258
Action spectrum, 579
Activated complex, 335
Activation energy, 26, 334
Active site, 364
Active transport, 178
Activity, 140, 160
Activity coefficient, 140
    Debye-Hückel theory, 164
    determination of, 246
    ionic, 162
    mean ionic, 162
    of electrolytes, 162
    of nonelectrolytes, 140
Adenine, 504
Adenosine diphosphate, see ADP
Adenosine triphosphate, see ATP
Adiabatic bomb calorimeter, 59
Adiabatic process, 56
ADP, 220
Aerobic process, 226
Alcohol dehydrogenase, 344, 375, 481

Allosteric interaction, 2, 385
    concerted model, 390
    sequential model, 391
Alpha decay, 424
Alpha (α) helix, 619
Amide group, 616
Amino acids, 288, 289t
    in proteins, 289
    pI values of, 289
Ammonia, 454
Anaerobic process, 226
Andrews, Thomas, 18
Anfinsen, Christian, 626
Angstrom (Å), 429
Angular momentum quantum number (l), 427
Anharmonicity constant, 530
Anode, 236
Antenna molecule, 581
Antibiotics, 180
Antibonding molecular orbital, 458
Apparent equilibrium constant, 201
Apparent solubility product, 168
Arrhenius equation, 333
Arrhenius, Svante, 333
Atmosphere
    ionic, 165
    unit of, 5
Atomic emission spectra, 407
Atomic mass unit, 5
Atomic orbitals, 428
Atomic radius, 437
ATP, 220
    NMR spectra of, 548
    structure of, 220
    synthesis of, 227, 252
Aufbau principle, 435
Average speed, 26
Avogadro's constant, 5
Avogadro's law, 10
Axon, 382
Azimuthal quantum number, 427

B

BAL, see British anti-Lewisite
Balmer series, 408
Bar (unit), 4

Barometer, 4
Bases, 267
    Brønsted, 267
    dissociation constants of, 274
    hard, 469
    Lewis, 268
    soft, 469
Bathorhodopsin, 589
Beer-Lambert law, 521
Beer, Wilhelm, 521
Belousov-Zhabotinskii reaction, 353
Benzene
    ESR spectrum of, 556
    molecular orbitals in, 467
    resonance structures of, 466
Beta (β) structure, 618
Bicarbonate-chloride shift, 296
Bimolecular reaction, 327
Binary collisions, 29
Binary liquid mixture, 134
    vapor-liquid equilibrium of, 135
Binding equilibria, 209
Binomial distribution, 545
Biochemists' standard state, 218
Bioelectrochemistry, 255
Bioenergetics, 217
Biological membranes, 175
Biological oxidation, 248
Biphasic, 273
2,3-Bisphosphoglycerate (BPG), 386
Blackbody radiation, 404
Blood
    pH of, 293
Bohr, Christian, 386
Bohr effect, 386
Bohr, Neil, 408
Bohr radius, 429
Bohr's theory of atomic spectra, 408
Boiling-point elevation, 143
Boltzmann constant, 24
Boltzmann distribution law, 52
    electronic, 575
    NMR, 542
Boltzmann equation, 84
Boltzmann, Ludwig, 24
Bomb calorimeter, 59

Bond
    covalent, 447
    dipole moment, 457
    dissociation energy, 70
    dissociation enthalpy, 71, 72t
    energy, 70
    enthalpy, 70, 72t
    hydrogen, 502, 617
    ionic, 490
    moment, 457
    order, 460
Bonding molecular orbital, 459
Born-Haber cycle, 492
Born, Max, 417
Bound state, 423
Boundary conditions, 418
Boundary surface diagram, 432
Bovine pancreatic trypsin inhibitor, 632
Boyle's law, 8
Briggs, George, 368
Briggs-Haldane kinetics, 368
British anti-Lewisite, 483
Brønsted, Johannes, 267
Brønsted theory of acids and bases, 267
Buffer capacity, 285
Buffer range, 286
Buffer solution, 280
Burk, Dean, 370
Butadiene, 423
B-Z reaction, 353

C

Cadmium, 482
Calomel electrode, 247
Calorie, 44
Calorimeter
    bomb, 59
    constant-pressure, 61
    constant-volume, 59
Capacitance, 183
Capacitor, 183
Carbaminohemoglobin, 295
Carbon-12, 5
Carbon dioxide
    binding to hemoglobin, 295
    critical state of, 18
    dry ice, 116
    fixation, 586

669

Carbon dioxide (*continued*)
  isotherm of, 18
  normal modes of, 531
  phase diagram of, 116
  standard enthalpy of
    formation of, 65
  supercritical fluid, 21
Carbon monoxide
  binding to hemoglobin,
    479
  Lewis structure of, 447
  molecular orbital
    diagram of, 464
  standard enthalpy of
    formation of, 67
Carbonic acid, 276, 293,
    371
Carbonic anhydrase, 294,
    371
Carbonyl sulfide, 526
Carboxyhemoglobin, 479
Carnot heat engine, 87
Carnot, Sadi, 87
Carotene, 581
Catalase, 372
Catalysis, 363
Catalytic rate constant
    (k_{cat}), 371
Cathode, 236
Cell
  concentration, 245
  electrochemical, 235
  membrane, 175
Cell constant, 250
Cell emf, 238
  and concentration, 242
  temperature dependence,
    244
Cell membrane, 175
Celsius temperature scale,
    10
Center of symmetry, 562
Centrifugal constant, 524
Chain reaction, 332
Charge-transfer spectra, 537
Charles, Jacques, 9
Charles' law, 9
Chemical equilibrium, 193
Chemical kinetics, 211
Chemical potential (μ), 131
  and partial molar Gibbs
    energy, 131
  effect of solute on, 141
  of ions, 161
Chemical reactions
  enthalpy change in, 65
  entropy change in, 98
  Gibbs energy change in,
    106

Chemical shift, 543
Chemiosmotic theory, 253
Chiralty, 562
Chlorophyll
  chlorophyll *a*, 581
  chlorophyll *b*, 581
  dimer, 584
Chloroplast, 580
Choline, 384
Christiansen, Jens, 326
Chromophore, 535
Chymotrypsin, 372
Chymotrypsinogen, 372
Circular birefringence, 564
Circular dichroism, 567
Circularly polarized light,
    563
Cis-trans isomerization,
    587
Citric acid cycle, 249
Clapeyron, Benoit, 113
Clapeyron equation, 112
Clausius-Clapeyron equa-
    tion, 113
Clausius, Rudolf, 113
Closed system, 7
Cobalt, 480
Coenzyme, 480
Coherence, 561
Collagen, 620
Colligative properties, 142
  of electrolyte solutions,
    170
  of nonelectrolyte solu-
    tions, 142
Collision diameter, 28
Collision frequency (Z_1), 28
Collision theory of chemical
    kinetics, 336
Color, 474
Competitive inhibition, 377
Complex conjugate, 417
Complexity, 322
Complex number, 417
Compressibility factor (Z),
    14
Concentration cell, 245
Concentration units, 127
Concerted model, 390
Condon, Edward, 534
Cone cells, 586
Configuration, 613
Conformation, 613
Consecutive reactions, 330
Constant
  anharmonicity, 530
  Avogadro, 5
  Boltzmann, 24
  centrifugal, 524

Constant (*continued*)
  critical, 19
  dielectric, 156, 183
  equilibrium, 195
  Faraday, 239
  force, 527
  gas, 11
  hyperfine splitting, 553
  Michaelis, 369
  normalization, 420
  Planck, 404
  rate, 312
  Rydberg, 407
Constructive interference,
    402
Cooperativity, 2, 385, 392
Coordination compound,
    469
Coordination number, 469
Copper
  electron configuration,
    435
  in bioinorganic com-
    pounds, 480
Corey, Robert, 616
Cotton, Aimé, 567
Cotton effect, 567
Coulomb, August, 156
Coulomb's law, 156, 490
Coupled reaction, 222
Covalent bond, 447
Covalent hydrolysis, 374
Covalent radius, 437
Crenation, 153
Crick, Francis H. C., 505
Critical constants, 19
Critical point, 19
Critical pressure, 19
Critical state, 18
Critical temperature, 19
Critical volume, 19
Crystal field splitting, 471
Crystal field stabilization
    energy, 471, 482
Crystal field theory, 470
Cycle
  citric acid, 249
Cyclohexane, 546
Cytochrome *c*, 480, 632
Cytochrome oxidase, 250
Cytosine, 504

D
d-d transitions, 537
Dalton, John, 12
Dalton's law of partial
    pressures, 12
Daniell cell, 236
Dark reaction, 580, 586

Davisson, Clinton, 411
de Broglie, Louis, 410
de Broglie relation, 410
Debye-Hückel limiting law,
    165
Debye, Peter, 96
  cube law, 96
  dipole moment, 457
Degeneracy
  atomic energy level, 430
  benzene, 467
Degenerate vibrations, 532
Degrees of freedom
  and molecular motion,
    530
  in phase equilibria, 117
Delocalized molecular
    orbital, 467
Density-gradient
    sedimentation, 607
Deoxyribonucleic acid, *see*
    DNA
Desalination, 154
Destructive interference,
    402
Dextrorotatory, 565
Dialysis, 213
  equilibrium, 213
Dielectric constant (ε), 156,
    183
Differential
  exact, 647
  inexact, 647
  partial, 645
  total, 645
Diffusion
  facilitated, 177
  gaseous, 30
  simple, 177
Diffusion coefficient, 348,
    606
  Fick's laws of, 605
Diffusion-controlled
    reactions, 348
Diisopropyl phosphofluori-
    date, 384
2,3-Dimercaptopropanol,
    626
Dipeptide, 616
Dipolar ion, 290
Dipole-dipole interaction,
    492
Dipole-induced dipole
    interaction, 495
Dipole moment, 457
  and molecular symmetry,
    457
  induced, 495
  transition, 520

Diprotic acid, 276
   titration of, 301
Dispersion curve, 566
Dispersion interaction, 497
Distribution, 99
Distribution diagram, 279
Disulfide bond, 624, 632
DNA, 621
   and radiation damage, 591
   hypochromism, 536
   melting temperature of, 536, 623
   renaturation of, 320
   semiconservative replication of, 623
   structure of, 622
   Watson-Crick model of, 504
Donnan effect, 172, 295
Donnan, Frederick George, 172
Doppler broadening, 516
Doppler, Christian, 516
Double-reciprocal plot, 370
Dye laser, 562

E
Eadie-Hofstee plot, 370
EDTA, 483
Effective nuclear charge, 434
Effector, 385
Efficiency, thermodynamic, 87, 227
Effusion, 30
Einstein, 405
   photoelectric effect, 405
   unit for photons, 576
Einstein coefficient
   of spontaneous emission, 518
   of stimulated absorption, 518
   of stimulated emission, 518
EKG, 261
Elastic collision, 21
Electric field, 182
Electric potential, 182
Electrical work, 104, 239
Electrocardiogram, see EKG
Electrocatalyst, 246
Electrochemical cell, 235
Electrochemistry, 235
Electrode
   calomel, 247
   gas, 236

Electrode (continued)
   glass, 247
   hydrogen, 236
Electrode potential, 236
   temperature dependence, 244
Electrolyte solutions, 154
Electromagnetic radiation, 402
Electromagnetic wave, 402
Electromotive force, 236
   temperature dependence, 244
Electron affinity, 439
Electron configuration
   of atoms, 433
   of molecules, 460
Electron delocalization, 467
Electron density, 429
Electron diffraction, 411
Electron microscope, 413
Electron spin, 431
Electron spin resonance spectroscopy, 552
Electronegativity, 455
Electronic spectra
   of amino acids, 536
   of DNA bases, 336
   relation with emission spectra, 555
   of tetracyanoethylene, 537
   of $Ti(H_2O)_6^{3+}$, 472
Electronic spectroscopy, 534
Electrophoresis, 608
Electrophoretic mobility, 609
Electrostatics, 182
Elementary step, 325
Emerson enhancement factor, 584
Emerson, Robert, 581
Emf, 236
Emission spectra, 555
End point, 287
Endergonic process, 102
Endothermic reaction, 64
Energy, 5
Energy level
   hydrogen atom, 433
   many-electron atoms, 433
Enthalpy (H), 46
   bond, 70
   chemical reactions, 65
   of activation, 340
   of formation, standard, 65

Enthalpy (H) (continued)
   temperature dependence, 69
Entropy (S), 83
   absolute, 96
   and spontaneity, 89
   and the second law of thermodynamics, 89
   and the third law of thermodynamics, 95
   Boltzmann equation, 84
   chemical reactions, 97
   gas expansion, 85
   of activation, 340
   of fusion, 91
   of heating, 92
   meaning of, 98
   of mixing, 90
   and probability, 84
   of rubber molecules, 119
   of vaporization, 92
   phase changes, 91
   statistical definition, 83
   standard molar, 97t
   thermodynamic definition, 86
Enzyme
   acetylcholinesterase, 260
   alcohol dehydrogenase, 344, 375
   carbonic anhydrase, 294, 371
   chymotrypsin, 372
   cytochrome oxidase, 250
   fumarse, 393
   hexokinase, 365
   inhibition, 377
   kinetics, 363
   peptidyl propyl isomerase, 631
   photolyase, 592
   protein disulfide isomerase, 631
   urease, 364
Enzyme kinetics, 363
   and pH, 393
Enzyme mechanism
   Briggs-Haldane, 368
   Michaelis-Menton, 367
   nonsequential, 376
   ordered sequential, 376
   "ping-pong," 376
   random sequential, 376
   sequential, 376
Equation
   Arrhenius, 333
   Clapeyron, 112
   Clausius-Clapeyron, 113
   diffusion, 605

Equation (continued)
   Gibbs-Helmholtz, 108, 205
   Goldman, 258
   Henderson-Hasselbalch, 281
   Hill, 387
   ideal gas, 11
   Nernst, 242
   van der Waals, 15
   van't Hoff, 206
   virial, 16
Equation of state, 11
   ideal gas, 11
   van der Waals, 15
   virial, 16
Equilibrium
   heterogenous, 203
   homogeneous, 193
   and Le Chatelier's principle, 206
   and transition-state theory, 340
   dialysis, 213
   phase, 110
Equilibrium constant
   apparent, 195
   and Gibbs energy change, 195
   in glycolysis, 303
   temperature dependence, 205
   thermodynamic, 201
Equilibrium dialysis, 213
Equilibrium isotope effect, 343
Equilibrium vapor pressure, 19
Equipartition of energy theorem, 530
Erythrocytes, see red blood cells
Ethane, 614
Ethanol
   NMR spectrum of, 542
   proton exchange, 542, 545
   with NAD, 45
Ethylene
   hybridization of, 453
   molecular orbitals in, 467
Ethylenediaminetetraacetate, see EDTA
Euler, Leonhard, 648
Euler's theorem, 648
Eutectic point, 229
Exact differential, 647
Excitation transfer, 583
Exergonic process, 102

Exothermic reaction, 64
Extensive properties, 8
Extent of reaction, 193
Eyring, Henry, 339

F

Facilitated diffusion, 285
Factorial, 640
Faraday constant (F), 256, 356
Faraday, Michael, 256
Fast reactions, 485
Fibrous protein, 620
Fick, Adolf, 605
Fick's laws of diffusion, 605
First law of thermodynamics, 44
First-order reaction, 315
Fischer, Emil, 365
Flow method, 349
Fluid mosaic model, 175
Fluorescence, 554
Fluorescence lifetime, 555
Fluorescence quantum yield, 555
Force, 3
Force constant (k), 527
Formal charge, 447
Fourier transform spectroscopy
NMR, 547
Franck-Condon principle, 534
Franck, James, 534
Free induction decay (FID), 549
Free-electron model, 423
Freezing-point depression, 146
Frequency (ν), 401
Frequency factor, 335
Frictional coefficient, 601
FT-NMR, 547
Fuel cell, 245
Fugacity (f), 200
Fugacity coefficient (γ), 200
Fundamental band, 530

G

g factor, 553
Galvanic cell, 236
Gamow, George, 424
Gas
diffusion, 30
effusion, 30
expansion, 40, 55
heat capacity, 50

Gas (*continued*)
ideal, 8
kinetic theory of, 21
partial pressure of, 13
real, 14
Gay-Lussac, Joseph, 9
Gas constant (R), 11
Gas expansions, 40, 55
adiabatic, 56
irreversible, 40
isothermal, 55
reversible, 41
Gel electrophoresis, 611
Geometric isomerization, 587
Germer, Lester, 411
Gibbs energy (G), 101, 103
of activation, 340
and electrical work, 104, 239
of formation, standard, 105
chemical reactions, 106
and spontaneity, 102
of mixing, 132
and phase equilibria, 110
pressure dependence, 108
temperature dependence, 107
Gibbs-Helmholtz equation, 108, 205
Gibbs phase rule, 117
Gibbs, Willard, 101
Glass electrode, 247
Globular protein, 620
Glucose
breakdown in glycolysis, 223
combustion, 107
Glucose-6-phosphate, 224, 376
Glycine
isoelectric point of, 290
titration curve of, 290
Glycolysis, 223
Goldman, David, 258
Goldman equation, 258
Goodyear, Charles, 117
Graham, Thomas, 31
Graham's law of diffusion, 31
Graham's law of effusion, 31
Gramicidin, 181
Gravitational acceleration constant, 40
Guanine, 504
Gyromagnetic ratio, 540

H

Haldane, John, 368
Half-cell reaction, 236
Half-life, 315
first order, 315
other orders, 315
second order, 318
Hard acid, 469
Hard base, 469
Harmonic oscillator, 528
Hartley, B. S., 373
Heat, 43
Heat capacity, 49, 51
and Debye cube law, 96
constant pressure, 50, 651
constant volume, 50
molecular interpretation, 51
Heat capacity ratio (γ), 57
Heat engine, 87
Heat of hydration, 158
Heat of reaction, 65
Heat of solution, 158
Heavy metals, 482
Heisenberg uncertainty principle, 414
and particle in a box, 420
and rotational motion, 524
and spectral line width, 515
and vibrational motion, 528
energy and time, 416, 515
in NMR, 549
Heisenberg, Werner, 414
Helium-neon laser, 558
Helmholtz, 102
Helmholtz energy (A), 102, 118
Heme proteins, 477
Hemoglobin
allosteric model of, 2, 385
and sickle-cell anemia, 2, 500
as buffer component, 294
binding of carbon dioxide, 295
binding of carbon monoxide, 479
binding of oxygen, 2, 385
structure of, 621
Hemolysis, 153
Henderson-Hasselbalch equation, 281
Henri, Victor, 367
Henry, William, 137

Henry's law, 137
Henry's law constant, 137
Hertz (Hz), 401
Hess, Germain, 66
Hess's law, 66
Heterogeneous equilibria, 203
Heterotropic effect, 385
High-spin complex, 473
Hill, Archibald, 389
Hill, Robert, 580
Hill coefficient (n), 389
Hill equation, 387
Histidine
as buffer component, 294
distal, 477
isoelectric point of, 291
proximal, 477
Hologram, 561
Holography, 561
Homogeneous equilibria, 193
Homotropic effect, 385
Hooke's law, 527
Hot band, 530
Hückel, Walter, 165
Hughes-Klotz plot, 213
Hund, Frederick, 435
Hund's rule, 435
Hybrid orbitals
dsp^3, 476
d^2sp^3, 476
sp, 454
sp^2, 453
sp^3, 451
Hybridization, 450
Hydration, 155
Hydration number, 155
Hydrogen (H_2)
bond dissociation energy, 70
bond dissociation enthalpy, 70
molecular orbital treatment, 458
potential energy curve, 70
valence bond treatment, 448
Hydrogen atom
Bohr's theory of, 407
emission spectra of, 407
energy of, 408
orbitals in, 428
wave functions of, 429
Hydrogen bond, 502, 617
Hydrogen-bromine chain reaction, 332

Hydrogen chloride, 532
Hydrogen electrode, 236
Hydrogen fluoride, 463
Hydrogen-oxygen fuel cell, 245
Hydrolysis
  covalent, 374
  p-nitrophenyl acetate, 374
  p-nitrotrimethyl acetate, 374
  salts, 275
Hydronium ion, 268
Hydrophobic interaction, 509, 625
Hydrostatic pressure, 14, 149
Hyperfine splitting constant, 553
Hypertonic solution, 153
Hypochromism, 536
Hypothermia, 54
Hypotonic solution, 153

I

Ice
  structure, 505
Ideal-dilute solution, 141
Ideal gas, 8
Ideal-gas equation, 11
Ideal solution, 135
Imidazole, 393
Indicator
  acid-base titrations, 287
Infrared active, 530
Infrared group frequencies, 533
Infrared spectroscopy, 527
Inhibition
  irreversible, 384
  reversible, 377
Initial rate, 366
Intensity of spectral lines, 518
Intensive properties, 8
Interference of waves, 402
Intermediate, 325
Intermolecular forces, 489
Internal energy (U), 45
International system of units (SI), 3
International unit, 371
Intersystem crossing, 593
Intrinsic dissociation constant, 212
Intrinsic viscosity, 608
Iodoacetate, 632
Iodopsin, 588

Ion
  ion pair, 170
  structure-breaking, 155
  structure-making, 155
  thermodynamics of, 157
Ion-dipole interaction, 494
Ion-induced dipole, 495
Ion pair, 170
Ion product of water, 269
Ionic activity, 162
Ionic atmosphere, 165
Ionic bond, 490
Ionic strength (I), 165
Ionization energy, 438
Ionophore, 180
Irreversible inhibition, 384
Isoelectric focusing, 612
Isoelectric point (pI), 174, 291, 610
Isoelectronic, 404
Isolated system, 7
Isolation method, 324
Isosbestic point, 538
Isotherm, 8
Isothermal gas expansion, 55
Isothermal process, 55
Isotonic solution, 153
Isotope effect, 343

J

Jablonski diagram, 557
Joule, 5

K

Kelvin, 9
Keratin, 620
Kilby, B. A., 373
Kinetic energy, 24
Kinetic isotope effect, 343
Kinetic salt effect, 347
Kinetic theory of gases, 21
Kirchhoff, Gustav, 69
Kirchhoff's law, 69
Kramers' theorem, 554

L

Lactate, 223
Lambert, Johann, 521
Landé g factor, 553
Lanthanides, 437
Larmor frequency, 540
Laser, 557
  dye, 562
  helium-neon, 558
  ruby, 557
  YAG, 559

Laser-induced fluorescence, 562
Lattice energy, 157
Law
  Beer-Lambert, 521
  Boyle's, 8
  Charles', 9
  Dalton's, 12
  Debye cube, 96
  Debye-Hückel limiting, 165
  Fick's, 605
  Graham's, 31
  Henry's, 137
  Hess's, 66
  Hooke's, 527
  Kirchhoff's, 69
  Newton's second, 22
  of thermodynamics, 8, 44, 88, 95
  Raoult's, 135
  rate, 312
  Stokes, 601
Lead, 482
Le Chatelier, Henry, 206
Le Chatelier's principle, 206
  and catalyst, 209
  and pressure, 208
  and temperature, 206
Lennard-Jones potential, 498
Lennard-Jones, John, 498
Levorotatory, 565
Lewis acid, 268
Lewis base, 268
Lewis, Gilbert, 165, 447
Lewis structure, 447
Lifetime broadening, 515
Ligand field theory, 475
Light
  intensity, laser, 562
  Maxwell's theory, 402
  measurement of, 578
  particle theory, 405
  speed, 403
  visible, 403
  wave theory, 402
Light intensity, 562, 578
Light reaction, 580
Lindermann, Alexander, 326
Line spectra, 407
Line width, 514
  and Doppler effect, 516
  and pressure effect, 516
  natural, 515

Linear combination of atomic orbitals (LCAO), 458
Linearly polarized light, *see* plane-polarized light
Lineweaver-Burk plot, 370
Lineweaver, H., 370
Lipid bilayer, 175
Liquid
  radial distribution function, 506
Liquid scintillation counting, 555
Lock-and-key theory, 365
London, Fritz, 497
London interaction, *see* dispersion interaction
Lone pair, 448
Low-spin complex, 473
Lummer, Otto, 403
Lyman series, 408

M

Maclaurin's theorem, 145
Macromolecules, 599
  binding of ligands, 209
  diffusion coefficients, 604
  electrophoresis of, 608
  molar mass of, 599
  sedimentation coefficient of, 601
  structure of, 613, 616
  viscosity of, 607
Macrostate, 99
Magnetic quantum number (m_l), 427
Magnetic resonance imaging, 551
Magnetogyric ratio, *see* gyromagnetic ratio
Maiman, Harold, 558
Manganese, 480
Manometer, 4
Many-electron atoms, 432
Maximum rate (V_{max}), 368
Maximum work, 41
Maxwell distribution of speeds, 25
Maxwell distribution of velocities, 25
Maxwell, James, 25
Maxwell's electromagnetic theory of radiation, 402
Mean activity coefficient, 162
Mean free path (λ), 29
Mean square velocity, 23

Melanin, 591
Membrane capacitance, 184
Membrane potential, 255
Membrane structure, 175
Membrane transport, 177
  active transport, 178
  facilitated diffusion, 177
  simple diffusion, 177
Menten, Maud, 367
β-Merceptoethanol, 626
Mercury, 482
Meselson, Matthew, 623
Metallothionein, 483
Methane
  bonding, 451
  combustion, 107
Methane hydrate, 509
8-Methoxypsoralen (8-MOP), 593
Michaelis constant (K_M), 369
Michaelis, Leonor, 367
Michaleis-Menton kinetics, 367
Microscopy
  electron, 413
  scanning tunneling, 426
Microstate, 99
Microwave spectroscopy, 522
Mitchell, Peter, 252
Mitochondria, 253
Molality (m), 128
Molar absorptivity, 522
Molar extinction coefficient, 522
Molar heat capacity, 49
Molar mass, 6
  comparison of experimental methods, 152
  from freezing-point depression, 147
  from osmotic pressure, 151
  from SDS-electrophoresis, 612
  from ultracentrifugation, 604
  from viscosity, 609
  number average, 599
  weight average, 599
Molar rotation, 565
Molarity (M), 128
Mole, 5
Mole fraction, 13, 128
Molecular chaperone, 631
Molecular collision, 29

Molecular orbital
  pi, 461
  sigma, 458
Molecular orbital energy level diagram
  amide group, 468
  C_2H_4, 467
  C_6H_6, 467
  CO, 464
  H_2, 458
  HF, 463
  homonuclear diatomic molecule, 460
Molecular orbital theory, 458, 475
Molecular symmetry, 562
Molecularity, 324
Molten globule, 630
Moment of inertia (I), 522
Monochromaticity, 561
Monochromator, 561
Monodispersity, 599
Most probable speed (c_{mp}), 26
Moving boundary method, 609
Mulliken, Robert, 537
Multiphoton transition, 560
Multisubstrate systems, 375
Myoglobin, 385

N

NAD^+, 224, 344
NADH, 224, 344
$NADP^+$, 585
Negative cooperativity, 391
Nernst equation, 242
Nernst, Walter, 242
Nerve gas, 384
Neuromuscular junction, 260
Neuron, 255
Neurotransmitter, 260
Newton (unit), 3
Newton, Issac, 3
Newton's second law of motion, 22
Nickel, 481
Nicotinamide adenine dinucleotide, *see* NAD^+
Nicotinamide adenine dinucleotide phosphate, *see* $NADP^+$
Nitric oxide
  electron configuration of, 465

Nitrogen (N_2)
  electron configuration of, 462
  nacorsis, 14
p-Nitrophenyl acetate, 373
p-Nitrophenyltrimethyl-acetate, 374
Nitroxide radical, 553
Noble gas core, 435
Node, 412, 421, 429
Noncompetitive inhibition, 380
Nonelectrolyte solutions, 127
Normalization constant, 420
Nuclear magnetic resonance (NMR) spectroscopy, 539
Nuclear spin quantum number, 539
Number-average molar mass, 599

O

Octahedral complex, 470
Octet rule, 447
Open system, 7
Operator, 417, 644
Opsin, 588
Optical activity, 562
Optical isomer, *see* enantiomers
Optical pumping, 558
Optical rotation, 564
Optical rotatory dispersion, 566
Orbital, 428
  atomic, 429
  d, 431
  delocalized, 467
  hybrid, 451
  LCAO, 458
  molecular, 458
  nonbonding, 463
  p, 431
  s, 430
Oscillating reaction, 353
Osmosis, 148
Osmotic pressure (π), 149
Ostwald viscometer, 607
Ostwald, Wolfgang, 607
Overtone, 530
Oxidative phosphorylation, 248
Oximeter, 522
Oxygen (O_2)
  binding to hemoglobin, 1, 385, 478

Oxygen (O_2) (*continued*)
  electron configuration of, 462
  heat capacity of, 51
  paramagnetism of, 462
Oxyhemoglobin, 294, 385
Ozone
  absorption of UV radiation, 591
  formal charges, 448

P

P branch, 533
Paramagnetism, 435, 462
Partial derivative, 645
Partial molar Gibbs energy, 131
Partial molar quantities, 129
Partial molar volume, 129
Partial pressure, 13
Partial specific volume ($\bar{v}$), 601
Particle in a one-dimensional box, 419
Particle-wave duality, 410
Pascal (unit), 3
Paschen series, 408
Passive transport, 171
Pauli exclusion principle, 434
Pauli, Wolfgang, 434
Pauling, Linus, 455, 616
Pearson, Ralph, 469
Peptide bond, 467
  orbitals of, 468
Peptidyl propyl isomerase, 631
Percent ionic character, 457
Percent dissociation, 271
Periodicity
  atomic radius, 437
  electronegativity, 455
  ionization energy, 438
Permittivity of the vacuum, 156, 491
Perpetual motion machine, 209
pH
  of blood, 293
  and enzyme kinetics, 393
  definition, 269
  measurement, 247
Phase, 110
Phase diagram, 115
  of carbon dioxide, 116
  of water, 115

Phase rule, 117
Phospholipids, 175
Phosphorescence, 556
Phosphorescence lifetime, 556
Phosphoric acid, 279
Photochemical reaction, 575
  primary processes, 576
  secondary processes, 576
Photodynamic therapy, 592
Photoelectric effect, 405
Photolyases, 592
Photomedicine, 592
Photon, 405
Photophosphorylation, 584
Photosensitizer, 592
Photosynthesis, 580
Photosynthetic unit, 583
Photosystem I, 583
Photosystem II, 583
Pi (π) bond, 450
pK_a, 275
pK_w, 270
Planck, Max, 404
Planck's constant, 404
Plane of symmetry, 562
Plane-polarized light, 563
pOH, 270
Polarimeter, 564
Polarizability (α), 495
Polarized light, 563
Poly-cis-isoprene, 117
Polydispersity, 599
Polyenes, 423
Polyethylene, 615
Polymer, see macro-
  molecules
Polypeptide, 616
Population inversion, 557
Porphyrin, 477
Potassium ferrioxalate, 578
Potential energy curve, 70
Potential energy surface, 335
Pre-exponential factor, 335
Pressure, 3
  critical, 19
  equilibrium vapor
    pressure, 19
  hydrostatic, 14, 149
  measurement of, 4
  partial, 13
  standard, 11
  units of, 3
Pre-steady-state kinetics, 368

Prigogine, Ilya, 353
Primary structure, 621
Principal quantum number
  (n), 427
Principle
  Aufbau, 435
  Franck-Condon, 534
  Heisenberg uncertainty, 414
  Le Chatelier's, 206
  microscopic reversibility, 329
  Pauli exclusion, 434
Pringsheim, Ernst, 404
Probability factor, 338
2-Propenenitrile, 533
Protein disulfide isomerase, 631
Proteins, 616
  and Donnan effect, 172, 295
  binding of ligands, 209
  denaturation, 62, 626
  fibrous, 620
  folding, 629
  globular, 620
  hydrophobic interactions
    in, 509, 625
  partial specific volume of, 601
  primary structure of, 621
  quaternary structure of, 621
  secondary structure of, 621
  tertiary structure of, 621
  thermodynamic stability
    of, 624
  two-state model for, 62, 628
Proton-decoupling, 547
Pseudo first-order reaction, 320
Pyruvate, 223
Pythagoras theorem, 22, 643

Q

Q branch, 533
Quanta, 404
Quantum mechanical
  tunneling, 424
Quantum mechanics, 417
Quantum numbers, 408, 427
Quantum yield (Φ), 555, 576
Quaternary structure, 620

R

Radial distribution
  function, 429, 506
Radiationless transition, 557
Radioactive decay, 424
Radius
  atomic, 437
  van der Waals, 500
Ramanchandran, G. N., 620
Ramanchandran plot, 620
Random-walk model, 614
Raoult, François, 135
Raoult's law, 135
  deviation from, 137
Rare earth metals, see
  lanthanides
Rate constant, 312
Rate-determining step, 325
Rate law, 312
Reaction
  chain, 332
  consecutive, 330
  diffusion-controlled, 485
  elementary, 325
Reaction center, 581
Reaction coordinate, 335
Reaction mechanism, 324
Reaction order, 312, 323
  determination of, 323
Reaction quotient, 199
Real gases, 14
Real solutions, 139
Red blood cells, 153, 500
Red drop, 583
Reduced mass (μ), 523
Relaxation kinetics, 349
Relaxation time
  in chemical kinetics, 349
  in Debye-Hückel theory, 274
  in NMR, 547
Repulsive forces, 498
Residue, 616
Resolution
  in spectroscopy, 517
Resonance, 466
Resonance structure, 466
Respiratory chain, 248
11-cis-Retinal, 814
Retrolental fibroplasia, 14
Reverse osmosis, 154
Reversible inhibition, 377
Reversible process, 41
Reversible reaction, 328
Rhodamine 6G, 562
Rhodopsin, 588

Ribonuclease, 626
Ribozyme, 364
Rigid rotor, 522
Rod cells, 586
Root-mean-square velocity, 25
Rotation
  about double bonds, 589
Rotation spectra, 525
Rotational constant, 525
Rotational degrees of
  freedom, 530
Rotational quantum
  number, 524
Rubber
  elasticity of, 117
  structure of, 117
Ruby laser, 558
Rule
  Hund's, 435
  phase, 117
  selection, 519
Rydberg formula, 407
Rydberg, Johannes, 407
Rydberg's constant, 407

S

Saddle point, 335
Salt bridge, 236
Salt hydrolysis, 275
Salting-in effect, 169
Salting-out effect, 169
Scalar quantity, 25
Scanning tunneling micro-
  scope (STM), 426
Scatchard, George, 213
Scatchard plot, 213
Schrödinger, Erwin, 416
Schrödinger wave equation, 416
  for hydrogen atom, 426
  for particle in a box, 418
Screening constant, 543
Scuba diving, 14
SDS-PAGE, 611
Second law of thermo-
  dynamics, 88
Second-order reaction, 318
Secondary structure, 620
Sedimentation
  coefficient (s), 601
  density gradient, 607
  equilibrium, 605
  moving boundary, 601
  velocity, 600
  zonal, 605
Selection rules, 519
  atomic, 520

Selection rules (*continued*)
ESR, 553
NMR, 540
rotational, 524
spin-forbidden, 519
symmetry-forbidden, 520
vibrational, 530
Self-consistent field method, 432
Semiconservative model, 623
Semipermeable membrane, 148
Sequential model, 391
Shell, 428
Shielding constant, 434
SI system, *see* International system of units
Sickle-cell anemia, 2, 500
Sigma (σ) bond, 450
Signal-to-noise ratio, 521
Silver chloride
salting-in effect, 167
solubility product, 168
Simple harmonic motion, 527
Singlet state, 520
Smoluckowski, Roman, 348
Sodium chloride
lattice energy of, 157
solution process, 157
Sodium dodecyl sulfate (SDS), 611
Sodium-potassium ATPase, 179
Sodium-potassium pump, 179
Soft acid, 469
Soft base, 469
Solubility of gases, 139
Solute
chemical potential, 141
standard state, 142
Solutions
colligative properties of, 142, 170
concentration units, 127
dilute ideal, 141
electrolyte, 154
ideal, 135
nonelectrolyte, 127
Solvent
cage, 346
chemical potential, 132
Specific activity, 371
Specific heat, 49
Specific rotation, 564
Specific viscosity, 608
Spectrochemical series, 472

Spectroscopy
atomic emission, 407
electronic, 534
ESR, 552
fluorescence, 554
IR, 527
microwave, 522
NMR, 539
phosphorescence, 554
Speed
average, 27
distribution of, 25
of light, 403
most probable, 26
root-mean-square, 25
Spherical polar coordinates, 427
Spin-decoupling, 547
Spin-forbidden transitions, 519
Spin multiplicity, 520
Spin quantum number (m_s), 432
Spin-spin coupling, 544
Spontaneous emission, 518
Spontaneous processes, 81
Square-planar complex, 473
Stahl, Franklin, 623
Standard enthalpy of formation, 64
of compounds, 103, 651
of ions, 265
Standard enthalpy of reaction, 99
Standard entropy
of inorganic compounds, 97, 651
of ions, 265
of organic compounds, 97, 654
Standard Gibbs energy of formation
of compounds, 106, 651
of ions, 159
Standard Gibbs energy of reaction, 106
Standard hydrogen electrode, 236
Standard reduction potential, 236
Standard state, 64
in biochemistry, 218
Standard temperature and pressure (STP), 11
Standing waves, 412
State
equations of, 11
standard, 64

State (*continued*)
standard, biochemists', 218
standard, physical chemists', 218
State function, 43
State of the system, 7
Steady state, 228
Steady-state approximation, 326, 368
Steric factor, 338
Stimulated absorption, 518
Stimulated emission, 518
Stokes, George, 601
Stokes law, 601
Stopped-flow kinetics, 349
Sublimation, 116
Subshell, 428
Substrate, 365
Subunits, 385
Succinic dehydrogenase, 380
Sulfur hexafluoride (SF_6), 20
Sumner, James, 365
Supercritical fluid, 21
Surroundings, 7
Svedberg (s), 601
Svedberg, Theodor, 601
Symmetry-forbidden transitions, 520
Synaptic junction, 260
System, 7
closed, 7
isolated, 7
open, 7

T
Temperature
absolute zero scale, 9
and kinetic theory, 24
operational definition, 8
Temperature-jump relaxation, 349
Terminal respiratory chain, 250
Termolecular reaction, 327
Tertiary structure, 620
Tesla (T), 541
Tesla, Nikola, 541
Tetracyanoethylene, 537
Tetrahedral complex, 473
Tetramethylsilane (TMS), 543
Thermal motion, 24
Thermal reaction, 575
Thermochemistry, 64
Thermodynamic efficiency, 87, 227

Thermodynamic equilibrium constant, 201
Thermodynamic hypothesis, 628
Thermodynamic solubility product, 168
Thermodynamics
first law, 44
second law, 88
third law, 95
zeroth law, 8
Thermodynamics of electrochemical cells, 238
Thermodynamics of mixing, 90, 132
Third-law entropies, 96
Third law of thermodynamics, 95
Thomson, George, 411
Thomson, William, 9
Threshold frequency, 405
Threshold potential, 259
Thymine, 504
Thymine dimer, 591
Tiselius, Arne, 610
Titration
acid-base, 286, 301
protein, 292
TMS, *see* tetramethylsilane
Total differential, 645
Toxic heavy metals, 338
Transition dipole moment, 520
Transition metals, 435
Transition state, *see* activated complex
Transition state theory, 340
thermodynamic formulation, 340
Transmittance, 522
Transpiration, 154
Tripeptide, 616
Triple point, 115
Triplet state, 520
Turnover number, 371
Two-photon spectroscopy, 560
Tyrosine, 538

U
Ultracentrifugation, 600
Ultraviolet catastrophe, 404
Uncompetitive inhibition, 381
Unimolecular reaction, 325
Units, SI, 2
Urea, 501
Urease, 364

UV radiation damage, 591
UV-A, 591
UV-B, 591
UV-C, 591

V

Valence bond theory, 448, 466, 476
Valence electron, 447
Valine, 2, 500
Valinomycin, 181
van der Waals constants, 16
van der Waals equation, 15
van der Waals forces, 498
van der Waals, Johannes, 15
van der Waals radii, 500
van't Hoff equation, 206
van't Hoff factor, 170
van't Hoff, Jacobus, 170
Vapor (comparison with gas), 111
Vapor pressure, 19
Vapor-pressure lowering, 143
Vector, 25

Vibration, normal modes, 530
Vibration-rotation spectra, 532
Vibrational degrees of freedom, 530
Vibrational quantum number, 528
Virial coefficients, 17
Virial equation of state, 16
Viscosity
intrinsic, 608
of liquids, 607
of macromolecular solutions, 607
reduced, 608
relative, 608
specific, 608
Vision, 586
Vitamin B_{12}, 480
Voltaic cell, 236
Vulcanization, 117

W

Wald, George, 586
Water
acid-base properties of, 268

Water (*continued*)
dielectric constant of, 156
hydrogen bonds in, 506
ion product of, 269
normal modes, 530
phase diagram of, 115
properties of, 507
radial distribution function for, 506
structure of, 507
surface tension of, 508
viscosity of, 508
Watson-Crick base pairs, 504
Watson, James, 504
Watt (unit), 560
Wave function
angular, 429
for particle in a box, 420
hybrid, 451, 453, 454
hydrogen atom, 429
hydrogen molecule, 458
normalization of, 420
radial, 429
unacceptable, 417
Wave mechanics, 417
Wave number (v), 514

Weight average molar mass, 600
Wien effect, 167
Wien, Wilhelm, 167
Work, 39
electrical, 104, 239
in gas expansions, 39, 55
maximum, 41
Work function, 405

Y

YAG laser, 559

Z

Z scheme, 585
Zero-order reaction, 313
Zero-point energy, 71
effect on rate, 343
harmonic oscillator, 528
particle in a box, 420
Zero-point vibration, 343
Zeroth law of thermodynamics, 8
Zinc, 481
Zinc finger, 482
Zonal sedimentation, 605
Zone electrophoresis, 610
Zwitterion, *see* dipolar ion

Values of Some Fundamental Constants

Constant	Value
Avogadro's constant (N_A)	6.0221367×10^{23} mol^{-1}
Bohr radius (a_o)	$5.29177249 \times 10^{-11}$ m
Boltzmann constant (k_B)	1.380658×10^{-23} J K^{-1}
Electron charge (e)	1.602177×10^{-19} C
Electron mass (m_e)	$9.1093897 \times 10^{-31}$ kg
Faraday constant (F)	96485.309 C mol^{-1}
Gas constant (R)	8.314510 J K^{-1} mol^{-1}
Neutron mass (m_N)	1.674928×10^{-27} kg
Permittivity of vacuum (ε_0)	8.854×10^{-12} C^2 N^{-1} m^{-2}
Planck constant (h)	6.626075×10^{-34} J s
Proton mass (m_P)	1.672623×10^{-27} kg
Rydberg constant (R_H)	109737.31534 cm^{-1}
Speed of light in vacuum (c)	299792458 m s^{-1}

Pressure of Water Vapor at Various Temperatures

Temperature/°C	Water Vapor Pressure/mmHg
0	4.58
5	6.54
10	9.21
15	12.79
20	17.54
25	23.76
30	31.82
35	42.18
40	55.32
45	71.88
50	92.51
55	118.04
60	149.38
65	187.54
70	233.7
75	289.1
80	355.1
85	433.6
90	525.76
95	633.90
100	760.00